THE EPIGENETICS OF AUTOIMMUNITY

Translational Epigenetics Series

Trygve O. Tollefsbol, Series Editor

Transgenerational Epigenetics
Edited by Trygve O. Tollefsbol, 2014

Personalized Epigenetics
Edited by Trygve O. Tollefsbol, 2015

Epigenetic Technological Applications
Edited by Y. George Zheng, 2015

Epigenetic Cancer Therapy
Edited by Steven G. Gray, 2015

DNA Methylation and Complex Human Disease
By Michel Neidhart, 2015

Epigenomics in Health and Disease
Edited by Mario F. Fraga and
Agustin F. F Fernández, 2015

Epigenetic Gene Expression and Regulation
Edited by Suming Huang, Michael Litt, and C. Ann Blakey, 2015

Epigenetic Biomarkers and Diagnostics
Edited by Jose Luis García-Giménez, 2015

Drug Discovery in Cancer Epigenetics
Edited by Gerda Egger and
Paola Barbara Arimondo, 2015

Medical Epigenetics
Edited by Trygve O. Tollefsbol, 2016

Chromatin Signaling
Edited by Olivier Binda and
Martin Fernandez-Zapico, 2016

Chromatin Regulation and Dynamics
Edited by Anita Göndör, 2016

Neuropsychiatric Disorders and Epigenetics
Edited by Dag H. Yasui, Jacob Peedicayil and
Dennis R. Grayson, 2016

Polycomb Group Proteins
Edited by Vincenzo Pirrotta, 2016

Epigenetics and Systems Biology
Edited by Leonie Ringrose, 2017

Cancer and Noncoding RNAs
Edited by Jayprokas Chakrabarti and
Sanga Mitra, 2017

Epigenetic Mechanisms in Cancer
Edited by Sabita Saldanha, 2017

THE EPIGENETICS OF AUTOIMMUNITY

Volume 5

Edited by

RONGXIN ZHANG

Department of Immunology, School of Basic Medical Sciences,
Tianjin Medical University, Tianjin, China

ELSEVIER

ACADEMIC PRESS

An imprint of Elsevier

Academic Press is an imprint of Elsevier
125 London Wall, London EC2Y 5AS, United Kingdom
525 B Street, Suite 1800, San Diego, CA 92101-4495, United States
50 Hampshire Street, 5th Floor, Cambridge, MA 02139, United States
The Boulevard, Langford Lane, Kidlington, Oxford OX5 1GB, United Kingdom

Library of Congress Cataloging-in-Publication Data
A catalog record for this book is available from the Library of Congress

British Library Cataloguing-in-Publication Data
A catalogue record for this book is available from the British Library

ISBN: 978-0-12-809912-4

For information on all Academic Press publications visit our website at https://www.elsevier.com/books-and-journals

Publisher: Mica Haley
Senior Acquisition Editor: Versteeg-Buschman
Editorial Project Manager: Megan Ashdown
Production Project Manager: Debasish Ghosh
Cover Designer: Victoria Pearson Esser

Typeset by Thomson Digital

Contents

List of Contributors

Alessandro Antonelli University of Pisa, Pisa, Italy

Saeed Aslani Rheumatology Research Center, Tehran University of Medical Sciences, Tehran, Iran

Teresa Ayuso Complejo Hospitalario de Navarra, Navarra Institute for Health Research (IdiSNA), Pamplona, Navarra, Spain

Patricia Aznar Complejo Hospitalario de Navarra, Navarra Institute for Health Research (IdiSNA), Pamplona, Navarra, Spain

Debarati Banik The George Washington University Cancer Center, The George Washington University Washington, DC, United States

Salvatore Benvenga University of Messina School of Medicine; University Hospital, Messina, Italy

Vanessa Beynon Ann Romney Center for Neurologic Diseases, Brigham and Women's Hospital and Harvard Medical School, Boston, MA, United States

Anne Bordron European University of Brittany, Brest, France

Wesley H. Brooks University of South Florida, Tampa, FL, United States

Luigi Cari University of Perugia, Perugia, Italy

Susan Carpenter University of California, Santa Cruz, CA, United States

Amandine Charras European University of Brittany, Brest, France

Katherine B. Chiappinelli The George Washington University Cancer Center, The George Washington University, Washington, DC, United States

Fabio Coppedè University of Pisa, Pisa, Italy

Patricia Costa-Reis Lisbon University, Santa Maria Hospital, Lisbon, Portugal

Sergio Covarrubias University of California, Santa Cruz, CA, United States

Giusy Elia University of Pisa, Pisa, Italy

Poupak Fallahi University of Pisa, Pisa, Italy

Silvia Martina Ferrari University of Pisa, Pisa, Italy

Lucien P. Garo Ann Romney Center for Neurologic Diseases, Brigham and Women's Hospital and Harvard Medical School, Boston, MA, United States

Steffen Gay Center of Experimental Rheumatology, University Hospital Zurich, Zurich, Switzerland

William T. Gerthoffer The Center for Lung Biology, University of South Alabama, Mobile, AL, United States

Pietro Invernizzi University of Milan-Bicocca, Milan, Italy

Sundararajan Jayaraman University of Illinois at Chicago, Chicago, IL, United States

Emmanuel Karouzakis Center of Experimental Rheumatology, University Hospital Zurich, Zurich, Switzerland

Kerstin Klein Center of Experimental Rheumatology, University Hospital Zurich, Zurich, Switzerland

Orsia D. Konsta European University of Brittany, Brest, France; National University of Athens, Greece

Christelle Le Dantec European University of Brittany, Brest, France

Yiu T. Leung Temple University School of Medicine, Philadelphia, PA, United States

Ana Lleo Liver Unit and Center for Autoimmune Liver Diseases, Humanitas Clinical and Research Center; Humanitas University, Rozzano,Milan, Italy

Hai Long The Second Xiangya Hospital, Central South University, Changsha, China

Qianjin Lu The Second Xiangya Hospital, Central South University, Changsha, China

Mahdi Mahmoudi Rheumatology Research Center, Tehran University of Medical Sciences, Tehran, Iran

Simona Marzorati Liver Unit and Center for Autoimmune Liver Diseases, Humanitas Clinical and Research Center, Rozzano, Milan, Italy

Maite Mendioroz Complejo Hospitalario de Navarra, Navarra Institute for Health Research (IdiSNA); Navarrabiomed-Navarra Institute for Health Research (IdiSNA), Pamplona, Navarra, Spain

Graziella Migliorati University of Perugia, Perugia, Italy

Gopal Murugaiyan Ann Romney Center for Neurologic Diseases, Brigham and Women's Hospital and Harvard Medical School, Boston, MA, United States

Giuseppe Nocentini University of Perugia, Perugia, Italy

Steven O'Reilly Northumbria University, Newcastle Upon Tyne, United Kingdom

Rab K. Prinjha GlaxoSmithKline, Medicines Research Centre, Stevenage, United Kingdom

Francesca Ragusa University of Pisa, Pisa, Italy

Sabrina Ramelli University of South Alabama, Mobile, AL, United States

Yves Renaudineau European University of Brittany; Brest University Medical School, CHU Morvan, Brest, France

Carlo Riccardi University of Perugia, Perugia, Italy

Inmaculada Rioja GlaxoSmithKline, Medicines Research Centre, Stevenage, United Kingdom

Takashi Sekiya Keio University School of Medicine, Shinjuku-ku, Tokyo; Section of Immune Response

Modification, The Research Center for Hepatitis and Immunology, National Center for Global Health and Medicine, Ichikawa, Chiba, Japan

Kathleen E. Sullivan The Children's Hospital of Philadelphia, University of Pennsylvania Perelman School of Medicine, Philadelphia, PA, United States

Paul-Peter Tak GlaxoSmithKline, Medicines Research Centre, Stevenage, United Kingdom

David F. Tough GlaxoSmithKline, Medicines Research Centre, Stevenage, United Kingdom

Alejandro Villagra The George Washington University Cancer Center, The George Washington University, Washington, DC, United States

Roberto Vita University of Messina School of Medicine, Messina, Italy

Ling Wang The Third Hospital of Changsha, Changsha, China

Zhi Yao School of Basic Medical Sciences, Tianjin Medical University, Tianjin, China

Heng Yin The Second Xiangya Hospital, Central South University, Changsha, China

Rongxin Zhang School of Basic Medical Sciences, Tianjin Medical University, Tianjin; School of Life Sciences and Biopharmaceutics, Guangdong Pharmaceutical University, Guangzhou, China

Zimu Zhang School of Basic Medical Sciences, Tianjin Medical University, Tianjin, China

About the Editor

Dr. Rongxin Zhang is Professor of Biology in the School of Life Sciences and Biopharmaceutics at the Guangdong Pharmaceutical University (Guangzhou, China), and is Professor of Immunology in the School of Basic Medical Sciences and Senior Scientist in the Research Center of Basic Medical Sciences at the Tianjin Medical University (Tianjin, China).

He trained as Postdoctoral Fellow and Instructor at Baylor College Medicine (Houston, Texas) and as Research Assistant Professor in the Department of Medicine at the University of Hong Kong. He earned doctorate degrees (Ph.D. in Genetics) from the Nankai University (Tianjin, China) and received his B.S. in Zoology from the School of Life Sciences at the Inner Mongolia University (Huhehot, China).

His research interests have covered a broad range of topics such as genetic and epigenetic mechanisms of pathogenesis of autoimmune central nervous system inflammation, cancer immunology and cancer immunotherapy, HIV vaccine, liver injury and immunotherapy, development of animal models for tumorigenesis and autoimmunity investigations. Dr. Zhang has published over 70 scientific papers and has been invited to speak on numerous international scientific conferences. Dr. Zhang also serves on the editorial boards of several relevant journals and has many patents in this area.

Preface

The Epigenetics of Autoimmunity is a topic directly linking the environmental and genetic factors to autoimmune responses. Via epigenetic mechanisms, a number of internal and external environmental risk factors, including smoking, nutrition, viral infection, and exposure to chemicals, could exert their influence on the pathogenesis of autoimmune diseases. Such factors could impact the epigenetic mechanisms, which, in turn, affect the regulation of gene expression, and eventually trigger immunologic events that result in instability of immune system. This book is a volume to be included in a series entitled "Translational Epigenetics" which is being edited by Trygve Tollefsbol. The book brings together an international group of basic and clinical scientists in autoimmunity to contribute to a state-of-the-art review of the underlying epigenetic regulation of autoimmune responses, pathogenic mechanisms of individual autoimmune disease, and epigenetic treatment strategies.

This volume is divided into four parts. The first part provides an overview of the epigenetics of autoimmunity and reviews the basics of autoimmunity epigenetics as it applies to medicine. The second part, Chapters 2–5, reviews the basic principles of the epigenetics of autoimmunity. The third part, Chapters 6–14, provides advances in widespread disorders or diseases such as systemic lupus erythematosus (SLE), rheumatoid arthritis (RA), multiple sclerosis (MS), type 1 diabetes (T1DM), systemic sclerosis (SSc), primary Sjögren's syndrome (pSS), autoimmune thyroid diseases (AITDs), primary biliary cirrhosis, and immune-mediated pulmonary diseases. The fourth part, Chapters 15–17, focuses on the therapy development of autoimmunity epigenetics. The last part discusses the future challenges and prospects for the epigenetics of autoimmunity.

In summary, this volume provides new insights into autoimmune diseases, raising great expectations among researchers and clinicians. This seminal book on this topic comprehensively covers the most recent advances in this exciting and rapidly developing new science.

This volume is written by 56 contributors representing 11 countries. Most of all, I would like thank these contributors for their writing of the chapters. I would also like to thank Linda Versteeg-Buschman, Halima Williams, Joslyn Paguio, Megan Ashdown, and other staff at Elsevier for their hard work. It has been a great pleasure to work with them on this book. I greatly appreciate the thoughtful help extended to me by Trygve Tollefsbol, the Series Editor. I am thankful to my graduate student Zimu Zhang for his assistance on the book.

Rongxin Zhang

Foreword

One of the major advances in clinic immunology, and indeed in all of medicine, was the discovery of the major histocompatibility complex. This event initiated the concept that autoimmunity would be due to genetic factors. Indeed, studies in both humans and inbred mice clearly indicated that the risk of developing an autoimmune disease was higher in families with an affected proband. However, it was equally clear that the concordance of autoimmunity in identical twins was not 100%. In some cases, such as rheumatoid arthritis or systemic lupus, it was only approximately 25%–30%. In other words, there was clearly a genetic influence, but environmental factors were also important.

The ability to rapidly sequence DNA and thence the development of high throughput assays led to an avalanche of studies involving genome-wide associations. With the increasing identification of SNPs, the ability to map and eventually the ability to recruit, define, and analyze large populations, it became widely assumed that the genetic contributions of such common diseases as type I diabetes, rheumatoid arthritis, and lupus, would be quickly defined. Indeed, there has likely been billions of research dollars spent throughout the world on genome-wide associations in many autoimmune diseases. Unfortunately, there has not been a single study, which has led to a result that can be translated to patients in the clinic. In other words, although there are clearly genetic influences, of which the contribution of the major histocompatibility complex is usually the most dominant, the data complexity has suggested that there are many genes involved, that these genes may not always be the same in every population and that our concept of a relatively simple genetic influence was incorrect.

These results in GWAS led naturally to the thesis that deep sequencing was required and that we were still missing clear and well-defined genetic influences. This concept of rare genetic influences is still a possibility, but it is even more likely that there are epigenetic contributions, that is environmentally induced changes in DNA. In this volume, Dr. Rongxin Zhang has emphasized these points and included a significant number of contributions by leaders in epigenetics. The chapters in this volume highlight the multitude of epigenetic modifications, which have been documented within diseases and within specific immune populations. They also emphasize the great need to translate these observations to functional data. Any time there is disease and, in particular, inflammation, there is likely to be epigenetic changes on tissues. It is our challenge to define the significance of these changes and whether they are the cause or the result of the disease. The papers in this volume emphasize these points and serve a bridge between our understanding of genetics and environment and subsequent loss of tolerance.

The future must define what these epigenetic influences are and how they function. Until then, our concept of autoimmunity can essentially be described as a situation of bad luck and bad genes. Our goal must be to refine this broad concept down to its molecular basis.

Merrill Eric Gershwin

University of California at Davis School of Medicine, Davis, CA, United States

CHAPTER

1

The Epigenetics of Autoimmunity: An Overview

Hai Long, Heng Yin*, Ling Wang**, Qianjin Lu**

***The Second Xiangya Hospital, Central South University, Changsha, China; **The Third Hospital of Changsha, Changsha, China**

OUTLINE

The Epigenetics of Autoimmunity. http://dx.doi.org/10.1016/B978-0-12-809912-4.00001-5

1.1 INTRODUCTION

Autoimmunity refers to the dysregulated immune state that causes loss of immunological tolerance to self-antigens and damage to normal cells and tissues. This aberrant state presents clinically as a wide spectrum of autoimmune disorders, which are characterized by the presence of autoreactive immune cells and (or) the development of autoantibodies. These disorders include not only organ-specific autoimmune diseases such as primary biliary cirrhosis (PBC), type 1 diabetes mellitus (T1DM), and Graves' disease, but also various systemic autoimmune diseases such as systemic lupus erythematosus (SLE), primary Sjögren's syndrome (pSS), and rheumatoid arthritis (RA).

The reasons for the loss of immunological tolerance to self-antigens, and the mechanisms underlying autoimmunity onset and development, remain largely unknown. Both genetic predisposition and environmental factors (i.e., nutrition, infection, and ultraviolet exposure) are considered to contribute to the pathogenesis of autoimmune diseases [1]; however, how each of these factors leads to autoimmunity remains a challenging question. The emergence of epigenetic studies in recent decades has provided us with a new perspective for understanding these complex mechanisms and has shed light on a new era of autoimmunity research.

Epigenetics is the study of potentially heritable changes in gene expression and function that do not involve alterations of the original DNA nucleotide sequence. Epigenetic regulation (i.e., DNA methylation, histone modification, and noncoding RNAs) plays a crucial role in various life processes, including cellular differentiation, growth, development, ageing, and the immune response [2]. Epigenetics provides a better understanding of how environmental triggers can alter gene expression and disturb immune homeostasis, and is implicated in the pathogenic mechanisms of many complex diseases (i.e., cancers and autoimmune diseases). For example, latent infection with Epstein-Barr virus, which is a possible trigger of autoimmune diseases such as SLE, RA, and multiple sclerosis, plays a role in the break of self-tolerance through the expression of latent viral oncoproteins and nontranslated RNAs under epigenetic control (e.g., DNA methylation and histone modification) in germinal center B cells and lymphoblastoid cells [3]. Another example is the protective effect of vitamin D supplementation, which reduces the risk of developing multiple sclerosis. Vitamin D affects multiple signaling and metabolic pathways that are critical for T-cell activation and Th1 and Th17 differentiation, partly via the modulation of epigenetic mechanisms such as DNA methylation, histone modification, and noncoding RNAs [4]. Unlike gene mutations and chromosome anomalies, epigenetic abnormalities are reversible and thus have the potential to be "corrected." The number of research publications regarding epigenetic modifications in autoimmune diseases has increased exponentially in recent decades, highlighting the attractiveness of epigenetic research in this field [5–8].

A brief interpretation of how genetic studies have failed to completely elucidate the pathogenesis of autoimmune diseases will be provided to explain the significance of epigenetics in autoimmunity. SLE will be used as a typical example, and an overview of the epigenetic mechanisms implicated in the pathogenesis of autoimmunity, and how these different epigenetic mechanisms interplay with each other and with genetic variants and environmental factors will be presented.

1.2 LIMITATIONS OF GENETICS IN AUTOIMMUNITY RESEARCH

1.2.1 Genetics of Autoimmunity: Advances and Disappointments

Genetic susceptibility to an autoimmune disease is generally attributable to multiple genetic variants, each of which has a moderate-to-low effect size; however, strong associations between human leukocyte antigen (HLA) genes and autoimmune diseases have been revealed [9]. Some non-HLA susceptibility genes (i.e., *IL23R*, *OLIG3/TNFAIP3*, *IL2RA* [10], *IRF8* [11,12], *IRF5*, *PTPN22*, *BANK1*, *ICAM3* [13], and *STAT4* [14]) are also known to be shared by different autoimmune diseases. These shared genetic loci may point to common pathways that are potentially implicated in the pathogenesis of autoimmune diseases.

A number of genome-wide association studies (GWAS), which compare all single nucleotide variants by sequencing large cohorts of individuals with and without the disease, have been performed for various autoimmune diseases [12]. For example, several GWAS for SLE variants have been reported in different ethnic populations, and at least 40 common risk loci have been definitively linked to SLE susceptibility in case–control studies [9,15,16]. Some of these studies have highlighted strong associations with genetic variants in HLA genes, while others have identified non-HLA susceptibility loci that are located within or near genes that are functionally implicated in the immune system (i.e., type I interferon (IFN) signaling, B- and T-cell signaling, or the clearance of cellular debris) [9].

However, one disappointing aspect of GWAS is that few have localized the actual causal variants [12] for diseases such as SLE, diabetes, and virtually all other autoimmune diseases [9,12,17–27]. Therefore the biological or functional effects of these genetic variants remain unidentified, undemonstrated, or merely putative using data from known canonical pathways. Except for a small number of genes that have been functionally characterized (i.e., *TNFAIP3*) [28,29], whether and how these polymorphisms contribute to the pathogenesis of autoimmune diseases remains unanswered.

Another defect of GWAS is that the identified risk variants, according to the rationale of GWAS analysis, are more likely to be associated with the variant frequency in the test population than in the disease itself [30]. Indeed, differences in the genetic variant findings are evident within distinct ethnic populations. For example, studies in type 2 diabetes have shown that several variants, or different variants at the same loci, are associated with the same disease in different ethnic populations, including Europeans, East Asians, South Asians, and Hispanics. Therefore it may be possible to use transethnic mapping of genetic variants as a strategy to overcome this limitation [31]. However, similar work is missing in the field of systemic autoimmune diseases, partly due to the rarity of these diseases in comparison to type 2 diabetes. In addition, efforts to determine the association between genetic variants and subphenotypes of autoimmune diseases have also been limited [32,33]. One reason why this is so difficult is the need for large sample sizes to account for the relatively rarity of these subphenotypes.

1.2.2 The Involvement of Nongenetic Factors in Autoimmunity

Twin studies have been successfully used to research many autoimmune diseases, including SLE. The concordance of disease occurrence in monozygotic twins, who have identical genetic information, is considerably below 100% for most autoimmune diseases [34,35]. This highlights an essential role for nongenetic factors (i.e., environmental influences and epigenetic mechanisms) in the pathogenesis of autoimmune diseases. Evidence suggests that epigenetic mechanisms are important intermediate regulators as the immune system responds to various environmental stimuli (i.e., ultraviolet exposure, drugs, and oxidative stress) [7]. Investigations into the roles of these epigenetic mechanisms in the pathogenesis of autoimmune diseases may provide a better understanding of these complex diseases.

1.3 EPIGENETICS AND ITS ROLE IN THE PATHOGENESIS OF AUTOIMMUNITY

The major epigenetic machinery (e.g., DNA methylation, histone modification, and noncoding RNAs) regulate the chromatin state, control gene expression activity, and are implicated in various life processes [36]. These epigenetic mechanisms are critical to the human immune system because of the high plasticity of the transcriptome in the various immune cells, each of which plays a role in the response to environmental changes and the protection of immune homeostasis [2]. The basics of these different epigenetic mechanisms will be introduced here, and an overview of their significance in the pathogenesis of autoimmune diseases will be presented using SLE as an example.

1.3.1 DNA Methylation

1.3.1.1 DNA Methylation and Gene Transcription

DNA methylation is the addition of a methyl group to the 5′ carbon in the pyrimidine ring of a cytosine residue, and it typically occurs in cytosine–guanine dinucleotides (CpG) in mammalian DNA. It is considered as a relatively stable epigenetic mark, and is present in many eukaryotic organisms. CpG dinucleotides are present throughout the whole DNA sequence and represent approximately 4% of the human genome; however, DNA methylation research has mainly focused on CpG sites located within the 5′ upstream promoter regions of genes. Methylation of CpG sites in a promoter region generally blocks the accessibility of transcriptional activators, thereby inhibiting gene transcription and serving as a repressive "lock." Conversely, an unmethylated state at the promoter permits transcription.

1.3.1.2 Reversible Regulations of DNA Methylation

Methylation is catalytically mediated by DNA methyltransferases (DNMTs) (Fig. 1.1), mainly DNMT1, DNMT3a, and DNMT3b. DNMT1 serves as a maintenance methyltransferase, recognizing and copying the preexisting methylation profile of a DNA strand to a new strand during DNA replication in the S phase of the cell cycle; DNMT3a and DNMT3b induce de novo methylation [7].

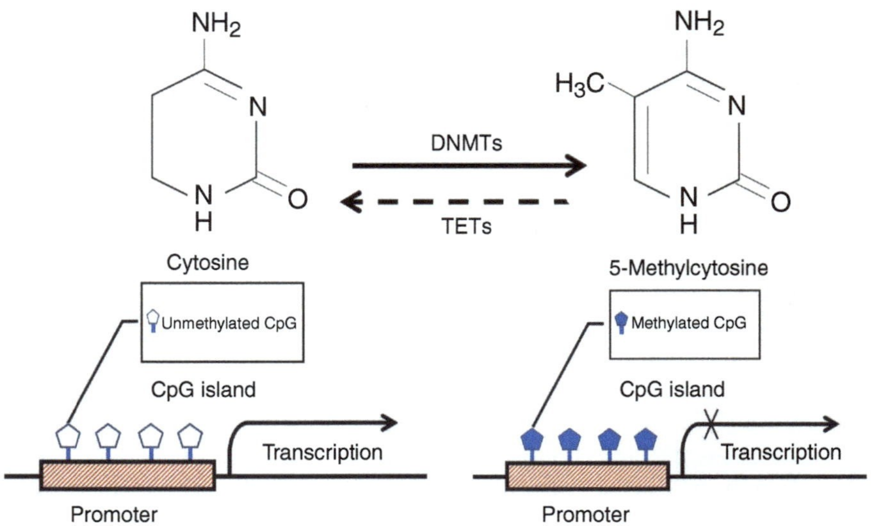

FIGURE 1.1 **DNA methylation and demethylation processes and their influences on gene expression.** In the methylation process, DNA methyltransferases (*DNMTs*) catalyze the addition of a methyl group to the 5′ carbon site of the cytosine within the CpG dinucleotide, while ten-eleven translocation (*TET*) family proteins mediate the demethylation process, which is more complex and not completely understood. Heavily methylated CpG islands in the promoter region of a gene generally lead to silencing of gene expression.

Although relatively stable, DNA methylation can be reversed by certain mechanisms (Fig. 1.1). It is generally accepted that demethylation (the removal of the methyl group from a methylated cytosine) can occur in either an active or passive manner [37]. Passive demethylation occurs during DNA replication when the maintenance methylation machinery (mediated by DNMT1) is halted and the copying of the methylation profile from the original methylated DNA strand fails. The methylation marks are thus "diluted" and lost from one cell division to the next. The mechanisms of active demethylation remained largely unknown until the identification of 5-hydroxymethylcytosine (5-hmC). Ten-eleven translocation (TET) family proteins, which are considered to be key enzymes in the facilitation of active demethylation, catalyze the transformation of 5-methylcytosine (5-mC) to 5-hmC, which then undergoes a series of steps to return it to an unmethylated cytosine and thus completes the active demethylation process [38]. The TET family members that may participate in active demethylation include TET1, TET2, and TET3.

1.3.1.3 DNA Hypomethylation in SLE

SLE has been used as a classical prototype of all systemic autoimmune diseases and has attracted intensive global research interest. Although the exact cause and pathogenesis of SLE remain largely unknown, cross-talk between genetic predisposition and environmental stimuli is generally considered to be implicated. Accumulating evidence has clearly shown the key role of DNA methylation changes in the onset and development of this autoimmune disorder [5] (Table 1.1).

One of the key epigenetic events in lupus is the hypomethylation at specific DNA regulatory regions, usually the upstream promoter regions. This leads to the overexpression of autoimmune-associated genes, which are normally suppressed by DNA methylation, in lupus $CD4^+$ T cells, and contributes to disease pathogenesis and progression. These methylation-sensitive immune-related genes include *CD40L*, a critical costimulatory molecule expressed on T cells to help B cells produce IgG [61]; *CD11a*, an integrin involved in costimulation and cellular adhesion and a key component of leukocyte function-associated antigen 1 (LFA-1, also known as integrin $\alpha L\beta 2$, or CD11a/CD18); *CD70*, a costimulatory molecule and a ligand of CD27, which contributes to B cell overstimulation for IgG production [49,50]; perforin, a cytolytic protein implicated in the disruption and lysis of target cell membranes [51–53]; killer cell immunoglobulin-like receptors [55,56]; and the catalytic subunit α isoform of protein phosphatase 2A (PP2Acα) [57]. Among these genes, *CD40L* is located on the X chromosome. Hypomethylation of this gene in the silenced X chromosome in female patients may lead to its reactivation, which at least partially explains the female predominance in the overall incidence of lupus [61].

The use of high-throughput sequencing technologies has further confirmed that hypomethylation is a dominant genome-wide feature in SLE $CD4^+$ T cells [41,47]. Interestingly, abundant hypomethylation events have also been identified in naive $CD4^+$ T cells from patients with SLE; however, most of these hypomethylated genes are not overexpressed in naive $CD4^+$ T cells but are epigenetically "poised" for rapid expression upon stimulation. A number of IFN-regulated genes account for the majority of these genes [58].

DNA methylation has been extensively investigated, and 5-mC is considered as "the fifth base" in the human genome; however, recent advances in epigenetic modification research have highlighted the widespread existence and biological significance of "the sixth base," 5-hmC, which is a key intermediate in active demethylation. The genome-wide landscape of 5-hmC in SLE $CD4^+$ T cells has also been depicted using next-generation sequencing technologies. Briefly, 2748 genes with increased 5-hmC levels in their promoter regions were identified in SLE $CD4^+$ T cells. These

TABLE 1.1 Changes of Global and Promoter DNA Methylation Identified in Different Cell Populations From Patients With SLE

Disease	Cell type	Changes in DNA methylation	References
SLE	Peripheral whole blood cells	IFN-induced protein 44-like (IFI44L) methylation ↓ in SLE	[39]
SLE	Peripheral whole blood cells	FOXP3 TSDR methylation ↑ in active SLE	[40]
SLE	Peripheral blood leukocytes	Global ↓	[41]
SLE	PBMCs	ERα ↓	[42]
SLE	T cells	X-chromosome genes (*CXCR3*, *OGT*, miR-98, let-7f-2*, miR 188-3p, miR-421, and miR-503) ↓	[43,44]
SLE	T cells	IL-4↓, IL-6 ↓	[45,46]
SLE	$CD4^+$ T cells	Global ↓	[47]
SLE	$CD4^+$ T cells	IFN-regulated genes ↓	[48]
SLE	$CD4^+$ T cells	CD11a ↓, CD70 ↓, CD40L ↓, perforin ↓	[49–54]
SLE	$CD4^+$ T cells	PP2Acα ↓ and KIR2DL4 ↓	[55–57]
SLE	$CD4^+$ T cells	IL-10 ↓, IL-13 ↓	[45,46]
SLE	Naïve $CD4^+$ T cells	IFN-regulated genes ↓	[58]
SLE	Naïve $CD4^+$ T cells	MIR886 ↓, TRIM69 ↓, RNF39 ↑ in SLE patients with a history of malar rash, and RHOJ ↓ in lupus patients with a history of discoid rash	[59]
SLE	Naïve $CD4^+$ T cells	CHST12 (locus: CG10152449) ↓ in SLE patients with renal involvement	[60]
SLE	B cells	IFN-regulated genes ↓	[48]
SLE	Monocytes	IFN-regulated genes ↓	[48]

IFN, Interferon; *PBMCs*, peripheral blood mononuclear cells; *SLE*, systemic lupus erythematosus; *TSDR*, T_{reg}-specific demethylated region.

genes were enriched in neurotrophin, WNT, MAPK, calcium, and mTOR signaling pathways [62]. When combining these data with gene expression data, 131 genes with increased promotor levels of 5-hmC were found to be overexpressed in SLE $CD4^+$ T cells. These included several immune-related genes (i.e., SOCS1, NR2F6, and IL15RA) [62], which were validated by chromatin immunoprecipitation (ChIP)-quantitative PCR.

1.3.1.4 Causative Mechanisms of DNA Hypomethylation in SLE

It remains largely unknown how widespread DNA hypomethylation develops in SLE $CD4^+$ T cells, although several studies have proposed possible mechanisms that could at least partially account for this phenomenon. One common end pathway of the different mechanisms that lead to DNA demethylation in lupus is the inhibition of DNMT1, which is the maintenance methyltransferase responsible for copying the methylation profile of a DNA strand to a new strand during replication. It can be inferred that the loss of DNMT1 function will stop the "heritability" of methylation marks during the cell cycle and thus result in a gradual "dilution" and eventual loss of the methylation marks. Indeed, DNMT1expression is decreased in lupus T cells [63], and inhibiting this enzyme in proliferating cells results in passive DNA hypomethylation [64].

Several mechanisms have been proposed to explain how DNMT1 is inhibited and DNA is demethylated in lupus CD4$^+$ T cells. Among the most well-documented is defective signaling via the T cell ERK pathway [63,65–67] and its upstream regulator, protein kinase C delta 43,68,69]. Other proposed mechanisms include: (1) inhibition of DNMT1 by the overexpression of microRNAs (i.e., miR-148a and miR-126) [70,71]; (2) dysregulation of transcription factors [i.e., regulatory factor X 1 (RFX1)], which can reduce the recruitment of DNMT1 and histone deacetylase 1 (HDAC1) to gene promoters [72,73]; and (3) enhanced expression of growth arrest and DNA damage-induced 45alpha, which promotes DNA demethylation [74,75]. In addition, TET family proteins, which catalyze the conversion of 5-mC to 5-hmC, are considered to be key enzymes in active DNA demethylation. Zhao et al. [62] observed increased 5-hmC levels in the genomic DNA of CD4$^+$ T cells from patients with SLE compared to healthy controls, which was also accompanied by an upregulation of *TET2* and *TET3*. Whether TET family protein-mediated active DNA demethylation contributes to DNA hypomethylation in SLE CD4$^+$ T cells warrants further study.

1.3.1.5 DNA Hypomethylation Contributes to Autoimmunity in SLE

Numerous studies have confirmed the fundamental role of T-cell DNA hypomethylation in the pathogenesis of lupus [76,77]. It was shown that experimentally inducing T-cell DNA demethylation could alter gene expression levels and cause T-cell autoreactivity in vitro, or a lupus-like autoimmune phenotype in vivo [78,79]. These features of demethylated T cells are similar to those seen in T cells from patients with SLE, which have decreased expression of *DNMT1*, impaired DNA methylation, and increased expression of lupus-associated genes (i.e., *CD11a*) that contribute to T-cell autoreactivity [78,79].

Three distinct and well-characterized DNA methyltransferase inhibitors (i.e., 5-azacytidine, procainamide, and hydralazine) were used in the abovementioned experimental studies. When treated with 5-azacytidine or procainamide, normal murine T lymphocytes gained augmented expression of the methylation-sensitive gene *LFA-1* (*CD11a/CD18*) and the ability to spontaneously lyse autologous macrophages, a similar property to a T-cell subset found in patients with active SLE [79,80]. Injecting these treated (but not untreated) T cells into syngeneic mice induced the production of anti-ssDNA, anti-dsDNA, and anti-histone autoantibodies. It also produced various clinical manifestations, including severe immune complex glomerulonephritis, pulmonary alveolitis, central nervous system abnormalities, and bile duct proliferation with periportal inflammatory-cell infiltration resembling PBC. These mouse model studies demonstrated that demethylation of normal T cells using DNA methyltransferase inhibitors was sufficient to cause T-cell autoreactivity and an autoimmune disease that resembled human lupus and autoimmune liver disease [80,81]. Consistently, treating normal human T cells with methyltransferase inhibitors also resulted in DNA demethylation and overexpression of lupus-associated genes, which induced T-cell autoreactivity and the overstimulation of autologous B cells [49,54,77].

Procainamide and hydralazine also cause drug-induced lupus. Similar murine studies have confirmed that procainamide and hydralazine, but not their structural analogs (i.e., N-acetylprocainamide, phthalazine, or hydroxyurea), can inhibit T-cell DNA methylation, increase *LFA-1* expression, and induce autoreactivity. Adoptive transfer of these drug-induced autoreactive T cells can also cause a lupus-like disease in recipient mice [82].

1.3.1.6 DNA Methylation Biomarkers in SLE

1.3.1.6.1 Biomarkers for the Diagnosis of SLE

SLE is a systemic autoimmune disease with significant heterogeneity in terms of both its clinical manifestation and its autoantibody profile, therefore biomarkers with good specificity and sensitivity are of great significance for the clinical diagnosis. However, the conventional serological biomarkers for SLE diagnosis have numerous limitations. For example, antinuclear antibody tests have a perfect sensitivity (almost 100%) for the diagnosis of SLE but an unsatisfactory specificity (65%) [83]; anti-Smith has a very high specificity (99%) but a low sensitivity (25%–40%) [84]; and antidouble-strand DNA (dsDNA) occurs in only 50%–60% of patients with lupus at some point during the course of their disease [85,86].

One advance is the successful identification of IFN-induced protein 44-like (IFI44L) promoter methylation in peripheral blood cells. This can now be used as a highly sensitive and specific biomarker for the diagnosis of SLE, and is superior to that of other currently available tests. Zhao et al. [39] performed this study in a Chinese population using a discovery set containing 377 patients with SLE, 358 healthy controls, and 353 patients with RA. These results were further validated in a Chinese cohort consisting of 529 patients with SLE, 569 healthy controls, 429 patients with RA, and 199 patients with pSS. Significant hypomethylation at two CpG sites within the IFI44L promoter (site one at chr1:79 085 222 and site two at chr1:79 085 250; cg06872964) was identified in peripheral blood cells from patients with SLE but not in patients with RA, pSS, or healthy controls. With the optimized cutoff methylation level for each site, methylation of sites one and two within the IFI44L promoter showed sensitivities of 93.6% and 94.1%, respectively, and specificities of 96.8% and 98.2%, respectively, thus suggesting that it had almost perfect properties for use as an SLE biomarker [39]. The IFI44L promoter methylation marker has been further validated in a European cohort consisting of 615 SLE patients and 781 healthy controls [39].

1.3.1.6.2 Biomarkers for the Evaluation of Disease Activity in SLE

The T_{reg}-specific demethylated region (TSDR) is a highly conserved locus on the *FOXP3* gene, which is fully demethylated in natural T_{reg} cells but methylated in effector T cells. It was reported that the use of high resolution melt-polymerase chain reaction to detect *FOXP3* TSDR methylation was a reliable and easy method to predict natural T_{reg} cell levels in peripheral blood samples from lupus patients [40]. The methylation level of *FOXP3* TSDR was significantly higher in patients with active SLE compared to healthy controls and patients with inactive SLE, and directly correlated with the disease activity [40].

The aforementioned IFI44L promoter methylation level may also act as a potential biomarker to predict SLE disease activity, as this gene promoter exhibits significantly increased methylation levels during remission than during the active stage of lupus [39]. In addition, GWAS have identified many differential-methylated loci in T and B lymphocytes from SLE patients compared to those from healthy individuals, with the methylation status of several genes (i.e., *RAB22A*, *STX1B2*, *LGALS3BP*, *DNASE1L1*, and *PREX1*) [47] correlating with disease activity; however, further selection and validation are needed.

1.3.1.6.3 Methylation Biomarkers Associated With Specific Organ Involvement of SLE

A recent study examining the individual DNA methylomes of $CD4^+$ T cells from 42 patients with SLE using high-throughput sequencing technologies identified significant differences in the methylation patterns within subphenotypes of lupus [87]. The aforementioned IFI44L promoter methylation also showed significantly lower levels in SLE patients with renal damage compared to those without [39].

To identify methylation loci as predictors of specific organ involvement in lupus, Coit et al. [60] examined genome-wide DNA methylation profiles in naive $CD4^+$ T cells from 28 SLE patients with renal involvement, 28 patients without a history of renal involvement, and 56 age-, sex-, and ethnicity-matched controls. Upon comparison, 191 CpG sites and 121 genes were differentially methylated in lupus patients with renal involvement compared to those without a history of renal involvement. Of the CpG sites only hypomethylated in patients with renal involvement, CG10152449 in *CHST12* had a sensitivity of 85.7% and a specificity of 64.3% for stratifying lupus patients for renal involvement [60]. Other differentially methylated genes included *TNK2*, which is involved in cell trafficking and tissue invasion; the phosphatase gene *DUSP5*, which dephosphorylates and inhibits the ERK signaling pathway; and the type-I IFN master regulator gene *IRF7* [60].

Epigenetic markers associated with skin lesions in SLE were also studied using a comparison of the DNA methylome in naive $CD4^+$ T cells [59]. The results showed that 36 and 37 unique differentially methylated regions were associated with malar rash and discoid rash, respectively. Hypomethylation of *MIR886* and *TRIM69*, and hypermethylation of *RNF39* were specific to lupus patients with a history of malar rash, whereas hypomethylation of *RHOJ* was specific to lupus patients with a history of discoid rash [59].

1.3.1.7 *DNA Hypomethylation as a Potential Therapeutic Target in SLE*

Corticosteroids and immunosuppressants have been used as the standard therapeutic treatments for lupus for many years. Advancements in biological agents for the therapeutic treatment of SLE remain satisfactory. Given that DNA hypomethylation plays a crucial role in the pathogenesis of SLE, it may be a potential therapeutic strategy to treat lupus by reversing the aberrant DNA demethylation. Certain nutrients (i.e., folate and vitamin B12) can act as methyl donors in one-carbon metabolism, a process that produces the universal methyl donor, S-adenosylmethionine. Thus this might provide a way to enhance DNA methylation in vivo. It was reported that increased dietary intake of folate elevated both global and p16 promoter methylation levels in the colon mucosa of aged mice [88]. Strickland et al. demonstrated that the addition of dietary micronutrients containing methyl donors or methyl cofactors (i.e., folate, vitamin B12, and choline) could increase DNA methylation levels and ameliorate lupus in a transgenic murine model [89]. Conversely, a reduced intake of folate was associated with global DNA hypomethylation in animal models [90].

1.3.2 Histone Modifications

1.3.2.1 *The Regulation of Histone Modifications*

Histones are highly conserved proteins that form nucelosomes with DNA molecules. Each nucleosome consists of a histone core containing two copies of histones H2A, H2B, H3, and H4, with 146 base pairs of genomic DNA wrapped around its outer surface. A histone modification is the posttranscriptional modification of a specific amino acid in the polypeptide sidechain of a histone that protrudes from the nucleosome. Histone modifications include acetylation, methylation, ubiquitination, and sumoylation, and each has a different function and biological significance. Acetylation and methylation are the most extensively studied histone modifications to date [91]. Histone modifications affect the local chromatin conformation, alter its accessibility, and influence gene transcription. Histone acetylation, which generally occurs at lysine residues within histone tails, is associated with an open chromatin state and increased gene transcription. This is largely because acetylation changes the overall charge of the histone tail from positive to neutral, thus decreasing the interaction of negatively charged DNA with histones and producing relaxed DNA that is accessible to transcription factors [92]. The impact of histone methylation on gene transcription depends upon the residues involved; for instance, the addition of three methyl groups to lysine four of histone H3 (H3K4me3) is a canonical chromatin mark of active transcription, while the trimethylation mark of histone H3 lysine 27 (H3K27me3) represses gene transcription [91].

Both acetylation and methylation are reversible and are catalyzed by enzymes with opposing functions [93]. Acetylation is catalyzed by histone acetyltransferases (HATs), including PCAF, Tip60, and p300/CBP. Histone deacetylation (the removal of acetyl groups from acetylated lysine residues in histone tails) is mediated by HDACs, including HDAC1 and sirtuins (i.e., SIRT1–7). Similarly, histone methylation and demethylation are catalyzed by histone methyltransferases (HMTs) and histone demethylases (HDMs), respectively. The reversibility of histone modifications provides an opportunity to manually modulate these epigenetic mechanisms.

1.3.2.2 *Aberrant Histone Acetylation in SLE*

Early studies of histone modifications in lupus T cells reported global site-specific histone H3 and H4 hypoacetylation in both splenocytes from MRL/lpr lupus mice [94] and $CD4^+$ T cells from SLE patients [95]. Histone H3K18 deacetylation, which is mediated by cAMP response element modulator alpha recruiting HDAC1 to the IL-2 promoter, has also been identified as a key epigenetic mechanism for silencing the *IL-2* gene in SLE T cells [6,96]. Several HDAC inhibitors (HDACi) (i.e., Trichostatin A (TSA), suberoylanilide hydroxamic acid (SAHA), and 4-phenylbutyric acid) have since been tested as therapies to reverse histone hypoacetylation and treat lupus in animal models [94,97,98]. Indeed, treating MRL/lpr mice with TSA or SAHA restored histone acetylation in vivo, inhibited the aberrant upregulation of proinflammatory cytokines, and significantly ameliorated the disease activity [94,97–99]. More recently, a specific class I and II HDACi (ITF2357) was shown to decrease serum inflammatory cytokines, reduce the Th17 phenotype, increase the percentage of T_{reg} cells, increase *Foxp3* acetylation, and ameliorate renal disease in lupus-prone NZB/W mice [100]. Administration of ITF2357 also favored β-cell survival during inflammatory conditions in a mouse model of T1DM [101].

Several enzymes that regulate histone acetylation and deacetylation (i.e., HATs and HDACs, respectively) were also shown to be abnormally expressed in $CD4^+$ T cells from patients with active SLE. Among these enzymes, SIRT1 was found to be overexpressed while CREBBP, P300, HDAC2, and HDAC7 were downregulated [95]. Inhibiting the aberrant overexpression of SIRT1 in MRL/lpr lupus-like mice using RNA interference transiently elevated global histone H3 and H4 acetylation levels in $CD4^+$ T cells, decreased serum anti-dsDNA antibodies, and reduced renal damage and IgG deposition [102]. It is plausible that histone hypoacetylation may be implicated in the pathogenesis of lupus, although the specific loci and detailed mechanisms remain to be elucidated.

However, there is also evidence that does not support the pathogenic role of histone hypoacetylation in lupus, although this data does reflect the acetylation status of $CD4^+$ T cells in lupus. Enhanced histone acetylation has been associated with the overexpression of cytokine genes (i.e., *TNFα* and *IL-17*) in lupus [103,104]. Furthermore, the administration of the triacetylated histone H4 peptide, but not the nonacetylated H4 peptide, into lupus-prone mice before the onset of the disease accelerated the disease progression and enhanced the mortality [105]. A lupus-associated antihistone autoantibody (KM-2), isolated from both lupus-prone mice and patients with SLE, reacted

TABLE 1.2 Histone Modifications Identified in Different Cell Populations From Patients With SLE

Disease	Cell type	Changes in histone modification	References
SLE	T cells	H3K18 deacetylation in the IL-2 promoter	[6,96]
SLE	T cells	Increased H3 acetylation at the IL-17 locus	[103,104]
SLE	$CD4^+$ T cells	Global site-specific H3 and H4 hypoacetylation and global H3K9 hypomethylation	[95]
SLE	Monocytes	Global H4 hyperacetylation	[107]
SLE	Monocytes	Increased histone acetylation of the *TNFα* gene	[103,104]
SLE	Monocytes	Narrow peaks of H3K4me3 enriched in housekeeping genes, and broader peaks of H3K4me3 enriched in immune-response genes	[108]
SLE	Apoptotic cells	Apoptosis-induced acetylation of histone H2B	[106]

SLE, Systemic lupus erythematosus.

more strongly with triacetylated H4 and hyperacetylated apoptotic histones than with the nonacetylated H4 peptide and normal histones [105]. Similarly, another lupus-derived autoantibody (LG11-2), which targets the N-terminal of histone H2B, exhibited enhanced reactivity with apoptotic and hyperacetylated H2B compared to normal H2B [106]. One possible explanation for these phenomena might be that apoptosis-induced acetylation of nucleosomes may act as important epitopes for autoantibodies in the early phase of the antichromatin autoimmune response of lupus.

A recent study that examined genome-wide histone H4 acetylation (H4ac) levels using chromatin immunoprecipitation arrays revealed significant hyperacetylation in SLE monocytes. Monocytes from SLE patients contained 179 genes with significantly elevated promoter H4ac levels compared to healthy controls. Network analysis of these genes identified ERK, p38, NFκB, CREB1, and IFNα as significant nodes in the three highest-scoring networks. Specifically, almost 63% of genes with increased H4ac levels had upstream IFN regulatory factor 1 (IRF1) binding sites and had the potential to be regulated by IRF1, which is consistent with previous knowledge of type-I IFN hyperresponsiveness in lupus [107]. The histone modifications implicated in the pathogenesis of lupus are summarized in Table 1.2.

1.3.2.3 *Abnormalities of Histone Methylation in SLE*

Altered histone methylation has been reported in both lupus patients and mouse models; for example, global histone H3K9 hypomethylation was reported in SLE $CD4^+$ T cells [95], and global site-specific histone hypermethylation was identified at histones H3 and H4 in splenocytes from MRL-lpr/lpr mice [94]. However, no difference in global H3K4 methylation (a dominant histone mark associated with active chromatin) was observed in samples from either lupus patients or lupus mouse models [94,95]. Understanding of the histone H3K4 methylation patterns in SLE has advanced following an integrative analysis that combined genome-wide H3K4me3 enrichment and gene transcription data, which were obtained by ChIP sequencing and RNA sequencing, respectively, from primary SLE monocytes. H3K4me3 enrichment was identified at 8399 transcription start sites (TSS), with narrow peaks of H3K4me3 enriched in housekeeping genes, and broader peaks of H3K4me3 expanding immediately upstream or downstream of TSS predominantly in immune response genes [108]. Furthermore, these data showed that the

H3K4me3 mark was consistently associated with gene transcription activity, with every 1% increase of H3K4me3 in the downstream region of TSS leading to an average increase of transcription of ~1.5% [108].

Aberrant expression of HMTs and HDMs may contribute to the imbalance of histone methylation and demethylation seen in the pathogenesis of lupus. Decreased expression of HMTs (i.e., SUV39H2 and EZH2) was observed in $CD4^+$ T cells from patients with SLE [95]. Several HDMs also showed altered expression levels in $CD4^+$ T cells of MRL/lpr lupus mice compared to controls, among which JMJD3 was found to be upregulated [109]. A ChIP-on-chip analysis identified hematopoietic progenitor kinase 1 (*HPK1*, also called *MAP4K1*) as a gene with increased levels of the repressive histone mark, H3K27me3, and decreased transcription in SLE $CD4^+$ T cells, which contributed to T cell overactivation and B cell overstimulation in lupus [110]. Further mechanistic studies revealed that reduced binding of the HDM, JMJD3, but not a change in the binding of the HMT, EZH2, to the HPK1 promoter caused the enhanced H3K27me3 enrichment and gene expression loss [110].

1.3.3 Noncoding RNA

Noncoding RNAs are also key epigenetic modulators. They are transcripts without the ability to code for proteins, and include microRNAs (miRNAs), long noncoding RNAs (lncRNAs), and circular RNAs. miRNAs are the most intensively studied noncoding RNAs. They are small (generally 21-25 nucleotides) single-stranded noncoding RNAs that negatively regulate gene expression by destabilizing or cleaving mRNA or inhibiting translation. Over 2000 miRNAs are registered in human miRNA databases, and approximately 50% of these regulate one-third of the transcriptome, thus controlling diverse biological processes (i.e., immune responses) [111,112]. Emerging evidence has revealed that these noncoding RNA molecules play crucial roles in the regulation of immunological functions and autoimmunity.

1.3.3.1 miRNAs and SLE

1.3.3.1.1 Aberrant Expression of miRNAs Implicated in Pathogenesis of SLE

Altered miRNA expression was identified in different immune cells in patients with SLE, and many of these cells have the potential to play a role in the disease pathogenesis. In peripheral blood mononuclear cells (PBMCs) from SLE patients, miR-155 and miR-146a expression was downregulated [113,114]. Interestingly, transfection of miR-155 and miR-146a into SLE PBMCs rescued IL-2 secretion [113] and inhibited the excessive production of IFN-α and IFN-β [114]. However, it remains unclear from which subset of PBMCs these two miRNAs originate and whether the ratio of cell subsets in PBMCs changes due to the two altered miRNAs.

Studies into specific immune cells have also found miRNA expression changes that are associated with SLE pathogenesis. For example, upregulation of miR-21, miR-148a, miR-126, and miR-29b in circulating $CD4^+$ T cells from SLE patients can directly or indirectly suppress *DNMT1* and thus contribute to DNA hypomethylation in these cells [70,71,115]. miR-21 also inhibited *PDCD4* expression and increased cell proliferation and *CD40L* and *IL-10* expression in lupus T cells [116]. In addition, miR-142 [117] and miR-31 [118] regulate T cell function by inhibiting *IL-4, IL-10, CD40L,* and *ICOS* expression and stimulating IL-2 production, respectively. Recent studies into the therapeutic mechanisms of mycophenolic acid as a treatment for SLE showed that it ameliorates T cell autoreactivity by upregulating the expression of miR-142-3P/5P and miR-146a [119].

TABLE 1.3 Dysregulated miRNAs, and Their Potential Target Genes, Identified in Different Cell Populations From Patients With SLE

Disease	Cell type	miRNAs	Target genes	References
SLE	PBMCs	Decreased expression of miR-155	*PP2Ac*	[113,114]
SLE	PBMCs	Decreased expression of miR-146a	*IFN-α* and *IFN-β*	[113,114]
SLE	T cells	Overexpression of miR-21	*PDCD4*	[116]
SLE	T cells	Decreased expression of miR-31	*RhoA*	[118]
SLE	$CD4^+$ T cells	Overexpression of miR-21, miR-148a, miR-126, and miR-29b	*DNMT1* (directly or indirectly)	[70,71,115]
SLE	$CD4^+$ T cells	Decreased expression of miRNA-142-3p/5p	*SAP*, *CD84*, and *IL-10*	[117]
SLE	B cells	Overexpression of miR-30a	*Lyn*	[120]
SLE	B cells	Decreased expression of miR-1246	*EBF1*	[123]

PBMCs, Peripheral blood mononuclear cells; *SLE*, systemic lupus erythematosus.

In lupus B cells, miR-30a expression was augmented and correlated negatively with Lyn, a key negative regulator of B cell activation [120]. miR-155 and miR-181b negatively modulate activation-induced cytidine deaminase, which regulates B cell antibody diversification [121,122]. Recent findings also revealed a role of miR-1246, which specifically targets *EBF1* mRNA, in the regulation of B cell activation in the pathogenesis of SLE [123]. The miRNAs implicated in the pathogenesis of lupus are summarized in Table 1.3.

1.3.3.1.2 miRNAs: Potential Biomarkers in Blood Cells, Serum, Urine, and Urinary Exosomes

Compared to DNA methylation and histone modifications, miRNAs are easier to detect in laboratory settings and may serve as ideal epigenetic biomarkers for autoimmune diseases. The expression level of miR-146a decreased in PBMCs from SLE patients, and was inversely correlated with disease activity [114]. miR-15a expression in regulatory B cells positively correlated with anti-dsDNA autoantibody serum levels in an IFN-accelerated lupus mouse model, although it remains to be validated in the human scenario [124]. Other studies observed aberrant serum levels of miR-200a, miR-200b, miR-200c, miR-429, miR-205, miR-192, miR-126, miR-16, miR-451, miR-223, miR-21, and miR-125a-3p [125,126] in patients with SLE, which correlated with the disease activity. These miRNAs in blood cells or serum may be useful as predictors of SLE disease activity.

There is also evidence that urinary levels of miR-146a and miR-155 may be used as noninvasive biomarkers for the diagnosis, disease activity, and therapeutic response of lupus. Specifically, urinary miR-146a and miR-155 levels significantly correlate with the estimated glomerular filtration rate and with proteinuria and the SLE disease activity index, respectively [127].

Recent studies have highlighted the clinical values of urinary exosomes, which are microvesicles released by epithelial cells facing the urinary space and contain the most urinary microRNA molecules. Examining exosomal miRNAs in the urine, which is also a noninvasive method, might be a promising technique for predicting renal dysfunction and injury in patients with autoimmune diseases. Evidence shows that the presence of exosomal miR-146a in the urine can help to discriminate active lupus nephritis [128]. Other studies show that miR-29c expression levels in urinary exosomes can predict the degree of chronicity of lupus nephritis with a high sensitivity and specificity (94% and 82%, respectively). This marker shows a strong negative correlation with the histological chronicity index and glomerular sclerosis but not with renal function, and thus has great potential to act as a noninvasive biomarker of early fibrosis in patients with lupus nephritis [129].

1.3.3.2 *LncRNAs and SLE*

LncRNAs are mRNA-like molecules that are larger than 200 nucleotides but do not contain functional open reading frames [130]. Studies of lncRNAs in patients with autoimmune diseases showed differential expression in diseases such as SLE, polymyositis and dermatomyositis, RA, T1DM, multiple sclerosis (MS), and autoimmune thyroid disease [131]. The growth arrest-specific transcript (also known as *GAS5*) is a lncRNA that is potentially implicated in the pathogenesis of SLE [131]. It plays essential roles in growth arrest, apoptosis, and the cell cycle both in T-cell lines and nontransformed lymphocytes [132]. Downregulation of *GAS5* may inhibit the cell cycle and apoptosis, and thus may contribute to the promotion of antigen exposure and the production of autoantibodies [133].

Large intergenic noncoding RNAs (lincRNAs), a specific type of lncRNAs, also regulate gene expression and are implicated in various physiological and pathogenic conditions. Two lincRNAs, linc0949 and linc0597, were found to be significantly downregulated in PBMCs from patients with SLE compared to patients with RA or controls [134]. Notably, decreased expression of linc0949 correlated with complement component C3 levels, the SLE disease activity index, and the presence of lupus nephritis and cumulative organ damage. Linc0949 also increased significantly after effective lupus treatment, thus suggesting its potential role as a biomarker for disease activity and the therapeutic response in SLE [134,135].

1.4 EPIGENETICS AND OTHER AUTOIMMUNE DISEASES

Epigenetic regulation abnormalities are also present in many other autoimmune diseases. Interestingly, promoter hypomethylation of *CD40L*, and the associated overexpression of the gene, were also identified in $CD4^+$ T cells from female patients with systemic sclerosis, RA, and PBC [136–138]. This phenomenon is similar to that revealed in SLE and provides insights into the female predominance of these autoimmune diseases. It also indicates the presence of common epigenetically regulated pathways among these diseases (Fig. 1.2).

1.5 REGULATORY NETWORKS OF EPIGENETIC MECHANISMS IN AUTOIMMUNITY

1.5.1 Interactive Regulation Between DNA Methylation and miRNAs

Integrating genome-wide DNA methylation data with global gene and miRNA expression data from SLE $CD4^+$ T cells identified a set of alternatively expressed genes that were modulated by the DNA methylation status. These included dozens of genes encoding miRNAs. Of these, 36 overexpressed miRNAs (i.e., miR-181) were located near CpG sites hypomethylated in SLE $CD4^+$ T cells, while 8 downregulated miRNAs were hypermethylated [87]. These findings were consistent with a previous study, which found that the DNA methylation status played an important role

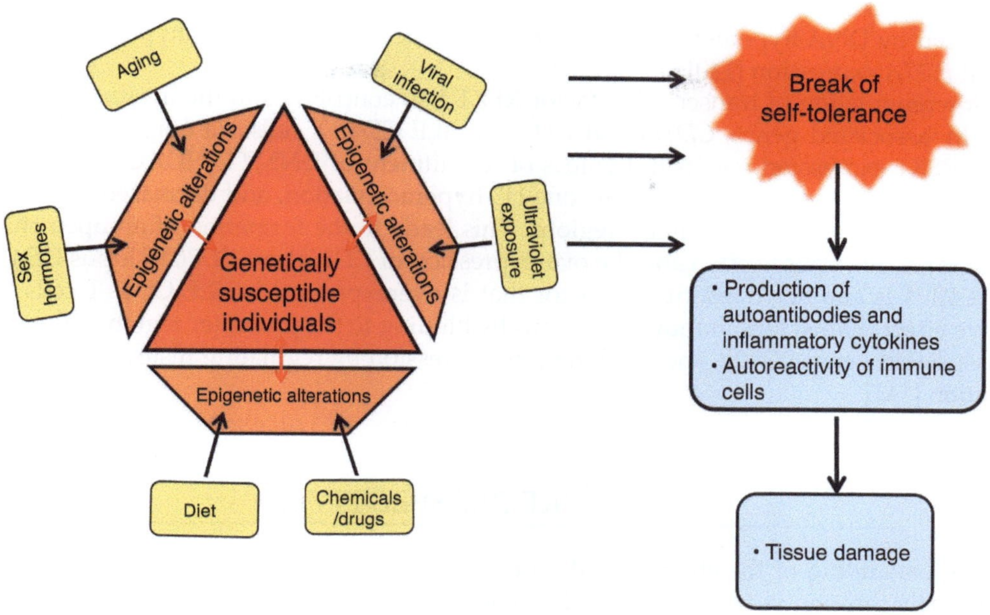

FIGURE 1.2 **Schematic representation of the different roles of genetics, epigenetics, and environmental factors in the pathogenesis of autoimmune diseases.** Environmental factors (i.e., viral infection, diet, ultraviolet exposure, and chemical or drug exposure), sex hormones, and aging all have the ability to affect the epigenome of different human cell subsets. These factors interact with genetic risk variants in genetically predisposed individuals, and this complex crosstalk may lead to the break of self-tolerance in the immune system and cause the clinical phenotypes of autoimmune diseases.

in regulating the expression of relevant miRNAs (i.e., miR-126 and miR-142-3p/5p) in SLE $CD4^+$ T cells [71,117].

More recent studies reported a set of upregulated miRNAs at the imprinted DLK1-Dio3 domain, which caused overproduction of lupus-relevant cytokines (i.e., IFNγ, IL-1β, IL-6, and IL-10) in the splenocytes of MRL/*lpr* lupus mice compared to control MRL mice. Global hypomethylation was also present in these affected splenocytes. A similarly augmented DLK1-Dio3 miRNA expression pattern was also induced by experimentally demethylating splenocytes from control MRL mice (but not MRL-*lpr* lupus mice) using 5-aza-2′-deoxycytidine. This suggests that there is a connection between DNA methylation, miRNA, and genomic imprinting in the pathogenesis of lupus [139].

Several miRNAs have also been identified as modulators of DNA methylation due to their direct or indirect targeting of DNMTs, particularly DNMT1. Classic examples include miR-29b [115], miR-126 [71], miR-21 [70], and miR-148a [70].

1.5.2 Regulatory Networks Involving DNA Methylation, Histone Modifications, and Transcriptional Factors

The transcription factor CREMα, which is overexpressed in T cells from SLE patients, physically binds to a cAMP-responsive element within the proximal human IL17A promoter and increases IL17A expression in SLE. It was suggested that CREMα decreased the recruitment of DNMT3a and

HDAC1 to this cAMP-responsive element site, and thus produced an accessible chromatin state characterized by DNA hypomethylation, H3K18 hyperacetylation, H3K27 hypomethylation, and enhanced *IL17A* expression [140].

The downregulation of transcription factor RFX1 also contributes to the overexpression of autoreactivity-associated genes, *CD11a* and *CD70*, in SLE $CD4^+$ T cells via epigenetic remodeling. Reduced RFX1 expression results in the loss of recruitment of DNMT1, HDAC1, and SUV39H1. It also causes DNA hypomethylation, histone H3 hyperacetylation, and decreased H3K9 trimethylation in *CD11a* and *CD70* promoter regions. This leads to the formation of an open chromatin structure for active transcription and the overexpression of *CD11a* and *CD70* in lupus $CD4^+$ T cells [72,73]. E4BP4 is another transcription factor that is overexpressed in SLE $CD4^+$ T cells. It has a protective effect against autoimmunity in lupus by binding to the promoter region of the autoreactivity-associated gene, *CD40L*, and inhibiting its expression by regulating histone acetylation and methylation [141].

1.6 FUTURE PERSPECTIVES

Our understanding of the etiology and pathogenesis of autoimmune diseases has progressed fundamentally through years of research. It is widely accepted that genetic susceptibility, epigenetic alterations, and environmental impact all contribute to the pathogenesis of autoimmunity; however, there is still a gap between what must be elucidated and what is known. For example, the exact role of epigenetics in the interplay with autoimmunity-risk genetic variants remains to be elucidated. Some genetic variations associated with lupus are believed to function by inducing epigenetic alterations that change the local chromatin conformation and regulate the expression of other genes [142]. Conversely, it remains to be determined whether disease-associated genetic variants originate from unstable genetic loci that are prone to mutation due to aberrant epigenetic modulation [7].

An important field that requires further investigation is the investigation of the causes or triggers (either genetic or environmental) of dysregulated DNA methylation, histone modification, or noncoding RNAs. This may direct help to direct the appropriate intervention or therapeutic measures for clinical practice. The recent upsurge of research in the field of the gut microbiome, and its interaction with immunity and epigenetics, has the potential to enhance our understanding of the impact of this "internal environmental factor" on epigenetic dysregulation in autoimmune diseases [143,144].

Examining epigenetic marks on a genome-wide level is ever more popular due to the maturation of high-throughput sequencing technologies and their decreasing costs. Integrating multiple layers of epigenetic data (i.e., DNA methylome data, genome-wide miRNA expression data, and histone modification marks) with transcriptome and proteome data from pivotal cell types in autoimmune diseases, will bring us much closer to uncovering the functional consequences of the disease-associated epigenetic changes. Furthermore, identifying the common epigenetic changes or pathogenic pathways in the same cell types in different autoimmune diseases would be extremely valuable.

Whether the autoimmunity-related epigenetic changes are partly attributed to aging remains to be answered. A recent study sought to define age-associated DNA methylation changes by investigating the DNA methylome of naive CD4+ T cells from 74 healthy individuals of varying ages.

Among the 11,431 age-associated CpG sites identified, 43% were hypomethylated and showed T-cell specific enrichment in active enhancers marked by H3K27ac and H3K4me1. These age-associated hypomethylation sites were enriched in immune-related pathways, including T-cell receptor signaling, FCγR-mediated phagocytosis, apoptosis, and the mTOR signaling pathway [145]. Our group attempted to compare the epigenomic differences in $CD4^+$ T cells between new-born, middle-aged, and long-lived individuals, and the findings suggested that methylation-dependent transcriptional regulation was involved in the development and function of T cells during aging [146]. However, more in-depth evidence of the ageing-associated epigenetic changes in autoimmune diseases is still lacking.

Epigenetic research into autoimmune diseases has accelerated the identification of potential biomarkers and therapeutic targets for these diseases. The development of epigenetic biomarkers for disease identification and disease activity are likely to become a reality and greatly facilitate the early diagnosis of autoimmune diseases and the prediction of flares, respectively. As for epigenetic therapies, the use of HDACi has shown promise in mouse models; however, the mechanisms must be clearly deciphered before further progress can be made. Although there are currently no drugs to reverse global DNA hypomethylation in $CD4^+$ T cells, a characteristic feature in most autoimmune diseases, alternative measures may be able to reverse site-specific hypomethylation, such as targeting key microRNAs that inhibit DNMT1, key molecules (i.e., PP2Ac), or key transcriptional factors (i.e., RFX1) [147,148].

Acknowledgments

This work was supported by the National Natural Science Foundation of China (No. 81220108017, No. 81430074, No. 81301357, No. 30972745, No. 81773334 and No. 81522038), the PhD Programs Foundation of Ministry of Education of China (No. 20120162130003), the Hunan Provincial Natural Science Foundation of China (13JJ4025 and 14JJ1009), the Programs of Science-Technology Commission of Hunan Province (2013FJ4202), and the National Key Clinical Specialty Construction Project of National Health and Family Planning Commission of the People's Republic of China.

The authors declare that they have no conflict of interest.

References

[1] Costenbader KH, Gay S, Alarcon-Riquelme ME, Iaccarino L, Doria A. Genes, epigenetic regulation and environmental factors: which is the most relevant in developing autoimmune diseases? Autoimmun Rev 2012;11:604–9.

[2] Zhao M, Wang Z, Yung S, Lu Q. Epigenetic dynamics in immunity and autoimmunity. Int J Biochem Cell Biol 2015;67:65–74.

[3] Niller HH, Wolf H, Ay E, Minarovits J. Epigenetic dysregulation of Epstein-Barr virus latency and development of autoimmune disease. Adv Exp Med Biol 2011;711:82–102.

[4] Zeitelhofer M, Adzemovic MZ, Gomez-Cabrero D, Bergman P, Hochmeister S, N'Diaye M, et al. Functional genomics analysis of vitamin D effects on $CD4^+$ T cells in vivo in experimental autoimmune encephalomyelitis. Proc Natl Acad Sci USA 2017;114:E1678–87.

[5] Zhao S, Long H, Lu Q. Epigenetic perspectives in systemic lupus erythematosus: pathogenesis, biomarkers, and therapeutic potentials. Clin Rev Allergy Immunol 2009;39:3–9.

[6] Hedrich CM, Tsokos GC. Epigenetic mechanisms in systemic lupus erythematosus and other autoimmune diseases. Trends Mol Med 2011;17:714–24.

[7] Jeffries MA, Sawalha AH. Autoimmune disease in the epigenetic era: how has epigenetics changed our understanding of disease and how can we expect the field to evolve? Expert Rev Clin Immunol 2015;11:45–58.

[8] Long H, Yin H, Wang L, Gershwin ME, Lu Q. The critical role of epigenetics in systemic lupus erythematosus and autoimmunity. J Autoimmun 2016;74:118–38.

[9] Ghodke-Puranik Y, Niewold TB. Immunogenetics of systemic lupus erythematosus: a comprehensive review. J Autoimmun 2015;64:125–36.

[10] Ramos PS, Criswell LA, Moser KL, Comeau ME, Williams AH, Pajewski NM, et al. A comprehensive analysis of shared loci between systemic lupus erythematosus (SLE) and sixteen autoimmune diseases reveals limited genetic overlap. PLoS Genet 2011;7:8.
[11] De Jager PL, Jia X, Wang J, de Bakker PI, Ottoboni L, Aggarwal NT, et al. Meta-analysis of genome scans and replication identify CD6, IRF8 and TNFRSF1A as new multiple sclerosis susceptibility loci. Nat Genet 2009;41:776–82.
[12] Lessard CJ, Ice JA, Adrianto I, Wiley GB, Kelly JA, Gaffney PM, et al. The genomics of autoimmune disease in the era of genome-wide association studies and beyond. Autoimmun Rev 2012;11:267–75.
[13] Zhernakova A, Withoff S, Wijmenga C. Clinical implications of shared genetics and pathogenesis in autoimmune diseases. Nat Rev Endocrinol 2013;9:646–59.
[14] Zheng J, Yin J, Huang R, Petersen F, Yu X. Meta-analysis reveals an association of STAT4 polymorphisms with systemic autoimmune disorders and anti-dsDNA antibody. Hum Immunol 2013;74:986–92.
[15] Deng Y, Tsao BP. Advances in lupus genetics and epigenetics. Curr Opin Rheumatol 2014;26:482–92.
[16] Ghodke-Puranik Y, Niewold TB. Genetics of the type I interferon pathway in systemic lupus erythematosus. Int J Clin Rheumtol 2013;8:58.
[17] Seldin MF. The genetics of human autoimmune disease: a perspective on progress in the field and future directions. J Autoimmun 2015;64:1–12.
[18] Niewold TB. Advances in lupus genetics. Curr Opin Rheumatol 2015;27:440–7.
[19] Bossini-Castillo L, Lopez-Isac E, Martin J. Immunogenetics of systemic sclerosis: defining heritability, functional variants and shared-autoimmunity pathways. J Autoimmun 2015;64:53–65.
[20] Hersh AO, Prahalad S. Immunogenetics of juvenile idiopathic arthritis: a comprehensive review. J Autoimmun 2015;64:113–24.
[21] Harden JL, Krueger JG, Bowcock AM. The immunogenetics of psoriasis: a comprehensive review. J Autoimmun 2015;64:66–73.
[22] Hollenbach JA, Oksenberg JR. The immunogenetics of multiple sclerosis: a comprehensive review. J Autoimmun 2015;64:13–25.
[23] Lee HJ, Li CW, Hammerstad SS, Stefan M, Tomer Y. Immunogenetics of autoimmune thyroid diseases: a comprehensive review. J Autoimmun 2015;64:82–90.
[24] Messemaker TC, Huizinga TW, Kurreeman F. Immunogenetics of rheumatoid arthritis: understanding functional implications. J Autoimmun 2015;64:74–81.
[25] Noble JA. Immunogenetics of type 1 diabetes: a comprehensive review. J Autoimmun 2015;64:101–12.
[26] Takeuchi M, Kastner DL, Remmers EF. The immunogenetics of Behcet's disease: a comprehensive review. J Autoimmun 2015;64:137–48.
[27] de Lange KM, Barrett JC. Understanding inflammatory bowel disease via immunogenetics. J Autoimmun 2015;64:91–100.
[28] Adrianto I, Wen F, Templeton A, Wiley G, King JB, Lessard CJ, et al. Association of a functional variant downstream of TNFAIP3 with systemic lupus erythematosus. Nat Genet 2011;43:253–8.
[29] Bates JS, Lessard CJ, Leon JM, Nguyen T, Battiest LJ, Rodgers J, et al. Meta-analysis and imputation identifies a 109 kb risk haplotype spanning TNFAIP3 associated with lupus nephritis and hematologic manifestations. Genes Immun 2009;10:470–7.
[30] Manolio TA, Collins FS, Cox NJ, Goldstein DB, Hindorff LA, Hunter DJ, et al. Finding the missing heritability of complex diseases. Nature 2009;461:747–53.
[31] Visscher PM, Brown MA, McCarthy MI, Yang J. Five years of GWAS discovery. Am J Hum Genet 2012;90:7–24.
[32] Yang W, Zhao M, Hirankarn N, Lau CS, Mok CC, Chan TM, et al. ITGAM is associated with disease susceptibility and renal nephritis of systemic lupus erythematosus in Hong Kong Chinese and Thai. Hum Mol Genet 2009;18:2063–70.
[33] Taylor KE, Chung SA, Graham RR, Ortmann WA, Lee AT, Langefeld CD, et al. Risk alleles for systemic lupus erythematosus in a large case-control collection and associations with clinical subphenotypes. PLoS Genet 2011;7:17.
[34] Ballestar E. Epigenetics lessons from twins: prospects for autoimmune disease. Clin Rev Allergy Immunol 2010;39:30–41.
[35] Selmi C, Lu Q, Humble MC. Heritability versus the role of the environment in autoimmunity. J Autoimmun 2012;39:249–52.
[36] Bird A. Perceptions of epigenetics. Nature 2007;447:396–8.
[37] Schomacher L. Mammalian DNA demethylation: multiple faces and upstream regulation. Epigenetics 2013;8:679–84.
[38] Guo JU, Su Y, Zhong C, Ming GL, Song H. Emerging roles of TET proteins and 5-hydroxymethylcytosines in active DNA demethylation and beyond. Cell Cycle 2011;10:2662–8.

[39] Zhao M, Zhou Y, Zhu B, Wan M, Jiang T, Tan Q, et al. IFI44L promoter methylation as a blood biomarker for systemic lupus erythematosus. Ann Rheum Dis 2016;75:1998–2006.
[40] Ngalamika O, Liang G, Zhao M, Yu X, Yang Y, Yin H, et al. Peripheral whole blood FOXP3 TSDR methylation: a potential marker in severity assessment of autoimmune diseases and chronic infections. Immunol Invest 2015;44:126–36.
[41] Javierre BM, Fernandez AF, Richter J, Al-Shahrour F, Martin-Subero JI, Rodriguez-Ubreva J, et al. Changes in the pattern of DNA methylation associate with twin discordance in systemic lupus erythematosus. Genome Res 2010;20:170–9.
[42] Liu HW, Lin HL, Yen JH, Tsai WC, Chiou SS, Chang JG, et al. Demethylation within the proximal promoter region of human estrogen receptor alpha gene correlates with its enhanced expression: implications for female bias in lupus. Mol Immunol 2014;61:28–37.
[43] Gorelik G, Sawalha AH, Patel D, Johnson K, Richardson B. T cell PKCδ kinase inactivation induces lupus-like autoimmunity in mice. Clin Immunol 2015;158:193–203.
[44] Hewagama A, Gorelik G, Patel D, Liyanarachchi P, McCune WJ, Somers E, et al. Overexpression of X-linked genes in T cells from women with lupus. J Autoimmun 2013;41:60–71.
[45] Mi XB, Zeng FQ. Hypomethylation of interleukin-4 and -6 promoters in T cells from systemic lupus erythematosus patients. Acta Pharmacol Sin 2008;29:105–12.
[46] Zhao M, Tang J, Gao F, Wu X, Liang Y, Yin H, et al. Hypomethylation of IL10 and IL13 promoters in $CD4^+$ T cells of patients with systemic lupus erythematosus. J Biomed Biotechnol 2010;2010:2.
[47] Jeffries MA, Dozmorov M, Tang Y, Merrill JT, Wren JD, Sawalha AH. Genome-wide DNA methylation patterns in $CD4^+$ T cells from patients with systemic lupus erythematosus. Epigenetics 2011;6:593–601.
[48] Absher DM, Li X, Waite LL, Gibson A, Roberts K, Edberg J, et al. Genome-wide DNA methylation analysis of systemic lupus erythematosus reveals persistent hypomethylation of interferon genes and compositional changes to $CD4^+$ T-cell populations. PLoS Genet 2013;9:e1003678.
[49] Oelke K, Lu Q, Richardson D, Wu A, Deng C, Hanash S, et al. Overexpression of CD70 and overstimulation of IgG synthesis by lupus T cells and T cells treated with DNA methylation inhibitors. Arthritis Rheum 2004;50:1850–60.
[50] Lu Q, Wu A, Richardson BC. Demethylation of the same promoter sequence increases CD70 expression in lupus T cells and T cells treated with lupus-inducing drugs. J Immunol 2005;174:6212–9.
[51] Kaplan MJ, Lu Q, Wu A, Attwood J, Richardson B. Demethylation of promoter regulatory elements contributes to perforin overexpression in $CD4^+$ lupus T cells. J Immunol 2004;172:3652–61.
[52] Kozlowska A, Hrycaj P, Lacki JK, Jagodzinski PP. Perforin level in $CD4^+$ T cells from patients with systemic lupus erythematosus. Rheumatol Int 2010;30:1627–33.
[53] Balada E, Castro-Marrero J, Felip L, Ordi-Ros J, Vilardell-Tarres M. Clinical and serological findings associated with the expression of ITGAL, PRF1, and CD70 in systemic lupus erythematosus. Clin Exp Rheumatol 2014;32:113–6.
[54] Lu Q, Kaplan M, Ray D, Zacharek S, Gutsch D, Richardson B. Demethylation of ITGAL (CD11a) regulatory sequences in systemic lupus erythematosus. Arthritis Rheum 2002;46:1282–91.
[55] Basu D, Liu Y, Wu A, Yarlagadda S, Gorelik GJ, Kaplan MJ, et al. Stimulatory and inhibitory killer Ig-like receptor molecules are expressed and functional on lupus T cells. J Immunol 2009;183:3481–7.
[56] Sawalha AH, Wang L, Nadig A, Somers EC, McCune WJ, Hughes T, et al. Sex-specific differences in the relationship between genetic susceptibility, T cell DNA demethylation and lupus flare severity. J Autoimmun 2012;38:3.
[57] Sunahori K, Juang YT, Kyttaris VC, Tsokos GC. Promoter hypomethylation results in increased expression of protein phosphatase 2A in T cells from patients with systemic lupus erythematosus. J Immunol 2011;186:4508–17.
[58] Coit P, Jeffries M, Altorok N, Dozmorov MG, Koelsch KA, Wren JD, et al. Genome-wide DNA methylation study suggests epigenetic accessibility and transcriptional poising of interferon-regulated genes in naive $CD4^+$ T cells from lupus patients. J Autoimmun 2013;43:78–84.
[59] Renauer P, Coit P, Jeffries MA, Merrill JT, McCune WJ, Maksimowicz-McKinnon K, et al. DNA methylation patterns in naive $CD4^+$ T cells identify epigenetic susceptibility loci for malar rash and discoid rash in systemic lupus erythematosus. Lupus Sci Med 2015;2:e000101.
[60] Coit P, Renauer P, Jeffries MA, Merrill JT, McCune WJ, Maksimowicz-McKinnon K, et al. Renal involvement in lupus is characterized by unique DNA methylation changes in naive $CD4^+$ T cells. J Autoimmun 2015;61:29–35.
[61] Lu Q, Wu A, Tesmer L, Ray D, Yousif N, Richardson B. Demethylation of CD40LG on the inactive X in T cells from women with lupus. J Immunol 2007;179:6352–8.
[62] Zhao M, Wang J, Liao W, Li D, Li M, Wu H, et al. Increased 5-hydroxymethylcytosine in $CD4^+$ T cells in systemic lupus erythematosus. J Autoimmun 2016;69:64–73.
[63] Deng C, Kaplan MJ, Yang J, Ray D, Zhang Z, McCune WJ, et al. Decreased Ras-mitogen-activated protein kinase signaling may cause DNA hypomethylation in T lymphocytes from lupus patients. Arthritis Rheum 2001;44:397–407.

[64] Jones PA, Taylor SM, Wilson VL. Inhibition of DNA methylation by 5-azacytidine. Recent Results Cancer Res 1983;84:202–11.
[65] Sawalha AH, Jeffries M, Webb R, Lu Q, Gorelik G, Ray D, et al. Defective T-cell ERK signaling induces interferon-regulated gene expression and overexpression of methylation-sensitive genes similar to lupus patients. Genes Immun 2008;9:368–78.
[66] Strickland FM, Li Y, Johnson K, Sun Z, Richardson BC. CD4+ T cells epigenetically modified by oxidative stress cause lupus-like autoimmunity in mice. J Autoimmun 2015;62:75–80.
[67] Sunahori K, Nagpal K, Hedrich CM, Mizui M, Fitzgerald LM, Tsokos GC. The catalytic subunit of protein phosphatase 2A (PP2Ac) promotes DNA hypomethylation by suppressing the phosphorylated mitogen-activated protein kinase/extracellular signal-regulated kinase (ERK) kinase (MEK)/phosphorylated ERK/DNMT1 protein pathway in T-cells from controls and systemic lupus erythematosus patients. J Biol Chem 2013;288:21936–44.
[68] Gorelik G, Fang JY, Wu A, Sawalha AH, Richardson B. Impaired T cell protein kinase C delta activation decreases ERK pathway signaling in idiopathic and hydralazine-induced lupus. J Immunol 2007;179:5553–63.
[69] Gorelik GJ, Yarlagadda S, Patel DR, Richardson BC. Protein kinase Cδ oxidation contributes to ERK inactivation in lupus T cells. Arthritis Rheum 2012;64:2964–74.
[70] Pan W, Zhu S, Yuan M, Cui H, Wang L, Luo X, et al. MicroRNA-21 and microRNA-148a contribute to DNA hypomethylation in lupus CD4+ T cells by directly and indirectly targeting DNA methyltransferase 1. J Immunol 2010;184:6773–81.
[71] Zhao S, Wang Y, Liang Y, Zhao M, Long H, Ding S, et al. MicroRNA-126 regulates DNA methylation in CD4+ T cells and contributes to systemic lupus erythematosus by targeting DNA methyltransferase 1. Arthritis Rheum 2011;63:1376–86.
[72] Zhao M, Wu X, Zhang Q, Luo S, Liang G, Su Y, et al. RFX1 regulates CD70 and CD11a expression in lupus T cells by recruiting the histone methyltransferase SUV39H1. Arthritis Res Ther 2010;12:30.
[73] Zhao M, Sun Y, Gao F, Wu X, Tang J, Yin H, et al. Epigenetics and SLE: RFX1 downregulation causes CD11a and CD70 overexpression by altering epigenetic modifications in lupus CD4+ T cells. J Autoimmun 2010;35:58–69.
[74] Li Y, Zhao M, Yin H, Gao F, Wu X, Luo Y, et al. Overexpression of the growth arrest and DNA damage-induced 45alpha gene contributes to autoimmunity by promoting DNA demethylation in lupus T cells. Arthritis Rheum 2010;62:1438–47.
[75] Li Y, Huang C, Zhao M, Liang G, Xiao R, Yung S, et al. A possible role of HMGB1 in DNA demethylation in CD4+ T cells from patients with systemic lupus erythematosus. Clin Dev Immunol 2013;2013. Article ID 206298.
[76] Yung RL, Richardson BC. Role of T cell DNA methylation in lupus syndromes. Lupus 1994;3:487–91.
[77] Zhou Y, Lu Q. DNA methylation in T cells from idiopathic lupus and drug-induced lupus patients. Autoimmun Rev 2008;7:376–83.
[78] Richardson B, Scheinbart L, Strahler J, Gross L, Hanash S, Johnson M. Evidence for impaired T cell DNA methylation in systemic lupus erythematosus and rheumatoid arthritis. Arthritis Rheum 1990;33:1665–73.
[79] Richardson BC, Strahler JR, Pivirotto TS, Quddus J, Bayliss GE, Gross LA, et al. Phenotypic and functional similarities between 5-azacytidine-treated T cells and a T cell subset in patients with active systemic lupus erythematosus. Arthritis Rheum 1992;35:647–62.
[80] Quddus J, Johnson KJ, Gavalchin J, Amento EP, Chrisp CE, Yung RL, et al. Treating activated CD4+ T cells with either of two distinct DNA methyltransferase inhibitors, 5-azacytidine or procainamide, is sufficient to cause a lupus-like disease in syngeneic mice. J Clin Invest 1993;92:38–53.
[81] Yung RL, Quddus J, Chrisp CE, Johnson KJ, Richardson BC. Mechanism of drug-induced lupus. I. Cloned Th2 cells modified with DNA methylation inhibitors in vitro cause autoimmunity in vivo. J Immunol 1995;154:3025–35.
[82] Yung R, Chang S, Hemati N, Johnson K, Richardson B. Mechanisms of drug-induced lupus. IV. Comparison of procainamide and hydralazine with analogs in vitro and in vivo. Arthritis Rheum 1997;40:1436–43.
[83] Solomon DH, Kavanaugh AJ, Schur PH. Evidence-based guidelines for the use of immunologic tests: antinuclear antibody testing. Arthritis Rheum 2002;47:434–44.
[84] Zieve GW, Khusial PR. The anti-Sm immune response in autoimmunity and cell biology. Autoimmun Rev 2003;2:235–40.
[85] Bizzaro N, Villalta D, Giavarina D, Tozzoli R. Are anti-nucleosome antibodies a better diagnostic marker than anti-dsDNA antibodies for systemic lupus erythematosus? A systematic review and a study of metanalysis. Autoimmun Rev 2012;12:97–106.
[86] Kurien BT, Scofield RH. Autoantibody determination in the diagnosis of systemic lupus erythematosus. Scand J Immunol 2006;64:227–35.
[87] Zhao M, Liu S, Luo S, Wu H, Tang M, Cheng W, et al. DNA methylation and mRNA and microRNA expression of SLE CD4+ T cells correlate with disease phenotype. J Autoimmun 2014;54:127–36.

[88] Keyes MK, Jang H, Mason JB, Liu Z, Crott JW, Smith DE, et al. Older age and dietary folate are determinants of genomic and p16-specific DNA methylation in mouse colon. J Nutr 2007;137:1713–7.

[89] Strickland FM, Hewagama A, Wu A, Sawalha AH, Delaney C, Hoeltzel MF, et al. Diet influences expression of autoimmune-associated genes and disease severity by epigenetic mechanisms in a transgenic mouse model of lupus. Arthritis Rheum 2013;65:1872–81.

[90] Sauer J, Jang H, Zimmerly EM, Kim KC, Liu Z, Chanson A, et al. Ageing, chronic alcohol consumption and folate are determinants of genomic DNA methylation, p16 promoter methylation and the expression of p16 in the mouse colon. Br J Nutr 2010;104:24–30.

[91] Black JC, Van Rechem C, Whetstine JR. Histone lysine methylation dynamics: establishment, regulation, and biological impact. Mol Cell 2012;48:491–507.

[92] Aslani S, Mahmoudi M, Karami J, Jamshidi AR, Malekshahi Z, Nicknam MH. Epigenetic alterations underlying autoimmune diseases. Autoimmunity 2016;49:69–83.

[93] Peserico A, Simone C. Physical and functional HAT/HDAC interplay regulates protein acetylation balance. J Biomed Biotechnol 2011;2011. Article ID 371832.

[94] Garcia BA, Busby SA, Shabanowitz J, Hunt DF, Mishra N. Resetting the epigenetic histone code in the MRL-lpr/lpr mouse model of lupus by histone deacetylase inhibition. J Proteome Res 2005;4:2032–42.

[95] Hu N, Qiu X, Luo Y, Yuan J, Li Y, Lei W, et al. Abnormal histone modification patterns in lupus $CD4^+$ T cells. J Rheumatol 2008;35:804–10.

[96] Tenbrock K, Juang YT, Leukert N, Roth J, Tsokos GC. The transcriptional repressor cAMP response element modulator alpha interacts with histone deacetylase 1 to repress promoter activity. J Immunol 2006;177:6159–64.

[97] Mishra N, Reilly CM, Brown DR, Ruiz P, Gilkeson GS. Histone deacetylase inhibitors modulate renal disease in the MRL-lpr/lpr mouse. J Clin Invest 2003;111:539–52.

[98] Reilly CM, Mishra N, Miller JM, Joshi D, Ruiz P, Richon VM, et al. Modulation of renal disease in MRL/lpr mice by suberoylanilide hydroxamic acid. J Immunol 2004;173:4171–8.

[99] Mishra N, Brown DR, Olorenshaw IM, Kammer GM. Trichostatin A reverses skewed expression of CD154, interleukin-10, and interferon-gamma gene and protein expression in lupus T cells. Proc Natl Acad Sci USA 2001;98:2628–33.

[100] Regna NL, Chafin CB, Hammond SE, Puthiyaveetil AG, Caudell DL, Reilly CM. Class I and II histone deacetylase inhibition by ITF2357 reduces SLE pathogenesis in vivo. Clin Immunol 2014;151:29–42.

[101] Lewis EC, Blaabjerg L, Storling J, Ronn SG, Mascagni P, Dinarello CA, et al. The oral histone deacetylase inhibitor ITF2357 reduces cytokines and protects islet beta cells in vivo and in vitro. Mol Med 2011;17:369–77.

[102] Hu N, Long H, Zhao M, Yin H, Lu Q. Aberrant expression pattern of histone acetylation modifiers and mitigation of lupus by SIRT1-siRNA in MRL/lpr mice. Scand J Rheumatol 2009;38:464–71.

[103] Sullivan KE, Suriano A, Dietzmann K, Lin J, Goldman D, Petri MA. The TNFα locus is altered in monocytes from patients with systemic lupus erythematosus. Clin Immunol 2007;123:74–81.

[104] Apostolidis SA, Rauen T, Hedrich CM, Tsokos GC, Crispin JC. Protein phosphatase 2A enables expression of interleukin 17 (IL-17) through chromatin remodeling. J Biol Chem 2013;288:26775–84.

[105] Dieker JW, Fransen JH, van Bavel CC, Briand JP, Jacobs CW, Muller S, et al. Apoptosis-induced acetylation of histones is pathogenic in systemic lupus erythematosus. Arthritis Rheum 2007;56:1921–33.

[106] van Bavel CC, Dieker J, Muller S, Briand JP, Monestier M, Berden JH, et al. Apoptosis-associated acetylation on histone H2B is an epitope for lupus autoantibodies. Mol Immunol 2009;47:511–6.

[107] Zhang Z, Song L, Maurer K, Petri MA, Sullivan KE. Global H4 acetylation analysis by ChIP-chip in systemic lupus erythematosus monocytes. Genes Immun 2010;11:124–33.

[108] Zhang Z, Shi L, Dawany N, Kelsen J, Petri MA, Sullivan KE. H3K4 tri-methylation breadth at transcription start sites impacts the transcriptome of systemic lupus erythematosus. Clin Epigenet 2016;8:14.

[109] Long H, Huang W, Yin H, Zhao S, Zhao M, Lu Q. Abnormal expression pattern of histone demethylases in $CD4^+$ T cells of MRL/lpr lupus-like mice. Lupus 2009;18:1327–8.

[110] Zhang Q, Long H, Liao J, Zhao M, Liang G, Wu X, et al. Inhibited expression of hematopoietic progenitor kinase 1 associated with loss of jumonji domain containing 3 promoter binding contributes to autoimmunity in systemic lupus erythematosus. J Autoimmun 2011;37:180–9.

[111] Shukla GC, Singh J, Barik S. MicroRNAs: processing, maturation, target recognition and regulatory functions. Mol Cell Pharmacol 2011;3:83–92.

[112] Baltimore D, Boldin MP, O'Connell RM, Rao DS, Taganov KD. MicroRNAs: new regulators of immune cell development and function. Nat Immunol 2008;9:839–45.

[113] Lashine YA, Salah S, Aboelenein HR, Abdelaziz AI. Correcting the expression of miRNA-155 represses PP2Ac and enhances the release of IL-2 in PBMCs of juvenile SLE patients. Lupus 2015;24:240–7.

[114] Tang Y, Luo X, Cui H, Ni X, Yuan M, Guo Y, et al. MicroRNA-146A contributes to abnormal activation of the type I interferon pathway in human lupus by targeting the key signaling proteins. Arthritis Rheum 2009;60:1065–75.

[115] Qin H, Zhu X, Liang J, Wu J, Yang Y, Wang S, et al. MicroRNA-29b contributes to DNA hypomethylation of $CD4^{+}$ T cells in systemic lupus erythematosus by indirectly targeting DNA methyltransferase 1. J Dermatol Sci 2013;69:61–7.

[116] Stagakis E, Bertsias G, Verginis P, Nakou M, Hatziapostolou M, Kritikos H, et al. Identification of novel microRNA signatures linked to human lupus disease activity and pathogenesis: miR-21 regulates aberrant T cell responses through regulation of PDCD4 expression. Ann Rheum Dis 2011;70:1496–506.

[117] Ding S, Liang Y, Zhao M, Liang G, Long H, Zhao S, et al. Decreased microRNA-142-3p/5p expression causes $CD4^{+}$ T cell activation and B cell hyperstimulation in systemic lupus erythematosus. Arthritis Rheum 2012;64:2953–63.

[118] Fan W, Liang D, Tang Y, Qu B, Cui H, Luo X, et al. Identification of microRNA-31 as a novel regulator contributing to impaired interleukin-2 production in T cells from patients with systemic lupus erythematosus. Arthritis Rheum 2012;64:3715–25.

[119] Tang Q, Yang Y, Zhao M, Liang G, Wu H, Liu Q, et al. Mycophenolic acid upregulates miR-142-3P/5P and miR-146a in lupus $CD4^{+}$ T cells. Lupus 2015;24:935–42.

[120] Liu Y, Dong J, Mu R, Gao Y, Tan X, Li Y, et al. MicroRNA-30a promotes B cell hyperactivity in patients with systemic lupus erythematosus by direct interaction with Lyn. Arthritis Rheum 2013;65:1603–11.

[121] Dorsett Y, McBride KM, Jankovic M, Gazumyan A, Thai TH, Robbiani DF, et al. MicroRNA-155 suppresses activation-induced cytidine deaminase-mediated Myc-Igh translocation. Immunity 2008;28:630–8.

[122] de Yebenes VG, Belver L, Pisano DG, Gonzalez S, Villasante A, Croce C, et al. miR-181b negatively regulates activation-induced cytidine deaminase in B cells. J Exp Med 2008;205:2199–206.

[123] Luo S, Liu Y, Liang G, Zhao M, Wu H, Liang Y, et al. The role of microRNA-1246 in the regulation of B cell activation and the pathogenesis of systemic lupus erythematosus. Clin Epigenet 2015;7:24.

[124] Yuan Y, Kasar S, Underbayev C, Vollenweider D, Salerno E, Kotenko SV, et al. Role of microRNA-15a in autoantibody production in interferon-augmented murine model of lupus. Mol Immunol 2012;52:61–70.

[125] Wang G, Tam LS, Li EK, Kwan BC, Chow KM, Luk CC, et al. Serum and urinary free microRNA level in patients with systemic lupus erythematosus. Lupus 2011;20:493–500.

[126] Wang H, Peng W, Ouyang X, Li W, Dai Y. Circulating microRNAs as candidate biomarkers in patients with systemic lupus erythematosus. Transl Res 2012;160:198–206.

[127] Wang G, Tam LS, Kwan BC, Li EK, Chow KM, Luk CC, et al. Expression of miR-146a and miR-155 in the urinary sediment of systemic lupus erythematosus. Clin Rheumatol 2012;31:435–40.

[128] Perez-Hernandez J, Forner MJ, Pinto C, Chaves FJ, Cortes R, Redon J. Increased urinary exosomal microRNAs in patients with systemic lupus erythematosus. PLoS One 2015;10:e0138618.

[129] Sole C, Cortes-Hernandez J, Felip ML, Vidal M, Ordi-Ros J. miR-29c in urinary exosomes as predictor of early renal fibrosis in lupus nephritis. Nephrol Dial Transplant 2015;30:1488–96.

[130] Li J, Tian H, Yang J, Gong Z. Long noncoding RNAs regulate cell growth, proliferation, and apoptosis. DNA Cell Biol 2016;35:459–70. May 23, 2016 [Epub ahead of print].

[131] Wu GC, Pan HF, Leng RX, Wang DG, Li XP, Li XM, et al. Emerging role of long noncoding RNAs in autoimmune diseases. Autoimmun Rev 2015;14:798–805.

[132] Mourtada-Maarabouni M, Hedge VL, Kirkham L, Farzaneh F, Williams GT. Growth arrest in human T-cells is controlled by the non-coding RNA growth-arrest-specific transcript 5 (GAS5). J Cell Sci 2008;121:939–46.

[133] Haywood ME, Rose SJ, Horswell S, Lees MJ, Fu G, Walport MJ, et al. Overlapping BXSB congenic intervals, in combination with microarray gene expression, reveal novel lupus candidate genes. Genes Immun 2006;7:250–63.

[134] Wu Y, Zhang F, Ma J, Zhang X, Wu L, Qu B, et al. Association of large intergenic noncoding RNA expression with disease activity and organ damage in systemic lupus erythematosus. Arthritis Res Ther 2015;17:131.

[135] Duarte JH. Connective tissue diseases: large intergenic noncoding RNA linked to disease activity and organ damage in SLE. Nat Rev Rheumatol 2015;11:9.

[136] Lian X, Xiao R, Hu X, Kanekura T, Jiang H, Li Y, et al. DNA demethylation of CD40l in $CD4^{+}$ T cells from women with systemic sclerosis: a possible explanation for female susceptibility. Arthritis Rheum 2012;64:2338–45.

[137] Liao J, Liang G, Xie S, Zhao H, Zuo X, Li F, et al. CD40L demethylation in $CD4^{+}$ T cells from women with rheumatoid arthritis. Clin Immunol 2012;145:13–8.

[138] Lleo A, Liao J, Invernizzi P, Zhao M, Bernuzzi F, Ma L, et al. Immunoglobulin M levels inversely correlate with CD40 ligand promoter methylation in patients with primary biliary cirrhosis. Hepatology 2012;55:153–60.

[139] Dai R, Lu R, Ahmed SA. The upregulation of genomic imprinted DLK1-Dio3 miRNAs in murine lupus is associated with global DNA hypomethylation. PLoS One 2016;11:e0153509.

[140] Rauen T, Hedrich CM, Juang YT, Tenbrock K, Tsokos GC. cAMP-responsive element modulator (CREM)alpha protein induces interleukin 17A expression and mediates epigenetic alterations at the interleukin-17A gene locus in patients with systemic lupus erythematosus. J Biol Chem 2011;286:43437–46.
[141] Zhao M, Liu Q, Liang G, Wang L, Luo S, Tang Q, et al. E4BP4 overexpression: a protective mechanism in CD4+ T cells from SLE patients. J Autoimmun 2013;41:152–60.
[142] Koelsch KA, Webb R, Jeffries M, Dozmorov MG, Frank MB, Guthridge JM, et al. Functional characterization of the MECP2/IRAK1 lupus risk haplotype in human T cells and a human MECP2 transgenic mouse. J Autoimmun 2013;41:168–74.
[143] Vieira SM, Pagovich OE, Kriegel MA. Diet, microbiota and autoimmune diseases. Lupus 2014;23:518–26.
[144] Zhang H, Liao X, Sparks JB, Luo XM. Dynamics of gut microbiota in autoimmune lupus. Appl Environ Microbiol 2014;80:7551–60.
[145] Dozmorov MG, Coit P, Maksimowicz-McKinnon K, Sawalha AH. Age-associated DNA methylation changes in naive CD4+ T cells suggest an evolving autoimmune epigenotype in aging T cells. Epigenomics 2017;9:429–45.
[146] Zhao M, Qin J, Yin H, Tan Y, Liao W, Liu Q, et al. Distinct epigenomes in CD4+ T cells of newborns, middle-ages and centenarians. Sci Rep 2016;6:38411.
[147] Zhang Y, Zhao M, Sawalha AH, Richardson B, Lu Q. Impaired DNA methylation and its mechanisms in CD4+ T cells of systemic lupus erythematosus. J Autoimmun 2013;41:92–9.
[148] Wu H, Zhao M, Tan L, Lu Q. The key culprit in the pathogenesis of systemic lupus erythematosus: aberrant DNA methylation. Autoimmun Rev 2016;15:684–9.

CHAPTER

2

Immune Cell Development and Epigenetics

*Takashi Sekiya**,**

*Keio University School of Medicine, Shinjuku-ku, Tokyo, Japan;
**Section of Immune Response Modification, The Research Center for Hepatitis and Immunology, National Center for Global Health and Medicine, Ichikawa, Chiba, Japan

OUTLINE

2.1 OVERVIEW OF EPIGENETIC REGULATION OF IMMUNE CELL DEVELOPMENT

Epigenetic mechanisms for cell development have been reported in an increasing number of studies. It reflects the importance of the covalent modification of DNA and histones, the remodeling of chromatin structures, and the noncoding RNA-mediated regulation of chromatid DNA in the correct patterning of diverse cell types. The transcriptional effects of epigenetic regulation are multidimensional (Fig. 2.1) as they not only include on/off regulation, but also include moderation of the kinetics and robustness, maintenance of transcriptional status, and responsiveness or unresponsiveness to external stimuli of transcription. Accordingly, epigenetic modifications coordinate

The Epigenetics of Autoimmunity. http://dx.doi.org/10.1016/B978-0-12-809912-4.00002-7

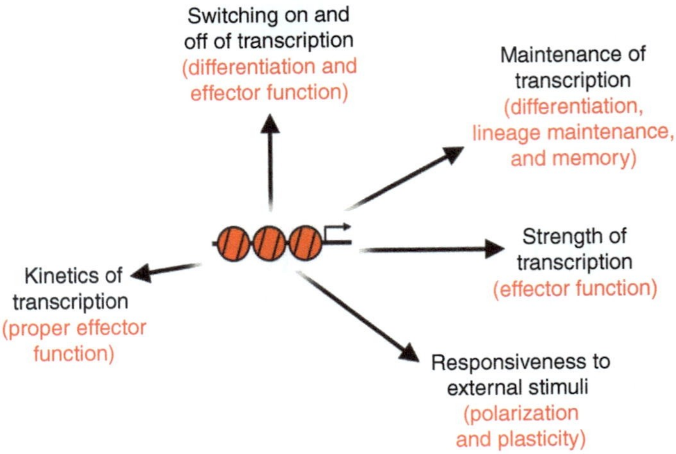

FIGURE 2.1 **"Multidimensional" regulation of gene activities by epigenetic mechanisms.** Gene activities regulated by epigenetic modifications (written in *black*) and their resulting cellular traits (written in *red*) are described. Epigenetic modifications not only regulate the switching on and off of transcription, but also the kinetics and robustness and the maintenance of the transcriptional status. Epigenetic modifications also define the responsiveness or unresponsiveness of genes to external stimuli.

various aspects of immune cell biological events, including initial fate decisions, maintenance of differentiation, effector functions, and plasticity (Fig. 2.1).

This chapter initially summarizes the basic epigenetic mechanisms with descriptions of the epigenetic modifications and their currently known effects and of the ensuing multidimensional regulation of cell development and function. To describe how epigenetic modifications regulate transcription and cell development, the present descriptions include various immune cells as well as other nonimmune cell types. Subsequent sections describe the epigenetic regulation of the development of individual immune cell subsets from both innate and adaptive immunity.

2.2 OUTLINE OF EPIGENETIC MODIFICATIONS AND THEIR OUTPUTS

The eukaryotic genome is packaged in chromatin structures that comprise the nucleosome unit, which is composed of 146 bp of DNA wrapped around octamers of histones H2A, H2B, H3, and H4. Epigenetic modifications include DNA methylation and more than a dozen different types of posttranslational modifications on the N-terminus of histone tails, including acetylation, methylation, ubiquitination, phosphorylation, sumoylation, and citrullination. Other contributors to epigenetic regulation include histone variants [1] and noncoding RNAs (i.e., long noncoding RNAs) [2], which will be described in depth in other chapters in this book.

The effects of each of these modifications have been extensively analyzed in a variety of cell types. As mentioned above, the modes of epigenetic transcriptional regulation range from dichotomous to continuous and can modulate the responsiveness of genes to external stimuli. The contributions of specific epigenetic parameters to gene regulation are discussed in the following paragraphs.

2.2.1 Epigenetic Modifications That Switch Transcription On and Off

Nucleosomal DNA is less accessible to the majority of transcription factors than naked DNA elements, and higher-order chromatin structures of condensed nucleosomal units are even more inaccessible to the transcriptional machinery. This mechanism fundamentally underlies the cell-specific and spatially- and temporally-regulated gene expression. Transcription is initiated by changes in repressive chromatin structures using methods such as nucleosome repositioning, loosening of condensed chromatin structures, and the removal of repressive histone modifications and repressive chromatin modifiers (Fig. 2.2A). Another important principle of epigenetics is that most known modifications are reversible [3], therefore the balance of incorporation and removal of certain epigenetic modifications is a critical determinant of transcription.

Histone acetyltransferases (HATs) acetylate the N-terminal lysine residues of histone core proteins in nucleosomes (histone acetylation), generally resulting in permissive chromatin structures and transcriptional activation [4,5]. However, the acetylation of histones can be reversed by histone deacetylases (HDACs), and many aspects of gene repression are mediated by these enzymes [6]. Thus the interplay between HATs and HDACs results in dynamic transitions of chromatin structure and results in the switching on and off of transcription.

In contrast to histone acetylation, histone methylation results in both transcriptional activation and repression, depending on the target residues. Methylation of H3K4 and H3K79 residues is commonly associated with gene activation. The histone lysine methyltransferases (KMTs) MLL and KMT4/DOT1L mediate H3K4 trimethylation and H3K79 methylation, respectively, and both factors are associated with a permissive chromatin status [7–11]. Modifications of H3K9, H3K27, and H4K20 residues are commonly associated with gene silencing. Specifically, H3K9 is di- and trimethylated by the KMTs G9a and SUV39H1, respectively, while H3K27 di- and trimethylation is mediated by the polycomb protein, Ezh2 [12–16]. Importantly, nearly all known histone methylations can be reversed by histone lysine demethylases. Although a H3K79 demethylase has not yet been identified, evidence suggests that H3K79 methylation is also reversible [17].

Chemical modifications of DNA bases, particularly methylation of the fifth position of cytosine in a symmetrical CpG context, are among the best-understood epigenetic mechanisms and play crucial roles in the regulation of gene expression. DNA methylation is generally repressive and is performed by DNA methyltransferase (DNMT) 1, DNMT3A, and DNMT3B, which deposit and maintain methyl groups on cytosine residues [18,19]. DNA methylation is also reversible through enzymatic removal ("active" process) or dilution through DNA replication ("passive" process). Enzymatic removal is mediated by the ten-eleven translocation 1–3 (TET1–3) family of enzymes, which convert 5-methylcytosine (5mC) to 5-hydroxymethylcytosine in an oxidation-driven reaction, which generates the intermediates 5-formylcytosine and 5-carboxylcytosine [20–22]. Subsequent enzymatic excision of these modified bases by DNA glycosylases leads to complete demethylation of DNA [23].

The two repressive epigenetic regulations, histone deacetylation and DNA methylation, are interconnected in many cases. It has been shown that a methylated CpG can recruit methyl CpG binding domain protein, Mbd2, which in turn recruits HDACs and results in histone deacetylation around the methylated CpG and transcriptional repression [24]. Interactions between histone modifications and DNA methylation in fungal and plant models indicated that the histone H3K9 methylation was required for DNA methylation [25,26]. Moreover, a counter interaction between activating histone modifications and DNA methylation has been described. Specifically, CXXC-type zinc finger protein 1 (CFP1) recruits activating KMTs and prevents DNA methylation

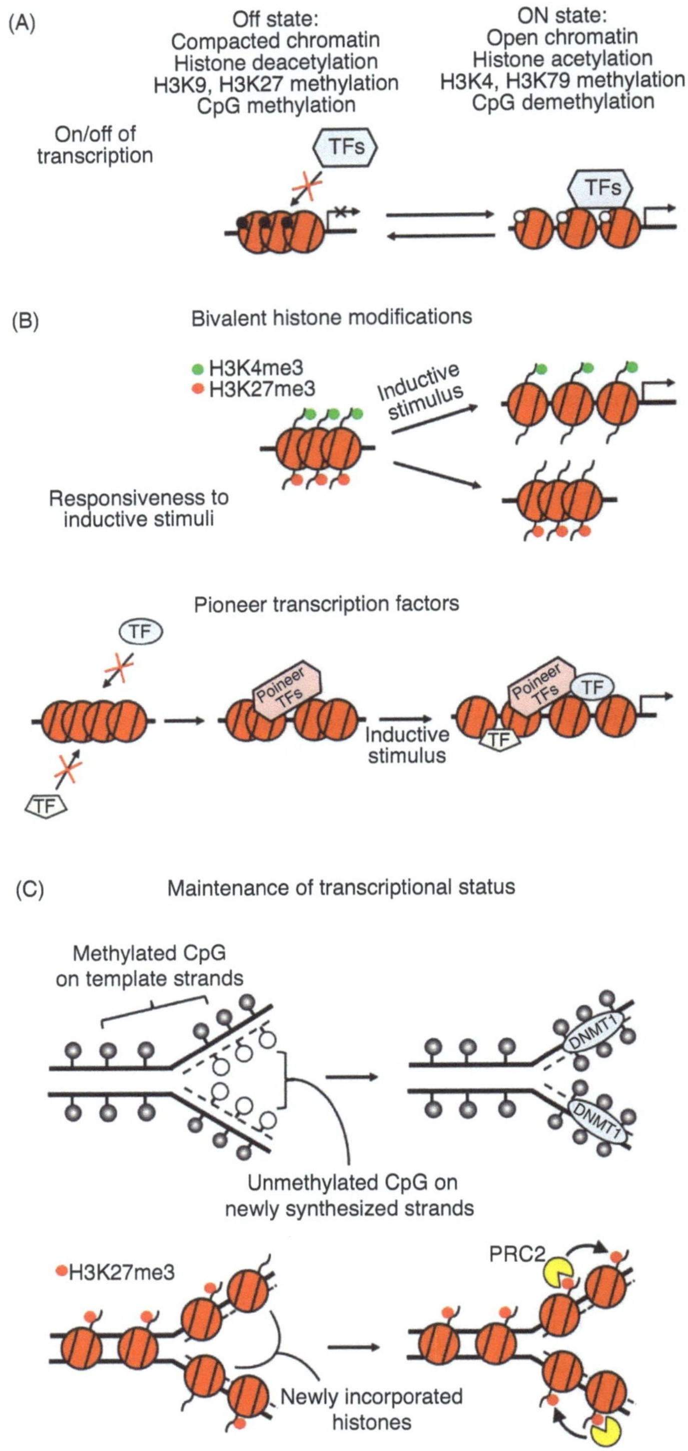

FIGURE 2.2 **Epigenetic regulation of gene activity.** (A) Examples of epigenetic mechanisms that switch on and off of transcription. The balance of incorporation and removal of a certain epigenetic modification determines the chromatin structure and the accessibility of transcriptional machineries. (B) Examples of epigenetic mechanisms that "poise" the gene to be activated by inductive stimuli, including bivalent modifications (upper) and pioneer transcription factors (lower). (C) Examples of epigenetic modification transmission during cell division, including transmission of DNA methylation by DNMT1 (upper) and H3K27me3 by the PRC2 complex. *TFs*, transcription factors.

at CpG-dense regions (CpG islands) [27,28]. In accordance, CpG islands that are prevalent at transcription start sites of housekeeping and developmental regulator genes are largely resistant to DNA methylation.

The epigenetic mechanisms described above, including histone modifications and DNA methylation, exert regulatory activities through direct biophysical consequences of the modification or through catalytic activities of epigenetic modifiers that recognize the modified histones and DNA residues. Biophysically, histone acetylation neutralizes basic charges on histones, leading to loosening of the nucleosomal DNA wound around histone proteins and thus improving access for transcription factors. Catalytic examples of regulation include the actions of ATP-dependent chromatin remodeling complexes [i.e., SWItch/sucrose nonfermentable (SWI/SNF) and nucleosome remodeling factor (NURF)] and ATP-dependent nucleosome remodeling complexes. ATP-dependent chromatin remodeling complexes slide nucleosomes and insert or remove histone molecules to regulate the accessibility of transcription factors. Many chromatin remodeling factors have domains that recognize modified histones in one or more of their subunits. Among these, bromodomain interacts specifically with acetylated histone tails to mediate communication between histone acetylation and chromatin remodeling [29]. In contrast, repressive histone modifications are recognized and translated by structural protein complexes that construct silent higher-order structures of chromatin. For example, H3K9me2/3 is recognized by the chromodomain of heterochromatin protein 1 (HP1) [30,31], which then propagates heterochromatin by recruiting the H3K9 trimethyltransferase Suv39h1, and by self-oligomerization [32,33]. The repressive histone modification H3K27me2/3 is catalyzed by Ezh2, which is an enzymatic subunit of the polycomb repressive complex 2 (PRC2) [13–16]. H3K27me3 is then recognized by the chromobox (CBX) component of PRC1, leading to the construction of a higher-order chromatin structure driven by the intrinsic ability of PRC1 to compact chromatin [34].

2.2.2 Epigenetic Modifications Poise Genes for Activation

Epigenetic modifications that endow potential for transcriptional initiation determine the cellular responsiveness or nonresponsiveness to external stimuli. One of the epigenetic mechanisms that endow activation potential is known as a "bivalent histone modification" (Fig. 2.2B). A"bivalent" modification represents an overlapping presence of both permissive H3K4me3 and repressive H3K27me3 histone modifications at the same locus [35]. These "bivalent chromatins" are usually seen at promoters of key developmental regulators in stem or progenitor cells. Although genes with this modification are not expressed due to the presence of H3K27me3, they are poised to be quickly activated upon loss of H3K27me3 during differentiation, due to the remaining presence of H3K4me3. Bivalent modifications were originally identified in embryonic stem cells, but have since been shown to play important roles in immune cell development as well.

Few known transcription factors have the ability to access their target sites buried in condensed chromatin and further potentiate the genes for transcription. Transcription factors with such an extraordinary ability are known as "pioneer transcription factors." Among these, the FoxA family of transcription factors are recognized as the prototype and were initially discovered due to their ability to stably bind to nucleosomal DNA and access target sites in highly compacted nucleosome arrays [36–39]. Initial binding of FoxA factors to enhancer regions of *alb1* does not induce transcription; however, it results in partial opening of the condensed chromatin structure and subsequent binding by other transcription factors, including GATA4 (Fig. 2.2B). Hence FoxA1 and FoxA2 play crucial roles in the developing endoderm for *alb1* gene expression in early hepatogenesis [40,41].

Although such pioneering transcriptional activity has not yet been well documented in immune cell development, recent technological advances are expected to detect such activities in the immune system.

2.2.3 Epigenetic Maintenance of the Transcriptional Status

Epigenetic modifications are often propagated faithfully at cell division, leading to continuity of identity and cellular memory. Maintenance of DNA methylation is largely mediated by DNMT1, which binds hemimethylated CpG sites after DNA replication and methylates cytosine residues on newly synthesized strands. In addition to DNA methylation, histone methylation patterns are also maintained during cellular replication. PRC2 complexes maintain histone modifications by associating with H3K27me3 during DNA replication and by methylating H3K27 on daughter strands during the incorporation of newly synthesized histones [42]. Furthermore, the embryonic endoderm development (EED) subunit of the PRC2 complex binds to H3K27me3 and allosterically activates its methyltransferase activity [43]. These mechanisms collectively enable the propagation and transmission of H3K27me3 signatures from mother to daughter cells. Inheritance of H3K9 methylation through multiple cell divisions was also demonstrated, even in the absence of cis-acting transcription factors and target DNA sequences [44]. The intrinsic affinity of the HP1 chromodomain for methylated H3K9 partially explains the inheritance of this histone modification through cell division [30] (Fig. 2.2C).

2.3 EPIGENETIC REGULATION OF IMMUNE CELL DEVELOPMENT

During immune cell development, multipotent precursors with various fates progress towards terminal differentiation with increasingly restricted developmental potential, simultaneously acquiring traits of specific cell lineages. Some common characteristics of transcriptional programs for the development of immune cell subsets include repression of precursor phenotypes, induction of developmental genes, and repression of genes of opposing cell lineages. The increasing limitation of fate choices during development, as well as the self-renewal capacity of multipotent progenitors, are mediated by cooperative actions of transcription factors and epigenetic modifiers, and these processes have been demonstrated for the development of both the innate and adaptive immune cells.

2.3.1 Epigenetic Regulation of Innate Immune Cells

Innate immune responses are the first line of defense against infection, comprising powerful activities that are harmful when temporally or spatially inappropriate. Hence the concerted actions of transcription factors and epigenetic modifiers are critical for generating the necessary cell lineages and context-specific gene expression in differing innate immune cell types, including neutrophils, macrophages, ILCs, basophils, mast cells, and dendritic cells. This section describes the epigenetic regulation of the development of macrophages and ILCs.

2.3.1.1 Macrophages

Macrophages are present in most tissues and they maintain tissue homeostasis by secreting trophic factors and clearing debris and dead cells, thus preventing excessive inflammation in

response to external stresses under physiological conditions [45,46]. Upon infection or tissue injury, activated macrophages kill microbes and produce inflammatory agents such as cytokines and chemokines. Upon activation, macrophages polarize towards specific functional phenotypes, depending on the pathogens and the cytokine profile of the microenvironment [47–49]. Accumulating evidence shows the important roles of epigenetics in macrophage development, polarization, and function [50].

2.3.1.1.1 Epigenetic Regulation of the Early Stages of Macrophage Development

Macrophages are a subset of myeloid cells that develop from hematopoietic stem cells (HSCs), and the roles of epigenetics in this multistep process are increasingly understood. Epigenetic modifications that characterize multipotent progenitor cells are removed during development and those that specify macrophages are established. Although HSCs do not express the myeloid-lineage specifying transcription factor CCAAT enhancer binding protein (C/EBP), the gene that encodes this transcription factor (*Cebpa*) is already potentiated to be expressed. This is done by a specific type of histone modification, "bivalent modification" [51] described above in Section 2.2.2 (Fig. 2.2B). After the onset of myeloid cell development, H3K27me3 modifications are removed from the *Cebpa* loci, while H3K4me3 modifications remain, and thus the expression of this gene is initiated (Fig. 2.3A).

DNA methylation also plays an important role in the early phase of myeloid development, and DNMT1 is critical for preventing premature differentiation of HSCs [52]. DNA methylation levels in cells from *Dnmt1*-hypomorphic mice were lower at signature myeloerythroid progenitor genes, including those encoding Gata1, inhibitor of DNA binding 2 (Id2), and C/EBPα, which led to their aberrantly high expression [52] (Fig. 2.3A). In contrast, the self-renewal and multipotency transcription factors, Runx1 and Gata3, were silenced by methylation of their corresponding gene loci by Dnmt3a and Dnmt3b [53].

After the initial specification of myeloid progenitors from HSCs, key transcription factors including C/EBPα, C/EBPβ, and PU.1 play crucial roles in the development of granulocyte–monocyte progenitors, and subsequent continuous expression of PU.1 specifies monocyte and macrophage lineages. The relative dosage of PU.1 versus C/EBP regulates the macrophage or neutrophil lineage decision, with high PU.1 favoring macrophage development and high C/EBP favoring neutrophil development [54]. As PU.1 has the ability to recruit both TET2 and DNMT3B to target sites, it is speculated that PU.1 regulates DNA methylation changes in both directions to facilitate or repress gene expression [55].

2.3.1.1.2 Epigenetic and Transcriptional Regulation of Polarization and Effector Function in Macrophages

After terminal differentiation, macrophages exert their function upon activation. It has been suggested that the polarization of macrophages into M1 or M2 subsets depends on the stimuli and microenvironmental variables [47–49]. "Classical" activation is mediated by proinflammatory cytokines (i.e., TNF-α) and microbial products (i.e., toll-like receptor (TLR) ligands) and skews cells towards the M1 phenotype, while "alternative" activation in the presence of type 2-associated immune mediators (i.e., IL-4 and IL-13) drives terminal differentiation towards the M2 phenotype [56]. M1 macrophages have host defense activities that result in pathogen death and are characterized by their high expression of proinflammatory molecules (i.e., cytokines TNFα, IL-1β, and IL-12), antimicrobial molecules, and reactive oxygen species. In contrast, M2 macrophages play important roles in tissue repair and the resolution of inflammation [56].

Extensive epigenetic investigations of macrophage polarization revealed substantial post-translational modifications of histones and CpG DNA motifs, and ATP-dependent nucleosome

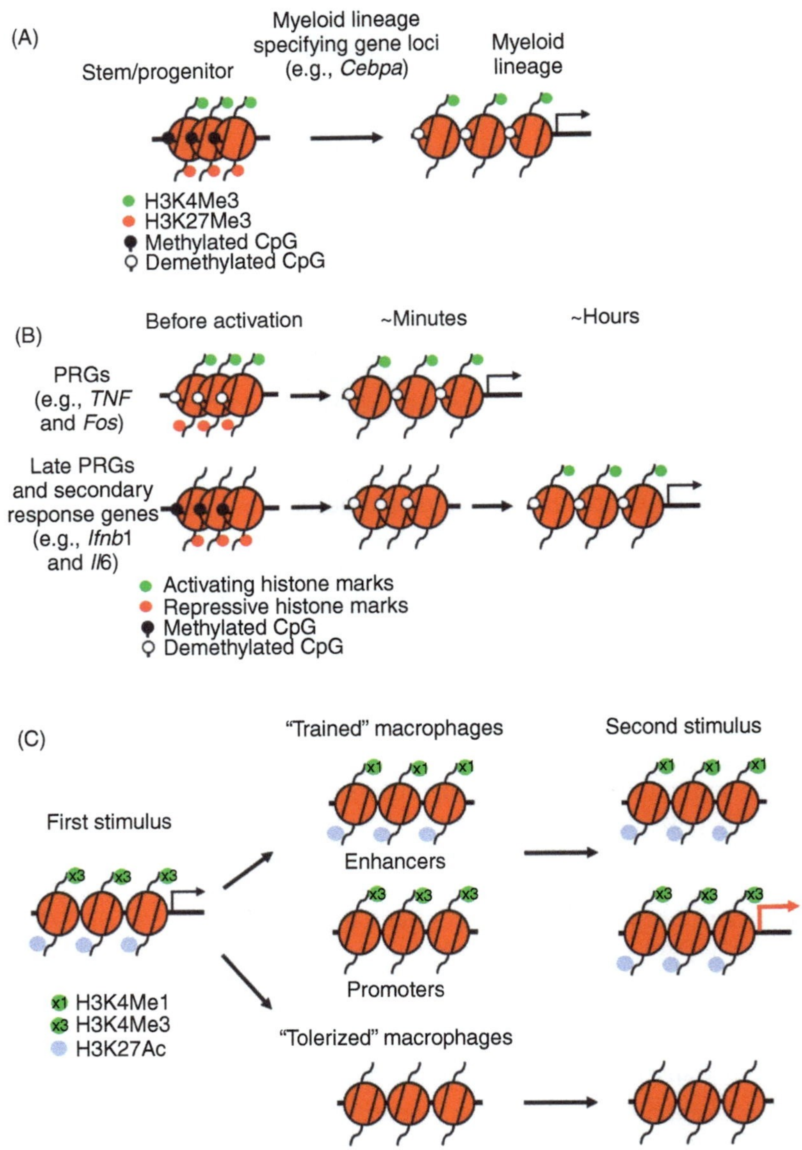

FIGURE 2.3 **Epigenetics of macrophage gene expression.** (A) Epigenetic regulation of some of the myeloid lineage-specifying genes (i.e., *Cebpa*). The gene expression remains silent but is potentiated with bivalent histone modifications in stem/progenitor cells. (B) Epigenetic regulations that explain the kinetic difference between primary response genes (PRGs), late PRGs, and secondary response genes. (C) Epigenetic regulations that establish "trained" and "tolerized" macrophage status. *Red bold* arrow shows an augmented transcription in the secondary stimulation of trained macrophages.

remodeling in this process. TLR signaling is considered to be the main driver of M1 polarization and function, and is mediated by mitogen-activated protein kinase (MAPK), NF-κB, and interferon-regulatory factors (IRFs), which induce inflammatory cytokines such as TNFα, IL-1β, IL-6, and IL-12p40. Important epigenetic observations show that the regulatory regions of TLR target genes are poised for induction by the master transcription factors of myeloid, monocyte, or macrophage cell development (i.e., PU.1 and C/EBPα) [57–60]. Even resting macrophages carry gene promoters with the permissive histone modifications, H3K4me3 and acetylated H3. Their enhancers are found in the open-chromatin state, as suggested by their hypersensitivity to nucleases, PU.1 binding, and H3K4me1 modification [61–64]. Prior to transcriptional activation, poised promoters are also further primed for rapid and strong responses to infectious challenges, in part by IFN-β, which maintains macrophages in the primed state via Jak/Stat signaling [65]. HDAC3 is required for the expression of *Ifnb*, potentially because it represses *Ptgs1*, which encodes a repressor of *Ifnb* [66]. However, as resting macrophages also have repressive histone modifications (i.e., H3K9me3, H3K27me3, H4K20me3, and deacetylated histones) at these loci, the transcription of M1 effector genes is limited in the absence of TLR signaling. The transcriptional repressor BCL-6 is responsible for incorporation of the repressive modifications by recruiting corepressor complexes, including HDACs and histone demethylases [57,67]. The mechanism through which TLR stimulation removes the repressive histone modifications and leads to the transcriptional initiation of M1 effector genes are described below.

First, TLR pathway activation removes repressive histone methylations by inducing demethylases (i.e., JMJD3, JMJD2d, AOF1, and PHF2) [68–71]. Second, the HAT p300 is recruited to the same loci upon lipopolysaccharide (LPS) stimulation, and overrides the repressive modifications to induce inflammatory genes. As these loci are enriched in binding sites for transcription factors such as NFκB, IRFs, AP-1, and PU-1, these factors likely recruit further epigenetic modifiers. Third, in addition to histone modifications, the ATP-dependent chromatin remodeling factors SWI/SNF induce genes including *Il6* and *Il12b* [64,72]. This cooperative function of permissive histone modifications and chromatin remodeling promotes the recruitment of transcription factors and RNA polymerase II to the promoters, leading to accelerated transcriptional initiation and elongation of M1 effector genes.

Promoters of TLR-signaling target genes also exhibit distinct types of CpG methylation patterns. Whereas promoters of *Tnf*, *Fos*, and *Nfkbia* contain CpG islands that are resistant to methylation, those of *Ifnb1*, *Il12b*, and *Il6* possess low-density CpG promoters that are often methylated before activation. Promoters for the latter class of genes also have H3K4me3 and H4Ac modifications at low levels [64,72]. These observations indicate that the collective differences in the chromatin status before cell stimulation can kinetically regulate target-gene expression, which is composed of two distinct phases of gene induction: "primary response genes" *Tnf*, *Fos*, and *Nfkbia* with lower CpG methylations are induced within minutes, whereas the "secondary response genes" *Il12b* and *Il6* with repressive epigenetic modifications are transcribed hours after cell stimulation (Fig. 2.3B).

The histone demethylase, JMJD3, plays an important role during M2 activation by removing repressive H3K27me3 from regulatory regions of M2-activating genes (i.e., *Irf4*) [73]. JMJD3 is induced by IL-4/Stat6 signaling and then bound the M2 genes *Chi3l3*, *Retnla*, and *Arg1* [74]. Several reports have shown an important role of JMJD3 in M2 polarization. In contrast, HDAC3, which is a positive regulator of M1 polarization, has been shown to repress M2 programs [75].

2.3.1.1.3 Epigenetic Insights into "Trained" and "Tolerized" States of Macrophages

As described above, much of the epigenetic landscape in macrophages is established during cellular development, and polarization occurs after exposure to pathogens. However, recent studies revealed that pathogens and other environmental stimuli also induce epigenetic modifications that can lead cells to a "tolerized" state or "train" macrophages for a more robust response upon a secondary challenge. Such tolerized or trained states are now recognized as immunological "memory" of the innate arm of immunity.

Several epigenetic analyses have reported the mechanisms for the "training" of macrophages. Among these, macrophages that entered a trained status by β-glucan challenge showed stable changes in H3K4me3 at promoters of proinflammatory cytokines (i.e., TNFα, IL-6, and IL-18) [76], whereas no changes were observed in H3K27me3. Another study revealed that 17% of dynamic promoters and 40% of enhancers gained H3K27 acetylation in β-glucan-trained macrophages [77]. H3K4me1 marks persist at enhancers after a pathogen challenge, and contribute to the faster and stronger induction of multiple genes upon restimulation in trained macrophages [78] (Fig. 2.3C). Epigenetic modifications are also suggested to underlie the metabolic switch required for monocyte activation and "training," as metabolic gene expression is defined by both histone methylation and acetylation patterns in β-glucan-trained macrophages [79]. Another study showed that regulation of the stress-response transcription factor ATF7 is important for imprinting memory in LPS-stimulated macrophages [80]. Specifically, ATF7 repressed genes encoding the innate immune factors TNF-α, CCL3, and CXCL2 in unstimulated macrophages by recruiting the histone H3K9 dimethyltransferase G9a. In contrast, LPS stimulation induced phosphorylation of ATF7 via p38, leading to ATF7 release from chromatin and the concomitant removal of repressive histone H3K9me2 modifications. This mechanism led to sustained basal expression of target genes to enhance resistance to pathogens in trained macrophages.

In contrast to trained macrophages, after initial stimulation of TLR4 with LPS, cells enter a long-term refractory state known as "LPS tolerance" or "endotoxin tolerance," in which proinflammatory responses to subsequent TLR4 stimulation are attenuated. Under these conditions, the induction of genes encoding proinflammatory mediators (i.e., TNFα, IL-6, IL-12, and CCL3) is limited after secondary challenge, whereas genes for antiinflammatory molecules (i.e., IL-10, TGF-β, and IL-1RA) are nontolerized and lead to increased or sustained expression following a secondary challenge [81–83]. During endotoxin tolerance, H3K4me3 modifications are incorporated into promoters of both tolerized and nontolerized genes; however, only nontolerized gene promoters maintain the H3K4me3 modification [81] (Fig. 2.3C).

2.3.1.2 *Innate Lymphoid Cells (ILCs)*

Innate lymphoid cells (ILCs) are categorized to the innate arm of the lymphoid lineage and play important roles in immune defense, inflammation, and tissue remodeling. ILCs are further categorized into three groups, ILC1, ILC2, and ILC3, which show some degree of similarity to Th1, Th2, and Th17 cells, respectively, in terms of their cytokine and transcription factor expression profiles. ILCs are also morphologically similar to B cells and T cells and their development can be traced back to the same origin, common lymphoid progenitors (CLPs). However, ILCs are distinguished from B cells and T cells by their lack of clonotypic antigen receptor expression. ILCs are also characterized by their quicker response to activation (hours) than the adaptive immune system (days to weeks for full differentiation).

2.3.1.2.1 Development of CLPs and Common Helper ILC Progenitor (CHILP) Lineages

Hematopoiesis is a continuous process that is characterized by progressive limits of cell-fate potential. During lymphopoiesis, erythroid, megakaryocytoid, and myeloid potentials are progressively lost, which results in the generation of CLPs that have restricted developmental potentials for lymphoid fates. Several transcription factors, including Ikaros, E2A, PU.1, and GFI1, are associated with CLP generation [84]. These transcription factors activate lymphoid-restricted transcriptional programs and concomitantly repress gene programs of other cell lineages, thus leading to the development of cell populations that have restricted potential for B cell, T cell, and ILC fates. Committed precursors for the ILC family, termed common helper ILC progenitors ("CHILPs"), were recently identified and can give rise to several ILC subsets. Several transcription factors (i.e., Id2, NFIL3, and GATA3) have been suggested to play roles in CHILPs [85–89]. Although deficiencies of these transcription factors disrupt all known ILC subsets, epigenetic modifications by these factors have not yet been revealed in depth.

2.3.1.2.2 Natural Killer Cell (NK) Cells

NK cells develop from CLPs and are categorized to the ILC1 subset. NK cells were identified 40 years ago as the first member of the ILC family [90,91]. They recognize and eliminate tumor cells and virus-infected cells by releasing toxic agents (i.e., perforin and granzyme B) and by secreting cytokines (i.e., IFN-γ and TNFα). NK cell activities are regulated by complex interactions of activating and inhibitory receptors on their surface. The activities of activating receptors (i.e., NKG2D and CD16) are mediated by immunodominant tyrosine-based activation motifs (ITAM) on their cytoplasmic tails, while all inhibitory receptors [i.e., killer cell immunoglobulin-like receptors (KIR), CD158], contain tyrosine-based inhibitory motifs (ITIM) on their cytoplasmic tails that mediate inhibitory signaling upon ligation with immune receptors [i.e., major histocompatibility complex (MHC) class I] [92]. Epigenetic modifications that regulate the development of NK cells remain poorly characterized [93,94]; however, the known epigenetic modifications and their functions in NK cell development are summarized below.

First, it has been shown that the expression of *KIR* genes, which encode the inhibitory receptor described above, are regulated by DNA methylation [95,96]. In contrast, the importance of histone acetylation in *KIR* gene regulation is enigmatic. Incubation of NK cells with inhibitors of HDAC (HDACi) did not affect *KIR* gene expression [97]. NK cell effector functions depended largely on epigenetic regulation of corresponding effector molecules. Combined functions of perforin and granzymes are central to the NK-mediated death of tumor cells and virus-infected cells. In NK cells, the Ets family transcription factor, myeloid Elf1-like factor, directly binds and transactivates *PRF1* promoters and leads to the production of Perforin [98]. Two upstream enhancers of the *PRF1* gene contain binding sites for Stat5, thus are responsive to IL-2 receptor-activated signal transduction. The importance of these upstream enhancers was indicated in analyses of IL2Rb and Stat5b knockout mice, which showed significantly lowered Perforin expression [99]. In contrast, the epigenetic regulation of granzyme expression has not been elucidated in depth yet in NK cells; however, some hints have been obtained from studies using $CD8^+$ T cells. Transcription factors including AP1, IKAROS, RUNX1, CREB, and Eomes facilitate the transcription of the *GZMB* gene in these cells, seemingly by constituting an open chromatin structure in their promoters as revealed by their extended nuclease hypersensitivities [100–102]. Acetylation of histone H3K9 was also correlated with a higher expression of *GZMB* [102]. As another effector molecule of NK cells, IFN-γ is rapidly produced upon stimulation of NK cells, which is mediated in part by the sustained histone acetylation of conserved noncoding sequences (CNS) in the *IFNG* locus. These acetylation modifications are established early during NK cell development and CNS-22, which is located 22 kb upstream of the transcriptional start site of mouse *Ifng*, was shown to open the gene locus and allow expression of *Ifng* in NK cells [103].

Although NK cells are innate lymphocytes, they have been shown to express memory-like characteristics following infection with cytomegalovirus [104]. NK cell memory facilitates self-renewal, a longer life, and a robust expansion upon subsequent viral challenges. "Memory NK" cells can be identified by their high expression of CD57 and NKG2C and the absence of FcR, SYC, DAB2, EAT-2, and the transcription factors PLZF and IKZF2 [105]. Some epigenetic modifications have been shown to be associated with memory characteristics of NK cells. For example, dense CpG methylation was observed at the FcεRγ and EAT-2 promoters in these cells, explaining their low levels of expression [106]. The development of memory NK cells is dependent on IL12R/Stat4-signaling but not on IFN-γ [107]. The roles of IL12R/Stat4-signaling in the constitution of memory NK cells are still largely unknown; however, their roles in the establishment of $CD8^+$ memory T cells may provide some hints, as the DNA methylation profile of memory NK cells is more similar to that of $CD8^+$ T cells than to canonical NK cells [106,108]. A previous comparison of memory NK and canonical NK cells revealed 2372 genomic regions that were differentially methylated, whereas only 61 regions were differentially methylated between memory NK cells and $CD8^+$ T cells [106]. As IL12R/Stat4 signaling enhances the expression of the antiapoptotic molecules Bcl2 and BclXL and increases the sensitivity to IL-7 and IL-15, similar mechanisms are likely to contribute to memory NK cell formation. Comprehensive analyses of epigenetic modifications in memory NK cells with chromatin immunoprecipitation sequencing (ChIP-seq) may further elucidate the characteristics and differentiation mechanisms in these cells.

2.3.1.2.3 ILC2 cells

Group 2 ILCs (ILC2) express Th2 type cytokines such as IL-5 and IL-13 [109–111]. ILC2s express receptors for IL-25, IL-33, and thymic stromal lymphopoietin (TSLP) and are therefore poised to respond rapidly to those cytokines from epithelial and hematopoietic cells during inflammation or infection [85,109–118]. Hence ILC2s are activated in response to helminth infections, viral infections, and allergic reactions [109–111,116,117,119]. Increased numbers of activated ILC2s have been observed in human inflammatory syndromes, including allergic asthma [120]. As this cell subset was identified relatively recently, comparatively little is known about the epigenetic mechanisms that regulate their differentiation.

Several transcription factors that regulate ILC2 differentiation have been identified, including GATA3, GFI1, TCF1, and Rorα [112,121–126]. Cell populations that express GATA3 at high levels are highly enriched in mature ILC2, thus GATA3 is considered to be a signature transcription factor in these cells [121–123,127]. GATA3 transactivates type 2 cytokines and is important for effector functions and survival in mature ILC2. TCF1 is also essential for ILC2 development [125,128]; for example, ILC2 cells do not develop in mice that lack the TCF1 gene, *Tcf7*, therefore these mice cannot mount an ILC2-mediated innate type 2 immune response. Expression of *Rora* was detected in several ILC subsets, and its absence resulted in the selective reduction of ILC2 numbers, particularly in the bone marrow [124,129]. GFI1 is also involved in the development of ILC2 and is implicated in the maintenance of mature ILC2. In the absence of GFI1, ILC2 cells aberrantly express RORγt and produce IL-17 [126].

Transcriptional and epigenetic programs that regulate ILC2 development have been extensively researched after the recent discovery of this cell subset. Although several transcription factors play crucial roles in the development of ILC2 (as described above), how they affect the expression of their targets remains largely undefined.

2.3.2 Epigenetic Regulation of Adaptive Immune Cells

Defense against infectious pathogens is mediated by coordinated interactions between innate and adaptive immunity. B and T cells are key cellular components of adaptive immune systems and share features that characterize adaptive immunity, such as the ability to recognize pathogen components via clonally expressed cell surface receptors and the ability to expand upon antigen recognition with simultaneous acquisition of cell lineage-specific immune functions. Memory is another key feature of adaptive immunity and it enables cells to respond more rapidly and vigorously to previously encountered infections. The following sections describe the epigenetic mechanisms that mediate development, effector function, lineage maintenance, and immunological memory in B and T cells.

2.3.2.1 *B Cells*

B cells play a central role in humoral immunity through the secretion of antibodies. B-cell development from HSCs starts in the bone marrow and proceeds via CLP, progenitor B (pro-B), precursor B (pre-B), and immature B-cell stages. B-cell development is characterized by the orderly expression of fate-defining marker proteins and the rearrangement of immunoglobulin H and L chain gene loci. B cells that encounter antigens undergo proliferation, immunoglobulin class switching, and somatic hypermutation. Some populations of activated B cells differentiate into long-lived plasma cells or memory B cells, cooperating to lower the probability of reinfection and to quickly and efficiently eliminate familiar pathogens. Epigenetic modifications play crucial roles in all these events [130,131].

2.3.2.1.1 Epigenetic Regulation of Early B Cell Development

During the transition of CLP into B-cell lineages via pre–pro-B cells, transcription factors E2A, early B cell factor 1 (EBF1), and PAX5 play crucial combined roles in a hierarchical manner. E2A induces the expression of EBF1 in CLPs, which induces PAX5. E2A and EBF1 specify the B-cell lineage by activating B-cell genes in pre–pro-B cells [132,133]. PAX5 subsequently regulates B-cell commitment at the pro-B-cell stage by restricting the developmental potential of lymphoid progenitors to the B-cell-developmental pathway [134]. This hierarchical dependence of PAX5 function on EBF and E2A in B-cell development is exemplified in their regulation of *mb-1* (CD79a; Fig. 2.4A). During the development of pro-B cells from HSC and CLP, the promoter region of the early B-cell specific *mb-1* gene is progressively demethylated [135] by the cooperative functions of EBF and E2A. This epigenetic event facilitates PAX5-mediated transactivation of *mb-1* expression [135]. ATP-dependent chromatin remodeling complexes also regulate transactivation of *mb-1* by EBF and PAX5. Accordingly, short hairpin RNA (shRNA)-mediated knockdown of the catalytic subunits of the SWI/SNF chromatin remodeling complex, Brg1 and Brm, inhibited transactivation of *mb-1* by EBF and PAX5 in B cells [136]. In contrast, the ATP-dependent chromatin remodeling complex Mi-2/NuRD repressed remodeling of the *mb-1* promoter. ShRNA-mediated knockdown of the Mi-2 subunit of the Mi-2/NuRD complex enhanced chromatin accessibility of EBF and PAX5 to the *mb-1* promoter, thus accelerating demethylation of the locus [136].

PAX5 is an essential transcription factor in B-cell differentiation and maintenance as it induces B-cell-specific genes while repressing genes of other lineages [134]. Gene expression profiling studies revealed that 110 genes were suppressed by PAX5 in wildtype pro-B cells [137,138]. Conversely, PAX5 was shown to activate 170 genes in committed pro-B cells [138,139]. Activating histone modifications (i.e., H3K9Ac, H3K4me2, and H3K4me3) are important molecular mechanisms that mediate transactivation by PAX5 [140–142]. Indeed, PAX5 interacts with the basal transcription factor TFIID, the chromatin remodeling BAF complex, the HAT CBP, and the PTIP protein, which is a known component of MLL-containing H3K4 methyltransferase complex to mediate permissive histone modifications (i.e., H3K9Ac and H3K4me3) at target sites [142]. Repressive activities of

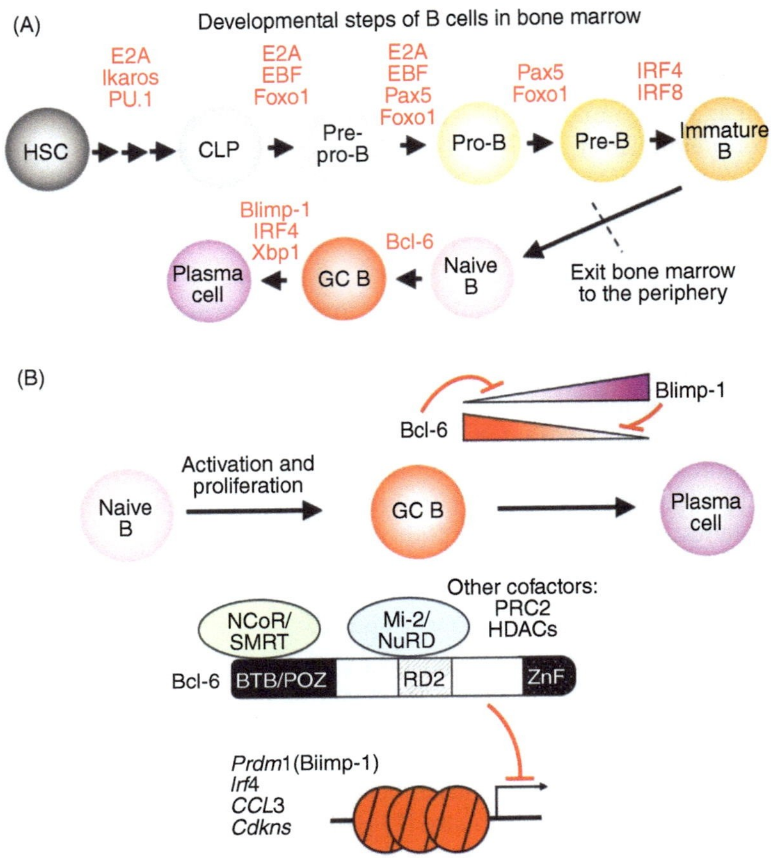

FIGURE 2.4 **Transcriptional control of B cell development.** (A) Schematic representation of B-cell development and the key transcription factors. The position of the transcription factors represents the stage in which they work. (B) Reciprocal roles of Bcl-6 and Blimp-1 in peripheral B-cell differentiation. Bcl-6 and Blimp-1 repress expression of each other. Bcl-6 recruits distinct repressive chromatin remodeling complexes via different domains. *CLP*, common lymphoid progenitor; *GC*, germinal center; *HSC*, hematopoietic stem cell.

PAX5 are mediated by histone deacetylation of the promoters [142]. PAX5 was also shown to recruit NCoR1 corepressor complexes that contain HDAC3 to their repressive targets, such as *Tmem66*, *Ccr2*, *Ncf4*, and *P2rx7* [142,143]. Less than 50% of PAX5 activated genes and less than 25% of PAX5-repressed genes were regulated by direct binding of PAX5 to their promoters or enhancers, thus indicating the presence of secondary effects by other transcription factors that are regulated by PAX5. Transcription factors such as Ebf1, Bach2, Irf4, Tcf4, Aiolos, Id3, and Irf8 have been revealed as direct targets of PAX5 and regulate various aspects of development and B-cell function [142].

Foxo1 is another transcription factor that plays roles in B-cell development, particularly in the early stages. For example, B cells do not develop further than the pro-B-cell stage in Foxo1-deficient mice. During the development from pre–pro-B to pro-B-cell stage, E2A and EBF1 bind to regulatory elements in the *Foxo1* locus, then E2A, EBF1, and Foxo1 are wired together to interact with putative enhancer elements of B cell-specifying genes, including *Pax5*. Foxo1 presumably allows the binding of other B-cell-associated transcription factors to the enhancers through a series of permissive epigenetic modifications [133]. As several forkhead box (Fox) transcription factors, including Foxo, have the ability to bind to condensed chromatin and enable subsequent binding of other transcription factors, the so-called "pioneer" activities described above [38,39,144,145] (Fig. 2.2B), it is suggested that some Foxo1 activities during early B-cell development may also be mediated by such pioneering activities.

Taken together, these studies indicate that early B-cell development is orchestrated by hierarchical functions of key transcription factors (i.e., E2A, EBF1, Foxo1, and PAX5) which, in conjunction with IRF4, IRF8, and PU.1, regulate the epigenome modifications at specific gene loci (Fig. 2.4A).

2.3.2.1.2 Epigenetic Regulation of B Cell Antigen Receptor Construction

B-cell development is characterized by the recombination of loci that encode antigen receptors. Precisely orchestrated assemblies of V, D, and J gene segments generate vast diversities of antigen receptors in developing B cells, and epigenetic modifications play pivotal roles in these V(D)J recombinations. Permissive histone modifications allow access to target sites by recombination complexes, therefore antigen receptor loci that are poised for V(D)J rearrangement are typically marked by permissive histone modifications, including H3Ac, H4Ac, H3K4me2, H3K4me3, and H3K79me2 [146–155]. H3K4me3 is directly recognized by the plant homeodomain (PHD) domain of RAG2. This interaction between H3K4me3 and the PHD domain enables the precise recruitment of RAG2 to the receptor loci to be recombined. Furthermore, interactions of the PHD domain with H3K4me3 stimulate the cleavage activity of RAG proteins [156]. Permissive histone modifications also enhance the transcription of noncoding RNA at the loci from both sense and antisense strands, known as "germline transcripts," further increasing the accessibility for advancing recombination reactions [157–160].

DNA methylation also regulates the rearrangement of antigen receptor loci. In vitro studies showed that specific methylation within heptamer recombination signal sequences markedly reduced V(D)J cleavage without affecting Rag1/Rag2–DNA complex formation [161]. In the IgH locus recombination, although D_H and J_H gene segments were densely methylated prior to recombination, the recombined DJ_H product was demethylated [162]. These data collectively indicate that epigenetic modifications play critical roles in the precise control of V(D)J rearrangements.

2.3.2.1.3 Epigenetic Regulation of Germinal Center Reactions That are Balanced by Bcl-6 and Blimp-1 Activities

The clearance of pathogens and the formation of effective immunity requires the activation of B cells and their subsequent differentiation into antibody-secreting plasma cells and memory cells. Following V(D)J recombination, B cells leave the bone marrow and enter peripheral lymphoid tissues, where they encounter antigens and receive appropriate T-cell mediated activation. Germinal centers (GC) are sites in peripheral lymphoid tissues where B cells proliferate, switch classes of antigen receptors, and increase their affinity to antigens. At GC, B cells are also selected to continue to divide or to differentiate to long-lived plasma or memory cells [163–165]. Activated B cells that show high affinity to the antigens and are selected to survive may differentiate into plasma cells or memory cells. Plasma cells continually secrete high-affinity antibodies and reside in the bone marrow, where extrinsic factors from niche cells maintain their survival. Memory cells play central roles in the secondary response, where pathogens are cleared more quickly than in the primary response. Both plasma cells and memory cells cooperate to lower the probability of reinfection and contribute to the rapid elimination of pathogens upon reinfection. Numerous gene expression changes occur during the differentiation of naive B cells to memory or plasma cells [166].

Furthermore, aberrant differentiation of B-cell subsets following disturbances to epigenetic modifiers have been reported, thus indicating the importance of the epigenetic regulation of transcriptional programs during peripheral B-cell differentiation.

Transcriptional mechanisms of peripheral B-cell differentiation are well characterized. Among these, the balance of Bcl-6 and Blimp-1 transcription factor activities is particularly important (Fig. 2.4B). These transcription factors exhibit mutually exclusive patterns of expression. For example, following antigen encounters in the GC, Bcl-6 is upregulated in some B cells to regulate the expression of genes that are important for GC formation and function [167–169]. Bcl-6 generally works as a transcriptional repressor by interacting with Mi-2/NuRD chromatin remodeling complexes in GC B cells and by recruiting HDAC1 and HDAC2 to target sites. Moreover, Bcl-6 recruits the Mi-2/NuRD complex to plasma cell-specific genes (i.e., *CCL3*) and represses their expression [170]. NCoR and SMRT also mediate the repressive transcriptional activity of Bcl-6 [171,172], and because these proteins use HDAC3 to repress transcription, Bcl-6 may also repress its target genes by transactivating this histone deacetylase [173–175]. Bcl-6 utilizes different domains to associate with Mi-2/NuRD and NCoR and SMRT. The N-terminal BTB-POZ domain of Bcl-6 associates with NCoR and SMRT, whereas the central region of the protein interacts with Mi-2/NuRD [170,171,176] (Fig. 2.4B).

As another mediator of the transcriptional repressive activities of Bcl-6, associations with the PRC2 component lead to incorporation of H3K27me3 at specific gene loci [177,178]. Bcl-6 shares some common target sites with EZH2, an enzymatic component of the PRC2 complex, in GC B cells [179]. ChIP-on-chip experiments with GC B cells revealed a specific EZH2 regulatory program, in which genes are preferentially modified with H3K27me3 and are repressed only in these cells. GC B-cell-specific EZH2 targets were enriched with BCL-6 binding, and approximately half of the genes that were bound by both PRC2 and BCL-6 in wildtype GC B cells were upregulated in EZH2-deficient cells. In contrast, EZH2 deficiency had little effect on the expression of BCL-6 target genes that lacked the H3K27me3 modification [179]. Hence repression targets of the Bcl-6/PRC2 complex contain various genes that are important for GC B cell differentiation and cell cycle arrest, including *Blimp1*, *Irf4*, *Cdkn1a*, *Cdkn1b*, and *Cdkn2a*. Accordingly, one of the main functions of Bcl-6/PRC2 complexes is to prevent terminal differentiation of B lymphocytes into plasma cells in the absence of appropriate signals.

Conversely, Blimp-1 is expressed in plasma cells and acts as a master regulator of plasma cell differentiation [169] (Fig. 2.4A and B), with targeted disruption of Blimp-1 impairing the generation of plasma cells [180]. Blimp-1 expression was repressed by Bcl-6 and was induced in cells with low Bcl-6 expression levels, where it drove plasma cell differentiation [169,181]. On the other hand, Blimp-1 reciprocally inhibits the expression of *Bcl6* in plasma cells [182]. In addition, Blimp-1 activity is required for stable repression of GC B-cell-associated genes including *CIITA*, *Blnk*, *Igb*, and *Syk* [182], that are mediated by recruitment of the histone methyltransferase, G9a [183].

2.3.2.1.4 Epigenetic Mechanisms for Sustained Lineage Commitment in B Cells

The transcription factors EBF1 and PAX5 are important for the sustained lineage commitment of B cells [134,184–186], and the absence of either after commitment of B cells results in the derepression of genes of alternative lineages (i.e., ILC and T cells). Although cells lacking either EBF1 or PAX5 adopt T-cell fates, these transcription factors have different target specificities; for example, whereas EBF1 mainly represses genes encoding transcription factors (i.e., *Tcf7* and *Id2*), PAX5 mainly represses genes encoding receptors for alternative cell fates (i.e., *Csf1r* and *Notch1*). Thus the cooperative actions of EBF1 and PAX5 hardwire the sustained lineage commitment of B cells.

2.3.2.2 *T Cells*

T cells are a type of lymphocyte that expresses the T-cell antigen receptor (TCR). T cells are subclassified as either αβ or γδ T cells, which rearrange and express alpha and beta chains or gamma and delta chains of the TCR, respectively. Further subsets of αβ T cells include $CD4^+$ and $CD8^+$ T cells, with $CD4^+$ T cells mediating immune reactions against diverse antigens after terminal differentiation. T helper 1 (Th1) cells are predominantly induced by intracellular bacterial infections and activate the bactericidal activity of macrophages. Th2 cells emerge upon infection with extracellular parasites and mediate humoral immunity by facilitating IgE production by B cells and subsequent activation of mast cells. Regulatory T (Treg) cells are a distinct subset of $CD4^+$ T cell that have immune-suppressive activities, preventing excessive inflammation and mediating tolerance against self-antigens and nonharmful environmental antigens. $CD8^+$ T cells are cytotoxic T cells with cell-killing activities that eliminate infected cells, particularly those infected with viruses. Epigenetic regulation of early T-cell development is common to all T cell subsets, and independent epigenetic mechanisms regulate $CD4^+$ T and $CD8^+$ T cells.

2.3.2.2.1 Epigenetics of Early T Cell Development

The development of T lymphocytes from multipotent precursors in the thymus occurs in 10 phenotypically distinct stages. Hematopoietic progenitor cells that enter the thymus do not initially express CD4 or CD8, and are referred to as double negative (DN) cells. Cells progress through phenotypically distinct stages of maturation (i.e., DN1, DN2a, DN2b, DN3a, and DN3b) before expressing both CD4 and CD8 in the double-positive (DP) stage. During the early stages of T-cell development, T-cell gene expression programs are mobilized to prepare cells for TCR expression and the developmental potentials for alternative lineages are gradually excluded. Cells subsequently permanently downregulate either CD4 or CD8 and exit the thymus as naive CD4 or CD8 single positive cells.

Initial commitment of hematopoietic precursors to the T-cell fate is triggered by Notch signaling after contact of the Notch receptor with delta-family Notch ligands expressed by thymic epithelial cells [187]. Notch signaling initiates and sustains the differentiation of T cells by activating T-cell-associated genes and by repressing genes of progenitor cells and of alternative lineages [188–191]. However, Notch signaling is not the only mechanism for the repression of progenitor cell- or alternative lineage-associated genes, as different genes are applied with diverse types of histone modifications, with different kinetics [192]. This implies that repression of the progenitor cell- or alternative lineage-associated genes is not mediated by a single switch, but rather that a variety of distinct repressive mechanisms are employed for their repression. For example, genes encoding the B-cell transcription factors, Pax5 and Ebf1, are repressed by H3K27me3 throughout T-cell development. Conversely, the myeloid regulatory gene, *Cebpa*, is kept silent by bivalent modification with H3K27me3 and H3K4me3 [192]. Among the erythroid genes, *EpoR* is repressed via H3K27me3, whereas *Gata1* is repressed independently of this histone mark. Some genes show H3K27me3 modification after their repression, suggesting that the likely role of this modification is the stabilization of repression [192]. In contrast to epigenetic modifications for gene repression, activation of T-cell-associated genes are strongly and temporally correlated with histone acetylation. Hence histone acetylation represents a rate-limiting step in the activation of T-cell gene transcription.

DNA methylation also regulates early T-cell development [193,194]. Demethylation has been observed in many essential T-cell regulators, including genes that encode TCR components (i.e., CD3 molecules) and key developmental genes (i.e., *Runx3*, *LEF1*, *RORC*, *IKAROS*, *RAG*, and *Lck*) [194,195]. De novo methylation events are less commonly associated with gene suppression during T-cell development and are mostly restricted to the early T-cell commitment stage [194]. Instead of acting as a sole component for gene repression, a main role of DNA methylation seems to be the reinforcement of other silencing mechanisms that are already in place (i.e., repressive histone modifications). The importance of DNA methylation in early T-cell development can also be seen in DNMT1-defficient mice, which show approximately 90% reductions in the numbers of DP T cells [196].

Transcription factor networks, including TCF-1, GATA-3, and Bcl-11b, regulate the T-cell lineage commitment [197]. Expression of these transcription factors is robustly activated in T-cell precursors in response to Notch signaling and is then sustained at various levels during T-cell development. Deletion of Bcl-11b temporally extends the potential for myeloid differentiation of T-cell precursors and forces cell fates towards NK cell precursors [198–200]. Stem- or progenitor-associated genes fail to be repressed in Bcl-11b mutant cells, whereas T-cell transcription factors GATA3 and TCF-1 are fully induced in those cells [198]. TCF-1 and GATA3 have long been recognized as essential transcription factors for T-cell development [201,202], and *Tcf7* is directly activated by Notch signaling through an enhancer located 31.5 kb upstream of the promoter. TCF-1 subsequently plays essential roles during the early T progenitor stage [203,204]. Forced expression of TCF-1 in prethymic precursors activates most T-cell identity programs, including those involving Gata3 and Bcl11b, even in the absence of Notch signaling [204]. Hence TCF-1 takes on many parts of the T-cell developmental programs that are elicited by Notch signaling. Moreover, GATA3 binds to the upstream enhancer of *Tcf7*, indicating that the interaction between TCF-1 and GATA3 is mutual [192]. The studies summarized here suggest that a feed-forward circuit among key transcription factors operates in early T cell development.

2.3.2.2.2 Epigenetics of T Cell Development to CD4 and CD8 Lineages

The lineage choice between CD4 and CD8 T cells follows the rearrangement of TCRα and TCRβ genes in thymocytes. Several transcription factors have been shown to control lineage determination of the mutually exclusive CD4 and CD8 lineages. The zinc finger protein, ThPOK, and the runt domain transcription factors, Runx1 and Runx3, are particularly important in the development of CD4 and CD8 lineages, respectively.

The ThPOK gene (*Zbtb7b*) was initially identified as a responsible gene for the helper deficient phenotype of spontaneous mutant mice that were deficient in CD4 commitment but showed normal maturation of the CD8 lineage [205]. Forced expression of ThPOK during thymocyte development caused redirection of MHC class I-restricted thymocytes, such as CD8-fated cells, to the CD4 lineage. Thus ThPOK activity is necessary and sufficient for CD4 lineage commitment. One of the important roles of ThPOK in CD4 T-cell specification is its positive regulation on the CD4 molecule. There is a "silencer" region on the *Cd4* gene locus that leads to repression of its gene expression in non-CD4 T lineages. ThPOK directly binds to the *Cd4* silencer and antagonizes its repressive function [206]. This counteracting function of ThPOK is abrogated in the presence of HDACi, suggesting that ThPOK also represses factors that mediate the repressive activities of the *Cd4* silencer [207]. Conversely, Runx3 is upregulated during differentiation of CD8 cells in the thymus [208], and although disruption of Runx3 reduced $CD8^+$ T cell numbers, deletion of both Runx1 and Runx3 or their common obligatory dimerization partner, Cbfβ, abrogated the development of CD8 cells [209–211]. Furthermore, Runx-deficient MHC class I-restricted thymocytes failed to develop into CD8 cells, and aberrantly developed into CD4 cells [209,211].

ThPOK and Runx factors are mutual repressors, and whereas Runx3 directly binds to a silencer element in the *Zbtb7b* locus [211,212], ThPOK seems to represses Runx3 expression indirectly

[213,214] by inducing suppressor of cytokine signalling (SOCS) family cytokine signaling suppressors that abolish the cytokine signaling required for Runx3 induction [215]. Taken together, CD4 and CD8 lineage distinctions are regulated by the key transcription factors ThPOK and Runx, respectively, and their mutual repression enables unambiguous lineage fate decisions.

2.3.2.2.3 Epigenetic Regulation of Terminal Differentiation and Effector Function of CD4$^+$and CD8$^+$ T Cells

After their terminal differentiation, CD4$^+$ T cells exert their function as either helper T cell subsets (Th1, Th2, Th17, and Tfh cells) or T_{reg} cells. This section describes how epigenetics regulates naive T (Tnaive) cells, which retain enormous proliferative potential and functional plasticity, to undergo a program of proliferation and functional differentiation upon activation. Specific sets of transcription factors have been identified in the polarization of each Th and T_{reg} cell subset (Fig. 2.5A).

ChIP-seq analysis of H3K4me3 and H3K27me3 in Tnaive, Th1, Th2, Th17, and T_{reg} cells demonstrated the presence of H3K4me3 in genes that encode signature transcription factors and cytokines for the corresponding cell subsets [216]. For example, H3K4me3 modifications were identified in *Tbx21* (encoding Tbet) and *Ifng* loci in Th1 cells, and in *Gata3* and *Il4* loci in Th2 cells; however, these loci were shown to carry repressive H3K27me3 modifications in opposing cell subsets (i.e., *Tbx21* and *Ifng* loci in Th2 cells and *Gata3* and *Il4* loci in Th1 cells; Fig. 2.5C). Hence histone modifications play crucial roles in the specification of CD4$^+$ T-cell subsets. Bivalent histone modifications with H3K4me3 and H3K27me3 were frequently observed at signature transcription factor loci in opposing cell subsets (i.e., *Tbx21* loci in T_{reg} cells), thus suggesting functional plasticity among Th and T_{reg} cell subsets. Accordingly, *Tbx21* and *Ifng* genes were induced in T_{reg} cells upon polarization under Th1-skewing culture conditions [216]. However, although some level of functional plasticity was observed among the CD4$^+$ T-cell subsets, fluctuations of cell subsets would be undesirable in most situations. Therefore various epigenetic mechanisms have been shown to stabilize the lineage commitments of each CD4$^+$ T-cell subset. For example, disruption of Suv39H1, which mediates repressive H3K9me3 modification, resulted in aberrant induction of *Ifng* in Th2 cells after reculture in Th1-skewing conditions. As the absence of Suv39H1 did not disturb Th2 cell differentiation per se, it was suggested that Suv39H1-mediated H3K9me3 modifications keep Th1 effector genes silent in Th2-committed cells [217].

An increasing number of transcription factors have been shown to play important roles in CD4$^+$ T-cell polarization and maintenance. For example, the Nr4a family of nuclear orphan receptors are important in T_{reg} cell development [218,219] and for maintaining the silence of Th2 and Tfh effector genes in mature T_{reg} cells [220]. Simultaneous disruption of Nr4a1, Nr4a2, and Nr4a3 in mature T_{reg} cells resulted in aberrant induction of Th2 and Tfh effector genes in T_{reg} cells and converted the cells into pathogenic Th2 and Tfh cells. Nr4a factors mediate T_{reg} cell development and maintenance by regulating histone modifications at key genetic loci for T_{reg} and Th subsets, including *Foxp3*, *Ikzf4*, *Il4*, *Il13*, and *Il21* [218–220].

DNA methylation also plays an important role in Th and T_{reg} differentiation and is critical for polarizing naive CD4$^+$ T cells to Th1 and Th2 cells [221,222]. *IFNG*, which encodes the Th1 effector cytokine IFN-γ, is hypermethylated in human naive CD4$^+$ T cells and is only demethylated during Th1 cell differentiation [223]. Conversely, *IL4* and *IL13*, which encode the Th2 effector cytokines IL-4 and IL-13, respectively, are partially demethylated only during Th2 differentiation [224]. Furthermore, human Th17 cells show DNA demethylation in *IL17A* and *RORC*, which encode the Th17 specifying transcription factor RORγt, and these changes were correlated with

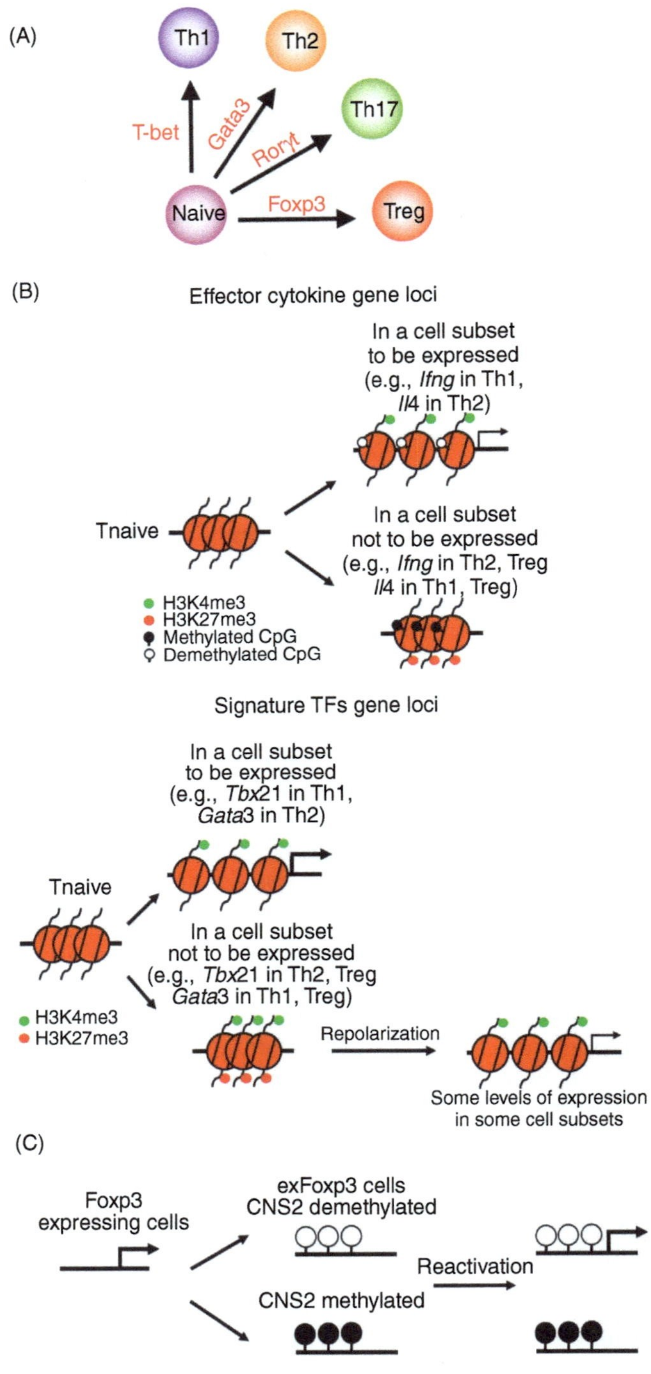

FIGURE 2.5 **Epigenetic regulation of T-cell development.** (A) Polarization of Th and T_{reg} subsets. In the differentiation of effector CD4$^+$ T cell subsets from naive cells, the lineage specifying transcription factors (*red*) play crucial roles. (B) Epigenetic events that occur at effector cytokine gene and lineage specifying transcription factor loci. Bivalent histone modification at genes that encode lineage-specifying transcription factors render some degree of plasticity. (C) Memorization of Foxp3 expression status by T_{reg} cells is mediated by demethylation of the *Foxp3* CNS2 enhancer region. *TF*, transcription factor.

specific expression of the genes [225,226]. DNA methylation at the CNS2 intronic enhancer region of *Foxp3* prevents aberrant expression of this gene in non-T_{reg} cell lineages. Thus in the absence of Dnmt1, TCR stimulation becomes sufficient to induce Foxp3 expression in Foxp3-negative conventional $CD4^+$ T cells, $CD8^+$ T cells, and NK cells [227,228]. In addition to the *Foxp3* CNS2 locus, T_{reg}-associated genes (i.e., *Ctla4*, *Il2ra*, *Tnfrsf18*, and *Ikzf4*) are also specifically demethylated in T_{reg} cells. Demethylation of these loci follows stimulation of the TCR during the development of T_{reg} cells and leads to their stable expression, thus contributing to the proper maintenance and function of T_{reg} cells [228]. Among these T_{reg} cell-specific demethylated loci, it is known that the *Foxp3* CNS2 locus is actively demethylated by the Tet family of demethylation factors [229–233].

Epigenetic modifications also play important roles in the regulation of $CD8^+$ T-cell effector functions. For example, heavily incorporated H3K27me3 modifications at the *Ifng* locus in naive $CD8^+$ T cells were removed upon activation and differentiation to effector cells, whereas the permissive histone modifications H3K9Ac and H3K4me3 were deposited to this locus. This epigenetic regulation enables spatial and temporal expressional control of *Ifng*, which encodes the $CD8^+$ T-cell effector cytokine, IFN-γ [234]. Furthermore, the *Ifng* locus had reduced levels of total histone H3 in activated $CD8^+$ T cells, suggesting that evacuation of nucleosomes from the locus allows transcriptional machinery to access the promoter. The *Gzmb* locus, which encodes another $CD8^+$ T-cell effector molecule, granzyme B, also showed similar epigenetic regulation during differentiation from naive to effector cells [235]. In addition to histone modifications, a recent genome-wide DNA methylation analysis in naive and effector $CD8^+$ T cells suggested the importance of this modification [236]. Approximately 650,000 regions showed differential methylation between naive and effector $CD8^+$ T cells in this study, and the abundance of DNA methylation was inversely correlated with gene expression.

2.3.2.2.4 Epigenetic Regulation of T Cell Memory

Memory T cells are characterized by their ability to rapidly express effector genes without the need for further differentiation. For example, activating histone modifications at the *Ifng* promoter of effector $CD8^+$ T cells were maintained as long-term memory modifications. Furthermore, RNA polymerase remained docked at the *Ifng* promoter until reactivation. These observations collectively suggest that the rapid production of IFN-γ following the reactivation of memory $CD8^+$ T cells reflects the maintenance of a transcriptionally permissive state of this promoter by a combination of histone modifications and transcriptional pausing [234,237].

DNA demethylation of the *Foxp3* CNS2 enhancer plays important roles in differentiated T_{reg} cell memory and provides an example of an epigenetic regulation of memory of immune suppression. Some T_{reg} cell populations lose the expression of Foxp3, and although this occurs at low levels under homeostatic conditions, it is accelerated under certain conditions, such as inflammatory stimulation and accelerated proliferation in lymphopenic mice [238]. T_{reg} cells that lose Foxp3 expression are referred to as "exFoxp3" or "exT_{reg}" cells, and approximately 30% of exT_{reg} cells memorize their previous *Foxp3* expression patterns and allow the rapid reacquisition of Foxp3 expression after activation. It is now known that the demethylation of the *Foxp3* CNS2 region endows the "memory" of Foxp3 expression in T_{reg} and exT_{reg} cells, as it was observed that the CNS2 region was demethylated in exT_{reg} cells which reacquired Foxp3 expression, but that this same region remained fully methylated in cells that could not reexpress Foxp3 [239] (Fig. 2.5C).

These studies collectively show that memory T cells are reconfigured by multiple epigenetic modifications that lead to more potent effector functions, a more effective and rapid control of secondary infections, and to tightly maintained immune homeostasis.

2.4 EPIGENETIC MISREGULATION IN IMMUNE-ASSOCIATED DISEASES

As this subject is robustly described in other chapters of this book, the present discussion is limited to the principles of pathogenic epigenetic misregulation of immune cell development and function. In general, mutations that interrupt developmental progression lead to uncontrolled growth of immature cells, thus causing tumorigenesis. Conversely, mutations that drive the uncontrolled expression of effector molecules (i.e., cytokines) lead to inflammatory diseases. Incorrect skewing to unwanted cell subsets might undermine the processes that eliminate pathogen and permit the expansion of infection. Epigenetic regulation plays crucial roles in all of these biological events, thus genetic mutations that alter epigenetic modifiers or transcription factors are responsible for various types of immune disorders.

2.5 CONCLUDING REMARKS

As epigenetic modifications regulate gene expression multidimensionally, improved understanding of their mechanisms and roles will substantially facilitate our knowledge of immune cell development. To this end, several key issues remain to be resolved. First, although the roles of transcription factors and epigenetic modifications have been abundantly described independently, few studies have sufficiently incorporated both processes. Transcription factors modify gene expression not only by interacting with the basic transcription machinery, but also by interacting with epigenetic regulators to mediate various modification at specific regulatory regions. Recent technical advances in mass spectrometry and proteomic analysis, particularly in terms of their high-throughput nature and their compatibility with scaling-down, will likely lead to the identification of further interactions between transcription factors and epigenetic modifiers.

Another important issue to be addressed is that limited information has been obtained for several cell subsets, specifically those with small populations. Although there have been difficulties when using epigenetic analyses for such small numbers of cells, recent advances in single-cell analysis are expected to unveil epigenetic mechanisms even in such minor cell subsets.

Finally, the combined use of next-generation sequencing and epigenetic analyses (i.e., ChIP and bisulfite sequencing) and the powerful in silico tools for analyzing their output will likely reveal global networks of epigenetic and transcriptional functions. Tremendous progress is expected to be made in the understanding of epigenetic contributions to immune cell development, which will in turn ensure the discovery of novel therapeutic agents for immune-associated diseases.

Acknowledgments

This work was supported by JSPS KAKENHI grant-in-aid for scientific research (16KT0114), Mochida Memorial Foundation for Medical and Pharmaceutical Research, SENSHIN Medical Research Foundation, and the Takeda Science Foundation.

References

[1] Sarai N, Nimura K, Tamura T, Kanno T, Patel MC, Heightman TD, et al. WHSC1 links transcription elongation to HIRA-mediated histone H3.3 deposition. EMBO J 2013;32(17):2392–406.

[2] Carpenter S, Aiello D, Atianand MK, Ricci EP, Gandhi P, Hall LL, et al. A long noncoding RNA mediates both activation and repression of immune response genes. Science 2013;341(6147):789–92.

[3] Allis CD, Jenuwein T. The molecular hallmarks of epigenetic control. Nat Rev Genet 2016;17(8):487–500.

[4] Brownell JE, Zhou J, Ranalli T, Kobayashi R, Edmondson DG, Roth SY, et al. Tetrahymena histone acetyltransferase A: a homolog to yeast Gcn5p linking histone acetylation to gene activation. Cell 1996;84(6):843–51.
[5] Kuo MH, Brownell JE, Sobel RE, Ranalli TA, Cook RG, Edmondson DG, et al. Transcription-linked acetylation by Gcn5p of histones H3 and H4 at specific lysines. Nature 1996;383(6597):269–72.
[6] Taunton J, Hassig CA, Schreiber SL. A mammalian histone deacetylase related to the yeast transcriptional regulator Rpd3p. Science 1996;272(5260):408–11.
[7] Krogan NJ, Dover J, Khorrami S, Greenblatt JF, Schneider J, Johnston M, et al. COMPASS, a histone H3 (Lysine 4) methyltransferase required for telomeric silencing of gene expression. J Biol Chem 2002;277(13):10753–5.
[8] Milne TA, Briggs SD, Brock HW, Martin ME, Gibbs D, Allis CD, et al. MLL targets SET domain methyltransferase activity to Hox gene promoters. Mol Cell 2002;10(5):1107–17.
[9] Nakamura T, Mori T, Tada S, Krajewski W, Rozovskaia T, Wassell R, et al. ALL-1 is a histone methyltransferase that assembles a supercomplex of proteins involved in transcriptional regulation. Mol Cell 2002;10(5):1119–28.
[10] Yokoyama A, Wang Z, Wysocka J, Sanyal M, Aufiero DJ, Kitabayashi I, et al. Leukemia proto-oncoprotein MLL forms a SET1-like histone methyltransferase complex with menin to regulate Hox gene expression. Mol Cell Biol 2004;24(13):5639–49.
[11] van Leeuwen F, Gafken PR, Gottschling DE. Dot1p modulates silencing in yeast by methylation of the nucleosome core. Cell 2002;109(6):745–56.
[12] Tachibana M, Sugimoto K, Fukushima T, Shinkai Y. Set domain-containing protein, G9a, is a novel lysine-preferring mammalian histone methyltransferase with hyperactivity and specific selectivity to lysines 9 and 27 of histone H3. J Biol Chem 2001;276(27):25309–17.
[13] Czermin B, Melfi R, McCabe D, Seitz V, Imhof A, Pirrotta V. Drosophila enhancer of Zeste/ESC complexes have a histone H3 methyltransferase activity that marks chromosomal polycomb sites. Cell 2002;111(2):185–96.
[14] Muller J, Hart CM, Francis NJ, Vargas ML, Sengupta A, Wild B, et al. Histone methyltransferase activity of a Drosophila polycomb group repressor complex. Cell 2002;111(2):197–208.
[15] Kuzmichev A, Nishioka K, Erdjument-Bromage H, Tempst P, Reinberg D. Histone methyltransferase activity associated with a human multiprotein complex containing the enhancer of zeste protein. Genes Dev 2002;16(22):2893–905.
[16] Cao R, Wang L, Wang H, Xia L, Erdjument-Bromage H, Tempst P, et al. Role of histone H3 lysine 27 methylation in polycomb-group silencing. Science 2002;298(5595):1039–43.
[17] Nguyen AT, Zhang Y. The diverse functions of Dot1 and H3K79 methylation. Genes Dev 2011;25(13):1345–58.
[18] Li E, Bestor TH, Jaenisch R. Targeted mutation of the DNA methyltransferase gene results in embryonic lethality. Cell 1992;69(6):915–26.
[19] Okano M, Bell DW, Haber DA, Li E. DNA methyltransferases Dnmt3a and Dnmt3b are essential for de novo methylation and mammalian development. Cell 1999;99(3):247–57.
[20] Kriaucionis S, Heintz N. The nuclear DNA base 5-hydroxymethylcytosine is present in Purkinje neurons and the brain. Science 2009;324(5929):929–30.
[21] Tahiliani M, Koh KP, Shen Y, Pastor WA, Bandukwala H, Brudno Y, et al. Conversion of 5-methylcytosine to 5-hydroxymethylcytosine in mammalian DNA by MLL partner TET1. Science 2009;324(5929):930–5.
[22] Ito S, D'Alessio AC, Taranova OV, Hong K, Sowers LC, Zhang Y. Role of Tet proteins in 5mC to 5hmC conversion, ES-cell self-renewal and inner cell mass specification. Nature 2010;466(7310):1129–33.
[23] He YF, Li BZ, Li Z, Liu P, Wang Y, Tang Q, et al. Tet-mediated formation of 5-carboxylcytosine and its excision by TDG in mammalian DNA. Science 2011;333(6047):1303–7.
[24] Ng HH, Bird A. DNA methylation and chromatin modification. Curr Opin Genet Dev 1999;9(2):158–63.
[25] Jackson JP, Lindroth AM, Cao X, Jacobsen SE. Control of CpNpG DNA methylation by the KRYPTONITE histone H3 methyltransferase. Nature 2002;416(6880):556–60.
[26] Tamaru H, Selker EU. A histone H3 methyltransferase controls DNA methylation in *Neurospora crassa*. Nature 2001;414(6861):277–83.
[27] Blackledge NP, Zhou JC, Tolstorukov MY, Farcas AM, Park PJ, Klose RJ. CpG islands recruit a histone H3 lysine 36 demethylase. Mol Cell 2010;38(2):179–90.
[28] Thomson JP, Skene PJ, Selfridge J, Clouaire T, Guy J, Webb S, et al. CpG islands influence chromatin structure via the CpG-binding protein Cfp1. Nature 2010;464(7291):1082–6.
[29] Syntichaki P, Topalidou I, Thireos G. The Gcn5 bromodomain co-ordinates nucleosome remodelling. Nature 2000;404(6776):414–7.
[30] Bannister AJ, Zegerman P, Partridge JF, Miska EA, Thomas JO, Allshire RC, et al. Selective recognition of methylated lysine 9 on histone H3 by the HP1 chromo domain. Nature 2001;410(6824):120–4.

[31] Lachner M, O'Carroll D, Rea S, Mechtler K, Jenuwein T. Methylation of histone H3 lysine 9 creates a binding site for HP1 proteins. Nature 2001;410(6824):116–20.
[32] Aagaard L, Laible G, Selenko P, Schmid M, Dorn R, Schotta G, et al. Functional mammalian homologues of the Drosophila PEV-modifier Su(var)3-9 encode centromere-associated proteins which complex with the heterochromatin component M31. EMBO J 1999;18(7):1923–38.
[33] Canzio D, Chang EY, Shankar S, Kuchenbecker KM, Simon MD, Madhani HD, et al. Chromodomain-mediated oligomerization of HP1 suggests a nucleosome-bridging mechanism for heterochromatin assembly. Mol Cell 2011;41(1):67–81.
[34] Francis NJ, Kingston RE, Woodcock CL. Chromatin compaction by a polycomb group protein complex. Science 2004;306(5701):1574–7.
[35] Bernstein BE, Mikkelsen TS, Xie X, Kamal M, Huebert DJ, Cuff J, et al. A bivalent chromatin structure marks key developmental genes in embryonic stem cells. Cell 2006;125(2):315–26.
[36] Chaya D, Hayamizu T, Bustin M, Zaret KS. Transcription factor FoxA (HNF3) on a nucleosome at an enhancer complex in liver chromatin. J Biol Chem 2001;276(48):44385–9.
[37] Cirillo LA, McPherson CE, Bossard P, Stevens K, Cherian S, Shim EY, et al. Binding of the winged-helix transcription factor HNF3 to a linker histone site on the nucleosome. EMBO J 1998;17(1):244–54.
[38] Sekiya T, Zaret KS. Repression by Groucho/TLE/Grg proteins: genomic site recruitment generates compacted chromatin in vitro and impairs activator binding in vivo. Mol Cell 2007;28(2):291–303.
[39] Sekiya T, Muthurajan UM, Luger K, Tulin AV, Zaret KS. Nucleosome-binding affinity as a primary determinant of the nuclear mobility of the pioneer transcription factor FoxA. Genes Dev 2009;23(7):804–9.
[40] Davidson EH, Erwin DH. Gene regulatory networks and the evolution of animal body plans. Science 2006;311(5762):796–800.
[41] Lee CS, Friedman JR, Fulmer JT, Kaestner KH. The initiation of liver development is dependent on Foxa transcription factors. Nature 2005;435(7044):944–7.
[42] Hansen KH, Bracken AP, Pasini D, Dietrich N, Gehani SS, Monrad A, et al. A model for transmission of the H3K27me3 epigenetic mark. Nat Cell Biol 2008;10(11):1291–300.
[43] Margueron R, Justin N, Ohno K, Sharpe ML, Son J, Drury WJ III, et al. Role of the polycomb protein EED in the propagation of repressive histone marks. Nature 2009;461(7265):762–7.
[44] Ragunathan K, Jih G, Moazed D. Epigenetics: epigenetic inheritance uncoupled from sequence-specific recruitment. Science 2015;348(6230):1258699.
[45] Murray PJ, Wynn TA. Obstacles and opportunities for understanding macrophage polarization. J Leukoc Biol 2011;89(4):557–63.
[46] Ivashkiv LB. Epigenetic regulation of macrophage polarization and function. Trends Immunol 2013;34(5):216–23.
[47] Gordon S, Martinez FO. Alternative activation of macrophages: mechanism and functions. Immunity 2010;32(5):593–604.
[48] Lawrence T, Natoli G. Transcriptional regulation of macrophage polarization: enabling diversity with identity. Nat Rev Immunol 2011;11(11):750–61.
[49] Mosser DM, Edwards JP. Exploring the full spectrum of macrophage activation. Nat Rev Immunol 2008;8(12):958–69.
[50] Alvarez-Errico D, Vento-Tormo R, Sieweke M, Ballestar E. Epigenetic control of myeloid cell differentiation, identity and function. Nat Rev Immunol 2015;15(1):7–17.
[51] Sun D, Luo M, Jeong M, Rodriguez B, Xia Z, Hannah R, et al. Epigenomic profiling of young and aged HSCs reveals concerted changes during aging that reinforce self-renewal. Cell Stem Cell 2014;14(5):673–88.
[52] Broske AM, Vockentanz L, Kharazi S, Huska MR, Mancini E, Scheller M, et al. DNA methylation protects hematopoietic stem cell multipotency from myeloerythroid restriction. Nat Genet 2009;41(11):1207–15.
[53] Challen GA, Sun D, Jeong M, Luo M, Jelinek J, Berg JS, et al. Dnmt3a is essential for hematopoietic stem cell differentiation. Nat Genet 2011;44(1):23–31.
[54] DeKoter RP, Singh H. Regulation of B lymphocyte and macrophage development by graded expression of PU.1. Science 2000;288(5470):1439–41.
[55] de la Rica L, Rodriguez-Ubreva J, Garcia M, Islam AB, Urquiza JM, Hernando H, et al. PU.1 target genes undergo Tet2-coupled demethylation and DNMT3b-mediated methylation in monocyte-to-osteoclast differentiation. Genome Biol 2013;14(9):R99.
[56] Biswas SK, Mantovani A. Macrophage plasticity and interaction with lymphocyte subsets: cancer as a paradigm. Nat Immunol 2010;11(10):889–96.
[57] Ghisletti S, Barozzi I, Mietton F, Polletti S, De Santa F, Venturini E, et al. Identification and characterization of enhancers controlling the inflammatory gene expression program in macrophages. Immunity 2010;32(3):317–28.

[58] Heinz S, Benner C, Spann N, Bertolino E, Lin YC, Laslo P, et al. Simple combinations of lineage-determining transcription factors prime cis-regulatory elements required for macrophage and B cell identities. Mol Cell 2010;38(4):576–89.
[59] Jin F, Li Y, Ren B, Natarajan R. PU.1 and C/EBP(alpha) synergistically program distinct response to NF-kappaB activation through establishing monocyte specific enhancers. Proc Natl Acad Sci U S A 2011;108(13):5290–5.
[60] Pham TH, Benner C, Lichtinger M, Schwarzfischer L, Hu Y, Andreesen R, et al. Dynamic epigenetic enhancer signatures reveal key transcription factors associated with monocytic differentiation states. Blood 2012;119(24): e161–71.
[61] Adelman K, Kennedy MA, Nechaev S, Gilchrist DA, Muse GW, Chinenov Y, et al. Immediate mediators of the inflammatory response are poised for gene activation through RNA polymerase II stalling. Proc Natl Acad Sci U S A 2009;106(43):18207–12.
[62] Escoubet-Lozach L, Benner C, Kaikkonen MU, Lozach J, Heinz S, Spann NJ, et al. Mechanisms establishing TLR4-responsive activation states of inflammatory response genes. PLoS Genet 2011;7(12):e1002401.
[63] Hargreaves DC, Horng T, Medzhitov R. Control of inducible gene expression by signal-dependent transcriptional elongation. Cell 2009;138(1):129–45.
[64] Ramirez-Carrozzi VR, Braas D, Bhatt DM, Cheng CS, Hong C, Doty KR, et al. A unifying model for the selective regulation of inducible transcription by CpG islands and nucleosome remodeling. Cell 2009;138(1):114–28.
[65] Gough DJ, Messina NL, Clarke CJ, Johnstone RW, Levy DE. Constitutive type I interferon modulates homeostatic balance through tonic signaling. Immunity 2012;36(2):166–74.
[66] Chen X, Barozzi I, Termanini A, Prosperini E, Recchiuti A, Dalli J, et al. Requirement for the histone deacetylase Hdac3 for the inflammatory gene expression program in macrophages. Proc Natl Acad Sci U S A 2012;109(42):E2865–74.
[67] Barish GD, Yu RT, Karunasiri M, Ocampo CB, Dixon J, Benner C, et al. Bcl-6 and NF-kappaB cistromes mediate opposing regulation of the innate immune response. Genes Dev 2010;24(24):2760–5.
[68] De Santa F, Narang V, Yap ZH, Tusi BK, Burgold T, Austenaa L, et al. Jmjd3 contributes to the control of gene expression in LPS-activated macrophages. EMBO J 2009;28(21):3341–52.
[69] Stender JD, Pascual G, Liu W, Kaikkonen MU, Do K, Spann NJ, et al. Control of proinflammatory gene programs by regulated trimethylation and demethylation of histone H4K20. Mol Cell 2012;48(1):28–38.
[70] van Essen D, Zhu Y, Saccani S. A feed-forward circuit controlling inducible NF-kappaB target gene activation by promoter histone demethylation. Mol Cell 2010;39(5):750–60.
[71] Zhu Y, van Essen D, Saccani S. Cell-type-specific control of enhancer activity by H3K9 trimethylation. Mol Cell 2012;46(4):408–23.
[72] Ramirez-Carrozzi VR, Nazarian AA, Li CC, Gore SL, Sridharan R, Imbalzano AN, et al. Selective and antagonistic functions of SWI/SNF and Mi-2beta nucleosome remodeling complexes during an inflammatory response. Genes Dev 2006;20(3):282–96.
[73] Satoh T, Takeuchi O, Vandenbon A, Yasuda K, Tanaka Y, Kumagai Y, et al. The Jmjd3-Irf4 axis regulates M2 macrophage polarization and host responses against helminth infection. Nat Immunol 2010;11(10):936–44.
[74] Ishii M, Wen H, Corsa CA, Liu T, Coelho AL, Allen RM, et al. Epigenetic regulation of the alternatively activated macrophage phenotype. Blood 2009;114(15):3244–54.
[75] Mullican SE, Gaddis CA, Alenghat T, Nair MG, Giacomin PR, Everett LJ, et al. Histone deacetylase 3 is an epigenomic brake in macrophage alternative activation. Genes Dev 2011;25(23):2480–8.
[76] Quintin J, Saeed S, Martens JH, Giamarellos-Bourboulis EJ, Ifrim DC, Logie C, et al. *Candida albicans* infection affords protection against reinfection via functional reprogramming of monocytes. Cell Host Microbe 2012;12(2):223–32.
[77] Saeed S, Quintin J, Kerstens HH, Rao NA, Aghajanirefah A, Matarese F, et al. Epigenetic programming of monocyte-to-macrophage differentiation and trained innate immunity. Science 2014;345(6204):1251086.
[78] Ostuni R, Piccolo V, Barozzi I, Polletti S, Termanini A, Bonifacio S, et al. Latent enhancers activated by stimulation in differentiated cells. Cell 2013;152(1–2):157–71.
[79] Cheng SC, Quintin J, Cramer RA, Shepardson KM, Saeed S, Kumar V, et al. mTOR- and HIF-1alpha-mediated aerobic glycolysis as metabolic basis for trained immunity. Science 2014;345(6204):1250684.
[80] Yoshida K, Maekawa T, Zhu Y, Renard-Guillet C, Chatton B, Inoue K, et al. The transcription factor ATF7 mediates lipopolysaccharide-induced epigenetic changes in macrophages involved in innate immunological memory. Nat Immunol 2015;16(10):1034–43.
[81] Foster SL, Hargreaves DC, Medzhitov R. Gene-specific control of inflammation by TLR-induced chromatin modifications. Nature 2007;447(7147):972–8.
[82] Mages J, Dietrich H, Lang R. A genome-wide analysis of LPS tolerance in macrophages. Immunobiology 2007;212 (9-10):723–37.

[83] del Fresno C, Garcia-Rio F, Gomez-Pina V, Soares-Schanoski A, Fernandez-Ruiz I, Jurado T, et al. Potent phagocytic activity with impaired antigen presentation identifying lipopolysaccharide-tolerant human monocytes: demonstration in isolated monocytes from cystic fibrosis patients. J Immunol 2009;182(10):6494–507.
[84] Ng SY, Yoshida T, Zhang J, Georgopoulos K. Genome-wide lineage-specific transcriptional networks underscore Ikaros-dependent lymphoid priming in hematopoietic stem cells. Immunity 2009;30(4):493–507.
[85] Bernink JH, Peters CP, Munneke M, te Velde AA, Meijer SL, Weijer K, et al. Human type 1 innate lymphoid cells accumulate in inflamed mucosal tissues. Nat Immunol 2013;14(3):221–9.
[86] Fuchs A, Vermi W, Lee JS, Lonardi S, Gilfillan S, Newberry RD, et al. Intraepithelial type 1 innate lymphoid cells are a unique subset of IL-12- and IL-15-responsive IFN-gamma-producing cells. Immunity 2013;38(4):769–81.
[87] Eberl G, Marmon S, Sunshine MJ, Rennert PD, Choi Y, Littman DR. An essential function for the nuclear receptor RORgamma(t) in the generation of fetal lymphoid tissue inducer cells. Nat Immunol 2004;5(1):64–73.
[88] Rothenberg EV. Transcriptional control of early T and B cell developmental choices. Annu Rev Immunol 2014;32: 283–321.
[89] Verykokakis M, Zook EC, Kee BL. ID'ing innate and innate-like lymphoid cells. Immunol Rev 2014;261(1):177–97.
[90] Di Santo JP. Natural killer cell developmental pathways: a question of balance. Annu Rev Immunol 2006;24:257–86.
[91] Shi FD, Ljunggren HG, La Cava A, Van Kaer L. Organ-specific features of natural killer cells. Nat Rev Immunol 2011;11(10):658–71.
[92] Lanier LL. Up on the tightrope: natural killer cell activation and inhibition. Nat Immunol 2008;9(5):495–502.
[93] Schenk A, Bloch W, Zimmer P. Natural killer cells: an epigenetic perspective of development and regulation. Int J Mol Sci 2016;17(3):326.
[94] Cichocki F, Miller JS, Anderson SK, Bryceson YT. Epigenetic regulation of NK cell differentiation and effector functions. Front Immunol 2013;4:55.
[95] Gao XN, Lin J, Wang LL, Yu L. Demethylating treatment suppresses natural killer cell cytolytic activity. Mol Immunol 2009;46(10):2064–70.
[96] Santourlidis S, Graffmann N, Christ J, Uhrberg M. Lineage-specific transition of histone signatures in the killer cell Ig-like receptor locus from hematopoietic progenitor to NK cells. J Immunol 2008;180(1):418–25.
[97] Ogbomo H, Michaelis M, Kreuter J, Doerr HW, Cinatl J Jr. Histone deacetylase inhibitors suppress natural killer cell cytolytic activity. FEBS Lett 2007;581(7):1317–22.
[98] Zhang J, Scordi I, Smyth MJ, Lichtenheld MG. Interleukin 2 receptor signaling regulates the perforin gene through signal transducer and activator of transcription (Stat)5 activation of two enhancers. J Exp Med 1999;190(9):1297–308.
[99] Imada K, Bloom ET, Nakajima H, Horvath-Arcidiacono JA, Udy GB, Davey HW, et al. Stat5b is essential for natural killer cell-mediated proliferation and cytolytic activity. J Exp Med 1998;188(11):2067–74.
[100] Wargnier A, Legros-Maida S, Bosselut R, Bourge JF, Lafaurie C, Ghysdael CJ, et al. Identification of human granzyme B promoter regulatory elements interacting with activated T-cell-specific proteins: implication of Ikaros and CBF binding sites in promoter activation. Proc Natl Acad Sci U S A 1995;92(15):6930–4.
[101] Babichuk CK, Duggan BL, Bleackley RC. In vivo regulation of murine granzyme B gene transcription in activated primary T cells. J Biol Chem 1996;271(28):16485–93.
[102] Araki Y, Fann M, Wersto R, Weng NP. Histone acetylation facilitates rapid and robust memory CD8 T cell response through differential expression of effector molecules (eomesodermin and its targets: perforin and granzyme B). J Immunol 2008;180(12):8102–8.
[103] Hatton RD, Harrington LE, Luther RJ, Wakefield T, Janowski KM, Oliver JR, et al. A distal conserved sequence element controls Ifng gene expression by T cells and NK cells. Immunity 2006;25(5):717–29.
[104] Sun JC, Beilke JN, Lanier LL. Adaptive immune features of natural killer cells. Nature 2009;457(7229):557–61.
[105] Lee J, Zhang T, Hwang I, Kim A, Nitschke L, Kim M, et al. Epigenetic modification and antibody-dependent expansion of memory-like NK cells in human cytomegalovirus-infected individuals. Immunity 2015;42(3):431–42.
[106] Schlums H, Cichocki F, Tesi B, Theorell J, Beziat V, Holmes TD, et al. Cytomegalovirus infection drives adaptive epigenetic diversification of NK cells with altered signaling and effector function. Immunity 2015;42(3):443–56.
[107] Sun JC, Madera S, Bezman NA, Beilke JN, Kaplan MH, Lanier LL. Proinflammatory cytokine signaling required for the generation of natural killer cell memory. J Exp Med 2012;209(5):947–54.
[108] Luetke-Eversloh M, Hammer Q, Durek P, Nordstrom K, Gasparoni G, Pink M, et al. Human cytomegalovirus drives epigenetic imprinting of the IFNG locus in NKG2Chi natural killer cells. PLoS Pathog 2014;10(10):e1004441.
[109] Moro K, Yamada T, Tanabe M, Takeuchi T, Ikawa T, Kawamoto H, et al. Innate production of T(H)2 cytokines by adipose tissue-associated c-Kit(+)Sca-1(+) lymphoid cells. Nature 2010;463(7280):540–4.
[110] Neill DR, Wong SH, Bellosi A, Flynn RJ, Daly M, Langford TK, et al. Nuocytes represent a new innate effector leukocyte that mediates type-2 immunity. Nature 2010;464(7293):1367–70.

[111] Price AE, Liang HE, Sullivan BM, Reinhardt RL, Eisley CJ, Erle DJ, et al. Systemically dispersed innate IL-13-expressing cells in type 2 immunity. Proc Natl Acad Sci U S A 2010;107(25):11489–94.
[112] Halim TY, Krauss RH, Sun AC, Takei F. Lung natural helper cells are a critical source of Th2 cell-type cytokines in protease allergen-induced airway inflammation. Immunity 2012;36(3):451–63.
[113] Ikutani M, Yanagibashi T, Ogasawara M, Tsuneyama K, Yamamoto S, Hattori Y, et al. Identification of innate IL-5-producing cells and their role in lung eosinophil regulation and antitumor immunity. J Immunol 2012;188(2):703–13.
[114] Fallon PG, Ballantyne SJ, Mangan NE, Barlow JL, Dasvarma A, Hewett DR, et al. Identification of an interleukin (IL)-25-dependent cell population that provides IL-4, IL-5, and IL-13 at the onset of helminth expulsion. J Exp Med 2006;203(4):1105–16.
[115] Bartemes KR, Iijima K, Kobayashi T, Kephart GM, McKenzie AN, Kita H. IL-33-responsive lineage- CD25+ CD44(hi) lymphoid cells mediate innate type 2 immunity and allergic inflammation in the lungs. J Immunol 2012;188(3):1503–13.
[116] Chang YJ, Kim HY, Albacker LA, Baumgarth N, McKenzie AN, Smith DE, et al. Innate lymphoid cells mediate influenza-induced airway hyper-reactivity independently of adaptive immunity. Nat Immunol 2011;12(7):631–8.
[117] Monticelli LA, Sonnenberg GF, Abt MC, Alenghat T, Ziegler CG, Doering TA, et al. Innate lymphoid cells promote lung-tissue homeostasis after infection with influenza virus. Nat Immunol 2011;12(11):1045–54.
[118] Serafini N, Vosshenrich CA, Di Santo JP. Transcriptional regulation of innate lymphoid cell fate. Nat Rev Immunol 2015;15(7):415–28.
[119] Barlow JL, Bellosi A, Hardman CS, Drynan LF, Wong SH, Cruickshank JP, et al. Innate IL-13-producing nuocytes arise during allergic lung inflammation and contribute to airways hyperreactivity. J Allergy Clin Immunol 2012;129(1). 191-198 e1-4.
[120] Bernink JH, Germar K, Spits H. The role of ILC2 in pathology of type 2 inflammatory diseases. Curr Opin Immunol 2014;31:115–20.
[121] Klein Wolterink RG, Serafini N, van Nimwegen M, Vosshenrich CA, de Bruijn MJ, Fonseca Pereira D, et al. Essential, dose-dependent role for the transcription factor Gata3 in the development of IL-5+ and IL-13+ type 2 innate lymphoid cells. Proc Natl Acad Sci U S A 2013;110(25):10240–5.
[122] Hoyler T, Klose CS, Souabni A, Turqueti-Neves A, Pfeifer D, Rawlins EL, et al. The transcription factor GATA-3 controls cell fate and maintenance of type 2 innate lymphoid cells. Immunity 2012;37(4):634–48.
[123] Yagi R, Zhong C, Northrup DL, Yu F, Bouladoux N, Spencer S, et al. The transcription factor GATA3 is critical for the development of all IL-7Ralpha-expressing innate lymphoid cells. Immunity 2014;40(3):378–88.
[124] Wong SH, Walker JA, Jolin HE, Drynan LF, Hams E, Camelo A, et al. Transcription factor RORalpha is critical for nuocyte development. Nat Immunol 2012;13(3):229–36.
[125] Yang Q, Monticelli LA, Saenz SA, Chi AW, Sonnenberg GF, Tang J, et al. T cell factor 1 is required for group 2 innate lymphoid cell generation. Immunity 2013;38(4):694–704.
[126] Spooner CJ, Lesch J, Yan D, Khan AA, Abbas A, Ramirez-Carrozzi V, et al. Specification of type 2 innate lymphocytes by the transcriptional determinant Gfi1. Nat Immunol 2013;14(12):1229–36.
[127] Serafini N, Klein Wolterink RG, Satoh-Takayama N, Xu W, Vosshenrich CA, Hendriks RW, et al. Gata3 drives development of RORgammat+ group 3 innate lymphoid cells. J Exp Med 2014;211(2):199–208.
[128] Mielke LA, Groom JR, Rankin LC, Seillet C, Masson F, Putoczki T, et al. TCF-1 controls ILC2 and NKp46+RORgammat+ innate lymphocyte differentiation and protection in intestinal inflammation. J Immunol 2013;191(8):4383–91.
[129] Halim TY, MacLaren A, Romanish MT, Gold MJ, McNagny KM, Takei F. Retinoic-acid-receptor-related orphan nuclear receptor alpha is required for natural helper cell development and allergic inflammation. Immunity 2012;37(3):463–74.
[130] Bao Y, Cao X. Epigenetic control of B cell development and B-cell-related immune disorders. Clin Rev Allergy Immunol 2016;50(3):301–11.
[131] Good-Jacobson KL. Regulation of germinal center, B-cell memory, and plasma cell formation by histone modifiers. Front Immunol 2014;5:596.
[132] Treiber N, Treiber T, Zocher G, Grosschedl R. Structure of an Ebf1:DNA complex reveals unusual DNA recognition and structural homology with Rel proteins. Genes Dev 2010;24(20):2270–5.
[133] Lin YC, Jhunjhunwala S, Benner C, Heinz S, Welinder E, Mansson R, et al. A global network of transcription factors, involving E2A, EBF1 and Foxo1, that orchestrates B cell fate. Nat Immunol 2010;11(7):635–43.
[134] Nutt SL, Heavey B, Rolink AG, Busslinger M. Commitment to the B-lymphoid lineage depends on the transcription factor Pax5. Nature 1999;401(6753):556–62.
[135] Maier H, Ostraat R, Gao H, Fields S, Shinton SA, Medina KL, et al. Early B cell factor cooperates with Runx1 and mediates epigenetic changes associated with mb-1 transcription. Nat Immunol 2004;5(10):1069–77.

[136] Gao H, Lukin K, Ramirez J, Fields S, Lopez D, Hagman J. Opposing effects of SWI/SNF and Mi-2/NuRD chromatin remodeling complexes on epigenetic reprogramming by EBF and Pax5. Proc Natl Acad Sci U S A 2009;106(27):11258–63.
[137] Delogu A, Schebesta A, Sun Q, Aschenbrenner K, Perlot T, Busslinger M. Gene repression by Pax5 in B cells is essential for blood cell homeostasis and is reversed in plasma cells. Immunity 2006;24(3):269–81.
[138] Pridans C, Holmes ML, Polli M, Wettenhall JM, Dakic A, Corcoran LM, et al. Identification of Pax5 target genes in early B cell differentiation. J Immunol 2008;180(3):1719–28.
[139] Schebesta A, McManus S, Salvagiotto G, Delogu A, Busslinger GA, Busslinger M. Transcription factor Pax5 activates the chromatin of key genes involved in B cell signaling, adhesion, migration, and immune function. Immunity 2007;27(1):49–63.
[140] Zouali M. The epigenetic landscape of B lymphocyte tolerance to self. FEBS Lett 2013;587(13):2067–73.
[141] Cho YW, Hong T, Hong S, Guo H, Yu H, Kim D, et al. PTIP associates with MLL3- and MLL4-containing histone H3 lysine 4 methyltransferase complex. J Biol Chem 2007;282(28):20395–406.
[142] McManus S, Ebert A, Salvagiotto G, Medvedovic J, Sun Q, Tamir I, et al. The transcription factor Pax5 regulates its target genes by recruiting chromatin-modifying proteins in committed B cells. EMBO J 2011;30(12):2388–404.
[143] Perissi V, Jepsen K, Glass CK, Rosenfeld MG. Deconstructing repression: evolving models of co-repressor action. Nat Rev Genet 2010;11(2):109–23.
[144] Cirillo LA, Lin FR, Cuesta I, Friedman D, Jarnik M, Zaret KS. Opening of compacted chromatin by early developmental transcription factors HNF3 (FoxA) and GATA-4. Mol Cell 2002;9(2):279–89.
[145] Hatta M, Cirillo LA. Chromatin opening and stable perturbation of core histone:DNA contacts by FoxO1. J Biol Chem 2007;282(49):35583–93.
[146] Xu CR, Feeney AJ. The epigenetic profile of Ig genes is dynamically regulated during B cell differentiation and is modulated by pre-B cell receptor signaling. J Immunol 2009;182(3):1362–9.
[147] McMurry MT, Krangel MS. A role for histone acetylation in the developmental regulation of VDJ recombination. Science 2000;287(5452):495–8.
[148] Chowdhury D, Sen R. Stepwise activation of the immunoglobulin mu heavy chain gene locus. EMBO J 2001;20(22):6394–403.
[149] Hesslein DG, Pflugh DL, Chowdhury D, Bothwell AL, Sen R, Schatz DG. Pax5 is required for recombination of transcribed, acetylated, 5′ IgH V gene segments. Genes Dev 2003;17(1):37–42.
[150] Johnson K, Angelin-Duclos C, Park S, Calame KL. Changes in histone acetylation are associated with differences in accessibility of V(H) gene segments to V-DJ recombination during B-cell ontogeny and development. Mol Cell Biol 2003;23(7):2438–50.
[151] Morshead KB, Ciccone DN, Taverna SD, Allis CD, Oettinger MA. Antigen receptor loci poised for V(D)J rearrangement are broadly associated with BRG1 and flanked by peaks of histone H3 dimethylated at lysine 4. Proc Natl Acad Sci U S A 2003;100(20):11577–82.
[152] Ng HH, Ciccone DN, Morshead KB, Oettinger MA, Struhl K. Lysine-79 of histone H3 is hypomethylated at silenced loci in yeast and mammalian cells: a potential mechanism for position-effect variegation. Proc Natl Acad Sci U S A 2003;100(4):1820–5.
[153] Perkins EJ, Kee BL, Ramsden DA. Histone 3 lysine 4 methylation during the pre-B to immature B-cell transition. Nucleic Acids Res 2004;32(6):1942–7.
[154] Goldmit M, Ji Y, Skok J, Roldan E, Jung S, Cedar H, et al. Epigenetic ontogeny of the Igk locus during B cell development. Nat Immunol 2005;6(2):198–203.
[155] Espinoza CR, Feeney AJ. The extent of histone acetylation correlates with the differential rearrangement frequency of individual VH genes in pro-B cells. J Immunol 2005;175(10):6668–75.
[156] Shimazaki N, Tsai AG, Lieber MR. H3K4me3 stimulates the V(D)J RAG complex for both nicking and hairpinning in trans in addition to tethering in cis: implications for translocations. Mol Cell 2009;34(5):535–44.
[157] Lennon GG, Perry RP. Cμ-containing transcripts initiate heterogeneously within the IgH enhancer region and contain a novel 5′-nontranslatable exon. Nature 1985;318(6045):475–8.
[158] Martin DJ, van Ness BG. Initiation and processing of two kappa immunoglobulin germ line transcripts in mouse B cells. Mol Cell Biol 1990;10(5):1950–8.
[159] Singh N, Bergman Y, Cedar H, Chess A. Biallelic germline transcription at the kappa immunoglobulin locus. J Exp Med 2003;197(6):743–50.
[160] Bolland DJ, Wood AL, Johnston CM, Bunting SF, Morgan G, Chakalova L, et al. Antisense intergenic transcription in V(D)J recombination. Nat Immunol 2004;5(6):630–7.

[161] Nakase H, Takahama Y, Akamatsu Y. Effect of CpG methylation on RAG1/RAG2 reactivity: implications of direct and indirect mechanisms for controlling V(D)J cleavage. EMBO Rep 2003;4(8):774–80.

[162] Selimyan R, Gerstein RM, Ivanova I, Precht P, Subrahmanyam R, Perlot T, et al. Localized DNA demethylation at recombination intermediates during immunoglobulin heavy chain gene assembly. PLoS Biol 2013;11(1):e1001475.

[163] Victora GD, Nussenzweig MC. Germinal centers. Annu Rev Immunol 2012;30:429–57.

[164] Victora GD, Schwickert TA, Fooksman DR, Kamphorst AO, Meyer-Hermann M, Dustin ML, et al. Germinal center dynamics revealed by multiphoton microscopy with a photoactivatable fluorescent reporter. Cell 2010;143(4):592–605.

[165] Meyer-Hermann M, Mohr E, Pelletier N, Zhang Y, Victora GD, Toellner KM. A theory of germinal center B cell selection, division, and exit. Cell Rep 2012;2(1):162–74.

[166] Klein U, Tu Y, Stolovitzky GA, Keller JL, Haddad J Jr, Miljkovic V, et al. Transcriptional analysis of the B cell germinal center reaction. Proc Natl Acad Sci U S A 2003;100(5):2639–44.

[167] Dent AL, Shaffer AL, Yu X, Allman D, Staudt LM. Control of inflammation, cytokine expression, and germinal center formation by BCL-6. Science 1997;276(5312):589–92.

[168] Ye BH, Cattoretti G, Shen Q, Zhang J, Hawe N, de Waard R, et al. The BCL-6 proto-oncogene controls germinal-centre formation and Th2-type inflammation. Nat Genet 1997;16(2):161–70.

[169] Shaffer AL, Yu X, He Y, Boldrick J, Chan EP, Staudt LM. BCL-6 represses genes that function in lymphocyte differentiation, inflammation, and cell cycle control. Immunity 2000;13(2):199–212.

[170] Fujita N, Jaye DL, Geigerman C, Akyildiz A, Mooney MR, Boss JM, et al. MTA3 and the Mi-2/NuRD complex regulate cell fate during B lymphocyte differentiation. Cell 2004;119(1):75–86.

[171] Huynh KD, Bardwell VJ. The BCL-6 POZ domain and other POZ domains interact with the co-repressors N-CoR and SMRT. Oncogene 1998;17(19):2473–84.

[172] Dhordain P, Albagli O, Lin RJ, Ansieau S, Quief S, Leutz A, et al. Corepressor SMRT binds the BTB/POZ repressing domain of the LAZ3/BCL6 oncoprotein. Proc Natl Acad Sci U S A 1997;94(20):10762–7.

[173] Urnov FD, Yee J, Sachs L, Collingwood TN, Bauer A, Beug H, et al. Targeting of N-CoR and histone deacetylase 3 by the oncoprotein v-erbA yields a chromatin infrastructure-dependent transcriptional repression pathway. EMBO J 2000;19(15):4074–90.

[174] Wen YD, Perissi V, Staszewski LM, Yang WM, Krones A, Glass CK, et al. The histone deacetylase-3 complex contains nuclear receptor corepressors. Proc Natl Acad Sci U S A 2000;97(13):7202–7.

[175]Li J, Wang J, Wang J, Nawaz Z, Liu JM, Qin J, et al. Both corepressor proteins SMRT and N-CoR exist in large protein complexes containing HDAC3. EMBO J 2000;19(16):4342–50.

[176] Ahmad KF, Melnick A, Lax S, Bouchard D, Liu J, Kiang CL, et al. Mechanism of SMRT corepressor recruitment by the BCL6 BTB domain. Mol Cell 2003;12(6):1551–64.

[177] Beguelin W, Popovic R, Teater M, Jiang Y, Bunting KL, Rosen M, et al. EZH2 is required for germinal center formation and somatic EZH2 mutations promote lymphoid transformation. Cancer Cell 2013;23(5):677–92.

[178] Caganova M, Carrisi C, Varano G, Mainoldi F, Zanardi F, Germain PL, et al. Germinal center dysregulation by histone methyltransferase EZH2 promotes lymphomagenesis. J Clin Invest 2013;123(12):5009–22.

[179] Velichutina I, Shaknovich R, Geng H, Johnson NA, Gascoyne RD, Melnick AM, et al. EZH2-mediated epigenetic silencing in germinal center B cells contributes to proliferation and lymphomagenesis. Blood 2010;116(24):5247–55.

[180] Shapiro-Shelef M, Lin KI, McHeyzer-Williams LJ, Liao J, McHeyzer-Williams MG, Calame K. Blimp-1 is required for the formation of immunoglobulin secreting plasma cells and pre-plasma memory B cells. Immunity 2003;19(4):607–20.

[181] Reljic R, Wagner SD, Peakman LJ, Fearon DT. Suppression of signal transducer and activator of transcription 3-dependent B lymphocyte terminal differentiation by BCL-6. J Exp Med 2000;192(12):1841–8.

[182] Shaffer AL, Lin KI, Kuo TC, Yu X, Hurt EM, Rosenwald A, et al. Blimp-1 orchestrates plasma cell differentiation by extinguishing the mature B cell gene expression program. Immunity 2002;17(1):51–62.

[183] Gyory I, Wu J, Fejer G, Seto E, Wright KL. PRDI-BF1 recruits the histone H3 methyltransferase G9a in transcriptional silencing. Nat Immunol 2004;5(3):299–308.

[184] Mikkola I, Heavey B, Horcher M, Busslinger M. Reversion of B cell commitment upon loss of Pax5 expression. Science 2002;297(5578):110–3.

[185] Rolink AG, Nutt SL, Melchers F, Busslinger M. Long-term in vivo reconstitution of T-cell development by Pax5-deficient B-cell progenitors. Nature 1999;401(6753):603–6.

[186]Nechanitzky R, Akbas D, Scherer S, Gyory I, Hoyler T, Ramamoorthy S, et al. Transcription factor EBF1 is essential for the maintenance of B cell identity and prevention of alternative fates in committed cells. Nat Immunol 2013;14(8):867–75.

[187] Radtke F, Wilson A, Stark G, Bauer M, van Meerwijk J, MacDonald HR, et al. Deficient T cell fate specification in mice with an induced inactivation of Notch1. Immunity 1999;10(5):547–58.

[188] Wilson A, MacDonald HR, Radtke F. Notch 1-deficient common lymphoid precursors adopt a B cell fate in the thymus. J Exp Med 2001;194(7):1003–12.

[189] Bell JJ, Bhandoola A. The earliest thymic progenitors for T cells possess myeloid lineage potential. Nature 2008;452(7188):764–7.

[190] Wada H, Masuda K, Satoh R, Kakugawa K, Ikawa T, Katsura Y, et al. Adult T-cell progenitors retain myeloid potential. Nature 2008;452(7188):768–72.

[191] Feyerabend TB, Terszowski G, Tietz A, Blum C, Luche H, Gossler A, et al. Deletion of Notch1 converts pro-T cells to dendritic cells and promotes thymic B cells by cell-extrinsic and cell-intrinsic mechanisms. Immunity 2009;30(1):67–79.

[192] Zhang JA, Mortazavi A, Williams BA, Wold BJ, Rothenberg EV. Dynamic transformations of genome-wide epigenetic marking and transcriptional control establish T cell identity. Cell 2012;149(2):467–82.

[193] Ji H, Ehrlich LI, Seita J, Murakami P, Doi A, Lindau P, et al. Comprehensive methylome map of lineage commitment from haematopoietic progenitors. Nature 2010;467(7313):338–42.

[194] Rodriguez RM, Suarez-Alvarez B, Mosen-Ansorena D, Garcia-Peydro M, Fuentes P, Garcia-Leon MJ, et al. Regulation of the transcriptional program by DNA methylation during human alphabeta T-cell development. Nucleic Acids Res 2015;43(2):760–74.

[195] Rodriguez RM, Lopez-Larrea C, Suarez-Alvarez B. Epigenetic dynamics during CD4(+) T cells lineage commitment. Int J Biochem Cell Biol 2015;67:75–85.

[196] Lee PP, Fitzpatrick DR, Beard C, Jessup HK, Lehar S, Makar KW, et al. A critical role for Dnmt1 and DNA methylation in T cell development, function, and survival. Immunity 2001;15(5):763–74.

[197] Rothenberg EV. Transcriptional drivers of the T-cell lineage program. Curr Opin Immunol 2012;24(2):132–8.

[198] Li L, Leid M, Rothenberg EV. An early T cell lineage commitment checkpoint dependent on the transcription factor Bcl11b. Science 2010;329(5987):89–93.

[199] Li P, Burke S, Wang J, Chen X, Ortiz M, Lee SC, et al. Reprogramming of T cells to natural killer-like cells upon Bcl11b deletion. Science 2010;329(5987):85–9.

[200] Ikawa T, Hirose S, Masuda K, Kakugawa K, Satoh R, Shibano-Satoh A, et al. An essential developmental checkpoint for production of the T cell lineage. Science 2010;329(5987):93–6.

[201] Verbeek S, Izon D, Hofhuis F, Robanus-Maandag E, te Riele H, van de Wetering M, et al. An HMG-box-containing T-cell factor required for thymocyte differentiation. Nature 1995;374(6517):70–4.

[202] Ting CN, Olson MC, Barton KP, Leiden JM. Transcription factor GATA-3 is required for development of the T-cell lineage. Nature 1996;384(6608):474–8.

[203] Germar K, Dose M, Konstantinou T, Zhang J, Wang H, Lobry C, et al. T-cell factor 1 is a gatekeeper for T-cell specification in response to Notch signaling. Proc Natl Acad Sci U S A 2011;108(50):20060–5.

[204] Weber BN, Chi AW, Chavez A, Yashiro-Ohtani Y, Yang Q, Shestova O, et al. A critical role for TCF-1 in T-lineage specification and differentiation. Nature 2011;476(7358):63–8.

[205] He X, He X, Dave VP, Zhang Y, Hua X, Nicolas E, et al. The zinc finger transcription factor Th-POK regulates CD4 versus CD8 T-cell lineage commitment. Nature 2005;433(7028):826–33.

[206] Muroi S, Naoe Y, Miyamoto C, Akiyama K, Ikawa T, Masuda K, et al. Cascading suppression of transcriptional silencers by ThPOK seals helper T cell fate. Nat Immunol 2008;9(10):1113–21.

[207] Wildt KF, Sun G, Grueter B, Fischer M, Zamisch M, Ehlers M, et al. The transcription factor Zbtb7b promotes CD4 expression by antagonizing Runx-mediated activation of the CD4 silencer. J Immunol 2007;179(7):4405–14.

[208] Woolf E, Xiao C, Fainaru O, Lotem J, Rosen D, Negreanu V, et al. Runx3 and Runx1 are required for CD8 T cell development during thymopoiesis. Proc Natl Acad Sci U S A 2003;100(13):7731–6.

[209] Egawa T, Littman DR. ThPOK acts late in specification of the helper T cell lineage and suppresses Runx-mediated commitment to the cytotoxic T cell lineage. Nat Immunol 2008;9(10):1131–9.

[210] Egawa T, Tillman RE, Naoe Y, Taniuchi I, Littman DR. The role of the Runx transcription factors in thymocyte differentiation and in homeostasis of naive T cells. J Exp Med 2007;204(8):1945–57.

[211] Setoguchi R, Tachibana M, Naoe Y, Muroi S, Akiyama K, Tezuka C, et al. Repression of the transcription factor Th-POK by Runx complexes in cytotoxic T cell development. Science 2008;319(5864):822–5.

[212] He X, Park K, Wang H, He X, Zhang Y, Hua X, et al. CD4-CD8 lineage commitment is regulated by a silencer element at the ThPOK transcription-factor locus. Immunity 2008;28(3):346–58.

[213] Taniuchi I, Ellmeier W. Transcriptional and epigenetic regulation of CD4/CD8 lineage choice. Adv Immunol 2011;110:71–110.

[214] Xiong Y, Bosselut R. The enigma of CD4-lineage specification. Eur J Immunol 2011;41(3):568–74.

[215] Luckey MA, Kimura MY, Waickman AT, Feigenbaum L, Singer A, Park JH. The transcription factor ThPOK suppresses Runx3 and imposes CD4(+) lineage fate by inducing the SOCS suppressors of cytokine signaling. Nat Immunol 2014;15(7):638–45.

[216] Wei G, Wei L, Zhu J, Zang C, Hu-Li J, Yao Z, et al. Global mapping of H3K4me3 and H3K27me3 reveals specificity and plasticity in lineage fate determination of differentiating $CD4^+$ T cells. Immunity 2009;30(1):155–67.

[217] Allan RS, Zueva E, Cammas F, Schreiber HA, Masson V, Belz GT, et al. An epigenetic silencing pathway controlling T helper 2 cell lineage commitment. Nature 2012;487(7406):249–53.

[218] Sekiya T, Kashiwagi I, Inoue N, Morita R, Hori S, Waldmann H, et al. The nuclear orphan receptor Nr4a2 induces Foxp3 and regulates differentiation of CD4+ T cells. Nat Commun 2011;2:269.

[219] Sekiya T, Kashiwagi I, Yoshida R, Fukaya T, Morita R, Kimura A, et al. Nr4a receptors are essential for thymic regulatory T cell development and immune homeostasis. Nat Immunol 2013;14(3):230–7.

[220] Sekiya T, Kondo T, Shichita T, Morita R, Ichinose H, Yoshimura A. Suppression of Th2 and Tfh immune reactions by Nr4a receptors in mature T reg cells. J Exp Med 2015;212(10):1623–40.

[221] Sanders VM. Epigenetic regulation of Th1 and Th2 cell development. Brain Behav Immun 2006;20(4):317–24.

[222] Wilson CB, Rowell E, Sekimata M. Epigenetic control of T-helper-cell differentiation. Nat Rev Immunol 2009;9(2): 91–105.

[223] Schoenborn JR, Dorschner MO, Sekimata M, Santer DM, Shnyreva M, Fitzpatrick DR, et al. Comprehensive epigenetic profiling identifies multiple distal regulatory elements directing transcription of the gene encoding interferon-gamma. Nat Immunol 2007;8(7):732–42.

[224] Santangelo S, Cousins DJ, Winkelmann N, Triantaphyllopoulos K, Staynov DZ. Chromatin structure and DNA methylation of the IL-4 gene in human T(H)2 cells. Chromosome Res 2009;17(4):485–96.

[225] Janson PC, Linton LB, Bergman EA, Marits P, Eberhardson M, Piehl F, et al. Profiling of CD4+ T cells with epigenetic immune lineage analysis. J Immunol 2011;186(1):92–102.

[226] Thomas RM, Sai H, Wells AD. Conserved intergenic elements and DNA methylation cooperate to regulate transcription at the IL17 locus. J Biol Chem 2012;287(30):25049–59.

[227] Josefowicz SZ, Wilson CB, Rudensky AY. Cutting edge: TCR stimulation is sufficient for induction of Foxp3 expression in the absence of DNA methyltransferase 1. J Immunol 2009;182(11):6648–52.

[228] Ohkura N, Hamaguchi M, Morikawa H, Sugimura K, Tanaka A, Ito Y, et al. T cell receptor stimulation-induced epigenetic changes and Foxp3 expression are independent and complementary events required for Treg cell development. Immunity 2012;37(5):785–99.

[229] Yue X, Trifari S, Aijo T, Tsagaratou A, Pastor WA, Zepeda-Martinez JA, et al. Control of Foxp3 stability through modulation of TET activity. J Exp Med 2016;213(3):377–97.

[230] Wang L, Liu Y, Han R, Beier UH, Thomas RM, Wells AD, et al. Mbd2 promotes foxp3 demethylation and T-regulatory-cell function. Mol Cell Biol 2013;33(20):4106–15.

[231] Toker A, Engelbert D, Garg G, Polansky JK, Floess S, Miyao T, et al. Active demethylation of the Foxp3 locus leads to the generation of stable regulatory T cells within the thymus. J Immunol 2013;190(7):3180–8.

[232] Nair VS, Oh KI. Down-regulation of Tet2 prevents TSDR demethylation in IL2 deficient regulatory T cells. Biochem Biophys Res Commun 2014;450(1):918–24.

[233] Sekiya T, Nakatsukasa H, Lu Q, Yoshimura A. Roles of transcription factors and epigenetic modifications in differentiation and maintenance of regulatory T cells. Microbes Infect 2016;18(6):378–86.

[234] Denton AE, Russ BE, Doherty PC, Rao S, Turner SJ. Differentiation-dependent functional and epigenetic landscapes for cytokine genes in virus-specific CD8+ T cells. Proc Natl Acad Sci U S A 2011;108(37):15306–11.

[235] Juelich T, Sutcliffe EL, Denton A, He Y, Doherty PC, Parish CR, et al. Interplay between chromatin remodeling and epigenetic changes during lineage-specific commitment to granzyme B expression. J Immunol 2009;183(11):7063–72.

[236] Scharer CD, Barwick BG, Youngblood BA, Ahmed R, Boss JM. Global DNA methylation remodeling accompanies CD8 T cell effector function. J Immunol 2013;191(6):3419–29.

[237] Zediak VP, Johnnidis JB, Wherry EJ, Berger SL. Cutting edge: persistently open chromatin at effector gene loci in resting memory CD8+ T cells independent of transcriptional status. J Immunol 2011;186(5):2705–9.

[238] Komatsu N, Mariotti-Ferrandiz ME, Wang Y, Malissen B, Waldmann H, Hori S. Heterogeneity of natural Foxp3+ T cells: a committed regulatory T-cell lineage and an uncommitted minor population retaining plasticity. Proc Natl Acad Sci U S A 2009;106(6):1903–8.

[239] Miyao T, Floess S, Setoguchi R, Luche H, Fehling HJ, Waldmann H, et al. Plasticity of Foxp3(+) T cells reflects promiscuous Foxp3 expression in conventional T cells but not reprogramming of regulatory T cells. Immunity 2012;36(2): 262–75.

CHAPTER

3

DNA Methylation and Histone Modifications in Autoimmunity

Debarati Banik, Katherine B. Chiappinelli, Alejandro Villagra

The George Washington University Cancer Center, The George Washington University, Washington, DC, United States

OUTLINE

The Epigenetics of Autoimmunity. http://dx.doi.org/10.1016/B978-0-12-809912-4.00003-9

3.1 INTRODUCTION

All cells in an organism share the same genetic information; however, cells get access to this information selectively, in a tissue- or cell-specific manner, during the process of development. In other words, specific genes are expressed at a given time in a given cell or tissue. The term "epigenetic landscape," was first coined by Conrad Waddington in 1942, and ingeniously encompasses the molecular mechanisms for translating genetic information into observable traits or phenotypes [1]. Epigenetics is defined as the study of a change in phenotype that does not arise from a change in the underlying DNA makeup or genotype. It is the nongenetic influence on the regulation of gene expression, which is heritable through generations of a species. Epigenetics is fundamental to a variety of processes that include, but is not limited to, gene expression, microRNA expression, genomic imprinting, transposable gene regulation, DNA–protein interactions, and X-chromosome inactivation. Epigenetic control is fundamental to cellular differentiation, beginning at embryogenesis and continuing until death [2]. Moreover, epigenetics plays a central role in determining the penetrance of certain pathologies within the population, such as cancer or autoimmune conditions.

The purpose of this chapter is to delineate the relationship between epigenetic regulation and the immune system in the context of the development, sustenance, and progression of autoimmune conditions. While the immune system is the decisive factor in combating pathogen-mediated invasion in a host, its misdirection may present host components as foreign bodies to the immune system and thus result in cell or tissue damage. This condition is collectively defined as autoimmunity, and can range from being mild to massively self-damaging. Mild autoimmune reactions are harmless in the subclinical range and are universal phenomena within vertebrates; however, the progression from benign to pathogenic self-reactive autoimmune conditions is accompanied by both genetic and environmental triggers [3]. Specific immune cells (e.g., T-lymphocytes [4], natural killer (NK) cells [5], and their subsets such as T-regulatory (T_{reg}) [6] and regulatory NK cells [7]) play a central role in the development of autoimmune conditions. The effector functions of cytotoxic cells are essentially regulated by the antigen-presenting cells, such as macrophages, dendritic cells (DCs), and B cells (Fig. 3.1). Therefore the involvement of these cells is critical in the biology of autoimmune conditions, as described in detail later in the chapter.

3.2 HISTONE DEACETYLASES AND DNA METHYLTRANSFERASES AT A GLANCE

Epigenetic modifications are crucial for controlling gene expression during mammalian cellular differentiation; however, they can also determine the differences between normal and diseased states. The main epigenetic mechanisms include covalent modifications of cytosine bases and histones as well as the repositioning of nucleosomes (Fig. 3.1). The modification of cytosine residues gives rise to methylated DNA; histone modifications produce acetylated, methylated, or phosphorylated histones; and the repositioning of nucleosomes functions during chromatin remodeling. DNA methylation is executed and maintained by de novo and maintenance DNA methyltransferases (DNMTs), respectively. Histone modifications are maintained by histone acetyltransferases (HATs), histone deacetylases (HDACs) (acetylation), histone methyltransferases (HMTs), histone demethylates (HDMs) (methylation), histone kinases, and histone phosphorylases (phosphorylation). Chromatin remodeling is executed by distinct families of enzymes including SWI/SNF, ISWI, Mi-2, and INO80 [2]. Due to the limited scope of this chapter, only the first two classes of histone modifications will be discussed, with focus on preclinical and clinical settings in autoimmunity.

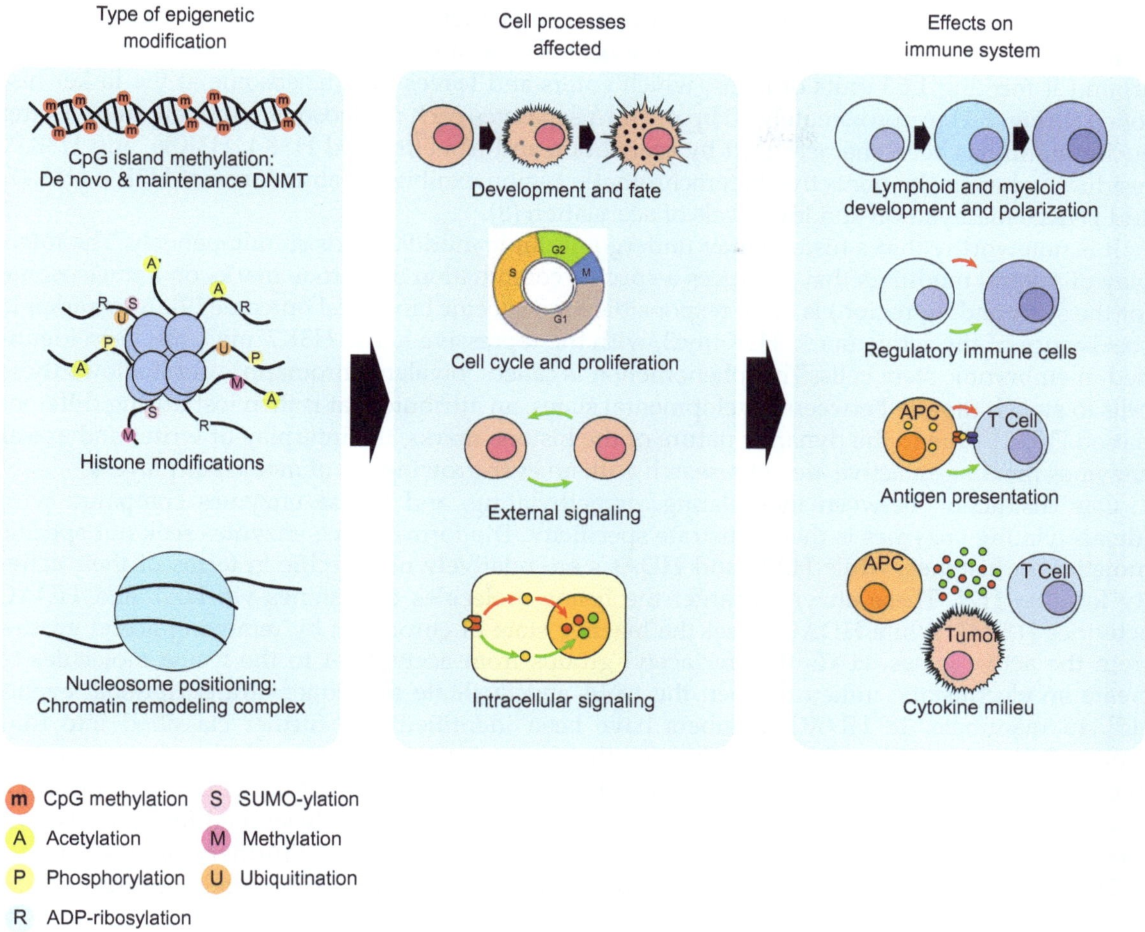

FIGURE 3.1 **Interplay of epigenetics in immunological outcomes.** Epigenetic mechanisms translate into various physiological functions that range from molecular and cellular levels to developmental stages. In multicellular complex life forms (i.e., mammals) the effects of such modifications are further translated into the immune system, which is composed of many cellular subsets with distinct functions. This figure segregates these different levels of functionality of epigenetic modification according to their overall effects on specific systems. The first tier of the figure summarizes the various mechanisms of epigenetic modification that a cell may acquire. The second tier represents the variety of molecular, cellular, and physiological phenomena influenced by the first tier. The development and fate of a cell is determined by phenomena such as transcriptional activation and repression, hematopoiesis, and cell differentiation. Cell cycle and proliferation involve growth factor-mediated signaling, DNA and protein synthesis, DNA repair, chaperone and structural functions, and nuclear transportation. Cell survival is balanced through the regulation of apoptosis, autophagy, ROS-induced death, senescence, and growth arrest. These activities are dependent upon a cascade of intra- and extracellular signaling pathways. Epigenetics also determine the changes that are induced at a macro level, such as X-chromosome inactivation, developmental stages of embryogenesis, and genomic imprinting. Most of these process are interdependent and overlapping in a complex network of temporal regulation. The third tier represents how the molecular functions of epigenetics are translated to immunological events. It involves the development and lineage selection within immune cells following hematopoiesis, maintenance of the plasticity of effector to regulatory cells and vice versa, and balancing innate and adaptive immunity by regulating the functionality of antigen presentation and cytokine milieu.

The nucleosome is central to posttranslational modifications and is constituted by two molecules of H2A, H2B, H3, and H4. This nucleosome octamer has a segment of 147 bp of DNA wrapped around it forming 1.65 units of turns, which enters and leaves the nucleosome at the linker histone H1 (Fig. 3.1). Approximately 50 bp of DNA separates each nucleosome. Actively transcribing euchromatin has been characterized by acetylation and trimethylated H3K4, H3K36, and H3K79 residues, whereas the nonactive heterochromatin region exhibits an abundance of H3K9, H3K27, and H4K20 methylation and low levels of acetylation [8].

It is noteworthy that a histone may undergo multiple modifications simultaneously. The interplay of several modifiers that produces a specific combination of histone marks on a nucleosome (or the surrounding region) is often responsible for a specific biological outcome [9]. A remarkable coexistence of the active mark, H3K4me3, with the repressive mark, H3K27me3, has been identified in embryonic stem cells. This phenomenon is called "bivalent chromatin" and it allows these cells to switch rapidly between developmental states, an attribute that is then lost during differentiation [10,11]. Due to the dynamic nature of the histone marks, the interplay of writer and eraser enzymes remains an active area of research with an ever-growing list of molecular players.

One distinction between methylating, demethylating, and kinase enzymes compared with (de)acetylating enzymes is their substrate specificity. The former three enzymes seek out specific moieties on histones, while HATs and HDACs are relatively nonspecific in terms of their activity location [12]. These enzymes target the lysine molecules of histones via HAT and HDAC activities [13,14]. While HDACs reset the inactive state of chromatin by removing acetyl groups from the active genes, HATs transfer acetyl groups from acetyl-CoA to the lysine molecules to create an electrostatic milieu to open the DNA and facilitate subsequent transcriptional events [15]. In mammals, 18 HDAC members have been identified and further classified into four groups [16]: class I, II, and IV use Zn^{2+} for their enzymatic activity and demonstrate a certain degree of sequence homology, while class III HDACs (sirtuins) are distinct in their use of NAD^+ as coenzymes. Class I enzymes (HCAC1, 2, 3, and 8) are predominantly studied for their histone-modifying and transcription-repressing properties; class II enzymes are further classified into IIa (HDAC4, 5, 7, and 9) and IIb (HDAC6 and 10); while HDAC11 is the sole member in the class IV family [17,18].

Although they were primarily identified for histone modifications, HDACs were later shown to modulate a wide array of nonhistone target proteins in distinct molecular and immunological pathways. Some of their substrates include critical transcription factors (e.g., p53, GATA1–4, NFκB, c-Myc, IRF-2, IRF-7, CREB, HIF1α, Rb, and STAT3), steroid receptors (e.g., androgen receptor, estrogen receptor α, and glucocorticoid receptor), DNA repair enzymes (e.g., Ku70, WRN, and TDG), inflammation mediators (e.g., HMGB1), nuclear import proteins (e.g., Rch1 and importin-α7), chaperone proteins (e.g., HSP90), structural proteins (e.g., α-tubulin), and viral proteins (e.g., E1A, L-HDAg, S-HDAg, T-antigen, and HIV-Tat) [19,20]. Therefore the molecular regulation exerted by HDACs regulates crucial pathways such as autophagy, extrinsic and intrinsic apoptosis, mitosis, angiogenesis, ROS-induced cell death, and senescence (Fig. 3.1). HDACs are mostly nonredundant and achieve their versatility partly through the unique localization profile of each class. For example, class I HDACs are ubiquitously expressed and localized in the nucleus whereas class II HDACs have tissue-specific expression and may show context-dependent shuttling between the cytoplasm and the nucleus [21]. HDAC inhibitors (HDACi) compounds were historically associated with modulating the carcinogenic properties of tumor cells due to their profound ability to regulate cell proliferation, differentiation, and death. Various pan-HDACi compounds (i.e., suberoylanilide hydroxamic acid (SAHA), givinostat, and panobinostat) have been successfully used in clinical trials as either standalone or combination therapies [22,23]. The use of these compounds in inflammatory and autoimmune diseases tends to capture an additional aspect of HDACi properties, such as the local or systemic modification of the cytokine milieu, which can be exploited to change the outcome of the ongoing inflammation (as discussed in greater length later in the chapter).

DNA methylation is a covalent modification added to cytosine bases in CpG dinucleotides by DNMT enzymes. These include the de novo enzymes, DNMT3A and DNMT3B, which add methyl groups to unmethylated DNA and thereby set up patterns of DNA methylation during development. The DNMT3L protein, which has no catalytic activity, shares homology with DNMT3A and DNMT3B and forms a complex with these proteins to assist in methylation. The DNMT1 protein (the maintenance methyltransferase) adds methyl groups to the new strand of DNA being synthesized from a methylated strand during cell division and is the main active methyltransferase in differentiated cells. Finally, the DNMT2 protein is a DNMT homologue that cannot methylate DNA, but it methylates cytosine 38 in the anticodon loop of aspartic acid transfer RNA [24,25].

DNA methylation at CpG dinucleotides in the promoter region of a gene generally represses transcription. DNA methylation can prevent transcription factors from binding; however, it can also recruit other proteins [i.e., methyl-CpG-binding domain proteins (MBDs)], which in turn recruit HDACs and other chromatin modifiers to establish a repressive chromatin state. This repressive chromatin state prevents transcription of the associated gene and may be inherited throughout mitotic cell divisions (through the DNMT1 maintenance enzyme). Approximately 50% of the human genome is susceptible to "silencing" by DNA methylation and this helps to prevent transcription of noncoding DNA that may encode transposable elements and threaten genome stability. Methylated cytosine residues spontaneously deaminate to become thymine residues, changing the genomic sequence. The promoter regions of many genes exhibit stretches of unmethylated CpG dinucleotides termed "CpG islands," which represent regions that are evolutionarily protected from DNA methylation so as to maintain sequence integrity and prevent silencing of the associated gene [24,25].

The first foray into the therapeutic inhibition of DNA methylation was in the context of cancer. Just as normal cells with different cell fates have different patterns of DNA methylation, cancer cells acquire strikingly different DNA methylation profiles to normal cells. Cancers lose DNA methylation on a global scale, specifically at repetitive elements, but they can also gain methylation in the promoter regions of tumor suppressors, which causes aberrant gene silencing [24,26]. Thus the inhibition of DNA methylation could reverse some of the molecular pathways that are altered in cancer. DNMT inhibitors (DNMTi) are cytidine analogs that incorporate into DNA during the S phase. DNMTi such as 5-azacytidine (azacytidine or "Aza") and 5-aza-2′-deoxycytidine (decitabine or "Dac") bind to and trap the DNMTs, block their catalytic action, and cause degradation of the enzymes [27]. The degradation of DNMTs can also disrupt the other repressive proteins that bind to DNMTs to create a repressive chromatin structure [28–30]. Azacytidine integrates into both RNA and DNA, while decitabine only integrates into DNA. Therefore DNMTi treatment can reverse the silencing of genes with promoter DNA methylation and cause reexpression of these genes [24,25]. Transient low-dose treatment of tumor cells (i.e., leukemia and breast cancer) with DNMTi promoted apoptosis, decreased the cell cycle activity, and decreased self-renewal [31]. DNMTi are efficacious for the treatment of hematologic neoplasms and are FDA approved for the preleukemic disorder, myelodysplasia (MDS), and acute myeloid leukemia (AML) [32–34]. A novel DNMTi, guadecitabine, which has a longer in vivo stability than Aza or Dac, has shown clinical activity in patients with MDS and AML [35]. As will be discussed further in this chapter, DNA methylation patterns at the promoters of immune genes are also found to be modified under autoimmune conditions such as rheumatoid arthritis, systemic lupus erythematosus (SLE), and asthma.

Significant information has been generated from tumor models to understand how the molecular mechanisms of epigenetic modulation affect the immune pathways. These findings have also accelerated insight into other pathologies related to the immune system, most notably autoimmune conditions, where the interplay of epigenetic mechanisms often influences the functionality of the immune components. However, before the specifics of this are described, an overview of the epigenetic control of specific molecular pathways is provided.

3.3 EFFECTS OF EPIGENETIC MODIFIERS ON IMMUNE PATHWAYS

This section addresses the molecular pathways that are influenced by epigenetics, beginning at the molecular level and gradually progressing into cellular functions. Fig. 3.1 summarizes the epigenetic modifications that are translated into various physiological functions at both the cellular and immune level.

Several pathways and transcription factors that play significant roles within the immune system are modulated by HDACs. Although the roles of HDACs in the cell cycle, apoptosis, differentiation, and growth have been well investigated, their contribution to immune cell function remains poorly understood. Microarray analysis of a trichostatin A (TSA)-treated human lymphoid cell line showed that it modulated the expression of only 2% of genes, whereas the modulation was limited to 1%–2% within the non-T cells [36–38]. These studies bolstered the notion that the post-translational modifications conferred by the HDACs through acetylation are beyond the scope of such mRNA-based studies.

3.3.1 Role of Acetylation and Deacetylation in Cell Survival, Differentiation, and Death

Acetylation induces conformational changes within various nonhistone transcription factors to facilitate sequence-specific DNA binding and confer cellular fate. For example, p53 is a regulator of the cell cycle and apoptosis and is acetylated by the HAT cofactor p300 at its C-terminus [39]. The same HAT is also able to acetylate lysine 685 of STAT3 to facilitate the stable dimerization of STAT3 for cytokine-stimulated DNA binding and transcriptional regulation [40]. GATA1, which is a significant determinant of hematopoietic lineage and terminal differentiation, is also modified by p300 to undergo DNA-binding [41]. P300 plays a critical role in the functions of E2F1–3 transcription factors, which regulate DNA replication and cell proliferation [42]. YY1 is another critical mediator of differentiation and development and is also subject to p300-mediated acetylation [43]. HAT enzymes (i.e., CBP and P/CAF) acetylate HMGI(Y), which is an essential component in the IFN-β pathway [44]. A continuous interplay between HATs and HDACs reversibly regulates the activation status for each of these transcription factors.

3.3.2 Proinflammatory Mediators Regulated by Epigenetic Modifiers

NFκB/IκB Pathway: A critical determinant of immune processes of early, acute-phase, and inflammatory responses is the multifunctional transcription factor NFκB, which can be either up- or downregulated by HDACi depending on the cellular context. IκBα, which is the regulator of NFκB, is induced by HDACi and leads to the proteolytic degradation of NFκB and prevents its nuclear translocation [45]. Conversely, the HAT enzyme CBP (p300/CREB-binding protein)-mediated acetylation of lysine 310 on p65 prevents its association to NFκB. By catalyzing the deacetylation of this lysine residue, HDAC3 promotes NFκB degradation, which can be reversed by HDAC3 inhibition [46–49].

JAK–STAT pathway: HDAC enzymes, especially the class I HDACs, are actively involved in determining the proinflammatory state of cells by regulating the JAK–STAT pathway. A significant example is the maintenance of acetylated versus deacetylated STAT1 through cytoplasmic and nuclear locations. Phosphorylation and dimerization of STAT1 is an important step in the induction of its nuclear transportation to transcriptionally activate several proinflammatory IFNγ-responsive genes. CBP recruits the tyrosine phosphatase, TCP45, to dephosphorylate STAT1, which also results in the disintegration of the STAT1 complex. The acetylated but dephosphorylated STAT1 exits the nucleus but cannot be activated unless it is deacetylated. HDAC3 plays a critical role in this process, regenerating STAT1 to continue the cycle [50,51] (Fig. 3.2). Pan-HDACi promote an antiinflammatory milieu in this scenario; however, selective HDAC6 inhibitors work in the opposite direction by abrogating antiinflammatory pathways [52,53].

3.3.3 Epigenetic Involvement in the Antigen Presentation Pathway: Signal 1, 2, and 3

HDAC enzymes regulate several key pathways in the process of antigen presentation, which is referred as the gateway of the immune response. The process of antigen presentation relies heavily on major histocompatibility complex (MHC) class I molecules, which are expressed in almost all cell types including tumors, and MHC class II molecules, which are only expressed in professional antigen presenting cells (APCs). Some of the pan-HDACi upregulate the expression of surface MHC class I molecules, as well as multiple components of the antigen presenting machinery (i.e., TAP-1, TAP-2, tapasin, calnexin, calreticulin, ERp57, LMP2, and LMP7) [54–56] (Fig. 3.1). MHC class I molecules are downregulated by a repressor complex consisting of HDAC1, 2, and 8, and thereby evade cytotoxic T-cell-mediated elimination [57]. HDAC6/Hsp90 also stabilizes the peptide structure to facilitate peptide presentation by MHC class I molecules [58]. Conversely, MHC class II molecules are induced by both pan- and class I HDACi molecules [52,59]. The master regulator of MHC class II molecules, CIITA, is also susceptible to this CBP/PCAF and HDAC1/2-mediated acetylation–deacetylation mode of regulation [60].

Involvement of both innate and adaptive immunity is required to establish an effective and durable immune response. The signaling cascade or sequential activation of the associated pathways is controlled by cytokines. APCs are the early initiators of the innate arm of immunity and convey danger signals from foreign or altered self-antigens to the adaptive arm of immunity. Apart from presenting the T cell receptor (TCR)–MHC mediated signal 1, APCs also determine the subsequent activation or repression of the effector cells via costimulatory signals (signal 2) and cytokine secretion (signal 3). CD80 and CD86 are the two costimulatory molecules of the B7 family that ligate to CD28 on T cells and lead to downstream activation of IL-2 production [61]. CD40 on the APC binds to CD40L on T cells and activates APCs, leading to the expression of digested peptides on the macrophage cell surface and antibody production in B cells. In this context, HDACi can reduce signals 2 and 3 from APCs (Figs. 3.2 and Figs. 3.3). Using both bone marrow-derived DCs (BMDC) and macrophages, pan-HDACi downregulate genes associated with these pathways; for example, TSA significantly diminishes the surface expression of CD40, CD80, CD86, and CCR7 in BMDCs along with the production of proinflammatory cytokines IL-6, TNFα, and IL-12 within endothelial cells [62]. HDACi can also modulate the expression of adhesion molecules; for

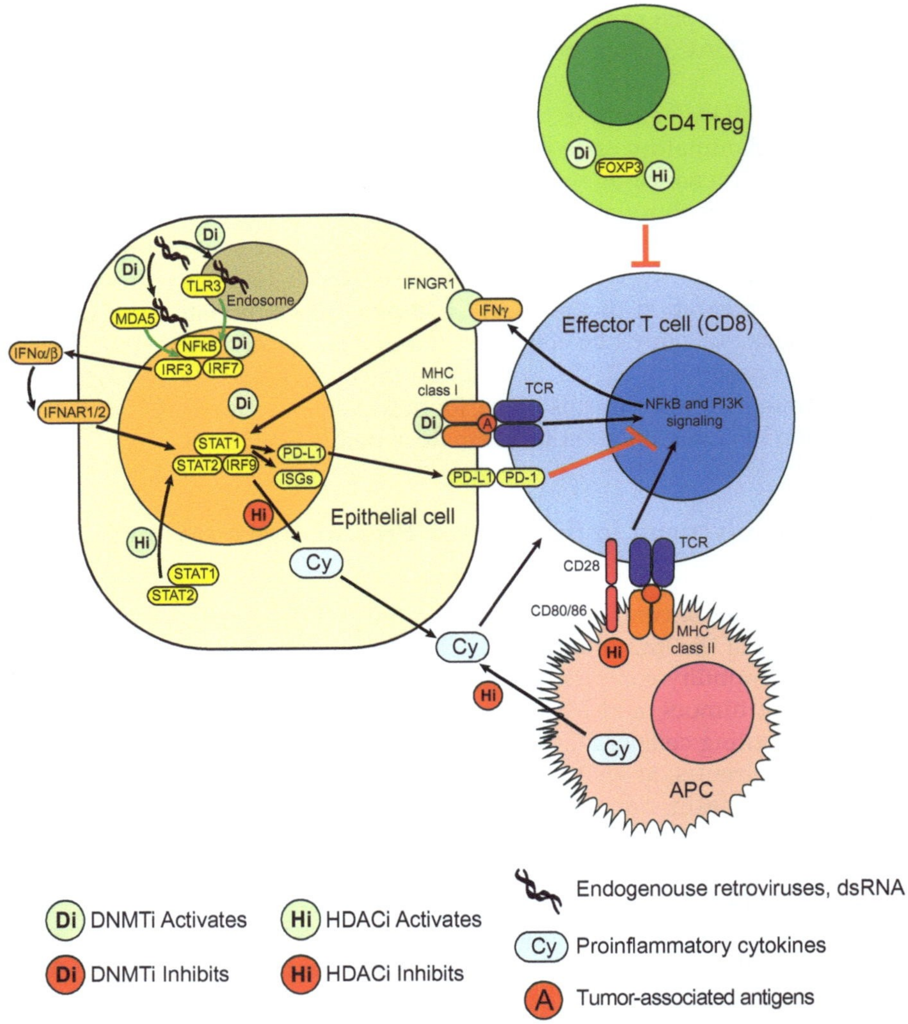

FIGURE 3.2 **Effects of DNMTi and HDACi on epithelial and immune cells.** DNMT inhibitors (DNMTi) remove methylation from double-stranded RNA species (i.e., endogenous retroviruses) which activate sensors (i.e., TLR3 and MDA5). This activates a type I interferon response, which may be boosted by DNMTi demethylation of the IRF7 promoter. STAT1 and STAT2 signaling mediate the transcription of interferon stimulated genes (ISGs), including the immune checkpoint ligand PD-L1. Downstream of the interferon response, antigen processing and presentation (MHC class I) are increased. DNMTi directly demethylate promoters of cancer testis antigens (CTAs), which may be presented on the cell surface and recognized by the T cell receptor (TCR). HDACi may also upregulate antigen presentation, but they predominantly have antiinflammatory effects. Specifically, they can induce acetylation of STAT1 and render it inactive. They reduce CD80/86 expression on antigen presenting cells (APCs) and cytokine secretion by these cells. Both DNMTi and HDACi increase the expression of FOXP3 in CD4 cells, promoting a regulatory T cell (T_{reg}) phenotype.

example, ICAM-1 is upregulated following HDACi treatment in tumor models [63]. ICAM-1 binds to LFA-1 molecules on the surface of T cells to strengthen the immunological synapse. Selective targeting of HDAC6 strengthens the immune synapse through cytoskeletal rearrangements [64]. HDAC6 is also involved in the regulation of several coregulatory molecules extremely important in both cancer and autoimmune diseases (i.e., PD-L1 and PD-L2), which suggests that targeting

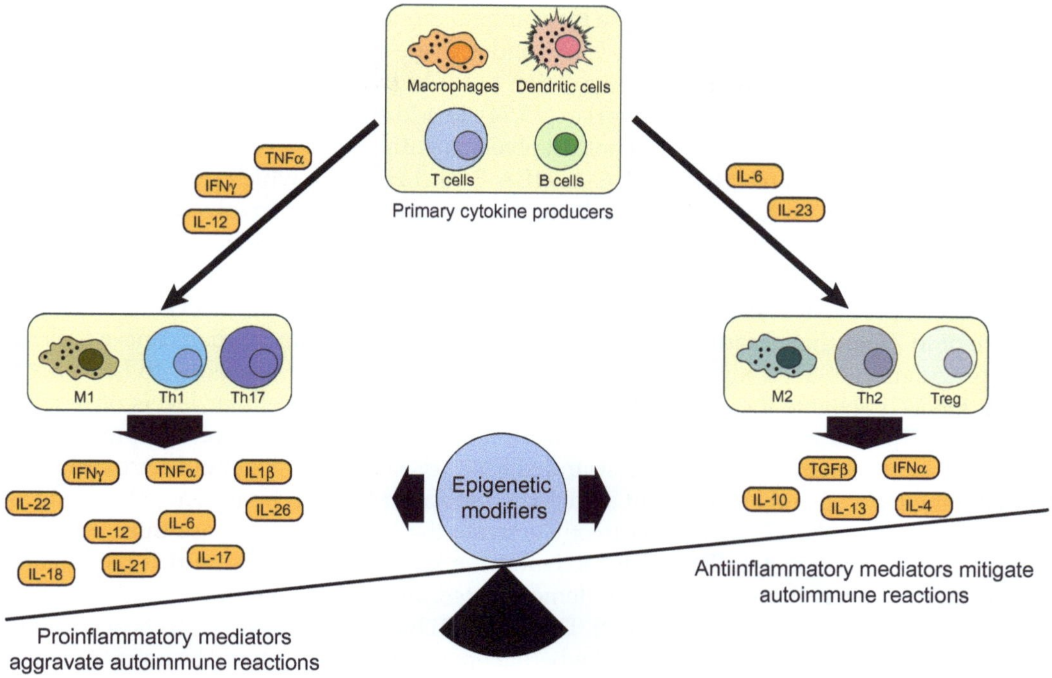

FIGURE 3.3 **Interaction between immune cells to regulate autoimmune conditions.** Various subsets of immune cells play distinct roles from the establishment to the resolution of autoimmune conditions. This is a simplistic representation of the hierarchy and interplay between the immune subsets. APCs activate specific effector cells and establish proinflammatory versus antiinflammatory milieu. Under autoimmune conditions, the resolution phase is critically dependent on the development of an antiinflammatory milieu thereby emphasizing the context-dependent use of epigenetic modifies.

this specific HDAC could be used to regulate the expression of these negative immune checkpoint pathways [65].

Genetic ablation of HDAC3 in a murine model is associated with a deficiency in the production of proinflammatory cytokines in response to the toll-like receptors (TLR) agonist, LPS. This observation was corroborated by a reduction in Ifnβ and the constitutive overexpression of Cox1, which may interfere with the proinflammatory cytokine pathways [66]. A similar trend was observed in humans in both in vitro and in vivo applications of HDACi. Human monocyte-derived DCs exhibited a reduced translocation of NFκB, RelB, IRF-3, and IRF-8 and a subsequent reduction in cytokine production upon treatment with pan-HDACi [67]. Ex vivo analysis of peripheral blood mononuclear cells (PBMCs) in givinostat-treated human subjects revealed a suppression of proinflammatory cytokine production upon stimulation [68]. These data collectively suggest that APCs depend upon HDACs to carry out their antigen-presenting functions. Fig. 3.2 summarizes the interdependence of molecular pathways converging on cellular outcomes in the presence of epigenetic modifiers.

3.3.4 Epigenetic Control of Immune Signaling in Nonimmune Cells

The immunomodulatory effects of DNMTi on cancer cells can inform our understanding of these epigenetic regulators in autoimmunity. These effects include the upregulation of cancer testis antigens (CTAs), which may be recognized by the host immune system and are currently being tested as vaccine antigens in clinical trials [69,70]. Data from animal models support using a combination of epigenetic and immune therapy to upregulate CTAs and activate the host immune system against them [71,72]. Other work has confirmed the upregulation of CTAs by DNMTi in multiple different cancer types, including lung, breast, colon, and ovarian [73,74]. In addition to the cancer antigens themselves, DNMTi also increase the expression of genes involved in antigen processing and presentation, such as the MHC class I proteins (i.e., B2M, HLA-A, HLA-B, and HLA-C), which present antigens on the surface of epithelial cells for host immune-cell recognition, and proteins involved in antigen processing by the proteasome (i.e., PSMB8, PSMB9, and TAP1) [74]. These genes lack DNA methylation in their promoters and can be activated by the interferon response and upstream STAT signaling [75]. The interferon response, antigen processing and presentation, and CTAs were all activated by Aza or Dac treatment in 77 epithelial cancer cell lines [76]. Tumor biopsies from ovarian cancer patients also showed upregulation of STAT signaling and cytokines when they were treated with Dac plus carboplatin [77]. DNMTi activate the interferon response by increasing the cellular levels of repetitive or viral RNAs [78,79]. This signaling inhibits colon cancer stem cells [80] and drives apoptosis in ovarian cancer [79]. One component of this double-stranded RNA (dsRNA) is endogenous retroviruses, which are normally methylated and silenced but are reexpressed after DNMTi treatment [79]. In addition to dsRNA upregulation, DNMTi may also directly activate the interferon response through demethylation of the transcription factor, IRF7. IRF7 activates the interferon response and is methylated and silenced in many cancers; however, IRF7 is demethylated and upregulated by DNMTi treatment in colon [80], lung [76], and ovarian cancers [79]. Thus DNMTi have a spectrum of effects on immune signaling in tumor cells, begging the question of whether epigenetic therapy could be used to increase the sensitivity to immune therapies in cancer. Preliminary data from mouse models shows that DNMTi sensitized the animals to treatment with the checkpoint inhibitor, anti-CTLA-4, in melanoma [79] and ovarian cancer models [81]. In the latter, Dac treatment increased the expression of chemokines that recruited NK and CD8 T cells and promoted their production of cytokines. In another mouse model of ovarian cancer, DNMTi treatment alongside an epigenetic modifier (EZH2 inhibitor) removed the repressive epigenetic marks from the chemokines, CXCL9 and CXCL10, on tumor cells, thus promoting effector T-cell tracking to the tumor microenvironment and sensitizing to treatment with the immune checkpoint inhibitor, anti-PD-1 [82]. Clinical trials to test these paradigms are currently being initiated or are under way. One trial in nonsmall cell lung cancer used a combined epigenetic therapy of a DNMTi (5-azacytidine) and a HDACi (entinostat) and allowed patients to move on to the therapy of their choice after progressing on the epigenetic therapy. The clinical trial was of the immune checkpoint inhibitor, anti-PD1, and 20% of nonsmall cell lung cancer patients responded to anti-PD1 alone; however, all five patients who had previously received epigenetic therapy reached 24 weeks without progression and three of the five developed durable Response Evaluation Criteria In Solid Tumors (RECIST) criteria responses [76]. Thus DNMTi increases immune signaling from tumors and may help to sensitize cancer patients who do not respond to immune therapies.

3.4 EFFECT OF EPIGENETICS ON EFFECTOR AND REGULATORY CELLS

Epigenetic modulation, especially the acetylation–deacetylation cycle, is a significant determinant of immune outcomes at the level of signaling pathways. The following sections will further discuss how these effects are translated to the functionalities of effector T cells, T_{reg} cells, NK cells, and B cells, which are the cornerstones of autoimmune outcomes (Fig. 3.1).

3.4.1 Dependence of T Cells on Epigenetics

Activation: During the initial stage of T-cell priming, APCs mediate signal 1 (MHC–TCR complex formation), signal 2 (costimulation), and signal 3 (cytokine stimulation). Low levels of IL-4 and IFN-γ are produced by naive T cells before the activation of any of the polarizing transcription factors (e.g., T-bet for Th1 or GATA3 for Th2). The timeline of lineage-specific gene expression implicates epigenetic regulation. This hypothesis is supported by the acetylation of histone 4 and the subsequent upregulation of IFN-γ gene expression following TSA treatment in Th0 cells. Pan-HDACi treatment also downregulates the expression of CD62L in naive T cells, supporting the notion that HDACi compounds may promote the early activation of $CD4^+$ T cells by modifying the gene loci involved in the Th1 or Th2 fate of Th0 cells [83].

Proliferation: Proliferation of naive T cells through mitosis is closely regulated by the posttranslational modification of histones; for example, both HDAC1 and HDAC2 are critical for promoting the proliferation stages past G1/S of mitosis through the inhibition of p21 [84]. Mitotic cells are found to be particularly enriched in the acetylated form of cell division-promoting proteins such as the RNA processing proteins eIF4G, RNA helicase A, APC1, anillin, and NudC [85]. Supporting this notion, Pan-HDACi as well as class I HDACi compounds exert antiproliferative actions on T cells, including their ability to produce IL-2 [86,87].

Polarization: Cellular environment, costimulatory signals, and cytokine milieu are the critical determinants of T-cell fate. Polarization towards type 1 (Th1) is dependent upon activation of IFN-γ, T-bet, and subsequent synthesis of IL-12, which in turn activates STAT4. A positive feedback loop promotes further IFN-γ production to ensure Th1 differentiation continues. Conversely, type 2 (Th2) polarization relies on IL-4 to activate the STAT6-mediated upregulation of GATA3 in Th0 cells, which turns on IL-5–and IL-13 gene expression to ensure Th2 fate [88]. Acetylation of some of the critical gene promoters are coordinated with these events, such as acetylation of the IFN-γ promoter before the production of IL-12 [89]. The balance of HAT and HDAC enzymes at different stages of differentiation appears to be critical for the activation status of specific regulators; for example, H3 and H4 acetylation were observed on IFN-γ and T-bet promoters in Th1 cells whereas the IL-4 promoter was acetylated within Th2 cells. Moreover, the lineage-associated transcription factors (i.e., GATA3, STAT1, and STAT4) are themselves the subjects of acetylation [20]. Conversely, DNA methylation, which is controlled specifically by the de novo methyltransferase DNMT3A, is necessary for IFN-γ silencing and polarization to Th2 [90]. IL-4 expression is also suppressed by DNA methylation as T cells polarize to Th1. DNMT3A is important in controlling the $CD4^+$ T cell fate and $CD4^+$ cells that lack DNMT3A simultaneously express IFN-γ and IL-4 after activation [91]. In conclusion, the reversible molecular signature defined by HAT, HDAC, and DNMT3A helps to dictate the phase-specific expression of molecular players (i.e., transcription factors and cytokines) in T cells (Fig. 3.2).

3.4.2 Reliance of T_{reg}-Functionality on Epigenetics

Several studies have emphasized that there is a correlation between HDAC activity and the generation and function of T_{regs} [92]. FOXP3 must be expressed for T_{regs} to properly modulate immune responses, and Foxp3 deficient mice fail to produce functional T_{regs} and die of autoimmune disease by 1 month of age [93,94]. An increased population of T cells from the thymus and peripheral blood show a propensity to acquire the $FOXP3^+$ T_{reg} phenotype under the influence of pan-HDACi compounds (i.e., TSA and SAHA). Pan-HDACi compounds also enhance the suppressive abilities of T_{regs} both in vitro and in vivo [95]. A complex interplay of HATs, HDACs, and DNMTs maintain the functional status of FOXP3, which is the master transcription factor for maintaining the suppressive functionality of T_{regs} [96]. In tumor models, the HAT, p300, maintains the acetylated state of FOXP3 [97], and this can be experimentally reversed by ablating HDAC6 [98], HDAC9 [95], sirtuin-1, and HDAC5, thereby leading to dysfunction of T_{regs} [99]. Treatment with a chemical inhibitor of HDAC6 can restore or increase T_{reg} function [100] and may be a future avenue to pursue autoimmune disease therapies.

Unfortunately, the interspecies differences in T_{reg} biology mean that the observations from murine and human models draw contrasting conclusions regarding the effect of HDACi on T_{reg} function. For example, while class II and III HDACi successfully augmented T_{reg} functions in murine models, experiments performed on human blood and rats showed minimal or no such effects [96]. Further investigations are needed to better understand how to harness the power of HDACi for T_{regs}.

The *FOXP3* gene in CD4-T cells is also regulated by DNA methylation. As mentioned previously, the *FOXP3* gene contains a CpG island that is methylated in most immune cells but is demethylated in human and mouse T_{regs} to promote *FOXP3* expression and proper T_{reg} functions [101,102]. Preventing DNA methylation with a DNMTi [103] increases Foxp3 expression in T cells and causes them to convert into T_{regs} (Fig. 3.2). However, conditional deletion of Dnmt1 in Foxp3^{+} T_{regs} impairs their development and promotes autoimmunity. Dnmt1 loss caused upregulation of inflammatory genes in T_{regs}, mononuclear cell infiltration of host tissues, and strong T- and B-cell responses due to T_{reg} dysfunction [103].

3.4.3 Epigenetics of NK Cells

NK cells represent a critical arm of effector functions in which the effector–target interaction is dependent upon the balance of activating and inhibitory signals experienced by the NK cells. Most of the information on NK cell biology and its HDAC-dependence have been elucidated using tumor models. Stress-induced activating ligands such as MHC class I-related chain A (MICA) and B molecules (MICB) and UL16-binding proteins are often expressed by tumor cells, and NK cells can recognize these via the activating NKG2D receptor [104]. Some HDACi compounds may increase the visibility of the target cells by increasing the activating signals. As opposed to nonmalignant primary hepatocytes, the HDAC class I inhibitor, valproic acid (VPA), enhances the activator signal expression on hepatocarcinoma cells [105]. Similar observations have been made for pan-HDACi molecules using various other cancer models [106]; however, other pan-HDACi molecules decrease NK cell activity by downregulating the B7-H6 molecule on target cells and eliminating the NKp30-dependent degranulation mechanism in NK cells [107]. Other in vitro studies have suggested a significant role of cytokine dependence on the overall outcome of NK cells in the presence of HDACi [106]. The variation of quantitative and qualitative features of NK cells under autoimmune conditions, as well as their ability to influence the cytokine milieu, suggests a strong association of NK cells with autoimmunity [5].

3.4.4 Epigenetics of Humoral Immunity (B Cells)

B cells stand at the critical junction of humoral and cell-mediated immunity, as they have the unique ability to produce antibodies and present antigens. Autoreactive antibodies produced by malfunctioning plasma cells also form the basis of numerous autoimmune conditions. Epigenetic modulators, especially HDACi, have been shown to modulate the differentiation, proliferation, and survival of B cells [108]. B cells are functionally activated through binding of foreign antigens to the B-cell receptor. With the help of helper cells and cytokines, activated B cells are then transformed into antibody-producing short-lived plasmablasts or undergo somatic hypermutation in the germinal center to produce high-affinity antibodies [109]. Memory B cells are generated from activated B cells to confer long-lasting protection against foreign invasion [110]. Failure to eliminate autoreactive plasma cells can accelerate the accumulation of self-antigen specific antibodies and result in multi-organ damage [111], as is seen in SLE [112]. Specific HDAC members (i.e., HDAC10) are associated with B-cell malignancies [113]. Pan-HDACi and class I specific HDACi have significantly reduced the autoantibody production and cytokine-mediated inflammation in preclinical murine models of B-cell mediated autoimmunity [114,115]. Fig. 3.3 summarizes the interaction of different immune cells in establishing or resolving inflammation and their relationship to autoimmune conditions.

3.5 ROLE OF EPIGENETIC MODIFIERS IN RESOLVING AUTOIMMUNE CONDITIONS

Epigenetic modulation as a treatment modality for diverse pathological conditions is gaining significant attention. This chapter will be limited to discussing autoimmune conditions such as the inflammatory conditions that arise due to genetic predisposition, environmental cues, or transplant-related interventions. Cytokine-driven inflammation plays a pivotal role in the development and maintenance of most of these diseases, therefore one of the main objectives is to target both the acute and chronic phases of inflammation through epigenetic modifications. Specific examples of autoimmune conditions that affect large numbers of the population will be discussed, including: (1) SLE, which may have both genetic and environmental pathological components, (2) lung inflammation associated with asthma and chronic obstructive pulmonary disease (COPD), which often arises from smoking-related behaviors, and (3) GVHD in the context of bone marrow transplantation, which is a dire consequence of an essential treatment modality for many debilitating diseases (i.e., cancer).

3.5.1 Systemic Lupus Erythematosus

SLE is an autoimmune disorder in which genetic predisposition coupled with environmental triggers may initiate autoimmunity. A major factor that causes morbidity in this disease is lupus nephritis, which affects a large proportion of SLE patients [116]. The concordance of SLE between monozygotic twins is approximately 24%, which points to a cause beyond genetics [117]. The disease presents classic autoimmune symptoms such as the loss of self-tolerance in T cells, a decrease in the T_{reg} population, an increase in the immune response, and a loss of B-cell suppression that leads to the deposition of glomerular immune complexes [118–120]. SLE patients have a lower cytotoxic T-cell activity alongside an increased B-cell activity and an increased autoantibody production [121,122]. Th2 cells are instrumental to humoral immunity and are responsible for secreting IL-4, IL-5, and IL-10 [123,124]. The Th1, Th2, and Th17 subpopulations of T cells tend to be enhanced in SLE subjects, which leads to the increased activation of inflammatory mediators and tissue damage [125]. Dysfunction or abrogation of T_{regs} in patients fuels the condition due to the lack of regulation over inflammatory mediators (i.e., IL-2, IFN-γ, and TNFα) [126–129]. HDACi compounds work at several levels to mitigate autoimmune reactions, such as enhancing T_{reg} function and controlling the systemic inflammatory milieu; for example, a low dose (1 mg/kg) of ITF2357, which is a class I and II HDACi, reduces the production of proinflammatory cytokines (i.e., IL-1β, TNFα, IL-6, and IFN-γ) without producing significant adverse cytotoxic effects [68,130–133]. Regna et al. used a preclinical murine model of lupus and demonstrated that ITF2357 was able to consistently reduce SLE markers by enhancing the T_{reg} phenotype and decreasing the Th17 phenotype within the $CD4^+$ subpopulation of T cells [134].

DNA methylation in $CD4^+$ lymphocytes is also believed to be responsible for some of the pathology of SLE. DNA methylation in $CD4^+$ lymphocytes is significantly lower in SLE patients than the general population [135], and methylation is often lost at the promoters of genes encoding perforin; immunoglobulin-like receptor; IL-4; IL-6; and the B-cell costimulatory molecules CD70, CD6 [136], and CD154. A specific subset of proinflammatory $CD4^+CD28^-$ cells often emerges with aging and autoimmune diseases. These cells overexpress the B-cell costimulatory molecule CD70 [137], the cytolytic perforin, and the killer-cell immunoglobulin-like receptor gene *KIRDL4*. These genes are demethylated within the $CD4^+CD28^-$ cells and their methylation and expression are regulated by DNMT1 and DNMT3A [138]. The cytokines IL-10 and IL-14 also have lower levels of promoter methylation and higher expression in SLE $CD4^+$ T cells relative to healthy controls [139]. These cytokines can push cells toward the Th2 response and the development of autoantibodies. The CD40LG costimulatory molecule, which is located on the X chromosome, is demethylated on the inactive X chromosome of women with SLE [140]. As SLE is approximately nine times more prevalent in women than men, and disproportionately affects men with Klinefelter's syndrome (XXY), the presence of an extra X chromosome and genes encoded on it (i.e., CD40LG) may be central to the disease development [140]. Genome-wide SLE studies have implicated hypomethylation near genes involved in the type I interferon response [141–143]. Thus a wide range of genes involved in autoimmunity are controlled by methylation in SLE.

B cells in patients with SLE also exhibit abnormal DNA methylation. B cells expressing the CD5 coreceptor produce antibodies that react to self-antigens and may be a significant source of autoantibodies [144]. DNMT1 activity is reduced in $CD5^+$ B cells and these cells express a specific exon of CD5 that is designated as CD5–E1B, which is a fusion between a human endogenous retrovirus (HERV) and the CD5 gene. HERVs are usually silenced by DNA methylation, therefore hypomethylation of this locus causes increased transcription of CD5–E1B. This encodes a truncated variant of CD5 that can no longer translocate to the cell membrane and promote anergy of autoreactive B cells [145–147]. However, it is unclear what causes the DNA hypomethylation in SLE. While DNMT1 and DNMT3A are decreased in $CD4^+$ T cells from SLE patients in some studies [148,149], other studies show no difference in the expression of DNMTs [150]. Two miRNAs (miR-21 and miR-148a), which target the RASGRP1 protein upstream of DNMT1 and downregulate DNMT1 expression, are known to be overexpressed in both SLE patients and the lupus-prone MRL/lpr mice. Inhibition of these miRNAs increased DNMT1 expression and reversed the hypomethylation of genes involved in autoimmunity (i.e., CD70) [151]. MBDs also play a role; for example, these proteins bind to methylated DNA and recruit repressive chromatin modifiers to repress transcription. MBD4 is upregulated in $CD4^+$ and $CD8^+$ T cells from SLE patients, but it is unclear how this increase affects DNA methylation [150,152]. Sunlight, particularly UVB radiation, can also reduce DNA methylation and is known to induce hypomethylation and the increased expression of immune genes involved in SLE, including LFA-1(CD11a) [153].

3.5.2 Asthma and Chronic Obstructive Pulmonary Disease (COPD)

A leading global cause of death is attributed to airway stress and the associated inflammation within susceptible populations. Asthma is a heterogeneous chronic inflammatory disease of the airways and shares some of its pathology with COPD. Both disease states are characterized by inflamed epithelial tissue and a therapy-resistant fibrotic component. Epigenetic modifications, including DNA methylation and histone modifications, have been implicated in the pathology of asthma. More specifically, hypomethylation of immune genes (i.e., IL-13, IL-4, and RUNX3) have been observed in children with asthma compared to healthy control subjects [154]; RUNX3 is involved in T-cell maturation and IL-13 and IL-4 are Th2 cytokines that drive autoimmunity [155].

High serum concentrations of immunoglobulin E (IgE), which drives the IgE allergic response, are also associated with the hypomethylation of a group of genes in eosinophils [156]. One study observed significant changes in the methylation status of asthma-associated genes within 24 hours of airway epithelial cells being treated with the cytokine IL-13 [157]. In murine models of asthma, $CD4^+$ T cells exhibit loss of methylation at the IL-4 promoter and a gain of methylation at the IFN-γ promoter, inducing a shift towards the Th2 phenotype [158]. As these changes involve hypomethylation of immune genes, DNMTi treatment has not been considered for asthma; however, treatment with HDACi, which can mediate antiinflammatory effects, seems promising.

Although corticosteroids and β-agonists are routinely used to ameliorate inflammation in asthma, they are not designed to target the fibrosis-induced structural damage to the airway. Moreover, the increasing incidence of steroid-resistance in these diseases means that the development of antiinflammatory drugs, specifically HDACi compounds, are of particular importance. HDAC1 and HDAC2 are reported to play critical roles during the aging-associated inflammation in lung tissue. While HDAC1 inhibits the inflammation-causing C/EBP-δ gene, both HDAC1 and HDAC2 play critical roles during airway development [159,160]. Pharmacological activation of HDAC2 has the potential to regain structural integrity in the airway epithelium and overcome steroid resistance in the COPD model [161,162]. Cigarette smokers show significant inhibition of HDAC activity in the pulmonary artery endothelial cells, implicating HDAC2 as a biomarker for COPD development [163,164]. Development of apoptotic resistance in fibrotic lung diseases remains a challenge that can be circumvented by inducing proapoptotic pathways through HDACi [165]. HDACi compounds also upregulate the expression of antimicrobial peptides in the lung epithelium regarding the pathology of asthma and COPD. Examples of such peptides include defensin, which is upregulated by HDAC1 inhibitors, and cathelicidin LL37, which is upregulated by TSA and sodium butyrate [166,167]. Pan-HDACi counter the pathology of lung airway-associated inflammatory events; for example, VPA inhibits the differentiation of naive B cells into plasmablasts by interfering with B-cell proliferation [168]. Treatment with TSA caused Ovalbumin (OVA)-challenged mice to exhibit reduced inflammation and reduced eosinophil cell counts in the bronchoalveolar lavage fluid and ameliorated the goblet cell metaplasia during subepithelial fibrosis in the airways [169]. In summary, although our understanding of the roles of individual HDACs in the development of asthma and COPD is in its rudimentary phase, the overall benefits of HDACi in tissue remodeling, epithelial defense, and fibrosis provide promise for future therapeutic implications [170].

3.5.3 Graft Versus Host Disease and Hematopoietic Transplantation

Allogenic hematopoietic cell transplant (HCT) remains a treatment option for several malignant and nonmalignant pathological conditions [171]. One of the complications resulting from HCT is the graft-versus-host immune reaction, which leads to severe tissue destruction in the host [172]. The pathophysiology of GVHD results from the graft T cells (referred as donor T cells) reacting to the major and minor histocompatible antigens of the host [173]. It arises from the interaction of the host's APCs with the donor T cells, which leads to the priming of $CD4^+$ and $CD8^+$ T cell activity and subsequent emergence of a proinflammatory milieu [174].

The antiinflammatory and immunomodulatory effects of epigenetic modifiers have been successfully harnessed to mitigate the severity of GVHD during the allogeneic transplantation of bone marrow [175]. The chronic and acute forms of GVHD are arbitrarily defined on the basis of the time taken for the pathological state to be established. This section will focus on the immunomodulatory effect of epigenetic modifiers in GVHD.

3.5.3.1 Biology of Establishing and Dampening GVHD

Innate immune mediators trigger adaptive immune components by signaling through DAMP-associated pathways to initiate signaling cascades (i.e., NFκB) and result in the secretion of proinflammatory cytokines (i.e., IL-6, IL-1β, and TNFα) [176]. T cell-mediated effector and regulatory functions also play a central role in establishing GVHD [177].

Proliferation of donor T cells: The alloreactive donor T cells contain several subtypes with different activation conditions and effector functions, therefore the in situ functionality and proliferation of specific subsets depends on the alloantigen composition. Unlike the donor effector memory ($CD62L^-$) cells, the naive T ($CD62L^+$) cells take an active part in alloreactivity and therefore drive the GVHD [177]. Th1 cytokines (i.e., TNFα, IL-2, and IFN-γ) play critical roles in establishing acute GVHD [178]. Conversely, IL-2 facilitates the generation and maintenance of T_{reg} cells, which mediate the resolution of autoimmune reactivity in GVHD [179].

Organ damage by effector molecules: Cytolytic effector cells and molecules such as $CD8^+$ cytotoxic T-lymphocytes (CTLs), NK cells, IFN-γ, IL-1β, TNFα, and ROS are critical mediators of local and systemic organ damage. Cell-to-cell contact involving the cellular counterpart initiates several cytolytic pathways, such as Fas-FasL, TRAIL-TNFR, perforin, granzyme, TWEAK, and LTβ-LIGHT [176].

Mediators that dampen GVHD: As observed under common autoimmune conditions, $CD4^+$ $CD25^+$ T_{regs} play a critical role in dampening GVHD, primarily due to their inherent functionality at suppressing effector T-cell proliferation and function [180]. Donor-derived NK cells may also play a significant protective role by directly attacking the host APCs, which prime donor T cells to destroy organ tissues [105].

Role of APCs in GVHD: By sensing the allodisparity and the resulting DAMP-mediated signaling cascade, host DC populations play critical roles in presenting antigens and conveying the second (costimulatory) and third (cytokine) signals for T-cell activation. Host and donor DCs may take consecutive turns during the activation and sustenance phases. Macrophages and monocytes carry out additional roles in exacerbating the autoimmune reactions by producing proinflammatory mediators (i.e., IL-1β, TNFα, and nitric oxide) [181].

3.5.3.2 Preclinical Applications of HDACi in Mitigating GVHD

HDACi compounds can significantly diminish the proinflammatory cytokine secretion by APCs. SAHA successfully restricts the production of TNFα, IL-1, and IFN-γ after allo-HCT, which correspond to an enhancement in H3-acetylation [182]. In vitro treatment of bone marrow-derived DC with SAHA and stimulation with TLR agonists was shown to significantly impair proinflammatory cytokine production, reduce phosphorylated STAT1 [183], and reduce allostimulatory capacity without disrupting the CTL functions [182]. HDACi compounds are also capable of affecting the metabolic arm of T-cell biology; for example, the treatment of DCs with SAHA resulted in the upregulation of the indoleamine 2,3-dioxygenase (IDO) enzyme, which in turn depleted the tryptophan availability and reduced the viability of T cells [184]. Acetylation of the H4 histone at the IDO promoter was also identified as a causal factor that was directly related to the ability of HDACi to dampen DC-mediated T-cell activation [185]. Sotomayor et al. made significant progress in understanding the association of HDAC11 and HDAC6 with the synthesis of IL-10 to dissect the involvement of specific HDAC members. The authors identified that while HDAC11 negatively regulates IL-10, HDAC6 may have an opposing effect [52,53,186]. In a fully MHC-mismatched model, vorinostat dampens inflammation and reduces mortality by targeting the activated allospecific T cells, leaving the donor-derived T-cell expansion and immune reconstitution intact [183,187].

HDACi compounds also augment T_{reg} and NK cell functionality to reduce the severity of GVHD. Zorn et al. reported a significant decrease in the total number of T_{reg} cells within GVHD-affected patients without a functional deterioration in the cells [188], which implicates this as a causal event in the pathology of GVHD. Tao et al. reported an increase in the number of these cells as well as increase in their function upon HDACi treatment, particularly the $CD4^{+}CD25^{+}FOXP3^{+}$ subset of T_{reg} cells [95]. In vivo application of HDACi simultaneously enhanced the expression of FOXP3, CTLA, and GITR while concomitantly decreasing IL-2 production and thus facilitating retardation of T-cell functions. HDAC6 inhibition by KA1010 also promotes T_{reg} functionality by acetylating FOXP3, cortactin, and HSP90 [189,190]. The cytotoxicity of NK cells, which can mitigate GVHD by destroying host-derived APCs, is also assisted by HDACi. Several investigators used tumor cells to confirm that HDACi (e.g., SAHA and VPA) treatment augmented NK cell-mediated killing through NKG2, member D (NKG2D) ligands, including MHC class I-related chain A and B (MICA/B), on target cells [105].

3.5.4 Rheumatoid Arthritis (RA)

RA is an autoimmune disease in which synovial fibroblasts interact with immune cells that end up attacking the joints of the body. These synovial fibroblasts (fibroblast-like synoviocytes) produce chemokines and proinflammatory cytokines, contribute to an inflamed environment, and promote the destruction of joint tissue by immune cells. In vitro experiments using broad-spectrum HDACi mitigated inflammation in the RA macrophage through decay of IL-6 mRNA [132]. Another study where enhanced activity of HDACs was identified in the PBMCs from RA patients found that the levels of IL-6 and TNFα were further diminished with both TSA and a selective HDAC3-inhibitor, MI192 [191]. The application of selective HDAC1 inhibitors, NW-21 and MS-275 in RA patients not only successfully reduced the levels of MCP-1 and MIP-1α but also reduced bone loss [192]. Moreover, methylation patterns in the T cells of RA patients are dysregulated [135] and LINE1 repetitive elements have also been shown to be hypermethylated in RA synovial tissues [193]. Differential methylation at the MHC region also stands out as a risk factor in RA patients [194]; however, the mechanism of how DNA methylation affects RA remains unclear.

3.5.5 Other Autoimmune Diseases

Other autoimmune diseases that are modulated by epigenetic mechanisms include those that have low concordance rates among monozygotic twins, including type 1 diabetes (13%–67.7%) [195] and multiple sclerosis (MS) (6%–30%) [196].

Type I diabetes: Patients with type 1 diabetes exhibit reduced methylation at seven CpG sites that are proximal to the insulin gene promoter [197] and decreased methylation and increased levels of the insulin-like growth factor binding protein-1 [198]. Monocytes isolated from patients with type 1 and 2 diabetes were shown to exhibit hyperacetylation of histone H3 in the promoters of TNFα and the inflammatory-associated enzyme, cyclooxygenase 2 [199]. The use of HDACi promoted β-cell development, proliferation, differentiation, and function; prevented β-cell inflammatory damage; improved insulin resistance; and positively affected late diabetic microvascular complications. This therefore indicates a promising avenue for future clinical interventions [200].

Multiple sclerosis (MS): The methylation status of 15 genes were shown to be different between healthy controls and patients exhibiting varying severities of MS [201]. Experimental trials using broad-spectrum HDACi for MS have shown variable results, partly due to the neuroprotective action of some HDACs (i.e., HDAC6) [202], which has left this field open to further research [203]. Additional autoimmune diseases that are potentially influenced by epigenetic mechanisms are listed in Table 3.1.

TABLE 3.1 Various Autoimmune Conditions and Their Potential Roots in Epigenetics

Disease name	Hallmarks of the disease	Associated epigenetic signature	Molecular pathways and cells involved	References
Alzheimer's disease	Oxidative stress and Aβ plaques	DNA methylation, DNA hydroxyl-methylation, and deacetylation	JNK and p38 MAPK, stress-related MAPK, and HAT p300/CREB	[204]
Antiphospholipid syndrome	Anti-β_2-glycoprotein I or anticardiolipin antibodies and oxidative stress	Not identified	TLRs, BAFF, tissue factor, NOS, and factor XI	[205]
Autoimmune hepatitis	Cirrhosis and liver failure	Not identified	Inherited	[206]
Behçet's disease	Inflammatory disease with multisystem involvement	DNA methylation and miRNA dysregulation	Microtubule-associated proteins, myosin genes, and actin-processing genes	[207,208]
Celiac disease	Reaction to gluten	Under investigation	HLA-DQ variant	[209]
Chagas disease	Cardiac alterations, digestive, neurological, or mixed alterations	Histone deacetylation in vector	*Trypanosoma cruzi* infection	[210]
Crohn's disease and IBD	Chronic relapsing intestinal inflammation	DNA methylation	NOD2, NFκB activation, and autophagy in dendritic cells	[211]
Hashimoto's thyroiditis and Graves' disease	Thyroid disorder and hyper- or hypothyroidism	X-chromosome inactivation	Amino acid substitution in the HLA-DR3 gene, CD40, CTLA-4, PTPN22, FOXP3, CD25 and thyroid-specific genes (thyroglobulin and TSHR)	[212]
SLE and glomerulonephritis	Nuclear-targeted autoantibody production	DNA methylation, deacetylation, and X-chromosome inactivation	ITGAL (CD11a), CD40LG (CD40L), TNFSF7 (CD70), KIR2DL4, PRF1 (perforin), and CD5	[213]
Guillain–Barre syndrome	Inflammatory demyelinating polyneuropathy	Not identified	Neuronal demyelination	[214]
Thalassemia and sickle cell anemia	Abnormal form of hemoglobin	DNA methylation	IGSF4 gene, fetal hemoglobin genes, and inherited traits	[215,216]
Juvenile idiopathic arthritis and rheumatoid arthritis	Chronic inflammation of joints, destruction, deformity, and swelling	DNA methylation	IL-32 in $CD4^+$ T cells Fibroblast-like synoviocytes	[217,218]
Kawasaki disease	Vasculitis	DNA methylation and miRNA dysregulation	FCGR2A, proinflammatory disease	[208]

Addison's disease	Destruction of endocrine cells of the adrenaline cortex	DNA methylation	$CD4^+$ T cells	[219]
Myasthenia gravis	Muscle weakness	DNA methylation Potentially linked through miRNA	Antibodies against acetylcholine receptor, muscle specific kinase, and low-density lipoprotein receptor-related protein(LRP4)	[220]
Multiple sclerosis	Inflammation and neurodegeneration	Deacetylation, methylation, and citrullination	Myelin proteins and progressive axonal loss. Risk factor is the leukocyte antigen-DRB1*15 allele, vitamin D deficiency, smoking, and Epstein–Barr virus	[221]
Psoriasis	Aberrant skin cell proliferation	CpG methylation	KYNU, OAS2, S100A12, and SERPINB3 genes	[222]
Sarcoidosis	Fluid homeostasis	miR-144	CFTR	[223]
Idiopathic thrombocytopenic purpura	Platelet destruction	DNA methylation	Polymorphisms of the DNA methyltransferase 3B (DNMT3B) gene	[224]
Sjogren's syndrome	Epithelitis in exocrine glands, mainly in the salivary and lacrimal glands	DNA methylation and X-chromosome inactivation	Dysfunctional $FoxP3^+$ T cells and $CD5^+$ B cells	[225]
ANCA-associated vasculitis (microscopic polyangiitis or, granulomatosis with polyangiitis, and eosinophilic granulomatosis with polyangiitis)	Inflammation of blood vessels and antineutrophil cytoplasmic antibodies	DNA methylation	MPO, RUNX3, and PR3	[208]
Scleroderma	Vasculopathy, immune activation, and extensive fibrosis of the skin and internal organs	DNA methylation in lymphocytes, deacetylation in fibroblasts, and miRNA	TGFβ, SMAD, FOXp3, COL1A1, and FN1	[226]

In addition to the aforementioned diseases, there are an abundance of clinically relevant autoimmune conditions with varying degrees of penetrance within the population. This table is a brief sampling of such conditions in search of their potential root into epigenetics of the immune system or other molecular pathways contributing to the etiology of the disease. As indicated, 75% of the total diseases have one or more epigenetic alterations as causal factors, emphasizing the relevance of the study of epigenetics to better understand and manage autoimmune diseases.
HAT, Histone acetyltransferases; *TLR*, toll-like receptors.

The role of DNMTi in the treatment of autoimmune diseases is yet to be established. Although the impact of their immunomodulatory roles has been explored and appreciated within the space of clinical cancer research, the field of autoimmunity has yet to capture their therapeutic benefits. The availability of resources and information has therefore skewed this discussion towards HDACi. The following concluding section is an effort to highlight the progress in applying the available knowledge towards therapy.

3.6 CURRENT PROGRESS ON CLINICAL TRIALS

The scope of this chapter meant that the discussions were limited to specific examples of inflammatory diseases with pronounced preclinical examination of epigenetic approaches. However, it is important to mention that several autoimmune or inflammatory conditions are successfully responding to epigenetic modifying compounds and therefore are progressing into phase I and II clinical trials for MS, type 2 diabetes [227], inflammatory bowel disease [228], RA [229], and juvenile idiopathic arthritis [230] (Table 3.2).

TABLE 3.2 Summary of Previous and Ongoing Clinical Trials Involving Epigenetic Modifiers as Therapeutic Options in Autoimmune Conditions and Diseases [231]

Agent	Targeted disease	Trial identifier	Phase	Status
Resveratrol	Chronic subclinical inflammation	NCT01492114	Phase 3	Completed
Resveratrol	Type 2 diabetes mellitus	NCT01038089	N/A	Completed
Resveratrol	Obesity, the metabolic syndrome, and diabetes	NCT00823381	N/A	Completed
Resveratrol	The metabolic syndrome and obesity	NCT01150955	N/A	Completed
Curcumin	Atopic asthma	NCT01179256	N/A	Completed
Curcumin	Irritable bowel syndrome	NCT00779493	Phase 4	Completed
Curcumin	Alzheimer's disease	NCT00164749	Phases 1 and 2	Completed
Curcumin and bioperine	Chronic obstructive pulmonary disease	NCT01514266	N/A	Completed
Curcumin C3 complex	Alzheimer's disease	NCT00099710	Phase 2	Completed
Curcumin C3 complex	Psoriasis	NCT00235625	Phase 2	Completed
Givinostat (ITF2357)	Systemic onset juvenile idiopathic arthritis	NCT00570661	Phase 2	Completed
HAT and HDAC levels in PBMC	Demyelinating autoimmune disease, clinically isolated syndrome	NCT02862301	N/A	Ongoing
Curcumin	Inflammatory bowel disease, ulcerative colitis, and Crohn's disease	NCT00889161	Phase 1	Completed

HAT, Histone acetyltransferases; *HDAC*, histone deacetylases; *N/A*, not applicable.

Acknowledgments

We acknowledge Dr. Jayakumar Nair and Tessa Knox for their help in reviewing the material.

LIST OF ABBREVIATIONS

AML	Acute myeloid leukemia
APC	Antigen presenting cells
CD	Cluster of differentiation
COPD	Chronic obstructive pulmonary disease
CTL	Cytotoxic T-lymphocytes
CTLA	Cytotoxic T-lymphocyte-associated protein
DAMP	Damage-associated molecular pattern
DC	Dendritic cell
DNMT	DNA methyltransferase
DNMTi	DNA methyltransferase inhibitors
FOXP3	Forkhead box P3
GITR	Glucocorticoid-induced TNFR-related protein
GVHD	Graft-versus host disease
HAT	Histone acetyltransferase
HCT	Hematopoietic cell transfer
HDAC	Histone deacetylases
HDACi	Histone deacetylases inhibitors
IDO	Indoleamine 2,3-dioxygenase
IFN	Interferon
IL	Interleukin
MHC	Major histocompatibility complex
NK	Natural killer cells
PAMP	Pathogen-associated molecular pattern
RA	Rheumatoid arthritis
ROS	Reactive oxygen species
SAHA	Suberoylanilide hydroxamic acid
SLE	Systemic lupus erythematosus
STAT	Signal transducer and activator of transcription
TCR	T cell receptor
TF	Transcription factor
TLR	Toll-like receptors
TNF	Tumor necrosis factor
TSA	Trichostatin A

References

[1] Waddington CH. An introduction to modern genetics. New York: The Macmillan Company; 1939. p. 460.
[2] Portela A, Esteller M. Epigenetic modifications and human disease. Nat Biotechnol 2010;28(10):1057–68.
[3] Rose NR, Bona C. Defining criteria for autoimmune diseases (Witebsky's postulates revisited). Immunol Today 1993;14(9):426–30.
[4] Zinkernagel RM, Pircher H, Ohashi PS, Hengartner H. T cells causing immunological disease. Springer Semin Immunopathol 1992;14(2):105–13.
[5] Schleinitz N, Vely F, Harle JR, Vivier E. Natural killer cells in human autoimmune diseases. Immunology 2010;131(4):451–8.

[6] Pesenacker AM, Cook L, Levings MK. The role of FOXP3 in autoimmunity. Curr Opin Immunol 2016;43:16–23.
[7] Tian Z, Gershwin ME, Zhang C. Regulatory NK cells in autoimmune disease. J Autoimmun 2012;39(3):206–15.
[8] Li B, Carey M, Workman JL. The role of chromatin during transcription. Cell 2007;128(4):707–19.
[9] Wang Z, Zang C, Rosenfeld JA, Schones DE, Barski A, Cuddapah S, et al. Combinatorial patterns of histone acetylations and methylations in the human genome. Nat Genet 2008;40(7):897–903.
[10] Mikkelsen TS, Ku M, Jaffe DB, Issac B, Lieberman E, Giannoukos G, et al. Genome-wide maps of chromatin state in pluripotent and lineage-committed cells. Nature 2007;448(7153):553–60.
[11] Bernstein BE, Mikkelsen TS, Xie X, Kamal M, Huebert DJ, Cuff J, et al. A bivalent chromatin structure marks key developmental genes in embryonic stem cells. Cell 2006;125(2):315–26.
[12] Kouzarides T. Chromatin modifications and their function. Cell 2007;128(4):693–705.
[13] Choudhary C, Kumar C, Gnad F, Nielsen ML, Rehman M, Walther TC, et al. Lysine acetylation targets protein complexes and co-regulates major cellular functions. Science 2009;325(5942):834–40.
[14] Villagra A, Sahakian E, Seto E. Preparation and biochemical analysis of classical histone deacetylases. Methods Enzymol 2016;573:161–81.
[15] Wang Z, Zang C, Cui K, Schones DE, Barski A, Peng W. Genome-wide mapping of HATs and HDACs reveals distinct functions in active and inactive genes. Cell 2009;138:1019–31.
[16] Seto E, Yoshida M. Erasers of histone acetylation: the histone deacetylase enzymes. Cold Spring Harb Perspect Biol 2014;6(4):a018713.
[17] Shakespear MR, Halili MA, Irvine KM, Fairlie DP, Sweet MJ. Histone deacetylases as regulators of inflammation and immunity. Trends Immunol 2011;32(7):335–43.
[18] Chen J, Sahakian E, Powers J, Lienlaf M, Perez-Villarroel P, Knox T, et al. Functional analysis of histone deacetylase 11 (HDAC11). Methods Mol Biol 2016;1436:147–65.
[19] Villagra A, Sotomayor EM, Seto E. Histone deacetylases and the immunological network: implications in cancer and inflammation. Oncogene 2010;29(2):157–73.
[20] Glozak MA, Sengupta N, Zhang X, Seto E. Acetylation and deacetylation of non-histone proteins. Gene 2005;363:15–23.
[21] Haberland M, Montgomery RL, Olson EN. The many roles of histone deacetylases in development and physiology: implications for disease and therapy. Nat Rev Genet 2009;10(1):32–42.
[22] West AC, Johnstone RW. New and emerging HDAC inhibitors for cancer treatment. J Clin Invest 2014;124(1):30–9.
[23] Mottamal M, Zheng S, Huang TL, Wang G. Histone deacetylase inhibitors in clinical studies as templates for new anticancer agents. Molecules 2015;20(3):3898–941.
[24] Baylin SB, Jones PA. A decade of exploring the cancer epigenome: biological and translational implications. Nat Rev Cancer 2011;11(10):726–34.
[25] Allis D, Caparros M-L, Jenuwein T, Reinberg D. Epigenetics. 2nd ed. Cold Spring Harbor Laboratory Press; 2015. p. 984.
[26] Ehrlich M. DNA methylation in cancer: too much, but also too little. Oncogene 2002;21(35):5400–13.
[27] Stresemann C, Brueckner B, Musch T, Stopper H, Lyko F. Functional diversity of DNA methyltransferase inhibitors in human cancer cell lines. Cancer Res 2006;66(5):2794–800.
[28] Rountree MR, Bachman KE, Baylin SB. DNMT1 binds HDAC2 and a new co-repressor, DMAP1, to form a complex at replication foci. Nat Genet 2000;25(3):269–77.
[29] Robertson KD, Wolffe AP. DNA methylation in health and disease. Nat Rev Genet 2000;1(1):11–9.
[30] Clements EG, Mohammad HP, Leadem BR, Easwaran H, Cai Y, Van Neste L, et al. DNMT1 modulates gene expression without its catalytic activity partially through its interactions with histone-modifying enzymes. Nucleic Acids Res 2012;40(10):4334–46.
[31] Tsai HC, Li H, Van Neste L, Cai Y, Robert C, Rassool FV, et al. Transient low doses of DNA-demethylating agents exert durable antitumor effects on hematological and epithelial tumor cells. Cancer Cell 2012;21(3):430–46.
[32] Kaminskas E, Farrell AT, Wang YC, Sridhara R, Pazdur R. FDA drug approval summary: azacitidine (5-azacytidine, Vidaza) for injectable suspension. Oncologist 2005;10(3):176–82.
[33] Kantarjian H, Giles F, Wunderle L, Bhalla K, O'Brien S, Wassmann B, et al. Nilotinib in imatinib resistant CML and Philadelphia chromosome-positive ALL. N Engl J Med 2006;354:2542.
[34] Kantarjian HM, O'Brien S, Cortes J, Giles FJ, Faderl S, Issa JP, et al. Results of decitabine (5-aza-2'deoxycytidine) therapy in 130 patients with chronic myelogenous leukemia. Cancer 2003;98(3):522–8.
[35] Issa JP, Roboz G, Rizzieri D, Jabbour E, Stock W, O'Connell C, et al. Safety and tolerability of guadecitabine (SGI-110) in patients with myelodysplastic syndrome and acute myeloid leukaemia: a multicentre, randomised, dose-escalation phase 1 study. Lancet Oncol 2015;16(9):1099–110.

[36] Moreira J, Scheipers P, Sorensen P. The histone deacetylase inhibitor trichostatin A modulates CD4⁺ T cell responses. BMC Cancer 2003;3(1):30.

[37] Van Lint C, Emiliani S, Verdin E. The expression of a small fraction of cellular genes is changed in response to histone hyperacetylation. Gene Expr 1996;5(4–5):245–53.

[38] Richon VM, Sandhoff TW, Rifkind RA, Marks PA. Histone deacetylase inhibitor selectively induces p21WAF1 expression and gene-associated histone acetylation. Proc Natl Acad Sci USA 2000;97(18):10014–9.

[39] Gu W, Roeder RG. Activation of p53 sequence-specific DNA binding by acetylation of the p53 C-terminal domain. Cell 1997;90(4):595–606.

[40] Yuan Z-L, Guan Y-J, Chatterjee D, Chin YE. Stat3 dimerization regulated by reversible acetylation of a single lysine residue. Science 2005;307(5707):269–73.

[41] Boyes J, Byfield P, Nakatani Y, Ogryzko V. Regulation of activity of the transcription factor GATA-1 by acetylation. Nature 1998;396(6711):594–8.

[42] Marzio G, Wagener C, Gutierrez MI, Cartwright P, Helin K, Giacca M. E2F family members are differentially regulated by reversible acetylation. J Biol Chem 2000;275(15):10887–92.

[43] Yao Y-L, Yang W-M, Seto E. Regulation of transcription factor YY1 by acetylation and deacetylation. Mol Cell Biol 2001;21(17):5979–91.

[44] Munshi N, Merika M, Yie J, Senger K, Chen G, Thanos D. Acetylation of HMG I(Y) by CBP turns off IFNβ expression by disrupting the enhanceosome. Mol Cell 1998;2(4):457–67.

[45] Yin L, Laevsky G, Giardina C. Butyrate suppression of colonocyte NF-kappa B activation and cellular proteasome activity. J Biol Chem 2001;276(48):44641–6.

[46] Chen L, Fischle W, Verdin E, Greene WC. Duration of nuclear NF-kappaB action regulated by reversible acetylation. Science 2001;293(5535):1653–7.

[47] Kiernan R, Brès V, Ng RWM, Coudart M-p, El Messaoudi S, Sardet C, et al. Post-activation turn-off of NF-κB-dependent transcription is regulated by acetylation of p65. J Biol Chem 2003;278(4):2758–66.

[48] Gloire G, Horion J, El Mjiyad N, Bex F, Chariot A, Dejardin E, et al. Promoter-dependent effect of IKKalpha on NF-kappaB/p65 DNA binding. J Biol Chem 2007;282(29):21308–18.

[49] Jung SB, Kim CS, Naqvi A, Yamamori T, Mattagajasingh I, Hoffman TA, et al. Histone deacetylase 3 antagonizes aspirin-stimulated endothelial nitric oxide production by reversing aspirin-induced lysine acetylation of endothelial nitric oxide synthase. Circ Res 2010;107(7):877–87.

[50] Krämer OH, Heinzel T. Phosphorylation–acetylation switch in the regulation of STAT1 signaling. Mol Cell Endocrinol 2010;315(1–2):40–8.

[51] Kramer OH, Knauer SK, Greiner G, Jandt E, Reichardt S, Gahrs K-h, et al. A phosphorylation-acetylation switch regulates STAT1 signaling. Genes Dev 2009;23(2):223–35.

[52] Cheng F, Lienlaf M, Wang HW, Perez-Villarroel P, Lee C, Woan K, et al. A novel role for histone deacetylase 6 in the regulation of the tolerogenic STAT3/IL-10 pathway in APCs. J Immunol 2014;193(6):2850–62.

[53] Cheng F, Lienlaf M, Perez-Villarroel P, Wang HW, Lee C, Woan K, et al. Divergent roles of histone deacetylase 6 (HDAC6) and histone deacetylase 11 (HDAC11) on the transcriptional regulation of IL10 in antigen presenting cells. Mol Immunol 2014;60(1):44–53.

[54] Khan AN, Gregorie CJ, Tomasi TB. Histone deacetylase inhibitors induce TAP, LMP, tapasin genes and MHC class I antigen presentation by melanoma cells. Cancer Immunol Immun 2008;57(5):647–54.

[55] Setiadi AF, Omilusik K, David MD, Seipp RP, Hartikainen J, Gopaul R, et al. Epigenetic enhancement of antigen processing and presentation promotes immune recognition of tumors. Cancer Res 2008;68(23):9601–7.

[56] Woan KV, Sahakian E, Sotomayor EM, Seto E, Villagra A. Modulation of antigen-presenting cells by HDAC inhibitors: implications in autoimmunity and cancer. Immunol Cell Biol 2012;90(1):55–65.

[57] Li H, Ou X, Xiong J, Wang T. HPV16E7 mediates HADC chromatin repression and downregulation of MHC class I genes in HPV16 tumorigenic cells through interaction with an MHC class I promoter. Biochem Biophys Res Commun 2006;349(4):1315–21.

[58] Lazaro E, Kadie C, Stamegna P, Zhang SC, Gourdain P, Lai NY, et al. Variable HIV peptide stability in human cytosol is critical to epitope presentation and immune escape. J Clin Invest 2011;121(6):2480–92.

[59] Khan AN, Magner WJ, Tomasi TB. An epigenetic vaccine model active in the prevention and treatment of melanoma. J Transl Med 2007;5:64.

[60] Zika E, Greer SF, Zhu XS, Ting JP. Histone deacetylase 1/mSin3A disrupts gamma interferon-induced CIITA function and major histocompatibility complex class II enhanceosome formation. Mol Cell Biol 2003;23(9):3091–102.

[61] Fuse S, Zhang W, Usherwood EJ. Control of memory CD8⁺ T cell differentiation by CD80/CD86-CD28 costimulation and restoration by IL-2 during the recall response. J Immunol 2008;180(2):1148–57.

[62] Roger T, Lugrin J, Le Roy D, Goy G, Mombelli M, Koessler T, et al. Histone deacetylase inhibitors impair innate immune responses to Toll-like receptor agonists and to infection. Blood 2011;117(4):12.
[63] Tomasi TB, Magner WJ, Khan AN. Epigenetic regulation of immune escape genes in cancer. Cancer Immunol Immun 2006;55(10):1159–84.
[64] Serrador JM, Cabrero JR, Sancho D, Mittelbrunn M, Urzainqui A, Sanchez-Madrid F. HDAC6 deacetylase activity links the tubulin cytoskeleton with immune synapse organization. Immunity 2004;20(4):417–28.
[65] Lienlaf M, Perez-Villarroel P, Knox T, Pabon M, Sahakian E, Powers J, et al. Essential role of HDAC6 in the regulation of PD-L1 in melanoma. Mol Oncol 2016;10(5):735–50.
[66] Chen X, Barozzi I, Termanini A, Prosperini E, Recchiuti A, Dalli J, et al. Requirement for the histone deacetylase Hdac3 for the inflammatory gene expression program in macrophages. Proc Natl Acad Sci USA 2012;109(42):E2865–74.
[67] Nencioni A, Beck J, Werth D, Grünebach F, Patrone F, Ballestrero A, et al. Histone deacetylase inhibitors affect dendritic cell differentiation and immunogenicity. Clin Cancer Res 2007;13(13):3933–41.
[68] Furlan A, Monzani V, Reznikov LL, Leoni F, Fossati G, Modena D, et al. Pharmacokinetics, safety and inducible cytokine responses during a phase 1 trial of the oral histone deacetylase inhibitor ITF2357 (givinostat). Mol Med 2011;17(5–6):353–62.
[69] Akers SN, Odunsi K, Karpf AR. Regulation of cancer germline antigen gene expression: implications for cancer immunotherapy. Future Oncol 2010;6(5):717–32.
[70] Vignali DA, Kuchroo VK. IL-12 family cytokines: immunological playmakers. Nat Immunol 2012;13(8):722–8.
[71] Guo ZS, Hong JA, Irvine KR, Chen GA, Spiess PJ, Liu Y, et al. De novo induction of a cancer/testis antigen by 5-aza-2′-deoxycytidine augments adoptive immunotherapy in a murine tumor model. Cancer Res 2006;66(2):1105–13.
[72] Srivastava P, Paluch BE, Matsuzaki J, James SR, Collamat-Lai G, Karbach J, et al. Immunomodulatory action of SGI-110, a hypomethylating agent, in acute myeloid leukemia cells and xenografts. Leuk Res 2014;38(11):1332–41.
[73] Karpf AR, Lasek AW, Ririe TO, Hanks AN, Grossman D, Jones DA. Limited gene activation in tumor and normal epithelial cells treated with the DNA methyltransferase inhibitor 5-aza-2′-deoxycytidine. Mol Pharmacol 2004;65(1):18–27.
[74] Li H, Chiappinelli KB, Guzzetta AA, Easwaran H, Yen RW, Vatapalli R, et al. Immune regulation by low doses of the DNA methyltransferase inhibitor 5-azacitidine in common human epithelial cancers. Oncotarget 2014;5(3):587–98.
[75] Karpf AR, Peterson PW, Rawlins JT, Dalley BK, Yang Q, Albertsen H, et al. Inhibition of DNA methyltransferase stimulates the expression of signal transducer and activator of transcription 1, 2, and 3 genes in colon tumor cells. Proc Natl Acad Sci USA 1999;96(24):14007–12.
[76] Wrangle J, Wang W, Koch A, Easwaran H, Mohammad HP, Vendetti F, et al. Alterations of immune response of non-small cell lung cancer with azacytidine. Oncotarget 2013;4(11):2067–79.
[77] Matei D, Fang F, Shen C, Schilder J, Arnold A, Zeng Y, et al. Epigenetic resensitization to platinum in ovarian cancer. Cancer Res 2012;72(9):2197–205.
[78] Leonova KI, Brodsky L, Lipchick B, Pal M, Novototskaya L, Chenchik AA, et al. P53 cooperates with DNA methylation and a suicidal interferon response to maintain epigenetic silencing of repeats and noncoding RNAs. Proc Natl Acad Sci USA 2013;110(1):E89–98.
[79] Chiappinelli KB, Strissel PL, Desrichard A, Li H, Henke C, Akman B, et al. Inhibiting DNA methylation causes an interferon response in cancer via dsRNA including endogenous retroviruses. Cell 2015;162(5):974–86.
[80] Roulois D, Loo Yau H, Singhania R, Wang Y, Danesh A, Shen SY, et al. DNA-demethylating agents target colorectal cancer cells by inducing viral mimicry by endogenous transcripts. Cell 2015;162(5):961–73.
[81] Wang L, Amoozgar Z, Huang J, Saleh MH, Xing D, Orsulic S, et al. Decitabine enhances lymphocyte migration and function and synergizes with CTLA-4 blockade in a murine ovarian cancer model. Cancer Immunol Res 2015;3(9):1030–41.
[82] Peng D, Kryczek I, Nagarsheth N, Zhao L, Wei S, Wang W, et al. Epigenetic silencing of TH1-type chemokines shapes tumour immunity and immunotherapy. Nature 2015;527(7577):249–53.
[83] Chang S, Collins PL, Aune TM. T-bet dependent removal of sin3A-histone deacetylase complexes at the Ifng locus drives Th1 differentiation. J Immunol 2008;181(12):8372–81.
[84] Yamaguchi T, Cubizolles F, Zhang Y, Reichert N, Kohler H, Seiser C, et al. Histone deacetylases 1 and 2 act in concert to promote the G1-to-S progression. Genes Dev 2010;24(5):455–69.
[85] Chuang C, Lin SH, Huang F, Pan J, Josic D, Yu-Lee LY. Acetylation of RNA processing proteins and cell cycle proteins in mitosis. J Proteome Res 2010;9(9):4554–64.
[86] Li N, Zhao D, Kirschbaum M, Zhang C, Lin C-l, Todorov I, et al. HDAC inhibitor reduces cytokine storm and facilitates induction of chimerism that reverses lupus in anti-CD3 conditioning regimen. Proc Natl Acad Sci USA 2008;105(12):4796–801.

[87] Skov S, Rieneck K, Bovin LF, Skak K, Tomra S, Michelsen BK, et al. Histone deacetylase inhibitors: a new class of immunosuppressors targeting a novel signal pathway essential for CD154 expression. Blood 2003;101(4):1430–8.
[88] Ansel KM, Lee DU, Rao A. An epigenetic view of helper T cell differentiation. Nat Immunol 2003;4:616.
[89] Morinobu A, Kanno Y, O'Shea JJ. Discrete roles for histone acetylation in human T helper 1 cell-specific gene expression. J Biol Chem 2004;279(39):40640–6.
[90] Jones B, Chen J. Inhibition of IFN-gamma transcription by site-specific methylation during T helper cell development. EMBO J 2006;25(11):2443–52.
[91] Gamper CJ, Agoston AT, Nelson WG, Powell JD. Identification of DNA methyltransferase 3a as a T cell receptor-induced regulator of Th1 and Th2 differentiation. J Immunol 2009;183(4):2267–76.
[92] Hancock WW, Akimova T, Beier UH, Liu Y, Wang L. HDAC inhibitor therapy in autoimmunity and transplantation. Ann Rheum Dis 2012;71(Suppl. 2):i46–54.
[93] Brunkow ME, Jeffery EW, Hjerrild KA, Paeper B, Clark LB, Yasayko SA, et al. Disruption of a new forkhead/winged-helix protein, scurfin, results in the fatal lymphoproliferative disorder of the scurfy mouse. Nat Genet 2001;27(1):68–73.
[94] Kim J, Lahl K, Hori S, Loddenkemper C, Chaudhry A, deRoos P, et al. Cutting edge: depletion of Foxp3+ cells leads to induction of autoimmunity by specific ablation of regulatory T cells in genetically targeted mice. J Immunol 2009;183(12):7631–4.
[95] Tao R, de Zoeten EF, Ozkaynak E, Chen C, Wang L, Porrett PM, et al. Deacetylase inhibition promotes the generation and function of regulatory T cells. Nat Med 2007;13(11):1299–307.
[96] Akimova T, Beier UH, Liu Y, Wang L, Hancock WW. Histone/protein deacetylases and T-cell immune responses. Blood 2012;119(11):2443–51.
[97] Liu Y, Wang L, Predina J, Han R, Beier UH, Wang LC, et al. Inhibition of p300 impairs Foxp3(+) T regulatory cell function and promotes antitumor immunity. Nat Med 2013;19(9):1173–7.
[98] Beier UH, Akimova T, Liu Y, Wang L, Hancock WW. Histone/protein deacetylases control Foxp3 expression and the heat shock response of T-regulatory cells. Curr Opin Immunol 2011;23(5):670–8.
[99] Xiao H, Jiao J, Wang L, O'Brien S, Newick K, Wang LC, et al. HDAC5 controls the functions of Foxp3(+) T-regulatory and CD8(+) T cells. Int J Cancer 2016;138(10):2477–86.
[100] Segretti MC, Vallerini GP, Brochier C, Langley B, Wang L, Hancock WW, et al. Thiol-based potent and selective HDAC6 inhibitors promote tubulin acetylation and T-regulatory cell suppressive function. ACS Med Chem Lett 2015;6(11):1156–61.
[101] Zheng Y, Josefowicz S, Chaudhry A, Peng XP, Forbush K, Rudensky AY. Role of conserved non-coding DNA elements in the Foxp3 gene in regulatory T-cell fate. Nature 2010;463(7282):808–12.
[102] Kim HP, Leonard WJ. CREB/ATF-dependent T cell receptor-induced FoxP3 gene expression: a role for DNA methylation. J Exp Med 2007;204(7):1543–51.
[103] Lal G, Zhang N, van der Touw W, Ding Y, Ju W, Bottinger EP, et al. Epigenetic regulation of Foxp3 expression in regulatory T cells by DNA methylation. J Immunol 2009;182(1):259–73.
[104] Vivier E, Tomasello E, Baratin M, Walzer T, Ugolini S. Functions of natural killer cells. Nat Immunol 2008;9(5):503–10.
[105] Armeanu S, Bitzer M, Lauer UM, Venturelli S, Pathil A, Krusch M, et al. Natural killer cell-mediated lysis of hepatoma cells via specific induction of NKG2D ligands by the histone deacetylase inhibitor sodium valproate. Cancer Res 2005;65(14):6321–9.
[106] Kroesen M, Gielen P, Brok IC, Armandari I, Hoogerbrugge PM, Adema GJ. HDAC inhibitors and immunotherapy: a double edged sword? Oncotarget 2014;5(16):6558–72.
[107] Fiegler N, Textor S, Arnold A, Rolle A, Oehme I, Breuhahn K, et al. Downregulation of the activating NKp30 ligand B7-H6 by HDAC inhibitors impairs tumor cell recognition by NK cells. Blood 2013;122(5):684–93.
[108] Waibel M, Christiansen AJ, Hibbs ML, Shortt J, Jones SA, Simpson I, et al. Manipulation of B-cell responses with histone deacetylase inhibitors. Nat Commun 2015;6:6838.
[109] Liu YJ, Zhang J, Lane PJ, Chan EY, MacLennan IC. Sites of specific B cell activation in primary and secondary responses to T cell-dependent and T cell-independent antigens. Eur J Immunol 1991;21(12):2951–62.
[110] Inamine A, Takahashi Y, Baba N, Miyake K, Tokuhisa T, Takemori T, et al. Two waves of memory B-cell generation in the primary immune response. Int Immunol 2005;17(5):581–9.
[111] Tsubata T. B cell abnormality and autoimmune disorders. Autoimmunity 2005;38(5):331–7.
[112] Kono DH, Theofilopoulos AN. Genetics of SLE in mice. Springer Semin Immunopathol 2006;28(2):83–96.
[113] Powers J, Lienlaf M, Perez-Villarroel P, Deng S, Knox T, Villagra A, et al. Expression and function of histone deacetylase 10 (HDAC10) in B cell malignancies. Methods Mol Biol 2016;1436:129–45.
[114] Mishra N, Reilly CM, Brown DR, Ruiz P, Gilkeson GS. Histone deacetylase inhibitors modulate renal disease in the MRL-lpr/lpr mouse. J Clin Invest 2003;111:539–52.

[115] Sweet MJ, Shakespear MR, Kamal NA, Fairlie DP. HDAC inhibitors: modulating leukocyte differentiation, survival, proliferation and inflammation. Immunol Cell Biol 2012;90(1):14–22.
[116] Aran AA, Putterman C. Treatment of lupus nephritis: facing the era of immunotherapy. Panminerva Med 2008;50(3):235–45.
[117] Renaudineau Y, Youinou P. Epigenetics and autoimmunity, with special emphasis on methylation. Keio J Med 2011;60(1):10–6.
[118] Mok CC, Lau CS. Pathogenesis of systemic lupus erythematosus. J Clin Pathol 2003;56(7):481–90.
[119] Kuhn A, Beissert S, Krammer PH. CD4(+)CD25 (+) regulatory T cells in human lupus erythematosus. Arch Dermatol Res 2009;301(1):71–81.
[120] Crispin JC, Liossis SN, Kis-Toth K, Lieberman LA, Kyttaris VC, Juang YT, et al. Pathogenesis of human systemic lupus erythematosus: recent advances. Trends Mol Med 2010;16(2):47–57.
[121] Puliaeva I, Puliaev R, Via CS. Therapeutic potential of CD8$^+$ cytotoxic T lymphocytes in SLE. Autoimmun Rev 2009;8(3):219–23.
[122] Crispin JC, Kyttaris VC, Terhorst C, Tsokos GC. T cells as therapeutic targets in SLE. Nat Rev Rheumatol 2010;6(6):317–25.
[123] Sabry A, Sheashaa H, El-Husseini A, Mahmoud K, Eldahshan KF, George SK, et al. Proinflammatory cytokines (TNF-alpha and IL-6) in Egyptian patients with SLE: its correlation with disease activity. Cytokine 2006;35(3–4):148–53.
[124] Apostolidis SA, Lieberman LA, Kis-Toth K, Crispin JC, Tsokos GC. The dysregulation of cytokine networks in systemic lupus erythematosus. J Interferon Cytokine Res 2011;31(10):769–79.
[125] Miyake K, Akahoshi M, Nakashima H. Th subset balance in lupus nephritis. J Biomed Biotechnol 2011;2011:980286.
[126] Liu MF, Lin LH, Weng CT, Weng MY. Decreased CD4+CD25+bright T cells in peripheral blood of patients with primary Sjogren's syndrome. Lupus 2008;17(1):34–9.
[127] Wahl SM, Chen W. Transforming growth factor-beta-induced regulatory T cells referee inflammatory and autoimmune diseases. Arthritis Res Ther 2005;7(2):62–8.
[128] Wenzel J, Henze S, Brahler S, Bieber T, Tuting T. The expression of human leukocyte antigen-DR and CD25 on circulating T cells in cutaneous lupus erythematosus and correlation with disease activity. Exp Dermatol 2005;14(6):454–9.
[129] Hayashi T, Hasegawa K, Adachi C. Elimination of CD4(+)CD25(+) T cell accelerates the development of glomerulonephritis during the preactive phase in autoimmune-prone female NZB x NZW F mice. Int J Exp Pathol 2005;86(5):289–96.
[130] Bodar EJ, Simon A, van der Meer JW. Effects of the histone deacetylase inhibitor ITF2357 in autoinflammatory syndromes. Mol Med 2011;17(5–6):363–8.
[131] Glauben R, Siegmund B. Inhibition of histone deacetylases in inflammatory bowel diseases. Mol Med 2011;17(5–6):426–33.
[132] Grabiec AM, Korchynskyi O, Tak PP, Reedquist KA. Histone deacetylase inhibitors suppress rheumatoid arthritis fibroblast-like synoviocyte and macrophage IL-6 production by accelerating mRNA decay. Ann Rheum Dis 2012;71(3):424–31.
[133] Leoni F, Fossati G, Lewis EC, Lee JK, Porro G, Pagani P, et al. The histone deacetylase inhibitor ITF2357 reduces production of pro-inflammatory cytokines in vitro and systemic inflammation in vivo. Mol Med 2005;11(1–12):1–15.
[134] Regna NL, Chafin CB, Hammond SE, Puthiyaveetil AG, Caudell DL, Reilly CM. Class I and II histone deacetylase inhibition by ITF2357 reduces SLE pathogenesis in vivo. Clin Immunol 2014;151(1):29–42.
[135] Richardson B, Scheinbart L, Strahler J, Gross L, Hanash S, Johnson M. Evidence for impaired T cell DNA methylation in systemic lupus erythematosus and rheumatoid arthritis. Arthritis Rheum 1990;33(11):1665–73.
[136] Singer NG, Richardson BC, Powers D, Hooper F, Lialios F, Endres J, et al. Role of the CD6 glycoprotein in antigen-specific and autoreactive responses of cloned human T lymphocytes. Immunology 1996;88(4):537–43.
[137] Lu Q, Wu A, Richardson BC. Demethylation of the same promoter sequence increases CD70 expression in lupus T cells and T cells treated with lupus-inducing drugs. J Immunol 2005;174(10):6212–9.
[138] Liu Y, Chen Y, Richardson B. Decreased DNA methyltransferase levels contribute to abnormal gene expression in "senescent" CD4(+)CD28(−) T cells. Clin Immunol 2009;132(2):257–65.
[139] Zhao M, Tang J, Gao F, Wu X, Liang Y, Yin H, et al. Hypomethylation of IL10 and IL13 promoters in CD4$^+$ T cells of patients with systemic lupus erythematosus. J Biomed Biotechnol 2010;2010:931018.
[140] Lu Q, Wu A, Tesmer L, Ray D, Yousif N, Richardson B. Demethylation of CD40LG on the inactive X in T cells from women with lupus. J Immunol 2007;179(9):6352–8.
[141] Jeffries MA, Dozmorov M, Tang Y, Merrill JT, Wren JD, Sawalha AH. Genome-wide DNA methylation patterns in CD4$^+$ T cells from patients with systemic lupus erythematosus. Epigenetics 2011;6(5):593–601.

[142] Lin SY, Hsieh SC, Lin YC, Lee CN, Tsai MH, Lai LC, et al. A whole genome methylation analysis of systemic lupus erythematosus: hypomethylation of the IL10 and IL1R2 promoters is associated with disease activity. Genes Immun 2012;13(3):214–20.

[143] Absher DM, Li X, Waite LL, Gibson A, Roberts K, Edberg J, et al. Genome-wide DNA methylation analysis of systemic lupus erythematosus reveals persistent hypomethylation of interferon genes and compositional changes to $CD4^+$ T-cell populations. PLoS Genet 2013;9(8):e1003678.

[144] Youinou P, Renaudineau Y. The paradox of CD5-expressing B cells in systemic lupus erythematosus. Autoimmun Rev 2007;7(2):149–54.

[145] Renaudineau Y, Vallet S, Le Dantec C, Hillion S, Saraux A, Youinou P. Characterization of the human CD5 endogenous retrovirus-E in B lymphocytes. Genes Immun 2005;6(8):663–71.

[146] Garaud S, Le Dantec C, Berthou C, Lydyard PM, Youinou P, Renaudineau Y. Selection of the alternative exon 1 from the cd5 gene down-regulates membrane level of the protein in B lymphocytes. J Immunol 2008;181(3):2010–8.

[147] Okada M, Ogasawara H, Kaneko H, Hishikawa T, Sekigawa I, Hashimoto H, et al. Role of DNA methylation in transcription of human endogenous retrovirus in the pathogenesis of systemic lupus erythematosus. J Rheumatol 2002;29(8):1678–82.

[148] Januchowski R, Wudarski M, Chwalinska-Sadowska H, Jagodzinski PP. Prevalence of ZAP-70, LAT, SLP-76, and DNA methyltransferase 1 expression in $CD4^+$ T cells of patients with systemic lupus erythematosus. Clin Rheumatol 2008;27(1):21–7.

[149] Lei W, Luo Y, Lei W, Luo Y, Yan K, Zhao S, et al. Abnormal DNA methylation in $CD4^+$ T cells from patients with systemic lupus erythematosus, systemic sclerosis, and dermatomyositis. Scand J Rheumatol 2009;38(5):369–74.

[150] Wang GS, Zhang M, Li XP, Zhang H, Chen W, Kan M, et al. Ultraviolet B exposure of peripheral blood mononuclear cells of patients with systemic lupus erythematosus inhibits DNA methylation. Lupus 2009;18(12):1037–44.

[151] Pan W, Zhu S, Yuan M, Cui H, Wang L, Luo X, et al. MicroRNA-21 and microRNA-148a contribute to DNA hypomethylation in lupus $CD4^+$ T cells by directly and indirectly targeting DNA methyltransferase 1. J Immunol 2010;184(12):6773–81.

[152] Balada E, Ordi-Ros J, Serrano-Acedo S, Martinez-Lostao L, Vilardell-Tarres M. Transcript overexpression of the MBD2 and MBD4 genes in $CD4^+$ T cells from systemic lupus erythematosus patients. J Leukoc Biol 2007;81(6):1609–16.

[153] Richardson B, Powers D, Hooper F, Yung RL, O'Rourke K. Lymphocyte function-associated antigen 1 overexpression and T cell autoreactivity. Arthritis Rheum 1994;37(9):1363–72.

[154] Yang IV, Pedersen BS, Liu A, O'Connor GT, Teach SJ, Kattan M, et al. DNA methylation and childhood asthma in the inner city. J Allergy Clin Immunol 2015;136(1):69–80.

[155] Grunig G, Warnock M, Wakil AE, Venkayya R, Brombacher F, Rennick DM, et al. Requirement for IL-13 independently of IL-4 in experimental asthma. Science 1998;282(5397):2261–3.

[156] Liang L, Willis-Owen SA, Laprise C, Wong KC, Davies GA, Hudson TJ, et al. An epigenome-wide association study of total serum immunoglobulin E concentration. Nature 2015;520(7549):670–4.

[157] Nicodemus-Johnson J, Naughton KA, Sudi J, Hogarth K, Naurekas ET, Nicolae DL, et al. Genome-wide methylation study identifies an IL-13-induced epigenetic signature in asthmatic airways. Am J Respir Crit Care Med 2016;193(4):376–85.

[158] Liu J, Ballaney M, Al-alem U, Quan C, Jin X, Perera F, et al. Combined inhaled diesel exhaust particles and allergen exposure alter methylation of T helper genes and IgE production in vivo. Toxicol Sci 2008;102(1):76–81.

[159] Sundar IK, Yao H, Rahman I. Oxidative stress and chromatin remodeling in chronic obstructive pulmonary disease and smoking-related diseases. Antioxid Redox Signal 2013;18(15):1956–71.

[160] Wang Y, Tian Y, Morley MP, Lu MM, Demayo FJ, Olson EN, et al. Development and regeneration of Sox2+ endoderm progenitors are regulated by a Hdac1/2-Bmp4/Rb1 regulatory pathway. Dev Cell 2013;24(4):345–58.

[161] Barnes PJ. Development of new drugs for COPD. Curr Med Chem 2013;20(12):1531–40.

[162] Barnes PJ. Corticosteroid resistance in patients with asthma and chronic obstructive pulmonary disease. J Allergy Clin Immunol 2013;131(3):636–45.

[163] Ortiz JL, Milara J, Lluch J, De Diego A, Sanz C, Cortijo J. Phosphodiesterase-4 inhibition improves corticosteroid insensitivity in pulmonary endothelial cells under oxidative stress. Allergy 2013;68(1):64–73.

[164] Qu Y, Yang Y, Ma D, He L, Xiao W. Expression level of histone deacetylase 2 correlates with occurring of chronic obstructive pulmonary diseases. Mol Biol Rep 2013;40(6):3995–4000.

[165] Huang SK, Scruggs AM, Donaghy J, Horowitz JC, Zaslona Z, Przybranowski S, et al. Histone modifications are responsible for decreased Fas expression and apoptosis resistance in fibrotic lung fibroblasts. Cell Death Dis 2013;4:e621.

[166] Kallsen K, Andresen E, Heine H. Histone deacetylase (HDAC) 1 controls the expression of beta defensin 1 in human lung epithelial cells. PLoS One 2012;7(11):e50000.

[167] Liu Q, Liu J, Roschmann KI, van Egmond D, Golebski K, Fokkens WJ, et al. Histone deacetylase inhibitors up-regulate LL-37 expression independent of toll-like receptor mediated signalling in airway epithelial cells. J Inflamm (Lond) 2013;10(1):15.
[168] Kienzler AK, Rizzi M, Reith M, Nutt SL, Eibel H. Inhibition of human B-cell development into plasmablasts by histone deacetylase inhibitor valproic acid. J Allergy Clin Immunol 2013;131(6):1695–9.
[169] Royce SG, Dang W, Yuan G, Tran J, El-Osta A, Karagiannis TC, et al. Effects of the histone deacetylase inhibitor, trichostatin A, in a chronic allergic airways disease model in mice. Arch Immunol Ther Exp 2012;60(4):295–306.
[170] Royce SG, Karagiannis TC. Histone deacetylases and their inhibitors: new implications for asthma and chronic respiratory conditions. Curr Opin Allergy Clin Immunol 2014;14(1):44–8.
[171] Atkins HL, Muraro PA, van Laar JM, Pavletic SZ. Autologous hematopoietic stem cell transplantation for autoimmune disease: is it now ready for prime time? Biol Blood Marrow Transplant 2012;18(Suppl. 1):S177–83.
[172] Kambham N, Higgins JP, Sundram U, Troxell ML. Hematopoietic stem cell transplantation: graft versus host disease and pathology of gastrointestinal tract, liver, and lung. Adv Anat Pathol 2014;21(5):301–20.
[173] Talmadge JE. Hematopoietic stem cell graft manipulation as a mechanism of immunotherapy. Int Immunopharmacol 2003;3(8):1121–43.
[174] Xia G, Truitt RL, Johnson BD. Graft-versus-leukemia and graft-versus-host reactions after donor lymphocyte infusion are initiated by host-type antigen-presenting cells and regulated by regulatory T cells in early and long-term chimeras. Biol Blood Marrow Transplant 2006;12(4):397–407.
[175] Sullivan KM, Muraro P, Tyndall A. Hematopoietic cell transplantation for autoimmune disease: updates from Europe and the United States. Biol Blood Marrow Transplant 2010;16(Suppl. 1):S48–56.
[176] Choi SW, Gatza E, Hou G, Sun Y, Whitfield J, Song Y, et al. Histone deacetylase inhibition regulates inflammation and enhances Tregs after allogeneic hematopoietic cell transplantation in humans. Blood 2015;125(5):815–9.
[177] Choi S, Reddy P. HDAC inhibition and graft versus host disease. Mol Med 2011;17(5–6):404–16.
[178] Ferrara JL, Krenger W. Graft-versus-host disease: the influence of type 1 and type 2 T cell cytokines. Transfus Med Rev 1998;12(1):1–17.
[179] Zeiser R, Nguyen VH, Beilhack A, Buess M, Schulz S, Baker J, et al. Inhibition of $CD4^+CD25^+$ regulatory T-cell function by calcineurin-dependent interleukin-2 production. Blood 2006;108(1):390–9.
[180] Heinrichs J, Bastian D, Veerapathran A, Anasetti C, Betts B, Yu XZ. Regulatory T-cell therapy for graft-versus-host disease. J Immunol Res Ther 2016;1(1):1–14.
[181] Medzhitov R, Janeway CA Jr. Decoding the patterns of self and nonself by the innate immune system. Science 2002;296(5566):298–300.
[182] Reddy P, Maeda Y, Hotary K, Liu C, Reznikov LL, Dinarello CA, et al. Histone deacetylase inhibitor suberoylanilide hydroxamic acid reduces acute graft-versus-host disease and preserves graft-versus-leukemia effect. Proc Natl Acad Sci USA 2004;101(11):3921–6.
[183] Leng C, Gries M, Ziegler J, Lokshin A, Mascagni P, Lentzsch S, et al. Reduction of graft-versus-host disease by histone deacetylase inhibitor suberonylanilide hydroxamic acid is associated with modulation of inflammatory cytokine milieu and involves inhibition of STAT1. Exp Hematol 2006;34(6):776–87.
[184] Mellor AL, Munn DH. IDO expression by dendritic cells: tolerance and tryptophan catabolism. Nat Rev Immunol 2004;4(10):762–74.
[185] Reddy P, Sun Y, Toubai T, Duran-Struuck R, Clouthier SG, Weisiger E, et al. Histone deacetylase inhibition modulates indoleamine 2,3-dioxygenase-dependent DC functions and regulates experimental graft-versus-host disease in mice. J Clin Invest 2008;118(7):2562–73.
[186] Villagra A, Cheng F, Wang HW, Suarez I, Glozak M, Maurin M, et al. The histone deacetylase HDAC11 regulates the expression of interleukin 10 and immune tolerance. Nat Immunol 2009;10(1):92–100.
[187] Hancock WW. Effects of histone deacetylase inhibitors on alloresponses. Lancet Oncol 2014;15(1):10–1.
[188] Zorn E, Kim HT, Lee SJ, Floyd BH, Litsa D, Arumugarajah S, et al. Reduced frequency of $FOXP3^+CD4^+CD25^+$ regulatory T cells in patients with chronic graft-versus-host disease. Blood 2005;106(8):2903–11.
[189] Ellis JD, Neil DA, Inston NG, Jenkinson E, Drayson MT, Hampson P, et al. Inhibition of histone deacetylase 6 reveals a potent immunosuppressant effect in models of transplantation. Transplantation 2016;100(8):1667–74.
[190] Hancock WW. Isoform-selective HDAC inhibitor therapy for transplantation: are we ready for HDAC6? Transplantation 2016;100(8):1597–8.
[191] Gillespie J, Savic S, Wong C, Hempshall A, Inman M, Emery P, et al. Histone deacetylases are dysregulated in rheumatoid arthritis and a novel histone deacetylase 3-selective inhibitor reduces interleukin-6 production by peripheral blood mononuclear cells from rheumatoid arthritis patients. Arthritis Rheum 2012;64(2):418–22.

[192] Cantley MD, Fairlie DP, Bartold PM, Marino V, Gupta PK, Haynes DR. Inhibiting histone deacetylase 1 suppresses both inflammation and bone loss in arthritis. Rheumatology (Oxford) 2015;54(9):1713–23.
[193] Neidhart M, Rethage J, Kuchen S, Kunzler P, Crowl RM, Billingham ME, et al. Retrotransposable L1 elements expressed in rheumatoid arthritis synovial tissue: association with genomic DNA hypomethylation and influence on gene expression. Arthritis Rheum 2000;43(12):2634–47.
[194] Gul FC, Turgut B, Dagli F, Ilhan N, Ozgen M. The comparison of the impact of ghrelin and tacrolimus on vitreous cytokine levels in an experimental uveitis model. Graefes Arch Clin Exp Ophthalmol 2013;251(4):1235–41.
[195] Huber A, Menconi F, Corathers S, Jacobson EM, Tomer Y. Joint genetic susceptibility to type 1 diabetes and autoimmune thyroiditis: from epidemiology to mechanisms. Endocr Rev 2008;29(6):697–725.
[196] Koch MW, Metz LM, Kovalchuk O. Epigenetic changes in patients with multiple sclerosis. Nat Rev Neurol 2013;9(1):35–43.
[197] Fradin D, Le Fur S, Mille C, Naoui N, Groves C, Zelenika D, et al. Association of the CpG methylation pattern of the proximal insulin gene promoter with type 1 diabetes. PLoS One 2012;7(5):e36278.
[198] Gu T, Falhammar H, Gu HF, Brismar K. Epigenetic analyses of the insulin-like growth factor binding protein 1 gene in type 1 diabetes and diabetic nephropathy. Clin Epigenetics 2014;6(1):10.
[199] Miao F, Gonzalo IG, Lanting L, Natarajan R. In vivo chromatin remodeling events leading to inflammatory gene transcription under diabetic conditions. J Biol Chem 2004;279(17):18091–7.
[200] Christensen DP, Dahllof M, Lundh M, Rasmussen DN, Nielsen MD, Billestrup N, et al. Histone deacetylase (HDAC) inhibition as a novel treatment for diabetes mellitus. Mol Med 2011;17(5–6):378–90.
[201] Liggett T, Melnikov A, Tilwalli S, Yi Q, Chen H, Replogle C, et al. Methylation patterns of cell-free plasma DNA in relapsing-remitting multiple sclerosis. J Neurol Sci 2010;290(1–2):16–21.
[202] Rivieccio MA, Brochier C, Willis DE, Walker BA, D'Annibale MA, McLaughlin K, et al. HDAC6 is a target for protection and regeneration following injury in the nervous system. Proc Natl Acad Sci USA 2009;106(46):19599–604.
[203] Faraco G, Cavone L, Chiarugi A. The therapeutic potential of HDAC inhibitors in the treatment of multiple sclerosis. Mol Med 2011;17(5–6):442–7.
[204] Zuo L, Hemmelgarn BT, Chuang CC, Best TM. The role of oxidative stress-induced epigenetic alterations in amyloid-beta production in Alzheimer's disease. Oxid Med Cell Longev 2015;2015:604658.
[205] Million M, Raoult D. The pathogenesis of the antiphospholipid syndrome. N Engl J Med 2013;368(24):2335.
[206] van Gerven NM, de Boer YS, Mulder CJ, van Nieuwkerk CM, Bouma G. Autoimmune hepatitis. World J Gastroenterol 2016;22(19):4651–61.
[207] Hughes T, Ture-Ozdemir F, Alibaz-Oner F, Coit P, Direskeneli H, Sawalha AH. Epigenome-wide scan identifies a treatment-responsive pattern of altered DNA methylation among cytoskeletal remodeling genes in monocytes and $CD4^+$ T cells from patients with Behcet's disease. Arthritis Rheumatol 2014;66(6):1648–58.
[208] Renauer P, Coit P, Sawalha AH. Epigenetics and vasculitis: a comprehensive review. Clin Rev Allergy Immunol 2016;50(3):357–66.
[209] Kocsis D, Beres N, Veres G, Szabo D, Muller KE, Arato A, et al. Genetic and epigenetic aspects of celiac disease. Orv Hetil 2014;155(3):83–8.
[210] Alvarenga EM, Rodrigues VL, Moraes AS, Naves LS, Mondin M, Felisbino MB, et al. Histone epigenetic marks in heterochromatin and euchromatin of the Chagas' disease vector, *Triatoma infestans*. Acta Histochem 2016;118(4):401–12.
[211] Loddo I, Romano C. Inflammatory bowel disease: genetics, epigenetics, and pathogenesis. Front Immunol 2015;6:551.
[212] Hasham A, Tomer Y. Genetic and epigenetic mechanisms in thyroid autoimmunity. Immunol Res 2012;54(1–3):204–13.
[213] Strickland FM, Hewagama A, Wu A, Sawalha AH, Delaney C, Hoeltzel MF, et al. Diet influences expression of autoimmune-associated genes and disease severity by epigenetic mechanisms in a transgenic mouse model of lupus. Arthritis Rheum 2013;65(7):1872–81.
[214] Rajabally YA, Hiew FL. Optimizing electrodiagnosis for Guillain-Barre syndrome: clues from clinical practice. Muscle Nerve 2016;55(5):748–51.
[215] Gao T, Nie Y, Hu H, Liang Z. Hypermethylation of IGSF4 gene for noninvasive prenatal diagnosis of thalassemia. Med Sci Monit 2012;18(1):BR33–40.
[216] Ginder GD. Epigenetic regulation of fetal globin gene expression in adult erythroid cells. Transl Res 2015;165(1):115–25.
[217] Meyer B, Chavez RA, Munro JE, Chiaroni-Clarke RC, Akikusa JD, Allen RC, et al. DNA methylation at IL32 in juvenile idiopathic arthritis. Sci Rep 2015;5:11063.
[218] Plant D, Wilson AG, Barton A. Genetic and epigenetic predictors of responsiveness to treatment in RA. Nat Rev Rheumatol 2014;10(6):329–37.

[219] Bjanesoy TE, Andreassen BK, Bratland E, Reiner A, Islam S, Husebye ES, et al. Altered DNA methylation profile in Norwegian patients with autoimmune Addison's disease. Mol Immunol 2014;59(2):208–16.
[220] Berrih-Aknin S, Le Panse R. Myasthenia gravis: a comprehensive review of immune dysregulation and etiological mechanisms. J Autoimmun 2014;52:90–100.
[221] Kucukali CI, Kurtuncu M, Coban A, Cebi M, Tuzun E. Epigenetics of multiple sclerosis: an updated review. Neuromolecular Med 2015;17(2):83–96.
[222] Roberson ED, Liu Y, Ryan C, Joyce CE, Duan S, Cao L, et al. A subset of methylated CpG sites differentiate psoriatic from normal skin. J Invest Dermatol 2012;132(3 Pt 1):583–92.
[223] Liu Y, Li H, Xiao T, Lu Q. Epigenetics in immune-mediated pulmonary diseases. Clin Rev Allergy Immunol 2013;45(3):314–30.
[224] Shaheen IA, Abukhalil RE, Ali DK, Afifi RA. DNMT3B promoter polymorphism and risk of immune thrombocytopenic purpura in pediatric Egyptians. Blood Coagul Fibrinolysis 2012;23(7):636–9.
[225] Le Dantec C, Varin MM, Brooks WH, Pers JO, Youinou P, Renaudineau Y. Epigenetics and Sjogren's syndrome. Curr Pharm Biotechnol 2012;13(10):2046–53.
[226] Broen JC, Radstake TR, Rossato M. The role of genetics and epigenetics in the pathogenesis of systemic sclerosis. Nat Rev Rheumatol 2014;10(11):671–81.
[227] Crunkhorn S. Metabolic disease: new role for HDACs in glucose homeostasis. Nat Rev Drug Discov 2011;10(7):492.
[228] Felice C, Lewis A, Armuzzi A, Lindsay JO, Silver A. Review article: selective histone deacetylase isoforms as potential therapeutic targets in inflammatory bowel diseases. Aliment Pharmacol Ther 2015;41(1):26–38.
[229] Buckland J. Rheumatoid arthritis: HDAC and HDACi: pathogenetic and mechanistic insights. Nat Rev Rheumatol 2011;7(12):682.
[230] Vojinovic J, Damjanov N. HDAC inhibition in rheumatoid arthritis and juvenile idiopathic arthritis. Mol Med 2011;17(5–6):397–403.
[231] Nebbioso A, Carafa V, Benedetti R, Altucci L. Trials with 'epigenetic' drugs: an update. Mol Oncol 2012;6(6):657–82.

CHAPTER

4

MicroRNA and T Helper Cell-Mediated Immune Responses

Lucien P. Garo, Vanessa Beynon, Gopal Murugaiyan

Ann Romney Center for Neurologic Diseases, Brigham and Women's Hospital and Harvard Medical School, Boston, MA, United States

OUTLINE

4.1 INTRODUCTION

4.1.1 T Helper Subsets and Functions

$CD4^+$ T helper cells are major regulators of adaptive immune responses. Naive $CD4^+$ T cells differentiate into distinct effector lineages (e.g., T helper (Th) type 1 (Th1), Th2, and IL-17-producing Th17 subsets) following antigen stimulation to exert specific effector functions [1] (Fig. 4.1). Th1 cells express the transcription factor T-bet and produce IFN-γ as their signature cytokine [2]. Although these cells play a critical role in protecting the host against intracellular pathogens, dysregulated Th1 responses to self-antigens can also lead to tissue inflammation and autoimmunity. IL-12 and IFN-γ are the major cytokines involved in the differentiation of Th1 cells [3]. IL-12 is mainly produced by antigen-presenting cells (APCs), such as dendritic cells (DCs) and macrophages, and

The Epigenetics of Autoimmunity. http://dx.doi.org/10.1016/B978-0-12-809912-4.00004-0

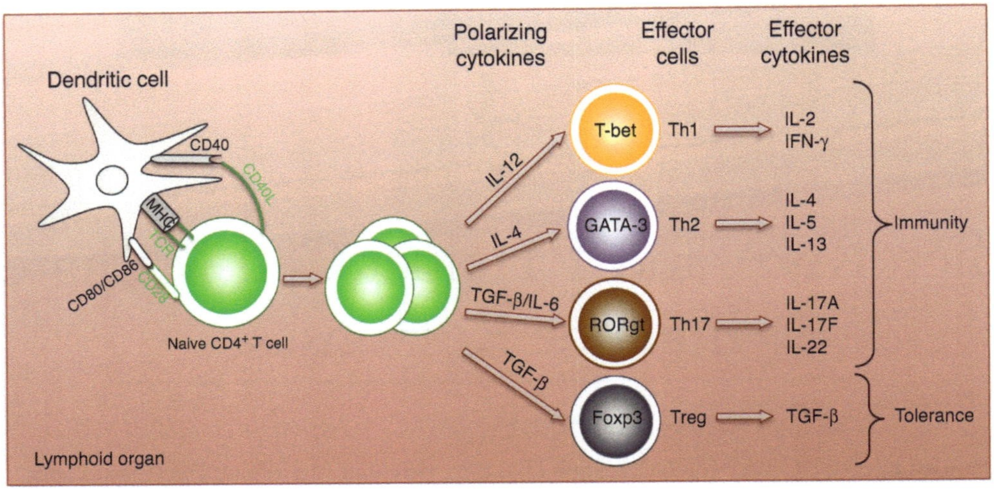

FIGURE 4.1 **Differentiation of T helper subsets.** Naive CD4⁺ T cells differentiate into functional effector cell subsets (i.e., Th1, Th2, and Th17 cells) and regulatory subsets (e.g., T_{reg} cells) depending on the cytokine milieu. In the presence of IL-12, naive CD4⁺ T cells differentiate into the Th1 cell phenotype, which are characterized by expression of the transcription factor, T-bet. IL-4 promotes differentiation to Th2 cells, which are marked by expression of GATA3, whereas TGF-β and IL-6 promote the differentiation of naive CD4⁺ T cells into the Th17 cell phenotype, which is characterized by expression of RORγt. Naive CD4⁺ T cells can also differentiate into Foxp3-expressing T_{reg} cells, for which TGF-β is crucial. While the effector Th1, Th2, and Th17 cells mediate immunity to infection, T_{reg} cells control overactive effector T cells and immune tolerance to self-antigens.

induces IFN-γ production in CD4⁺ T cells through activation of signal transducer and activator of transcription 4 (STAT4). IFN-γ then activates STAT1 to induce T-bet, which is a Th1-specific transcription factor. T-bet can directly upregulate IFN-γ further in activated CD4⁺ T cells, thereby functioning as a crucial mediator of Th1 differentiation [4].

While Th1 cells fight against intracellular pathogens, Th2 cells protect against extracellular pathogens [5]; however, excessive Th2 responses have been linked to allergies and asthma. Th2 cells produce cytokines (e.g., IL-4, IL-5, and IL-13) that are responsible for antibody production and eosinophil activation. IL-4 is a critical inducer of Th2 cells and it signals through STAT6 to upregulate the transcription factor GATA3, which is the master regulator of Th2 differentiation [6]. Repression of this activity results in the developmental failure of Th2 cells; however, IL-4/STAT6-independent Th2 cell differentiation has been observed in vivo, as Th2 cell differentiation can also occur through dendritic cell-derived thymic stromal lymphopoietin and other cytokines (i.e., IL-25 and IL-33) [7–9].

Another subset of effector CD4⁺ T cells, Th17 cells, has also been described. Th17 cells play a key role in the host's defense against extracellular bacteria and fungi [10]. The proinflammatory cytokine IL-6, together with TGF-β, drives initial Th17 differentiation, including expression of the Th17-specific transcription factor, RAR-related orphan receptor gamma (RORγt), which induces the effector cytokine IL-17A [11–13]. The induction of RORγt is dependent on STAT3, which is preferentially activated by IL-6 [14]. In addition to IL-17A, Th17 cells also secrete a range of inflammatory mediators, including IL-17F, IL-21, IL-22, and granulocyte-macrophage colony-stimulating factor (GM-CSF). Other transcription factors (e.g., RORα, IRF4, HIF1α, and BATF) are also involved in the control of Th17 lineage commitment [15–18]. IL-23 and IL-1β play a critical role in stabilizing the Th17 cell phenotype after initial Th17 differentiation, as they increase expression of IL-23R and IL-1R and endow Th17 cells with inflammatory effector functions [19–21]. Dysregulated Th17 responses have been implicated in the pathogenesis of several inflammatory and autoimmune diseases, including rheumatoid arthritis (RA), inflammatory bowel diseases (IBDs), asthma, multiple sclerosis (MS), and psoriasis [10].

Naive CD4$^+$ T cells can also develop into regulatory T (T_{reg}) cells, which help to maintain immunological unresponsiveness to self-antigens and counterbalance effector T-cell responses that are deleterious to the host [22]. Defects in the frequency and function of T_{reg} cells have also been associated with several inflammatory and autoimmune disorders, including MS, RA, type 1 diabetes, and systemic lupus erythematosus (SLE) [22]. T_{reg} cells are characterized by higher levels of the IL-2Rα chain (CD25) and the T_{reg}-specific transcription factor, forkhead box P3 (Foxp3) [23]. Naturally occurring T_{reg} cells are primed with antigens and generated in the thymus; however, induced T_{reg} cells can also be generated in peripheral lymphoid organs where they collaborate with natural T_{reg} cells to maintain immunological self-tolerance and prevent autoimmunity. Although several factors promote T_{reg} cell differentiation, TGF-β is critical for the induction of Foxp3 [24]. Furthermore, IL-2 and the transcription factor STAT5 are indispensable for stabilizing the expression of Foxp3 in T_{reg} cells [25]. The combined deficiency of TGF-β and IL-2 almost completely abolishes thymic T_{reg} cell development [26,27]. All-trans retinoic acid, a vitamin A metabolite, has also been shown to promote T_{reg} cell differentiation in synergy with TGF-β [28]. Most recently, IL-33, a member of the IL-1 family of cytokines, has been shown to induce T_{reg} cells as well [29].

As CD4$^+$ T cells differentiate towards one lineage, other lineage fates are usually suppressed. The transcription factors and cytokines specific to a T-cell subset can inhibit the differentiation of other helper T-cell subsets; for example, the Th1 transcription factor, T-bet, suppresses the generation of Th2 cells by blocking the expression of IL-4 and GATA3 [4,30]. T-bet also suppresses RORγt expression by interacting with Runx1, a transcription factor important for the induction of RORγt expression during Th17 differentiation [31]. During Th2 differentiation, GATA3 downregulates the expression of STAT4, thereby limiting IL-12-mediated Th1 differentiation [32,33]. STAT5 is critical for maintaining GATA3 expression in differentiated Th2 cells. A constitutively active form of STAT5 also inhibits T-bet to promote Th2 differentiation. Th17 cells and T_{reg} cells must also cross-regulate each other. TGF-β promotes the differentiation of both Th17 and T_{reg} cells [12]. At an intermediate stage of differentiation, transcription factors for both Th17 (i.e., RORγt) and T_{reg} (i.e., Foxp3) lineages are induced. Higher concentrations of TGF-β cause Foxp3 to interact with RORγt to block its function and thus inhibit Th17 differentiation; however, lower concentrations of TGF-β, in combination with IL-6, induce RORγt [34]. The relative expression of these transcription factors is controlled by a delicate cytokine balance, which ultimately determines whether the Th17 or T_{reg} fate is adopted.

4.1.2 Epigenetic Control of T Helper Responses

The differentiation of naive CD4$^+$ T helper cells into effector T helper cells depends on the coordinated action of lineage-specific transcription factors. However, the accessibility of these transcription factors to promoter binding sites and the subsequent induction of gene expression is highly regulated by epigenetic modifications. Epigenetics refers to the study of potentially heritable changes in gene expression that do not stem from a change in the DNA sequence. Two epigenetic mechanisms that have profound effects on gene expression are histone modifications and DNA methylation [35–37].

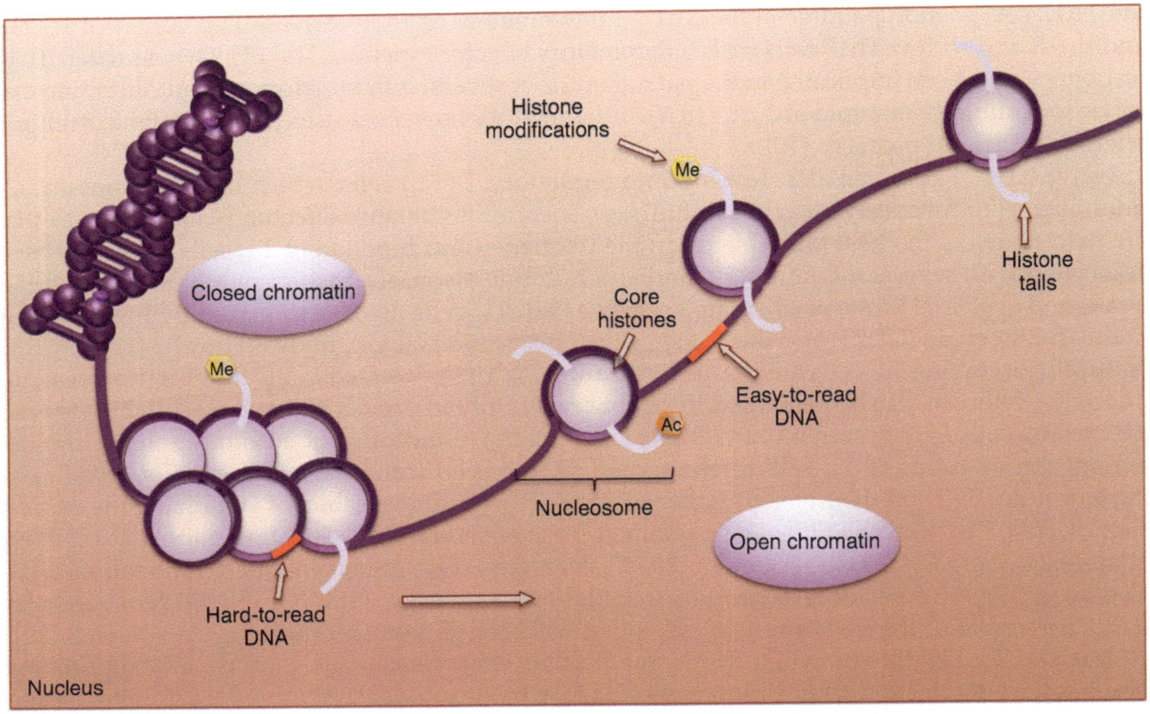

FIGURE 4.2 **Simplified schematic representation of nucleosome and core histone modifications.** Histones provide the basis for the nucleosome, which is the basic unit of chromatin structure. The nucleosome consists of hundreds of base pairs of DNA wrapped around a histone octamer. Nucleosome packaging of DNA and methylation (Me)-induced changes in the DNA structure present a physical barrier to the initiation of transcription, making the DNA difficult to read. However, the posttranslational covalent modifications of histones or histone tails by acetylation (Ac) can have a profound effect on gene expression as it opens the closed chromatin and allows efficient transcription.

Histones are proteins that condense and organize the DNA of eukaryotic cell nuclei into units called nucleosomes. Nucleosomes are the basic structural units of chromatin and they consist of two copies of each of the core histones H2A, H2B, H3, and H4. Each nucleosome is encircled by DNA and interconnected by linker DNA. The nucleosome packaging of DNA presents a physical barrier to the initiation of transcription (Fig. 4.2). Tight association between DNA and histones forms a closed nucleosomal configuration in which RNA polymerase is prevented from binding to the transcription start site and transcription factors are precluded from interacting with their cognate binding sites in gene regulatory regions. However, nucleosomes can adopt an open configuration in response to enzymatic modification of specific histone residues, which allows RNA polymerases and transcription factors to access their respective binding sites. A range of posttranslational histone modifications have been described; for example, the N-terminal tails of histone proteins may be covalently modified through acetylation, methylation, phosphorylation, ubiquitination, and sumoylation. The balance of histone acetylation and deacetylation plays a critical role in the regulation of gene expression. Histone acetylation, induced by histone acetyltransferases (HACs), is associated with gene transcription, while histone deacetylation, induced by histone deacetylase (HDAC) activity, is associated with gene silencing [35–37]; for example, acetylation of lysine residues 9, 14, and 27 in histone H3 tails (H3K9ac, H3K14ac, and H3K27ac, respectively) is indicative of active gene expression. Histone methylation can promote or inhibit the transcriptional activation of a gene depending upon the location of the methylated lysine residue and the number of methyl moieties; for example, methylation of lysine 4 of histone H3 (H3K4) correlates with transcriptional activation, while methylation of lysine 9 or 27 of histone H3 (H3K9 and H3K27, respectively) correlates with transcriptional repression [35–37].

DNA transcription is also epigenetically controlled by changing the structure of DNA itself via the addition and removal of methyl groups [38]. DNA methylation refers to the covalent addition of a methyl group at the 5′ carbon of a cytosine ring to produce 5-methylcytosine. This process is carried out by a family of enzymes called DNA methyltransferases (DNMTs). DNMT1, DNMT3a, and DNMT3b are required for the establishment and maintenance of DNA methylation patterns. DNA methylation can regulate the promoter and enhancer activity of genes through the disruption of transcription factor binding. [38,39].

Several lines of evidence indicate that histone modifications and DNA methylation have a profound impact on T helper cell differentiation [39–41]; for example, the treatment of $CD4^+$ T cells with HDAC inhibitors leads to enhanced Th1 and Th2 cytokine production [42,43]. Mice that are haplosufficient for H3K4 methyltransferase also show defects in maintaining IL-4, IL-5, and GATA3 levels in Th2 cells [44]. In addition, mice that lack MEL18 (a polycomb repressor complex 1 (PRC1) protein that binds to trimethylated H3K27) present with impaired Th2 differentiation. Conversely, the conditional deletion of DNMT1 increases the expression of Th1 and Th2 cytokines and increases the inability of Th1 and Th2 cells to silence the genes associated with opposing lineages [45].

Studies focusing on individual loci within T cells have revealed more specific roles for epigenetic modifications in T-cell differentiation; for example, exposure to the Th1-polarizing cytokine, IL-12, causes remodeling of the IFN-γ locus in naive $CD4^+$ T cells to a permissive epigenetic signature that reinforces and maintains IFN-γ expression [46,47]. The IL-4 and IL-17 loci in Th1 cells accordingly present with repressive epigenetic modifications, thus resulting in the inhibition of IL-4 and IL-17 expression [46–48]. The IFN-γ promoter conversely shows repressive histone marks (i.e., H3K27me3) in Th2 cells [48]. Stimulation with the Th2-polarizing cytokine, IL-4, causes the IL-4 locus in naive $CD4^+$ T cells to acquire permissive epigenetic marks that reinforce IL-4 expression while maintaining a repressive epigenetic signature that inhibits IFN-γ expression [49]. Genes encoding the key transcriptional factors that drive lineage fate also exhibit either repressive or accessible epigenetic marks depending on the T helper subset; for example, the T-bet promoter contains permissive H3K4me3 marks in Th1 cells [48] and repressive H3K27me3 marks in Th2 and Th17 cells. In addition, a significant reduction of repressive H3K27me3 and an increase of permissive H3K4me3 marks at the GATA3 regulatory regions have been observed under Th2 differentiation conditions [48].

The Foxp3 locus has also been associated with strong permissive epigenetic marks in T_{reg} cells and repressive marks in Th1 and Th2 cells [48,50,51]. No H3K27me3 repressive marks were observed in the Foxp3 locus in Th17 cells, which agrees with the finding that Foxp3 isexpressed at lower levels in TGF-β-driven Th17 cells [34]. Th17 cells also show permissive H3K4me3 marks at the IL-17A and IL-17F loci, which is consistent with their ability to produce high levels of IL-17A and IL-17F [52]. Interestingly, Th17 cells have been observed with significant remodeling of distal enhancers in the IFN-γ locus, even in the absence of Th1-polarizing cytokines. The Th1-promoting T-bet locus within Th17 cells has also been shown to present with both permissive (H3K4me3) and repressive (H3K27me3) histone marks, and is thus termed a bivalent locus [53]. These findings support reports that there is considerable intrinsic instability in Th17 cells compared to Th1 and Th2 cells [53–55]; for example, adoptive transfer of Th17 cells results in the loss of RORγt expression and IL-17 production and the acquisition of IFN-γ production and a Th1-like phenotype [55], and it is the exposure of Th17 cells to IL-12 that can induce such changes. Interestingly, the addition of IL-12 also enhances the generation of IFN-γ^+IL-17^+ double producers, while inhibiting IL-17^+ single producers [53,54]. This "Th1 permissiveness" in Th17 cells contrasts with the absence of permissive histone modifications at the IL-17A and IL-17F loci in Th1 cells, which are correspondingly resistant to the developmental plasticity that characterizes Th17 cells [53].

4.1.3 MicroRNAs and Epigenetics

Epigenetic mechanisms not only regulate the expression of protein-coding genes, but also the expression of microRNAs (miRNAs). MiRNAs are a family of small noncoding RNAs that posttranscriptionally downregulate gene expression in a sequence-specific manner [56]. The pathways involved in the biogenesis of miRNAs are reviewed in detail elsewhere [57]. Briefly, miRNA genes are generally transcribed from transcription start sites by RNA polymerase II to form primary transcripts (pri-miRNAs) that are capped with 7-methylguanosine and polyadenylated. The nuclear RNase III enzyme, Drosha, and its cofactor, DiGeorge syndrome critical region gene-8 (DGCR8), mediate the processing of pri-miRNAs into precursor miRNAs (pre-miRNAs) to generate stem-loop structures. The pre-miRNAs are then transported into the cytoplasm by exportin 5 and are subsequently processed into mature miRNAs through the actions of the endoribonuclease Dicer. Mature miRNAs are finally loaded into an RNA-induced silencing complex (RISC). Argonaute (AGO) proteins directly interact with miRNAs and are key factors in the assembly and function of this complex. The miRNA–RISC complex downregulates specific gene products by repressing translation (by binding to complementary sequences in the 3′ untranslated regions (UTR) of target mRNAs) or by degrading mRNA (by binding to complementary sequences) (Fig. 4.3).

Most miRNA genes are transcribed in a similar manner to coding genes. Unsurprisingly, several miRNAs are also known to be regulated by the same epigenetic mechanisms that regulate coding genes, including DNA methylation and histone modifications [58–60]. For example, upregulation of miRNAs has been observed in cancer cells treated with a DNA demethylating agent (5-aza-2′-deoxycytidine) and a HDAC inhibitor (4-phenylbutyric acid) [61,62]. DNA methylation and histone methylation have also been shown to contribute to the loss of miR-31 expression in breast cancer and melanoma [63,64]. This is significant as miR-31 is believed to act as a tumor suppressor by regulating proliferation, invasion, and metastasis in several tumors. In fact, the parent gene for miR-31 is frequently deleted in melanoma and other malignancies. A separate study in prostate cancer cell lines identified that a loss of DNA methylation, and a gain of histone acetylation, induces miR-615 [65]. MiR-155 is one of the most well-characterized, immunologically relevant, and oncogenic miRNAs (oncomiRs) and is repressed by deacetylation of histone H3 and H2A in breast cancer [66]. Researchers have also identified several other miRNAs that are regulated through epigenetic pathways in various cancer models (reviewed in [63]).

Evidence also suggests that miRNAs are themselves capable of targeting genes that control epigenetic pathways [60]. This presents a new mechanism, in addition to the RNA interference (RNAi) pathway, through which miRNAs can control gene expression. Several miRNAs control the expression of important epigenetic regulators, including DNMTs, HDACs, and polycomb group proteins. DNMT3A and DNMT3B are key DNA methylation enzymes that are involved in the silencing of target genes. The miR-29 family (29a, 29b, and 29c) can directly target DNMT3A and DNMT3B, and exogenous miR-29 has been shown to reactivate methylation-silenced tumor suppressor genes by restoring normal DNA methylation patterns in cancer cells [67]. Recent studies have also highlighted a role for miRNAs in the establishment and maintenance of histone acetylation. MiR-449

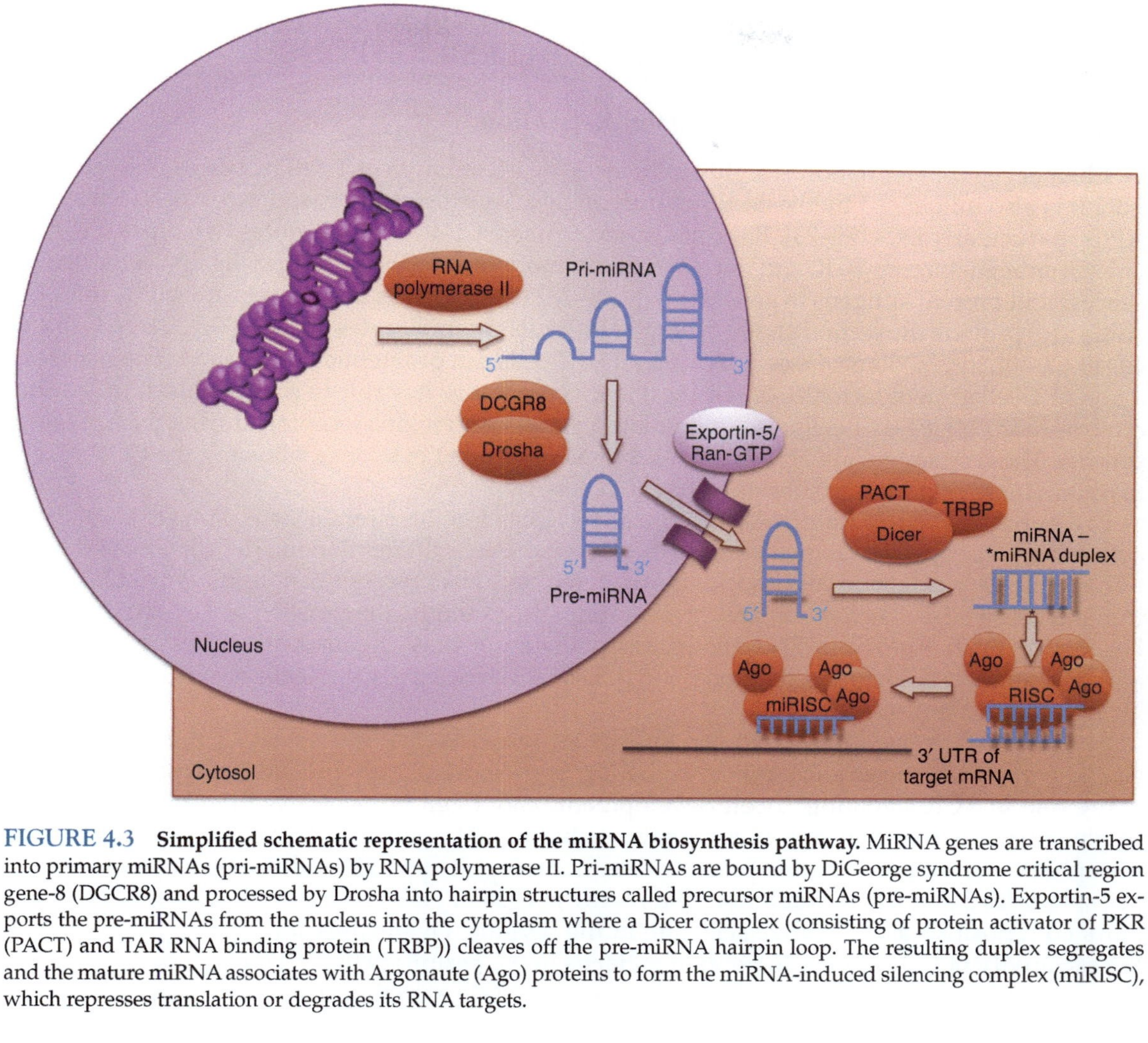

FIGURE 4.3 **Simplified schematic representation of the miRNA biosynthesis pathway.** MiRNA genes are transcribed into primary miRNAs (pri-miRNAs) by RNA polymerase II. Pri-miRNAs are bound by DiGeorge syndrome critical region gene-8 (DGCR8) and processed by Drosha into hairpin structures called precursor miRNAs (pre-miRNAs). Exportin-5 exports the pre-miRNAs from the nucleus into the cytoplasm where a Dicer complex (consisting of protein activator of PKR (PACT) and TAR RNA binding protein (TRBP)) cleaves off the pre-miRNA hairpin loop. The resulting duplex segregates and the mature miRNA associates with Argonaute (Ago) proteins to form the miRNA-induced silencing complex (miRISC), which represses translation or degrades its RNA targets.

regulates the expression of HDAC1 and miR-1 regulates HDAC4 [68,69]. Dysregulation of these miRNAs can contribute to the overexpression of HDACs in cancer cells; for example, downregulation of miR-449a leads to HDAC1 overexpression and abnormal epigenetic patterns in prostate cancer [68]. MiRNAs have also been implicated in the regulation of polycomb group proteins, which are transcriptional repressors that regulate cellular differentiation and maintain cell fates. There are two PRCs: PRC1, whose core complex contains the genes *Cbx*, *Mph*, *Ring*, *Bmi-1*, and *Me118*; and PRC2, which consists of *Ezh2*, *Suz12*, and *Eed*. Several miRNAs (e.g., miR-26a, miR-101, miR-205, and miR-214) have been shown to control the expression of *Ezh2* [60], while miR-128 and miR-203 have been shown to regulate *Bmi-1* [60]. This evidence suggests that posttranscriptional regulation by miRNAs and transcriptional control by epigenetic mechanisms cooperate to organize gene expression profiles and maintain physiological functions. Aberrations in miRNA expression that have been linked to epigenetic pathways, and epigenetic pathways that are known to control miRNA expression, have both been associated with altered immune cell function and immunologic diseases, including autoimmunity and cancer [39,60].

4.2 MicroRNA CONTROL OF INFLAMMATORY AND REGULATORY T HELPER RESPONSES

4.2.1 MicroRNA Control of T Helper Activation

T helper cell differentiation is highly influenced by the strength of T-cell receptor (TCR) stimulation. It is also influenced by the costimulatory molecules expressed by, and cytokines secreted by, APCs, particularly DCs [70–73]. Basic activation of naive $CD4^+$ T cells requires two types of signal: one via antigen-specific TCR, and one via costimulatory molecules. To trigger the first signal, APCs process and present antigens to generate peptide–MHC complexes that are recognized by the TCR. APCs also provide the second signal through costimulatory molecules (i.e., CD80/CD86-CD28 and CD40-CD40L interactions). Once activated, $CD4^+$ T cells can proliferate and expand. The release of a potent T-cell growth factor, IL-2, is critical to this process as it is able to act upon itself in an autocrine fashion [74]. $CD4^+$ T cells also upregulate a variety of cytokine receptors during the activation process. These receptors are able to respond to APC-secreted cytokines, allowing the $CD4^+$ T cells to differentiate into specific subsets of effector cells [1].

Evidence suggests that miRNAs are critically involved in these initial steps of naive $CD4^+$ T-cell activation, which are fundamental to further specific T-cell differentiation [75–78]. Global disruption of miRNA production (via knockout of different miRNA processing molecules) in $CD4^+$ T cells leads to altered T-cell activation and proliferation; for example, the deletion of key enzymes (i.e., Dicer and Drosha) in the miRNA biogenesis pathway in $CD4^+$ T cells leads to reduced survival and proliferation following activation [79–81]. DGCR8-deficient $CD4^+$ T cells also display aberrant T-cell activation and survival [82].

Several studies have identified the individual miRNAs that are involved in these basic T-cell activation processes such asTCR signaling; for example, miR-181 has been shown to play a key role in TCR signaling, T-cell development, and T-cell homeostasis through its targeting of phosphatases that dampen TCR signals, including SH2 domain-containing protein tyrosine phosphatase 2, protein tyrosine phosphatase, nonreceptor type 22, and dual-specificity protein phosphatases 5 and 6 [83]. Inhibition of miR-181 in $CD4^+$ T cells reduces TCR sensitivity and impairs both positive and negative selection in the thymus. Overexpression of miR-181 augments the sensitivity of $CD4^+$ T cells to peptide antigens in the periphery [84]. The phosphatidylinositol 3-kinase (PI3K) pathway downstream of TCR and CD28 signaling has also emerged as a key molecular regulator of T-cell proliferation and survival [85]. While PI3K signaling positively influences T-cell proliferation and survival, phosphatase and tensin homolog deleted on chromosome 10 (PTEN) and the proapoptotic protein, BIM, negatively regulate T-cell survival. Several miRNAs have been shown to alter the balance of PI3K and PTEN expression. One miRNA cluster that is strongly upregulated during T-cell activation is the miR-17-92 cluster, which is comprised of six miRNAs from four families. miR-17-92 promotes the proliferation of $CD4^+$ T cells upon TCR stimulation and inhibits activation-induced cell death. Two of the miR-17-92 cluster members, miR-17 and miR-92, enhance $CD4^+$ T-cell proliferation in the absence of other miRNA members, while another member of this cluster member, miR-19, positively regulates the survival of activated $CD4^+$ T cells [86]. These miRNAs mediate their effects by directly targeting several genes, including PTEN and BIM. MiR-214 also targets PTEN and promotes PI3K signaling and $CD4^+$ T-cell proliferation [87].

MiRNAs can also favor costimulation. Cytotoxic T-lymphocyte-associated protein 4 (CTLA4) serves as an inhibitory T-cell coreceptor due to its high affinity for CD80/CD86 [88]. CTLA4 competes with CD28 and blocks costimulatory signaling provided by CD28 cross-linking. miR-155 is strongly upregulated upon TCR stimulation and promotes CD28 signaling and $CD4^+$ T-cell proliferation by targeting CTLA4 [89]. Inducible T-cell costimulator (ICOS) is another costimulatory molecule that is expressed on the surface of T cells. It provides a survival signal to activated T cells by increasing the activation of PI3K [90]. MiR-101 has been shown to repress ICOS expression in association with Roquin (a ubiquitin ligase that limits ICOS expression by binding to the carboxy-terminal region of its 3′ UTR and promoting mRNA degradation) [91]. This carboxy-terminal region also has a binding site for miR-101, and disruption of this site reverses ICOS repression by Roquin, thus suggesting that Roquin-mediated repression of ICOS requires miR-101.

MiRNA-IL-2 signaling also plays a crucial role in T-cell proliferation and clonal expansion [74]. Several miRNAs affect T-cell growth by directly affecting IL-2 levels or by regulating molecules associated with IL-2 signaling pathways; for example, miR-181c (a member of the miR-181 family) directly targets IL-2 in human $CD4^+$ T cells and leads to a reduction of IL-2 levels and an inhibition of $CD4^+$ T-cell activation [92]. While miR-181c directly targets IL-2 mRNA, miR-184 and miR-31 regulate a transcription factor that mediates IL-2 transcription, NFAT [93,94]. Following activation by antigenic stimulation and costimulatory molecules, naive $CD4^+$ T cells switch from resting to clonal expansion. This switch requires inactivation of forkhead box O1 (FOXO1), a proliferation-suppressing transcription factor that is expressed in resting helper T cells [95]. MiR-182 enhances clonal T-cell expansion by inhibiting FOXO1 [96]. The miRNAs described above regulate the initial events in T-cell activation and survival; however, many miRNAs regulate the polarization of naive $CD4^+$ T cells towards specific effector-cell subsets [76–78].

4.2.2 MicroRNA Control of Inflammatory T Helper Responses

4.2.2.1 *Th1*

The first evidence of miRNA regulation of Th1 cells came from studies in Dicer-deficient mice. These mice exhibit increased differentiation toward effector cells expressing the Th1-specific transcription factor, T-bet, and the Th1 signature cytokine, IFN-γ [79,80]. $CD4^+$ T cells lacking Drosha or DGCR8 also display a similar phenotype of aberrant Th1 differentiation, thus suggesting that miRNA processing genes, and miRNAs in general, within $CD4^+$ T cells are important for Th1 development [81,82]. Some of these Th1 defects can now be attributed to specific miRNAs. Several miRNAs modulate Th1 differentiation by affecting the Th1-polarizing cytokine, IL-12, and its downstream signaling pathways; for example, a functional screen in DGCR8-deficient $CD4^+$ T cells identified miR-29a and miR-29b as the most potent inhibitors of Th1 differentiation [82]. miR-29 inhibits Th1 development by targeting the IL-12-induced Th1-specific transcription factor, T-bet, as well as Eomes, a transcription factor implicated in IFNγ synthesis. MiR-29 also directly targets IFN-γ mRNA, thereby further inhibiting Th1 development. In vivo inhibition of miR-29 isassociated with enhanced Th1 differentiation and increased resistance to intracellular bacterial pathogens (e.g., *Mycobacterium tuberculosis* and *Listeria monocytogenes*). MiR-29b levels decrease significantly during T-cell activation in MS patients, which suggests that miR-29 dysregulation may contribute to tissue inflammation and autoimmunity [97]. MiR-21 also inhibits Th1 differentiation by inhibiting the Th1-polarizing cytokine, IL-12, from macrophages and DCs [98,99]. MiR-21 specifically targets the p35 subunit of IL-12. Initial studies with miR-146a found that this miRNA also regulates Th1 differentiation, albeit indirectly through the control of T_{reg} cell responses [100]; however, recent research suggests that miR-146a can also directly regulate Th1 differentiation by targeting protein kinase C epsilon (PRKCε), which is part of a functional complex consisting of PRKCε and STAT4. PRKCε promotes STAT4 phosphorylation and Th1 differentiation [101].

In contrast to the miRNAs mentioned above, miR-155 positively regulate Th1 differentiation [102–104]. MiR-155 within DCs and macrophages targets suppressor of cytokine signaling 1 (SOCS1) and enhances IL-12 secretion to promote Th1 development. It also inhibits src homology 2 domain-containing inositol-5-phosphatase 1, a major negative regulator of IL-12 in DCs and macrophages [105]. Consistent with these observations, depletion of miR-155 leads to defective Th1 responses in vivo and miR-155-deficient mice are susceptible to *Salmonella typhimurium, M. tuberculosis*, and other bacterial and viral infections [106]. Defective Th1 responses in these mice also contribute to resistance to tissue inflammation and autoimmunity [102–104]. The miR-17-92 cluster is also important for Th1 differentiation [107,108]. MiR-17-92-deficient $CD4^+$ T cells produce less IFN-γ and express less T-bet in response to Th1-polarizing stimulation. Reconstituting miR-17-92-deficient $CD4^+$ T cells with individual miRNAs revealed that miR-19b was the only miRNA from this cluster that isessential for promoting IFN-γ production [108].

4.2.2.2 *Th2*

$CD4^+$ T cells lacking DGCR8 also present with aberrant Th2 differentiation, implicating an important role for miRNAs [82]. Dysregulated Th2 differentiation is also observed in $CD4^+$ T cells with reduced expression of AGO, the core effector protein of the miRNA silencing complex [109]. Functional screening in DGCR8-deficient $CD4^+$ T cells determined that miR-24 and miR-27 areamong the most potent Th2 inhibitors [110,111]. These studies also revealed that miR-24 and miR-27 areco-expressed from two genomic clusters (Mirc11 and Mirc22), although each functions independently to limit Th2 development. MiR-27 inhibits Th2 cells by targeting the Th2-specific transcription factor, GATA3, while miR-24 blocks cytokine production by these effector cells. Mice lacking these miRNAs in $CD4^+$ T cells exhibit Th2 bias and show enhanced allergic airway hypersensitivity. MiR-155 was one of the first miRNAs to be implicated in the regulation of Th2 differentiation [106,112]. This miRNA specifically suppresses c-Maf (a transactivator of IL-4) and inhibits Th2 differentiation. Thus the absence of miR-155 leads to excessive Th2 responses, and ageing miR-155-deficient mice have been shown to develop Th2-mediated spontaneous lung inflammation [106].

While miR-155, miR-24, and miR-27 inhibit Th2 differentiation, miR-126 promotes differentiation by targeting POU domain class 2 associating factor 1 (POU2F1). POU2F1 is an activator of the transcription factor, PU.1, which in turn inhibits the DNA-binding activity of the Th2 transcription factor, GATA3 [113]. MiR-126 levels correlate with Th2 cells in a mouse model of asthma, and the in vivo inhibition of miR-126 ameliorates Th2 responses and asthmatic phenotypes. Another miRNA implicated in Th2 development is miR-21. $CD4^+$ T-cell-intrinsic miR-21 promotes GATA3 expression and Th2 differentiation by directly targeting Sprouty1, a mitogen-activated protein kinase (MAPK) pathway inhibitor [114]. Overexpression of miR-21 promotes the differentiation of Th2 cells in vitro, and miR-21-deficient mice show reduced allergic inflammation in the lungs after allergen challenge [99]. MiR-21 has been shown to be one of the most highly upregulated miRNAs in murine models of asthma and human eosinophilic esophagitis [115].

4.2.2.3 *Th17*

MiR-326 was the first miRNA to be implicated in the differentiation and function of Th17 cells [78,116]. MiRNA profiling in $CD4^+$ T cells from patients with MS and mice with experimental autoimmune encephalomyelitis (EAE) revealed a connection between miR-326 and Th17 cells in central nervous system autoimmunity. The expression of Th17 miR-326 correlates with disease severity in both MS and EAE. MiR-326 ispreferentially increased in murineTh17 cells and regulates Th17 differentiation *in vitro* and *in vivo*. Overexpression of miR-326 increases Th17 differentiation and worsens EAE symptoms, while silencing miR-326 inhibits Th17 cells and ameliorates EAE. MiR-326 is now known to promote Th17 cells by targeting the transcription factor Ets1, a negative regulator of Th17 differentiation [116].

MiR-155, which promotes Th17 differentiation, is one of the most extensively studied immune-related miRNAs [78]. T-cell miR-155 expression is strongly upregulated in response to Th17-polarizing conditions, and miR-155 promotes the Th17 phenotype by targeting Ets1 [102,103,117]. MiR-155 also promotes Th17 development by regulating epigenetic mechanisms, specifically through suppression of the inhibitor Jarid2, a DNA-binding protein that recruits PRC2 to chromatin [118]. MiR-155 appears to promote responsiveness to the Th17-stabilizing cytokine, IL-23, to maintain the Th17 phenotype, as miR-155-deficient Th17 cells are hyporesponsive to IL-23 stimulation [117]. The promotion of Th17 cells by miR-155 has been implicated in multiple autoimmune diseases; for example, miR-155 was shown to be increased in brain lesions in MS [119]. MiR-155 is also elevated in T helper cells in EAE, and miR-155-deficient mice are resistant to EAE symptoms and central nervous system inflammation [102,103], which is associated with reduced Th17 activity. MiR-155 regulation of Th17 cells also appears to mediate arthritis. Defective Th17 cells in miR-155-deficient mice are thought to protect against collagen-induced arthritis [120,121], and synovial tissue and T cells in RA patients show elevated miR-155 levels [121]. Furthermore, impaired Th17 responses in miR-155-deficient $CD4^+$ T cells have also been observed in a mouse model of Th17-driven chronic colitis [122].

MiR-21, a well-known oncomiR, also modulates Th17 cells (reviewed in [81]). It is elevated in Th17 cells and promotes Th17 differentiation by inhibiting Smad7, a negative regulator of TGF-β signaling [123,124]. MiR-21-deficient mice present with a striking defect in Th17 differentiation, which is associated with protection from EAE symptoms. Elevated miR-21 expression levels have also been identified in peripheral blood mononuclear cells from MS patients [125]. Overexpression and silencing of miR-21 in a mouse model of myocarditis elevates and reduces Th17 frequency, respectively [126]. MiR-21-deficiency also confers protection and inhibits Th17 cells in a chronic graft-versus-host disease model of SLE, [127], as does miR-21 silencing using pharmacological therapies [128]. A major bioactive cannabinoid derived from the *Cannabis sativa* plant, Δ-9-tetrahydrocannabinol, has also been shown to suppress Th17-mediated delayed-type hypersensitivity through miR-21 regulation [129].

Several other miRNAs have been implicated in Th17 defects and Th17-mediated diseases, specifically EAE [78]. MiR-301a is enriched in CNS-infiltrating Th17 cells [130] where it promotes Th17 development by targeting protein inhibitor of activated STAT3, an inhibitor of the STAT3 pathway. Silencing miR-301a in myelin antigen-specific (2D2) $CD4^+$ T cells ameliorates EAE, while overexpressing miR-301a exacerbates EAE. MiR-132/212 cluster-deficient mice also present with defective Th17 differentiation and are resistant to EAE [131]. The miR-17-92 cluster has also been found to promote Th17 cells and the development of EAE by targeting PTEN and Ikaros family zinc finger 4. It induces inflammation via other targets as well, including the proapoptotic protein, Bim. Mice overexpressing miR-17-92 eventually develop fatal systemic autoimmunity [132].

In contrast to the numerous well-characterized miRNAs that promote Th17 development and Th17-mediated inflammation, relatively little is known about miRNAs that negatively regulate Th17 cells [78]. However, miR-23b has been shown to ameliorate multiple autoimmune diseases by suppressing Th17-mediated signaling pathways. More recent work established that miR-23b levels are downregulated in lesions associated with lupus and RA in both animal models and human patients

[133]. MiR-210 also inhibits IL-17 secretion from Th17 cells and prevents Th17-associated inflammation in a mouse model of IBD [134]. MiR-10a can inhibit Th17 differentiation [135]; however, this effect depends on the presence of retinoic acid, which induces both miR-10a and the Th1 transcription factor, T-bet. Suppression of miR-10a targets, B-cell lymphoma 6 (BCL-6) or nuclear receptor corepressor 2 (NCOR2), using RNA interference limits Th17 differentiation in wildtype but not in T-bet-deficient $CD4^+$ T cells. As BCL-6 and NCOR2 modulate the transcriptional activity of T-bet, and T-bet inhibits the expression of the Th17 transcription factor RORγt, miR-10a and T-bet may cooperate to balance Th1 and Th17 differentiation in retinoic acid-rich microenvironments. In line with these observations, miR-10a overexpression in myelin oligodendrocyte glycoprotein-specific $CD4^+$ T cells delays neurological disease in EAE.

4.2.3 MicroRNA Control of Regulatory T Helper Responses

4.2.3.1 T_{reg} Cells

MiRNAs regulate the development, homeostasis, and immunosuppressive functions of natural and induced T_{reg} cells. $CD4^+$ T cells without miRNAs fail to differentiate into T_{reg} cells in the thymus and periphery. Conditional deletion of the entire miRNA repertoire (via Dicer or Drosha knockout) within $Foxp3^+$ T_{reg} cells results in fatal autoimmunity that is indistinguishable from autoimmunity in Foxp3-deficient mice [136]. This suggests a critical role for miRNAs in the preservation of T_{reg} function, and several miRNAs are now known to specifically contribute to T_{reg} development and efficacy. The earliest example came from studies with miR-155 [78,137,138]. Expression of miR-155 is preferentially elevated in T_{regs}, where it is known to target SOCS-1 [138]. miR-155-targetting of SOCS-1 promotes IL-2-induced STAT5 signaling and differentiation of T_{reg} cells. Although miR-155 is important for T_{reg} cell differentiation, and miR-155-deficient T_{regs} present with increased SOCS1, the immunosuppressive abilities of miR-155-deficient T_{regs} remain intact [137].

MiR-146a is also naturally enriched in T_{reg} cells, where it promotes differentiation and has suppressive functions. It downregulates STAT1, a key transcription factor implicated in Th1 differentiation, and is essential for T_{reg} cells to prevent self-reactive pathogenic Th1 responses [139]. T_{reg} cells from patients with RA show a diminished upregulation of miR-146a in response to T-cell stimulation; decreased miR-146a expression is also seen in patients with active disease and correlates with joint inflammation [140]. MiR-10a also substantially impacts T_{reg} differentiation and function [78,141]. Expression levels of miR-10a are higher in natural T_{reg} cells, and can be upregulated by a combination of retinoic acid and TGF-β in inducible T_{reg} cells [141]. Blockade of either the retinoic acid- or the TGF-β-signaling pathway prevents miR-10a induction. MiR-10a limits the conversion of T_{reg} cells into inflammatory T cells by targeting the transcriptional repressor BCL6 and the corepressor NCOR2. MiR-10a levels in T_{reg} cells have therefore been inversely correlated with autoimmune susceptibility; for example, the genetic susceptibility for autoimmune diabetes in nonobese diabetic mice is linked to lower T_{reg}-specific miR-10a expression [142].

Although most of the miRNA-based regulation of T_{reg} cells primarily modulates their differentiation, some miRNAs have also been shown to affect their suppressive functions; for example, miR-125a can stabilize both the differentiation and immunosuppressive capacity of T_{reg} cells [143]. A genome-wide analysis revealed that miR-125a promotes T_{reg} differentiation by targeting negative regulators of this process (i.e., Stat3, IFN-γ, and IL-13). More recent studies also suggest that miR-125a-5p decreases the sensitivity of T_{reg} cells toward IL-6-mediated pathogenic conversion by inhibiting IL-6R and STAT3 expression, thus enhancing T_{reg} stability and suppressive functions [144]. MiR-125a deficiency results in severe colitis and heightened EAE pathogenesis, and a chemically synthesized miR-125a analogue has been shown to inhibit established EAE. Downregulation of miR-125a in humans is associated with enhanced inflammatory mediators in peripheral CD4$^+$ T cells in various autoimmune diseases, including SLE [145].

Several miRNAs have also been implicated in the negative regulation of T_{reg} cells, such as miR-31 [146]. MiR-31 is induced in CD4$^+$ T cells during TCR signaling, and conditional deletion of miR-31 in CD4$^+$ T cells promotes T_{reg} development. MiR-31 appears to control T_{reg} differentiation by targeting Gprc5a, a recently identified regulator of T_{reg} differentiation. MiR-31-deficient mice present with resistance to EAE, which is associated with increased T_{reg} frequency. Other miRNAs, including the miR-17-92 cluster, have been more closely linked with the inhibition of T_{reg} suppressive abilities [147,148]; however, the relative contributions of the individual miRNAs to the regulation of T helper development remain to be elucidated. MiR-17 has been shown to inhibit T_{reg}-mediated suppression by targeting the Foxp3 coregulators Eos, Satb1, and IRF4 [149]. Overexpression of miR-17, both in vitro and in vivo, diminishes T_{reg} suppression functions, an effect that is mitigated following coexpression with an Eos mutant lacking miR-17 target sites. MiR-27 has also been shown to impair T_{reg} development [150]. Mice with T-cell specific overexpression of miR-27 harbor dysregulated Th1 responses and develop an autoimmune pathology. The severe autoimmune phenotype in these mice is not driven by miR-27 in effector T cells, but from miR-27 in T_{reg} cells perturbing T_{reg} differentiation and suppressive functions.

4.3 CONCLUDING REMARKS

Research regarding the miRNA-mediated regulation of T helper cell differentiation and function has recently gained interest, as has the epigenetic regulation of adaptive immunity. The discovery of multidirectional cross talk between epigenetic mechanisms and miRNA targeting suggests that epigenetic–miRNA interactions may underlie T helper cell development. Many informative studies exploring epigenetic–miRNA interactions have been performed within the context of cancer [58–60]; however, very little work examining epigenetic–miRNA regulation has been performed in T cells [118]. The presence of powerful epigenetic and miRNA regulatory networks makes further investigations within T cells, and within T-cell-dependent cancer immunotherapeutic strategies, promising lines of research.

References

[1] Zhu J, Yamane H, Paul WE. Differentiation of effector CD4 T cell populations. Annu Rev Immunol 2010;28:445–89.
[2] Mosmann TR, Coffman RL. TH1 and TH2 cells: different patterns of lymphokine secretion lead to different functional properties. Annu Rev Immunol 1989;7:145–73.
[3] Trinchieri G. Interleukin-12: a proinflammatory cytokine with immunoregulatory functions that bridge innate resistance and antigen-specific adaptive immunity. Annu Rev Immunol 1995;13:251–76.
[4] Lazarevic V, Glimcher LH, Lord GM. T-bet: a bridge between innate and adaptive immunity. Nat Rev Immunol 2013;13(11):777–89.
[5] Nakayama T, Hirahara K, Onodera A, Endo Y, Hosokawa H, Shinoda K, et al. Th2 cells in health and disease. Annu Rev Immunol 2017;26:53–84. doi: 10.1146/annurev-immunol-051116-052350.
[6] Zheng W, Flavell RA. The transcription factor GATA-3 is necessary and sufficient for Th2 cytokine gene expression in CD4 T cells. Cell 1997;89:587–96.
[7] Zhang Y, Zhou X, Zhou B. DC-derived TSLP promotes Th2 polarization in LPS-primed allergic airway inflammation. Eur J Immunol 2012;42:1735–43.
[8] Wang YH, Angkasekwinai P, Lu N, Voo KS, Arima K, Hanabuchi S, et al. IL-25 augments type 2 immune responses by enhancing the expansion and functions of TSLP-DC-activated Th2 memory cells. J Exp Med 2007;204:1837–47.

[9] Kurowska-Stolarska M, Kewin P, Murphy G, Russo RC, Stolarski B, Garcia CC, et al. IL-33 induces antigen-specific IL-5+ T cells and promotes allergic-induced airway inflammation independent of IL-4. J Immunol 2008;181:4780–90.
[10] Korn T, Bettelli E, Oukka M, Kuchroo VK. IL-17 and Th17 cells. Annu Rev Immunol 2009;27:485–517.
[11] Mangan PR, Harrington LE, O'Quinn DB, Helms WS, Bullard DC, Elson CO, et al. Transforming growth factor-β induces development of the T(H)17 lineage. Nature 2006;441:231–4.
[12] Bettelli E, Carrier Y, Gao W, Korn T, Strom TB, Oukka M, et al. Reciprocal developmental pathways for the generation of pathogenic effector TH17 and regulatory T cells. Nature 2006;441:235–8.
[13] Ivanov II, McKenzie BS, Zhou L, Tadokoro CE, Lepelley A, Lafaille JJ, et al. The orphan nuclear receptor RORγt directs the differentiation program of proinflammatory IL-17+ T helper cells. Cell 2006;126:1121–33.
[14] Yang XO, Panopoulos AD, Nurieva R, Chang SH, Wang D, Watowich SS, et al. STAT3 regulates cytokine-mediated generation of inflammatory helper T cells. J Biol Chem 2007;282:9358–63.
[15] Yang XO, Pappu BP, Nurieva R, Akimzhanov A, Kang HS, Chung Y, et al. T helper 17 lineage differentiation is programmed by orphan nuclear receptors RORα and RORγ. Immunity 2008;28(1):29–39.
[16] Huber M, Brüstle A, Reinhard K, Guralnik A, Walter G, Mahiny A, et al. IRF4 is essential for IL-21-mediated induction, amplification, and stabilization of the Th17 phenotype. Proc Natl Acad Sci USA 2008;105(52):20846–51.
[17] Schraml BU, Hildner K, Ise W, Lee WL, Smith WA, Solomon B, et al. The AP-1 transcription factor Batf controls T(H)17 differentiation. Nature 2009;460:405–9.
[18] Dang EV, Barbi J, Yang HY, Jinasena D, Yu H, Zheng Y, et al. Control of T(H)17/T(reg) balance by hypoxia-inducible factor 1. Cell 2011;146(5):772–84.
[19] McGeachy MJ, Chen Y, Tato CM, Laurence A, Joyce-Shaikh B, Blumenschein WM, et al. The interleukin 23 receptor is essential for the terminal differentiation of interleukin 17-producing effector T helper cells in vivo. Nat Immunol 2009;10:314–24.
[20] Ghoreschi K, Laurence A, Yang XP, Tato CM, McGeachy MJ, Konkel JE, et al. Generation of pathogenic T(H)17 cells in the absence of TGF-β signalling. Nature 2010;467:967–71.
[21] El-Behi M, Ciric B, Dai H, Yan Y, Cullimore M, Safavi F, et al. The encephalitogenicity of T(H)17 cells is dependent on IL-1- and IL-23-induced production of the cytokine GM-CSF. Nat Immunol 2011;12:568–75.
[22] Wing K, Sakaguchi S. Regulatory T cells exert checks and balances on self tolerance and autoimmunity. Nat Immunol 2010;11:7–13.
[23] Hori S, Nomura T, Sakaguchi S. Control of regulatory T cell development by the transcription factor Foxp3. Science 2003;299:1057–61.
[24] Chen Q, Kim YC, Laurence A, Punkosdy GA, Shevach EM. IL-2 controls the stability of Foxp3 expression in TGF-β-induced Foxp3+ T cells in vivo. J Immunol 2011;186:6329–37.
[25] Davidson TS, DiPaolo RJ, Andersson J, Shevach EM. Cutting edge: IL-2 is essential for TGF-β-mediated induction of Foxp3+ T regulatory cells. J Immunol 2007;178:4022–6.
[26] Marie JC, Liggitt D, Rudensky AY. Cellular mechanisms of fatal early-onset autoimmunity in mice with the T cell-specific targeting of transforming growth factor-β receptor. Immunity 2006;25:441–54.
[27] Fontenot JD, Rasmussen JP, Gavin MA, Rudensky AY. A function for interleukin 2 in Foxp3-expressing regulatory T cells. Nat Immunol 2005;6:1142–51.
[28] Coombes JL, Siddiqui KR, Arancibia-Cárcamo CV, Hall J, Sun CM, Belkaid Y, et al. A functionally specialized population of mucosal CD103+ DCs induces Foxp3+ regulatory T cells via a TGF-β and retinoic acid-dependent mechanism. J Exp Med 2007;204:1757–64.
[29] Schiering C, Krausgruber T, Chomka A, Fröhlich A, Adelmann K, Wohlfert EA, et al. The alarmin IL-33 promotes regulatory T-cell function in the intestine. Nature 2014;513(7519):564–8.
[30] Usui T, Preiss JC, Kanno Y, Yao ZJ, Bream JH, O'Shea JJ, et al. T-bet regulates Th1 responses through essential effects on GATA-3 function rather than on *IFNG* gene acetylation and transcription. J Exp Med 2006;203(3):755–66.
[31] Lazarevic V, Chen X, Shim JH, Hwang ES, Jang E, Bolm AN, et al. T-bet represses T(H)17 differentiation by preventing Runx1-mediated activation of the gene encoding RORγt. Nat Immunol 2011;12:96–104.
[32] Usui T, Nishikomori R, Kitani A, Strober W. GATA-3 suppresses Th1 development by downregulation of Stat4 and not through effects on IL-12Rβ2 chain or T-bet. Immunity 2003;18(3):415–28.
[33] Zhu J, Yamane H, Cote-Sierra J, Guo L, Paul WE. GATA-3 promotes Th2 responses through three different mechanisms: induction of Th2 cytokine production, selective growth of Th2 cells and inhibition of Th1 cell-specific factors. Cell Res 2006;16:3–10.
[34] Zhou L, Lopes JE, Chong MM, Ivanov II, Min R, Victora GD, et al. TGF-β-induced Foxp3 inhibits T(H)17 cell differentiation by antagonizing RORγt function. Nature 2008;453:236–40.
[35] Berger SL. The complex language of chromatin regulation during transcription. Nature 2007;447:407–12.

[36] Tessarz P, Kouzarides T. Histone core modifications regulating nucleosome structure and dynamics. Nat Rev Mol Cell Biol 2014;15:703–8.

[37] Zentner GE, Henikoff S. Regulation of nucleosome dynamics by histone modifications. Nat Struct Mol Biol 2013;20:259–66.

[38] Smith ZD, Meissner A. DNA methylation: roles in mammalian development. Nat Rev Genet 2013;14:204–20.

[39] Kanno Y, Vahedi G, Hirahara K, Singleton K, O'Shea JJ. Transcriptional and epigenetic control of T helper cell specification: molecular mechanisms underlying commitment and plasticity. Annu Rev Immunol 2012;30:707–31.

[40] Wilson CB, Rowell E, Sekimata M. Epigenetic control of T-helper-cell differentiation. Nat Rev Immunol 2009;9:91–105.

[41] Ansel KM, Lee DU, Rao A. An epigenetic view of helper T cell differentiation. Nat Immunol 2003;4:616–23.

[42] Valapour M, Guo J, Schroeder JT, Keen J, Cianferoni A, Casolaro V, et al. Histone deacetylation inhibits IL4 gene expression in T cells. J Allergy Clin Immunol 2002;109:238–45.

[43] Akimova T, Beier UH, Liu Y, Wang L, Hancock WW. Histone/protein deacetylases and T-cell immune responses. Blood 2012;119:2443–51.

[44] Yamashita M, Hirahara K, Shinnakasu R, Hosokawa H, Norikane S, Kimura MY, et al. Crucial role of MLL for the maintenance of memory T helper type 2 cell responses. Immunity 2006;24:611–22.

[45] Lee PP, Fitzpatrick DR, Beard C, Jessup HK, Lehar S, Makar KW, et al. A critical role for Dnmt1 and DNA methylation in T cell development, function, and survival. Immunity 2001;15:763–74.

[46] Schoenborn JR, Dorschner MO, Sekimata M, Santer DM, Shnyreva M, Fitzpatrick DR, et al. Comprehensive epigenetic profiling identifies multiple distal regulatory elements directing transcription of the gene encoding interferon-γ. Nat Immunol 2007;8:732–42.

[47] Hatton RD, Harrington LE, Luther RJ, Wakefield T, Janowski KM, Oliver JR, et al. A distal conserved sequence element controls *Ifng* gene expression by T cells and NK cells. Immunity 2006;25:717–29.

[48] Wei G, Wei L, Zhu J, Zang C, Hu-Li J, Yao Z, et al. Global mapping of H3K4me3 and H3K27me3 reveals specificity and plasticity in lineage fate determination of differentiating $CD4^+$ T cells. Immunity 2009;30:155–67.

[49] Ansel KM, Djuretic I, Tanasa B, Rao A. Regulation of Th2 differentiation and Il4 locus accessibility. Annu Rev Immunol 2006;24:607–56.

[50] Floess S, Freyer J, Siewert C, Baron U, Olek S, Polansky J, et al. Epigenetic control of the *foxp3* locus in regulatory T cells. PLoS Biol 2007;5:e38.

[51] Janson PC, Winerdal ME, Marits P, Thörn M, Ohlsson R, Winqvist O. *FOXP3* promoter demethylation reveals the committed Treg population in humans. PLoS One 2008;3:e1612.

[52] Akimzhanov AM, Yang XO, Dong C. Chromatin remodeling of interleukin-17 (IL-17)-IL-17F cytokine gene locus during inflammatory helper T cell differentiation. J Biol Chem 2007;282:5969–72.

[53] Mukasa R, Balasubramani A, Lee YK, Whitley SK, Weaver BT, Shibata Y, et al. Epigenetic instability of cytokine and transcription factor gene loci underlies plasticity of the T helper 17 cell lineage. Immunity 2010;32:616–27.

[54] Shi G, Cox CA, Vistica BP, Tan C, Wawrousek EF, Gery I. Phenotype switching by inflammation-inducing polarized Th17 cells, but not by Th1 cells. J Immunol 2008;181:7205–13.

[55] Bending D, De la Peña H, Veldhoen M, Phillips JM, Uyttenhove C, Stockinger B, et al. Highly purified Th17 cells from BDC2.5NOD mice convert into Th1-like cells in NOD/SCID recipient mice. J Clin Invest 2009;119:565–72.

[56] Ambros V. The functions of animal microRNAs. Nature 2004;431:350–5.

[57] Ha M, Kim VN. Regulation of microRNA biogenesis. Nat Rev Mol Cell Biol 2014;15:509–24.

[58] Chuang JC, Jones PA. Epigenetics and microRNAs. Pediatr Res 2007;61:24R–9R.

[59] Fabbri M, Calin GA. Epigenetics and miRNAs in human cancer. Adv Genet 2010;70:87–99.

[60] Sato F, Tsuchiya S, Meltzer SJ, Shimizu K. MicroRNAs and epigenetics. FEBS J 2011;278:1598–609.

[61] Saito Y, Liang G, Egger G, Friedman JM, Chuang JC, Coetzee GA, et al. Specific activation of microRNA-127 with downregulation of the proto-oncogene BCL6 by chromatin-modifying drugs in human cancer cells. Cancer Cell 2006;9:435–43.

[62] Scott GK, Mattie MD, Berger CE, Benz SC, Benz CC. Rapid alteration of microRNA levels by histone deacetylase inhibition. Cancer Res 2006;66:1277–81.

[63] Asangani IA, Harms PW, Dodson L, Pandhi M, Kunju LP, Maher CA, et al. Genetic and epigenetic loss of microRNA-31 leads to feed-forward expression of EZH2 in melanoma. Oncotarget 2012;3:1011–25.

[64] Augoff K, McCue B, Plow EF, Sossey-Alaoui K. miR-31 and its host gene lncRNA LOC554202 are regulated by promoter hypermethylation in triple-negative breast cancer. Mol Cancer 2012;11:5.

[65] Gao W, Gu Y, Li Z, Cai H, Peng Q, Tu M, et al. miR-615-5p is epigenetically inactivated and functions as a tumor suppressor in pancreatic ductal adenocarcinoma. Oncogene 2015;34:1629–40.

[66] Chang S, Wang RH, Akagi K, Kim KA, Martin BK, Cavallone L, et al. Tumor suppressor BRCA1 epigenetically controls oncogenic microRNA-155. Nat Med 2011;17:1275–82.
[67] Fabbri M, Garzon R, Cimmino A, Liu Z, Zanesi N, Callegari E, et al. MicroRNA-29 family reverts aberrant methylation in lung cancer by targeting DNA methyltransferases 3A and 3B. Proc Natl Acad Sci USA 2007;104:15805–10.
[68] Noonan EJ, Place RF, Pookot D, Basak S, Whitson JM, Hirata H, et al. miR-449a targets HDAC-1 and induces growth arrest in prostate cancer. Oncogene 2009;28:1714–24.
[69] Chen JF, Mandel EM, Thomson JM, Wu Q, Callis TE, Hammond SM, et al. The role of microRNA-1 and microRNA-133 in skeletal muscle proliferation and differentiation. Nat Genet 2006;38(2):228–33.
[70] Steinman RM, Young JW. Signals arising from antigen-presenting cells. Curr Opin Immunol 1991;3:361–72.
[71] Germain RN, Stefanová I. The dynamics of T cell receptor signaling: complex orchestration and the key roles of tempo and cooperation. Annu Rev Immunol 1999;17:467–522.
[72] Quezada SA, Jarvinen LZ, Lind EF, Noelle RJ. CD40/CD154 interactions at the interface of tolerance and immunity. Annu Rev Immunol 2004;22:307–28.
[73] Murugaiyan G, Agrawal R, Mishra GC, Mitra D, Saha B. Functional dichotomy in CD40 reciprocally regulates effector T cell functions. J Immunol 2006;177(10):6642–9.
[74] Boyman O, Sprent J. The role of interleukin-2 during homeostasis and activation of the immune system. Nat Rev Immunol 2012;12:180–90.
[75] O'Connell RM, Rao DS, Chaudhuri AA, Baltimore D. Physiological and pathological roles for microRNAs in the immune system. Nat Rev Immunol 2010;10:111–22.
[76] Baumjohann D, Ansel KM. MicroRNA-mediated regulation of T helper cell differentiation and plasticity. Nat Rev Immunol 2013;13:666–78.
[77] Pagani M, Rossetti G, Panzeri I, de Candia P, Bonnal RJ, Rossi RL, et al. Role of microRNAs and long-non-coding RNAs in CD4(+) T-cell differentiation. Immunol Rev 2013;253:82–96.
[78] Garo LP, Murugaiyan G. Contribution of MicroRNAs to autoimmune diseases. Cell Mol Life Sci 2016;73(10): 2041–51.
[79] Cobb BS, Nesterova TB, Thompson E, Hertweck A, O'Connor E, Godwin J, et al. T cell lineage choice and differentiation in the absence of the RNase III enzyme Dicer. J Exp Med 2005;201(9):1367–73.
[80] Muljo SA, Ansel KM, Kanellopoulou C, Livingston DM, Rao A, Rajewsky K. Aberrant T cell differentiation in the absence of Dicer. J Exp Med 2005;202(2):261–9.
[81] Chong MM, Rasmussen JP, Rudensky AY, Littman DR. The RNAse III enzyme Drosha is critical in T cells for preventing lethal inflammatory disease. J Exp Med 2008;205(9):2005–17.
[82] Steiner DF, Thomas MF, Hu JK, Yang Z, Babiarz JE, Allen CD, et al. MicroRNA-29 regulates T-box transcription factors and interferon-γ production in helper T cells. Immunity 2011;35(2):169–81.
[83] Li QJ, Chau J, Ebert PJ, Sylvester G, Min H, Liu G, et al. miR-181a is an intrinsic modulator of T cell sensitivity and selection. Cell 2007;129:147–61.
[84] Ebert PJ, Jiang S, Xie J, Li QJ, Davis MM. An endogenous positively selecting peptide enhances mature T cell responses and becomes an autoantigen in the absence of microRNA miR-181a. Nat Immunol 2009;10:1162–9.
[85] Okkenhaug K, Turner M, Gold MR. PI3K signaling in B cell and T cell biology. Front Immunol 2014;5:557.
[86] Xiao C, Srinivasan L, Calado DP, Patterson HC, Zhang B, Wang J, et al. Lymphoproliferative disease and autoimmunity in mice with increased miR-17-92 expression in lymphocytes. Nat Immunol 2008;9:405–14.
[87] Jindra PT, Bagley J, Godwin JG, Iacomini J. Costimulation-dependent expression of microRNA-214 increases the ability of T cells to proliferate by targeting *Pten*. J Immunol 2010;185:990–7.
[88] Krummel MF, Allison JP. CD28 and CTLA-4 have opposing effects on the response of T cells to stimulation. J Exp Med 1995;182(2):459–65.
[89] Sonkoly E, Janson P, Majuri ML, Savinko T, Fyhrquist N, Eidsmo L, et al. miR-155 is overexpressed in patients with atopic dermatitis and modulates T-cell proliferative responses by targeting cytotoxic T lymphocyte-associated antigen 4. J Allergy Clin Immunol 2010;126. 581-9.e1-20.
[90] Simpson TR, Quezada SA, Allison JP. Regulation of CD4 T cell activation and effector function by inducible costimulator (ICOS). Curr Opin Immunol 2010;22:326–32.
[91] Yu D, Tan AH, Hu X, Athanasopoulos V, Simpson N, Silva DG, et al. Roquin represses autoimmunity by limiting inducible T-cell co-stimulator messenger RNA. Nature 2007;450:299–303.
[92] Xue Q, Guo ZY, Li W, Wen WH, Meng YL, Jia LT, et al. Human activated CD4(+) T lymphocytes increase IL-2 expression by downregulating microRNA-181c. Mol Immunol 2011;48:592–9.
[93] Weitzel RP, Lesniewski ML, Haviernik P, Kadereit S, Leahy P, Greco NJ, et al. MicroRNA 184 regulates expression of NFAT1 in umbilical cord blood $CD4^+$ T cells. Blood 2009;113(26):6648–57.

[94] Xue F, Li H, Zhang J, Lu J, Xia Y, Xia Q. miR-31 regulates interleukin 2 and kinase suppressor of ras 2 during T cell activation. Genes Immun 2013;14:127–31.

[95] Hedrick SM, Hess Michelini R, Doedens AL, Goldrath AW, Stone EL. FOXO transcription factors throughout T cell biology. Nat Rev Immunol 2012;12(9):649–61.

[96] Stittrich AB, Haftmann C, Sgouroudis E, Kühl AA, Hegazy AN, Panse I, et al. The microRNA miR-182 is induced by IL-2 and promotes clonal expansion of activated helper T lymphocytes. Nat Immunol 2010;11:1057–62.

[97] Smith KM, Guerau-de-Arellano M, Costinean S, Williams JL, Bottoni A, Mavrikis Cox G, et al. miR-29ab1 deficiency identifies a negative feedback loop controlling Th1 bias that is dysregulated in multiple sclerosis. J Immunol 2012;189:1567–76.

[98] Lu TX, Hartner J, Lim EJ, Fabry V, Mingler MK, Cole ET, et al. MicroRNA-21 limits in vivo immune response-mediated activation of the IL-12/IFN-γ pathway, Th1 polarization, and the severity of delayed-type hypersensitivity. J Immunol 2011;187(6):3362–73.

[99] Lu TX, Munitz A, Rothenberg ME. MicroRNA-21 is up-regulated in allergic airway inflammation and regulates IL-12p35 expression. J Immunol 2009;182(8):4994–5002.

[100] Lu LF, Boldin MP, Chaudhry A, Lin LL, Taganov KD, Hanada T, et al. Function of miR-146a in controlling Treg cell-mediated regulation of Th1 responses. Cell 2010;142(6):914–29.

[101] Möhnle P, Schütz SV, van der Heide V, Hübner M, Luchting B, Sedlbauer J, et al. MicroRNA-146a controls Th1-cell differentiation of human $CD4^+$ T lymphocytes by targeting PRKCε. Eur J Immunol 2015;45:260–72.

[102] O'Connell RM, Kahn D, Gibson WS, Round JL, Scholz RL, Chaudhuri AA, et al. MicroRNA-155 promotes autoimmune inflammation by enhancing inflammatory T cell development. Immunity 2010;33(4):607–19.

[103] Murugaiyan G, Beynon V, Mittal A, Joller N, Weiner HL. Silencing microRNA-155 ameliorates experimental autoimmune encephalomyelitis. J Immunol 2011;187(5):2213–21.

[104] Zhang A, Wang K, Zhou C, Gan Z, Ma D, Ye P, Sun Y, Wu J, Huang X, Ren L, Deng P, Wu C, Yue Z, Ding X, Chen J, Xia J. Knockout of microRNA-155 ameliorates the Th1/Th17 immune response and tissue injury in chronic rejection. J Heart Lung Transplant. 2017; 36; 175-84.

[105] O'Connell RM, Chaudhuri AA, Rao DS, Baltimore D. Inositol phosphatase SHIP1 is a primary target of miR-155. Proc Natl Acad Sci USA 2009;106(17):7113–8.

[106] Rodriguez A, Vigorito E, Clare S, Warren MV, Couttet P, Soond DR, et al. Requirement of bic/microRNA-155 for normal immune function. Science 2007;316(5824):608–11.

[107] Wu T, Wieland A, Lee J, Hale JS, Han JH, Xu X, et al. Cutting edge: miR-17-92 is required for both CD4 Th1 and T follicular helper cell responses during viral infection. J Immunol 2015;195(6):2515–9.

[108] Jiang S, Li C, Olive V, Lykken E, Feng F, Sevilla J, et al. Molecular dissection of the miR-17-92 cluster's critical dual roles in promoting Th1 responses and preventing inducible Treg differentiation. Blood 2011;118(20):5487–97.

[109] Bronevetsky Y, Villarino AV, Eisley CJ, Barbeau R, Barczak AJ, Heinz GA, et al. T cell activation induces proteasomal degradation of Argonaute and rapid remodeling of the microRNA repertoire. J Exp Med 2013;210:417–32.

[110] Pua HH, Steiner DF, Patel S, Gonzalez JR, Ortiz-Carpena JF, Kageyama R, et al. MicroRNAs 24 and 27 suppress allergic inflammation and target a network of regulators of T helper 2 cell-associated cytokine production. Immunity 2016;44:821–32.

[111] Cho S, Wu CJ, Yasuda T, Cruz LO, Khan AA, Lin LL, et al. miR-23∼27∼24 clusters control effector T cell differentiation and function. J Exp Med 2016;213(2):235–49.

[112] Malmhäll C, Alawieh S, Lu Y, Sjöstrand M, Bossios A, Eldh M, et al. MicroRNA-155 is essential for T(H)2-mediated allergen-induced eosinophilic inflammation in the lung. J Allergy Clin Immunol 2014;133(5). 1429-38, 1438.e1-7.

[113] Mattes J, Collison A, Plank M, Phipps S, Foster PS. Antagonism of microRNA-126 suppresses the effector function of TH2 cells and the development of allergic airways disease. Proc Natl Acad Sci USA 2009;106(44):18704–9.

[114] Sawant DV, Wu H, Kaplan MH, Dent AL. The Bcl6 target gene microRNA-21 promotes Th2 differentiation by a T cell intrinsic pathway. Mol Immunol 2013;54(3–4):435–42.

[115] Chen RF, Huang HC, Ou CY, Hsu TY, Chuang H, Chang JC, et al. MicroRNA-21 expression in neonatal blood associated with antenatal immunoglobulin E production and development of allergic rhinitis. Clin Exp Allergy 2010;40:1482–90.

[116] Du C, Liu C, Kang J, Zhao G, Ye Z, Huang S, et al. MicroRNA miR-326 regulates TH-17 differentiation and is associated with the pathogenesis of multiple sclerosis. Nat Immunol 2009;10:1252–9.

[117] Hu R, Huffaker TB, Kagele DA, Runtsch MC, Bake E, Chaudhuri AA, et al. MicroRNA-155 confers encephalogenic potential to Th17 cells by promoting effector gene expression. J Immunol 2013;190(12):5972–80.

[118] Escobar TM, Kanellopoulou C, Kugler DG, Kilaru G, Nguyen CK, Nagarajan V, et al. miR-155 activates cytokine gene expression in Th17 cells by regulating the DNA-binding protein Jarid2 to relieve polycomb-mediated repression. Immunity 2014;40:865–79.

[119] Junker A, Krumbholz M, Eisele S, Mohan H, Augstein F, Bittner R, et al. MicroRNA profiling of multiple sclerosis lesions identifies modulators of the regulatory protein CD47. Brain 2009;132:3342–52.
[120] Blüml S, Bonelli M, Niederreiter B, Puchner A, Mayr G, Hayer S, et al. Essential role of microRNA-155 in the pathogenesis of autoimmune arthritis in mice. Arthritis Rheum 2011;63:1281–8.
[121] Kurowska-Stolarska M, Alivernini S, Ballantine LE, Asquith DL, Millar NL, Gilchrist DS, et al. MicroRNA-155 as a proinflammatory regulator in clinical and experimental arthritis. Proc Natl Acad Sci USA 2011;108:11193–8.
[122] Singh UP, Murphy AE, Enos RT, Shamran HA, Singh NP, Guan H, et al. miR-155 deficiency protects mice from experimental colitis by reducing T helper type 1/type 17 responses. Immunology 2014;143(3):478–89.
[123] Murugaiyan G, da Cunha AP, Ajay AK, Joller N, Garo LP, Kumaradevan S, et al. MicroRNA-21 promotes Th17 differentiation and mediates experimental autoimmune encephalomyelitis. J Clin Invest 2015;125:1069–80.
[124] Murugaiyan G, Garo LP, Weiner HL. MicroRNA-21, T helper lineage and autoimmunity. Oncotarget 2015;6:9644–5.
[125] Fenoglio C, Cantoni C, De Riz M, Ridolfi E, Cortini F, Serpente M, et al. Expression and genetic analysis of miRNAs involved in $CD4^+$ cell activation in patients with multiple sclerosis. Neurosci Lett 2011;504:9–12.
[126] Liu YL, Wu W, Xue Y, Gao M, Yan Y, Kong Q, et al. MicroRNA-21 and -146b are involved in the pathogenesis of murine viral myocarditis by regulating TH-17 differentiation. Arch Virol 2013;158:1953–63.
[127] Garchow B, Kiriakidou M. MicroRNA-21 deficiency protects from lupus-like autoimmunity in the chronic graft-versus-host disease model of systemic lupus erythematosus. Clin Immunol 2016;162:100–6.
[128] Garchow BG, Bartulos Encinas O, Leung YT, Tsao PY, Eisenberg RA, Caricchio R, et al. Silencing of microRNA-21 in vivo ameliorates autoimmune splenomegaly in lupus mice. EMBO Mol Med 2011;3:605–15.
[129] Sido JM, Jackson AR, Nagarkatti PS, Nagarkatti M. Marijuana-derived Δ-9-tetrahydrocannabinol suppresses Th1/Th17 cell-mediated delayed-type hypersensitivity through microRNA regulation. J Mol Med (Berl) 2016;94(9):1039–51. doi: 10.1007/s00109-016-1404-5.
[130] Mycko MP, Cichalewska M, Machlanska A, Cwiklinska H, Mariasiewicz M, Selmaj KW. MicroRNA-301a regulation of a T-helper 17 immune response controls autoimmune demyelination. Proc Natl Acad Sci USA 2012;109:E1248–57.
[131] Nakahama T, Hanieh H, Nguyen NT, Chinen I, Ripley B, Millrine D, et al. Aryl hydrocarbon receptor-mediated induction of the microRNA-132/212 cluster promotes interleukin-17-producing T-helper cell differentiation. Proc Natl Acad Sci USA 2013;110:11964–9.
[132] Liu SQ, Jiang S, Li C, Zhang B, Li QJ. miR-17-92 cluster targets phosphatase and tensin homology and Ikaros family zinc finger 4 to promote TH17-mediated inflammation. J Biol Chem 2014;289:12446–56.
[133] Zhu S, Pan W, Song X, Liu Y, Shao X, Tang Y, et al. The microRNA miR-23b suppresses IL-17-associated autoimmune inflammation by targeting TAB2, TAB3 and IKK-α. Nat Med 2012;18:1077–86.
[134] Wang H, Flach H, Onizawa M, Wei L, McManus MT, Weiss A. Negative regulation of *Hif1a* expression and TH17 differentiation by the hypoxia-regulated microRNA miR-210. Nat Immunol 2014;15:393–401.
[135] Takahashi H, Kanno T, Nakayamada S, Hirahara K, Sciumè G, Muljo SA, et al. TGF-β and retinoic acid induce the microRNA miR-10a, which targets Bcl-6 and constrains the plasticity of helper T cells. Nat Immunol 2012;13(6):587–95.
[136] Liston A, Lu LF, O'Carroll D, Tarakhovsky A, Rudensky AY. Dicer-dependent microRNA pathway safeguards regulatory T cell function. J Exp Med 2008;205:1993–2004.
[137] Kohlhaas S, Garden OA, Scudamore C, Turner M, Okkenhaug K, Vigorito E. Cutting edge: the Foxp3 target miR-155 contributes to the development of regulatory T cells. J Immunol 2009;182:2578–82.
[138] Lu LF, Thai TH, Calado DP, Chaudhry A, Kubo M, Tanaka K, et al. Foxp3-dependent microRNA155 confers competitive fitness to regulatory T cells by targeting SOCS1 protein. Immunity 2009;30:80–91.
[139] Lu LF, Boldin MP, Chaudhry A, Lin LL, Taganov KD, Hanada T, et al. Function of miR-146a in controlling Treg cell-mediated regulation of Th1 responses. Cell 2010;142:914–29.
[140] Zhou Q, Haupt S, Kreuzer JT, Hammitzsch A, Proft F, Neumann C, et al. Decreased expression of miR-146a and miR-155 contributes to an abnormal Treg phenotype in patients with rheumatoid arthritis. Ann Rheum Dis 2015;74:1265–74.
[141] Takahashi H, Kanno T, Nakayamada S, Hirahara K, Sciumè G, Muljo SA, et al. TGF-β and retinoic acid induce the microRNA miR-10a, which targets Bcl-6 and constrains the plasticity of helper T cells. Nat Immunol 2012;13:587–95.
[142] Jeker LT, Zhou X, Gershberg K, de Kouchkovsky D, Morar MM, Stadthagen G, et al. MicroRNA 10a marks regulatory T cells. PLoS One 2012;7:e36684.
[143] Pan W, Zhu S, Dai D, Liu Z, Li D, Li B, et al. miR-125a targets effector programs to stabilize Treg-mediated immune homeostasis. Nat Commun 2015;6:7096.
[144] Li D, Kong C, Tsun A, Chen C, Song H, Shi G, et al. miR-125a-5p decreases the sensitivity of Treg cells toward IL-6-mediated conversion by inhibiting IL-6R and STAT3 expression. Sci Rep 2015;5:14615.
[145] Zhao X, Tang Y, Qu B, Cui H, Wang S, Wang L, et al. MicroRNA-125a contributes to elevated inflammatory chemokine RANTES levels via targeting KLF13 in systemic lupus erythematosus. Arthritis Rheum 2010;62:3425–35.

[146] Zhang L, Ke F, Liu Z, Bai J, Liu J, Yan S, et al. MicroRNA-31 negatively regulates peripherally derived regulatory T-cell generation by repressing retinoic acid-inducible protein 3. Nat Commun 2015;6:7639.
[147] Jiang S, Li C, Olive V, Lykken E, Feng F, Sevilla J, et al. Molecular dissection of the miR-17-92 cluster's critical dual roles in promoting Th1 responses and preventing inducible Treg differentiation. Blood 2011;118:5487–97.
[148] de Kouchkovsky D, Esensten JH, Rosenthal WL, Morar MM, Bluestone JA, Jeker LT. MicroRNA-17-92 regulates IL-10 production by regulatory T cells and control of experimental autoimmune encephalomyelitis. J Immunol 2013;191:1594–605.
[149] Yang HY, Barbi J, Wu CY, Zheng Y, Vignali PD, Wu X, et al. MicroRNA-17 modulates regulatory T cell function by targeting co-regulators of the Foxp3 transcription factor. Immunity 2016;45:83–93.
[150] Cruz LO, Hashemifar SS, Wu CJ, Cho S, Nguyen DT, Lin LL, et al. Excessive expression of miR-27 impairs Treg-mediated immunological tolerance. J Clin Invest 2017;127:530–42.

Further Readings

[151] Chen W, Jin W, Hardegen N, Lei KJ, Li L, Marinos N, et al. Conversion of peripheral CD4+CD25- naive T cells to CD4+CD25+ regulatory T cells by TGF-β induction of transcription factor Foxp3. J Exp Med 2003;198:1875–86.
[152] Kimura M, Koseki Y, Yamashita M, Watanabe N, Shimizu C, Katsumoto T, et al. Regulation of Th2 cell differentiation by mel-18, a mammalian polycomb group gene. Immunity 2001;15:275–87.
[153] Hirahara K, Vahedi G, Ghoreschi K, Yang XP, Nakayamada S, Kanno Y, et al. Helper T-cell differentiation and plasticity: insights from epigenetics. Immunology 2011;134:235–45.
[154] Cobb BS, Hertweck A, Smith J, O'Connor E, Graf D, Cook T, et al. A role for Dicer in immune regulation. J Exp Med 2006;203:2519–27.

CHAPTER

5

Long Noncoding RNAs in the Immune Response

Sergio Covarrubias, Susan Carpenter

University of California, Santa Cruz, CA, United States

OUTLINE

The Epigenetics of Autoimmunity. http://dx.doi.org/10.1016/B978-0-12-809912-4.00005-2

5.1 INTRODUCTION

One of the most profound discoveries from the sequencing of the human genome is that over 85% of the genome is transcribed, yet less than 2% encodes protein-coding genes [1]. Large consortiums such as ENCODE and FANTOM have embarked on characterizing all functional coding and noncoding elements in the genome and have compiled important regulatory data for these elements [2–4]. Long noncoding RNAs (lncRNAs) represent the largest group of RNAs produced from the genome. LncRNAs are defined as transcripts greater than 200 nucleotides in length that do not have a protein-coding capacity. In the most recent GENCODE 26 release, 15,787 annotated lncRNAs were documented in the human genome (https://www.gencodegenes.org/). However, only 1% of annotated lncRNAs have ascribed functions, hence this class of RNA is in need of further investigation [4]. Evidence from those that have been characterized indicates that lncRNAs are able to function through a variety of mechanisms to regulate gene expression, both at the transcriptional and posttranscriptional levels [5].

Comparative genomics studies noted that the number of lncRNAs in the genome has increased with evolutionary complexity [6], with the greatest increase of lncRNAs correlating with the evolution of multicellularity [7]. Although yeast and human genomes contain similar numbers of protein-coding genes [8], the genome of single-cellular yeast contains ~25% noncoding DNA while the human genome contains ~97% noncoding sequence [6]. Increased biological complexity likely involves increased levels of regulation (provided by lncRNAs) rather than an increase in the number of coding genes [8]. LncRNAs show more cell-type specific expression compared to coding genes, which may suggest that they play a role in driving lineage-specific gene programs [4,9]. In addition, gene regulation by alternative splicing, enhancers, and posttranscriptional mechanisms are also major contributors of increased levels of overall regulation.

LncRNAs are currently classified from a "coding-gene-centric" point of view and are often named based on their nearest coding gene. The different types of lncRNAs include pseudogenes, antisense, bidirectional, enhancer-associated, and intergenic (lincRNAs). The specific type of lncRNA can give significant information on its mechanism of action; for example, lncRNAs near coding genes often coregulate the transcription of that gene. LncRNAs can also mediate their functions through interactions with proteins, RNA, DNA, or a combination of these. Furthermore, the function of lncRNAs is often dictated by their localization, sequence, or secondary structure. Some lncRNAs have also been demonstrated to encode small peptides, which adds another layer of complexity to lncRNA biology.

While the role of coding genes in immune-cell function has been well characterized, the role of lncRNAs in these processes is just beginning to emerge [10]. Hematopoiesis gives rise to both the innate and adaptive branches of the immune system, with each branch consisting of unique cell lineages with specific immune-related functions [10]. The innate immune system functions as a rapid general response against specific pathogens, while also promoting the activation and development of the adaptive immune system. The adaptive immune system is a slower response and takes days to develop; however, it allows for specific antigen-dependent recognition and elimination of pathogens. Macrophages and dendritic cells (DCs) are important innate immune cells that initiate the immune response through recognition of specific pathogen-associated molecular patterns (PAMPs) using their germline-encoded pattern recognition receptors. These receptors couple pathogen sensing to activation of downstream signaling cascades to produce upregulation of numerous inflammatory pathways [11]. The inflammatory program functions to activate macrophages and DCs, which endows them with the ability to activate specific T cells and initiate the adaptive immune response. In this way, coordination between innate and adaptive branches leads to amplification of the immune response and allows specific targeting and elimination of the pathogen. While a robust immune response is crucial for eliminating pathogens, prolonged activation can be detrimental to the surrounding tissue [12]. Unsurprisingly, many aspects of the immune response are tightly regulated at both the transcriptional and posttranscriptional level, thus permitting a transient antimicrobial response while subsequently promoting a return to homeostasis [13]. Perturbation in this regulation can have significant consequences that can manifest in diseases such as arthritis [14], multiple sclerosis (MS) [15], and cancer [12,16].

Before the discovery of microRNAs, the paradigm was that protein-coding genes were the key drivers of all processes within molecular biology. However, a more complete picture of a human genome is now beginning to emerge, which is one that is composed of thousands of lncRNAs, microRNAs, enhancers, repetitive elements, and proteins. LncRNAs represent the largest group of RNAs uncovered by large-scale transcriptome analysis [4]. It is now evident that lncRNAs are critical regulators of the immune response and that their dysregulation may be implicated in various inflammation-related autoimmune diseases. This chapter aims to summarize lncRNA-mediated gene regulation of the immune system and describe the roles of lncRNAs in autoimmune diseases.

5.2 ROLE OF LncRNAs IN INNATE IMMUNITY

5.2.1 Toll-like Receptor Signaling and Downstream Pathways

Toll-like receptors (TLRs) are type 1 transmembrane proteins that possess a ligand-binding leucine-rich domain and a cytosolic "adaptor domain" called the Toll–interleukin (IL)-1 receptor [17]. There are 10 different TLRs in the human genome and 13 in mice, each binding a different PAMP [17]. TLR4 was the first TLR to be discovered and has been the most extensively studied: it recognizes lipopolysaccharides found in the outer membrane of Gram-negative bacteria [11].

TLRs play a critical role in the innate immune response where they recognize specific PAMPs and initiate a complex signaling cascade that activates both inflammatory- and interferon (IFN)-related pathways [18]. The proinflammatory gene-activation branch of the signaling pathway involves the transcription factor (TF), nuclear factor kappa-B (NFκB), which activates genes including *IL-6* and tumor necrosis factor alpha (*TNFα*). The IFN branch of the pathway requires IFN-response factors 3 and 7 (IRF3 and 7), which activate a series of IFN-response genes needed for the antiviral response. These TFs function in concert with others to regulate the coordinated activation of specific genes [19].

5.2.2 LncRNA-Mediated Regulation of Inflammatory Pathways

RNA-sequencing (RNA-seq) studies on human and murine macrophages post-TLR stimulation have identified hundreds of differentially expressed lncRNAs. Our group identified the first example of a TLR-stimulated lncRNA, lincRNA-Cox2, which was capable of positively and negatively regulating distinct types of innate immune genes [20]. Knockdown of lincRNA-Cox2 resulted in impaired production of proinflammatory genes (i.e., *IL-6*), while IFN-related genes were hyperactivated in the absence of lincRNA-Cox2 [20]. While the inhibitory functions of lincRNA-Cox2 on IFN genes are thought to involve interactions with heterogeneous nuclear ribonucleoprotein (hnRNP)-A/B and A2/B1, the activation of lincRNA-Cox2 appears to involve interactions with the SWItch/sucrose nonfermentable (SWI/SNF) chromatin remodeling complex [21] (Fig. 5.1).

Li et al. stimulated a human monocytic cell line (THP1) with a TLR2 ligand (Pam3CSK4) and identified 156 differentially expressed lncRNAs [22]. From this study, they characterized the first

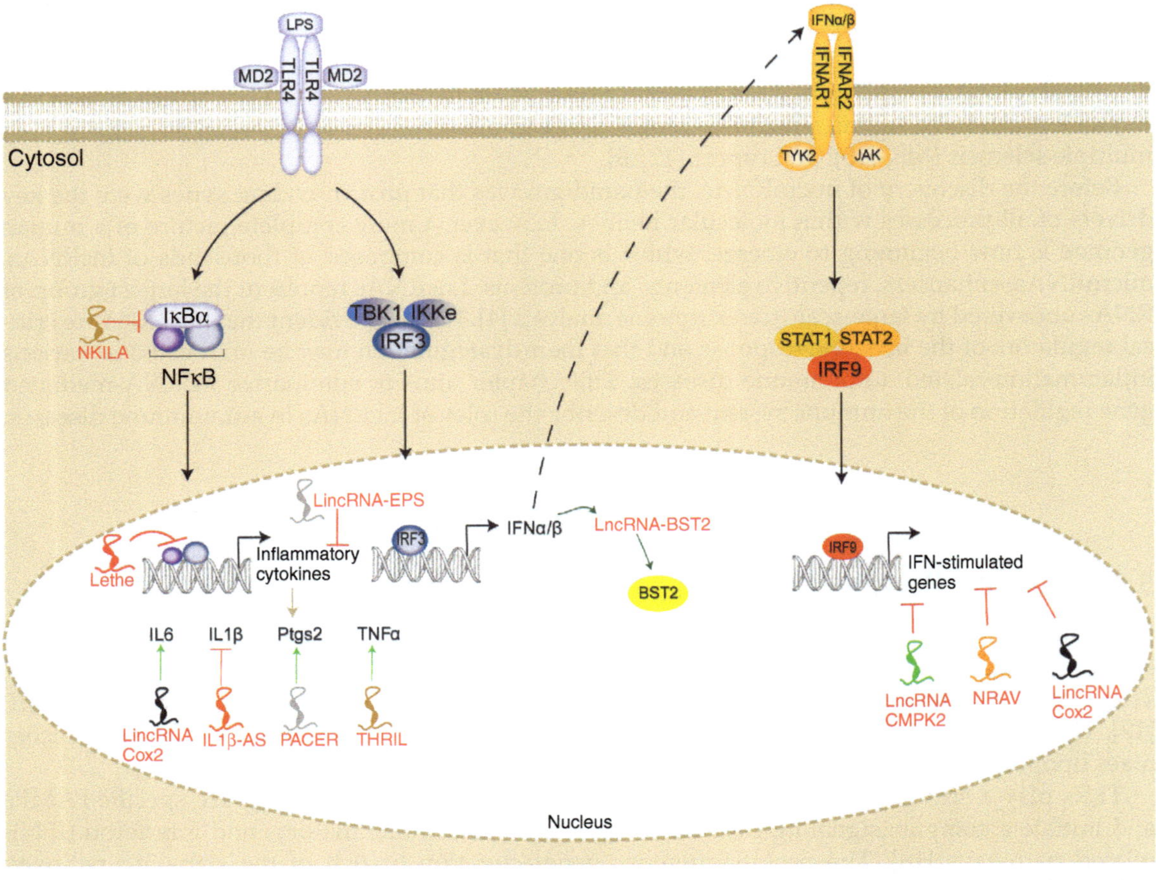

FIGURE 5.1 **LncRNAs that regulate innate immune signaling pathways.** Lipopolysaccharide (LPS) stimulation of the toll-like receptor 4 (TLR4) induces activation of the transcription factors, NFκB and IRF3, resulting in the production of proinflammatory cytokines and type I interferons (IFNα and β). IFNα and β subsequently induce the transcription factor IRF9 downstream of the IFN receptor, resulting in activation of IFN-stimulated genes. The lncRNAs involved in promoting (*green arrows*) or inhibiting (*red bars*) the production of proinflammatory cytokines or IFN-stimulated genes are highlighted. LncRNA names (*red*) and related coding genes (*black*) are also shown.

human TLR-activated lncRNA, TNFα and hnRNPL related immunoregulatory lincRNA (THRIL) [22]. THRIL regulates TNFα expression through interactions with hnRNPL. RNA-seq analysis on THRIL-depleted THP1 cells revealed predominantly (98%) decreased expression of hundreds of immune genes, suggesting that THRIL likely functions as a positive regulator of these genes [22]; however, the exact mechanism of action remains unclear.

NFκB is an important TF downstream of TLR-signaling, which has proven to be a major point of regulation by lncRNAs. NFκB is part of a family of TFs that includes RelA (p65), RelB, c-Rel, NFKB1 (p105/p50), and NFKB2 (p100/52) [23]. Inducible activation of NFκB depends upon proteosomal degradation of the inhibitor of NFκB (IκB) proteins, which retain inactive NFκB dimers in the cytosol. Following TLR stimulation, IκBs are phosphorylated and subsequently degraded, thus allowing nuclear relocalization of NFκB [23]. Given its importance in activating inflammatory-response genes, it is not surprising that lncRNAs have been identified that regulate NFκB function. The NFκB-interacting lncRNA (NKILA) acts as a direct inhibitor of NFκB signaling by binding to the NFκB–IκB complex and inhibiting phosphorylation and subsequent degradation of IκB [24]. In this way, NKILA functions to broadly inhibit NFκB function (Fig. 5.1).

Rapicavoli et al. identified Lethe, a nuclear-localized lncRNA that interacts with NFκB (RelA) and prevents it from binding to promoter regions of target genes [25] (Fig. 5.1). Lethe represents a class of lncRNAs called pseudogenes, which have lost their protein-coding ability and have been repurposed by the cell for other functions [26]. It should be noted that most pseudogenes do not have ascribed functions, and therefore Lethe could be the exception to the rule. Lethe is induced by TNFα stimulation, and TNFα itself is induced by NFκB. Hence Lethe completes a negative feedback loop, which ultimately results in decreased TNFα [25]. Interestingly, Lethe expression levels decrease dramatically with age, which is likely due to aberrant NFκB activity.

Regulation of NFκB by lncRNAs can also occur in a gene-specific manner. The p50-associated COX-2 antisense lncRNA (PACER) is produced from a bidirectional promoter that transcribes PACER [27] alongside the protein coding gene, Ptgs2 (Cox2), a gene with well-established roles in inflammation and immune responses (reviewed in Ref. [28]). One of the functions of PACER is to promote the expression of Cox2 by sequestering p50 (the repressive subunit of the NFκB complex) and preventing it binding to the promoter of Ptgs2 (Fig. 5.1). PACER also functions to recruit p300 and RNA polymerase II complexes to promote the induction of Cox2. PACERs interaction with p50 permits access to the activating subunits of NFκB (p50-p65) at the promoter of Cox2 following inflammatory stimulation [27].

Chronic inflammation is a major contributor to the development of autoimmune diseases [29]. In rheumatoid arthritis (RA), the inflammatory cytokine, TNFα, has established roles in the disease pathogenesis [30]. Given that lncRNAs have been shown to be integral regulators of the inflammatory response, it will not be surprising if they prove to be important contributors to the long-term development of autoimmunity.

5.2.3 LncRNAs and IFN Signaling

The IFN pathway is an ancient evolutionary branch of the innate immune response that is crucial for antiviral defense. Every human virus that causes disease is known to possess at least one mechanism for inhibiting IFN activation, which highlights how important the IFN response is for limiting viral replication [31]. The IFN pathway is divided into type I (IFNα and β), type II (IFNγ), and type III, with type I being the main pathway involved in the innate immune response [32]. The type II IFN pathway will be described in the context of the adaptive immune system. The IFN III pathway is less well characterized but appears to be most important in mucosal epithelial cells during viral infection [33]. The IFN α/β response is initiated following the recognition of nucleic acid PAMPs by TLRs (TLR3, 4, 7, 8, or 9) or by cytoplasmic localized sensors [i.e., the cyclic GMP-AMP synthase (cGAS), retinoic acid-inducible gene I (RIGI), or melanoma differentiation-associated protein 5 (MDA5) pathways] [34]. The TLR-induced IFN pathway results in the production of IFNα and IFNβ, which can then bind to membrane-bound IFN receptors, trigger the "secondary" janus kinase and signal transducer and activator of transcription (JAK–STAT) signaling pathway, and stimulate the transcriptional activation of IFN-stimulated genes (ISGs) [32] (Fig. 5.1).

Many ISGs are directly involved in antiviral defense [32]. A few studies have demonstrated that IFN signaling results in differential expression of a number of lncRNAs [35]. The MONOCLdb database is a bioinformatic tool that contains comprehensive lists of lncRNAs regulated by IFNα stimulation and viral infection [36]. There is some evidence that IFN-induced lncRNAs can function to regulate IFN pathways; however, few studies have provided extensive mechanistic understanding of how these lncRNAs function, particularly in the host defense. This section will summarize IFN-regulated lncRNAs and cover the experimental data supporting their roles in these pathways.

LncRNA-cytidine/uridine monophosphate kinase 2 (CMPK2) was first identified because of its elevated expression levels in the livers of patients with hepatitis C [37] (Fig. 5.1). This lncRNA was shown to be an IFNα-induced polyadenylated spliced nuclear lncRNA that was conserved in both humans and mouse. Kambara et al. demonstrated that knockdown of this lncRNAs resulted in the upregulation of ISGs, as well as a decrease in viral replication. The mechanism by which this lncRNA mediates its inhibitory effects remains unclear.

Stimulation of signaling can also result in the downregulation of lncRNAs; for example, the negative regulator of antiviral response (NRAV) lncRNA is dramatically downregulated following viral infection (Fig. 5.1). Overexpression of human NRAV negatively regulated ISGs at the epigenetic level and resulted in increased influenza virus replication [38], while small interfering ribonucleic acid (siRNA) knockdown of NRAV resulted in reduced influenza infection. Downregulation of NRAV was also observed to be beneficial for other DNA and RNA viruses tested, thus suggesting that it may be a useful therapeutic target for universal antiviral therapy [38].

LncRNAs can have dynamic expression kinetics during viral infection due to their direct manipulation by viral proteins; for example, the lncRNA noncoding repressor or nuclear (NRON) factor of activated T cells (NFAT) is actively downregulated during early infection with HIV, but is later upregulated during late stages of HIV viral replication [39]. NRON was shown to be a cytosolic lncRNA that can directly bind to the TF, NFAT, and repress its activation [39]. SiRNA knockdown of NRON resulted in enhanced HIV replication due to increased NFAT binding at the viral long terminal repeats (LTR) [39]. A recent study showed that NRON can bind to the HIV transactivator of transcription viral protein and induce its degradation, thereby promoting viral latency [40]. It will be interesting to determine if RNA interference (RNAi) depletion of NRON can be used as a method to reactivate and remove the latent reservoir of virus that persists in HIV-infected individuals.

Some lncRNAs have also been shown to exert effects on specific ISGs. The IFNα-inducible lncRNA, bone marrow stromal cell antigen 2 (BST2) IFN-stimulated positive regulator (BISPR), is transcribed from a bidirectional promoter that also produces the coding gene BST2 [37,41,42] (Fig. 5.1). BST2 is a well-characterized antiviral protein that directly inhibits viral particle release and acts on a variety of enveloped viruses [43]. SiRNA-mediated knockdown of BISPR resulted in decreased BST2 protein expression levels, suggesting a cis enhancer effect [41,42]. It is unclear if any of the BISPR sequence is important for mediating its function or whether RNA polymerase II recruitment is the main mechanism of action.

Numerous studies have now implicated the type I IFN pathway in the pathogenesis of multiple autoimmune diseases, thus suggesting the presence of a common underlying mechanism [44,45]. IFN signaling is correlated with increased pathogenicity in systemic lupus [44,46], while IFNβ is used as a therapy to reduce symptom relapses in MS. This highlights the similar connection of both diseases to the IFN pathway [47].

5.2.4 LncRNAs Involved in Innate Immune-Cell Differentiation

Regulation of cell differentiation is critical for generating cell lineages with specific roles. In the innate immune system, DCs and macrophages are both derived from a common precursor cell called a monocyte. This process is stimulated by cytokines, including the macrophage colony stimulation factor (MCSF) and granulocyte MCSF (GM-CSF), which activate specific TFs to drive the differentiation process. Given the critical role of differentiation in the generation of mature immune cells, it is not surprising that lncRNAs have been demonstrated to be integral components of this process. In DCs, long noncoding RNA DC (lnc-DC) was shown to be upregulated during differentiation (Fig. 5.2). Knockdown of lnc-DC inhibited the differentiation of monocytes into mature DCs

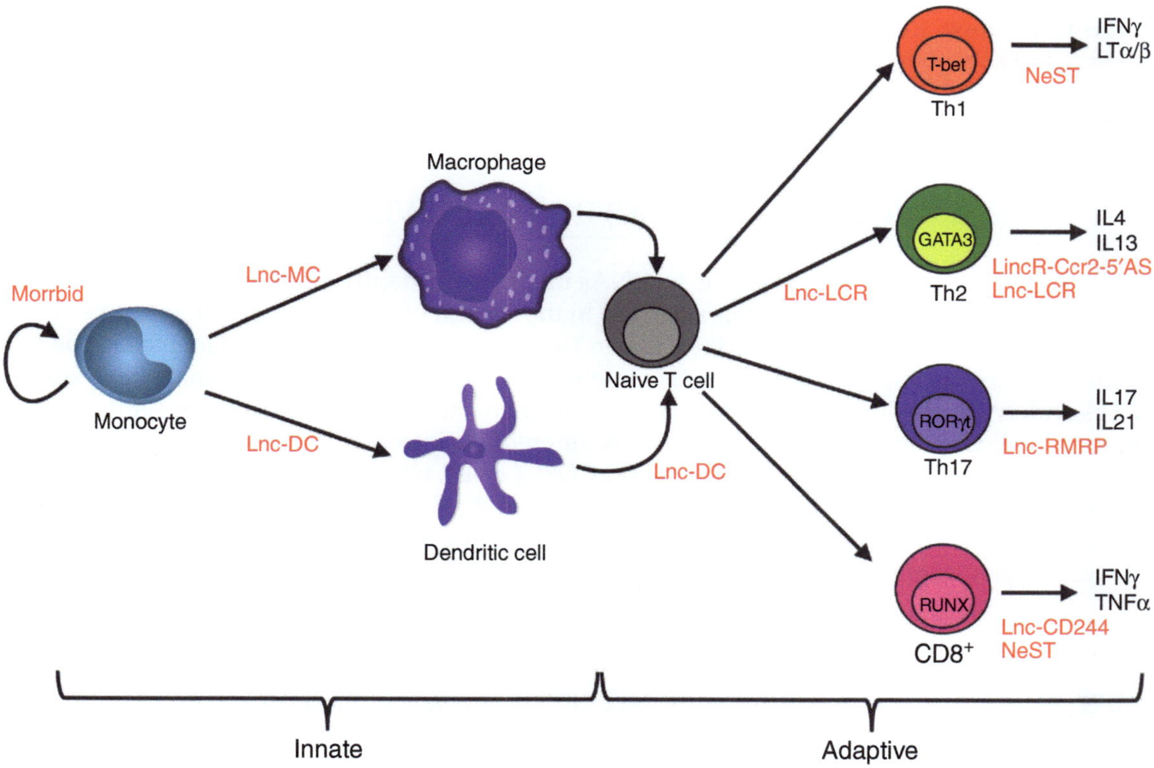

FIGURE 5.2 **Role of lncRNAs in distinct cell types of both innate and adaptive immune systems.** Macrophages and dendritic cells are derived from monocytes and function to stimulate T-cell activation and differentiation. Cells of the adaptive immune system are derived from lymphoid progenitor cells, which give rise to naive T cells. These cells undergo further differentiation into specialized subsets of T cells including T-helper 1 (Th1), T-helper 2 (Th2), T-helper 17 (Th17), and $CD8^+$ T ($CD8^+$) cells. Transcription factors that drive differentiation are shown in the nuclei of each cell type. Each cell type is characterized by the expression of specific cytokines. LncRNAs have roles in cytokine secretion and T-cell differentiation and specialization. Specific lncRNA names (*red*) and coding genes names (*black*) are given.

and significantly affected the ability of DCs to stimulate T cells [48]. Lnc-DC was shown to function by binding to the STAT3 TF in the cytoplasm and preventing the Src homology region 2 domain-containing phosphatase-1 (SHP1) phosphatase from dephosphorylating STAT3. Phosphorylation promotes STAT3 dimerization and relocation to the nucleus to initiate its transcriptional program [48]. The long noncoding RNA macrophage cell (lnc-MC) was the first lncRNA shown to be involved in macrophage differentiation (Fig. 5.2). Lnc-MC was shown to function as a competitive endogenous RNA, specifically sponging microRNA:miR-199a-5p away from its target (activin A receptor type 1B), which is a known regulator of macrophage differentiation [49].

Maintaining appropriate numbers of immune cells requires the generation of new cells and cell types via differentiation and the elimination of cells that are no longer needed. Kotzin et al. identified a novel lncRNA, myeloid RNA regulator of Bim-induced death (Morrbid), which they showed to be crucial for regulating the life span of myeloid cells (Fig. 5.2). The authors demonstrated that the removal of the Morrbid locus resulted in a dramatic decrease in circulating myeloid cells in the blood [50]. They also showed that Morrbid is a chromatin-associated lncRNA that functions in cis to suppress the expression of its neighboring proapoptotic gene *Bcl2l11* (Bim) by recruiting the polycomb repressive complex (PRC2) to the *Bcl2l11* promoter in short-lived myeloid cells [50].

It is well known that GM-CSF levels can be regulated by TLR stimulation, thus linking it to the inflammatory response [51,52]. An anti-GM-CSF antibody-based therapy has proven effective at treating both MS and RA in phase I–II trails, suggesting that targeting this pathway could be a successful strategy for treating autoimmune diseases [51].

5.3 ROLE OF LNCRNAS IN ADAPTIVE IMMUNITY

This section presents a review on the lncRNAs involved in adaptive immunity, focusing mostly on those lncRNAs that have been characterized in the context of T-cell biology. These studies have demonstrated that lncRNAs play a critical role in T-cell lineage development and in lineage-specific functions. Naive T-helper cells differentiate into effector cells through the coordinated expression of TFs and cytokines. TFs orchestrate complex epigenetic alterations to activate or repress genes required for lineage specification. Naive T cells differentiate into effector T subsets (i.e., T-helper 1 (Th1), T-helper 2 (Th2), T-helper 17 (Th17), and $CD8^+$ T cells), which are classified based on the cytokines they express: IFNγ (Th1); IL-4, IL-5, and IL-13 (Th2); IL-17 (Th17); or IFNγ and TNFα ($CD8^+$ T cells) (Fig. 5.2). This differentiation step is essential for the adaptive immune system to provide protection against infection.

5.3.1 LncRNA Regulators of Adaptive Immune-Cell Function

IFNγ is an important component of the type II IFN pathway and has roles in immune-cell differentiation and function [53]. LncRNA NeST (also known as TMEVPG1 and LincR-*Ifng*-3′AS) was identified as a Th1 lineage-specific lncRNA that was critical for controlling Theiler's virus infection in mice (Fig. 5.2). Gomez et al. showed that NeST positively controlled the expression of IFNγ through its interaction with a histone 3 methylation complex, which promoted the methylation and activation of the *Ifnγ* locus in $CD8^+$ T cells [54]. Another lncRNA that regulates IFNγ is lncRNA-CD244 (GSTT1-AS1), antisense to the gene CD244, which was shown to be highly overexpressed in $CD8^+$ T cells of patients infected with tuberculosis [55] (Fig. 5.2). Expression of lncRNA-CD244 was shown to be induced by CD244 signaling and promote silencing of the IFNγ and TNFα loci. Furthermore, inhibition of IFNγ and TNFα expression by this lncRNA involves interactions with enhancer of zeste homolog 2, a component of the PRC2 complex [55]. Given the critical role for IFNγ signaling in viral and bacterial infections, it is likely that other lncRNAs that also regulate other aspects of this pathway will be discovered.

Th17 cells have been implicated in a variety of autoimmune diseases, including MS and lupus [56]. Th17 cells are specified by *r*etinoic acid receptor-related *o*rphan nuclear *r*eceptor gamma t (RORγt) and basic leucine zipper ATF-like transcription factor (BATF) TFs, and they produce IL-17, IL-21, and IL-22 [57] (Fig. 5.2). IL-17 promotes inflammation and various clinical trials are now focused on this cytokine [58]. Huang et al. identified a role for lncRNA-RNA Component Of Mitochondrial RNA Processing Endoribonuclease (RMRP) in controlling gene expression (i.e., IL-17 expression) programs in Th17 cells. The authors identified a lncRNA-RMRP-dependent interaction between DEAD-box helicase 5 (DDX5) and TF RORγt, which was critical for controlling the transcription of target genes in Th17 cells (Fig. 5.2). RMRP is highly conserved, and mutations within this lncRNA are linked to cartilage hair hyperplasia (CHH), a genetic bone growth disorder that leads to short stature, skeletal abnormalities, and immune deficiencies. This was the first genetic disorder shown to be caused by a mutation within a lncRNA. This discovery provides new insights in the regulation and complexity of Th17 cell functions.

5.3.2 LncRNAs Involved in Adaptive Immune-Cell Differentiation

Th2-cell differentiation is regulated by the TFs, STAT6 and GATA-binding protein 3 (GATA3), which activate many Th2-specific genes. Hu et al. demonstrated that GATA3 also regulated ~28% of Th2-specific lncRNAs (e.g., lincR-Ccr2-5′AS) [59]. The authors showed that lincR-Ccr2-5′AS regulated the migration of Th2 cells to the lung. Knockdown of this lncRNA using short hairpin RNAs (shRNAs) did not decrease the expression of the Th2 cytokine, IL-4, but did decrease the expression of its neighboring protein-coding genes (i.e., *Ccr1-Ccr5*) (Fig. 5.2). Another Th2 cell-specific lncRNA, Lnc-LCR, was shown to be important for promoting the expression of the Th2-specific cytokines (i.e., IL-4, IL-5, and IL-13) [60] (Fig. 5.2). The current mechanism of TH2-LCR-mediated gene regulation is proposed to be through its recruitment of tryptophan-aspartic acid repeat protein 5 (WDR5)-containing complexes that regulate H3K4me3 patterns at the affected target gene loci.

LncRNAs have proven to be integral in the developmental and functional regulation of various cells of the adaptive immune system. Regulation of the immune system is critical for limiting excess inflammation and self-damage. Autoimmunity may be the result of decades of inability to tightly regulate specific immune-related pathways [61]; however, this representation of autoimmunity is likely to be an oversimplification as many known environmental and genetic factors also contribute to these diseases [61]. Nevertheless, some specific cell types (i.e., Th17 cells) have been proven to have direct roles in the pathogenesis of specific autoimmune diseases [62]. IL-17 is critical for Th17 function, and monoclonal antibodies targeting this cytokine have been shown to reduce Th17-specific tissue damage in clinical trials [63].

5.4 EPIGENETIC GENE REGULATION BY LncRNAs

5.4.1 Epigenetic Regulation by X-Inactivation Specific Transcript and Implications for Autoimmune Diseases

One of the best examples of epigenetic regulation by a lncRNA comes from the X-inactivation specific transcript (Xist) lncRNA, which is involved in dosage compensation and X-chromosome inactivation in females [64]. The exact mechanism of Xist remained elusive until recently [65] when McHugh et al. conclusively demonstrated that the inactivation function of Xist was directly dependent on its binding to the SMRT- and HDAC-associated repressor protein (SHARP, also called SPEN) [66]. The binding of Xist to SHARP recruits the SMRT corepressor, activates HDAC3-mediated silencing, and excludes RNA polymerase II from the X chromosome [66]. Although previous studies have shown that Xist can bind to PRC2, this interaction proved to be dispensable for Xist-mediated silencing [65].

Some autoimmune diseases are known to disproportionately affect women, which highlights the effects of sex-related differences in disease risk (www.cdc.gov/). Xist is a female-specific lncRNA that is responsible for silencing one of the X chromosomes to match the expression levels present in males [65]. However, Xist-based inactivation is now known to be reversible in T and B cells [67]. Wang et al. showed that many immune-related genes can escape X-inactivation in female lymphocytes and can be biallelically expressed as a result of reduced Xist expression during certain developmental stages [67]. Immune genes residing in the X chromosome include TLR7 and CXCR3, and they are interestingly shown to be overexpressed in lupus [68,69]. Wang et al. concluded that the partial activation of the second X chromosome in female lymphocytes could explain why women display enhanced immunity and higher rates of autoimmunity compared to men [67].

5.4.2 Other Mechanisms of LncRNA-Mediated Epigenetic Regulation

LncRNAs can regulate the epigenetic landscape directly, through modulation of the recruitment of chromatin-modifying factors to a specific genomic locus, or indirectly, as a consequence of their transcription [70]. TF binding can affect the epigenetic landscape and promote the activation of specific genes [71,72]. The various ways in which lncRNAs can regulate both NFκB binding and activation in the context of inflammation have already been described. Given that TFs are a major point of regulation for specific programs of cellular differentiation and effector functions, it is not surprising that lncRNA–TF regulation is a repeated theme seen in various other biological contexts [73]. LncRNAs often function as scaffolds to facilitate the recruitment of chromatin-modifying factors to specific genomic loci in nearby genes [74]. Numerous lncRNAs have been shown to recruit the PRC2 complex to mediate gene-specific silencing [75]; however, the role of PRC2 in gene repression is not always clear [66]. Changes in the methylation state have been implicated in various autoimmune diseases (i.e., lupus and RA) [76]. The muscle-specific lncRNA, lncRNA-Dum, was recently shown to modulate methylation at the promoter of its neighboring gene, *Dppa2*, in cis through the recruitment of DNA methyltransferases [77]. Although epigenetic modulation can result in gene expression changes, they can also result in changes in alternative splicing (reviewed in Ref. [78]). In some instances, the promoter of a lncRNA, not the RNA product itself, is the critical regulatory component needed to maintain an active chromatin state [79]. The three-dimensional organization of the genome has recently been shown to be important for regulating gene expression [80]. Rinn et al. described a lncRNA, FIRRE, that was involved in the topological organization of specific genome loci within the X chromosome. This three-dimensional organization was critical for efficient expression of numerous adipocyte-specific genes (discussed later) [81].

5.4.3 Genetic Factors in LncRNA Dysregulation

Genome-wide association studies (GWAS) have revolutionized the study of complex diseases by quantitating disease association at thousands of genetic loci [82]. Most GWAS studies performed to date have focused on protein-coding genes, although 90% of disease-associated single nucleotide polymorphism (SNPs) lie within noncoding regions of the genome [83]. However, one recent GWAS study focused on immune-cell specific SNPs in noncoding regions and found that ~60% were located within enhancer elements [84]. Furthermore, because of the cell-type specific activity of enhancers, they could pinpoint the exact cell types that were likely to be involved in each autoimmune disease [84]. A noncoding SNP can have direct consequences on gene expression through various mechanisms. The most characterized mechanism of SNP-mediated alterations in gene expression involve the disruption of TF binding motifs, which can result in positive [85] or negative [86] transcriptional effects [87]. Other polymorphisms can lead to disruptions in splicing [88,89] or to changes in polyadenylation [90]. A few studies have provided mechanistic insights into how SNPs can modulate lncRNA expression in cancer; however, no similar studies have been performed for autoimmunity (reviewed in Ref. [91]).

Guo et al. studied disease-associated SNPs that could potentially alter lncRNA expression [92], and identified a risk variant that increased binding of the one cut homeobox 2 TF and increased expression of the lncRNA PCAT1 [92]. Another study examining autoimmune thyroid disease (AITD) identified a SNP that increased the risk of developing the disease [93]. This SNP was subsequently shown to be located within the promoter of an antisense lncRNA, SAS-ZFAT, which affected its ability to regulate its neighboring coding gene, zinc finger gene in AITD [93].

SNPs can also alter the sequence of a lncRNA and affect its function. As mentioned previously, lncRNA-RMRP contains a SNP that was shown to be the cause of a genetic disorder CHH (reviewed in [94]). Interestingly, the specific mutation within lncRNA-RMRP was demonstrated to interfere with its binding to DDX5, which is critical for Th17-cell function [95]. GWAS studies have illustrated that most autoimmune conditions are not monogenic disorders, but involve the dysregulation of hundreds of genes.

5.4.4 LncRNAs and Autoimmune Diseases

Most of the studies exploring the role of lncRNAs in autoimmune diseases have been descriptive and have demonstrated dysregulated lncRNA expression in specific diseases [96,97]; for example, Müller et al. isolated $CD14^+$ monocytes from a small cohort of RA patients and identified TNFα-regulated lncRNAs [98]. Another study focused on fibroblast-like cells from the synovium and found 135 dysregulated lncRNAs in RA patients [99]. The role of lincRNA-Cox2 in mediating interactions with hnRNP A2B1 (A2B1) has already been described earlier in the text. A2B1, also known as RA-33 (RA of 33kD), is a known target of nuclear autoantibodies in both RA and lupus patients [100].

Systemic lupus erythematosus (SLE) is characterized by the presence of antinuclear antibodies and elevated IFN signaling ("IFN signature") [101]. The observation that numerous lncRNAs are involved in regulating the IFN pathway has already been discussed; however, it is yet to be determined whether one or more of these lncRNAs is dysregulated in SLE. Shi et al. compared nine female SLE patients to eight healthy controls and identified several differentially expressed lncRNAs [102]; interestingly, twice as many lncRNAs were downregulated than upregulated [102].

Psoriasis is a chronic inflammatory disease characterized by hyperproliferation of skin cells. Sonkoly et al. identified a psoriasis-associated lncRNA, which they named psoriasis susceptibility-related RNA gene induced by stress (PRINS) [103]. The authors demonstrated that PRINS was regulated by the proliferation and differentiation of keratinocytes, was highly inducible by stress (i.e., UV irradiation), and functioned to regulate the expression of G1P3, an antiapoptotic keratinocyte protein [103,104].

Alteration in lncRNA expression can clearly promote epigenetic defects, while defects in the chromatin state can promote aberrant expression of lncRNAs. Which comes first? It is likely that the development of a disease represents an accumulation of defects in both epigenetic dysregulation and lncRNA expression, but the question remains about how best to study these disease-associated lncRNAs?

5.5 FACTORS TO CONSIDER WHEN STUDYING LncRNA FUNCTION

5.5.1 Understanding Conservation Within LncRNAs

Unlike lncRNAs, coding genes are highly conserved across distal and related species. The requirement of coding genes to encode functional peptides likely constrains variation within the open reading frame (ORF) [105]. LncRNAs are not translated and often have poor sequence conservation across related species, therefore inferring their function based on sequence similarity is a challenge (reviewed in Ref. [106]). A more useful metric is to assess whether the lncRNA has conservation of synteny (location relative to flanking coding genes within the chromosome) alongside expression conservation [107]. The LNCipedia database (http://www.lncipedia.org/) has a user-friendly interface and allows the assessment of both conservation of synteny and sequence. Expression conservation can also be useful when assessing whether the specific lncRNA has the same biology in similar and divergent species [108]. Conservation of lncRNA expression can also indicate conservation of regulatory regions (i.e., TF binding sites within promoters) [109]; however, conservation of expression does not necessarily mean that the RNA product is important for lncRNA function. For example, enhancer RNAs (eRNAs) are thought to function by creating a localized active transcriptional state that can activate neighboring genes [19]; however, it is unclear whether the specific RNA sequence for eRNAs is important for their function [19]. Additionally, many lncRNAs display species-specific expression. In summary, understanding what feature of the lncRNA is conserved will allow for a better assessment of the likelihood of functional conservation across species.

5.5.2 Function of Repetitive Elements Within LncRNAs

Sequence conservation of a genetic element does not always equal function [110], as numerous nonconserved repetitive elements are known to be functionally important [111]. More than 50% of the human genome is composed of repetitive elements, and certain types of repeats have been shown to be enriched in lncRNAs [81,112,113]. Hacisuleyman et al. identified that the lncRNA, FIRRE, was essential for adipocyte differentiation [114]. The authors showed that a 156 bp repeat sequence within the lncRNA was essential for its interaction with hnRNP U and allowed it to act in the topological organization of genome loci within the X chromosome [81]. Another lncRNA, 1/2-sbsRNA, was found to contain Alu-repeats that were important for promoting Staufen 1-mediated degradation of mRNAs [115]. Therefore it is clear that repetitive elements represent functional regions within lncRNAs that can significantly contribute to their function.

5.5.3 Upregulated Versus Downregulated LncRNAs

RNA-seq studies in immune cells clearly show that similar numbers of lncRNAs are up- and downregulated during inflammatory stimulation [13]. Nevertheless, most lncRNAs studied in immunity and other fields have focused on upregulated lncRNAs (e.g., lncRNAs that are upregulated following TLR stimulation). The reasoning is not necessarily that upregulated lncRNAs are more interesting or biologically relevant than downregulated lncRNAs, but is likely due to the use of traditional methods of ablation when functionally studying genes (as discussed later). LincRNA-EPS, which is expressed in both macrophages and DCs, is an example of a lincRNA that is downregulated upon TLR stimulation. Deletion of lincRNA-EPS in mice resulted in elevated expression of immune response genes, thus providing evidence that it functions as a negative regulator of inflammation [116]. Furthermore, some of the function of lincRNA-EPS was demonstrated to involve interactions with hnRNPL via a *CANACA* motif in its 3′ end [116]. The identification that lincRNA-EPS acts as a repressor of inflammation serves as an example of how a downregulated lincRNA (often overlooked in preference for upregulated lncRNAs) can have an integral role in regulating a major pathway.

5.5.4 Assessing Whether LncRNAs Are Translated

A significant challenge in studying lncRNAs is that the transcript is noncoding. It is important to note that the categorization of lncRNAs as "noncoding" is mostly determined bioinformatically using an arbitrary cutoff of <100 codons. Therefore it is possible that a number of lncRNAs could produce small peptides. A gene annotated as a putative lncRNA (FBgn0087003) in drosophila was eventually shown to encode multiple small (~10 amino acid) peptides that were critical for development [117]. Several bioinformatic tools have been developed to predict small ORFs (smORF); however, the predictive ability of smaller ORF size is generally very poor [118]. PhyloCSF uses the codon substitution frequency and species conservation to provide a score metric that can be used to determine the presence of a conserved ORF [105]. Other groups have sought to identify novel small peptides using mass spectrometry; however, a major challenge is determining whether the peptides identified correspond to a novel smORF or are degradation intermediates of larger proteins [119].

The development of ribosomal footprinting and next generation sequencing (ribosome profiling) has provided a high-throughput method that has revealed the complexity of the mammalian translatome [120–122]. Ribosome profiling allows ribosome nuclease-protected RNA fragments to be mapped to transcripts to enable quantitative measurement of the translational efficiency for a transcript based on read counts [120]. Ingolia et al. also used a drug that stalls ribosomes at ATG start codons to globally map translation start sites. This study revealed significant levels of noncanonical translation initiation from CTG codons [120]. One of the surprising findings from various ribosome profiling studies has been the observation that many lncRNAs appear to be translated, which raises the possibility that some of these regions could be producing small peptides [123].

Guttman et al. developed the ribosome-release score metric as part of their ribosome profiling analysis pipeline as a way to more accurately predict translation efficiency. Their findings showed that ribosome occupancy of lncRNAs and 5′ UTRs does not always equate to translation [124], and the authors concluded that the majority of noncoding transcripts are not translated into peptides. Several other studies have also examined lncRNA binding to ribosomes and have concluded that ribosome binding may not be functional and may serve as a quality control process to degrade transcripts with low coding potential via the nonsense mediated decay pathway [125,126].

However, one study has demonstrated the production of a small peptide from a lncRNA, thus raising the possibility that other lncRNAs may also produce functional peptides [127]. Myoregulin (MLN) is a peptide encoded from a skeletal muscle-specific lncRNA. MLN was shown to bind and regulate the sarcoendoplasmic reticulum Ca^{2+}-ATPase pump, which is critical for muscle-cell function [127]. The discovery of MLN may open the door to the discovery of other novel regulatory peptides that could have significant roles in various biological contexts.

5.6 SYSTEMS FOR ABLATING THE EXPRESSION OF LncRNAs

One of the first steps when studying lncRNAs is to knockdown its expression using the following ablation methods:

5.6.1 RNA Interference

The laborious process of functionally characterizing lncRNAs has been a major limitation to the discovery of lncRNAs; for example, while 16,000 lncRNAs have been identified in the human genome, only 1% have been ascribed a function. RNAi has been the method of choice for loss of function studies due to its ease and versatility. RNAi uses short complementary RNAs, which when transfected into cells can utilize the endogenous cellular machinery to target specific transcripts [128,129] (Fig. 5.3A). In addition, shRNAs can also be expressed in a lentiviral context to accomplish stable RNAi [130]. RNAi is most active in the cytoplasm, which makes it most useful for targeting mRNAs and cytoplasmically localized lncRNAs [131]. Success in knocking down some lncRNAs, such as lincRNA-Cox2, has been attributed to the fact that this lncRNA is expressed in both the cytoplasm and the nucleus (perhaps cycling between both compartments) and hence is susceptible to RNAi [20]. However, many lncRNAs are thought to be nuclear restricted where RNAi has been demonstrated to have limited efficiency [132].

5.6.2 Antisense Oligonucleotides

Antisense oligonucleotides (ASOs) can be RNA- or DNA-based and can be used to target complementary sequences within a transcript. Unlike RNAi, ASOs do not engage the cellular RNAi machinery [133]. Instead, they function by hybridizing to the target RNA and inhibiting its function by steric inhibition or by inducing the RNase H pathway. RNase H normally functions to resolve unwanted DNA–RNA interactions that occur during replication or transcription [134]. One of the most widely used ASOs for targeting lncRNAs are gapmers. Gapmers contain a "hybrid" modified–unmodified configuration consisting of a ~10 nucleotide DNA core flanked by 2′-O-methyl- or locked nucleic acid (LNA)-modified synthetic nucleotides (Fig. 5.2). Gapmers possess stable modifications with reduced toxicity while still allowing engagement of the RNase H pathway (Fig. 5.3B).

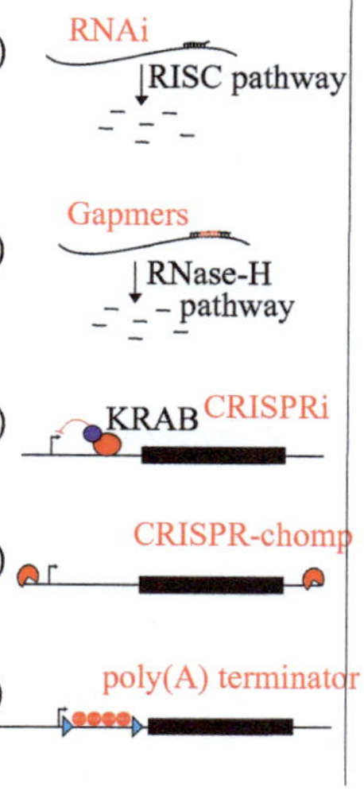

	Target	Reversible loss of function	Type of phenotype	Knockdown of nuclear lncRNAs	Time to phenotype
(A)	Transcript	Yes	Hypomorphic	No (some)	Short
(B)	Transcript	Yes	Hypomorphic	Yes	Short
(C)	gDNA (TSS)	Yes	Hypomorphic	Yes	Medium
(D)	gDNA	No	Null (complete)	Yes	Medium-to-long
(E)	gDNA	Yes (loxP)	Hypomorphic	Yes	Medium-to-long

FIGURE 5.3 **The repertoire of tools used to disrupt lncRNA expression.** (A) RNA interference (RNAi) acts posttranslationally to degrade transcripts in the cytoplasm. (B) Gapmers are composed of RNA and a core stretch of ~10 nucleotides of DNA, which are critical for engaging the RNase H pathway. (C) CRISPRi uses a catalytically inactive version of Cas9 fused to a Krüppel-associated box domain (KRAB). The CRISPRi system can target transcription start sites (TSS) to induce heterochromatin-based transcriptional silencing. (D) Clustered regularly interspaced short palindromic repeats (CRISPR)/Cas9 can be used to delete specific regions of lncRNA loci using two flanking guide RNAs (gRNAs). (E) Polyadenylation [poly(A)] signals [3–5] can be inserted downstream of the TSS, which result in premature termination of transcription. These poly(A) signals can be flanked by loxP sites to allow their removal. The different tools are compared using the criteria described in the headings of each column: Target describes what aspects of the lncRNAs gene expression are targeted [e.g., transcriptional (CRISPRi) or posttranscriptional silencing (RNAi)]; Reversible loss-of-function describes if the silencing effect of this tool is reversible in a cell. Only the nuclease active Cas9 activity is generally irreversible since the regions of the loci are actually deleted; Type of phenotype refers to whether the tool completely shuts off expression of the lncRNAs. Hypomorphic refers to the fact that some level of expression remains despite the level of knockdown, and only compete removal of the locus completely abolishes its expression; Knockdown of nuclear lncRNAs touches on the observation that RNAi is not very active in the nucleus; and Time to phenotype refers to the amount of time from planning the experiment to obtaining the phenotype. *Short*, requires transfection only; *Medium*, requires design and cloning in addition to the generation of a functional CRISPRi cell line; and *Medium-to-Long*, requires screening for cells and making single-cell clones.

Efficient depletion of nuclear lncRNAs has been demonstrated using modified DNA ASOs [131]; for example, Ilott et al. used gapmers to successfully knockdown TLR-induced eRNAs after they were unable to knockdown these eRNAs with four different siRNAs [135]. In vivo applications for DNA ASOs have also proven successful in the treatment of a neural degenerative disease, Angelman's disease, which is caused by aberrant expression of a specific lncRNA [136].

5.6.3 Cas9/CRISPR Technology

The introduction of CRISPR/Cas9 technology has revolutionized the field of functional genomics by providing a novel tool for interrogating gene function. CRISPR/Cas9 is a deoxyribonuclease that can be specifically targeted to genomic regions via a guide RNA (gRNA) [137,138]. Targeting of Cas9 to a region results in a blunt double-stranded DNA break that engages the cellular nonhomologous end joining DNA repair pathway. This promotes imprecise repair and produces small deletions in the repaired sequence. These small deletions then result in a frameshift that disrupts the ORF within coding genes and disrupts protein synthesis. LncRNAs do not contain ORFs and they tend to span tens of kilobases of sequence, therefore targeting them with a single gRNA is thought to be insufficient to disrupt their function [139]. An alternative application of Cas9 that has proven effective for targeting lncRNAs involves using two gRNAs that flank the lncRNA region of interest and induce a deletion of the entire locus (Fig. 5.3D). The main advantage of deleting lncRNAs is obtaining complete loss-of-function, as demonstrated in a recent study that performed a CRISPR-mediated deletion screen to identify lncRNAs that positively and negatively regulate cancer growth [140]. However, deleting the entire DNA sequence may result in an inability to resolve whether a phenotype is due to the loss of lncRNA production or loss of DNA sequence [141].

The ease with which Cas9 can be targeted to a specific genomic region sparked the development of a modified (catalytically inactivated) version of the protein fused to the Krüppel-associated box chromatin-silencing domain (CRISPRi) [142,143] (Fig. 5.3C). CRISPRi can be used to target both coding and noncoding genes and triggers localized heterochromatin silencing at the transcription start site (TSS) [144]. The transcription-based inhibition by CRISPRi offers the ability to efficiently target lncRNAs regardless of their localization in the cell. Gilbert et al. showed maximum knockdown efficiency when targeting regions +50 to +500 nucleotides relative to the TSS [142]. However, CRISPRi is limited by the availability of an accurately annotated TSS. Incorporating CAGE-seq, Gro-seq, or ChIP-seq data into gRNA design may improve targeting of a lncRNA of interest [145]. Nevertheless, Liu et al. used CRISPRi to target 16,401 lncRNA loci in 7 diverse cell lines and identified hundreds of lncRNAs required for cell growth [145]. Therefore despite its caveats, it seems that CRISPRi can be a useful and powerful tool for interrogating lncRNA biology.

5.6.4 Dissecting the Complex Functions of LncRNAs

LncRNAs can span large stretches of DNA and can contain important regulatory sequences (i.e., enhancers) that are functionally independent from the lncRNA product. Nevertheless, in vivo assessment of lncRNA function has predominantly relied on assessing the consequence of deleting the entire lncRNA locus [141]. The role of the lncRNA, Fendrr, in embryonic development has been established using two different ablation methods: locus deletion [146] (Fig. 5.3D) and the insertion of a polyadenylation signal (transcription terminator) [147] (Fig. 5.3E). Both methods of ablation led to embryonically lethal phenotypes; however, while deletion of Fendrr led to defects in lung development, transcriptional termination caused heart-related defects [9,146]. Furthermore, only the transcriptional- termination mutant of Fendrr could be rescued by a transgene, thus highlighting an important distinction between inhibiting transcript production and deleting the locus [141]. Numerous studies that have deleted lncRNA loci have been unable to rescue the phenotype using a transgene approach, therefore making it difficult to attribute the phenotype to the lncRNA product itself [148,149]. Deletion of an entire lncRNA loci also makes it difficult to identify the lncRNA component that is important for function [141]. Therefore to minimize disruption to the locus, recent mouse studies have opted for either promoter deletions or polyadenylation termination signal insertions to disrupt lncRNA transcription [150–152]. LncRNA promoters have also been proposed to function as enhancers and promote the recruitment of transcription-activating factors that can affect the local nucleosome environment and the expression of neighboring genes [19,153]. Hence the promoter itself could be the functional unit of a lncRNA locus [70]. Engreitz et al. demonstrated that a significant portion of lncRNAs had cis effects on their neighboring genes. Interestingly, the cis effect generally did not require production of the lncRNA transcripts themselves, but required the processes associated with their production (i.e., promoter activity and splicing) [70].

5.7 THERAPEUTIC TARGETING OF LncRNAs

Inflammatory cytokines have been the major target for therapeutic interventions in a number of inflammatory conditions [154–156]. A significant advancement in the treatment of RA came with the development of anti-TNFα therapies, including antibody-based formulations [157]. Recombinant IFNβ therapies have also been used with some success in the treatment of MS [47]. While these cytokine-centric drugs have had some success in the treatment of autoimmune diseases, it is important to note that a significant percentage of patients fail to respond to therapy [158,159]; for example, in some studies a staggering 20%–40% of patients failed to respond to anti-TNFα treatment [159]. Therefore there is an urgent need to develop other therapeutic options.

In addition to their efficacy, another important attribute of therapeutic agents is their specificity. LncRNAs have been shown to have strict cell-specific expression, therefore the targeting of disease-associated lncRNAs using ASOs or small-molecule inhibitors has become an appealing prospect for several companies. IONIS pharmaceuticals (formerly ISIS pharmaceuticals) has pioneered the development of ASO therapy for the treatment of human diseases, and most recently targeted a lncRNA using ASOs for the treatment of Angelman's syndrome. This disorder is characterized by developmental delays, seizures, and ataxia [160] and has very few treatment options available. Angelman's syndrome is a monogenic disorder caused by mutations in E3 ubiquitin ligase 3A (*UBE3A*) that lead to inactivation of this gene [136,160]. As *UBE3A* is a paternally imprinted gene, a mutation in the maternal allele is sufficient to cause the disease [136,160]. Paternal imprinting of *UBE3A* requires expression of an antisense lncRNA called UBE3A-ATS. Hence the strategy is to target the paternal lncRNAs UBE3A-ATS to derepress expression of paternal *UBE3A* and rescue the defective maternal copy [136]. In a mouse model, even partial restoration of UBE3A protein expression ameliorated some of the cognitive deficits associated with the disease; however, further phenotypic characterization is still needed [136]. One of the main challenges of oligo-based therapeutics is the efficient and specific delivery to organs or tissues of interest.

RaNA Therapeutics has also focused on diseases caused by the downregulation of specific protein-coding genes. They specifically aim to target the lncRNAs that mediate this repression through their interaction with repressor complexes (i.e., the PRC2 complex). By targeting these specific lncRNAs they can block PRC2-mediated coding-gene silencing. RaNA recently received a patent to develop strategies to target these PRC2-associated noncoding RNAs as a way to unsilence specific coding genes (United States Patent No. 9,328,346).

Another exciting application for disease-associated lncRNAs is their use as biomarkers. LncRNA-PCA3 is the first lncRNA approved by the FDA for use as a biomarker in prostate cancer, which opens the door for the discovery of other disease-specific lncRNA biomarkers [136a]. Diseases such as RA and osteoarthritis can take decades to develop, therefore the discovery of a lncRNA-based biomarker could provide a snapshot of early disease progression and provide an opportunity for early diagnosis and intervention.

5.8 CONCLUSION AND FUTURE PERSPECTIVES

In 1975, King and Wilson observed that the vast differences between evolutionarily distinct species could not solely be attributed to coding genes alone and likely involved differences in how these genes were regulated [161]. LncRNAs are now known to represent a significant component of gene regulation, functioning as the "fine-tuners" of this process. LncRNAs mediate specific functions through a combination of protein, RNA, or DNA interactions [162]. Their expression is highly cell-type, stimuli, and time-point specific and their regulatory role in a broad spectrum of biological processes has been established [162]. Functional characterization of specific lncRNAs has revealed interactions with "nonspecific" ubiquitously expressed proteins (i.e., DEAD-box helicases) [95,163,164] and heterologous nucleus ribonuclearproteins (i.e., hnRNPs) [20,114,165]. However, the question remains as to why a cell-type specific lncRNA would interact with a noncell-type specific protein? The modular nature of lncRNAs combined with their cell-type specific expression could allow exact interactions between proteins, RNA, or DNA to occur only under specific cellular conditions (including during disease conditions) [166]. In this manner, lncRNAs may allow each cell type to utilize the same cellular components for cell-type specific purposes. Similarly, aberrant expression of lncRNAs in autoimmune conditions could allow unwanted molecular interactions to occur that may have detrimental consequences to the host.

The novel and exciting single-cell transcriptome analysis methods have provided rich datasets that illustrate the genetic and epigenetic wiring of individual cells [167]. These technologies have allowed transcriptomic interrogation of large numbers of single cells using next-generation sequencing [167,168]. Single-cell sequencing has allowed investigators to identify new types of stem cells in the gut that are only present at low abundance [169]. Identification and isolation of gut stem cells would be particularly important for the treatment of inflammatory disorders (i.e., inflammatory bowel disease) that result in extensive damage to the gut lining [170]. The immune system depends on the coordinated functions of an immense number of heterogeneous cell types. Single-cell sequencing offers the opportunity to analyze, at a granular level, the specific cell-types that may be contributing to beneficial versus detrimental immune responses [171]. Given that lncRNAs are thought to be lineage specific, it will be interesting to profile their expression at a single-cell level. Fluorescent in situ hybridization studies have provided some evidence of drastic cell-to-cell heterogeneity in lncRNA expression [172]; for example, Kim et al. used single-cell sequencing to demonstrate significant and dramatic heterogeneity in lncRNA-expression during induced pluripotent stem cell development [173].

This chapter has summarized the roles of lncRNAs in the regulation of the immune system. It has specifically highlighted the many mechanisms used by lncRNAs to regulate gene expression and the ways in which dysregulation can occur. LncRNA expression profiling in disease contexts continues to emerge at a blistering pace; however, the functional characterization of these lncRNAs has lagged. The important factors to consider when studying lncRNAs have also been outlined and common tools used for ablating lncRNA expression have been described. Combining the approaches outlined in this review will enable the identification of functionally relevant lncRNAs in autoimmune diseases and permit the design of better tools for diagnosis and targets for therapeutic interventions in the near future.

References

[1] Hangauer MJ, Vaughn IW, McManus MT. Pervasive transcription of the human genome produces thousands of previously unidentified long intergenic noncoding RNAs. Rinn JL, editor. PLoS Genet. Public Library of Science; June 2013;9(6):e1003569.

[2] Carninci P, Kasukawa T, Katayama S, Gough J, Frith MC, Maeda N, et al. The transcriptional landscape of the mammalian genome. Science. American Association for the Advancement of Science September 2, 2005;309(5740):1559–63.

[3] Forrest ARR, Kawaji H, Rehli M, Baillie JK, de Hoon MJL, FANTOM Consortium and the RIKEN PMI and CLST (DGT). et al. A promoter-level mammalian expression atlas. Nature March 27, 2014;507(7493):462–70.

[4] Derrien T, Johnson R, Bussotti G, Tanzer A, Djebali S, Tilgner H, et al. The GENCODE v7 catalog of human long noncoding RNAs: analysis of their gene structure, evolution, and expression. Genome Res Cold Spring Harbor Lab September 2012;22(9):1775–89.

[5] Atianand MK, Fitzgerald KA. Long non-coding RNAs and control of gene expression in the immune system. Trends Mol Med. Elsevier Ltd November 1, 2014;20(11):623–31.

[6] Kapusta A, Feschotte C. Volatile evolution of long noncoding RNA repertoires: mechanisms and biological implications. Trends Genet. Elsevier October 2014;30(10):439–52.

[7] Mattick JS. The central role of RNA in human development and cognition. FEBS Lett June 6, 2011;585(11):1600–16.

[8] Claverie JM. Gene number. What if there are only 30,000 human genes? Science February 16, 2001;291(5507):1255–7.

[9] Grote P, Herrmann BG. Long noncoding RNAs in organogenesis: making the difference. Trends Genet. Elsevier June 2015;31(6):329–35.

[10] Satpathy AT, Chang HY. Long noncoding RNA in hematopoiesis and immunity. Immunity. Elsevier Inc. May 19, 2015;42(5):792–804.

[11] Medzhitov R, Janeway C. Innate immune recognition: mechanisms and pathways. Immunol Rev February 2000;173:89–97.

[12] Dajon M, Iribarren K, Cremer I. Toll-like receptor stimulation in cancer: a pro- and anti-tumor double-edged sword. Immunobiology January 2017;222(1):89–100.

[13] Carpenter S. Long noncoding RNA: novel links between gene expression and innate immunity. Virus Res. Elsevier B.V. September 15, 2015;1–9.

[14] Scanzello CR. Role of low-grade inflammation in osteoarthritis. Curr Opin Rheumatol January 2017;29(1):79–85.

[15] Miranda-Hernandez S, Baxter AG. Role of toll-like receptors in multiple sclerosis. Am J Clin Exp Immunol. e-Century Publishing Corporation 2013;2(1):75–93.

[16] Wunderlich R, Ruehle P-F, Deloch L, Unger K, Hess J, Zitzelsberger H, et al. Interconnection between DNA damage, senescence, inflammation, and cancer. Front Biosci (Landmark Ed) January 1, 2017;22:348–69.

[17] Gao D, Li W. Structures and recognition modes of toll-like receptors. Proteins October 3, 2016;85:3–9.

[18] Carpenter S, O'Neill LAJ. Recent insights into the structure of toll-like receptors and post-translational modifications of their associated signalling proteins. Biochem J August 15, 2009;422(1):1–10.

[19] Lam MTY, Li W, Rosenfeld MG, Glass CK. Enhancer RNAs and regulated transcriptional programs. Trends Biochem Sci. Elsevier April 2014;39(4):170–82.

[20] Carpenter S, Aiello D, Atianand MK, Ricci EP, Gandhi P, Hall LL, et al. A long noncoding RNA mediates both activation and repression of immune response genes. Science. American Association for the Advancement of Science August 16, 2013;341(6147):789–92.

[21] Tong Q, Gong A-Y, Zhang X-T, Lin C, Ma S, Chen J, et al. LincRNA-Cox2 modulates TNF-α-induced transcription of Il12b gene in intestinal epithelial cells through regulation of Mi-2/NuRD-mediated epigenetic histone modifications. FASEB J March 2016;30(3):1187–97.

[22] Li Z, Chao T-C, Chang K-Y, Lin N, Patil VS, Shimizu C, et al. The long noncoding RNA THRIL regulates TNFα expression through its interaction with hnRNPL. Proc Natl Acad Sci USA. National Acad Sciences January 21, 2014;111(3):1002–7.

[23] Oeckinghaus A, Ghosh S. The NF-kappaB family of transcription factors and its regulation. Cold Spring Harb Perspect Biol. Cold Spring Harbor Lab October 2009;1(4):a000034.

[24] Liu B, Sun L, Liu Q, Gong C, Yao Y, Lv X, et al. A cytoplasmic NF-κB interacting long noncoding RNA blocks IκB phosphorylation and suppresses breast cancer metastasis. Cancer Cell March 9, 2015;27(3):370–81.

[25] Rapicavoli NA, Qu K, Zhang J, Mikhail M, Laberge R-M, Chang HY. A mammalian pseudogene lncRNA at the interface of inflammation and anti-inflammatory therapeutics. elife. eLife Sciences Publications Limited July 23, 2013;2:e00762.

[26] Roberts TC, Morris KV. Not so pseudo anymore: pseudogenes as therapeutic targets. Pharmacogenomics. London, UK: Future Medicine Ltd December 2013;14(16):2023–34.

[27] Krawczyk M, Emerson BM. P50-associated COX-2 extragenic RNA (PACER) activates COX-2 gene expression by occluding repressive NF-κB complexes. elife. eLife Sciences Publications Ltd April 29, 2014;3:e01776.

[28] Minghetti L. Role of COX-2 in inflammatory and degenerative brain diseases. Subcell Biochem 2007;42:127–41.

[29] Yu MB, Langridge WHR. The function of myeloid dendritic cells in rheumatoid arthritis. Rheumatol Int. Springer Berlin Heidelberg July 2017;37(7):1043–51.

[30] McInnes IB, Schett G. Cytokines in the pathogenesis of rheumatoid arthritis. Nat Rev Immunol June 2007;7(6):429–42.

[31] Devasthanam AS. Mechanisms underlying the inhibition of interferon signaling by viruses. Virulence. Taylor & Francis February 15, 2014;5(2):270–7.

[32] Schneider WM, Chevillotte MD, Rice CM. Interferon-stimulated genes: a complex web of host defenses. Annu Rev Immunol 2014;32(1):513–45.

[33] Wack A, Terczyńska-Dyla E, Hartmann R. Guarding the frontiers: the biology of type III interferons. Nat Immunol August 2015;16(8):802–9.

[34] Uematsu S, Akira S. Toll-like receptors and type I interferons. J Biol Chem. American Society for Biochemistry and Molecular Biology May 25, 2007;282(21):15319–23.

[35] Valadkhan S, Gunawardane LS. LncRNA-mediated regulation of the interferon response. Virus Res. Elsevier B.V. January 2, 2016;212:127–36.

[36] Josset L, Tchitchek N, Gralinski LE, Ferris MT, Eisfeld AJ, Green RR, et al. Annotation of long non-coding RNAs expressed in collaborative cross founder mice in response to respiratory virus infection reveals a new class of interferon-stimulated transcripts. RNA Biol. Taylor & Francis 2014;11(7):875–90.

[37] Kambara H, Niazi F, Kostadinova L, Moonka DK, Siegel CT, Post AB, et al. Negative regulation of the interferon response by an interferon-induced long non-coding RNA. Nucleic Acids Res. Oxford University Press 2014;42(16):10668–80.

[38] Ouyang J, Zhu X, Chen Y, Wei H, Chen Q, Chi X, et al. NRAV, a long noncoding RNA, modulates antiviral responses through suppression of interferon-stimulated gene transcription. Cell Host Microbe. Elsevier November 12, 2014;16(5):616–26.

[39] Imam H, Bano AS, Patel P, Holla P, Jameel S. The lncRNA NRON modulates HIV-1 replication in a NFAT-dependent manner and is differentially regulated by early and late viral proteins. Sci Rep. Nature Publishing Group March 2, 2015;5:8639.

[40] Li J, Chen C, Ma X, Geng G, Liu B, Zhang Y, et al. Long noncoding RNA NRON contributes to HIV-1 latency by specifically inducing tat protein degradation. Nat Commun. Nature Publishing Group June 13, 2016;7:11730.

[41] Barriocanal M, Carnero E, Segura V, Fortes P. Long non-coding RNA BST2/BISPR is induced by IFN and regulates the expression of the antiviral factor tetherin. Front Immunol. Frontiers 2014;5(7):655.

[42] Kambara H, Gunawardane L, Zebrowski E, Kostadinova L, Jobava R, Krokowski D, et al. Regulation of interferon-stimulated gene BST2 by a lncRNA transcribed from a shared bidirectional promoter. Front Immunol. Frontiers 2014;5(Suppl. 1):676.

[43] Neil SJD. The antiviral activities of tetherin. Curr Top Microbiol Immunol. Berlin, Heidelberg: Springer Berlin Heidelberg 2013;371:67–104. Chapter 3.

[44] Niewold TB. Type I interferon in human autoimmunity. Front Immunol. Frontiers 2014;5:306.

[45] Di Domizio J, Cao W. Fueling autoimmunity: type I interferon in autoimmune diseases. Expert Rev Clin Immunol March 2013;9(3):201–10.

[46] Weidenbusch M, Kulkarni OP, Anders H-J. The innate immune system in human systemic lupus erythematosus. Clin Sci April 25, 2017;131(8):625–34.

[47] Plosker GL. Interferon-β-1b: a review of its use in multiple sclerosis. CNS Drugs. Springer International Publishing January 2011;25(1):67–88.

[48] Wang P, Xue Y, Han Y, Lin L, Wu C, Xu S, et al. The STAT3-binding long noncoding RNA lnc-DC controls human dendritic cell differentiation. Science April 18, 2014;344(6181):310–3.

[49] Chen M-T, Lin H-S, Shen C, Ma Y-N, Wang F, Zhao H-L, et al. PU.1-regulated long noncoding RNA lnc-MC controls human monocyte/macrophage differentiation through interaction with microRNA 199a-5p. Mol Cell Biol September 2015;35(18):3212–24.

[50] Kotzin JJ, Spencer SP, McCright SJ, Kumar DBU, Collet MA, Mowel WK, et al. The long non-coding RNA Morrbid regulates Bim and short-lived myeloid cell lifespan. Nature August 15, 2016;537(7619):239–43.

[51] Hamilton JA. GM-CSF as a target in inflammatory/autoimmune disease: current evidence and future therapeutic potential. Expert Rev Clin Immunol April 2015;11(4):457–65.

[52] Shiomi A, Usui T. Pivotal roles of GM-CSF in autoimmunity and inflammation. Mediators Inflamm. Hindawi 2015;2015(6):568543–613.

[53] Schoenborn JR, Wilson CB. Regulation of interferon-gamma during innate and adaptive immune responses. Adv Immunol. Elsevier 2007;96:41–101.

[54] Gomez JA, Wapinski OL, Yang YW, Bureau J-F, Gopinath S, Monack DM, et al. The NeST long ncRNA controls microbial susceptibility and epigenetic activation of the interferon-γ locus. Cell. Elsevier February 14, 2013;152(4):743–54.

[55] Wang Y, Zhong H, Xie X, Chen CY, Huang D, Shen L, et al. Long noncoding RNA derived from CD244 signaling epigenetically controls $CD8^+$ T-cell immune responses in tuberculosis infection. Proc Natl Acad Sci USA. National Acad Sciences July 21, 2015;112(29):E3883–92.

[56] Burkett PR, Meyer zu Horste G, Kuchroo VK. Pouring fuel on the fire: Th17 cells, the environment, and autoimmunity. J Clin Invest. American Society for Clinical Investigation June 2015;125(6):2211–9.

[57] Crome SQ, Wang AY, Levings MK. Translational mini-review series on Th17 cells: function and regulation of human T helper 17 cells in health and disease. Clin Exp Immunol. Blackwell Publishing Ltd February 2010;159(2):109–19.

[58] Tabarkiewicz J, Pogoda K, Karczmarczyk A, Pozarowski P, Giannopoulos K. The role of IL-17 and Th17 lymphocytes in autoimmune diseases. Arch Immunol Ther Exp (Warsz). Springer Basel December 2015;63(6):435–49.

[59] Hu G, Tang Q, Sharma S, Yu F, Escobar TM, Muljo SA, et al. Expression and regulation of intergenic long noncoding RNAs during T cell development and differentiation. Nat Immunol November 2013;14(11):1190–8.

[60] Spurlock CF, Tossberg JT, Guo Y, Collier SP, Crooke PS, Aune TM. Expression and functions of long noncoding RNAs during human T helper cell differentiation. Nat Commun. Nature Publishing Group April 23, 2015;6:6932.

[61] Rosenblum MD, Remedios KA, Abbas AK. Mechanisms of human autoimmunity. J Clin Invest. American Society for Clinical Investigation June 2015;125(6):2228–33.

[62] Singh RP, Hasan S, Sharma S, Nagra S, Yamaguchi DT, Wong DTW, et al. Th17 cells in inflammation and autoimmunity. Autoimmun Rev December 2014;13(12):1174–81.

[63] Robinson KM, Manni ML, Biswas PS, Alcorn JF. Clinical consequences of targeting IL-17 and TH17 in autoimmune and allergic disorders. Curr Allergy Asthma Rep December 2013;13(6):587–95.

[64] Lessing D, Anguera MC, Lee JT. X chromosome inactivation and epigenetic responses to cellular reprogramming. Annu Rev Genomics Hum Genet. Annual Reviews 2013;14(1):85–110.

[65] Engreitz JM, Ollikainen N, Guttman M. Long non-coding RNAs: spatial amplifiers that control nuclear structure and gene expression. Nat Rev Mol Cell Biol. Nature Research December 2016;17(12):756–70.

[66] McHugh CA, Chen C-K, Chow A, Surka CF, Tran C, McDonel P, et al. The Xist lncRNA interacts directly with SHARP to silence transcription through HDAC3. Nature May 14, 2015;521(7551):232–6.

[67] Wang J, Syrett CM, Kramer MC, Basu A, Atchison ML, Anguera MC. Unusual maintenance of X chromosome inactivation predisposes female lymphocytes for increased expression from the inactive X. Proc Natl Acad Sci USA. National Acad Sciences April 5, 2016;113(14):E2029–38.

[68] Hwang S-H, Lee H, Yamamoto M, Jones LA, Dayalan J, Hopkins R, et al. B cell TLR7 expression drives anti-RNA autoantibody production and exacerbates disease in systemic lupus erythematosus-prone mice. J Immunol. American Association of Immunologists December 15, 2012;189(12):5786–96.

[69] Moser K, Kalies K, Szyska M, Humrich JY, Amann K, Manz RA. CXCR3 promotes the production of IgG1 autoantibodies but is not essential for the development of lupus nephritis in NZB/NZW mice. Arthritis Rheum. Wiley Subscription Services, Inc., A Wiley Company April 2012;64(4):1237–46.

[70] Engreitz JM, Haines JE, Perez EM, Munson G, Chen J, Kane M, et al. Local regulation of gene expression by lncRNA promoters, transcription and splicing. Nature. Nature Research November 17, 2016;539(7629):452–5.

[71] Lee J-E, Ge K. Transcriptional and epigenetic regulation of PPARγ expression during adipogenesis. Cell Biosci. BioMed Central 2014;4(1):29.

[72] Brettingham-Moore KH, Taberlay PC, Holloway AF. Interplay between transcription factors and the epigenome: insight from the role of RUNX1 in leukemia. Front Immunol. Frontiers 2015;6(Spec No. 1):499.

[73] Geisler S, Coller J. RNA in unexpected places: long non-coding RNA functions in diverse cellular contexts. Nat Rev Mol Cell Biol November 2013;14(11):699–712.

[74] Chen J, Xue Y. Emerging roles of non-coding RNAs in epigenetic regulation. Sci China Life Sci January 29, 2016;59(3):227–35.

[75] Davidovich C, Cech TR. The recruitment of chromatin modifiers by long noncoding RNAs: lessons from PRC2. RNA. Cold Spring Harbor Lab December 2015;21(12):2007–222.

[76] Richardson B, Scheinbart L, Strahler J, Gross L, Hanash S, Johnson M. Evidence for impaired T cell DNA methylation in systemic lupus erythematosus and rheumatoid arthritis. Arthritis Rheum November 1990;33(11):1665–73.

[77] Wang L, Zhao Y, Bao X, Zhu X, Kwok YK-Y, Sun K, et al. LncRNA Dum interacts with Dnmts to regulate Dppa2 expression during myogenic differentiation and muscle regeneration. Cell Res. Nature Publishing Group February 17, 2015;25(3):335–50.
[78] Luco RF, Allo M, Schor IE, Kornblihtt AR, Misteli T. Epigenetics in alternative pre-mRNA splicing. Cell. Elsevier January 7, 2011;144(1):16–26.
[79] Li W, Notani D, Rosenfeld MG. Enhancers as non-coding RNA transcription units: recent insights and future perspectives. Nat Rev Genet April 2016;17(4):207–23.
[80] Bonev B, Cavalli G. Organization and function of the 3D genome. Nat Rev Genet October 14, 2016;17(11):661–78.
[81] Hacisuleyman E, Shukla CJ, Weiner CL, Rinn JL. Function and evolution of local repeats in the Firre locus. Nat Commun. Nature Publishing Group March 24, 2016;7:11021.
[82] Bush WS, Moore JH. Chapter 11 Genome-wide association studies. Lewitter F, Kann M, editors. PLoS Comput Biol. Public Library of Science; 2012;8(12):e1002822.
[83] Kumar V, Westra H-J, Karjalainen J, Zhernakova DV, Esko T, Hrdlickova B, et al. Human disease-associated genetic variation impacts large intergenic non-coding RNA expression. Cheung VG, editor. PLoS Genet. Public Library of Science; 2013;9(1):e1003201.
[84] Farh KK-H, Marson A, Zhu J, Kleinewietfeld M, Housley WJ, Beik S, et al. Genetic and epigenetic fine mapping of causal autoimmune disease variants. Nature. Nature Research February 19, 2015;518(7539):337–43.
[85] Ongen H, Andersen CL, Bramsen JB, Oster B, Rasmussen MH, Ferreira PG, et al. Putative cis-regulatory drivers in colorectal cancer. Nature August 7, 2014;512(7512):87–90.
[86] Hazelett DJ, Rhie SK, Gaddis M, Yan C, Lakeland DL, Coetzee SG, et al. Comprehensive functional annotation of 77 prostate cancer risk loci. Cheung VG, editor. PLoS Genet. Public Library of Science; January 2014;10(1):e1004102.
[87] Buckland PR. Allele-specific gene expression differences in humans. Hum Mol Genet. Oxford University Press October 1, 2004;13(Suppl. 2):R255–60.
[88] Wang D, Sadee W. CYP3A4 intronic SNP rs35599367 (CYP3A4*22) alters RNA splicing. Pharmacogenet Genomics January 2016;26(1):40–3.
[89] Agrawal A, Hamvas A, Cole FS, Wambach JA, Wegner D, Coghill C, et al. An intronic ABCA3 mutation that is responsible for respiratory disease. Pediatr Res. Springer Nature June 2012;71(6):633–7.
[90] Thomas LF, Sætrom P. Single nucleotide polymorphisms can create alternative polyadenylation signals and affect gene expression through loss of microRNA-regulation. Guigó R, editor. PLoS Comput Biol. Public Library of Science; 2012;8(8):e1002621.
[91] Stone L. Prostate cancer: risk SNPs modulate lncRNAs. Nat Rev Urol October 2016;13(10):565.
[92] Guo H, Ahmed M, Zhang F, Yao CQ, Li S, Liang Y, et al. Modulation of long noncoding RNAs by risk SNPs underlying genetic predispositions to prostate cancer. Nat Genet October 2016;48(10):1142–50.
[93] Shirasawa S, Harada H, Furugaki K, Akamizu T, Ishikawa N, Ito K, et al. SNPs in the promoter of a B cell-specific antisense transcript, SAS-ZFAT, determine susceptibility to autoimmune thyroid disease. Hum Mol Genet. Oxford University Press October 1, 2004;13(19):2221–31.
[94] Pagon RA, Adam MP, Ardinger HH, Wallace SE, Amemiya A, Bean LJ, et al. Cartilage-hair hypoplasia – anauxetic dysplasia spectrum disorders. Seattle (WA): University of Washington; 1993.
[95] Huang W, Thomas B, Flynn RA, Gavzy SJ, Wu L, Kim SV, et al. DDX5 and its associated lncRNA Rmrp modulate TH17 cell effector functions. Nature December 24, 2015;528(7583):517–22.
[96] Sigdel KR, Cheng A, Wang Y, Duan L, Zhang Y. The emerging functions of long noncoding RNA in immune cells: autoimmune diseases. J Immunol Res. Hindawi Publishing Corporation May 18, 2015;1–9.
[97] Wu G-C, Pan H-F, Leng R-X, Wang D-G, Li X-P, Li X-M, et al. Emerging role of long noncoding RNAs in autoimmune diseases. Autoimmun Rev September 2015;14(9):798–805.
[98] Müller N, Döring F, Klapper M, Neumann K, Schulte DM, Türk K, et al. Interleukin-6 and tumour necrosis factor-α differentially regulate lincRNA transcripts in cells of the innate immune system in vivo in human subjects with rheumatoid arthritis. Cytokine July 2014;68(1):65–8.
[99] Zhang Y, Xu Y-Z, Sun N, Liu J-H, Chen F-F, Guan X-L, et al. Long noncoding RNA expression profile in fibroblast-like synoviocytes from patients with rheumatoid arthritis. Arthritis Res Ther. BioMed Central October 6, 2016;18(1):227.
[100] Mediwake R, Isenberg DA, Schellekens GA, van Venrooij WJ. Use of anti-citrullinated peptide and anti-RA33 antibodies in distinguishing erosive arthritis in patients with systemic lupus erythematosus and rheumatoid arthritis. Ann Rheum Dis. BMJ Publishing Group January 2001;60(1):67–8.
[101] Bennett L, Palucka AK, Arce E, Cantrell V, Borvak J, Banchereau J, et al. Interferon and granulopoiesis signatures in systemic lupus erythematosus blood. J Exp Med. Rockefeller University Press March 17, 2003;197(6):711–23.

[102] Shi L, Zhang Z, Yu AM, Wang W, Wei Z, Akhter E, et al. The SLE transcriptome exhibits evidence of chronic endotoxin exposure and has widespread dysregulation of non-coding and coding RNAs. Prokunina-Olsson L, editor. PLoS One. Public Library of Science; 2014;9(5):e93846.

[103] Sonkoly E, Bata-Csorgo Z, Pivarcsi A, Polyanka H, Kenderessy-Szabo A, Molnar G, et al. Identification and characterization of a novel, psoriasis susceptibility-related noncoding RNA gene, PRINS. J Biol Chem. American Society for Biochemistry and Molecular Biology June 24, 2005;280(25):24159–67.

[104] Szegedi K, Sonkoly E, Nagy N, Németh IB, Bata-Csorgo Z, Kemeny L, et al. The anti-apoptotic protein G1P3 is overexpressed in psoriasis and regulated by the non-coding RNA PRINS. Exp Dermatol. Blackwell Publishing Ltd March 2010;19(3):269–78.

[105] Lin MF, Jungreis I, Kellis M. PhyloCSF: a comparative genomics method to distinguish protein coding and noncoding regions. Bioinformatics. Oxford University Press July 1, 2011;27(13):i275–82.

[106] Johnsson P, Lipovich L, Grandér D, Morris KV. Evolutionary conservation of long non-coding RNAs; sequence, structure, function. Biochim Biophys Acta March 2014;1840(3):1063–71.

[107] Quinn JJ, Chang HY. Unique features of long non-coding RNA biogenesis and function. Nat Rev Genet January 2016;17(1):47–62.

[108] Roberts TC, Morris KV, Weinberg MS. Perspectives on the mechanism of transcriptional regulation by long noncoding RNAs. Epigenetics January 2014;9(1):13–20.

[109] Nitsche A, Stadler PF. Evolutionary clues in lncRNAs. WIREs RNA. John Wiley & Sons, Inc. July 19, 2016;8(1).

[110] Dermitzakis ET, Reymond A, Lyle R, Scamuffa N, Ucla C, Deutsch S, et al. Numerous potentially functional but non-genic conserved sequences on human chromosome 21. Nature. Nature Publishing Group December 5, 2002;420(6915):578–82.

[111] Ule J. Alu elements: at the crossroads between disease and evolution. Biochem Soc Trans. Portland Press Limited December 2013;41(6):1532–5.

[112] de Koning APJ, Gu W, Castoe TA, Batzer MA, Pollock DD. Repetitive elements may comprise over two-thirds of the human genome. Copenhaver GP, editor. PLoS Genet. Public Library of Science; December 2011;7(12):e1002384.

[113] Wicker T, Sabot F, Hua-Van A, Bennetzen JL, Capy P, Chalhoub B, et al. A unified classification system for eukaryotic transposable elements. Nat Rev Genet December 2007;8(12):973–82.

[114] Hacisuleyman E, Goff LA, Trapnell C, Williams A, Henao-Mejia J, Sun L, et al. Topological organization of multichromosomal regions by the long intergenic noncoding RNA Firre. Nat Struct Mol Biol February 2014;21(2):198–206.

[115] Gong C, Maquat LE. LncRNAs transactivate STAU1-mediated mRNA decay by duplexing with 3′UTRs via Alu elements. Nature February 10, 2011;470(7333):284–8.

[116] Atianand MK, Hu W, Satpathy AT, Shen Y, Ricci EP, Alvarez-Dominguez JR, et al. A long noncoding RNA lincRNA-EPS acts as a transcriptional brake to restrain inflammation. Cell June 16, 2016;165(7):1672–85.

[117] Galindo MI, Pueyo JI, Fouix S, Bishop SA, Couso JP. Peptides encoded by short ORFs control development and define a new eukaryotic gene family. Martinez Arias A, editor. PLoS Biol. Public Library of Science; May 2007;5(5):e106.

[118] Cheng H, Chan WS, Li Z, Wang D, Liu S, Zhou Y. Small open reading frames: current prediction techniques and future prospect. Curr Protein Pept Sci. NIH Public Access September 2011;12(6):503–7.

[119] Slavoff SA, Mitchell AJ, Schwaid AG, Cabili MN, Ma J, Levin JZ, et al. Peptidomic discovery of short open reading frame-encoded peptides in human cells. Nat Chem Biol January 2013;9(1):59–64.

[120] Ingolia NT, Ghaemmaghami S, Newman JRS, Weissman JS. Genome-wide analysis in vivo of translation with nucleotide resolution using ribosome profiling. Science. American Association for the Advancement of Science April 10, 2009;324(5924):218–23.

[121] Ingolia NT. Genome-wide translational profiling by ribosome footprinting. Meth Enzymol. Elsevier 2010;470:119–42.

[122] Ingolia NT. Ribosome footprint profiling of translation throughout the genome. Cell. Elsevier March 24, 2016;165(1):22–33.

[123] Ingolia NT, Brar GA, Stern-Ginossar N, Harris MS, Talhouarne GJS, Jackson SE, et al. Ribosome profiling reveals pervasive translation outside of annotated protein-coding genes. Cell Rep. Elsevier September 11, 2014;8(5):1365–79.

[124] Guttman M, Russell P, Ingolia NT, Weissman JS, Lander ES. Ribosome profiling provides evidence that large noncoding RNAs do not encode proteins. Cell. Elsevier July 3, 2013;154(1):240–51.

[125] Carlevaro-Fita J, Rahim A, Guigó R, Vardy LA, Johnson R. Cytoplasmic long noncoding RNAs are frequently bound to and degraded at ribosomes in human cells. RNA. Cold Spring Harbor Lab June 2016;22(6):867–82.

[126] Chew G-l, Pauli A, Rinn JL, Regev A, Schier AF, Valen E. Ribosome profiling reveals resemblance between long noncoding RNAs and 5′ leaders of coding RNAs. Development.Oxford University Press for The Company of Biologists Limited July 2013;140(13):2828–34.

[127] Anderson DM, Anderson KM, Chang C-l, Makarewich CA, Nelson BR, McAnally JR, et al. A micropeptide encoded by a putative long noncoding RNA regulates muscle performance. Cell. Elsevier February 12, 2015;160(4):595–606.
[128] Boettcher M, McManus MT. Choosing the right tool for the job: RNAi, TALEN, or CRISPR. Mol Cell May 21, 2015;58(4):575–85.
[129] Kurreck J. RNA interference: from basic research to therapeutic applications. Angew Chem Int Ed Engl 2009;48(8):1378–98.
[130] Lambeth LS, Smith CA. Short hairpin RNA-mediated gene silencing. Methods Mol Biol. Totowa (NJ): Humana Press 2013;942:205–32. Chapter 12.
[131] Lennox KA, Behlke MA. Cellular localization of long non-coding RNAs affects silencing by RNAi more than by antisense oligonucleotides. Nucleic Acids Res. Oxford University Press January 29, 2016;44(2):863–77.
[132] Zong X, Huang L, Tripathi V, Peralta R, Freier SM, Guo S, et al. Knockdown of nuclear-retained long noncoding RNAs using modified DNA antisense oligonucleotides. Methods Mol Biol. New York (NY): Springer New York 2015;1262(Chapter 20): 321–31.
[133] Deleavey GF, Damha MJ. Designing chemically modified oligonucleotides for targeted gene silencing. Chem Biol August 24, 2012;19(8):937–54.
[134] Cerritelli SM, Crouch RJ. Ribonuclease H: the enzymes in eukaryotes. FEBS J. Blackwell Publishing Ltd March 2009;276(6):1494–505.
[135] Ilott NE, Heward JA, Roux B, Tsitsiou E, Fenwick PS, Lenzi L, et al. Long non-coding RNAs and enhancer RNAs regulate the lipopolysaccharide-induced inflammatory response in human monocytes. Nat Commun June 9, 2014;5:1–14.
[136] Meng L, Ward AJ, Chun S, Bennett CF, Beaudet AL, Rigo F. Towards a therapy for Angelman syndrome by targeting a long non-coding RNA. Nature February 19, 2015;518(7539):409–12.
[136a] de Kok JB, Verhaegh GW, Roelofs RW, Hessels D, Kiemeney LA, Aalders TW, et al. DD3(PCA3), a very sensitive and specific marker to detect prostate tumors. Cancer Res. 2002;62(9):2695–8.
[137] Hochstrasser ML, Doudna JA. Cutting it close: CRISPR-associated endoribonuclease structure and function. Trends Biochem Sci January 2015;40(1):58–66.
[138] Sternberg SH, Doudna JA. Expanding the biologist's toolkit with CRISPR-Cas9. Mol Cell May 21, 2015;58(4):568–74.
[139] Ho T-T, Zhou N, Huang J, Koirala P, Xu M, Fung R, et al. Targeting non-coding RNAs with the CRISPR/Cas9 system in human cell lines. Nucleic Acids Res. Oxford University Press February 18, 2015;43(3):e17.
[140] Zhu S, Li W, Liu J, Chen C-H, Liao Q, Xu P, et al. Genome-scale deletion screening of human long non-coding RNAs using a paired-guide RNA CRISPR–Cas9 library. Nat Biotechnol October 31, 2016;34(12):1279–86.
[141] Bassett AR, Akhtar A, Barlow DP, Bird AP, Brockdorff N, Duboule D, et al. Considerations when investigating lncRNA function in vivo elife. eLife Sciences Publications Ltd August 14, 2014;3:e03058.
[142] Gilbert LA, Horlbeck MA, Adamson B, Villalta JE, Chen Y, Whitehead EH, et al. Genome-scale CRISPR-mediated control of gene repression and activation. Cell October 23, 2014;159(3):647–61.
[143] Larson MH, Gilbert LA, Wang X, Lim WA, Weissman JS, Qi LS. CRISPR interference (CRISPRi) for sequence-specific control of gene expression. Nat Protoc November 2013;8(11):2180–96.
[144] Ying Y, Yang X, Zhao K, Mao J, Kuang Y, Wang Z, et al. The Krüppel-associated box repressor domain induces reversible and irreversible regulation of endogenous mouse genes by mediating different chromatin states. Nucleic Acids Res. Oxford University Press February 18, 2015;43(3):1549–61.
[145] Liu SJ, Horlbeck MA, Cho SW, Birk HS, Malatesta M, He D, et al. CRISPRi-based genome-scale identification of functional long noncoding RNA loci in human cells. Science. American Association for the Advancement of Science December 15, 2016;aah7111.
[146] Sauvageau M, Goff LA, Lodato S, Bonev B, Groff AF, Gerhardinger C, et al. Multiple knockout mouse models reveal lincRNAs are required for life and brain development elife. eLife Sciences Publications Limited December 31, 2013;2:e01749.
[147] Grote P, Wittler L, Hendrix D, Koch F, Währisch S, Beisaw A, et al. The tissue-specific lncRNA Fendrr is an essential regulator of heart and body wall development in the mouse. Dev Cell. Elsevier January 28, 2013;24(2):206–14.
[148] Marahrens Y, Panning B, Dausman J, Strauss W, Jaenisch R. Xist-deficient mice are defective in dosage compensation but not spermatogenesis. Genes Dev January 15, 1997;11(2):156–66.
[149] Ripoche MA, Kress C, Poirier F, Dandolo L. Deletion of the H19 transcription unit reveals the existence of a putative imprinting control element. Genes Dev June 15, 1997;11(12):1596–604.
[150] Fitzpatrick GV, Soloway PD, Higgins MJ. Regional loss of imprinting and growth deficiency in mice with a targeted deletion of KvDMR1. Nat Genet November 2002;32(3):426–31.
[151] Nakagawa S, Ip JY, Shioi G, Tripathi V, Zong X, Hirose T, et al. Malat1 is not an essential component of nuclear speckles in mice. RNA. Cold Spring Harbor Lab August 2012;18(8):1487–99.

[152] Sleutels F, Zwart R, Barlow DP. The non-coding Air RNA is required for silencing autosomal imprinted genes. Nature February 14, 2002;415(6873):810–3.

[153] Paralkar VR, Taborda CC, Huang P, Yao Y, Kossenkov AV, Prasad R, et al. Unlinking an lncRNA from its associated cis element. Mol Cell. Elsevier April 7, 2016;62(1):104–10.

[154] Semerano L, Minichiello E, Bessis N, Boissier M-c. Novel immunotherapeutic avenues for rheumatoid arthritis. Trends Mol Med. Elsevier March 2016;22(3):214–29.

[155] Korkmaz B, Lesner A, Guarino C, Wysocka M, Kellenberger C, Watier H, et al. Inhibitors and antibody fragments as potential anti-inflammatory therapeutics targeting neutrophil proteinase 3 in human disease. Sibley DR, editor. Pharmacol Rev. American Society for Pharmacology and Experimental Therapeutics; July 2016;68(3):603–30.

[156] Jacob N, Targan SR, Shih DQ. Cytokine and anti-cytokine therapies in prevention or treatment of fibrosis in IBD. United Eur Gastroenterol J. London (England, UK): SAGE Publications August 2016;4(4):531–40.

[157] Moreland LW, Baumgartner SW, Schiff MH, Tindall EA, Fleischmann RM, Weaver AL, et al. Treatment of rheumatoid arthritis with a recombinant human tumor necrosis factor receptor (p75)-Fc fusion protein. N Engl J Med. Massachusetts Medical Society July 17, 1997;337(3):141–7.

[158] Sormani MP, Gasperini C, Romeo M, Rio J, Calabrese M, Cocco E, et al. Assessing response to interferon-β in a multicenter dataset of patients with MS. Neurology. Lippincott Williams & Wilkins July 12, 2016;87(2):134–40.

[159] Rubbert-Roth A, Finckh A. Treatment options in patients with rheumatoid arthritis failing initial TNF inhibitor therapy: a critical review. Arthritis Res Ther. BioMed Central 2009;11(Suppl. 1):S1.

[160] Margolis SS, Sell GL, Zbinden MA, Bird LM. Angelman syndrome. Neurotherapeutics. Springer US July 2015;12(3):641–50.

[161] King MC, Wilson AC. Evolution at two levels in humans and chimpanzees. Science April 11, 1975;188(4184):107–16.

[162] Rinn JL, Chang HY. Genome regulation by long noncoding RNAs. Annu Rev Biochem. Annual Reviews 2012;81(1):145–66.

[163] Booy EP, McRae EKS, Howard R, Deo SR, Ariyo EO, Dzananovic E, et al. RNA helicase associated with AU-rich element (RHAU/DHX36) interacts with the 3′-tail of the long non-coding RNA BC200 (BCYRN1). J Biol Chem. American Society for Biochemistry and Molecular Biology March 4, 2016;291(10):5355–72.

[164] Marchese FP, Grossi E, Marín-Béjar O, Bharti SK, Raimondi I, González J, et al. A long noncoding RNA regulates sister chromatid cohesion. Mol Cell August 4, 2016;63(3):397–407.

[165] Atianand MK, Hu W, Satpathy AT, Shen Y, Ricci EP, Alvarez-Dominguez JR, et al. A long noncoding RNA lincRNA-EPS acts as a transcriptional brake to restrain inflammation. Cell. Elsevier June 16, 2016;165(7):1672–85.

[166] Guttman M, Rinn JL. Modular regulatory principles of large non-coding RNAs. Nature February 15, 2012;482(7385):339–46.

[167] Gawad C, Koh W, Quake SR. Single-cell genome sequencing: current state of the science. Nat Rev Genet. Nature Research March 2016;17(3):175–88.

[168] Zhang X, Marjani SL, Hu Z, Weissman SM, Pan X, Wu S. Single-cell sequencing for precise cancer research: progress and prospects. Cancer Res. American Association for Cancer Research March 15, 2016;76(6):1305–12.

[169] Grün D, Lyubimova A, Kester L, Wiebrands K, Basak O, Sasaki N, et al. Single-cell messenger RNA sequencing reveals rare intestinal cell types. Nature September 10, 2015;525(7568):251–5.

[170] Hawkey CJ, Snowden JA, Lobo A, Beglinger C, Tyndall A. Stem cell transplantation for inflammatory bowel disease: practical and ethical issues. Gut. BMJ Group June 2000;46(6):869–72.

[171] Proserpio V, Mahata B. Single-cell technologies to study the immune system. Immunology February 2016;147(2):133–40.

[172] Cabili MN, Dunagin MC, McClanahan PD, Biaesch A, Padovan-Merhar O, Regev A, et al. Localization and abundance analysis of human lncRNAs at single-cell and single-molecule resolution. Genome Biol. BioMed Central January 29, 2015;16(1):20.

[173] Kim DH, Marinov GK, Pepke S, Singer ZS, He P, Williams B, et al. Single-cell transcriptome analysis reveals dynamic changes in lncRNA expression during reprogramming. Cell Stem Cell. Elsevier January 8, 2015;16(1):88–101.

CHAPTER

6

Epigenetics of Systemic Lupus Erythematosus

Patricia Costa-Reis, Yiu T. Leung**, Kathleen E. Sullivan†*

*Lisbon University, Santa Maria Hospital, Lisbon, Portugal;
**Temple University School of Medicine, Philadelphia, PA, United States;
†The Children's Hospital of Philadelphia, University of Pennsylvania Perelman School of Medicine, Philadelphia, PA, United States

OUTLINE

6.1 INTRODUCTION

Systemic lupus erythematosus (SLE) is the quintessential systemic autoimmune disease. It is a chronic inflammatory disease that has an estimated population frequency of 1:1000 (affecting predominantly young women) and significant long-term morbidity [1]. The disease has three cardinal immunologic features: (1) immune complex deposition in end organs and inflammation across multiple tissues; (2) immune cell dysfunction; and (3) autoantibodies directed at nuclear antigens. Although the cause(s) of SLE are unknown, the model for its development includes genetic susceptibility and environmental triggers (Fig. 6.1). The aggregation of genetic susceptibility and diverse triggers leads to a gradual loss of self-tolerance. Autoreactive T and B cells emerge and over time this leads to high

The Epigenetics of Autoimmunity. http://dx.doi.org/10.1016/B978-0-12-809912-4.00006-4

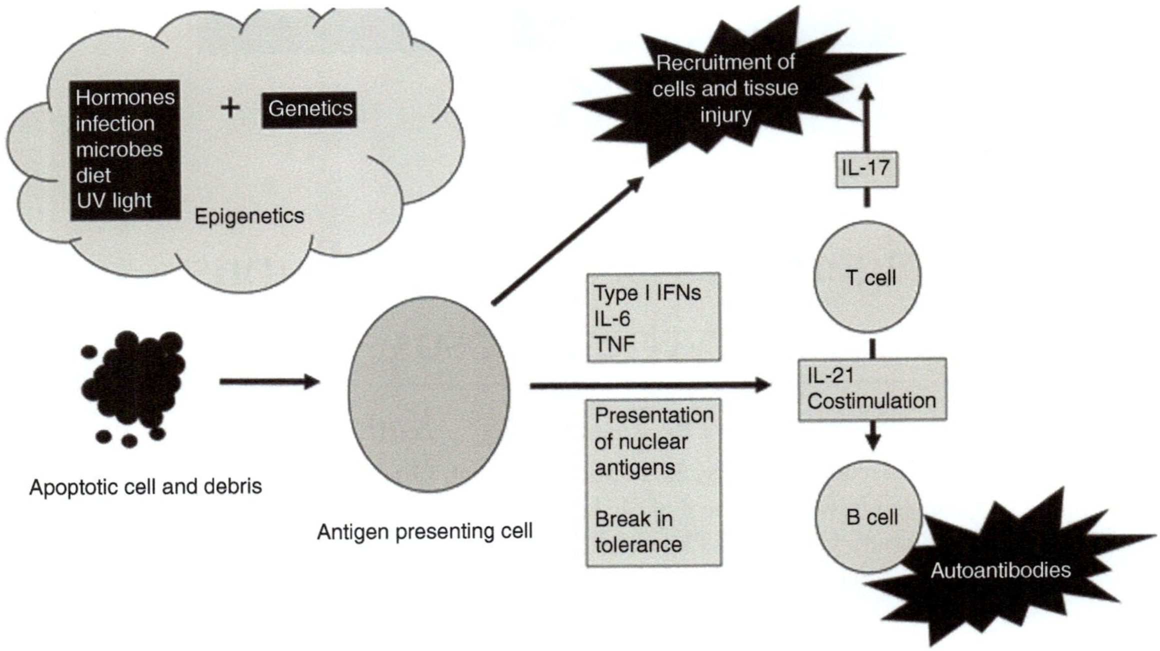

FIGURE 6.1 **Model for the development of SLE.** Both environmental and genetic factors contribute to the susceptibility state. These contributors may be filtered through a "lens" of epigenetic changes. Once the overall susceptibility state has been established and promoted through the accrual of environmental contributors, the features implicated in the progression to an active disease state are the imbalance of apoptotic cell production and clearance. Residual apoptotic debris contains nucleic acids, which stimulates dendritic cells and other myeloid cells to produce inflammatory cytokines. These cytokines in turn contribute to B- and T-cell activation and promote the loss of tolerance and nucleic acid responses. Additional cytokines promote inflammation and the recruitment of cells into end organs.

levels of autoantibodies. The autoantibodies bind to self-antigens, which are initially released from the nucleus during apoptosis or neutrophil extracellular trap (NET)osis, and are deposited in multiple organs where they trigger inflammation. Skin, joint, kidney, and blood cells are typically involved. An imbalance between apoptosis and the clearance of apoptotic debris is thought to be central to the development of SLE. Apoptotic debris is recognized as a viral pattern by the innate immune system and triggers the production of type I interferon. The genes implicated in SLE (as defined by genome-wide association studies) reflect the cardinal immunologic features of the disease, and include MHC class II genes, immune complex clearance genes, apoptosis-regulating genes, and interferon-response genes. Gene expression studies have been instrumental in driving the field forward and have exposed the interferon pathway as being critical in SLE [2–4]; however, the etiopathogenesis is still not fully understood despite there being over 20,000 published studies. The role of epigenetics in the pathophysiology of SLE has recently begun to be examined.

The etiology of most autoimmune diseases remains completely unknown, and SLE is no exception. Recent data have demonstrated the genetic interrelationships between most autoimmune diseases [5], and it is reasonable to believe that there are also common epigenetic underpinnings to autoimmune diseases. This theory is supported by the observation that multiple autoimmune diseases epidemiologically cosegregate in families. A common immunologic feature in multiple autoimmune diseases is the type I interferon signature [2–4]. This signature reflects multiple inputs but is common to a variety of autoimmune diseases [6–11]; a type I interferon signature can even be identified in unaffected relatives who are at risk of developing an autoimmune disease [12]. These commonalities have made it difficult to understand why lupus develops in some individuals while rheumatoid arthritis develops in others. Gene expression studies have more finely tuned the signatures associated with different autoimmune diseases and support the involvement of distinct pathways. Gene expression patterns have encouragingly been reported within distinct SLE subsets; for example, disease activity, as measured by the SLE Disease Activity Index, was associated with a plasmablast signature while nephritis was associated with a neutrophil signature [13]. These gene expression signatures reflect transcriptional dysregulation, which may reflect alterations to the epigenome.

Certain disease characteristics are relatively specific to lupus; for example, exposure to ultraviolet light has long been associated with exacerbations of lupus, and photosensitivity is one of the American College of Rheumatology's SLE diagnostic criteria [14]. Most patients with lupus will report photosensitivity. Sunlight exposure can induce systemic disease activity in addition to the more common cutaneous flares. This is thought to be due to the activation of the NFκB pathway, which is central to the production of most proinflammatory cytokines. Keratinocytes directly exposed to ultraviolet light mount an inflammatory response to the DNA damage, which results in apoptosis of the cells. Ultraviolet radiation has been shown to drive DNA demethylation during subsequent rounds of replication, linking the epidemiologic association to epigenetic changes [15]. A second epidemiologic feature that is not unique to SLE, but is most pronounced in SLE compared to other autoimmune diseases, is the strong female preponderance. A recent transgenic mouse model demonstrated that accelerated estrogen receptor signaling could induce a lupus-like disease associated with aberrant DNA methylation of the CD40L promoter [16]. This resulted in increased somatic hypermutations and class switch recombination.

The female preponderance of SLE has also been examined from the perspective of prolactin. Hyperprolactinemia is more common in SLE patients (15%–33%) than in the general population (0.4%–3%) [17–20], and males with SLE also have higher prolactin levels [21]. Disease activity and autoantibody levels have been correlated with prolactin levels in several studies [22–30]. Bromocriptine, a dopamine agonist that inhibits prolactin secretion, has proven therapeutic effects in SLE, thus confirming the pivotal role for this hormone in the disease pathogenesis [31–34]. Murine models have also confirmed the disease-causing or disease-accelerating effects of prolactin [35–37]. Prolactin is known to increase histone acetylation, which may suggest that it exerts its effects via an epigenetic mechanism [38]. Therefore some of the most strongly associated features of SLE appear to be partly mediated through epigenetics.

Epigenetics is often defined as stable and heritable changes in gene expression without alterations in the DNA sequence. Epigenetics controls gene expression patterns throughout the development of a cell, in response to environmental cues, and according to the cell program. Therefore epigenetics represents a critical interface between health and disease. Heritability, as defined by twin concordance or sibling recurrence risk, is moderately high in most autoimmune diseases; however, genome-wide association studies have only identified loci with low relative risks. This has led to the hypothesis that epigenetic factors majorly contribute to the heritability of autoimmune disorders. Epidemiologic studies implicating diet, toxins, hormones, and sunlight also support a potential role for epigenetics in the overall susceptibility to autoimmune diseases. Although the hypothesis that DNA methylation and histone modifications integrate environmental risk factors and classical genetic heritability is attractive, there are few studies that have directly examined this claim.

This chapter will review the epigenetic associations with SLE. There are some facets of epigenetic associations that are common to multiple autoimmune diseases, some that appear to be relatively specific to SLE, and some that require additional study. The mechanisms of DNA methylation, histone modifications, microRNAs (miRNAs), and long noncoding RNAs (lncRNAs) will also be discussed in this chapter. These are not all strictly in line with the formal definition of epigenetics; however, they act to regulate gene expression and the effects can be durable, therefore they are conceptually functioning as epigenetic regulatory process and are mechanistically linked.

6.2 DNA METHYLATION

DNA methylation is the addition of a methyl group to a cytosine at the 5′ position of a CpG dinucleotide, which results in the conversion of the cytosine to methylcytosine. CpG dinucleotides tend to cluster in regions called CpG islands, which are defined as regions of more than 200 bases with a GC content of at least 50% [39]. Approximately 60% of promoters are associated with CpG islands and they are normally unmethylated. DNA methylation typically represses gene expression through several mechanisms; for example, methyl CpG binding domain proteins can recruit histone-modifying complexes that drive repression [40]. DNA methylation can also directly inhibit transcription by interfering with the recruitment of transcription factors [41]. DNA methylation is mediated by DNA methyltransferases (DNMT); DNMT1 maintains the methylation status during cell replication while DNMT3A and DNMT3B induce de novo methylation. In contrast, DNA demethylation is regulated by the ten-eleven translocation (TET) family of demethylases [42]. DNA methylation is a fascinating topic of study in SLE as many of the epidemiologic features of the disease (i.e., diet, smoking, ultraviolet light, and toxins) are known to influence DNA methylation. Therefore this may be a mechanism through which these epidemiologic variables can contribute to the susceptibility [43–45]. It is generally believed that DNA methylation is tightly regulated; however, expression levels of pivotal DNA methyltransferases are aberrant in SLE [46].

DNA methylation was the first epigenetic change to be identified in human SLE patients. A key research finding was that drugs known to induce lupus-like features acted by inhibiting DNA methylation [47]. Transferring T cells from drug-treated mice into control mice induced lupus in the control population [48]. This group also demonstrated that 5-azacytidine-induced demethylation caused autoreactivity in otherwise healthy human CD4 T cells, leading to characteristics of SLE T cells [49]. Therefore this murine model was instrumental in advancing the understanding of DNA methylation in SLE.

T cells from patients with active lupus show global DNA hypomethylation [50], whereas patients with nephritis have more robust DNA demethylation [51]. DNA hypomethylation has been found in SLE CD4 T cells but not in CD8 T cells. CD4 T cells from patients with active lupus are more extensively demethylated and overexpress key adhesion molecules that mediate direct cytotoxicity. When examined on a genome-wide scale, many interferon-response genes were shown to be specifically hypomethylated [51,52]. One surprising observation was that the effect seen during a flare was also present in patients with quiescent disease in the postflare period, thus supporting a durable effect that can set the character of the disease in each patient [52]. Patients tend to recapitulate their original presentation during flares, and the epigenetic echoes of their previous disease may influence the subsequent course. Identifying effects in interferon-responsive genes has been interpreted as transcriptional poising of genes in naive CD4 T cells in SLE. The observation that interferon-responsive genes are specifically hypomethylated has been confirmed and is consistent with the known key pathways in SLE [53]. It is not known if the poising of interferon-response genes is causal, or if it occurs as a consequence of the disease. The DNA methylation status of the IFI44L promoter was found to be a sensitive marker for SLE, supporting a disease-specific effect [54]. Critical efforts to link DNA hypomethylation and gene expression data have begun and have so far demonstrated a strong concordance [55]. The DNA methylation profile observed in patients is likely to reflect both contributors that predate the onset of lupus as well as effects from the disease process.

Additional transcriptional consequences of altered DNA methylation in SLE have also been described; for example, helper 2 (Th2) cytokines (i.e., IL-4 and IL-13) were found to be derepressed [52]. Genes involved in T-cell activation were shown to be derepressed in early SLE studies, which supports the emerging evidence and suggests that these changes are critical for disease activity, progression, or development [56–58]. Although initially performed in T cells, these studies have now been extended, and B cells, monocytes, and neutrophils all exhibit demethylation of interferon genes, although there is little concordance between cell types [59]. Not all dysregulated gene expression in SLE is associated with altered DNA methylation, and active research is being performed to understand which gene sets are regulated by which strategy. Gene expression is regulated by multiple epigenetic mechanisms, and some poised genes are primarily regulated by transcription factor loading and the release of RNA polymerase II from a paused or stalled position [60]; for example, *TNFAIP3*, which is involved in the negative regulation of the NFκB pathway, is a known susceptibility gene for SLE. It has normal CpG DNA methylation in SLE but has been shown to be dysregulated due to altered histone modifications [61].

The robust association of DNA methylation changes with SLE has driven intense mechanistic studies. Oxidative stress, which is an umbrella term that refers to increased levels of superoxide or nitric oxide within cells, is often cited as a potential trigger for demethylation. Oxidative stress may be caused by ultraviolet light, smoking, infections, toxins, or air pollution [43–45], and it has been shown to decrease DNMT1 [62]. Sera from patients with SLE demonstrated signs of oxidative stress, and the level of nitration in the sera was related to the disease activity [63,64]. Thus one potential contribution to the development of SLE could be DNA demethylation secondary to oxidative stress. It is presumed that it is the presence of multiple aggregated variables that drives the disease susceptibility, and dietary factors are known to interact with other triggers [65]. Maternal methyl-donor supplementation in mice clearly shapes the epigenome of the developing embryo; for example, dietary methyl supplementation with folic acid, vitamin B12, choline, and betaine in the agouti mouse was shown to alter the phenotype of the offspring by increasing CpG methylation at the agouti gene locus [66]. In transgenic mice already predisposed to developing autoimmunity, dietary supplementation or the restriction of methyl-donor groups was shown to ameliorate or exacerbate the disease activity, respectively [67]. These pivotal murine studies have therefore emphasized the potential impact of dietary factors.

The oxidative derivative of 5-methylcytosine (5-mc), 5-hydroxymethylcytosine (5-hmC), was initially identified in mouse embryonic stem cells. Expression of 5-hmC was shown to be dependent on the TET family of proteins, which enzymatically convert 5-mC to 5-hmC. Subsequently, 5-hmC was found in certain terminally differentiated cells. The differentiation of hematopoietic progenitors, specifically effector T cells, led to the induction of TET2 and an increase in the 5-hmC content [68,69]. This correlated with increased binding of master transcriptional regulators and the induction of cytokines related to T-cell effector functions. One study investigated the level of 5-hmC in SLE CD4 T cells using a combined analysis of differential DNA 5-hmC and gene expression data. The authors identified that 131 genes differed from healthy controls, and this gene set was enriched for immunologically reactive genes [70]. Therefore changes in 5-hmC methylation may contribute to aberrant gene expression and disease pathogenesis; however, limited investigation of this modification has been performed mechanistically in SLE.

In SLE, demethylation is believed to alter the immunologic reactivity of DNA, regulate gene expression, and be implicated in the development of autoantibodies. In a murine model, activated lymphocyte-derived DNA induced a lupus-like disease phenotype [71]. The effects of the DNA in disease induction were related to apoptosis, as DNA derived from necrotic cells was unable to induce the disease [71]. The authors further demonstrated that the effects were mediated by the cytoplasmic DNA sensors, DAI and AIM2 [72,73]. This group also highlighted the importance of macrophages in DNA sensing, although macrophages have been less intensively studies in SLE than other cell types. Comparable studies are not possible in humans; however, one study from 1966 identified circulating DNA in the plasma of patients with SLE [74], and subsequent studies suggested that this was likely derived from apoptotic cells [75]. Next generation sequencing techniques have revealed that the level of hypomethylated DNA in the serum correlates with disease activity. Furthermore, anti-double stranded DNA (dsDNA) antibodies (which are characteristic of SLE) show enhanced binding to hypomethylated DNA compared to normally methylated DNA [76]. The presence of anti-dsDNA antibodies skews the types of free DNA in the serum and may alter the immune reactivity in the process. Hypomethylated DNA, specifically CpG dinucleotides, can drive both B-cell and myeloid-cell hyperactivity [72,73,77].

6.3 HISTONE MODIFICATIONS

Histone modifications are epigenetic marks that have been less well studied due to the technical complexity and expense of the methods. Nevertheless, epigenetic therapies have made progress in treating aberrant histone methylation and acetylation disorders, therefore there are compelling reasons to define SLE-specific histone modifications. DNA is wrapped around the core histones, which have their N-terminal tails protruding. Posttranslational modifications have been detected on over 60 different residues in these histone tails [78]. Histone modifications are dynamic and each individual modification provides a code for the cell to interpret. One complication in the study of posttranslational histone modifications lies in the diverse roles of the targeting enzymes; for example, many histone-modifying enzymes also modify nonhistone substrates and therefore attributing cause is often complicated by changes occurring on other signaling proteins. There are at least eight distinct types of posttranslational modifications found on histone tails: acetylation and methylation are the most characterized and evaluated global approaches; however, phosphorylation, ubiquitination, sumoylation, ribosylation, deamination, and proline isomerization also occur. Posttranslational histone modifications are generally described as part of the transcriptional regulatory process; however, they are also implicated in the regulation of DNA repair and chromatin condensation. Each modification is regulated by a "writer" and an "eraser" enzyme, and interpreted by a "reader" enzyme. There are histone acetyltransferases, histone methyltransferases, and corresponding eraser enzymes such as histone deacetylases (HDACs) and histone demethylases. Alterations to the histone modification profile correspondingly alter the cellular phenotype due to the profound role of these modifications on gene expression.

Very little attention has been paid to histone modifications in SLE, partly due to technical reasons. The technology used to study histone modifications requires large numbers of viable cells, and the sensitivity of the techniques have only recently increased to a point where studies are feasible. Histone modifications were initially studied in murine lupus models, and two HDAC inhibitors were specifically found to improve disease features in these animals [79,80]. These agents were strongly immunosuppressive and increased the production of regulatory T cells; however, these studies were the first epigenetic therapies to be tested for lupus and only provide a critical proof of principle.

Two strategies have been used to study histone modifications in SLE: (1) Measuring aggregate histone modifications. This approach was used to identify aberrant histone modifications in T cells from SLE patients, and mycophenolate mofetil was used to correct the defect [81]; (2) Next generation sequencing. This technique has been used to define genome-wide changes in histone modifications, and both chromatin immunoprecipitation sequencing (ChIP-seq) and assay for transposase-accessible chromatin sequencing (ATAC-seq) have been utilized frequently in human samples of limiting quantity. ChIP-seq measures the enrichment of a specific histone modification while ATAC-seq measures chromatin accessibility. In an early study, global H4 acetylation was demonstrated to be increased in SLE monocytes [82], which was concordant with DNA methylation studies. Both observations would be predicted to drive overexpression of the target genes. IRF1 binding sites were identified within the altered peaks of H4 acetylation, and IRF1 binding was subsequently shown to be increased in SLE monocytes [83,84]. IRF1 induces, and is induced by, type I interferons, and this finding conceptually connects the histone modifications to the type I interferon signature [85]. In vitro overexpression of IRF1 increases global H4 acetylation and IRF1 directly interacts with histone-modifying enzymes. IRF1 overexpression was also associated with increased recruitment of histone acetyltransferase enzyme P300 to IRF1 target genes [86]. These studies provide a conceptual connection between overexpression of type I interferons and changes to histone modifications at interferon-responsive genes.

Enhancer and promoter H3K4me3 levels have also been evaluated. Enhancers were globally altered in SLE monocytes and undoubtedly dictated the altered cell behavior [87]. Increased H3K4me3 levels were found in monocyte-specific enhancers where one would anticipate pioneer transcription factors to establish the basal enhancer activity. Position weight matrix analysis was used to identify transcription factors related to the altered peaks of H3K4me3. In SLE subjects, promoters with increased H3K4me3 levels were enriched for CEBPB and STAT1 binding sites. CEBPB is a monocyte pioneer transcription factor, which supports the specificity of the findings [88], and STAT1 is induced by type I interferons, which is consistent with the type I interferon effects. Promoters with decreased H3K4me3 levels were also enriched for STAT1 and CEBPB, but they were also enriched for PU.1, SP1, IRF1, BLIMP1, IRF8, and STAT6 binding sites. PU.1 is another pioneer transcription factor; STAT1, IRF1, and IRF8 are related to interferon pathways; and SP1 is a ubiquitous transcription factor involved in cell cycling [89]. BLIMP1 was previously associated with increased H4 acetylation peaks in SLE [82]. This study supports a model where promoter H3K4me3 levels are altered by convergent signaling pathways, including interferon effects. A similar finding was seen at enhancers, although the enrichment was less substantial. These studies provide insights into potential disease mechanisms and suggest possible druggable targets.

Epigenetic therapeutics represent an untapped but promising approach for autoimmune diseases [90]. Pan-HDAC inhibitors (i.e., panobinostat, dacinostat, vorinostat, and trichostatin A) are known to inhibit the production of inflammatory cytokines, inhibit B-cell activation, and prevent the T-cell production of cytokines [91,92]. Selective HDAC inhibitors (i.e., valproic acid, ITF3056, and tubastatin) also have immunomodulatory effects that could be useful in autoimmune diseases [93–95]; for example, a phase I trial of givinostat in autoinflammatory diseases demonstrated antiinflammatory activity and induced partial remission of Schnitzler syndrome [96,97]. One of the newer therapeutic efforts directed at the epigenome utilizes inhibitors to BRD4, a member of the bromodomain and extraterminal domain (BET) family that recognize acetylated lysine residues on histones [98]. This protein is integral for the upregulation of inflammatory genes as it binds to acetylated histones and influences RNA polymerase II elongation [99]. BRD4 inhibitors are being developed to target malignancies [100], and one of these has demonstrated efficacy in a murine lupus model [101]. BRD4 plays a complex role in the regulation of transcription, but within enhancers it appears to regulate the production of noncoding enhancer RNAs. These RNAs are essential for enhancer function for reasons that remain to be clarified.

Histone modifications appear to play a unique role in SLE independent of gene expression. Apoptotic cell modifications have been identified as autoantigens in SLE. Antibodies directed at nuclear antigens are the cornerstone of SLE. Murine lupus models express antibodies to H4K8ac, H4K12ac, and H4K16ac, which are apoptosis-associated modifications [102,103]. H2BK12ac and H3K27me3 have also been identified as autoantibody targets in SLE [104,105]. A key concept in the current SLE etiopathogenesis model is the imbalance between apoptotic cell production and clearance, with the uncleared debris stimulating the production of type I interferon. It has been suggested that lupus sera recognize apoptosis-modified histones more avidly than unmodified histones, which supports this etiopathogenesis model [102,104,105]. The histone modifications associated with apoptosis have also been described in NETosis, which is a phenomenon whereby neutrophils extrude their nuclear material to kill pathogens [106]. A new approach to confirm these findings has involved the use of microarrays of modified histone peptides. This approach confirmed the early studies and demonstrated higher reactivity in SLE patients with a high interferon signature, as predicted by the model [107]. It was also established that the ratio of H2Bac:H2B antibodies was strongly associated with renal involvement [108]. This phenomena of histone modifications participating in the disease process as an autoantibody target is unusual and unique in human disease, but is conceptually similar to the impact of hypomethylated DNA as an antigen in SLE.

6.4 MICRORNAs

miRNAs are single-stranded noncoding RNAs that posttranscriptionally regulate gene expression. miRNAs are derived from pri-miRNAs by sequential processing using the endonucleases Drosha and Dicer. miRNAs guide the RNA-induced silencing complex to mRNA targets, generally in the 3′ untranslated region, by imperfect base pairing. The targeted mRNAs are either translationally repressed or are targeted for degradation. Each miRNA can regulate hundreds of targets [109] and each mRNA can be regulated by more than one miRNA. This confounding issue has complicated the study of miRNAs.

The study of miRNAs in SLE has generally focused on known pathways, including innate immunity, type I IFN signaling, inflammation, and dysregulation of apoptosis; for example, miR-146a was found to be underexpressed in SLE, which increased type I IFN signaling [110]. miR-125a was also found to be downregulated in SLE, and this was associated with an increase in RANTES and inflammation [111]. Additional studies have attempted to link miRNA dysregulation with other epigenetic control mechanisms; for example, miR-126 has been found to be overexpressed in CD4 T cells of SLE patients, which leads to a downregulating of DNMT1 expression and contributes to T- and B-cell hyperactivity [112]. Conversely, miR-21 has been shown to be upregulated in SLE patients and also represses DNMT1, albeit indirectly through the downregulation of RasGRP1 [113]. miRNA effects can therefore be related to altered gene expression in SLE.

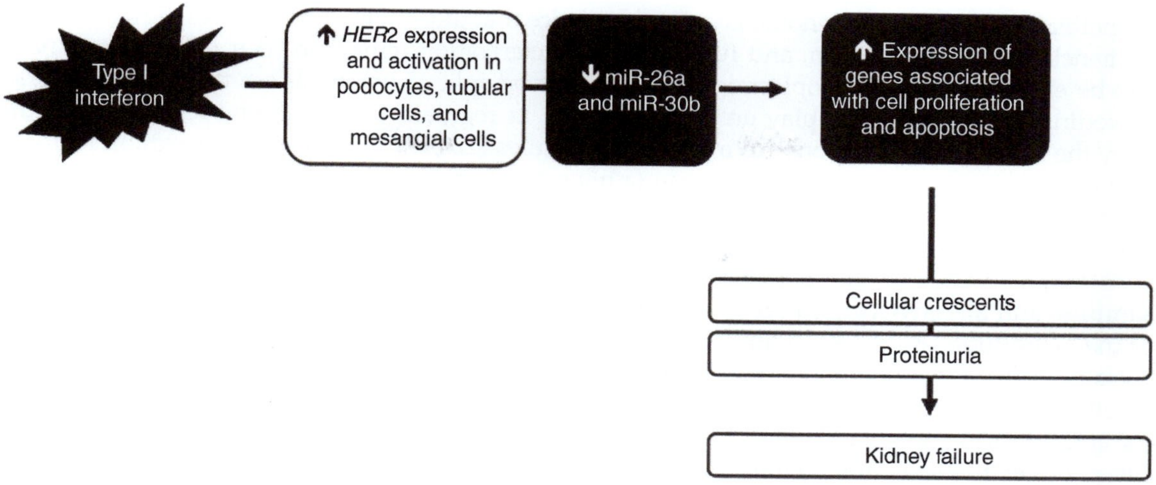

FIGURE 6.2 **Model for dysregulated *HER2* expression in lupus nephritis.** Interferon stimulates the expression of the orphan epidermal growth factor receptor, *HER2*. This in turn downregulates two miRNAs integrally involved in cell cycle regulation. Proliferation is central to the observed pathology in lupus nephritis and leads to the generation of glomerular crescents, which impede renal function.

Changes in miRNAs have also been identified in blood and renal tissue from patients with SLE [114,115]. Several of the miRNAs appear to have effects that are central to the lupus process and impact toll-like receptor signaling and type I interferon responses. An unbiased approach was used to define the renal miRNA profile in pediatric SLE nephritis [116]. A lupus nephritis miRNA signature was identified that reflected cell proliferation, and the miRNA pattern was different from that seen with poststreptococcal glomerulonephritis thus confirming disease specificity. The signature showed a significant decrease of miR-26a, miR-30b, and miR-4286, which are miRNAs that regulate the expression of genes involved in the cell cycle. This observation was consistent with the oncology literature, where lower levels of these miRNAs have been identified in states of increased cell proliferation. Indeed, diminished miR26a and miR30b were found to be secondary to increased *HER2* expression levels in breast cancer, and the beneficial effect of Trastuzumab treatment in breast cancer was mediated through modulation of these miRNAs [117]. *HER2* levels were shown to be dramatically increased in lupus nephritis, even spilling into the urine where it may act as a promising biomarker for disease activity [116]. The increased *HER2* expression could be related back to interferon since type I interferon increased the expression of *HER2* in a mesangial cell line (Fig. 6.2).

6.5 LONG NONCODING RNAs

The rapid increase in our understanding of the complexity of the human genome has revealed that, in addition to protein-coding genes, there is also broad transcription of noncoding RNAs [118,119]. LncRNAs can be processed in the same manner as protein-coding mRNAs, and increasing evidence suggests that these lncRNAs are involved in diverse biological processes. LncRNAs were originally studied in the context of genomic imprinting and malignancy; however, current research suggests that several lncRNAs may be implicated in immunologic function. The number of lncRNAs in the human genome is more than twice that of protein-coding genes; however, their nomenclature, categorization, and functional characterization remain in their infancy. LncRNAs can be either nuclear or cytoplasmic and they may be spliced or unspliced: they are as diverse as coding genes. LncRNAs play an important role in regulating the immune response, specifically the inflammatory response to microbial molecules [120,121]. Lupus patients exhibit features that are consistent with chronic endotoxin (a microbial byproduct) exposure, and it could be reasonably proposed that dysregulation of lncRNAs could alter this pathway. LncRNAs have also been demonstrated to regulate the development of dendritic cells, granulocytes, and lymphocytes [122–124]. However, studies performed to date have all utilized murine models or limited human samples, and all have focused on specific lncRNAs. Therefore there is a knowledge gap in terms of the magnitude and extent of lncRNA dysregulation in SLE. Interestingly, one genome-wide association locus identified in SLE falls within a lncRNA named *GAS5* [125]. This lncRNA regulates T-cell function and cell proliferation [126–128], and dysregulated *GAS5* expression was found in the BXSB murine model of lupus [129]. Nevertheless, although lncRNAs represent an important future target, there are limited data to support a role for them in SLE at this time.

Transcripts that formally comprise a distinct subset of (usually) noncoding RNAs are derived from retroviruses. Virus-like repeat elements comprise nearly 50% of the human genome. Retroviruses that integrated more than 25 million years ago have mutated, duplicated, and undergone gene conversion so that they now exhibit a range of functions [130]. Long interspersed nuclear element-1 (L1) is a member of an autonomous family of retroelements that have become activated in mammalian genomes. There are approximately 500,000 copies of the L1 genome. As these elements have retained their activity, they are able to generate RNA transcripts and in some cases can produce viral proteins. L1 expression is highly regulated and the DNA is usually fully methylated to prevent expression. L1 may be expressed during germ cell maturation but somatic tissues rarely express it [131,132]. However, L1 levels were recently shown to be upregulated in renal tissue from patients with lupus nephritis [133]. The increase in L1 was specifically associated with increased type I interferon. These data suggest that aberrant expression of virus-like repeat elements can participate directly in the type I interferon signature, which has been observed in a variety of autoimmune diseases. Aberrant expression of these transcripts could induce type I interferons through a nucleic acid recognition pathway. Supporting this hypothesis was the finding that salivary gland tissue in patients with Sjogren's syndrome also exhibits increased L1 levels corresponding with type I interferon [133].

6.6 CONCLUDING REMARKS

Substantial progress has been made in the study of epigenetics in SLE, and the mounting body of literature clearly supports the importance of epigenetic processes in SLE pathogenesis. Epidemiologic drivers of disease can reasonably be associated with epigenetic changes, and disease processes provide another link. The interactions between genetic and epigenetic factors significantly contribute to disease. Epigenetic biomarkers show some potential as diagnostic and prognostic variables for SLE. Furthermore, the potential therapeutic strategies based upon the targeting of site-specific or global aberrant epigenetic marks may bring about a durable, or perhaps even curative, therapy for this challenging and complex disease. New efforts to develop epigenetic therapeutics that target specific enzymes (i.e., EZH2, DOT1L, LSD1, and BET) have progressed and promise to selectively manipulate the epigenome [90]. Gene-specific targeting of relevant transcription factors (i.e., STAT3 or IRF1) is feasible but will require improved drug delivery approaches [83,134]. The role of epigenetic changes in the antigenicity of the nucleic acid structure is a novel feature of epigenetic changes, specifically in SLE.

References

[1] Danchenko N, Satia JA, Anthony MS. Epidemiology of systemic lupus erythematosus: a comparison of worldwide disease burden. Lupus 2006;15(5):308–18.

[2] Baechler EC, Batliwalla FM, Karypis G, Gaffney PM, Ortmann WA, Espe KJ, et al. Interferon-inducible gene expression signature in peripheral blood cells of patients with severe lupus. Proc Natl Acad Sci USA 2003;100(5):2610–5.

[3] Kirou KA, Lee C, George S, Louca K, Papagiannis IG, Peterson MG, et al. Coordinate overexpression of interferon-alpha-induced genes in systemic lupus erythematosus. Arthritis Rheum 2004;50(12):3958–67.

[4] Bennett L, Palucka AK, Arce E, Cantrell V, Borvak J, Banchereau J, et al. Interferon and granulopoiesis signatures in systemic lupus erythematosus blood. J Exp Med 2003;197(6):711–23.

[5] Li YR, Li J, Zhao SD, Bradfield JP, Mentch FD, Maggadottir SM, et al. Meta-analysis of shared genetic architecture across ten pediatric autoimmune diseases. Nat Med 2015;21(9):1018–27. doi: 10.1038/nm.3933.

[6] Baechler EC, Bauer JW, Slattery CA, Ortmann WA, Espe KJ, Novitzke J, et al. An interferon signature in the peripheral blood of dermatomyositis patients is associated with disease activity. Mol Med 2007;13(1–2):59–68.

[7] Greenberg SA, Pinkus JL, Pinkus GS, Burleson T, Sanoudou D, Tawil R, et al. Interferon-alpha/beta-mediated innate immune mechanisms in dermatomyositis. Ann Neurol 2005;57(5):664–78. doi: 10.1002/ana.20464.

[8] Walsh RJ, Kong SW, Yao Y, Jallal B, Kiener PA, Pinkus JL, et al. Type I interferon-inducible gene expression in blood is present and reflects disease activity in dermatomyositis and polymyositis. Arthritis Rheum 2007;56(11):3784–92. doi: 10.1002/art.22928.

[9] Lubbers J, Brink M, van de Stadt LA, Vosslamber S, Wesseling JG, van Schaardenburg D, et al. The type I IFN signature as a biomarker of preclinical rheumatoid arthritis. Ann Rheum Dis 2013;72(5):776–80. doi: 10.1136/annrheumdis-2012-202753.

[10] Gordon RA, Grigoriev G, Lee A, Kalliolias GD, Ivashkiv LB. The interferon signature and STAT1 expression in rheumatoid arthritis synovial fluid macrophages are induced by tumor necrosis factor alpha and counter-regulated by the synovial fluid microenvironment. Arthritis Rheum 2012;64(10):3119–28. doi: 10.1002/art.34544.

[11] van der Pouw Kraan TC, Wijbrandts CA, van Baarsen LG, Voskuyl AE, Rustenburg F, Baggen JM, et al. Rheumatoid arthritis subtypes identified by genomic profiling of peripheral blood cells: assignment of a type I interferon signature in a subpopulation of patients. Ann Rheum Dis 2007;66(8):1008–14. doi: 10.1136/ard.2006.063412.

[12] Munroe ME, Lu R, Zhao YD, Fife DA, Robertson JM, Guthridge JM, et al. Altered type II interferon precedes autoantibody accrual and elevated type I interferon activity prior to systemic lupus erythematosus classification. Ann Rheum Dis 2016;75:2014–21. doi: 10.1136/annrheumdis-2015-208140.

[13] Berthier CC, Kretzler M, Davidson A. A systems approach to renal inflammation in SLE. Clin Immunol 2016;185:109–18. doi: 10.1016/j.clim.2016.08.015.

[14] Tan EM, Cohen AS, Fries JF, Masi AT, McShane DJ, Rothfield NF, et al. The 1982 revised criteria for the classification of systemic lupus erythematosus. Arthritis Rheum 1982;25(11):1271–7.

[15] Gronniger E, Weber B, Heil O, Peters N, Stab F, Wenck H, et al. Aging and chronic sun exposure cause distinct epigenetic changes in human skin. PLoS Genet 2010;6(5):e1000971. doi: 10.1371/journal.pgen.1000971.

[16] Strickland FM, Hewagama A, Lu Q, Wu A, Hinderer R, Webb R, et al. Environmental exposure, estrogen and two X chromosomes are required for disease development in an epigenetic model of lupus. J Autoimmun 2012;38(2–3):J135–43. doi: 10.1016/j.jaut.2011.11.001.

[17] Biller BM, Luciano A, Crosignani PG, Molitch M, Olive D, Rebar R, et al. Guidelines for the diagnosis and treatment of hyperprolactinemia. J Reprod Med 1999;44(Suppl. 12):1075–84.

[18] El Miedany YM, Ahmed I, Moustafa H, El Baddini M. Hyperprolactinemia in Sjogren's syndrome: a patient subset or a disease manifestation? Joint Bone Spine 2004;71(3):203–8. doi: 10.1016/S1297-319X(03)00151-9.

[19] Buskila D, Lorber M, Neumann L, Flusser D, Shoenfeld Y. No correlation between prolactin levels and clinical activity in patients with systemic lupus erythematosus. J Rheumatol 1996;23(4):629–32.

[20] Buskila D, Shoenfeld Y. Prolactin, bromocriptine and autoimmune diseases. Isr J Med Sci 1996;32(1):23–7.

[21] Mok CC, Lau CS, Lee KW, Wong RW. Hyperprolactinemia in males with systemic lupus erythematosus. J Rheumatol 1998;25(12):2357–63.

[22] Miranda JM, Prieto RE, Paniagua R, Garcia G, Amato D, Barile L, et al. Clinical significance of serum and urine prolactin levels in lupus glomerulonephritis. Lupus 1998;7(6):387–91.

[23] Leanos-Miranda A, Cardenas-Mondragon G. Serum free prolactin concentrations in patients with systemic lupus erythematosus are associated with lupus activity. Rheumatology (Oxford) 2006;45(1):97–101. doi: 10.1093/rheumatology/kei115.
[24] Pacilio M, Migliaresi S, Meli R, Ambrosone L, Bigliardo B, Di Carlo R. Elevated bioactive prolactin levels in systemic lupus erythematosus: association with disease activity. J Rheumatol 2001;28(10):2216–21.
[25] Jacobi AM, Rohde W, Ventz M, Riemekasten G, Burmester GR, Hiepe F. Enhanced serum prolactin (PRL) in patients with systemic lupus erythematosus: PRL levels are related to the disease activity. Lupus 2001;10(8):554–61.
[26] Cardenas-Mondragon G, Ulloa-Aguirre A, Isordia-Salas I, Goffin V, Leanos-Miranda A. Elevated serum bioactive prolactin concentrations in patients with systemic lupus erythematosus are associated with disease activity as disclosed by homologous receptor bioassays. J Rheumatol 2007;34(7):1514–21.
[27] Ugarte-Gil MF, Gamboa-Cardenas RV, Zevallos F, Medina M, Cucho-Venegas JM, Perich-Campos RA, et al. High prolactin levels are independently associated with damage accrual in systemic lupus erythematosus patients. Lupus 2014;23(10):969–74. doi: 10.1177/0961203314531083.
[28] Karimifar M, Tahmasebi A, Bonakdar ZS, Purajam S. Correlation of serum prolactin levels and disease activity in systematic lupus erythematosus. Rheumatol Int 2013;33(2):511–6. doi: 10.1007/s00296-011-2211-5.
[29] Orbach H, Zandman-Goddard G, Boaz M, Agmon-Levin N, Amital H, Szekanecz Z, et al. Prolactin and autoimmunity: hyperprolactinemia correlates with serositis and anemia in SLE patients. Clin Rev Allergy Immunol 2012;42(2):189–98. doi: 10.1007/s12016-011-8256-0.
[30] Yang J, Li Q, Yang X, Li M. Increased serum level of prolactin is related to autoantibody production in systemic lupus erythematosus. Lupus 2016;25(5):513–9. doi: 10.1177/0961203315622276.
[31] Walker SE. Bromocriptine treatment of systemic lupus erythematosus. Lupus 2001;10(10):762–8.
[32] McMurray RW, Weidensaul D, Allen SH, Walker SE. Efficacy of bromocriptine in an open label therapeutic trial for systemic lupus erythematosus. J Rheumatol 1995;22(11):2084–91.
[33] Jara LJ, Cruz-Cruz P, Saavedra MA, Medina G, Garcia-Flores A, Angeles U, et al. Bromocriptine during pregnancy in systemic lupus erythematosus: a pilot clinical trial. Ann N Y Acad Sci 2007;1110:297–304. doi: 10.1196/annals.1423.031.
[34] Qian Q, Liuqin L, Hao L, Shiwen Y, Zhongping Z, Dongying C, et al. The effects of bromocriptine on preventing postpartum flare in systemic lupus erythematosus patients from South China. J Immunol Res 2015;2015:316965. doi: 10.1155/2015/316965.
[35] McMurray R, Keisler D, Kanuckel K, Izui S, Walker SE. Prolactin influences autoimmune disease activity in the female B/W mouse. J Immunol 1991;147(11):3780–7.
[36] Saha S, Tieng A, Pepeljugoski KP, Zandamn-Goddard G, Peeva E. Prolactin, systemic lupus erythematosus, and autoreactive B cells: lessons learnt from murine models. Clin Rev Allergy Immunol 2011;40(1):8–15. doi: 10.1007/s12016-009-8182-6.
[37] Peeva E, Michael D, Cleary J, Rice J, Chen X, Diamond B. Prolactin modulates the naive B cell repertoire. J Clin Invest 2003;111(2):275–83. doi: 10.1172/JCI16530.
[38] Leung Y, Sullivan K, Maurer K, Song L, Shi L. Prolactin induces interferon regulatory factor 1 activation and histone H4 hyperacetylation in primary monocytes comparable to changes seen in monocytes from systemic lupus erythematosus patients. Arthritis Rheumatol 2015;(Suppl. 10). [Abstract 804].
[39] Gardiner-Garden M, Frommer M. CpG islands in vertebrate genomes. J Mol Biol 1987;196(2):261–82.
[40] Matarazzo MR, De Bonis ML, Strazzullo M, Cerase A, Ferraro M, Vastarelli P, et al. Multiple binding of methyl-CpG and polycomb proteins in long-term gene silencing events. J Cell Physiol 2007;210(3):711–9. doi: 10.1002/jcp.20879.
[41] Kuroda A, Rauch TA, Todorov I, Ku HT, Al-Abdullah IH, Kandeel F, et al. Insulin gene expression is regulated by DNA methylation. PLoS One 2009;4(9):e6953. doi: 10.1371/journal.pone.0006953.
[42] Denis H, Ndlovu MN, Fuks F. Regulation of mammalian DNA methyltransferases: a route to new mechanisms. EMBO Rep 2011;12(7):647–56. doi: 10.1038/embor.2011.110.
[43] Yara S, Lavoie JC, Levy E. Oxidative stress and DNA methylation regulation in the metabolic syndrome. Epigenomics 2015;7(2):283–300. doi: 10.2217/epi.14.84.
[44] Campos AC, Molognoni F, Melo FH, Galdieri LC, Carneiro CR, D'Almeida V, et al. Oxidative stress modulates DNA methylation during melanocyte anchorage blockade associated with malignant transformation. Neoplasia 2007;9(12):1111–21.
[45] Menezo YJ, Silvestris E, Dale B, Elder K. Oxidative stress and alterations in DNA methylation: two sides of the same coin in reproduction. Reprod Biomed Online 2016;33(6):668–83. doi: 10.1016/j.rbmo.2016.09.006.
[46] Nawrocki MJ, Majewski D, Puszczewicz M, Jagodzinski PP. Decreased mRNA expression levels of DNA methyltransferases type 1 and 3A in systemic lupus erythematosus. Rheumatol Int 2017;37(5):775–83. doi: 10.1007/s00296-017-3711-8.

[47] Richardson B, Scheinbart L, Strahler J, Gross L, Hanash S, Johnson M. Evidence for impaired T cell DNA methylation in systemic lupus erythematosus and rheumatoid arthritis. Arthritis Rheum 1990;33(11):1665–73.

[48] Kaplan MJ, Lu Q, Wu A, Attwood J, Richardson B. Demethylation of promoter regulatory elements contributes to perforin overexpression in CD4+ lupus T cells. J Immunol 2004;172(6):3652–61.

[49] Richardson BC, Strahler JR, Pivirotto TS, Quddus J, Bayliss GE, Gross LA, et al. Phenotypic and functional similarities between 5-azacytidine-treated T cells and a T cell subset in patients with active systemic lupus erythematosus. Arthritis Rheum 1992;35(6):647–62.

[50] Jeffries M, Bruner G, Glenn S, Sadanandan P, Carson CW, Harley JB, et al. Sulpha allergy in lupus patients: a clinical perspective. Lupus 2008;17(3):202–5. doi: 10.1177/0961203307086640.

[51] Renauer PA, Coit P, Sawalha AH. The DNA methylation signature of human TCRalphabeta+CD4-CD8- double negative T cells reveals CG demethylation and a unique epigenetic architecture permissive to a broad stimulatory immune response. Clin Immunol 2015;156(1):19–27. doi: 10.1016/j.clim.2014.10.007.

[52] Absher DM, Li X, Waite LL, Gibson A, Roberts K, Edberg J, et al. Genome-wide DNA methylation analysis of systemic lupus erythematosus reveals persistent hypomethylation of interferon genes and compositional changes to CD4+ T-cell populations. PLoS Genet 2013;9(8):e1003678. doi: 10.1371/journal.pgen.1003678.

[53] Coit P, Jeffries M, Altorok N, Dozmorov MG, Koelsch KA, Wren JD, et al. Genome-wide DNA methylation study suggests epigenetic accessibility and transcriptional poising of interferon-regulated genes in naive CD4+ T cells from lupus patients. J Autoimmun 2013;43:78–84. doi: 10.1016/j.jaut.2013.04.003.

[54] Zhao M, Zhou Y, Zhu B, Wan M, Jiang T, Tan Q, et al. IFI44L promoter methylation as a blood biomarker for systemic lupus erythematosus. Ann Rheum Dis 2016;75(11):1998–2006. doi: 10.1136/annrheumdis-2015-208410.

[55] Zhu H, Mi W, Luo H, Chen T, Liu S, Raman I, et al. Whole-genome transcription and DNA methylation analysis of peripheral blood mononuclear cells identified aberrant gene regulation pathways in systemic lupus erythematosus. Arthritis Res Ther 2016;18:162. doi: 10.1186/s13075-016-1050-x.

[56] Oelke K, Lu Q, Richardson D, Wu A, Deng C, Hanash S, et al. Overexpression of CD70 and overstimulation of IgG synthesis by lupus T cells and T cells treated with DNA methylation inhibitors. Arthritis Rheum 2004;50(6):1850–60. doi: 10.1002/art.20255.

[57] Zhang Y, Zhao M, Sawalha AH, Richardson B, Lu Q. Impaired DNA methylation and its mechanisms in CD4(+)T cells of systemic lupus erythematosus. J Autoimmun 2013;41:92–9. doi: 10.1016/j.jaut.2013.01.005.

[58] Hedrich CM, Crispin JC, Tsokos GC. Epigenetic regulation of cytokine expression in systemic lupus erythematosus with special focus on T cells. Autoimmunity 2014;47(4):234–41. doi: 10.3109/08916934.2013.801462.

[59] Teruel M, Sawalha AH. Epigenetic variability in systemic lupus erythematosus: what we learned from genome-wide DNA methylation studies. Curr Rheumatol Rep 2017;19(6):32. doi: 10.1007/s11926-017-0657-5.

[60] Patel MC, Debrosse M, Smith M, Dey A, Huynh W, Sarai N, et al. BRD4 coordinates recruitment of pause release factor P-TEFb and the pausing complex NELF/DSIF to regulate transcription elongation of interferon-stimulated genes. Mol Cell Biol 2013;33(12):2497–507. doi: 10.1128/MCB.01180-12.

[61] Zhao H, Wang L, Luo H, Li QZ, Zuo X. TNFAIP3 downregulation mediated by histone modification contributes to T-cell dysfunction in systemic lupus erythematosus. Rheumatology (Oxford) 2017;56(5):835–43. doi: 10.1093/rheumatology/kew508.

[62] Kang KA, Zhang R, Kim GY, Bae SC, Hyun JW. Epigenetic changes induced by oxidative stress in colorectal cancer cells: methylation of tumor suppressor RUNX3. Tumour Biol 2012;33(2):403–12. doi: 10.1007/s13277-012-0322-6.

[63] Khan F, Siddiqui AA, Ali R. Measurement and significance of 3-nitrotyrosine in systemic lupus erythematosus. Scand J Immunol 2006;64(5):507–14. doi: 10.1111/j.1365-3083.2006.01794.x.

[64] Oates JC, Gilkeson GS. The biology of nitric oxide and other reactive intermediates in systemic lupus erythematosus. Clin Immunol 2006;121(3):243–50. doi: 10.1016/ j.clim.2006.06.001.

[65] Somers EC, Richardson BC. Environmental exposures, epigenetic changes and the risk of lupus. Lupus 2014;23(6):568–76. doi: 10.1177/0961203313499419.

[66] Waterland RA, Jirtle RL. Transposable elements: targets for early nutritional effects on epigenetic gene regulation. Mol Cell Biol 2003;23(15):5293–300.

[67] Strickland FM, Hewagama A, Wu A, Sawalha AH, Delaney C, Hoeltzel MF, et al. Diet influences expression of autoimmune-associated genes and disease severity by epigenetic mechanisms in a transgenic mouse model of lupus. Arthritis Rheum 2013;65(7):1872–81. doi: 10.1002/art.37967.

[68] Madzo J, Liu H, Rodriguez A, Vasanthakumar A, Sundaravel S, Caces DB, et al. Hydroxymethylation at gene regulatory regions directs stem/early progenitor cell commitment during erythropoiesis. Cell Rep 2014;6(1):231–44. doi: 10.1016/j.celrep.2013.11.044.

[69] Ichiyama K, Chen T, Wang X, Yan X, Kim BS, Tanaka S, et al. The methylcytosine dioxygenase Tet2 promotes DNA demethylation and activation of cytokine gene expression in T cells. Immunity 2015;42(4):613–26. doi: 10.1016/j.immuni.2015.03.005.

[70] Zhao M, Wang J, Liao W, Li D, Li M, Wu H, et al. Increased 5-hydroxymethylcytosine in CD4(+) T cells in systemic lupus erythematosus. J Autoimmun 2016;69:64–73. doi: 10.1016/j.jaut.2016.03.001.

[71] Wen ZK, Xu W, Xu L, Cao QH, Wang Y, Chu YW, et al. DNA hypomethylation is crucial for apoptotic DNA to induce systemic lupus erythematosus-like autoimmune disease in SLE-non-susceptible mice. Rheumatology (Oxford) 2007;46(12):1796–803. doi: 10.1093/rheumatology/kem275.

[72] Zhang W, Cai Y, Xu W, Yin Z, Gao X, Xiong S. AIM2 facilitates the apoptotic DNA-induced systemic lupus erythematosus via arbitrating macrophage functional maturation. J Clin Immunol 2013;33(5):925–37. doi: 10.1007/s10875-013-9881-6.

[73] Zhang W, Zhou Q, Xu W, Cai Y, Yin Z, Gao X, et al. DNA-dependent activator of interferon-regulatory factors (DAI) promotes lupus nephritis by activating the calcium pathway. J Biol Chem 2013;288(19):13534–50. doi: 10.1074/jbc.M113.457218.

[74] Tan EM, Schur PH, Carr RI, Kunkel HG. Deoxybonucleic acid (DNA) and antibodies to DNA in the serum of patients with systemic lupus erythematosus. J Clin Invest 1966;45(11):1732–40. doi: 10.1172/JCI105479.

[75] Rumore PM, Steinman CR. Endogenous circulating DNA in systemic lupus erythematosus. Occurrence as multimeric complexes bound to histone. J Clin Invest 1990;86(1):69–74. doi: 10.1172/JCI114716.

[76] Sun K, Jiang P, Chan KC, Wong J, Cheng YK, Liang RH, et al. Plasma DNA tissue mapping by genome-wide methylation sequencing for noninvasive prenatal, cancer, and transplantation assessments. Proc Natl Acad Sci USA 2015;112(40):E5503–12. doi: 10.1073/pnas.1508736112.

[77] Uccellini MB, Busconi L, Green NM, Busto P, Christensen SR, Shlomchik MJ, et al. Autoreactive B cells discriminate CpG-rich and CpG-poor DNA and this response is modulated by IFN-alpha. J Immunol 2008;181(9):5875–84.

[78] Kouzarides T. Chromatin modifications and their function. Cell 2007;128(4):693–705. doi: 10.1016/j.cell.2007.02.005.

[79] Mishra N, Reilly CM, Brown DR, Ruiz P, Gilkeson GS. Histone deacetylase inhibitors modulate renal disease in the MRL-lpr/lpr mouse. J Clin Invest 2003;111(4):539–52. doi: 10.1172/JCI16153.

[80] Garcia BA, Busby SA, Shabanowitz J, Hunt DF, Mishra N. Resetting the epigenetic histone code in the MRL-lpr/lpr mouse model of lupus by histone deacetylase inhibition. J Proteome Res 2005;4(6):2032–42.

[81] Hu N, Qiu X, Luo Y, Yuan J, Li Y, Lei W, et al. Abnormal histone modification patterns in lupus $CD4^+$ T cells. J Rheumatol 2008;35(5):804–10.

[82] Zhang Z, Song L, Maurer K, Petri MA, Sullivan KE. Global H4 acetylation analysis by ChIP-chip in systemic lupus erythematosus monocytes. Genes Immun 2010;11(2):124–33. doi: 10.1038/gene.2009.66. gene200966 [pii].

[83] Zhang Z, Shi L, Song L, Ephrem E, Petri M, Sullivan KE. Interferon regulatory factor 1 marks activated genes and can induce target gene expression in systemic lupus erythematosus. Arthritis Rheumatol 2015;67(3):785–96. doi: 10.1002/art.38964.

[84] Shi L, Perin JC, Leipzig J, Zhang Z, Sullivan KE. Genome-wide analysis of interferon regulatory factor I binding in primary human monocytes. Gene 2011;487(1):21–8. doi: 10.1016/j.gene.2011.07.004. S0378-1119(11)00325-8 [pii].

[85] Yarilina A, Park-Min KH, Antoniv T, Hu X, Ivashkiv LB. TNF activates an IRF1-dependent autocrine loop leading to sustained expression of chemokines and STAT1-dependent type I interferon-response genes. Nat Immunol 2008;9(4):378–87.

[86] Leung YT, Shi L, Maurer K, Song L, Zhang Z, Petri M, et al. Interferon regulatory factor 1 and histone H4 acetylation in systemic lupus erythematosus. Epigenetics 2015;10(3):191–9. doi: 10.1080/15592294.2015.1009764.

[87] Shi L, Zhang Z, Song L, Leung YT, Petri MA, Sullivan KE. Monocyte enhancers are highly altered in systemic lupus erythematosus. Epigenomics 2015;7(6):921–35. doi: 10.2217/epi.15.47.

[88] Zaret KS, Carroll JS. Pioneer transcription factors: establishing competence for gene expression. Genes Dev 2011;25(21):2227–41. doi: 10.1101/gad.176826.111.

[89] Deniaud E, Baguet J, Chalard R, Blanquier B, Brinza L, Meunier J, et al. Overexpression of transcription factor Sp1 leads to gene expression perturbations and cell cycle inhibition. PLoS One 2009;4(9):e7035. doi: 10.1371/journal.pone.0007035.

[90] Tough DF, Tak PP, Tarakhovsky A, Prinjha RK. Epigenetic drug discovery: breaking through the immune barrier. Nat Rev Drug Discov 2016;15(12):835–53. doi: 10.1038/nrd.2016.185.

[91] Leoni F, Zaliani A, Bertolini G, Porro G, Pagani P, Pozzi P, et al. The antitumor histone deacetylase inhibitor suberoylanilide hydroxamic acid exhibits antiinflammatory properties via suppression of cytokines. Proc Natl Acad Sci USA 2002;99(5):2995–3000. doi: 10.1073/pnas.052702999.

[92] Leoni F, Fossati G, Lewis EC, Lee JK, Porro G, Pagani P, et al. The histone deacetylase inhibitor ITF2357 reduces production of pro-inflammatory cytokines in vitro and systemic inflammation in vivo. Mol Med 2005;11(1–12):1–15. doi: 10.2119/2006-00005.Dinarello.

[93] Wu C, Li A, Leng Y, Li Y, Kang J. Histone deacetylase inhibition by sodium valproate regulates polarization of macrophage subsets. DNA Cell Biol 2012;31(4):592–9. doi: 10.1089/dna.2011.1401.

[94] Li S, Fossati G, Marchetti C, Modena D, Pozzi P, Reznikov LL, et al. Specific inhibition of histone deacetylase 8 reduces gene expression and production of proinflammatory cytokines in vitro and in vivo. J Biol Chem 2015;290(4):2368–78. doi: 10.1074/jbc.M114.618454.

[95] Li Y, Zhao T, Liu B, Halaweish I, Mazitschek R, Duan X, et al. Inhibition of histone deacetylase 6 improves long-term survival in a lethal septic model. J Trauma Acute Care Surg 2015;78(2):378–85. doi: 10.1097/TA.0000000000000510.

[96] Furlan A, Monzani V, Reznikov LL, Leoni F, Fossati G, Modena D, et al. Pharmacokinetics, safety and inducible cytokine responses during a phase 1 trial of the oral histone deacetylase inhibitor ITF2357 (givinostat). Mol Med 2011;17(5–6):353–62. doi: 10.2119/molmed.2011.00020.

[97] Bodar EJ, Simon A, van der Meer JW. Effects of the histone deacetylase inhibitor ITF2357 in autoinflammatory syndromes. Mol Med 2011;17(5–6):363–8. doi: 10.2119/molmed.2011.00039.

[98] Gonzales-Cope M, Sidoli S, Bhanu NV, Won KJ, Garcia BA. Histone H4 acetylation and the epigenetic reader Brd4 are critical regulators of pluripotency in embryonic stem cells. BMC Genomics 2016;17:95. doi: 10.1186/s12864-016-2414-y.

[99] Liu W, Ma Q, Wong K, Li W, Ohgi K, Zhang J, et al. Brd4 and JMJD6-associated anti-pause enhancers in regulation of transcriptional pause release. Cell 2013;155(7):1581–95. doi: 10.1016/j.cell.2013.10.056.

[100] Filippakopoulos P, Qi J, Picaud S, Shen Y, Smith WB, Fedorov O, et al. Selective inhibition of BET bromodomains. Nature 2010;468(7327):1067–73. doi: 10.1038/nature09504.

[101] Wei S, Sun Y, Sha H. Therapeutic targeting of BET protein BRD4 delays murine lupus. Int Immunopharmacol 2015;29(2):314–9. doi: 10.1016/j.intimp.2015.10.036.

[102] Dieker JW, Fransen JH, van Bavel CC, Briand JP, Jacobs CW, Muller S, et al. Apoptosis-induced acetylation of histones is pathogenic in systemic lupus erythematosus. Arthritis Rheum 2007;56(6):1921–33. doi: 10.1002/art.22646.

[103] Fullgrabe J, Hajji N, Joseph B. Cracking the death code: apoptosis-related histone modifications. Cell Death Differ 2010;17(8):1238–43. doi: 10.1038/cdd.2010.58.

[104] van Bavel CC, Dieker J, Muller S, Briand JP, Monestier M, Berden JH, et al. Apoptosis-associated acetylation on histone H2B is an epitope for lupus autoantibodies. Mol Immunol 2009;47(2–3):511–6. doi: 10.1016/j.molimm.2009.08.009.

[105] van Bavel CC, Dieker JW, Kroeze Y, Tamboer WP, Voll R, Muller S, et al. Apoptosis-induced histone H3 methylation is targeted by autoantibodies in systemic lupus erythematosus. Ann Rheum Dis 2011;70(1):201–7. doi: 10.1136/ard.2010.129320.

[106] Pieterse E, Hofstra J, Berden J, Herrmann M, Dieker J, van der Vlag J. Acetylated histones contribute to the immunostimulatory potential of neutrophil extracellular traps in systemic lupus erythematosus. Clin Exp Immunol 2015;179(1):68–74. doi: 10.1111/cei.12359.

[107] Price JV, Tangsombatvisit S, Xu G, Yu J, Levy D, Baechler EC, et al. On silico peptide microarrays for high-resolution mapping of antibody epitopes and diverse protein-protein interactions. Nat Med 2012;18(9):1434–40. doi: 10.1038/nm.2913.

[108] Dieker J, Berden JH, Bakker M, Briand JP, Muller S, Voll R, et al. Autoantibodies against modified histone peptides in SLE patients are associated with disease activity and lupus nephritis. PLoS One 2016;11(10):e0165373. doi: 10.1371/journal.pone.0165373.

[109] Krek A, Grun D, Poy MN, Wolf R, Rosenberg L, Epstein EJ, et al. Combinatorial microRNA target predictions. Nat Genet 2005;37(5):495–500. doi: 10.1038/ng1536.

[110] Tang Y, Luo X, Cui H, Ni X, Yuan M, Guo Y, et al. MicroRNA-146A contributes to abnormal activation of the type I interferon pathway in human lupus by targeting the key signaling proteins. Arthritis Rheum 2009;60(4):1065–75. doi: 10.1002/art.24436.

[111] Zhao X, Tang Y, Qu B, Cui H, Wang S, Wang L, et al. MicroRNA-125a contributes to elevated inflammatory chemokine RANTES levels via targeting KLF13 in systemic lupus erythematosus. Arthritis Rheum 2010;62(11):3425–35. doi: 10.1002/art.27632.

[112] Zhao S, Wang Y, Liang Y, Zhao M, Long H, Ding S, et al. MicroRNA-126 regulates DNA methylation in $CD4^+$ T cells and contributes to systemic lupus erythematosus by targeting DNA methyltransferase 1. Arthritis Rheum 2011;63(5):1376–86. doi: 10.1002/art.30196.

[113] Stagakis E, Bertsias G, Verginis P, Nakou M, Hatziapostolou M, Kritikos H, et al. Identification of novel microRNA signatures linked to human lupus disease activity and pathogenesis: miR-21 regulates aberrant T cell responses through regulation of PDCD4 expression. Ann Rheum Dis 2011;70(8):1496–506. doi: 10.1136/ard.2010.139857.
[114] Dai J, Liu B, Cua DJ, Li Z. Essential roles of IL-12 and dendritic cells but not IL-23 and macrophages in lupus-like diseases initiated by cell surface HSP gp96. Eur J Immunol 2007;37(3):706–15.
[115] Dai P, Jeong SY, Yu Y, Leng T, Wu W, Xie L, et al. Modulation of TLR signaling by multiple MyD88-interacting partners including leucine-rich repeat Fli-I-interacting proteins. J Immunol 2009;182:3450–60.
[116] Costa-Reis P, Russo PA, Zhang Z, Colonna L, Maurer K, Gallucci S, et al. The role of microRNAs and human epidermal growth factor receptor 2 in proliferative lupus nephritis. Arthritis Rheumatol 2015;67(9):2415–26. doi: 10.1002/art.39219.
[117] Ichikawa T, Sato F, Terasawa K, Tsuchiya S, Toi M, Tsujimoto G, et al. Trastuzumab produces therapeutic actions by upregulating miR-26a and miR-30b in breast cancer cells. PLoS One 2012;7(2):e31422. doi: 10.1371/journal.pone.0031422.
[118] Djebali S, Davis CA, Merkel A, Dobin A, Lassmann T, Mortazavi A, et al. Landscape of transcription in human cells. Nature 2012;489(7414):101–8. doi: 10.1038/nature11233.
[119] Bernstein BE, Birney E, Dunham I, Green ED, Gunter C, Snyder M. An integrated encyclopedia of DNA elements in the human genome. Nature 2012;489(7414):57–74. doi: 10.1038/nature11247.
[120] Carpenter S, Aiello D, Atianand MK, Ricci EP, Gandhi P, Hall LL, et al. A long noncoding RNA mediates both activation and repression of immune response genes. Science 2013;341(6147):789–92. doi: 10.1126/science.1240925.
[121] Guttman M, Amit I, Garber M, French C, Lin MF, Feldser D, et al. Chromatin signature reveals over a thousand highly conserved large non-coding RNAs in mammals. Nature 2009;458(7235):223–7. doi: 10.1038/nature07672.
[122] Wang P, Xue Y, Han Y, Lin L, Wu C, Xu S, et al. The STAT3-binding long noncoding RNA lnc-DC controls human dendritic cell differentiation. Science 2014;344(6181):310–3. doi: 10.1126/science.1251456.
[123] Zhang X, Lian Z, Padden C, Gerstein MB, Rozowsky J, Snyder M, et al. A myelopoiesis-associated regulatory intergenic noncoding RNA transcript within the human HOXA cluster. Blood 2009;113(11):2526–34. doi: 10.1182/blood-2008-06-162164.
[124] Hu G, Tang Q, Sharma S, Yu F, Escobar TM, Muljo SA, et al. Expression and regulation of intergenic long noncoding RNAs during T cell development and differentiation. Nat Immunol 2013;14(11):1190–8. doi: 10.1038/ni.2712.
[125] Suarez-Gestal M, Calaza M, Endreffy E, Pullmann R, Ordi-Ros J, Sebastiani GD, et al. Replication of recently identified systemic lupus erythematosus genetic associations: a case-control study. Arthritis Res Ther 2009;11(3):R69. doi: 10.1186/ar2698.
[126] Kino T, Hurt DE, Ichijo T, Nader N, Chrousos GP. Noncoding RNA gas5 is a growth arrest- and starvation-associated repressor of the glucocorticoid receptor. Sci Signal 2010;3(107):ra8. doi: 10.1126/scisignal.2000568.
[127] Mourtada-Maarabouni M, Hasan AM, Farzaneh F, Williams GT. Inhibition of human T-cell proliferation by mammalian target of rapamycin (mTOR) antagonists requires noncoding RNA growth-arrest-specific transcript 5 (GAS5). Mol Pharmacol 2010;78(1):19–28. doi: 10.1124/mol.110.064055.
[128] Williams GT, Mourtada-Maarabouni M, Farzaneh F. A critical role for non-coding RNA GAS5 in growth arrest and rapamycin inhibition in human T-lymphocytes. Biochem Soc Trans 2011;39(2):482–6. doi: 10.1042/BST0390482.
[129] Haywood ME, Rose SJ, Horswell S, Lees MJ, Fu G, Walport MJ, et al. Overlapping BXSB congenic intervals, in combination with microarray gene expression, reveal novel lupus candidate genes. Genes Immun 2006;7(3):250–63. doi: 10.1038/sj.gene.6364294.
[130] Hancks DC, Kazazian HH Jr. Active human retrotransposons: variation and disease. Curr Opin Genet Dev 2012;22(3):191–203. doi: 10.1016/j.gde.2012.02.006.
[131] Guo H, Zhu P, Yan L, Li R, Hu B, Lian Y, et al. The DNA methylation landscape of human early embryos. Nature 2014;511(7511):606–10. doi: 10.1038/nature13544.
[132] Ergun S, Buschmann C, Heukeshoven J, Dammann K, Schnieders F, Lauke H, et al. Cell type-specific expression of LINE-1 open reading frames 1 and 2 in fetal and adult human tissues. J Biol Chem 2004;279(26):27753–63. doi: 10.1074/jbc.M312985200.
[133] Mavragani CP, Sagalovskiy I, Guo Q, Nezos A, Kapsogeorgou EK, Lu P, et al. Expression of long interspersed nuclear element 1 retroelements and induction of type I interferon in patients with systemic autoimmune disease. Arthritis Rheumatol 2016;68(11):2686–96. doi: 10.1002/art.39795.
[134] Sen M, Thomas SM, Kim S, Yeh JI, Ferris RL, Johnson JT, et al. First-in-human trial of a STAT3 decoy oligonucleotide in head and neck tumors: implications for cancer therapy. Cancer Discov 2012;2(8):694–705. doi: 10.1158/2159-8290.CD-12-0191.

CHAPTER

7

Rheumatoid Arthritis and Epigenetics

Kerstin Klein, Emmanuel Karouzakis, Steffen Gay

Center of Experimental Rheumatology, University Hospital Zurich, Zurich, Switzerland

OUTLINE

7.1 INTRODUCTION

Rheumatoid arthritis (RA) is a chronic inflammatory joint disease characterized by joint swelling and pain, cartilage damage, bone erosion, severe joint deformation, disability, and premature mortality. The main pathogenic mechanisms occur symmetrically in the synovial tissues of the small joints in hands and feet; however, other joints may also be affected [1]. RA may also include extraarticular manifestations, such as rheumatoid nodules, pulmonary involvement and vasculitis, and systemic comorbidities [2,3]. The incidence of RA is approximately 0.5%–1%, and it has a heterogenous pathophysiology. The major subclassification is based on the presence or absence of anticitrullinated peptide antibodies (ACPAs). Seropositive RA is associated with a more aggressive disease course, an increased frequency of extraarticular manifestations, and increased mortality [4]. The etiology of RA is still not fully understood but a preclinical phase and ACPA formation may precede clinical symptoms for several years [5]. In the current concept of the RA pathogenesis, environmental insults (i.e., smoking or silica exposure) in a genetically predisposed individual, together with changes in the oral, bronchial, or gut microbiome or periodontitis, leads to epigenetic and posttranslational modifications that produce loss of self-tolerance, asymptomatic synovitis, and overt arthritis [1]. The heritability of RA is estimated to be 40%–65% for ACPA$^+$ and 20% for ACPA$^-$ [1], with the human leukocyte antigen (HLA) DRB1 alleles ("shared epitope") being the

The Epigenetics of Autoimmunity. http://dx.doi.org/10.1016/B978-0-12-809912-4.00007-6

most common genetic risk factors [6]. Genome-wide association studies (GWAS) have identified more than 100 risk loci associated with RA [7–10], which collectively explain approximately 50% of the genetic variance in RA [8,11]. Most of the GWAS-identified loci map to the noncoding genome and have yet to be functionally analyzed [8].

Epigenetic factors can play a role in the pathogenesis of RA at different levels. For example, cell differentiation in the adaptive and innate immune systems is controlled by epigenetic mechanisms (i.e., DNA methylation and histone modifications), therefore epigenetic factors may contribute to autoimmunity by affecting cell differentiation [12–21]. DNA methylation also contributes to the gender bias in RA by regulating the inactivation of the second X chromosome in females. Women are two-to-three times more likely to develop RA than men. Female sex hormones can also affect RA development and severity [22]. Female RA patients have an increased skewed X-chromosome inactivation pattern compared to controls [23], and the promoter of the X-chromosome encoded gene, *CD40L*, was shown to be demethylated in $CD4^+$ T cells in female (but not male) patients, leading to the overexpression of this gene [24]. Finally, epigenetic factors were also shown to be altered in specific cell types in the disease [25,26]. Whether these changes predispose individuals to disease risk or were inflammation-induced is incompletely understood and remains to be elucidated.

Epigenetic factors have the potential to act as intermediaries to confer environmental risk to altered gene expression levels in a cell-specific manner. Recent research efforts in RA have integrated genetic susceptibility loci with expression data, cis-regulatory element annotations, and cell-specific epigenetic changes to identify causal single nucleotide polymorphisms (SNPs), causal cell types, disease relevant pathways, and potential new therapeutic targets [8,25,27–29]. Since many approved drugs overlap with RA risk genes, the use of genetic data in drug development is of clear benefit [8,29]. In addition to known drug targets, these studies have the potential to discover novel and unexpected possibilities for drug development in RA [28,29]. Trynka et al. examined 15 chromatin marks in 38 different cell types and identified that trimethylation of histone 3 lysine 4 (H3K4me3), acetylation of H3K9 (H3K9ac), acetylation of DNase I hypersensitivity sites, hallmarks of open chromatin states, and active gene regulation were the most phenotypically cell-specific marks [26]. By overlaying 31 risk SNPs for RA with H3K4me3 marks, the authors identified a significant association with $CD4^+$ T cells, particularly with $CD4^+$ regulatory T cell (T_{reg}). Although the study was limited to certain cell types and lacked cells resident within joints, it highlighted the power of integrating genetic and epigenetic data [26]. Farh et al. developed a fine-mapping algorithm to identify candidate causal variants from genotyping data for 21 autoimmune diseases, including RA. The authors used an integrated analysis of genetic and transcriptional data sets with cis-regulatory annotations, and performed a series of epigenetic analyses in different subsets of resting and stimulated primary immune cells and different types of nonimmune cells. They mapped ~60% of likely causal variants to stimulus- and cell-specific enhancers, which were enriched for H3K27ac histone marks and noncoding RNAs [25]. A new powerful approach for identifying complex long-range interactions between distant genomic regions, particularly SNPs in enhancer regions, and their corresponding target genes, is the use of chromosome confirmation methods such as capture Hi-C [30] or the further developed promoter capture Hi-C (PCHi-C) [31]. Javierre et al. performed a PCHi-C analysis in 17 human primary hematopoietic cell types, providing cell- and lineage-specific promoter interactomes, and suggested that higher order genome structures undergo widespread and coordinated remodeling during lineage specification [31]. The authors showed that promoter-interacting regions were specifically enriched for histone marks associated with active enhancers, such as H3K27ac and H3K4me1, or H3K4me3 and H3K36me3, which are marks associated with active promoters and transcribed regions, respectively, which is consistent with noncoding transcription of regulatory regions. In the final step of their analysis, PCHi-C data were integrated with GWAS data for 8 autoimmune diseases (including RA) and 23 additional disease-related traits to build up the "core autoimmune disease network" containing the 421 highest-scoring genes prioritized for autoimmune disease [31].

7.2 EPIGENETIC PROTEINS AS POTENTIAL THERAPEUTIC TARGETS IN RA

Epigenetic proteins belong to three different classes of enzymes, and are the subjects of new epigenetic drug developments for RA [13,32,33]. Epigenetic modifications are attached to histone side chains or DNA by "writer" proteins. These modifications then influence the chromatin structure by making it denser or looser, thus blocking or enabling transcriptional regulation, respectively. Histone modifications are recognized by "reader" proteins, which serve as docking platforms for additional protein complexes involved in transcriptional regulation, DNA replication, or DNA damage repair. Epigenetic modifications are subject to endogenous and environmental influences and can be removed by "eraser" proteins [34]. In addition to DNA methylation and posttranslational histone modifications, noncoding RNA molecules add another level of complexity to the epigenetic regulation of gene expression [35].

Histone acetylation states are controlled by the antagonistic functions of histone acetyltransferase, which are positively correlated with gene expression, and histone deacetylases (HDAC), which reset chromatin to an inactive state by removing acetylation residues [36]. The most characterized and functionally studied group of epigenetic enzymes in RA are the eraser proteins, which are HDAC. Numerous studies have described the beneficial effects of HDAC inhibitors in vitro and in vivo in mouse models of arthritis [37]. The first isoform-specific HDAC inhibitors have recently been developed [32]. In addition to the epigenetic effects, HDAC inhibitors also affect acetylated cytoplasmic enzymes [38] and mRNA stability [39]. An unexpected epigenetic mechanism of HDAC inhibitor action was recently identified, whereby HDAC inhibitors were shown to deplete specific bromodomain reader protein levels within the nucleus and thus block transcription [40].

The human genome encodes 18 HDACs, which are subclassified into four groups based on their structural and functional properties. Class I HDACs (HDAC1, 2, 3, and 8) are widely expressed and are located in the nucleus. Class II HDACs (HDAC 4, 5, 6, 7, and 9) exhibit a more specific expression pattern and are located both in the nucleus and the cytoplasm. Sirtuins (SIRT1–7) are NAD^+-dependent enzymes of the class III HDACs, and HDAC11 is the sole member of the class IV HDACs [41]. The first studies to address the function of the individual HDAC isoforms has pointed to pro- and antiinflammatory roles for these proteins.

Silencing of HDAC1 in RA synovial fibroblasts (SF), which are the main resident cells in joints that are involved in promoting inflammation and cartilage destruction [42], reduced proliferation, migration, and invasion. Transcriptome analysis followed by pathway enrichment prediction revealed that HDAC1 regulates genes involved in apoptosis, regulation of glucose metabolism, circulatory system processes, DNA metabolism and repair, regulation of transcription factor activity, regulation of the extracellular matrix, cell proliferation, and cell migration. Furthermore, small interfering RNA (siRNA)-mediated silencing of HDAC1 in collagen-induced arthritis (CIA) mice attenuated paw swelling, reduced cartilage damage and bone loss, and decreased the expression of proinflammatory cytokines, most notably TNF-α [43]. Angiolilli et al. identified HDAC3 as a major factor required for the production of type I interferon (IFN) and the subsequent activation of signal transducer and activator of transcription 1 (STAT1) following IL-1β stimulation in RA SF [44]. Conversely, silencing of HDAC5 in RA SF increased the IL1-β-induced nuclear accumulation of IFN regulatory factor 1 (IRF1), and increased the transcription levels of type I IFN genes [45]. In addition to studies on class I and II HDACs, experiments on the functional role of SIRTs in RA have also been conducted. SIRT6 levels were found to be increased in the joint tissues of CIA mice, and functional studies revealed that SIRT6 overexpression attenuated the severity of arthritis by reducing both the inflammatory response and tissue destruction [46]. Smoking and TNF-α exposure were shown to increase SIRT6 levels in vivo and in vitro, as SIRT6 acts as part of a counterregulatory mechanism to attenuate the levels of matrix metalloproteinase 1 (MMP1) [47]. SIRT1 levels were also shown to be increased in synovial tissues and SF from RA patients compared to osteoarthritis (OA) patients, as well as in monocytes from RA patients compared to healthy individuals [48]. Furthermore, SIRT1 levels were increased in synovial tissues of RA smokers compared to nonsmokers and correlated with leukocytic infiltrates [49]. In contrast to the function of SIRT6, overexpression of SIRT1 contributed to apoptosis resistance, proliferation, and inflammatory cytokine production in RA SF, and TNF-α secretion in monocytes [48,49]. Detailed studies on the function of HDACs in other cell types implicated in the pathogenesis of RA, as well as studies on other members of the HDAC family, would be important for drug development and evaluation of the potential of HDAC inhibitors, as current data clearly show that they are implicated in distinct pathways and cellular functions. A clinical trial (NCT00570661) on the safety and efficacy of the orally active HDAC inhibitor, givinostat, in systemic-onset juvenile idiopathic arthritis (JIA) revealed benefits in patients after 12 weeks of treatment [50]. However, the lack of efficacy of givinostat in JIA in a subsequent dose-finding study (DSC08/2357/36) led to the termination of a phase II clinical trial (NCT01557452) on long-term safety (1 year). Another clinical trial (NCT01261624) on the safety and efficacy of givinostat in JIA was terminated due to a lack of enrolment. None of the studies were terminated due to tolerability concerns (www.ClinicalTrials.gov).

Epigenetic reader proteins, most importantly the bromodomain protein family, have become the focus of epigenetic drug developers due to their good drugability [51]. Bromodomain proteins recognize ε-N-acetylated histone side chains, and their function is closely related to transcriptional activation [52]. A first hint that the bromodomain and extraterminal (BET) proteins (BRD2, BRD3, BRD4, and BRDT) may be involved in the pathogenesis of RA was provided by Mahdi et al., who identified three SNPs in the *BRD2* locus that were associated with RA risk in a subset of patients positive for autoantibodies against cyclic citrullinated peptides and citrullinated α-enolase peptide 1 [53]. Although initially developed for the treatment of malignant diseases, pan-inhibitors against the BET protein family were also shown to have antiinflammatory and antidestructive properties in vitro [54] and in CIA mice in vivo [33,55,56]. The effects of BET inhibitors in CIA were attributed to the suppression of Th17 cell differentiation and function [33]. Effects on other cell types in vivo are also likely, as the BET-mediated regulation of cytokines, chemokines, and MMPs was described in RA SF [54], and BET protein inhibition had antiinflammatory properties in bone marrow-derived macrophages in mice [57]. Readers of other epigenetic marks, including selected methyl lysine binding-site readers, were also predicted to have good drugability and are now the subjects of inhibitor development programs [58].

7.3 DNA METHYLATION IN RA

Early DNA methylation studies showed that repetitive genomic sequences [i.e., long interspersed nuclear element-1 (LINE-1) retrotransposons] were expressed in RA SF but not in OA SF [59]. 5′-C-phosphate-G-3′ (CpG) sites of repetitive sequences are usually repressed by DNA methylation in the human genome, therefore they are commonly used in the field of cancer as a marker of global DNA hypomethylation [60]. The CpG island of LINE-1 was shown to be hypomethylated in RA SF in comparison to OA SF [61]. In addition, lower levels of 5-methylocytosine (5-mC) were also observed in RA SF than OA SF. These experiments revealed that the genome of RA SF is globally hypomethylated [61]. Global hypomethylation was also observed in T cells, B cells, and monocytes in early RA patients [62]. In contrast, Rhead et al. used Illumina HumanMethylation 450 bead chips and identified 1056 hypermethylated CpG sites in peripheral blood-derived $CD4^+$ naive T cells from RA patients compared to unaffected controls. These hypermethylated sites had also previously been shown to be hypermethylated in RA SF [63].

The mechanism of global hypomethylation was analyzed in RA SF. The expression of DNA methyltransferase 1 (DNMT1) was shown to be reduced in RA SF, and the proinflammatory cytokines, TNF-α and IL-1, reduced 5-mC levels and DNMT expression in these cells [61]. DNMT1 catalyzes the synthesis of the 5-mC modification by consuming S-adenosyl methionine (SAM) [64]. SAM levels were decreased in RA SF, and further investigation revealed that the polyamine pathway was responsible for the high consumption of SAM in these cells. Protein expression analysis of the polyamine metabolism revealed that enzymes (i.e., SSAT1), proteins, and carrier proteins involved in this pathway were highly expressed in RA SF and contributed to faster polyamine recycling and consumption of SAM (Fig. 7.1) [65]. The treatment of RA SF with the SSAT1 inhibitor, diminazene aceturate, reduced the invasiveness of these cells (by 70%) in the severe combined immunodeficient mouse model [66]. Therefore this strategy has been proposed as a potential future therapy for targeting RA SF.

Several studies have analyzed the methylation status of single promoter regions using bisulfite sequencing, and performed genome-wide methylation studies using methylation promoter arrays,

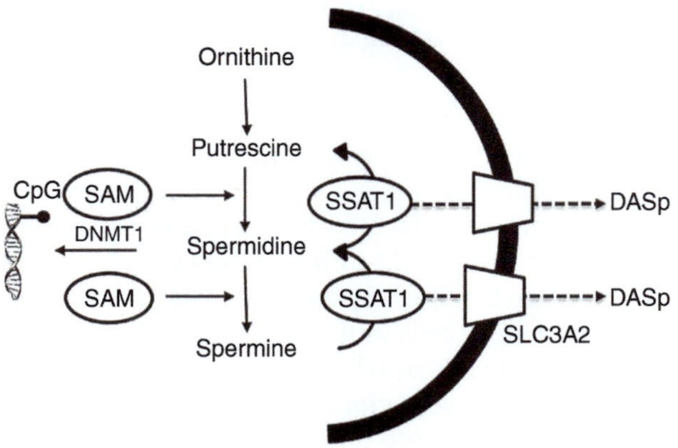

FIGURE 7.1 **Activation of the polyamine pathway reduces S-adenosyl methionine (SAM) and causes global hypomethylation in RA SF.** *DASp*, diacetylpolyamines; *DNMT1*, DNA methyl transferase 1; *SLC3A2*, solute carrier family 3 member 2; SSAT1, spermidine/spermine N1-acetyltransferase.

TABLE 7.1 Identification of Differentially Methylated Genes in Cell Types in RA

Genes	Hypomethylated	Hypermethylated	Methods	Tissues	References
CXCL12	✓	—	BS	RA SF	[67]
EBF3 and *IRX1*	—	✓	MeDIP and pyro	RA SF	[68]
TBX5	✓	—	MeDIP and pyro	RA SF	[69]
CHI3L1, *CASP1*, *STAT3*, *MAP3K5*, *MEFV*, and *WISP3*	✓	—	450K array	RA SF	[70]
TGFBR2 and *FOXO1*	—	✓	450K array	RA SF	[70]
IL6R, *CAPN8*, *HOXA11*, and *S100A14*	✓	—	450K array	RA SF	[71]
CCR6, *DPP4*, *HOXC4*, and *BCL6*	—	✓	450K array	RA SF	[71]
CD40L	✓	—	BS	$CD4^+$ T cells	[24]
DR3	—	✓	BS	RA SF	[72]
IL6	—	✓	BS	PBMC	[73]
GSTA2	✓	—	450K array	PBMC	[27]
HLADRB1	✓	—	450K array	PBMC	[27]
ELMO1	✓	—	450K array and GWAS	RA SF	[29]
LBH	✓	—	450K array, GWAS, and pyro	RA SF	[74]

BS, Bisulfite sequencing; *GWAS*, genome-wide association studies: *MeDIP*, methylation immunoprecipitation; *PBMC*, peripheral blood mononuclear cells; *pyro*, pyrosequencing; *RA*, rheumatoid arthritis; *SF*, synovial fibroblasts.

in SF, $CD4^+$ T cells, and macrophages. They have provided important insights into the pathogenesis of RA by contributing to the understanding of gene expression regulation of known genes and identifying novel gene targets (Table 7.1).

For example, *CXCL12* is known to be elevated in RA and expressed at higher levels in RA SF than OA SF [75]. The upregulation of *CXCL12* in RA SF was shown to be epigenetically regulated by promoter hypomethylation. High *CXCL12* expression levels have functional effects in RA as they increase the infiltration of inflammatory cells in the synovium. Another study showed that the promoter of *TBX5* was hypomethylated in RA SF, which resulted in the upregulation of *TBX5* [69]. *TBX5* increases the expression of proinflammatory cytokines (i.e., IL8, CXCL2, and CCL20) and contributes to SFs ability to attract inflammatory cells into the synovium. Novel candidate genes [i.e., limb bud and heart development (*LBH*) and engulfment and cell motility 1 (*ELMO1*)] have been identified in RA through the combination of GWAS and methylation studies [29,74]. A SNP associated with RA and a hypomethylated CpG site was found to be located on an *LBH* enhancer region in RA SF, and DNA methylation was responsible for the expression of *LBH* in these cells. Another study also showed that LBH promoted the cell growth of SF in RA [76]. *ELMO1* is another novel gene that is strongly expressed in RA SF and the RA synovium [29], and it has been found to regulate cell invasion, migration, and Rac1 signaling in RA SF.

Methotrexate is the most commonly used disease-modifying antirheumatic drug for the treatment of RA. It inhibits folate metabolism by inhibiting dihydrofolate reductase, and since folate is required to synthesize SAM and 5-mC during DNA replication, methotrexate was studied as a potential RA therapy. Interestingly, methotrexate reverts the global hypomethylation profile back towards hypermethylation in T cells, B cells, and monocytes after 1 month of treatment [62]. In RA, the T_{regs} response fails to control the activity of T helper cells. FOXP3 is the most important transcription factor for the differentiation of naive T cells to T_{regs}, and DNA methylation plays an important role in *FOXP3* expression [77]. Methotrexate is able to restore the suppressive function of T_{regs} by demethylating the FOXP3 promoter and upregulating the expression of *CTLA4* [78].

Biological drug therapies have improved the treatment of RA [79]; however, only 30% of patients respond well to this treatment [80]. A biomarker that could predict the treatment response would be extremely useful for patient care. A recent study in RA used differentially methylated genes as biomarkers of the treatment response to etanercept [81]. A group of responders was compared to nonresponders after 3 months of treatment, and the low-density lipoprotein receptor-related protein 1 gene was shown to be differentially methylated in those that responded to the treatment.

The risk of developing RA is associated with environmental exposure and genetic influence, particularly with smoking and the HLA-DRB1 genotype [82]. An early sign of RA is the presence of ACPA, which targets self-tissues in the joints. Therefore the impact of environmental factors in $ACPA^{+}$ RA is of interest. Monozygotic (MZ) twins who are discordant for a disease are powerful tools in which to study the influence of environmental factors on disease pathogenesis. DNA methylation is a dynamic and reversible process, therefore the environment can have a direct impact on this molecular mechanism. One study used total blood samples from MZ twins who were discordant for $ACPA^{+}$ RA, and identified that the differentially methylated region was associated with *EXOSC1* and protocadherin beta-14 [83]. Future studies using larger cohorts will hopefully provide valuable information on the functional role of these genes during disease progression.

7.4 NONCODING RNA IN RA

Noncoding RNAs are a versatile group of molecules that regulate a broad spectrum of cellular processes, including transcription, RNA processing, and translation [35]. Different classes of noncoding RNAs have been studied in RA, with microRNAs (miR) being the most investigated group, followed by long noncoding RNAs (lncRNA) and PIWI-associated RNA [84–88]. Recent data suggest that the role of the noncoding transcriptome has been underestimated in arthritis susceptibility and severity; for example, most of the GWAS hits in RA are located outside of protein coding sequences, but the likely causal variants are enriched for the transcription of noncoding variants [25]. In line with this, a SNP (rs2833522) in the lncRNA, AP000255.6-001, was significantly associated with the progression of joint destruction in a GWAS in $ACPA^{-}$ RA patients [89]. Another indication of the profound role of the noncoding transcriptome in RA is based on the finding that SF from different anatomic sites in RA and OA patients show improved clustering when based on the noncoding, and not the coding, transcriptome; clustering reached perfection when based on the expression of miRs [90,91].

miR are short noncoding RNAs that are approximately 22 nucleotides in length. Binding of a miR to the 5′ untranslated region of its target mRNA leads to degradation of the miR–mRNA duplex or inhibition of translational [92]. A major challenge in studying miR function in RA is the complexity of miRNA target-gene networks [28]. Several miR have been studied in various cell types in RA, and their aberrant expression has been shown to be involved in numerous processes involved in the pathogenesis of RA (i.e., inflammation, bone homeostasis, cell survival, and proliferation) [84]. Okada et al. developed a novel analytical method to analyze the enrichment of GWAS signals in miR target-gene networks, and provided a list the miR and their target genes that had excess genetic association signals in RA. The authors suggested that miR-4728-5p, which down-regulates peptidyl arginine deaminase 2 (PADI2), was a potential therapeutic target in RA [28]. The descriptions of miR involved in the pathogenesis of RA in this chapter have been restricted to those miRs with confirmed regulation of specific targets or in vivo evidence of a functional role in arthritis (Table 7.2).

On one hand, miRs are discussed as future therapeutic targets for RA, while on the other hand they are suggested as diagnosis tools and biomarkers for predicting therapy response and disease progression. miRs are stable in body fluids (i.e., whole blood, serum, plasma, and synovial fluid) and are relatively easy to measure; however, analyzing the data is challenging due to the lack of appropriate reference levels [86]. Studies on RA-specific changes of miR expression levels are mostly lacking and most miR have only been compared to expression levels in healthy individuals and not to other autoimmune diseases [109]. However, circulating miR-125b levels were suggested as a marker for rituximab therapy response in RA patients [110], and a combination of low miR-22 and high miR-886.3p levels were associated with a good response to adalimumab in early RA [111].

TABLE 7.2 Expression and Function of microRNAs in RA

miRNAs	Expressions	Validated targets	References
miR-27a	↓ in RA/healthy (serum, synovial tissue and SF)	FSTL1	[93]
miR-29b	↑ by betaine treatment (RA SF)	DNMT3a	[94]
miR-34a*	↓ in RA/OA (SF)	XIAP	[95]
miR-124a	↓ in RA/OA (SF)	CDK2, MCP1, and NFATc1	[96,97]
miR-146a	↑ in RA/OA (synovial tissues, SF, peripheral blood, and IL17-producing T cells) ↑ in RA/healthy ($CD4^+$ T cells and PBMC)	TRAF6, IRAK1, and STAT1	[98–100]
miR-155	↑ in RA/OA (synovial tissues, synovial fluid, SF, and PBMC) ↑ in RA/healthy (PBMC and B cells)	SHIP1, SOCS, and PU.1	[98,101,102]
miR-223	↑ in RA/OA (synovial tissues, synovial fluid, SF, and $CD14^+$ cells) ↑ in RA/healthy (naive $CD4^+$ T cells)	NF-1A and ARNT	[98,103–108]

OA, Osteoarthritis; *PBMC*, peripheral blood mononuclear cells; *SF*, synovial fibroblasts.
microRNAs were selected based on functional studies identifying target genes.

Changes in plasma miR-23 and miR-223 were suggested as biomarkers of the response to anti-TNF-α in RA [112]; however, these results were not confirmed by a similar study [113]. To date, no miR can consistently predict the response to anti-TNF-α, and validation for other therapies in independent cohorts is still missing.

The best-studied miR in RA are miR-146a and miR-155, which were originally identified in a screen of differentially expressed miR between RA and OA SF [114]. These miR have since been shown to be increased in all measured tissues and cells (i.e., synovial tissues, synovial fluid, SF, peripheral blood mononuclear cells (PBMCs), and specific blood-cell subtypes) [98,103,114,115] and in SF from the human TNF transgenic mouse model (TghuTNF, Tg197) [116]. miR-146a and miR-155 are considered as the "yin and yang" of inflammation as they are implicated in negative and positive regulation of inflammatory pathways, respectively [85]. The definitive role of miR-155 in the pathogenesis of RA was determined in an experimental mouse models of arthritis [101,117]. miR-155 deficient mice were resistant to CIA and had no detectable inflammatory cell infiltrates in the synovial membrane. These effects were attributed to higher expression levels of Src homology 2-containing inositol phosphatase 1 (SHIP1), which is a direct target of miR-155 [101]. Furthermore, the lack of miR-155 interfered with late-stage osteoclast differentiation in the K/BxN serum-transfer arthritis model, leading to reduced levels of bone erosions and no impact on joint inflammation [117]. In contrast, the enforced expression of miR-155 in RA SF reduced the expression levels of MMP1 and MMP3, thus pointing to a protective effect of miR-155 in this cell type [114]. The enforced expression of miR-155 in RA monocytes increased the expression levels of chemokines (i.e., CCL3, CCL4, CCL5, and CCL8) and increased the levels of the chemokine receptor, CCR7 [118]. miR-155 was also shown to be overexpressed in B cells, particularly memory B cells, that were already in an early phase of RA. The authors showed that high levels of miR-155 were paralleled by decreased levels of the transcription factor PU.1, which is crucial for B cell differentiation. Inhibition of miR-155 in B cells restored PU.1 expression levels and inhibited B-cell receptor crosslinking-induced antibody production [102]. MiR-155 and miR-146a were described as characteristically altered miR in naive and memory T_{reg} in peripheral blood from RA patients compared to healthy controls [119]. Conversely, the upregulation of miR-146a and miR-155 in response to T-cell stimulation was diminished in T_{reg} of RA patients compared to healthy controls. This resulted in a proinflammatory T_{reg} phenotype in RA, which was due to the augmented expression and activation of STAT1 (a direct target of miR-146a) and characterized by inflammatory cytokine expression [99]. miR-146a also acts as an antiapoptotic factor in T cells [84]. miR-146a was shown to be induced by several inflammatory stimuli, and functional studies in monocytes and macrophages suggested a counterregulatory role of miR-146a through the restriction of inflammation by targeting components of the Toll-like receptor pathway (i.e., IRAK1 and TRAF6) and the type I IFN pathway (STAT1) [120,121].

Increased miR-223 expression levels have also been identified in many cell types (i.e., SF, naive $CD4^+$ T cells, and $CD14^+$ monocytes) in RA compared to OA or healthy controls, as well as in the synovial fluid and serum [98,103–108]. Lentiviral silencing of miR-223 in CIA mice reduced the disease severity (characterized by reduced histological and arthritis scores), osteoclastogenesis, and bone erosions. Li et al. identified nuclear factor IA as a direct target of miR-223, which led to increased osteoclastogenesis [104]; however, Shibuya et al. described a suppressing effect of miR-223 on osteoclastogenesis [122], thus making it difficult to interpret the role of this miR in osteoclastogenesis. miR-223 was also identified as a negative regulator of the aryl hydrocarbon receptor (AHR)/AHR nuclear translocator (ARNT) pathway in macrophages from RA patients. This miR targets ARNT, and reduced levels of ARNT prevent the inhibition of inflammatory cytokine production [105]. In contrast, miR-223 was described to exert a proinflammatory function in T cells through the upregulation of LPS-induced IFN-γ production and reduction IL-17a synthesis [123,124]. Furthermore, overexpression of miR-223 in activated T cells suppressed the expression of insulin-like growth factor-1 receptor and impaired IL-10 production [107].

Another miR that plays a functional role in arthritis is miR-124a [96], which was shown to be downregulated in RA SF compared to OA SF [97]. Injection of pre-miR-124 (miR-124, the rat analogue of human miR-124a) into the ankle joints of adjuvant-induced arthritis (AIA) rats suppressed arthritis, decreased synoviocyte proliferation, reduced leucocyte infiltration, and suppressed cartilage and bone destruction. miR-124 suppressed the differentiation of osteoclasts by targeting nuclear factor of activated T cells cytoplasmic 1 (NFATc1) [96]. miR-124a was also shown to suppress expression levels of cyclin dependent kinase 2 (CDK2) and monocyte chemoattractant protein 1 (MCP1) and to decrease proliferation in RA SF [97].

Several miR are also regulated by DNA methylation [94,95,125]. DNA methylation downregulates the proapoptotic miR-34a* in RA SF, which results in increased expression of its target, X-linked inhibitor of apoptosis protein (XIAP), and contributes to the apoptosis-resistant phenotype of RA SF [95]. Conversely, treatment of RA SF with the methyl-donor, betaine, increased miR-29levels, which is a miR that targets DNMT3a and thereby limits the remethylation of globally hypomethylated SF [94].

Whereas miR have been widely studied in the field of inflammation and RA, studies looking at the roles of lncRNAs have only recently started to be conducted. LncRNAs are autonomously transcribed RNAs that are over 200 nucleotides in length and do not encode a protein [35]. LncRNAs exert their function either in the cell's cytoplasm or in the nucleus. In the cytoplasm they regulate mRNA stability, promote or inhibit translation of mRNAs, serve as precursors for miRNAs, or

TABLE 7.3 Expression and Function of lncRNAs in RA

lncRNA	Expression	Effects	References
HOTAIR	↑ in RA/OA PBMC ↑ in SF lower/upper extremities	Enforced expression in osteoclasts → ↓ of MMP2 MMP3 Silencing in SF → ↑ basal and TNF-α-induced production of MMP1, 3, 13, 14	[127]
C17orf76-AS1	↑ in RA/healthy SF from knees	Correlation with CRP, SDAI	[128]
RP11-534G203	↓ in RA/healthy SF from knees	Unknown	[128]
AK096159	↓ in RA/healthy SF from knees	Unknown	[128]
RP11-57I11.2	↓ in RA/healthy SF from knees	Unknown	[128]
LOC100652951	↑ in RA/healthy T cells	Unknown	[129]
LOC100506036	↑ in RA/healthy T cells	Silencing in Jurkat cells → ↓ of SMPD1, NFAT1, and IFN-γ production	[129]
GAS5	↓ in RA/healthy $CD4^+$ T cells and B cells	Unknown	[130]
lincRNA-p21	↓ in RA/healthy PBMC	Decreased lincRNA-p21→ ↑ levels of p-RelA	[131]

CRP, C-reactive protein; *IFN*, interferon; *MMP*, matrix metalloproteinase, *OA*, osteoarthritis, *PBMC*, peripheral blood mononuclear cells; *SDAI*, simplified disease activity index; *SF*, synovial fibroblasts.

compete for miRNA-mediated inhibition and thus increase mRNA expression levels. In the nucleus they guide chromatin modifying complexes to specific genomic loci to regulate gene expression, act as coactivators or corepressors and change the activity of transcription factors, or regulate alternative splicing by binding to splicing regulatory proteins [126] (Table 7.3).

The first lncRNA found to be upregulated in PBMCs in RA was HOX transcript antisense RNA (HOTAIR) [127], a lncRNA that interacts with polycomb repressive complex 2 (PRC2) to regulate levels of the repressive histone mark, H3K27me3, and chromatin, thus downregulating target genes [132]. Enforced expression of HOTAIR in osteoclasts decreased expression levels of MMP2 and MMP13 [127]. HOTAIR was also identified as one of the most differentially expressed long noncoding transcripts in SF derived from upper and lower extremities. Interestingly, silencing HOTAIR in knee SF, which exhibit high endogenous levels of HOTAIR, increased basal and TNFα-induced levels of MMP1, MMP3, MMP13, and MMP14, indicating a role of this lncRNA in the matrix-destructive behavior of lower-extremity SF and pointing to a general role of this lncRNA in regulating matrix degradation [90,91]. Zhang et al. were the first to screen differentially expressed lncRNAs in knee SF from RA patients compared to healthy individuals with trauma. Levels of C17orf76-AS1 were increased in RA SF compared to healthy SF and correlated with clinical characteristics [i.e., C-reactive protein (CRP) and simplified disease activity index (SDAI)], while levels of RP11-534G20.3, AK096159, and RP11-573I11.2 were significantly reduced in RA SF [128]. The functional role of these lncRNAs remains to be unraveled. The lncRNAs, LOC100652951 and LOC100506036, were increased in T cells from RA patients compared to healthy controls. The expression levels of these lncRNAs were downregulated by treatment with biologic agents or by gender (female), respectively. Silencing of LOC100506036 in activated Jurkat cells decreased nuclear factor of activated T cells 1 (NFAT1) levels, inhibited IFN-γ production, and reduced sphingomyelin phosphodiesterase 1 (SMPD1) levels [129]. The lncRNA, growth arrest specific 5 (GAS5), was also downregulated in $CD4^+$ T cells and B cells from RA and SLE patients. Mayama et al. suggested a potential role for GAS5 in the regulation of inflammatory pathways as GAS5 expression levels were also altered in other inflammatory conditions [130]. Another lncRNA associated with an inflammatory response was lincRNA-p21, which was reduced in PBMCs of RA but not SLE patients, thus resulting in increased levels of phosphorylated p65 (a marker of NF-κB activation). Spurlock et al. showed that methotrexate treatment induced levels of lincRNA-p21 [131]. Current knowledge regarding the functional role of lncRNAs in the regulation of inflammatory pathways has been reviewed by Pearson and Jones [87].

A novel group of lncRNAs are circular RNAs (circRNAs), which are characterized by a covalently closed loop without 5′–3′ polarity and without a polyA tail. CircRNAs regulate gene expression at the transcriptional or posttranscriptional level by binding to miR or other molecules. Recent studies suggest that circRNAs may be used as a novel biomarker for cancer, and that they may play a functional role in cancer and other human diseases [133]. Studies of circRNAs in RA have not yet been conducted.

7.5 SUMMARY

Epigenetic research in RA and other autoimmune diseases is an evolving field that has high expectations. Epigenetic changes are discussed as future markers of therapy response and disease progression, and epigenetic enzymes are currently the subjects for some drug development trials; however, further studies are needed to address the potential of epigenetic factors to act as biomarkers and drug targets. High-throughput technologies for interrogating genetic and epigenetic data sets have provided new insights into disease pathogenesis and have identified disease-relevant pathways in distinct cell types. The inclusion of further cell types (i.e., joint resident cells) is highly desirable and will likely lead to a better understanding of the disease pathogenesis.

LIST OF ABBREVIATIONS

ACPA	anticitrullinated protein antibodies
AHR	aryl hydrocarbon receptor
AIA	adjuvant-induced arthritis
ARNT	AHR nuclear translocator
BET	bromodomain and extraterminal domain
CIA	collagen-induced arthritis
circRNAs	circular RNAs
CpG	5′-C-phosphate-G-3′
CRP	C-reactive protein
DHS	DNase I hypersensitivity site
DMARD	disease-modifying anti-rheumatic drug
DNMT	DNA methyl transferase
ELMO1	engulfment and cell motility protein-1
GAS5	growth arrest specific 5
GWAS	genome-wide association study
HAT	histone acetyl transferase
HDAC	histone deacetylase
H3K4me3	histone 3 lysine 4 trimethylation
H3K9ac	histone 3 lysine 9 acetylation
HLA	human leukocyte antigen
HOTAIR	HOX transcript antisense RNA
IFN	interferon
JIA	juvenile idiopathic arthritis
LBH	limb and heart development
LINE-1	Long interspersed nuclear element
lncRNA	long-non coding RNA
miR	microRNA
5-mC	5-methyl cytosine
MMP	matrix metalloproteinase
OA	osteoarthritis
PBMC	peripheral blood mononuclear cells
PCHi-C	promoter capture Hi-C
PRC2	polycomb repressive complex 2
RA	rheumatoid arthritis
siRNA	small interfering RNA
SIRT	sirtuin
SDAI	simplified disease activity index
SLE	systemic lupus erythematosus
SNP	single nucleotide polymorphism
SF	synovial fibroblast
STAT	signal transducer and activator of transcription
T_{reg}	regulatory T cells

References

[1] Smolen JS, Aletaha D, McInnes IB. Rheumatoid arthritis. Lancet 2016;388(10055):2023–38.

[2] Van Doornum S, McColl G, Wicks IP. Accelerated atherosclerosis: an extraarticular feature of rheumatoid arthritis? Arthritis Rheum 2002;46(4):862–73.

[3] Cojocaru M, Cojocaru IM, Silosi I, Vrabie CD, Tanasescu R. Extra-articular manifestations in rheumatoid arthritis. Maedica (Buchar) 2010;5(4):286–91.

[4] van Zeben D, Hazes JM, Zwinderman AH, Cats A, van der Voort EA, Breedveld FC. Clinical significance of rheumatoid factors in early rheumatoid arthritis: results of a follow up study. Ann Rheum Dis 1992;51(9):1029–35.

[5] Gerlag DM, Raza K, van Baarsen LG, Brouwer E, Buckley CD, Burmester GR, et al. EULAR recommendations for terminology and research in individuals at risk of rheumatoid arthritis: report from the study group for risk factors for rheumatoid arthritis. Ann Rheum Dis 2012;71(5):638–41.

[6] Gregersen PK, Silver J, Winchester RJ. The shared epitope hypothesis. an approach to understanding the molecular genetics of susceptibility to rheumatoid arthritis. Arthritis Rheumatol 1987;30(11):1205–13.

[7] Stahl EA, Raychaudhuri S, Remmers EF, Xie G, Eyre S, Thomson BP, et al. Genome-wide association study meta-analysis identifies seven new rheumatoid arthritis risk loci. Nat Genet 2010;42(6):508–14.

[8] Okada Y, Wu D, Trynka G, Raj T, Terao C, Ikari K, et al. Genetics of rheumatoid arthritis contributes to biology and drug discovery. Nature 2014;506(7488):376–81.

[9] Okada Y, Terao C, Ikari K, Kochi Y, Ohmura K, Suzuki A, et al. Meta-analysis identifies nine new loci associated with rheumatoid arthritis in the Japanese population. Nat Genet 2012;44(5):511–6.

[10] Eyre S, Bowes J, Diogo D, Lee A, Barton A, Martin P, et al. High-density genetic mapping identifies new susceptibility loci for rheumatoid arthritis. Nat Genet 2012;44(12):1336–40.

[11] Viatte S, Plant D, Raychaudhuri S. Genetics and epigenetics of rheumatoid arthritis. Nat Rev Rheumatol 2013;9(3):141–53.

[12] Amaravadi L, Klemsz MJ. DNA methylation and chromatin structure regulate PU.1 expression. DNA Cell Biol 1999;18(12):875–84.

[13] Rodriguez RM, Lopez-Larrea C, Suarez-Alvarez B. Epigenetic dynamics during CD4(+) T cells lineage commitment. Int J Biochem Cell Biol 2015;67:75–85.

[14] Huehn J, Beyer M. Epigenetic and transcriptional control of FOXP3+ regulatory T cells. Semin Immunol 2015;27(1):10–8.

[15] Danbara M, Kameyama K, Higashihara M, Takagaki Y. DNA methylation dominates transcriptional silencing of PAX5 in terminally differentiated B cell lines. Mol Immunol 2002;38(15):1161–6.

[16] Maier H, Colbert J, Fitzsimmons D, Clark DR, Hagman J. Activation of the early B-cell-specific MB-1 (IG-alpha) gene by PAX-5 is dependent on an unmethylated ETS binding site. Mol Cell Biol 2003;23(6):1946–60.

[17] Schwab J, Illges H. Regulation of CD21 expression by DNA methylation and histone deacetylation. Int Immunol 2001;13(5):705–10.

[18] Xu CR, Feeney AJ. The epigenetic profile of IG genes is dynamically regulated during B cell differentiation and is modulated by pre-B cell receptor signaling. J Immunol 2009;182(3):1362–9.

[19] Lee SC, Bottaro A, Insel RA. Activation of terminal B cell differentiation by inhibition of histone deacetylation. Mol Immunol 2003;39(15):923–32.

[20] Kapellos TS, Iqbal AJ. Epigenetic control of macrophage polarisation and soluble mediator gene expression during inflammation. Mediators Inflamm 2016;2016:6591703.

[21] Suarez-Alvarez B, Baragano Raneros A, Ortega F, Lopez-Larrea C. Epigenetic modulation of the immune function: a potential target for tolerance. Epigenetics 2013;8(7):694–702.

[22] Gabay CMJN, van Laar JM. Rheumatoid arthritis: pathogenesis and clinical features. EULAR textbook on rheumatic diseases. United Kingdom: BMJ; 2015. 229–264.

[23] Chabchoub G, Uz E, Maalej A, Mustafa CA, Rebai A, Mnif M, et al. Analysis of skewed X-chromosome inactivation in females with rheumatoid arthritis and autoimmune thyroid diseases. Arthritis Res Ther 2009;11(4):R106.

[24] Liao J, Liang G, Xie S, Zhao H, Zuo X, Li F, et al. CD40l demethylation in CD4(+) T cells from women with rheumatoid arthritis. Clin Immunol 2012;145(1):13–8.

[25] Farh KK, Marson A, Zhu J, Kleinewietfeld M, Housley WJ, Beik S, et al. Genetic and epigenetic fine mapping of causal autoimmune disease variants. Nature 2015;518(7539):337–43.

[26] Trynka G, Sandor C, Han B, Xu H, Stranger BE, Liu XS, et al. Chromatin marks identify critical cell types for fine mapping complex trait variants. Nat Genet 2013;45(2):124–30.

[27] Liu Y, Aryee MJ, Padyukov L, Fallin MD, Hesselberg E, Runarsson A, et al. Epigenome-wide association data implicate DNA methylation as an intermediary of genetic risk in rheumatoid arthritis. Nat Biotechnol 2013;31(2):142–7.

[28] Okada Y, Muramatsu T, Suita N, Kanai M, Kawakami E, Iotchkova V, et al. Significant impact of miRNA-target gene networks on genetics of human complex traits. Sci Rep 2016;6:22223.

[29] Whitaker JW, Boyle DL, Bartok B, Ball ST, Gay S, Wang W, et al. Integrative omics analysis of rheumatoid arthritis identifies non-obvious therapeutic targets. PLoS One 2015;10(4):e0124254.

[30] Rudan MV, Hadjur S, Sexton T. Detecting spatial chromatin organization by chromosome conformation capture II: genome-wide profiling by HI-C. Methods Mol Biol 2016;47–74.

[31] Javierre BM, Burren OS, Wilder SP, Kreuzhuber R, Hill SM, Sewitz S, et al. Lineage-specific genome architecture links enhancers and non-coding disease variants to target gene promoters. Cell 2016;167(5):1369–84.

[32] Regna NL, Reillya CM. Isoform-selective HDAC inhibition in autoimmune disease. J Clin Cell Immunol 2014;5(2):206–13.

[33] Mele DA, Salmeron A, Ghosh S, Huang HR, Bryant BM, Lora JM. BET bromodomain inhibition suppresses TH17-mediated pathology. J Exp Med 2013;210(11):2181–90.

[34] Musselman CA, Lalonde ME, Cote J, Kutateladze TG. Perceiving the epigenetic landscape through histone readers. Nat Struct Mol Biol 2012;19(12):1218–27.

[35] Cech TR, Steitz JA. The noncoding RNA revolution-trashing old rules to forge new ones. Cell 2014;157(1):77–94.

[36] Wang Z, Zang C, Cui K, Schones DE, Barski A, Peng W, et al. Genome-wide mapping of hats and HDACs reveals distinct functions in active and inactive genes. Cell 2009;138(5):1019–31.

[37] Klein K, Gay S. Epigenetics in rheumatoid arthritis. Curr Opin Rheumatol 2015;27(1):76–82.

[38] Singh BN, Zhang G, Hwa YL, Li J, Dowdy SC, Jiang SW. Nonhistone protein acetylation as cancer therapy targets. Expert Rev Anticancer Ther 2010;10(6):935–54.

[39] Grabiec AM, Korchynskyi O, Tak PP, Reedquist KA. Histone deacetylase inhibitors suppress rheumatoid arthritis fibroblast-like synoviocyte and macrophage IL-6 production by accelerating mRNA decay. Ann Rheum Dis 2012;71(3):424–31.

[40] Mackmull MT, Iskar M, Parca L, Singer S, Bork P, Ori A, et al. Histone deacetylase inhibitors (HDACi) cause the selective depletion of bromodomain containing proteins (BCPs). Mol Cell Proteomics 2015;14(5):1350–60.

[41] McGee-Lawrence ME, Westendorf JJ. Histone deacetylases in skeletal development and bone mass maintenance. Gene 2011;474(1–2):1–11.

[42] Bottini N, Firestein GS. Duality of fibroblast-like synoviocytes in RA: passive responders and imprinted aggressors. Nat Rev Rheumatol 2013;9(1):24–33.

[43] Hawtree S, Muthana M, Wilkinson JM, Akil M, Wilson AG. Histone deacetylase 1 regulates tissue destruction in rheumatoid arthritis. Hum Mol Genet 2015;24(19):5367–77.

[44] Angiolilli C, Kabala PA, Grabiec AM, Van Baarsen IM, Ferguson BS, Garcia S, et al. Histone deacetylase 3 regulates the inflammatory gene expression programme of rheumatoid arthritis fibroblast-like synoviocytes. Ann Rheum Dis 2016;.

[45] Angiolilli C, Grabiec AM, Ferguson BS, Ospelt C, Malvar Fernandez B, van Es IE, et al. Inflammatory cytokines epigenetically regulate rheumatoid arthritis fibroblast-like synoviocyte activation by suppressing HDAC5 expression. Ann Rheum Dis 2016;75(2):430–8.

[46] Lee HS, Ka SO, Lee SM, Lee SI, Park JW, Park BH. Overexpression of sirtuin 6 suppresses inflammatory responses and bone destruction in mice with collagen-induced arthritis. Arthritis Rheum 2013;65(7):1776–85.

[47] Engler A, Niederer F, Klein K, Gay RE, Kyburz D, Camici GG, et al. Sirt6 regulates the cigarette smoke-induced signalling in rheumatoid arthritis synovial fibroblasts. J Mol Med 2014;92(7):757–67.

[48] Niederer F, Ospelt C, Brentano F, Hottiger MO, Gay RE, Gay S, et al. Sirt1 overexpression in the rheumatoid arthritis synovium contributes to proinflammatory cytokine production and apoptosis resistance. Ann Rheum Dis 2011;70(10):1866–73.

[49] Engler A, Tange C, Frank-Bertoncelj M, Gay RE, Gay S, Ospelt C. Regulation and function of SIRT1 in rheumatoid arthritis synovial fibroblasts. J Mol Med 2016;94(2):173–82.

[50] Vojinovic J, Damjanov N, D'Urzo C, Furlan A, Susic G, Pasic S, et al. Safety and efficacy of an oral histone deacetylase inhibitor in systemic-onset juvenile idiopathic arthritis. Arthritis Rheum 2011;63(5):1452–8.

[51] Gallenkamp D, Gelato KA, Haendler B, Weinmann H. Bromodomains and their pharmacological inhibitors. ChemMedChem 2014;9(3):438–64.

[52] Sanchez R, Zhou MM. The role of human bromodomains in chromatin biology and gene transcription. Curr Opin Drug Discov Devel 2009;12(5):659–65.

[53] Mahdi H, Fisher BA, Kallberg H, Plant D, Malmstrom V, Ronnelid J, et al. Specific interaction between genotype, smoking and autoimmunity to citrullinated alpha-enolase in the etiology of rheumatoid arthritis. Nat Genet 2009;41(12):1319–24.

[54] Klein K, Kabala PA, Grabiec AM, Gay RE, Kolling C, Lin LL, et al. The bromodomain protein inhibitor I-BET151 suppresses expression of inflammatory genes and matrix degrading enzymes in rheumatoid arthritis synovial fibroblasts. Ann Rheum Dis 2016;75(2):422–9.

[55] Zhang QG, Qian J, Zhu YC. Targeting bromodomain-containing protein 4 (BRD4) benefits rheumatoid arthritis. Immunol Lett 2015;166(2):103–8.

[56] Xiao Y, Liang L, Huang M, Qiu Q, Zeng S, Shi M, et al. Bromodomain and extra-terminal domain bromodomain inhibition prevents synovial inflammation via blocking ikappab kinase-dependent nf-kappab activation in rheumatoid fibroblast-like synoviocytes. Rheumatology 2016;55(1):173–84.

[57] Nicodeme E, Jeffrey KL, Schaefer U, Beinke S, Dewell S, Chung CW, et al. Suppression of inflammation by a synthetic histone mimic. Nature 2010;468(7327):1119–23.

[58] Santiago C, Nguyen K, Schapira M. Drugability of methyl-lysine binding sites. J Comput Aided Mol Des 2011;25(12):1171–8.

[59] Kuchen S, Seemayer CA, Rethage J, von Knoch R, Kuenzler P, Beat AM, et al. The L1 retroelement-related p40 protein induces p38delta map kinase. Autoimmunity 2004;37(1):57–65.

[60] Kazazian HH Jr, Goodier JL. Line drive: retrotransposition and genome instability. Cell 2002;110(3):277–80.

[61] Karouzakis E, Gay RE, Michel BA, Gay S, Neidhart M. DNA hypomethylation in rheumatoid arthritis synovial fibroblasts. Arthritis Rheum 2009;60(12):3613–22.

[62] de Andres MC, Perez-Pampin E, Calaza M, Santaclara FJ, Ortea I, Gomez-Reino JJ, et al. Assessment of global DNA methylation in peripheral blood cell subpopulations of early rheumatoid arthritis before and after methotrexate. Arthritis Res Ther 2015;17:233.

[63] Rhead B, Holingue C, Cole M, Shao X, Quach HL, Quach D, et al. Rheumatoid arthritis naive T cells share hypermethylation sites with synoviocytes. Arthritis Rheumatol 2017;69(3):550–9.

[64] Portela A, Esteller M. Epigenetic modifications and human disease. Nat Biotechnol 2010;28(10):1057–68.

[65] Karouzakis E, Gay RE, Gay S, Neidhart M. Increased recycling of polyamines is associated with global DNA hypomethylation in rheumatoid arthritis synovial fibroblasts. Arthritis Rheumatol 2012;64(6):1809–17.

[66] Neidhart M, Karouzakis E, Jungel A, Gay RE, Gay S. Inhibition of spermidine/spermine N1-acetyltransferase activity: a new therapeutic concept in rheumatoid arthritis. Arthritis Rheumatol 2014;66(7):1723–33.

[67] Karouzakis E, Rengel Y, Jungel A, Kolling C, Gay RE, Michel BA, et al. DNA methylation regulates the expression of CXCL12 in rheumatoid arthritis synovial fibroblasts. Genes Immun 2011;12(8):643–52.

[68] Park SH, Kim SK, Choe JY, Moon Y, An S, Park MJ, et al. Hypermethylation of EBF3 and IRX1 genes in synovial fibroblasts of patients with rheumatoid arthritis. Mol Cells 2013;35(4):298–304.

[69] Karouzakis E, Trenkmann M, Gay RE, Michel BA, Gay S, Neidhart M. Epigenome analysis reveals TBX5 as a novel transcription factor involved in the activation of rheumatoid arthritis synovial fibroblasts. J Immunol 2014;193(10):4945–51.

[70] Nakano K, Whitaker JW, Boyle DL, Wang W, Firestein GS. DNA methylome signature in rheumatoid arthritis. Ann Rheum Dis 2013;72(1):110–7.

[71] de la Rica L, Urquiza JM, Gomez-Cabrero D, Islam AB, Lopez-Bigas N, Tegner J, et al. Identification of novel markers in rheumatoid arthritis through integrated analysis of DNA methylation and microRNA expression. J Autoimmun 2013;41:6–16.

[72] Takami N, Osawa K, Miura Y, Komai K, Taniguchi M, Shiraishi M, et al. Hypermethylated promoter region of DR3, the death receptor 3 gene, in rheumatoid arthritis synovial cells. Arthritis Rheum 2006;54(3):779–87.

[73] Nile CJ, Read RC, Akil M, Duff GW, Wilson AG. Methylation status of a single CpG site in the IL6 promoter is related to IL6 messenger RNA levels and rheumatoid arthritis. Arthritis Rheum 2008;58(9):2686–93.

[74] Hammaker D, Whitaker JW, Maeshima K, Boyle DL, Ekwall AH, Wang W, et al. LBH gene transcription regulation by the interplay of an enhancer risk allele and DNA methylation in rheumatoid arthritis. Arthritis Rheumatol 2016;68(11):2637–45.

[75] Watanabe N, Ando K, Yoshida S, Inuzuka S, Kobayashi M, Matsui N, et al. Gene expression profile analysis of rheumatoid synovial fibroblast cultures revealing the overexpression of genes responsible for tumor-like growth of rheumatoid synovium. Biochem Biophys Res Commun 2002;294(5):1121–9.

[76] Ekwall AK, Whitaker JW, Hammaker D, Bugbee WD, Wang W, Firestein GS. The rheumatoid arthritis risk gene LBH regulates growth in fibroblast-like synoviocytes. Arthritis Rheumatol 2015;67(5):1193–202.

[77] Baron U, Floess S, Wieczorek G, Baumann K, Grutzkau A, Dong J, et al. DNA demethylation in the human FOXP3 locus discriminates regulatory T cells from activated FOXP3(+) conventional T cells. Eur J Immunol 2007;37(9):2378–89.
[78] Cribbs AP, Kennedy A, Penn H, Amjadi P, Green P, Read JE, et al. Methotrexate restores regulatory t cell function through demethylation of the FOXP3 upstream enhancer in patients with rheumatoid arthritis. Arthritis Rheumatol 2015;67(5):1182–92.
[79] Gaffo A, Saag KG, Curtis JR. Treatment of rheumatoid arthritis. Am J Health Syst Pharm 2006;63(24):2451–65.
[80] Soubrier M, Lukas C, Sibilia J, Fautrel B, Roux F, Gossec L, et al. Disease activity score-driven therapy versus routine care in patients with recent-onset active rheumatoid arthritis: data from the GUEPARD trial and ESPOIR cohort. Ann Rheum Dis 2011;70(4):611–5.
[81] Plant D, Webster A, Nair N, Oliver J, Smith SL, Eyre S, et al. Differential methylation as a biomarker of response to etanercept in patients with rheumatoid arthritis. Arthritis Rheumatol 2016;68(6):1353–60.
[82] Malmstrom V, Catrina AI, Klareskog L. The immunopathogenesis of seropositive rheumatoid arthritis: from triggering to targeting. Nat Rev Immunol 2017;17:60–75.
[83] Gomez-Cabrero D, Almgren M, Sjoholm LK, Hensvold AH, Ringh MV, Tryggvadottir R, et al. High-specificity bioinformatics framework for epigenomic profiling of discordant twins reveals specific and shared markers for ACPA and ACPA-positive rheumatoid arthritis. Genome Med 2016;8(1):124.
[84] Ammari M, Jorgensen C, Apparailly F. Impact of microRNAs on the understanding and treatment of rheumatoid arthritis. Curr Opin Rheumatol 2013;25(2):225–33.
[85] Vicente R, Noel D, Pers YM, Apparailly F, Jorgensen C. Deregulation and therapeutic potential of microRNAs in arthritic diseases. Nat Rev Rheumatol 2016;12(8):496.
[86] Churov AV, Oleinik EK, Knip M. MicroRNAs in rheumatoid arthritis: altered expression and diagnostic potential. Autoimmun Rev 2015;14(11):1029–37.
[87] Pearson MJ, Jones SW. Review: long noncoding RNAs in the regulation of inflammatory pathways in rheumatoid arthritis and osteoarthritis. Arthritis Rheumatol 2016;68(11):2575–83.
[88] Plestilova L, Neidhart M, Russo G, Frank-Bertoncelj M, Ospelt C, Ciurea A, et al. Expression and regulation of PIWIL-proteins and PIWI-interacting RNAs in rheumatoid arthritis. PLoS One 2016;11(11):e0166920.
[89] de Rooy DP, Tsonaka R, Andersson ML, Forslind K, Zhernakova A, Frank-Bertoncelj M, et al. Genetic factors for the severity of ACPA-negative rheumatoid arthritis in 2 cohorts of early disease: a genome-wide study. J Rheumatol 2015;42(8):1383–91.
[90] Frank-Bertoncelj M, Trenkmann M, Armaka M, Russo G, Bratus A, Kolling C, et al. Positional coding and noncoding transcriptomes of synovial fibroblasts in joint specific patterns of arthritis. Ann Rheum Dis 2015;74(Suppl 2):95.
[91] Frank-Bertoncelj M, Trenkmann M, Klein K, Karouzakis E, Rehrauer H, Bratus A, et al. Epigenetically-driven anatomical diversity of synovial fibroblasts guides joint-specific fibroblast functions. Nat Commun 2017;8:14852.
[92] Bartel DP. MicroRNAs: genomics, biogenesis, mechanism, and function. Cell 2004;116(2):281–97.
[93] Shi DL, Shi GR, Xie J, Du XZ, Yang H. MicroRNA-27a inhibits cell migration and invasion of fibroblast-like synoviocytes by targeting follistatin-like protein 1 in rheumatoid arthritis. Mol Cells 2016;39(8):611–8.
[94] Gaur N, Karouzakis E, Gluck S, Bagdonas E, Jungel A, Michel BA, et al. MicroRNAs interfere with DNA methylation in rheumatoid arthritis synovial fibroblasts. RMD Open 2016;2(2):e000299.
[95] Niederer F, Trenkmann M, Ospelt C, Karouzakis E, Neidhart M, Stanczyk J, et al. Down-regulation of microRNA-34a* in rheumatoid arthritis synovial fibroblasts promotes apoptosis resistance. Arthritis Rheum 2012;64(6):1771–9.
[96] Nakamachi Y, Ohnuma K, Uto K, Noguchi Y, Saegusa J, Kawano S. MicroRNA-124 inhibits the progression of adjuvant-induced arthritis in rats. Ann Rheum Dis 2016;75(3):601–8.
[97] Nakamachi Y, Kawano S, Takenokuchi M, Nishimura K, Sakai Y, Chin T, et al. MicroRNA-124a is a key regulator of proliferation and monocyte chemoattractant protein 1 secretion in fibroblast-like synoviocytes from patients with rheumatoid arthritis. Arthritis Rheum 2009;60(5):1294–304.
[98] Kriegsmann M, Randau TM, Gravius S, Lisenko K, Altmann C, Arens N, et al. Expression of miR-146a, miR-155, and miR-223 in formalin-fixed paraffin-embedded synovial tissues of patients with rheumatoid arthritis and osteoarthritis. Virchows Arch 2016;469(1):93–100.
[99] Zhou Q, Haupt S, Kreuzer JT, Hammitzsch A, Proft F, Neumann C, et al. Decreased expression of miR-146a and miR-155 contributes to an abnormal Treg phenotype in patients with rheumatoid arthritis. Ann Rheum Dis 2015;74(6):1265–74.
[100] Niimoto T, Nakasa T, Ishikawa M, Okuhara A, Izumi B, Deie M, et al. MicroRNA-146a expresses in interleukin-17 producing T cells in rheumatoid arthritis patients. BMC Musculoskelet Disord 2010;11:209.
[101] Kurowska-Stolarska M, Alivernini S, Ballantine LE, Asquith DL, Millar NL, Gilchrist DS, et al. MicroRNA-155 as a proinflammatory regulator in clinical and experimental arthritis. Proc Natl Acad Sci U S A 2011;108(27):11193–8.

[102] Alivernini S, Kurowska-Stolarska M, Tolusso B, Benvenuto R, Elmesmari A, Canestri S, et al. MicroRNA-155 influences B-cell function through PU.1 in rheumatoid arthritis. Nat Commun 2016;7:12970.

[103] Murata K, Yoshitomi H, Tanida S, Ishikawa M, Nishitani K, Ito H, et al. Plasma and synovial fluid microRNAs as potential biomarkers of rheumatoid arthritis and osteoarthritis. Arthritis Res Ther 2010;12(3):R86.

[104] Li YT, Chen SY, Wang CR, Liu MF, Lin CC, Jou IM, et al. Brief report: amelioration of collagen-induced arthritis in mice by lentivirus-mediated silencing of microRNA-223. Arthritis Rheum 2012;64(10):3240–5.

[105] Ogando J, Tardaguila M, Diaz-Alderete A, Usategui A, Miranda-Ramos V, Martinez-Herrera DJ, et al. Notch-regulated miR-223 targets the aryl hydrocarbon receptor pathway and increases cytokine production in macrophages from rheumatoid arthritis patients. Sci Rep 2016;6:20223.

[106] Fulci V, Scappucci G, Sebastiani GD, Giannitti C, Franceschini D, Meloni F, et al. MiR-223 is overexpressed in T-lymphocytes of patients affected by rheumatoid arthritis. Hum Immunol 2010;71(2):206–11.

[107] Lu MC, Yu CL, Chen HC, Yu HC, Huang HB, Lai NS. Increased miR-223 expression in T cells from patients with rheumatoid arthritis leads to decreased insulin-like growth factor-1-mediated interleukin-10 production. Clin Exp Immunol 2014;177(3):641–51.

[108] Wang H, Peng W, Ouyang X, Li W, Dai Y. Circulating microRNAs as candidate biomarkers in patients with systemic lupus erythematosus. Transl Res 2012;160(3):198–206.

[109] Ospelt C. Epigenetic biomarkers in rheumatology: the future? Swiss Med Wkly 2016;146:w14312.

[110] Duroux-Richard I, Pers YM, Fabre S, Ammari M, Baeten D, Cartron G, et al. Circulating miRNA-125b is a potential biomarker predicting response to rituximab in rheumatoid arthritis. Mediators Inflamm 2014;2014:342524.

[111] Krintel SB, Dehlendorff C, Hetland ML, Horslev-Petersen K, Andersen KK, Junker P, et al. Prediction of treatment response to adalimumab: a double-blind placebo-controlled study of circulating microRNA in patients with early rheumatoid arthritis. Pharmacogenomics J 2016;16(2):141–6.

[112] Castro-Villegas C, Perez-Sanchez C, Escudero A, Filipescu I, Verdu M, Ruiz-Limon P, et al. Circulating miRNAs as potential biomarkers of therapy effectiveness in rheumatoid arthritis patients treated with anti-TNFalpha. Arthritis Res Ther 2015;17:49.

[113] Cuppen BV, Rossato M, Fritsch-Stork RD, Concepcion AN, Schenk Y, Bijlsma JW, et al. Can baseline serum microRNAs predict response to TNF-alpha inhibitors in rheumatoid arthritis? Arthritis Res Ther 2016;18:189.

[114] Stanczyk J, Pedrioli DM, Brentano F, Sanchez-Pernaute O, Kolling C, Gay RE, et al. Altered expression of microRNA in synovial fibroblasts and synovial tissue in rheumatoid arthritis. Arthritis Rheum 2008;58(4):1001–9.

[115] Pauley KM, Satoh M, Chan AL, Bubb MR, Reeves WH, Chan EK. Upregulated miR-146a expression in peripheral blood mononuclear cells from rheumatoid arthritis patients. Arthritis Res Ther 2008;10(4):R101.

[116] Pandis I, Ospelt C, Karagianni N, Denis MC, Reczko M, Camps C, et al. Identification of microRNA-221/222 and microRNA-323-3p association with rheumatoid arthritis via predictions using the human tumour necrosis factor transgenic mouse model. Ann Rheum Dis 2012;71(10):1716–23.

[117] Bluml S, Bonelli M, Niederreiter B, Puchner A, Mayr G, Hayer S, et al. Essential role of microRNA-155 in the pathogenesis of autoimmune arthritis in mice. Arthritis Rheum 2011;63(5):1281–8.

[118] Elmesmari A, Fraser AR, Wood C, Gilchrist D, Vaughan D, Stewart L, et al. MicroRNA-155 regulates monocyte chemokine and chemokine receptor expression in rheumatoid arthritis. Rheumatology 2016;55(11):2056–65.

[119] Smigielska-Czepiel K, van den Berg A, Jellema P, van der Lei RJ, Bijzet J, Kluiver J, et al. Comprehensive analysis of miRNA expression in T-cell subsets of rheumatoid arthritis patients reveals defined signatures of naive and memory Tregs. Genes Immun 2014;15(2):115–25.

[120] Taganov KD, Boldin MP, Chang KJ, Baltimore D. Nf-kappaB-dependent induction of microRNA miR-146, an inhibitor targeted to signaling proteins of innate immune responses. Proc Natl Acad Sci U S A 2006;103(33):12481–6.

[121] Hou J, Wang P, Lin L, Liu X, Ma F, An H, et al. MicroRNA-146a feedback inhibits RIG-I-dependent type I IFN production in macrophages by targeting TRAF6, IRAK1, and IRAK2. J Immunol 2009;183(3):2150–8.

[122] Shibuya H, Nakasa T, Adachi N, Nagata Y, Ishikawa M, Deie M, et al. Overexpression of microRNA-223 in rheumatoid arthritis synovium controls osteoclast differentiation. Mod Rheumatol 2013;23(4):674–85.

[123] Dai R, Phillips RA, Zhang Y, Khan D, Crasta O, Ahmed SA. Suppression of LPS-induced interferon-gamma and nitric oxide in splenic lymphocytes by select estrogen-regulated microRNAs: a novel mechanism of immune modulation. Blood 2008;112(12):4591–7.

[124] Schaefer JS, Montufar-Solis D, Vigneswaran N, Klein JR. Selective upregulation of microRNA expression in peripheral blood leukocytes in IL-10-/- mice precedes expression in the colon. J Immunol 2011;187(11):5834–41.

[125] Stanczyk J, Ospelt C, Karouzakis E, Filer A, Raza K, Kolling C, et al. Altered expression of microRNA-203 in rheumatoid arthritis synovial fibroblasts and its role in fibroblast activation. Arthritis Rheum 2011;63(2):373–81.

[126] Zhang K, Shi ZM, Chang YN, Hu ZM, Qi HX, Hong W. The ways of action of long non-coding RNAs in cytoplasm and nucleus. Gene 2014;547(1):1–9.

[127] Song J, Kim D, Han J, Kim Y, Lee M, Jin EJ. PBMC and exosome-derived HOTAIR is a critical regulator and potent marker for rheumatoid arthritis. Clin Exp Med 2015;15(1):121–6.

[128] Zhang Y, Xu YZ, Sun N, Liu JH, Chen FF, Guan XL, et al. Long noncoding RNA expression profile in fibroblast-like synoviocytes from patients with rheumatoid arthritis. Arthritis Res Ther 2016;18(1):227.

[129] Lu MC, Yu HC, Yu CL, Huang HB, Koo M, Tung CH, et al. Increased expression of long noncoding RNAs loc100652951 and loc100506036 in T cells from patients with rheumatoid arthritis facilitates the inflammatory responses. Immunol Res 2016;64(2):576–83.

[130] Mayama T, Marr AK, Kino T. Differential expression of glucocorticoid receptor noncoding RNA repressor GAS5 in autoimmune and inflammatory diseases. Horm Metab Res 2016;48(8):550–7.

[131] Spurlock CF III, Tossberg JT, Matlock BK, Olsen NJ, Aune TM. Methotrexate inhibits NF-kappaB activity via long intergenic (noncoding) RNA-p21 induction. Arthritis Rheumatol 2014;66(11):2947–57.

[132] Gupta RA, Shah N, Wang KC, Kim J, Horlings HM, Wong DJ, et al. Long non-coding RNA HOTAIR reprograms chromatin state to promote cancer metastasis. Nature 2010;464(7291):1071–6.

[133] Zhang HD, Jiang LH, Sun DW, Hou JC, Ji ZL. CircRNA: a novel type of biomarker for cancer. Breast Cancer 2017;doi: 10.1007/s12282-017-0793-9.

CHAPTER

8

Epigenetic Modifications in Multiple Sclerosis Pathophysiology: Potential Diagnostic and Therapeutic Applications

*Patricia Aznar**, *Teresa Ayuso**, *Maite Mendióroz**,**

*Complejo Hospitalario de Navarra, Navarra Institute for Health Research (IdiSNA), Pamplona, Navarra, Spain; **Navarrabiomed-Navarra Institute for Health Research (IdiSNA), Pamplona, Navarra, Spain

OUTLINE

8.1 INTRODUCTION

Aberrant epigenetic modifications are associated with the development of many diseases, including neuropsychiatric and neurodegenerative diseases. Epigenetic studies have opened new doors, helped to expand knowledge in this field, and assessed the feasibility of developing new biomarkers and therapeutic targets [1].

Multiple sclerosis (MS) is an autoimmune disease of the central nervous system (CNS) that is characterized by inflammatory lesions, demyelination, and axonal loss. MS affects more than

The Epigenetics of Autoimmunity. http://dx.doi.org/10.1016/B978-0-12-809912-4.00008-8

2 million people worldwide and has an incidence rate of approximately 5–6 per 100,000 population per year in the United States [2]. The prevalence in Europe is 83 per 100,000 [3]. Interestingly, there are higher rates of MS in the northern countries (related with risk factors), and the disease mainly affects young people, being the most important cause of nontraumatic disability [4].

MS is caused by an autoimmune response against myelin proteins in neurons. Two mechanisms may act together to trigger MS: (1) molecular mimicry, whereby the immune system acts against infectious agents that are similar to self-antigens; and (2) bystander activation, which is the initial activation of antigen presenting cells in the CNS that starts the autoimmune response by activating preprimed autoreactive T cells. As in other autoimmune diseases, there is a potentiation of T helper 1 lymphocytes (Th1)-IFNγ and Th17-IL-17 immunity versus Th2-IL-4 and T_{reg} immune responses. However, there are many other immunological mechanisms (both cellular and humoral) that are involved in the pathophysiology of the disease [5].

MS is also characterized by the abnormal entry of inflammatory cells into the CNS. Under normal conditions, the blood-brain barrier (BBB) functions to prevent the entrance of inflammatory cells, the expression levels of histocompatibility leukocyte antigen (HLA) molecules are very low, and the cerebrospinal fluid (CSF) contains soluble factors that are responsible for neutralizing or counteracting the effects of proinflammatory cytokines (CKs). The altered inflammatory reaction in MS may cause demyelination and axonal loss. These three are the three—*altered inflammatory reaction, demyelination and axonal loss-* are the main pathophysiological characteristics of the disease. Under physiological conditions, demyelination and axonal loss could be repaired by oligodendrocytes, which are the brain cells responsible for myelin synthesis and reparation; however, chronic inflammation also affects oligodendrocyte function and differentiation [6]. The triggers and factors that keep this autoimmune response active remain unclear; however, some relevant self-antigens to myelin, myelin-associated proteins, and lipoproteins have been identified. Some of these self-antigens are associated with MS risk factors although their presence does not always cause the disease [7].

There is a genetic component to MS regarding the presence of certain histocompatibility alleles (i.e., HLADRB1*1501 and DRB1*1503) [8]; however, these HLA susceptibility alleles vary in different global populations [9], their presence is not always responsible for the disease, and their concordance between monozygotic twins (although high at 6%–30%) is not complete [10]. Genome-wide association studies (GWAS) have been performed to identify all the single-nucleotide polymorphisms (SNPs) associated with MS [11,12], and although approximately 1000 SNPs have been reported, they collectively explain only one-third of the recurrence in relatives [13]. This suggests the involvement of other etiological factors; for example, environmental factors such as Epstein–Barr virus (EBV) infection, vitamin D deficiency, and tobacco smoking have been shown to increase the risk of developing MS. These factors increase the risk of developing the disease and are associated with a worse prognosis. All of these factors are important mediators of epigenetic modifications, both in autoimmune and oncological diseases. EBV induces miRNA-155 expression, which is overexpressed in CD4 lymphocytes and demyelinating lesions in MS patients [14,15]; vitamin D ameliorates Th17 autoimmunity by recruiting histone deacetylases (HDAC) and decreasing IL-17A thus slowing disease progression in experimental autoimmune encephalomyelitis (EAE), the animal model of MS [16]; and, tobacco contains more than 500 chemical compounds that can influence most of the pathological mechanisms in MS [17]. Tobacco is also a powerful inducer of epigenetic changes in tumor and immune-related genes as it causes hypomethylation of IL-4 and IFNγ genes, which are important in MS pathogenesis [18].

As mentioned previously, concordance between monozygotic twins is not complete but it is well known that epigenetic modifications change during development and are potentially influenced by environmental factors [19]. Therefore it is likely that the different epigenetic changes that affect and regulate the immune system and CNS play a key role in the development and evolution of MS.

8.2 EPIGENETIC CHANGES IN MULTIPLE SCLEROSIS

Many epigenetic modifications have been described in MS patients compared to healthy controls. This wide variety of epigenetic changes may be detected at different disease stages and in different body tissues and cells. Epigenome-wide differences and single-gene epigenetic changes that affect different components of the pathological process (i.e., autoimmunity, neuroinflammation, demyelination, and neurodegeneration) have been reported. These epigenetic mechanisms can be subclassified in three groups: DNA methylation, histone modifications, and noncoding RNA.

8.3 DNA METHYLATION

DNA methylation is the process in which DNA methyltransferase enzymes (DNMTs) add methyl groups to cytosine residues in cytosine–guanine (CpG) dinucleotides. Methylation or hypermethylation of a gene promoter region decreases or inhibits gene transcription at this locus. Two families of proteins regulate gene expression after methylation: (1) methyl-binding domain proteins (MBDs), which form repressor complexes that silence gene expression by recruiting HDAC and DNMT repressors; and (2) ten-eleven translocation (TET) family proteins, which catalyze the transformation of 5-methylcytosine to 5-hydroxymethylcytosine and promote DNA demethylation [20].

TET enzymes are involved in the inflammatory response through their role in the regulation of stem cells and CK synthesis [21]. TET2 and DNMT1 are reportedly downregulated in peripheral blood mononuclear cell (PBMC) from MS patients compared to controls, and this downregulation is correlated to decreased expression of their target genes. TET1 and TET2 are involved in the transcriptional regulation of IL-1β, which is increased in MS patients [22]. Other researchers have similarly reported that the *TET3, MBD2,* and *MBD4* genes are downregulated in PBMCs of secondary progressive MS patients. TET3 has subsequently been correlated with the expression levels of three genes (*APLA2, SLC25A11,* and *ATP6AP2*) involved in the migration, differentiation, and metabolism of neural stem cells [23].

MeCP2 is another member of the MBD family that prevents DNA transcription when the CpG sites within a promoter are methylated. Interestingly, MeCP2 suppresses the expression of brain neurotrophic factor, which is necessary for myelin repair. Lysine-acetyl transferase and HDAC inhibitors (HDACi) may decrease MeCP2 expression and thus favor remyelination, and these compounds have been proposed as potential treatments for MS [24]. Serine hydroxymethyltransferase 1 (SHMT1) catalyzes the transfer of a carbon unit to the folate cycle, which is essential for methylation regulation and for the maintenance of epigenetic signatures. One GWAS in a German population identified this as a potential MS susceptibility locus [25].

One of the genetic susceptibility loci in MS is the major histocompatibility complex (MHC), specifically the HLA-DRB1*1501 allele. Methylation patterns of these genes have been assessed in

PBMCs from patients with different disease severities; however, no differential methylation patterns have been found [26].

Peptidylarginine deiminase enzymes (PAD)-2 is an enzyme that causes a pathological citrullination of an arginine in the myelin basic protein (MBP) of MS patients. Citrullination of MBP is one of the proposed antigens that is wrongly identified by T lymphocytes in MS. PAD-2 is overexpressed in PBMCs from patients with MS, which could be explained by decreased methylation levels of its promoter [27]. Remarkably, PAD-2 overexpression is also observed in the white matter of MS patients, where its promoter methylation is reduced (~33%) compared to healthy controls. PAD-2 demethylation is not observed in the thymus gland of MS patients or in the white matter of patients with other neurological diseases, which suggests that this demethylation is a tissue-specific epigenetic change that is characteristic of MS. Nevertheless, the PAD-2 concentration in peripheral blood does not correlate with MS duration, evolution, or disability [28]. PAD-4 is also upregulated in MS, both in the brain and peripheral blood; however, a differential methylation status of its promoter has not yet been identified in MS patients [27].

Forkhead box P3 (FOXP3) is a protein involved in many autoimmune diseases due to its role in regulating Th1, Th2, and Th17 differentiation. The *FOXP3* promoter has been found to be hypermethylated in CD4 lymphocytes of relapsing–remitting MS (RRMS) patients [29]. *FOXP3* downregulation consequently modifies the Th1–Th2 immune balance and favors the differentiation of T_{reg} and Th17 cells [30]. One of the methylation-modifying drugs tested in EAE in mice was decitabine (5-aza-2′-deoxicytidine), which is a DNA methylation inhibitor that induces *FOXP3* expression. Decitabine was shown to decrease spinal infiltration in an animal model and ameliorate the disease progression [31]. Furthermore, IL-17a production and further Th17 differentiation are also increased following hypomethylation of IL-17a gene promoter, which is observed in T cells from patients with MS [32].

Other interesting protein involved in the pathophysiology of MS is CD44, which is a membrane protein involved in lymphocyte proliferation, differentiation, function, and migration through the BBB. CD44 also modifies DNA methylation patterns in key genes; for example, the binding of CD44 to its ligand causes hypomethylation of the IFNγ and IL-17 genes and promotes differentiation towards Th1 and Th17 cells [33]. CD44 deficiency not only decreases Th1 and Th17 differentiation but it also promotes Th2 differentiation via hypomethylation of the IL-4 promoter. This may explain why CD44-deficient mice are protected against EAE [34].

The tyrosine phosphatase, SHP1, is a negative regulator of proinflammatory signaling and its expression has been shown to be reduced in PBMCs from MS patients [35]. SHP1 deficiency may be caused by increased methylation of its promoter in approximately 33% of patients, which may contribute to leukocyte-mediated inflammation; however, no association has been described between *SHP1* methylation and the disease type, progression, or disability [36].

In addition to altered methylation patterns described in single genes (Table 8.1), a number of studies have also profiled genome-wide (GW) DNA methylation changes in MS patients [37].

One GW DNA methylation study performed in PBMCs paid attention to the clinical course differences, comparing RRMS with primary-progressive multiples sclerosis (PPMS) patients. A larger number of differentially methylated CpG regions (DMRs) were identified in PPMS and RRMS patients compared to healthy controls; however, PPMS patients showed a shift towards hypermethylation while RRMS patients shifted towards hypomethylation. This study demonstrated that methylation profiles can change during MS development and could explain different processes that determine the MS phenotype [38].

TABLE 8.1 DNA Methylation in MS

(A) Identified by Candidate Gene Studies

Gene	Status in MS	Tissue or cells	Function	References
TET2 and *DNMT1*	Downregulated	PBMC	Demethylation	Calabrese et al. [22]
TET3	Downregulated	PBMC–SPMS	Demethylation	Fagone et al. [23]
MECP2 (MBD)	Downregulated	CNS	Avoids DNA when methylated	Khorshid Ahmad et al. [24]
SHMT1	Upregulated	GWAS	Hydroxymethyltransferase	Andlauer et al. [25]
*HLA-DRB1*1501* and *HLA-DRB5*	No difference	PBMC	Susceptibility genes	Handel et al. [26]
PAD2	Hypomethylated	PBMC	Citrullination of MBP	Calabrese et al. [27]
PAD2	Hypomethylated	WM	Citrullination of MBP	Mastronardi et al. [28]
FOXP3	Hypermethylated	CD4 T lymphocytes	Th1, Th2, Th17, and T_{reg} immunity	Janson et al. [32]
IL17a	Hypomethylated	T cells	IL-17 synthesis	Koch et al. [29]
CD44	Its deficiency protects mice against EAE	Lymphocytes	Hypomethylation of IFN and IL-17 genes	Guan et al. [34]
SHP-1	Hypomethylated	PBMC	Inhibition of proinflammatory signaling	Christophi et al. [35]
SHP-1	Hypomethylated	Leukocytes	Inhibition of proinflammatory signaling	Kumagai et al. [36]
MHC2TA	No difference	PBMC from monozygotic twins	MHC class II transactivator	Ramagopalan et al. [46]

(Continued)

TABLE 8.1 DNA Methylation in MS (*cont.*)

(B) Identified by Genome-Wide Studies

Tissues or cells	**Results and affected genes**	**References**
PBMC	Hypermethylation tendency in patients with PPMS Hypomethylation tendency in patients with RRMS	Kulakova et al. [38]
CD4 T cells	HLA-DRB1 *TCRα, ARSB*, and *KCTD11*	Graves et al. [39]
CD8 T cells, CD4 T cells, and WB cells	*TMEM48, APC2, and DNHD1* hypermethylation tendency in CD8 T cells	Bos et al. [40]
CD8 versus CD4 T cells	*FTL, ERG*, and *DCAF4*	Maltby et al. [41]
Brain	Hypermethylated *SOX8, BCL2L*, and *NDRG1*; and hypomethylated *LGMN* and *CTSZ*	Huynh et al. [42]
CFPDNA	Increased CFPDNA in RRMS. No differences between relapse or remitting phase CDKN2B	Liggett et al. [43]
WB, repetitive DNA sequences	Hypermethylation of repetitive DNA sequences and hypomethylation of Alu-disease severity	Neven et al. [44]
CD4 lymphocytes in monozygotic twins	No methylation differences	Baranzini et al. [45]

CFPDNA, Cell-free plasma DNA; *GWAS*, genome-wide association study; *MBP*, myelin basic protein; *PBMC*, peripheral blood mononuclear cell; *PPMS*, progressive multiples sclerosis; *RRMS*, relapsing–remitting MS; *WB*, whole blood; *SPMS*, Secondary progressive multiple sclerosis; *WM*, white matter.

A GW DNA methylation study in CD4 T lymphocytes also showed differential methylation patterns in patients with MS. The largest difference in DNA methylation levels was observed in HLA-DRB1, the histocompatibility allele classically related to MS. Outside of the MHC region, 55 genes showed differential methylation in MS patients, with *TCRα, ARSB*, and *KCTD11* being the most relevant [39]. However, another GW DNA methylation study focusing on CD8 T cells did not report the same methylation patterns found in CD4 T cells. In fact, this study found differences in the DNA methylation profiles in CD8 T cells, CD4 T cells, and whole blood (WB) from MS patients comparing to controls. CD8 T cells showed a trend towards hypermethylation at a larger number of CpG sites, particularly in individuals with a disease duration of less than 7 years. The most significantly methylated sites in CD8 T cells were *TMEM48, APC2*, and *DNHD1* [40]. Another study also reported that the methylation profile of CD8 T cells was distinct from CD4 T cells in MS. In this study, the methylation profile of CD8 T cells was strikingly distinct from CD4 T cells even in the *HLA-DRB1* locus, which was not differentially methylation in CD8 T cells of MS patients. The authors identified 79 CpG sites outside of the MHC region that were differentially methylated in CD8 T cells. The most outstanding genes were the ferritin light chain gene (*FLT*), the ETS-related gene (*ERG*), and the CUL4-associated factor 4 (*DCAF4*). *FTL*, a component of ferritin, was hypomethylated in CD8 T cells from MS patients and dysregulation of this gene could be related to iron deposits within the gray matter of RRMS patients; *ERG* was hypermethylated in MS but normally plays an important role in cell proliferation, apoptosis, and inflammation; and *DCAF4* was also hypomethylated in MS patients and could be involved in neurodegeneration due to its influence on leucocyte telomere length [41].

A remarkable GW DNA methylation study performed in brain samples from MS patients and healthy controls reported methylation differences in oligodendrocyte survival genes. These differences were found in pathology-free areas of MS brains called normal-appearing white matter (NAWM). The authors identified 220 hypomethylated and 319 hypermethylated DMRs. Hypermethylated DMRs (and decreased transcript levels) were detected for oligodendrocyte-specific genes (i.e., *MBP* and *SOX8*) and in genes regulating neuronal survival (i.e., *BCL2L* and *NDRG1*), while hypomethylated DMRs (and increased transcript levels) were found for genes related to proteolytic processing (i.e., *LGMN* and *CTSZ*) [42].

Cell-free plasma DNA (CFPDNA) has been also analyzed. This DNA is believed to be released from dying cells and could reflect the turnover of cell populations, and use of this sample has been proposed as a biomarker in genetic diseases and cancer. The concentration of CFPDNA in patients with RRMS was found to be several fold higher than in healthy controls; however, no differences were observed between different disease states. The authors observed that 15 genes were differentially methylated, with the most differentially methylated, *CDKN2B*, being hypermethylated in patients and showing a sensitivity and specificity of ~75% for discriminating patients from controls [43].

Repetitive DNA sequences are fundamental to genome stability and function because of their ability to form nucleoprotein complexes. They include Alu repeats, long interspersed nuclear elements (i.e., *LINE1*), and alpha satellite DNA sequences (i.e., *SATα*), which are found in the centromere. The methylation status of these sequences may be considered as a good estimate for global DNA methylation levels. Remarkably, MS patients showed hypermethylation in all these repetitive elements when compared to control subjects. Reduced *LINE1* methylation was associated with a low expanded disability status scale (EDSS) score, while hypomethylation of Alu was related to an increased EDSS score [44].

A GWAS has also been performed on CD4 lymphocytes from three pairs of MS-discordant monozygotic twins to identify potential epigenomic differences; however, no evidence of genetic, epigenetic, or transcriptomic differences were found. This is likely due to the small sample size and the need to include other disease mechanisms in this extensive analysis [45]. A separate twin study also reported that there were no epigenetic differences, but did report differences in the MHC class II transactivator gene (*MHC2TA*) in PBMCs from fifty pairs of MS-discordant monozygotic twins [46].

In summary, DNA methylation is clearly involved in immune-gene regulation and therefore in the pathogenesis of MS. Methylation changes are distinctly related to disease onset and progression and therefore these changes could be used as biomarkers or therapeutic targets. Several drugs related to methylation (i.e., decitabine) could be assessed in preclinical studies of MS; however, whether these drugs could be used in MS patients and how they target specific genes remains to be explored [47].

8.4 HISTONE MODIFICATIONS

Histones are proteins around which DNA is wrapped to form a nucleosome, which is the basic unit of the chromatin. The tails of histones are rich in lysine and arginine residues, which can be modified by extracellular signals to change their nucleosomal organization and thus influence gene expression. Epigenetic modifications that regulate the arrangement of histones and affect gene expression have also been suggested to play a role in the pathophysiology of MS. There are at least three types of histone modifications: (1) modification in the composition of nucleosomal octamers (histone variants), (2) rapid sliding movements using ATP hydrolysis to expose specific DNA regions, and (3) posttranslational modifications of amino acids in histone tails (i.e., acetylation, methylation, citrullination, sumoylation, and ubiquitylation). Histone acetylation of lysine residues is catalyzed by histone acetyltransferases (HATs) whereas deacetylation is catalyzed by HDACs. There are five groups of HATs and three classes of HDACs [48].

Sirtuins (SIRTs) are a group of HDACs that may be nuclear, cytoplasmic, or mitochondrial. They control cell survival by regulating transcriptional activity and they have been associated with aging and inflammation; for example, the recruitment of SIRTs (and other HDACs) was shown to be decreased in oligodendrocytes of aged brains, which increased the expression of inhibitors of myelin gene expression and oligodendrocyte differentiation [49]. SIRT1 has also been implicated in autoimmune demyelination and neurodegeneration [50] and its expression levels have been assessed in brain tissue and PBMCs from patients with RRMS. SIRT1 was shown to be deposited in demyelinating lesions, NAWM, and normal-appearing gray matter in subjects with RRMS, but was not expressed in healthy tissue. In PBMCs, SIRT1 mRNA and protein expression levels were decreased during a relapse when compared to stable MS patients and controls. This was also reflected in an increased level of H3 lysine 9 acetylation in PBMCs during relapse. Therefore it has been proposed that SIRT1 expression could be used as an activity marker in MS [51]. SIRT1 has also been proposed as a therapeutic target. Gene activators (SRT647 and SRT501) have been tested in animal models of optic neuritis, and intravitreal injections have been shown to decrease the retinal-cell loss but not the inflammation or clinical signs. This suggests that SIRT1 activation prevents axonal degeneration and may have a positive effect on neuronal damage [52]. Resveratrol (SRT501) is a SIRT activator that has shown promising results when tested in mice with EAE. It prevents neuronal loss during optic neuritis and in spinal cord, and has a secondary benefit in clinical dysfunction. These results have also been shown using other SIRT1 activators (i.e., SRT1720); however, SIRT1 activators did not prevent inflammatory cell infiltration in these tissues [53].

PAD-4 is a peptidyl arginine deiminase that catalyzes the citrullination of arginine residues in the MBP. As mentioned previously, PAD-4 is upregulated in both the brain and peripheral blood of MS subjects. Analysis in NAWM in MS patients and animal model demonstrated a relationship between H3 histone citrullination and PAD-4 expression and activity levels, thus suggesting that PAD-4 also catalyzes histone citrullination in nucleosomal histones. Unlike other PADs, PAD-4 can translocate into nucleus and act, and its translocation may be attributable to elevated tumor necrosis factor-α (TNF-α) levels, which are also known to be increased in the NAWM of MS patients. Overexpression of TNF-α in animal models is associated with increased PAD-4 and citrullinated histone levels in the nucleus before demyelination occurs. Chronic exposure of oligodendrocyte cells to TNF-α in vitro also increases PAD-4 levels [54].

Histone modifications control many inflammatory mechanisms that are involved in autoimmunity, such as IL-17 production. Therefore it has been suggested that IL-17 synthesis is regulated by transcription cell factor 1 (TCF1) through the acetylation of histones. TCF1 knockout mice showed hyperacetylation of histones in the IL-17 promoter and were more susceptible to EAE induction [55].

Histone acetylation is negatively associated with oligodendrocyte differentiation in vitro and in vivo, probably due to increased levels of transcriptional inhibitors. Studies in animal models of demyelination showed that HDAC1 and HDAC2 are needed for oligodendrocyte differentiation. Their inhibition promotes β-catenin signaling, activates the expression of oligodendrocyte differentiation inhibitors, represses transcriptional factors, and suppresses myelin gene expression. This HDAC regulation is attained through competition with β-catenin on TCF7L2 activity, which is a transcriptional repressor of oligodendrocyte differentiation. This function of HDAC1 and HDAC2 has not been noted in the development of other brain cells (i.e., neurons or astrocytes) [56]. HDACs are not only involved in oligodendrocyte differentiation through the promotion of other transcriptional activators (i.e., *ASCL1* or *SOX10*) but they are also related to myelin synthesis; for example, oligodendrocyte Notch signaling can repress MBP expression by forming complexes with HDACs [57].

Histone modifications in the myelination process have also been reported in the brains of MS patients; for example, the NAWM of chronic MS patients shows a shift towards histone acetylation, which reflects the percentage of acetyl-H3 histone nuclei over total cells. Increases in H3 histone acetylation are associated with high levels of oligodendrocyte-differentiation inhibitors (i.e., *TCF7L2* and *SOX2*). However, the H3 histone acetylation is significantly reduced in early MS lesions compared to perilesional white matter. This data suggest that histone deacetylation occurs during the early stages of the disease and that its efficiency decreases with disease progression, thus reflecting distinct stages of oligodendrocyte remyelination. In summary, increased histone acetylation may be correlated with the disease duration [58].

There is little evidence relating to the epigenetic mechanisms that regulate neuronal degeneration in MS. However, one interesting study sought to compare brain-volume loss in MS patients with SNPs. The authors identified four SNPs that could function as prognostic markers of cerebral atrophy, which mapped to genes from the histone deacetylase superfamily (i.e., *HDAC11*, *HDAC9*, *SIRT4*, and *SIRT5*). SIRT4 and SIRT5 modulate numerous proteins related to axonal degeneration and apoptosis, but they have also been associated with mitochondrial dysfunction [59].

Some therapeutic treatments that modify histone acetylation have been tested in MS patients and animal models (Table 8.2). HDACi have potential therapeutic value in MS because of their anti-inflammatory and neuroprotective effects both in in vitro and in vivo experiments. Sodium phenylbutyrate (SPB) was the first HDACi to be used in animal models of MS and it was able to prevent EAE development. Its metabolite, sodium phenylacetate (SPA), suppress neurological impairment

TABLE 8.2 Histone Modifying Therapies Used in MS Patients and Animal Models

Treatment	Treatment group	Effect	References
SIRT activators (SRT647 and SRT501) SIRT activators (SRT501 and SRT1720)	Animal models EAE mice	Prevention of axonal degeneration Clinical benefits	Shindler et al. [52] Shindler et al. [53]
SPB	EAE mice	Prevention of EAE development and milder neurological symptoms	Faraco et al. [60]
VPA, lithium, and TSA	Astrocytes culture	Nrf2 antioxidant effects	Correa et al. [61]
VPA	EAE	Decreased CK synthesis and inflammatory infiltration	Zhang et al. [62]
TSA	EAE	Decreased inflammation, cell infiltration, demyelination, and neuronal loss and improved disability	Camelo et al. [63]
TSA	PBMC	Regulation of cell cycle	Zhang et al. [64]
Largazole	EAE	Decreased CK synthesis and inflammatory infiltration, and improved disability	Chen et al. [65]
Vorinostat	EAE	Reduced Th1- and Th17- mediated inflammation	Ge et al. [66]
Curcumin	EAE	Decreased CK synthesis and inflammatory infiltration, and decreased MBP-reactive lymphocytes	Xie et al. [67]

CK, Cytokine; *EAE*, encephalomyelitis; *PBMC*, peripheral blood mononuclear cell; *SIRT*, sirtuin; *SPB*, sodium phenylbutyrate; *TSA*, trichostatin A; VPA, valproic acid.

in mice pretreated with MBP–primed T-cells [60]. Well-known drugs (i.e., valproic acid (VPA) and lithium) can also inhibit HDAC function as they restore the antioxidant effects of nuclear factor-erythroid 2-related factor 2 (Nrf2) and protect astrocytes from stress-induced death through the normalization of histone acetylation levels. VPA improves the Nrf2-inducible defense, and lithium restores histone acetylation and Nrf2-related transcription through inhibition of GSK3β [61]. VPA has also been shown to reduce the severity and duration of EAE in rats by regulating systemic and local inflammation through a decrease in proinflammatory CKs (i.e., IFNγ, TNF-α, and IL-17), an increase in IL-4 levels, and a decrease of macrophage and lymphocyte infiltration into the spinal cord [62]. Trichostatin A (TSA) is another HDACi that has been tested in mice. When TSA was injected after immunization it was shown to reduce inflammation, cell infiltration, demyelination, and neuronal loss in the spinal cord and ameliorate the disability of EAE relapse [63]. However, it has been reported that T cells from MS patients are relatively resistant to TSA-induced apoptosis via p53 and related genes, possibly due to the increased expression of HDAC3. Indeed, TSA is used to evaluate p53 activation (a key transcription factor that regulates apoptosis genes and cell cycle) in human cells from MS patients because of its regulation by HDAC [64].

Largazole is an oral HDACi with a strong antiproliferative activity. When tested in MS animal models it was shown to decrease IL-17 and IFNγ production, reduce CNS inflammatory infiltrates, and produce a clinical effect on the incidence, severity, and disability scores [65]. Vorinostat is a

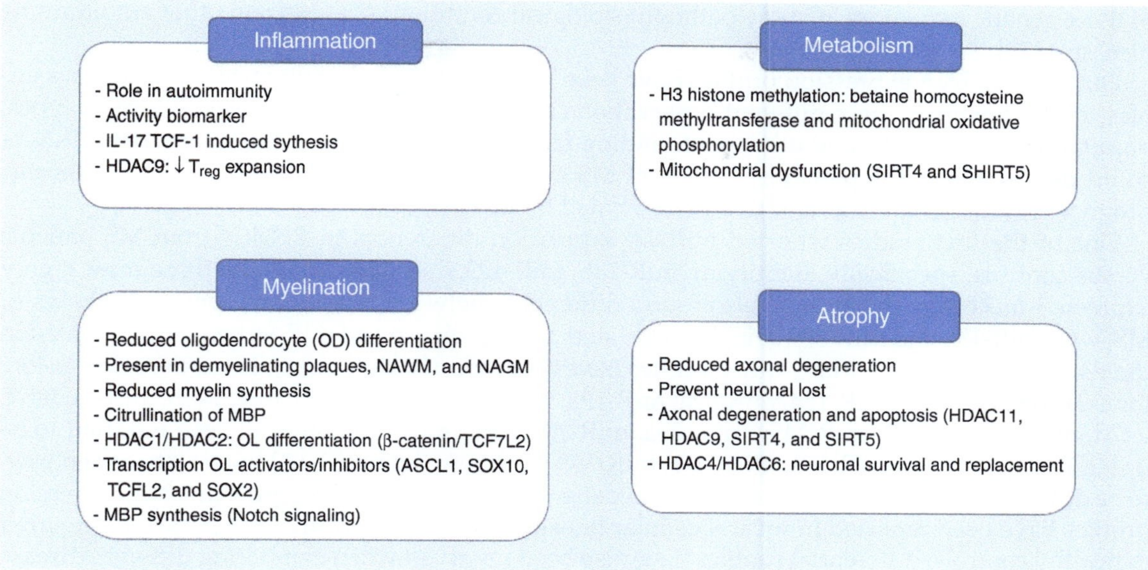

FIGURE 8.1 **Histone modifications in multiple sclerosis.**

HDACi that prevents human CD14 monocyte-derived dendritic cell differentiation and reduces Th1- and Th17-mediated inflammation and demyelination in the CNS of EAE mice [66]. Curcumin is a natural polyphenolic phytochemical agent derived from the rhizome of the medicinal plant, *Curcuma longa.* It inhibits enzymes with lysine acetyltransferase activity (i.e., HATs) and has widely demonstrated epigenetic effects. When tested in mice with EAE, curcumin was able to reduce CK synthesis (i.e., IL-17, TGFβ, IL-6, IL-21, STAT3, and RORγ) and inflammatory-cell infiltration into the spinal cord, thus leading to a secondary reduction in clinical severity. Curcumin has also been shown to cause a dose-dependent reduction in MBP-reactive lymphocyte proliferation [67,68].

In conclusion, numerous research studies have described and demonstrated the key roles of histone modifications in the development and progression of MS due to their influence on inflammation, degeneration, and metabolism (Fig. 8.1). Histone modifying drugs are promising MS therapies as their HDACi properties can modulate the overactive immune system and trigger antioxidant and neuroprotective pathways to prevent CNS damage. However, further studies are needed to verify whether these beneficial effects are seen in MS patients [69].

8.5 REGULATORY EFFECTS OF NONCODING RNAs

MicroRNAs (miRNAs) are endogenous noncoding single-stranded short (18–23 nucleotides) RNAs that regulate gene expression posttranscriptionally. They exert their function through two pathways: (1) by inducing degradation of their target mRNA when the miRNA sequence completely matches the 3′ untranslated region (UTR) sequence of the target mRNA, and (2) by modulating DNA methylation of promoter sites and histone modifications. Numerous miRNAs have been linked to autoimmunity, neuroinflammation, and myelination, thus highlighting the importance of these genetic regulators in many pathophysiological conditions (i.e., MS and other autoimmune diseases) [70].

Different miRNA expression profiles have been reported in WB, lymphocytes, and serum samples in MS; however, these different observations may be caused by differences in sample types, populations, techniques, and other confounding factors [71]. However, it is evident that miRNAs could play a role in the pathophysiology of MS and so they have been proposed as diagnostic, prognostic, process-specific, and treatment-related biomarkers [72].

One of the first studies reported miRNA expression differences in PBMCs from MS patients versus controls, specifically identifying miR-18b, miR-493, and miR-599 as the three most highly expressed miRNAs. The authors also found differences between relapse and remitting phases of MS, with miR-18b and miR-599 being upregulated during relapse and miR-96 being upregulated in the remitting phase [73]. Many other studies have also reported differences in miRNA expression; for example, miR-326, miR-155, miR-92a, miR-19a, miR-142-3p, miR-146a, miR-146b, miR-21, let7f, let7d, miR-744, miR-93, miR-145, miR-125a, miR-200c, and miR-146b have all been reported to be upregulated, whereas miR-15a, miR-16-1, miR-363, miR-199a, and miR-152 have all been reported to be downregulated in PBMCs isolated from patients with MS [74]. Differential miRNA expression profiles have been reported in several cellular lineages and in the serum of MS patients compared to healthy controls; differential profiles have also been observed when comparing different disease states [75–80] (Table 8.3).

miRNA expression profiles are also different between active brain lesions and inactive lesions and NAWM in MS patients, and between MS and healthy brains. At least 20 miRNAs are overexpressed within MS brain lesions, and these miRNAs play roles in different mechanisms of neuronal inflammation. The three most highly expressed miRNAs in MS plaques (miR-155, miR-326, and miR-34a) have all been associated with inflammatory pathways that could participate in the pathogenesis of MS. Interestingly, all three miRNAs were found to be differentially expressed in MS patients and animal models and active in different cell lineages that are implicated in the disease. They target the 3′ UTR of the CD47 protein and decrease its expression and activity, and CD47 has been shown to be reduced in active lesions and in the core of chronic inactive lesions [81,102]. CD47 is a cellular marker of macrophage inhibition and is also expressed on astrocytes and oligodendrocytes. Its reduction facilitates CK secretion, inflammatory activation, and myelin phagocytosis. Decreased CD47 expression in brain plaques could also reflect a decreased remyelination process [103]. MiR-155 and miR-326 also regulate the Th1–Th2 and Treg–T17 balance. MiR-155 is associated with a wide variety of immunological targets and seems to play a critical role in autoimmune disorders by promoting Th17 cell formation and activity [104].

miR-155 has also been shown to be associated with several other neuroinflammatory processes, including antigen presenting cell function, CK synthesis, immunoglobulin diversification, and neurosteroid synthesis. miR-338 and miR491 were also overexpressed in the white matter of MS patients and their expression levels were associated with the suppression of neurosteroids [82]. In mouse models of MS, miR-155 knockdown decreased the Th1 and Th17 populations and ameliorated EAE, while miR-155 overexpression enhanced the Th1 and Th17 cells and caused a severe phenotype [83,84].

miR-326, which has been shown to be upregulated in PBMCs, lymphocytes, and MS lesions, has been proposed as a biomarker to discriminate between relapsing and remitting phases of MS [85]. This miRNA participates in autoimmune regulation through its action on C-ets-1, a negative regulator of Th17 differentiation, and its expression is correlated with disease severity in animal models and MS patients [86].

TABLE 8.3 MicroRNAs Implicated in MS

miRNA	Sample	Conclusion	References
miR-18b, miR.493, and miR599	PBMC	Discriminates between relapsing or remitting phases	Otaegui et al. [73]
miR-326, miR-155, miR-92a, miR-19a, miR-146a, miR-146b, miR-21, let7f, let7d, miR-744, miR-93, miR-145, miR-125a, miR-200c, and miR-146b	PBMC	Upregulated in MS patients	Aslani et al. [74]
miR-363, miR-199a, miR-15a, miR16-1, and miR-152		Downregulated in MS patients	Aslani et al. [74]
miR-128, miR-27b, and miR-340	CD4 lymphocytes	Upregulated	Guerau-de-Arellano et al. [75]
miR-15b, miR-23a, and miR-223	Serum	Downregulated	Fenoglio et al. [76]
miR-614, miR-572, miR-648, miR-422a, miR-1826, miR-22, and miR-1979	Serum	Upregulated Downregulated	Siegel et al. [77]
miR-145 miR-1979	Serum	Upregulated Downregulated	Sondergaard et al. [78]
miR-16, miR-155, and miR-142-3 miR-15a and miR-16-1	CD4 and CD8	Upregulated Downregulated	Arruda et al. [79] Lorenzi et al. [80]
miR-155, miR-650, miR-155, miR-326, and miR -34a	Brain, microglia, and astrocytes	Upregulated in demyelinating plaques	Junker et al. [81]
miR-155, miR-338, and miR-491	WM	Upregulated in MS patients and associated with neurosteroids levels	Noorbakhsh et al. [82]
miR-155	EAE	Associated with Th1 and Th17 activity and disease severity	Hu et al. [83] Zhang et al. [84]
miR-326 and miR-26a	Lymphocytes	Discriminates between relapsing or remitting phases	Honardoost et al. [85]
miR-326	EAE	Correlation with disease severity	Du et al. [86]
miR-124	Hippocampus EAE	Upregulated in MS patients Improve memory tests and reduced myelin-specific T cells	Dutta et al. [87] Ponomarev et al. [88]
miR-17	WB	Downregulated in MS and regulation of T cell activation	Cox et al. [89]

(Continued)

TABLE 8.3 MicroRNAs Implicated in MS (*cont.*)

miRNA	Sample	Conclusion	References
miR-17-5p. miR-485-3p, miR-376, miR-497, miR-193a, and miR-126	CD4 lymphocytes	Upregulated in MS	Lindberg et al. [90]
MiR-34a		Downregulated	
miR-17	CD4 lymphocytes	Downregulated in patients following natalizumab treatment	Meira et al. [91]
miR-21, miR-142-3p, miR-146a, miR-146b, and miR-155	Brain and spinal cord cells from marmosets and mice with EAE	Upregulated	Lescher et al. [92]
miR-23b	EAE	Downregulated and autoimmune regulation	Zhu et al. [93]
miR-155-5p	EAE urine, plasma, and spinal cord cells	Increased before disease onset	Singh et al. [94]
miR-145	WB	Discriminates MS from controls	Keller et al. [95]
miR-20a-5p and miR-7-1-3p	WB	Discriminates MS from controls, including CIS patients without treatment	Keller et al. [96]
miR-27a miR-214	CD4 lymphocytes	Upregulated in relapsing phase Downregulated in relapsing phase	Ahmadian-Elmi et al. [97]
Let7a and miR-648	Serum	Discriminates between relapsing or remitting phases	Kacperska et al. [98]
miR-140-5p	PBMC	Correlated with disease severity and discriminates between relapsing or remitting phases	Guan et al. [99]
miR-92a-1, miR-145, and let-7	Serum	Upregulated in progressive MS	Gandhi et al. [100]
miR-181c, miR-633, and miR-922 miR-644 and miR-16	CSF	Unregulated Downregulated in SP compared with RR	Haghikia et al. [101]

CSF, Cerebrospinal fluid; *MS*, multiple sclerosis; *RR*, relapsing remitting; *SP*, secondary progressive; *WB*, whole blood; *WM*, white matter.

miR-124 is also found to be overexpressed in the demyelinated hippocampi of MS patients. This miRNA regulates 26 neuronal proteins that are crucial to hippocampal function, and its levels have been correlated with reduced memory functions in rodents [87]. It is also expressed in macrophages and microglia where it regulates motility and phagocytosis, and miR-124 overexpression in microglia decreases the cell function [105]. These findings have also been evidenced in EAE mice where miR-124 is downregulated in activated microglia. Moreover, the peripheral administration of miR-124 has been shown to reduce the activation of macrophages and myelin-specific T cells and decrease the disease severity [88].

Two additional miRNAs (miR-17 and miR-126) have also been associated with the pathogenesis of MS. These miRNAs help to regulate Intercellular Adhesion Molecule (ICAM) and vascular cell adhesion molecule (VCAM) expression, which are endothelial adhesion molecules that are required for the entry of inflammatory cells into the CNS, a key pathological process in MS. miR-17 has been reported to be decreased in WB [89], while miR-17-5p has been shown to be upregulated in CD4 lymphocytes from MS patients alongside concomitant alterations in the expression of its target genes [90]. Interestingly, this could be another mechanism of action of natalizumab, one of the second-line therapies for MS. Natalizumab is a recombinant humanized monoclonal antibody that prevents leucocytes interacting with VCAM by binding to surface integrins. MiR-17 was reportedly downregulated in CD4 lymphocytes from patients following natalizumab treatment compared with untreated patients, thus supporting its role in MS pathogenesis [91].

Numerous miRNAs have also been studied in animal models of MS; for example, miR-21, miR-142-3p, miR-146a, miR-146b, and miR-155 were all found to be upregulated in the spinal cord and brain of mice and marmoset with EAE [92]. Conversely, miR-23b was found to be downregulated in mouse models of MS. This miRNA decreases IL-17, TNF-α, IL-1β, and other proinflammatory factors, and thus its downregulation is essential for autoimmune pathogenesis [93].

Many lines of evidence support the idea that the detection of miRNAs in other tissues (i.e., peripheral blood or urine) could be used as biomarkers for MS. For example, it has been observed that miR-155-5p levels increase in the urine, plasma, and spinal cord of EAE mice at least 6 days before the disease onset, therefore suggesting that urine miRNAs could be used as early biomarkers in MS [94]. In addition, levels of miR-145, miR-201-5p, and miR7-1-3p in WB samples have been used to discriminate patients from controls with a high specificity and sensitivity [95,96]. miRNA detection can also be used to discriminate the disease activity; for example, the detection of miR-326 and miR-26a in lymphocytes is highly specific and sensitive for differentiating between relapsing and remitting phases of MS [85]. In addition, miR-27a is upregulated and miR-214 is downregulated in CD4 lymphocytes in the relapsing phase, with possible secondary effects on TGF-β and mTOR signaling [97]. Serum levels of Let-7a and miR-648 have also been proposed to discriminate between relapsing and remitting states of the disease [98]. miRNA detection could also be related to the disease severity; for example, miR-140-5p expression in PBMCs correlates negatively with STAT1, which is a transcription factor for Th1 differentiation [99]. Three miRNA were also shown to be upregulated in the serum of patients with progressive MS (miR-92a-1, miR-145, and let-7) [100] and an additional two miRNAs were upregulated in the CSF of these subjects (miR-633 and miR-181c). These miRNAs have since been proposed as biomarkers for differentiating different disease states because they are downregulated in the CSF of patients with secondary progressive disease compared to those with RR [101].

miRNA-modulating drugs have been proposed as potential MS treatments. Considering their relative specificity in modulating protein expression, they seem easy to monitor and evaluate. Suppressing miR-155 in mice decreased EAE development and severity [106], and repressing miR-326 also resulted in reduced numbers of Th17 cells and improved disease severity [107]. MiR-132 expression could be enhanced by 2,3,4,8-tetrachlorodibenzo-p-dioxin through the promotion of cholinergic antiinflammation and an improved EAE phenotype [108]. MiR-128, miR-27b, and miR-340 are known to be increased in CD4 cells from patients with MS and promote a shift towards Th1 differentiation and an inhibition of Th2 development. In fact, transferring T cells with these miRNAs into mice was shown to enhance the incidence and severity of EAE. Treating T cells from MS patients with oligonucleotide miRNA inhibitors can restore Th2 CK production and this may also decrease the differentiation of autoimmune T cells [75].

Numerous studies have evidenced the role of miRNAs in the posttranscriptional control of gene expression and disease development in autoimmune and neurodegenerative diseases. An increasing number of publications have highlighted the relationship between miRNAs and MS pathophysiology by showing differential mRNA expression in MS patients, the possible use pf miRNAs as diagnostic and prognostic biomarkers, and the potential use of miRNAs as therapeutic targets.

8.6 CONCLUSIONS

The pathophysiology of MS raises many questions regarding the risk of developing the disease and the factors that modify its progression. Evidence from genetic susceptibility studies, clinical variability in the presentation and progression of the disease, and the sensitivity to environmental factors clearly highlight that epigenetic modifications play a clear role in the development of MS. These factors are shared by other autoimmune diseases in which multiple epigenetic processes have been widely described.

This chapter has described the relationship between the different epigenetic modifications and pathophysiological factors of MS (i.e., inflammation, demyelination, and cerebral atrophy). There is a growing body of research, which when published in the coming years, will surely provide more details about the influence of these mechanisms in MS and other autoimmune diseases.

Research results are currently in their infancy, and there is still a long way to go before this knowledge can be applied in the clinical practice. Nevertheless, this new path of research has provided hope for increasing the knowledge of the pathophysiological mechanisms of MS and for developing new therapeutic targets. Epigenetic modifications will surely be used as biomarkers or therapeutic targets in MS; however, it will be necessary to define their functions in disease development and the consequences of modifying gene expression levels across different tissues.

References

[1] Gangisetty O, Murugan S. Epigenetic modifications in neurological diseases: natural products as epigenetic modulators as a treatment strategy. Adv Neurobiol 2016;12:1–25.

[2] Goodin DS. The epidemiology of multiple sclerosis: insights to disease pathogenesis. Handb Clin Neurol 2014;122: 231–66.

[3] Pugliatti M, Rosati G, Carton H, Riise T, Drulovic J, Vecsei L, et al. The epidemiology of multiple sclerosis in Europe. Eur J Neurol 2006;13:700–22.

[4] Ma VY, Chan L, Carruthers KJ. Incidence, prevalence, costs, and impact on disability of common conditions requiring rehabilitation in the United States: stroke, spinal cord injury, traumatic brain injury, multiple sclerosis, osteoarthritis, rheumatoid arthritis, limb loss, and back pain. Arch Phys Med Rehabil 2014;95(5):986–95.

[5] Mallucci F, Peruzzotti-Jametti L, Bernstocj JD, Pluchino S. The role of immune cells, glia and neurons in white and gray matter pathology in multiple sclerosis. Prog Neurobiol 2015;127–128:1–22.

[6] Deshmukh VA, Tardif V, Lyssiotis CA, Green CC, Kerman B, Kim HJ, et al. A regenerative approach to the treatment of multiple sclerosis. Nature 2013;502(7471):327–32.
[7] McCarthy DP, Richards MH, Miller SD. Mouse models of multiple sclerosis: experimental autoimmune encephalomyelitis and Theiler's virus-induced demyelinating disease. Methods Mol Biol 2012;900:381–401.
[8] Ebers GC, Sadovnick AD, Dyment DA, Yee IM, Willer CJ, Risch N. Parent-of-origin effect in multiple sclerosis: observations in half-siblings. Lancet 2004;363(9423):1773–4.
[9] Hollenbach AJ, Oksenberg JR. The immunogenetics of multiple sclerosis a comprehensive review. J Autoimmun 2015;64:13–25.
[10] Küçükali CI, Kürtüncü M, Coban A, Cebi M, Tüzün E. Epigenetics of multiple sclerosis: an updated review. Neuromolecular Med 2015;17(2):83–96.
[11] Hafler DA, Comston A, IMSCG. et al. Risk alleles for multiple sclerosis identified by a genome-wide study. N Engl J Med 2007;357(9):851–62.
[12] Sawcer S, Hellenthal G, IMSGC. et al. Genetic risk and a primary role for cell-mediated immune mechanisms in multiple sclerosis. Nature 2011;476(7359):214–9.
[13] Beecham AH, Patsopoulos NA, IMSGC. et al. Analysis of immune-related loci identifies 48 new susceptibility variants for multiple sclerosis. Nat Genet 2013;45(11):1353–60.
[14] Yin Q, Wang X, Roberts C, Flemington EK, Lasky JA. Methylation status and AP1 elements are involved in EBV-mediated miR-155 expression in EBV positive lymphoma cells. Virology 2016;494:158–67.
[15] Ma X, Zhou J, Zhong Y, Jiang L, Mu P, Li Y, et al. Expression, regulation and function of microRNA in multiple sclerosis. Int J Med Sci 2014;11(8):810–8.
[16] Joshi S, Pantalena LC, Liu XK, Gaffen SL, Liu H, Rohowsky-Kochan C, et al. 1,25-Dihydroxyvitamin D (3) ameliorates Th17 autoimmunity via transcriptional modulation of interleukin-17A. Mol Cell Biol 2011;31(17):3653–69.
[17] Sopori ML, Kozak W. Immunomodulatory effects of cigarette smoke. J Neuroimmunol 1998;83:148–56.
[18] Klingbeil EC, Hew KM, Nygaard UC, Nadeau KC. Polycyclic aromatic hydrocarbons, tobacco smoke, and epigenetic remodeling in asthma. Immunol Res 2014;58:369–73.
[19] Fraga MF, Ballestar E, Paz MF, Ropero S, Setien F, Ballestar ML, et al. Epigenetic differences arise during the lifetime of monozygotic twins. Proc Natl Acad Sci USA 2005;102(30):10604–9.
[20] Li X, Xiao B, Chen XS. DNA methylation: a new player in multiple sclerosis. Mol Nerobiol 2017;54(6):4049–59.
[21] Neves-Costa A, Moita LF. TET1 is a negative transcriptional regulator of IL-1β in the THP-1 cell line. Mol Immunol 2013;54(3–4):264–70.
[22] Calabrese R, Valentin E, Ciccarone F, Guastaferro T, et al. TET2 gene expression and 5-hydroxymethylcytosine level in multiple sclerosis peripheral blood cells. Biochim Biophys Acta 2014;1842(7):1130–6.
[23] Fagone P, Mangano K, Di Marco R, Touil-Boukoffa C, Chikovani T, Signorelli S, et al. Expression of DNA methylation genes in secondary progressive multiple sclerosis. J Neuroimmunol 2016;290:66–9.
[24] Khorshid Ahmad T, Acosta C, Cortes C, Lakowski TM, Gangadaran S, Namaka M. Transcriptional regulation of brain-derived neurotrophic factor (BDNF) by methyl CpG binding protein 2 (MeCP2): a novel mechanism for re-myelination and/or myelin repair involved in the treatment of multiple sclerosis. Mol Neurobiol 2016;53(2):1092–107.
[25] Andlauer TF, Buck D, Antony G, Bayas A, Bechmann L, Berthele A, et al. Novel multiple sclerosis susceptibility loci implicated in epigenetic regulation. Sci Adv 2016;172(6).
[26] Handel AE, De Luca GC, Morahan J, Handunnetthi L, Sadovnick AD, Ebers GC, et al. No evidence for an effect of DNA methylation on multiple sclerosis severity at HLADRB1*15 or HLA-DRB5. J Neuroimmunol 2010;223:120–3.
[27] Calabrese R, Zampieri M, Mechelli R, Annibali V, Guastafierro T, Ciccarone F, et al. Methylation-dependent PAD2 upregulaton in multiple sclerosis peripheral blood. Mult Scler 2012;18(3):299–304.
[28] Mastronardi FG, Noor A, Wood DD, Paton T, Moscarello MA, et al. Peptidyl argininedeiminase 2 CpG island in multiple sclerosis white matter is hypomethylated. J Neurosci Res 2007;85(9):2006–16.
[29] Koch MW, Metz LM, Kovalchuk O. Epigenetics changes in patients with multiple sclerosis. Nat Rev Neurol 2012;9:35–43.
[30] Brooks WH. Epigenetics and autoimmunity. J Autoimmun 2010;34(3):J207–19.
[31] Mangano K, Fagone P, Bendtzen K, Meroni PL, Quattrocchi C, Mammana S, et al. Hypomethylating agent 5-aza-2′-deoxycytidine (DAC) ameliorates multiple sclerosis in mouse models. J Cell Physiol 2014;229(12):1918–25.
[32] Janson PC, Linton LB, Bergman EA, Maritis P, et al. Profiling $CD4^+$ T cells with epigenetic immune lineage analysis. J Immunol 2011;186(1):92–102.
[33] Naor D, Sopmpv RV, Ish-Shalom D. CD44: structure, function and association with the malignant process. Adv Cancer Res 1997;71:241–319.

[34] Guan H, Nagarkatti PS, Nagarkatti M. CD44 reciprocally regulates the differentiation of encephalitogenic Th1/Th17 and Th2/T_{reg} cells through epigenetic modulation involving DNA methylation of cytokine gene promoters thereby controlling the development of experimental autoimmune encephalomyelitis. J Immunol 2011;186(12):6955–64.

[35] Christophi GP, Hudson CA, Gruber RC, Christophi CP, Mihai C, Mejico LJ, et al. SHP-1 deficiency and increased inflammatory gene expression in PBMCs of multiple sclerosis patients. Lab Invest 2008;88:243–55.

[36] Kumagai C, Kalman B, Middleton FA, Vyshkina T, Massa PT, et al. Increased promoter methylation of the immune regulatory gene SHP-1 in leukocytes of multiple sclerosis subjects. J Neuroimmunol 2012;246(1–2):51–7.

[37] Sokratous M, Dardiotis E, Tsouris Z, Bellou E, Michalopoulou A, Siokas V, et al. Deciphering the role of DNA methylation in multiple sclerosis: emerging issues. Auto Immun Highlights 2016;7(1):12.

[38] Kulakova OG, Kabilov MR, Danilova LV, Popova EV, Baturina OA, Tsareva EY, et al. Whole-genome DNA methylation analysis of peripheral blood mononuclear cells in multiple sclerosis patients with different disease courses. Acta Naturae 2016;8(3):103–10.

[39] Graves MC, Benton M, Lea RA, Boyle M, Tajouri L, Macartney-Coxson D, et al. Methylation differences at the HLA-DRB1 locus in $CD4^+$ T cells are associated with multiple sclerosis. Mult Scler 2014;20(8):1033–41.

[40] Bos SD, Page CM, Andreassen BK, Elboudwarej E, Gustavsen MW, Briggs F, et al. Genome-wide DNA methylation profiles indicate $CD8^+$ T cell hypermethylation in multiple sclerosis. PLoS One 2015;10(3). 2015.

[41] Maltby VE, Graves MC, Lea RA, Benton MC, Sanders KA, Tajouri L, et al. Genome-wide DNA methylation profiling of $CD8^+$ T cells shows a distinct epigenetic signature to $CD4^+$ T cells in multiple sclerosis patients. Clin Epigenet 2015;7118.

[42] Huynh JL, Garg P, Thin TH, Yoo S, Dutta R, Trapp BD, et al. Epigenome-wide differences in pathology-free regions of multiple sclerosis-affected brains. Nat Neurosci 2014;17(1):121–30.

[43] Liggett T, Melnikov A, Tilwalli S, Yi Q, Chen H, Replogle C, et al. Methylation patterns of cell-free plasma DNA in relapsing-remitting multiple sclerosis. J Neurol Sci 2010;290(1–2):16–21.

[44] Neven KY, Piola M, Angelici L, Cortini F, Fenoglio C, Galimberti D, et al. Repetitive element hypermethylation in multiple sclerosis patients. BMC Genet 2016;17.

[45] Baranzini SE, Mudge J, van Velkinburgh JC, Khankhanian P, Khrebtukova I, Miller NA, et al. Genome, epigenome and RNA sequences of monozygotic twins discordant for multiple sclerosis. Nature 2010;464:1351–6.

[46] Ramagopalan SV, Dyment DA, Morrison KM, Herrera BM, Deluca GC, Lincoln MR, et al. Methylation of class II transactivator gene promoter IV is not associated with susceptibility to multiple sclerosis. BMC Med Genet 2008;63.

[47] Iridoy Zulet M, Pulido Fontes L, Ayuso Blanco T, Lacruz Bescos F, Mendioroz Iriarte M. Modificaciones epigenéticas en neurología: alteraciones en la metilación del ADN en la esclerosis múltiple. Neurología 2015;11.

[48] Huynh JL. Epigenetic mechanisms in multiple sclerosis: implications for pathogenesis and treatment. Lancet Neurol 2013;12(2):195–206.

[49] Shen S, Sandoval J, Swiss VA, Li J, Dupree J, Franklin RJ, et al. Age-dependent epigenetic control of differentiation inhibitors is critical for remyelination efficiency. Nat Neurosci 2008;11(9):1024–34.

[50] Martin A, Tegla CA, Cudrici CD, Kruszewski AM, Azimzadeh P, Boodhoo D, et al. Role of SIRT1 in autoimmune demyelination and neurodegeneration. Immunol Res 2015;61(3):187–97.

[51] Tegla CA, Azimzadeh P, Andrian-Albescu M, Martin A, Cudrici CD, Trippe R, et al. SIRT1 is decreased during relapses in patients with multiple sclerosis. Exp Mol Pathol 2014;96:139–48.

[52] Shindler KS, Ventura E, Rex TS, Elliott P, Rostami A. SIRT1 activation confers neuroprotection in experimental optic neuritis. Invest Ophthalmol Visual Sci 2007;48:3602–9.

[53] Shindler KS, Ventura E, Dutt M, Elliott P, Fitzgerald DC, Rostami A. Oral reservatrol reduces neuronal damage in a model of multiple sclerosis. J Neuroophthalmol 2010;30(4):328–39.

[54] Mastronardi FG, Wood D, Mei J, Raijmakers R, Tseveleki V, Dosch HM, et al. Increased citrullination of histone H3 in multiple sclerosis brain and animal models of demyelination: a role for tumor necrosis factor-induced peptidylarginine deiminase 4 translocation. J Neurosci 2006;26:11387–96.

[55] Ma J, Wang R, Fang X, Ding Y, Sun Z. Critical role of TCF-1 in repression of the IL-17 gene. PLoS One 2011;6(9). e24768.

[56] Ye F, Chen Y, Hoang T, Montgomery RL, Zhao XH, Bu H, et al. HDAC1 and HDAC2 regulate oligodendrocyte differentiation by disrupting the b-catenin–TCF interaction. Nat Neurosci 2009;12(7):829–38.

[57] Li H, He Y, Richardson WD, Casaccia P. Two-tier transcriptional control of oligodendrocyte differentiation. Curr Opin Neurobiol 2009;19(5):479–85.

[58] Pedre X, Mastronardi F, Bruck W, Lopez-Rodas G, Kuhlmann T, Casaccia P. Changed histone acetylation patterns in normal-appearing white matter and early multiple sclerosis lesions. J Neurosci 2011;31:3435–45.

[59] Inkster B, Strijbis EM, Vounou M, Kappos L, Radue EW, Matthews PM, et al. Histone deacetylase gene variants predict brain volume changes in multiple sclerosis. Neurobiol Aging 2013;34:238–47.

[60] Faraco G, Cavone L, Chiarugi A. The therapeutic potential of HDAC inhibitors in the treatment of multiple sclerosis. Mol Med 2011;17(5–6):442–7.

[61] Correa F, Mallard C, Nilsson M, Sandberg M, et al. Activated microglia decrease histone acetylation and Nrf2-inducible anti-oxidant defence in astrocytes: restoring effects of inhibitors of HDACs, p38 MAPK and GSK3β. Neurobiol Dis 2011;44(1):142–51.

[62] Zhang Z, Zhang ZY, Wu Y, Schluesener H. Valproic acid ameliorates inflammation in experimental autoimmune encephalomyelitis rats. Neuroscience 2012;221:140–50.

[63] Camelo S, Iglesias AH, Hwang D, Due B, Ryu H, Smith K, et al. Transcriptional therapy with the histone deacetylase inhibitor trichostatin A ameliorates experimental autoimmune encephalomyelitis. J Neuroimmunol 2005;164(1–2): 10–21.

[64] Zhang F, Shi Y, Wang L, Sriram S. Role of HDAC3 on p53 expression and apoptosis in T cells of patients with multiple sclerosis. PLoS One 2011;6(2):e16795.

[65] Chen F, Chai H, Su MB, Zhang YM, Li J, Xie X, et al. Potent and orally efficacious bisthiazole-based histone deacetylase inhibitors. ACS Med Chem Lett 2014;5(6):628–33.

[66] Ge Z, Da Y, Xue Z, Zhang K, Zhuang H, Peng M, et al. Vorinostat, a histone deacetylase inhibitor, suppresses dendritic cell function and ameliorates experimental autoimmune encephalomyelitis. Exp Neurol 2013;241:56–66.

[67] Xie L, Li XK, et al. Amelioration of experimental autoimmune encephalomyelitis by curcumin treatment through inhibition of IL-17 production. Int Immunopharmacol 2009;9(5):575–81.

[68] Feng J, Tao T, Yan W, Chen CS, Qin X. Curcumin inhibits mitochondrial injury and apoptosis from the early stage in EAE mice. Oxid Med Cell Longev 2014;2014. 728751.

[69] Gray SG, Dangond F. Rationale for the use of histone deacetylase inhibitors as a dual therapeutic modality in multiple sclerosis. Epigenetics 2006;12:67–75.

[70] Ksiazek-Winiarek DJ, Kacperska MJ, Glabinski A. MicroRNAs as novel regulators of neuroinflammation. Mediators Inflamm 2013;2013. 172351.

[71] Ma X, Zhou J, Zhong Y, Jiang L, Mu P, Li Y, et al. Expression, regulation and function of microRNAs in multiple sclerosis. Int J Med Sci 2014;11(8):810–8.

[72] Gandhi R. miRNA in multiple sclerosis: search for novel biomarkers. Mult Scler 2015;21(9):1095–103.

[73] Otaegui D, Baranzini SE, Armañanzas R, Calvo B, Muñoz-Culla M, et al. Differential microRNA expression in PBMC from multiple sclerosis patients. PLoS One 2009;4(7). e6309.

[74] Aslani S, Jafari N, Javan MR, Karami J, Ahmadi M, Jafarnejad M. Epigenetic modifications and therapy in multiple sclerosis. Neuromolecular Med 2017;19(1):11–23.

[75] Guerau-de-Arellano M, Smith KM, Godlewski J, Liu Y, Winger R, Lawler SE, et al. Micro-RNA dysregulation in multiple sclerosis favours pro-inflammatory T-cell-mediated autoimmunity. Brain 2011;134:3578–89.

[76] Fenoglio C, Ridolfi E, Cantoni C, De Riz M, Bonsi R, Serpente M, et al. Decreased circulating miRNA levels in patients with primary progressive multiple sclerosis. Mult Scler 2013;19:1938–42.

[77] Siegel SR, Mackenzie J, Chaplin G, Jablonski NG, Griffiths L. Circulating microRNA involved in multiple sclerosis. Mol Biol Rep 2012;39(5):6219–25.

[78] Sondergaard HB, Hesse D, Krakauer M, Sorensen PS, Sellebjerg F. Differential microRNA expression in blood in multiple sclerosis. Mult Scler 2013;19(14):1849–57.

[79] Arruda LC, Lorenzi JC, Sousa AP, Zanette DL, Palma PV, et al. Autologous hematopoietic SCT normalizes miR-16-155 and -142-3p expression in multiple sclerosis. Bone Marrow Transplant 2015;50(30):380–9.

[80] Lorenzi JC, Brum DG, Zanette DL, de Paula Alves Souza A, Barbuzan FC, et al. miR-15a and 16-1 are downregulated in $CD4^+$ T cells of multiple sclerosis relapsing patients. Int J Neurosci 2012;122(8):466–71.

[81] Junker A, Krumbholz M, Eisele S, Mohan H, Augstein F, Bittner R, et al. MicroRNA profiling of multiple sclerosis lesions identifies modulators of the regulatory protein CD47. Brain 2009;132:3342–52.

[82] Noorbakhsh F, Ellestad KK, Maingat F, Warren KG, Han MH, Steinman L, et al. Impaired neurosteroid synthesis in multiple sclerosis. Brain 2011;134(9):2703–21.

[83] Hu R, Huffaker TB, Kagele DA, Runtsch MC, Bake E, Chaudhuri AA, et al. MicroRNA-155 confers encephalogenic potential to Th17 cells by promoting effector gene expression. J Immunol 2013;15190(12):5972–80.

[84] Zhang J, Cheng Y, Cui W, Li M, Li B. MicroRNA-155 modulates Th1 and Th17 cell differentiation and is associated with multiple sclerosis and experimental autoimmune encephalomyelitis. J Neuroimmunol 2014;266(1–2):56–63.

[85] Honardoost MA. miR-326 and miR-26a, two potential markers for diagnosis of relapse and remission phases in patient with relapsing-remitting multiple sclerosis. Gene 2014;544(2):128–33.

[86] Du C, Liu C, Kang J, Zhao G, Ye Z, Huang S, et al. MicroRNA miR-326 regulates TH-17 differentiation and is associated with the pathogenesis of multiple sclerosis. Nat Immunol 2009;10:1252–9.

[87] Dutta R, Chomyk AM, Chang A, Ribaudo MV, Deckard SA, Doud MK, et al. Hippocampal demyelination and memory dysfunction are associated with increased levels of the neuronal microRNA miR-124 and reduced AMPA receptors. Ann Neurol 2013;73:637–45.
[88] Ponomarev ED, Veremeyko T, Barteneva N, Krichevsky AM, Weiner HL. MicroRNA-124 promotes microglia quiescence and suppresses EAE by deactivating macrophages via the C/EBP-α-PU 1 pathway. Nat Med 2011;17(1):64–70.
[89] Cox MB, Cairns MJ, Gandhi KS, Carroll AP, Moscovis S, Stewart GJ, et al. MicroRNA miR-17 and miR-20a inhibit T cell activation genes and are under-expressed in MS whole blood. PLoS One 2010;115(8). e12132.
[90] Lindberg RL, Hoffmann F, Mehling M, Kuhle J, Kappos L. Altered expression of miR-17-5p in $CD4^+$ lymphocytes of relapsing-remitting multiple sclerosis patients. Eur J Immunol 2010;40(3):888–98.
[91] Meira M, Sievers C, Hoffmann F, Rasenack M, Kuhle J, Derfuss T, et al. Unraveling natalizumab effects on deregulated miR-17 expression in $CD4^+$ T cells of patients with relapsing-remitting multiple sclerosis. J Immunol Res 2014;2014. 897249.
[92] Lescher J, Paap F, Schultz V, Redenbach L, Scheidt U, Rosewich H, et al. MicroRNA regulation in experimental autoimmune encephalomyelitis in mice and marmosets resembles regulation in human multiple sclerosis lesions. J Neuroimmunol 2012;246:27–33.
[93] Zhu S, Pan W, Song X, Liu Y, Shao X, Tang Y, et al. The microRNA miR-23b suppresses IL-17 associated autoimmune inflammation by targeting TAB2 TAB3and IKK-α. Nat Med 2012;18(7):1077–86.
[94] Singh J, Deshpande M, Suhail H, Rattan R, Giri S. Targeted stage-specific inflammatory microRNA profiling in urine during disease progression in experimental autoimmune encephalomyelitis: markers of disease progression and drug response. J Neuroimmune Pharmacol 2016;11(1):84–97.
[95] Keller A, Leidinger P, Lange J, Borries A, Schroers H, Scheffler M, et al. Multiple sclerosis: microRNA expression profiles accurately differentiate patients with relapsing-remitting disease from healthy controls. PLoS One 2009;4:e7440.
[96] Keller A, Leidinger P, Steinmeyer F, Stähler C, Franke A, Hemmrich-Stanisak GF, et al. Comprehensive analysis of microRNA profiles in multiple sclerosis including next-generation sequencing. Mult Scler 2014;20:295–303.
[97] Ahmadian-Elmi M, Bidmeshki Pour A, Naghavian R, Ghaedi K, Tanhaei S, Izadi T, et al. miR-27a and miR-214 exert opposite regulatory roles in Th17 differentiation via mediating different signaling pathways in peripheral blood $CD4^+$ T lymphocytes of patients with relapsing-remitting multiple sclerosis. Immunogenetics 2016;68(1):43–54.
[98] Kacperska MJ, Jastrzebski K, Tomasik B, Walenczak J, Konarska-Krol M, Glabinski A. Selected extracellular microRNA as potential biomarkers of multiple sclerosis: a preliminary study. J Mol Neurosci 2015;56(1):154–63.
[99] Guan H, Singh UP, Rao R, Mrelashvili D, Sen S, Hao H, et al. Inverse correlation of expression of microRNA-140-5p progression of multiple sclerosis and differentiation of encephalitogenic T helper type 1 cells. Immunology 2016;147(4):488–98.
[100] Gandhi R, Healy B, Gholipour T, Egorova S, Musallam A, Hussain MS, et al. Circulating microRNAs as biomarkers for disease staging in multiple sclerosis. Ann Neurol 2013;73:729–40.
[101] Haghikia A, Haghikia A, Hellwig K, Baraniskin A, Holzmann A, Décard BF, et al. Regulated microRNAs in the CSF of patients with multiple sclerosis: a case-control study. Neurology 2012;79(22):2166–70.
[102] Koning N, Bo L, Hoek RM, Huntinga I. Downregulation of macrophage inhibitory molecules in multiple sclerosis lesions. Ann Neurol 2007;62:504–14.
[103] Gao Q, Zhang Y, Han C, Hu X, Zhang H, et al. Blockade of CD47 ameliorates autoimmune inflammation in CNS by suppressing IL-1-triggered infiltration of pathogenic Th17 cells. J Autoimmun 2016;69:74–85.
[104] O'Connell RM, Kahn D, Gibson WS, Round JL, Scholz RL, Chaudhuri AA, et al. MicroRNA-155 promotes autoimmune inflammation by enhancing inflammatory T cell development. Immunity 2010;33(4):607–19.
[105] Svahn AJ, Giacomotto J, Graeber MB, Rinkwitz S, Becker TS. miR-124 contributes to the functional maturity of microglia. Dev Neurobiol 2016;76(5):507–18.
[106] Murugaiyan G, Beynon V, Mittal A, Joller N, Weiner HL. Silencing microRNA-155 ameliorates experimental autoimmune encephalomyelitis. J Immunol 2011;187:2213–21.
[107] Junker A. Pathophysiology of translational regulation by microRNAs in multiple sclerosis. FEBS Lett 2011;585: 3738–46.
[108] Hanieh H, Alzahrani A. MicroRNA-132 suppresses autoimmune encephalomyelitis by inducing cholinergic anti-inflammation: a new Ahr-based exploration. Eur J Immunol 2013;43(10):2771–82.

CHAPTER

9

Type 1 Diabetes and Epigenetics

Sundararajan Jayaraman
University of Illinois at Chicago, Chicago, IL, United States

OUTLINE

The Epigenetics of Autoimmunity. http://dx.doi.org/10.1016/B978-0-12-809912-4.00009-X

9.1 INTRODUCTION

In 1901, Eugene Opie observed that the loss of the pancreatic islets, either partly or wholly, led to diabetes [1]. Type 1 diabetes (T1D) is the most common childhood autoimmune disease and can occur after a variable asymptomatic period [2]. It is prevalent among European descendants carrying the major histocompatibility complex class II loci, human leukocyte antigen *(HLA)-DRB1* and *HLA-DQB1* (Fig. 9.1) [3]. Type 1B (idiopathic or fulminant diabetes) is characterized by an abrupt onset, high serum pancreatic enzyme levels, and almost complete loss of islets without diabetes-related autoantibodies [4]. Type 2 diabetes, which accounts for ~90% of cases, generally affects individuals over 45 years of age and is a heterogeneous disease characterized by relative loss of β-cell mass, intrinsic β-cell secretory defects, and insulin resistance without evidence of autoimmune reactivity [5]. Approximately 11% of patients with type 2 diabetes are also positive for one of the islet autoantibodies and share genetic determinants that are common to both T1D and type 2 diabetes. These individuals are said to have latent autoimmune diabetes in adults or type 1.5 diabetes [6]. Type 3 diabetes was recently characterized by insulin resistance in the central nervous system of Alzheimer's patients [7]. This review focuses on the epigenetic mechanisms involved in the modulation of spontaneously occurring T1D in a nonobese diabetic (NOD) mouse model and virally induced diabetes in rodents, both of which are similar to autoimmune diabetes in humans. Direct chemical injury to β-cells (i.e., streptozotocin induced) without involving autoimmune mechanisms and studies using pharmacological agents that do not represent histone deacetylases (HDAC) inhibitors are beyond the scope of this review.

9.2 ENVIRONMENTAL INFLUENCE ON T1D

The risk of developing T1D is 15-fold higher in first-degree relatives than in the general population [8]. The Finnish Twin Cohort study, involving 44 monozygotic and 183 dizygotic twin pairs, reported a concordance rate for diabetes susceptibility of 33% in monozygotic compared to 3% in dizygotic twins [9]. It was suggested that additive genetic effects contribute to 88% of the phenotypic variance in T1D while 12% is attributed to environmental factors [10]. In developed countries, the incidence of T1D is approximately doubling every 20 years despite the decrease in T1D susceptibility HLA genes [10,11], which suggests a crucial role for environmental factors. The "hygiene hypothesis" postulated that decreased childhood exposure to pathogens could contribute to enhanced susceptibility to diseases such as allergy, asthma, and diabetes [12]; however, the mechanisms underlying this contention are poorly understood.

Congenital rubella and enteroviruses (especially Coxsackie B virus) have been implicated in the onset of T1D (Fig. 9.1) [13–15]. A recent study reported that sera of newly diagnosed T1D patients recognized the Coxsackie virus B4 VP 1 peptide, which is homologous to β-cell associated phogrin and voltage-gated L-type calcium channels involved in the regulation of β-cell apoptosis [16]. Release of self-peptides during a viral infection may trigger an immune response that recognizes both viral antigens and self-peptides by virtue of cross-reactivity ("molecular mimicry"), leading to autoimmune diseases such as T1D [17]. Although these studies provide circumstantial evidence [15,16,18], systematic studies are needed to provide convincing evidence to link β-cell destruction and viral infections. In addition to viral infections, exposure to drugs, intestinal microbial dysbiosis, dietary elements (i.e., bovine milk), and vitamin D3 deficiency have also been implicated in triggering autoimmune diabetes in genetically susceptible individuals (Fig. 9.1). This review highlights the current understanding of the prominent epigenetic mechanisms involved in T1D.

9.3 DNA METHYLATION AND T1D

Epigenetics is defined as heritable changes in gene expression that do not alter the DNA sequence [19]. Epigenetics provides a mechanism through which external factors (i.e., drugs, infections, and dietary elements) can produce a variety of phenotypic variations with identical genotypes. Epigenetic mechanisms such as DNA methylation and histone acetylation have been intensely studied for more than two decades, primarily in the context of cancer [20]. Methylation of cytosine residues at the carbon 5′ position is a common epigenetic mark found in sequences such as cytosine-guanosine dinucleotide (CpG) or CpHpG (H = A, T or C). The loss of normal DNA methylation patterns is the most understood epigenetic cause of cancer [20]. Whereas a methylated cluster or island of CpG pairs in a gene promoter is usually a repressive mark, unmethylated

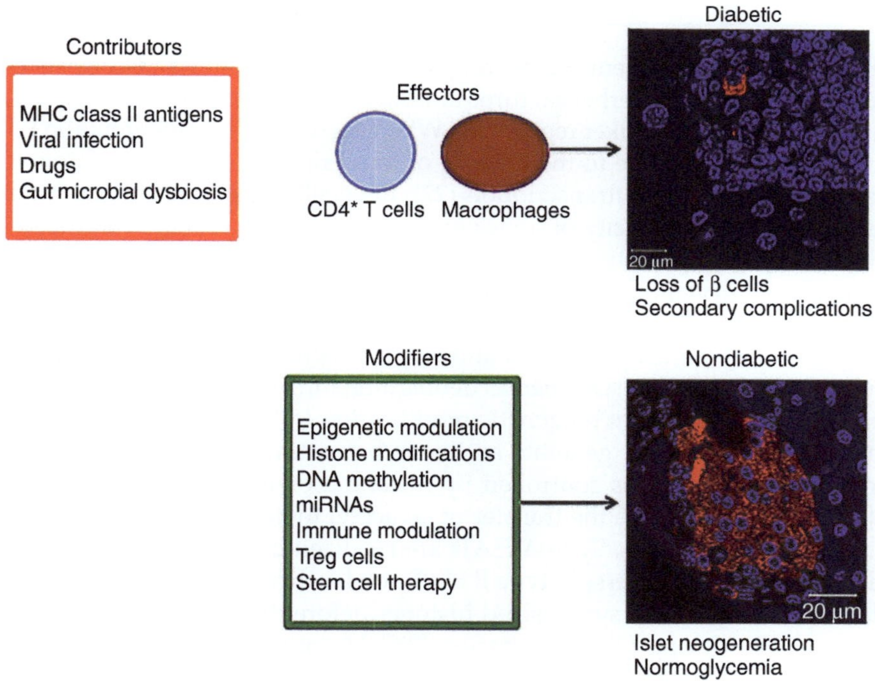

FIGURE 9.1 **A model depicting events during diabetes and their blockade by TSA treatment.** Autoimmune diabetes in humans is influenced by certain HLA genes as well as environmental factors including viral infections, drugs, and dysbiosis of the gut microbiota. Activated $CD4^+$ T lymphocytes engage macrophages to destroy β-cells (containing intracellular insulin visualized as *red* granules by confocal microscopy). Epigenetic modulation by TSA treatment can substantially impact the diabetogenesis via curtailing effector cells and/or facilitating neogeneration of β-cells. The role of other epigenetic mechanisms including miRNA-mediated immunoregulation and DNA methylation in prevention of diabetes is less understood. Immunotherapeutic approaches and islet and stem cell transplantation may also provide transient protection against diabetes.

cytosine residues are associated with transcription permissiveness. The methyl-CpG-binding domain proteins (i.e., MeCP2, MBD1, MBD2, MBD3, and MBD4) mediate the transcriptional repression of genes [21]. Although demethylation of the genome has a significant impact on the regulation of transposons, imprinted genes, and inactive X-chromosome genes, the underlying process remains poorly understood. A role for altered DNA methylation has been implicated in type 2 diabetes [22]; however, limited data are available to attest a role for DNA methylation in autoimmune diabetes (Fig. 9.1). DNA methylation profiling identified 19 prospective CpG sites that were associated with the risk of nephropathy in T1D patients [23]. Interestingly, a subset of genes showed significantly increased demethylation of lysine 9 of histone H3 in lymphocytes, but not monocytes, derived from long-standing T1D patients [24]. In β-cells, methylation of *Ins*1 (exon 2) and *Ins*2 (exon 1) changed over time in NOD mice as a consequence of changes in cytokine transcription [25]. However, specific DNA methylation alterations that could influence T1D pathogenesis are yet to be delineated. Accumulating evidence indicates that, unlike DNA methylation, histone hyperacetylation has a profound influence on both the immune system and endocrine cells.

9.4 ROLES OF HISTONE ACETYLASES AND DEACETYLASES IN T1D

The nucleosome is the fundamental repeating subunit of chromatin and consists of ~147 base pairs of DNA wrapped in two superhelical turns around an octamer of core histones (H2A, H2B, H3, and H4) with H1 comprising the linker region [26]. Whereas compact heterochromatin is inaccessible, relaxed euchromatin is permissible to the binding of transcription factors and chromatin-remodeling complexes that are critical for transcription [27]. Reversible posttranslational modifications of proteins are essential for the plasticity of transient biological signals and are controlled by contrasting enzymatic activities. Modification of the N-terminal tails of core histones occurs by methylation, acetylation, phosphorylation, sumoylation, and ubiquitination [27]. Phosphate groups, which are essential for protein function, are added and deleted by kinases and phosphatases, respectively. Similarly, methylases and demethylases add and remove methyl groups, respectively. Ubiquitin is added to proteins by ubiquitin ligases whereas deubiquitinating enzymes remove ubiquitin.

Acetylation is the most frequent epigenetic modification to histone tails and functions to neutralize the charge of histones and generate an open chromatin configuration [27,28]. The degree of acetylation and deacetylation is controlled by histone acetyltransferases (HATs) and HDACs, respectively [28,29]. HATs catalyze the transfer of an acetyl moiety from acetyl coenzyme A to the $\varepsilon-NH3^{+}$ groups of lysine residues. Type A HATs are localized in the nuclei and acetylate histones H2A, H2B, H3, and H4 [28]. Conversely, type B HATs are localized in cytoplasmic fractions and are responsible for acetylating newly synthesized histones before their translocation into the nucleus for chromatin assembly during DNA replication [28]. Mutations in HATs such as CREBBP and EP300 have been associated with Rubinstein–Taybi syndrome [30]; however, it is unknown whether similar genetic mutations in HATs can contribute to the pathogenesis of autoimmune diseases.

Based on molecular analysis of the primary structure, the 18 HDACs have been classified into five classes [31]: class I (HDAC1, 2, 3, and 8), class IIa (HDAC4, 5, 7, and 9), class IIb (HDAC6 and 10), and class IV (HDAC11) are all zinc-dependent metalloproteins; class III HDACs (sirtuins 1–7) are NAD^{+}-dependent and form a structurally- and mechanistically-distinct class [32]. The roles of class I, IIa, IIb, and IV HDACs in diabetogenesis are beginning to be appreciated and are discussed later in this review.

The combination of different posttranslational modifications appearing on different histones at different times in order to regulate cellular processes is termed the "histone code" [27]. This epigenetic marking system represents a fundamental regulatory mechanism that impacts cell-cycle progression, DNA replication, DNA damage repair, chromatin stability, X-chromosome inactivation, imprinting of maternal and paternal genes, stem cell plasticity, cellular memory, aging, and cancer. Therefore understanding the nature of the histone code and its translation under healthy and pathological conditions is essential to modulate the epigenome to mitigate autoimmune diseases including T1D. Although the past two decades have witnessed substantial progress in understanding the epigenetics of cancers [33,34], the field remains in its infancy in other human pathological conditions (i.e., T1D) (Fig. 9.1).

9.5 SELECTIVITY OF HDAC INHIBITORS

The connection between overall chromatin effects and cellular functions has been elucidated with the use of small-molecule HDAC inhibitors [33–36], which fall into four chemical classes: (1) hydroxamates, such as trichostatin A (TSA) and suberonylanilide hydroxamic acid (SAHA); (2) depsipeptides (i.e., FK-228); (3) short-chain fatty acids (i.e., n-butyrate and valproic acid); and (4) benzamides (i.e., MS-275). TSA derived from *Streptomyces platensis* [37] is the most potent of all HDAC inhibitors [38]. Although TSA and SAHA were previously described to be nonselective (based on their inhibition of enzymatic substrates in vitro) [34], the use of a newly synthesized tripeptide substrate revealed that both TSA and SAHA were more potent at inhibiting HDAC class I (HDAC1, 2, 3 and 8) and class IIb (HDAC6) enzymes than class IIa enzymes (HDAC4, 5, 7 and 9); however, the enzymatic activities of HDAC10 (class II) and HDAC11 (class IV) could not be determined using this new substrate [39]. Unpublished data from our laboratory indicate the efficacy of TSA at inhibiting the transcription of *Hdac10* and *Hdac11* in NOD mice in vivo. Several HDAC inhibitors, including short-chain fatty acids (i.e., butyrate and valproic acid), cyclic tetrapeptide-depsipeptides (i.e., FR901228), and benzamides (i.e., MS-275), are selective for class I HDACs [38,39]. Thus these pharmacological inhibitors have selectivity for HDAC enzymatic substrates that could be exploited to regulate pathological conditions in which certain HDACs are overexpressed.

It is conceivable that particular HDACs may represent risk factors that participate in the induction phase of a disease, while others may be crucial for disease manifestation and may be involved in the effector phase. The induction and manifestation of autoimmune diabetes requires the participation of a triad of neutrophils, B1a cells, and plasmacytoid dendritic cells to trigger T cell activation and the subsequent demise of β-cells (Fig. 9.1) [40]. It is possible that the levels of selected HDACs may differ in these distinct cellular compartments in NOD mice when compared to those of nonautoimmune mice. In support of this hypothesis, varying representations of HDACs have been identified in lymphoid organs compared to the pancreas in NOD mice. Importantly, some of these HDACs are transcriptionally regulated by TSA treatment. Selection of appropriate HDAC inhibitors that can modulate aberrantly expressed HDACs in all tissues participating in T1D pathogenesis would assure mitigation of this common complex disease. The following sections discuss the susceptibility of HDACs expressed in the pancreas and immune cells to pharmacological inhibition.

9.6 SUSCEPTIBILITY OF THE PANCREAS TO HDAC INHIBITION

Among the 18 HDACs described so far, class I and class IIa HDACs have been particularly analyzed for their function [41]. The class I HDAC enzymes are known to participate in transcriptional regulation and are dubbed as "gene repressors." Treatment of the mouse NIT1 cell line and the rat INS-1E cell line with a cocktail of cytotoxic cytokines (i.e., IL-1α, TNF-α and IFN-γ) reduced viability and insulin secretion [42,43]. Treatment of the β-cell line with broad-spectrum HDAC inhibitors (i.e., TSA, SAHA), and the class I selective MS-275 prevented the cytokine-mediated decrease in cellular ATP levels; however, caspase-3 activity was only reduced by MS-275 [43]. Treatment with MS-275, but not SAHA, restored glucose-stimulated insulin secretion, thus indicating the involvement of MS-275-sensitive class I HDACs in insulinoma cell biology [43] (Table 9.1). Although TSA could decrease nitrite production, and caspase-3 activation and restore cellular ATP levels in the cytokine-exposed INS-1E insulinoma cell line, its ability to restore glucose-stimulated insulin secretion remains unknown [43]. The insulinoma cell line expressed all 11 *Hdac* genes, as assessed by quantitative reverse transcriptase polymerase chain reaction (qRT-PCR). In response to in vitro treatment with cytotoxic cytokines, the levels of *Hdac1, 2, 3, 6,* and *8* transcripts were selectively decreased [42]. Although *HDAC1, 2, 6,* and *11* transcripts were detected in two isolated human islet preparations, in vitro cytokine treatment reduced their levels to varying degrees [42]. Importantly, cytokine treatment reduced the level of *HDAC11* in normal human islets; however, *Hdac11* expression was enhanced in insulinoma cells treated with the same cocktail of cytokines [42]. This suggests that HDACs can be differentially expressed in normal versus transformed cells and possibly subject to differential regulation by HDAC inhibitors. It is well established that tumor cells are more vulnerable to HDAC inhibitor (i.e., TSA)-induced cell death than untransformed cells [33–37,44]. Additional studies are required to characterize the expression of HDACs in normal islets and their responsiveness to HDAC inhibitors. Nevertheless, these data indicate that cytokine exposure leads to transcriptional silencing of selected HDAC members in transformed and normal endocrine tissues. Treatment of insulinoma cells with cytokines resulted in decreased cell viability and diminished insulin release in response to a high glucose challenge in vitro [43]. Transfection with a *Hdac3*-specific small-interfering RNA restored the insulin release function in cytokine-treated rat INS-1E insulinoma cells, suggesting a role for *Hdac3* in β-cell function [43]. Paradoxically, cytokine treatment alone was able to decrease *Hdac3* transcription in the same cell line [42]. Thus a role for HDAC3, or any other HDAC, in untransformed β-cell biology remains unclear. As indicated above, primary islets may behave differently from transformed insulinoma cells, hence it is important that data generated using insulinoma cell lines in vitro be replicated in primary islets in vivo. This knowledge is essential for the translation of these findings into clinical practice.

Research from our own laboratory demonstrated that the administration of TSA during the prediabetic stage provided robust protection against T1D in female NOD mice, but was not potent at reversing overt diabetes [45] (Table 9.1). Interestingly, treatment with the depsipeptide, FR901228, reversed overt diabetes in a small number (three out of four) of NOD mice [44]. This finding remains to be validated in larger numbers and the underlying mechanisms need to be elucidated. In contrast, administration of the hydroxamate, SAHA, which is less potent than TSA [38], provided modest protection against T1D in NOD mice in one study [48] but not in another study [49]. The dipeptidyl peptidase-IV inhibitor, MK-626, was not efficient at reversing T1D in NOD mice by itself or in combination with SAHA [49] despite reduced insulitis (Table 9.1). Givinostat (ITF2357) inhibited the development of spontaneously occurring T1D in NOD mice

TABLE 9.1 Effects of Epigenetic Drugs on T1D in Preclinical Models

Inhibitors	Class	Selectivity	Effects
Trichostatin A	Hydroxamate [38,39]	HDAC class I, IIa [39], and IV[a]	Reduced T1D onset [45–47] Minimal effects in overt T1D [45] Restored normoglycemia [45–47] Increased histone H3 acetylation in the spleen and pancreas [45] Decreased inflammatory response in islets [45] Increased frequency of $CD4^+CD62L^+$ cell [45] No effect on $CD4^+FoxP3^+$ T_{reg} cell population [45] Increased neogeneration of islets [45] Decreased selective HDACs[a] Decreased macrophage-specific *Cel* and *Cela3b* expression [46] Reduced cytokine-induced caspase-3 activation and nitrite production, and restored cellular ATP level in the rat INS-1E cell line. Did not affect insulin secretion and its effect on islets remains unknown [43]
Vorinostat (SAHA)	Hydroxamate [38,39]	HDAC class I and IIa [39]	Decreased cytokine-induced caspase-3 activation and restored cellular ATP without influencing insulin secretion in insulinoma cells [43] Reversed T1D by inhibiting lysine deacetylase but not histone deacetylase activity [48] No effect on the incidence of T1D, but in combination with the dipeptidyl peptidase-IV inhibitor, MK-626, reduced insulitis [49]
Givinostat (ITF2357)	Hydroxamate [38,39]	Pan-lysine deacetylases [48]	Inhibited T1D without H4 acetylation in the spleen, pancreas, or pancreatic lymph nodes [48] Reduced Kilham rat virus induced T1D, islet infiltration, and IL-12/IL-23 expression in rats [50]
FR901228	Depsipeptide [38,39]	HDAC class I [38,39]	Reversed diabetes development in a small number of NOD mice, although the underlying mechanisms are unknown [44]
MS-275	Benzamide	HDAC class I [38,39]	Prevented cytokine-induced caspase-3 activation, nitrite production, and cell death, and restored insulin secretion in insulinoma cells but its impact on T1D is unknown [43]

[a]*Unpublished data.*

without inducing H4 acetylation in the spleen, pancreas, or pancreatic lymph nodes [48]. Givinostat was also efficient at blocking Kilham rat virus-induced T1D in rats and demonstrated reduced islet infiltration and IL-12 and IL-23 expression [50]. TSA treatment protected against T1D, reduced islet inflammation [45], and abrogated islet infiltration with $CD4^+$ T cells, $CD8^+$ T cells, $Ly6C^+$ macrophages, and $Ly6G^+$ neutrophils (unpublished data). Whereas overtly diabetic mice contained large inflamed islets, TSA treatment resulted in numerous newly formed smaller islets without inflammatory-cell infiltration, which is indicative of islet regeneration [45]. It remains to be determined whether TSA administration can also produce selective transcriptional regulation of *Hdac* genes in the pancreas and promote islet neogeneration. In the NOD mouse model of T1D, adoptive transfer of preactivated T cells resulted in diabetes via macrophage mediation [51]; however, the mechanisms involved in the final act of β-cell destruction have not been fully elucidated. Whereas neutralization of IFN-γ protected mice from diabetes [52], the role of IL-17 remains contentious [53,54]. Although IL-1α is thought to be cytotoxic to β-cells, a clinical trial using IL-1α antagnoist failed to provide protection against T1D [55]. It is unclear whether soluble mediators derived from diabetogenic T cells (i.e., IFN-γ and IL-17) are directly responsible for β-cell destruction in T1D patients. Importantly, it remains to be delineated whether epigenetic regulation can impact β-cell destruction in T1D patients.

Exposure of mouse insulinoma 6 (MIN6) cells to high glucose concentrations led to hyperacetylation of histone H4 at the insulin and glucose transporter 2 gene promoters [56]. Treatment with TSA and sodium butyrate increased histone H4 acetylation and enhanced insulin gene transcription in the absence of high glucose concentrations [56], thus suggesting a role for epigenetic regulations in insulin gene expression. TSA treatment ameliorated T1D and induced histone H3 acetylation in the pancreas and spleen after 2 weeks of treatment [45]. Other genes within the spleen and pancreas were also differentially regulated. These data are consistent with the notion that TSA can act as a powerful HDAC inhibitor and lead to altered gene expression and amelioration of T1D. It is possible that TSA may induce hyperacetylation of nonhistone proteins, as only 61 of 3600 acetylation sites were detected on histones [57]. The transcription factors E2F, p53, STAT3, GATA3, c-MYC, E2F/Rb, NF-kB, SHP, GATA 1/3, HIF-1α, and Smad7 can also be acetylated [58]. Thus TSA can influence gene expression primarily via acetylation of histones although acetylation of transcription factors cannot be excluded.

9.7 SELECTIVE ROLES OF HDACs IN AUTOIMMUNE DIABETES PATHOGENESIS

The first-line HDAC inhibitors (i.e., TSA and SAHA) were originally thought to be nonselective and hence were described as pan-HDAC inhibitors [34]. It was not clear whether HDACs performed redundant or nonoverlapping functions; however, recent work based on genetic ablation in mice has revealed that HDACs are not redundant and that each member has a specific function during embryonic and postnatal life. Interestingly, insulin gene expression was decreased in the MIN6 cell line at low glucose concentrations through recruitment of HDAC1 and HDAC2 to the insulin promoter by the pancreatic duodenal homeobox-1 (*Pdx1*) transcription factor, which is required for β-cell development [59]. Treatment of mice with the HDAC inhibitor, TSA, may lead to the formation of neoislets (Fig. 9.1) [45]; however, this may not be due to the relief of *Pdx1* repression by TSA-mediated blockade of HDAC1 and 2, as the mRNA levels of *HDAC1* and *HDAC2* were not repressed in the pancreas' of TSA-treated mice (unpublished data). This inconsistency could be due to the inherent differences between normal and transformed insulin-producing β-cells, as discussed previously. In vitro treatment with TSA promoted the differentiation of α, β, δ and PP (pancreatic peptide)-cell lineages during pancreatic development [60]. It was also found that TSA could promote the differentiation of bone marrow cells into insulin-producing cells [61]. Whereas genetic ablation of *Hdac3* is embryonically lethal [62], HDAC4 inhibits cell cycle progression and is implicated in neuronal protection [63]. In addition, *Glut4* transcription is inhibited by HDAC1, HDAC4, and HDAC5 [64]. *Hdac7* deletion is embryonically lethal; however, *Hdac6* and *Hdac9* augment the suppressive function of T_{reg} (T regulatory) cells [65]. *Hdac9* deficiency affords colitis resistance and reduces autoantibody production in MRL/lpr (Murphy Roths Large/lymphoproliferation) mice [66]. In contrast, *Hdac5* deletion weakens the suppressive function and de novo induction of T_{reg} cells [67]. Taken together, these data indicate that different HDACs may be involved in a variety of cellular functions.

Although the roles of class I, IIa, IIb, and IV HDACs have not been studied thoroughly under diabetic conditions, preliminary data indicate that T1D prevention by TSA treatment in NOD mice is accompanied by selective transcriptional repression of a few class I and class IIa *Hdac* genes in the spleen with no concomitant effect on class IIb or class IV genes. However, the modulation of *Hdac* genes by TSA treatment in the pancreas of NOD mice remains unknown. This knowledge will be helpful in determining whether inhibitors with selectivity to certain HDACs will be more suited for T1D treatment than the isoform-specific HDAC inhibitors proposed.

9.8 HDAC INHIBITION LEADS TO THE AMELIORATION OF T1D

HDAC inhibitors have mostly been tested on selected tumor cell lines in vitro [35,68], although they have shown efficacy in some animal models of disease. Treatment with TSA has been shown to ameliorate autoimmune lupus [69,70], experimental colitis [66,71], airway hyperreactivity [72], and experimental autoimmune encephalomyelitis [73,74]. The prediabetic female NOD mouse has also been shown to be amenable to manipulation by TSA [45,46]. Weekly injections of TSA between 18 and 24 weeks of age (prediabetic stage) substantially prevented the onset of diabetes in this mouse model and was accompanied by reduced cellular infiltration of the pancreas, normoglycemia, and normalized glucose-induced insulin release from β-cells [45] (Table 9.1). When protection was fully stabilized (28–36 weeks of age), histone H3 acetylation in the pancreas and spleen was minimal, as detected by western blotting [45], which is consistent with the reversible nature of histone acetylation induced by HDAC inhibitors [33–35]; however, when the mice were analyzed 2 weeks after the last injection of TSA, detectable levels of H3 histone acetylation were still evident in the spleen and pancreas [45], indicating a possible role for histone acetylation in diabetes intervention. Importantly, administering TSA during the transition from peri-insulitis to full-blown diabetes prevented invasive cellular infiltration into the pancreas (Fig. 9.1). Although this period corresponds to check point two, which is thought to include a second wave of immune responses culminating in full-blown diabetes [75], some mice became diabetic despite continuous drug administration [45] (Table 9.1). This suggests that inhibition of the HDAC enzyme activity by TSA does not occur uniformly in all mice for unknown reasons. In addition, initiation of the drug treatment in overtly diabetic mice failed to reverse the disease [45], suggesting that inhibition of the functional activity of HDACs must occur during the early stage of the disease and not at the point of no return.

The beneficial effects of TSA treatment observed in prediabetic mice could be due to the cumulative effects of many events. In the advanced stage of the disease there is simply not enough time for these events to be completely reversed by the drug. This implies that disease intervention by HDAC inhibition is more likely to produce a desirable clinical outcome in newly diagnosed T1D patients than in those with overt diabetes. This is probably because more than 40% of the islet mass is already lost at the time of diabetes diagnosis [76], and any intervention at this stage may not sufficiently compensate for the loss of islets or block the ongoing inflammatory processes. It is possible that adjunct therapies or more vigorous treatment with TSA may provide reasonable benefits in advanced disease, although further work is necessary to assess these possibilities.

9.9 HISTONE ACETYLATION IMPINGES ON T LYMPHOCYTES

The most potent and best-characterized HDAC inhibitor, TSA, has been shown to repress several immune-response genes in T lymphocytes derived from mouse and human peripheral blood T lymphocytes in vitro [47,77,78]. Most investigations strive to understand the impact of dysregulated immune regulation by T_{reg} cells on the pathogenesis of diseases (i.e., T1D). In animal models, TSA has been reported to upregulate the expression of the *Foxp3* gene, which is required for T_{reg} cell development, as well as the number and function of FoxP3-expressing T_{reg} cells, resulting in a reduction of inflammatory bowel disease and augmentation of rapamycin-induced allograft tolerance in short-term observations [66]. However, TSA treatment failed to increase the frequency of the $CD4^+Foxp3^+$ T_{reg} cell population in NOD mice [45] (Table 9.1). Interestingly, treatment with the HDAC class I-specific inhibitor, entinostat, downregulated *Foxp3* transcription and suppressed the function of T_{reg} cells while augmenting the antitumor response [79]. The frequency of $CD4^+CD62L^+$ cells, but not $CD4^+Foxp3^+$ T_{reg} cells, was enhanced in the spleens of NOD mice that were protected from T1D when analyzed at a much later time point (32–36 weeks of age) [45] (Table 9.1). This indicates that one long-term effect of TSA may be the replenishment of $CD4^+CD62L^+$ T_{reg} cells, which are proposed to protect against T1D [80], rather than $CD4^+Foxp3^+$ T_{reg} cells [45].

The role of $CD4^+CD25^+$ T_{reg} cells in T1D patients has been intensely studied because of the implications for disease treatment by T_{reg} cell therapy. One study identified a lower frequency of $CD4^+CD25^+$ T_{reg} cells in T1D patients [81]; however, this observation was not replicated in long-standing T1D patients [82]. Another study demonstrated that the $CD4^+CD25^+$ T_{reg} cells did not differ numerically or in their suppressive function [83], although only a depressed suppressive function of T_{reg} cells was noted in T1D patients [84]. Thus the status of T_{reg} cells in long-standing T1D patients remains highly contentious. Additional complexities inherent in T_{reg} cell studies include the lack of clear phenotypic markers to distinguish between T_{reg} cells and activated conventional T-cells, the conversion of $CD25^-$ $T_{effector}$ cells into $CD25^+Foxp3^+$ T_{reg} cells [85], the conversion of T_{reg} cells into IL-17-producing Th17 (T helper type 17) cells [86], and Th17 transdifferentiation into T_{reg} cells [87]. Understanding the role of histone hyperacetylation in processes associated with the plasticity of $T_{effector}$ and T_{reg} cells requires further investigation.

As T lymphocytes have been implicated in T1D, they are the prime targets for disease manipulation. Early studies suggested that abnormalities in T-cell function were implicated in T1D, and it has recently been established that T-cells from T1D patients are defective in undergoing normal programmed cell death, which could account for the perpetuation of pathogenic T-cells [82]. Several attempts have been made to restore T-cell tolerance and minimize the adverse effects of autoreactive T-cells. These include the oral administration of insulin [88], the injection of glutamic acid decarboxylase in an alum adjuvant [89], and treatment with nonactivating anti-CD3 antibody [90]. These approaches have not provided adequate protection against T1D in patients, indicating that insulin and glutamic acid decarboxylase are less likely to serve as the molecular targets for disease manipulation. Instead of nonspecific anti-CD3 antibody treatment, antigen-specific manipulation could provide robust protection against T1D without causing nonspecific immunosuppression; however, the identity of self-antigens involved in T1D remains elusive. Chromogranin A, which is widely expressed in endocrine tissues, neuroendocrine tissues, the central nervous system, and the colonic mucosa, was unexpectedly identified as the target antigen in a highly diabetogenic mouse $CD4^+$ T cell clone, indicating that autoreactive T cells recognize unusually presented self-peptides [91]; however, the underlying mechanisms through which chromogranin A-reactive T cells manifest autoimmune diabetes remain unknown. Determining the specificity of the diabetogenic T cells in T1D patients will pave the way for antigen-specific immunotherapy; however, in the absence of this critical information, other strategies may be utilized to quell autoimmune diabetes.

TSA can simultaneously up- and downregulate a small number of genes (~20%) in cell lines in vitro [92]. Genes encoding immune response-related molecules (i.e., IL-2, IFN-γ, CD25, and CD154) were modulated in human peripheral blood T cells and in mouse splenic T cells following TSA treatment in vitro [77,78]. TSA treatment also modulated the expression of a number of adhesion and costimulatory molecules in both T cells and non-T cells [78]. Thus the observed modulatory effects of TSA could be due to the cumulative effects of the drug on both the innate and adaptive immune system.

The molecular mechanisms that underpin HDAC inhibitor-mediated protection against T1D in vivo are beginning to be understood. One study showed that the adoptive transfer of IL-17-producing Th17 cells into histocompatible immunodeficient NOD.*scid* (severe combined immunodeficiency) mice resulted in diabetes, and that the transferred Th17 cells were converted into IFN-γ-producing Th1-like cells [93]. Histone hyperacetylation has also been shown to repress gene expression levels of the Th17-specific *Il17a* gene, its transcription factor *Rorgt* [45], and *Il6* and *Il23*, which are involved in Th17 cell generation (unpublished data). Taken together, these data suggest that the conversion of Th17 cells to a Th1 cell type may accompany diabetogenesis. Although the treatment of NOD mice with TSA did not affect the ability of T cells to differentiate into Th1 and Th17 cells in vitro, it abrogated the ability of polarized Th1 and Th17 cells to manifest diabetes upon adoptive transfer into immunodeficient recipients (unpublished data). Transcriptome profiles from diabetes-causing Th1 and Th17 cells and from those treated with TSA and deprived of diabetogenicity will provide significant insights into the molecular basis of T1D pathogenesis.

Although the participation of HDACs in the transcriptional regulation of both pro- and antiinflammatory cytokines has been described in a variety of cellular models, information on the role of HDACs on cytokine expression in the immune system is limited [37,78,92,94,95]. HDAC1 is involved in the transcriptional control of *IL1, IL2, IL4, IL5*, and *IL12*, while HDAC2 is implicated in *IFNG* and *GMCSF* repression. HDAC3 controls the expression of the transcription factors *STAT1, STAT3, GATA1, GATA2, GATA3*, and *NFKB*, whereas HDAC5 regulates *IL8* transcription. It is noteworthy that more than one HDAC is involved in the regulation of *IL2, IL4, IFNG, STAT3*, and *GATA1*. However, a direct link between altered expression of HDACs and T1D manifestation has not been established. Histone hyperacetylation induced by TSA treatment was shown to increase the transcription of Th1-specific *Ifng* and its transcription factor *Tbet/Tbx21* [45]. This was similar to the upregulation of *Ifng* seen in the spleen of NOD mice treated with the disease-protective glutamic acid decarboxylase peptide [96]. The upregulation of *Ifng* observed under nondiabetic conditions [45,96] appears to contradict the ability of IFN-γ-producing Th1 cells to transfer diabetes [93]. It is possible that IFN-γ per se may not bestow protection against T1D or cause the disease. Unknown characteristics shared by both Th1 and Th17 cells may contribute to diabetogenesis. Nevertheless, these results suggest that histone hyperacetylation can impact the expression of pathogenic genes regardless of the functional status of T cells.

Previously activated T lymphocytes undergo apoptosis in response to further T-cell receptor ligation, a phenomenon termed activation-induced cell death [97]. This has implications for the control of immune responses, including those directed against autoantigens, and therefore represents an important peripheral tolerance mechanism. Polyclonally activated splenic $CD4^+$ and $CD8^+$ T cells derived from NOD mice displayed a reduced ability to undergo activation-induced cell death through the T cell receptor in vitro [98]. T lymphocytes derived from long-standing T1D patients were also refractory to antigen receptor-mediated apoptosis [82]; however, they did undergo apoptosis in response to stimulation through the Fas receptor. This is consistent with the notion that although Fas receptor expression is linked to T-cell receptor stimulation, it may be coupled to proximal apoptotic pathways that are distinct from that of the T-cell receptor complex. Further work is necessary to delineate whether self-tolerance can be restored through epigenetic modulation in T1D patients, which could lead to the diminished frequency of autoreactive T lymphocytes.

9.10 EPIGENETIC REGULATION OF THE INNATE IMMUNE SYSTEM FOR DIABETES TREATMENT

In addition to abnormalities in the adaptive immune system, the innate immune system, which provides the first-line defense against infections and injury, may also be impaired in T1D. NOD mice have been shown to harbor defects in plasmacytoid dendritic cells, which are involved in tolerance induction; natural killer (NK) cells, which are important for eliminating target cells; and NK-T cells, which recognize the glycolipid antigens presented by the nonmajor histocompatibility complex encoded by the CD1d molecule [99–101]. Numeric reduction in NK-T cells was also reported in long-standing T1D patients [102]. The complex interactions between cells of the innate and adaptive immune systems that impinge upon insulin-producing β-cells are only partially understood. Manipulations that result in the abrogation of autoimmune T1D are likely to intersect several molecular pathways that are engaged by both the innate and adaptive immune system. Epigenetic regulation of the innate immune system in T1D remains to be elucidated.

Genomic tools are useful for analyzing the correlation between genome-wide alterations in gene expression and phenotypic changes. Genome-wide expression analysis is a nonhypothesis-driven approach that is ideal for simultaneously analyzing the expression levels of thousands of genes without bias. The spleen contains T-cells that can mediate diabetes and macrophages that act as final effectors of disease manifestation. Therefore it is an ideal organ in which to interrogate the transcriptome under pathological and healthy conditions. Gene expression analysis in untreated nondiabetic mice, overtly diabetic mice, and TSA-treated mice yielded 164 differentially expressed genes [46]. Bioinformatic analysis revealed the upregulation of a distinct set of 17 genes (encoding lipases, peptidases, and elastases) in the uninduced splenocytes of overtly diabetic mice. Further analysis revealed that these proinflammatory genes were colocalized and interacted with each other. Treatment with TSA significantly suppressed the splenic expression of four of the nine genes tested by qRT-PCR: *Cela3b* (chymotrypsin-like elastase family, member 3B), *Cpb1* (carboxypeptidase B1), *Pnlip* (pancreatic lipase), and *Cel* (carboxyl ester lipase). Importantly, *Cela3b*, *Pnlip*, *Cel*, and *Pnliprp-1* (pancreatic lipase-related protein 1) were also repressed in the pancreas following TSA treatment [46]. Previous microarray analyses had already documented the overexpression of the pancreatic exocrine gene, *Cel*, in uninduced splenocytes of diabetes-prone NOD mice in comparison to diabetes-resistant mice [103,104]. Although the expression patterns of these genes in various tissues are not fully understood, *Cel*, for instance, is expressed in macrophages and associated with atherosclerosis [105]. Since monocytes are predominantly found in the inflammatory infiltrate in islets [106], the proinflammatory genes identified in the above study are likely to be expressed by monocytes localized in the pancreas under diabetic condition. Indeed, data indicate that *Cela3b*, which was downregulated by TSA treatment, is highly expressed in the spleen and pancreas of diabetic mice [46]. GeneMANIA predicts that *Cela3b* acts as a hub gene and interacts with functionally related proinflammatory genes [46]. These results demonstrate the power of genomics approaches for gleaning unprecedented insights into pathological mechanisms.

Previous studies indicated that TSA treatment downregulated the gene encoding the migration inhibition factor (*Mif*) in cell lines, primary cells, and in vivo models, resulting in increased antitumor activity [107]. Although the steady-state level of *Mif* was higher in splenocytes from drug-treated NOD mice, the ability to express *Mif* in response to T-cell receptor stimulation was compromised [46]. MIF was the first cytokine to be described [108], is implicated in a number of pathological situations (i.e., allograft rejection), and is phylogenically conserved [109]. Although TSA induced the deacetylation of histones H3 and H4 at the *MIF* promoter region, it did not affect the chromatin accessibility; however, it was associated with the impaired recruitment of RNA polymerase II, Sp1, and CREB transcription factors, which are necessary for *MIF* transcription [107]. GeneMANIA predicts that *Mif* interacts with proapoptotic genes such as *Bnipl* and *Bcl2l11* [46]. Thus chromatin remodeling by TSA treatment is associated with the repression of genes such as *Cela3b* and *Mif*, which are possibly involved in diabetogenesis. Further studies are required to elucidate whether MIF can cause β-cell demise or influence other aspects of T1D pathogenesis either directly or indirectly through interaction with proapoptotic molecules.

9.11 HDAC INHIBITION INCREASES GENE EXPRESSION

Early studies indicated that the treatment of tumor cell lines with TSA affected the expression of a small number (~20%) of genes, thus indicating the selective action of the drug [92]. Notably, drug treatments can simultaneously decrease as well as increase the expression of genes. Gene expression analysis in splenocytes and pancreas samples from TSA-treated uninduced NOD mice revealed the upregulation of *Ermap* [46], which codes for erythroid adhesion/receptor transmembrane protein in reticulocytes and circulating erythroblasts [110]. The *Lycat* gene, which is expressed in hematopoietic and endothelial progenitor cells [111], was also enhanced in the spleen and pancreas of TSA-treated mice [46]. TSA treatment also enhanced the expression of autophagy related genes (i.e., *Atg4*), and antioxidant genes (i.e., *Prdx1*, *Prdx6*, *Gpx1*, and *H2afz*, which replaces the canonical histone H2A in a subset of nucleosomes and therefore can modify gene expression) [112]. These data are consistent with the notion that protection afforded by epigenetic mechanisms (i.e., histone hyperacetylation) involve the up- and down-regulation of genes.

9.12 CONCLUSION

Epigenetic mechanisms represent a paradigm shift in our view of gene expression under diabetic conditions. The incidence of T1D in developed countries will double in children under 5 years of age by 2020 [113], and as of now there is no way of stopping this pandemic. Although genome-wide association studies have implicated several HLA and non-HLA associated genes s diabetes, it is difficult to predict how many of these genes, and what combination, could trigger diabetes. The NOD mouse faithfully recapitulates many of the pathological events of human T1D and facilitates the scrutiny of the intricate mechanisms involved in naturally occurring chronic autoimmune diabetes. While the contributions of major histocompatibility genes have been mostly reproduced in mice, no consensus has been reached on the mechanisms responsible for T1D pathology or the regulatory mechanisms that may contain this disease (Fig. 9.1). While >200 modalities are reported to "delay" T1D in NOD mice [114], none are capable of reversing overt T1D. Importantly, clinical trials based on previous findings in NOD mice have so far not translated into successful treatments for patients with T1D. This is partly due to the differences in the immune systems of mice and humans, as well as to variations in the etiology and clinical history of the disease. Data generated in our laboratory using TSA, which are reviewed herein, indicate that the epigenetic regulation of the genome could concurrently alter both the adaptive and innate immune systems to prevent diabetes in genetically susceptible female NOD mice [45–47] (Fig. 9.1). Evidence regarding the susceptibility of endocrine β-cells to epigenetic manipulation in vivo is relatively modest. Thus a holistic approach is necessary for a thorough understanding of the epigenomic landscape of autoimmune prone mice. This knowledge will pave the way for better manipulation of human T1D in the future.

9.13 FUTURE PERSPECTIVE

Although improved insulin formats help the management of T1D, micro- and macrovascular complications and end-stage renal disease resulting from previous hyperglycemic episodes (legacy effect or metabolic memory) are inevitable in T1D patients [115]. The administration of exogenous insulin is not a cure as it does not prevent the ongoing pathogenesis or replenish the lost β-cells. Although transplantation of islets from cadaveric allogeneic pancreata may provide transient benefits in T1D recipients, transplanted allogeneic β-cells are subject to the same immune destruction as the residual insulin producers (Fig. 9.1). Therefore the robust control of autoimmune responses against insulin-producing β-cells (restoration of self-tolerance) is an important aspect of T1D treatment. Due to the lack of understanding of the specificity of T lymphocytes involved in T1D, restitution of tolerance to self-antigens is not an option. Chromatin remodeling, which is known to modulate gene expression under a variety of pathological conditions including T1D, may provide an effective strategy to combat T1D via immunomodulation and possible β-cell effects. HDAC inhibitors are attractive drug candidates for treating cancer, inflammatory diseases, neuronal disorders, and autoimmune diseases. Isoform-specific HDAC inhibitors may provide precise tools for manipulating pathological conditions in which the restricted numbers of HDACs are aberrantly expressed. However, caution should be exercised when trying to implement this strategy for the treatment of polygenic diseases such as T1D; for example, HDACs play different important roles during embryonic development and in adult life. The pattern of expression of HDACs in normal tissues and their probable variations under pathological conditions have not yet been fully elucidated. Complex autoimmune disease such as T1D probably harbor multiple abnormally expressed HDACs in various tissues that require simultaneous repression and enhancement to prevent the disease. The fact that TSA can ameliorate T1D and unrelated diseases [33–35,45–47,66,68–74,94,95] suggests the involvement of overlapping and yet distinct HDACs in these diseases. Treatment of these diseases is likely to benefit from inhibitors that have selectivity to more than one HDAC isoforms, but further work must be performed to test this hypothesis.

Acknowledgments

The author acknowledges the contribution of laboratory members to the work discussed in this article. Arathi Jayaraman is acknowledged for critical reading of the manuscript.

References

[1] Opie EL. The relation of diabetes mellitus to lesions of the pancreas: hyaline degeneration of the islands of Langerhans. J Exp Med 1901;25:527–40.
[2] Gale EA. The rise of childhood type 1 diabetes in the 20th century. Diabetes 2002;51:3353–61.
[3] Redondo MJ, Eisenbarth GS. Genetic control of autoimmunity in type 1 diabetes and associated disorders. Diabetologia 2002;45:605–22.
[4] Imagawa A, Hanafusa T. Fulminant type 1 diabetes mellitus. Endocr J 2006;53:577–84.
[5] Haffner SM. Epidemiology of type 2 diabetes risk factors. Diabetes Care 1998;21(Suppl. 3):3C–6C.
[6] Naik RG, Brooks-Worrell BM, Palmer JP. Latent autoimmune diabetes in adults. J Clin Endocrinol Metab 2009;94:4635–44.
[7] Steen E, Terry BM, Rivera EJ, Cannon JL, Neely TR, Tavares R, et al. Impaired insulin and insulin-like growth factor expression and signaling mechanisms in Alzheimer's disease: is this type 3 diabetes? J Alzheimers Dis 2005;7:63–80.
[8] Baschal EE, Eisenbarth GS. Extreme genetic risk for type 1A diabetes in the post-genome era. J Autoimmun 2008;31:1–6.
[9] Hyttinen V, Kaprio J, Kinnunen L, Koskenvuo M, Tuomilehto J. Genetic liability of type 1 diabetes and the onset age among 22,650 young Finnish twin pairs: a nationwide follow-up study. Diabetes 2003;52:1052–5.
[10] EURODIAB Substudy 2 Study Group. Rapid early growth is associated with increased risk of childhood type 1 diabetes in various European populations. Diabetes Care 2002;25:1755–60.
[11] Vehik K, Hamman RF, Lezotte D, Norris JM, Klingensmith G, Bloch C, et al. Increasing incidence of type 1 diabetes in 0- to 17-year-old Colorado youth. Diabetes Care 2007;30:503–9.
[12] Bach JF. The effect of infections on susceptibility to autoimmune and allergic diseases. N Engl J Med 2003;347:911–20.
[13] Gale EA. Congenital rubella: citation virus or viral cause of type 1 diabetes? Diabetologia 2008;51:1559–66.
[14] Mäkelä M, Vaarala O, Hermann R, Salminen K, Vahlberg T, Veijola R, et al. Enteral virus infections in early childhood and an enhanced type 1 diabetes-associated antibody response to dietary insulin. J Autoimmun 2006;27:54–61.
[15] Hober D, Sauter P. Pathogenesis of type 1 diabetes mellitus: interplay between enterovirus and host. Nat Rev Endocrinol 2010;6:279–89.
[16] Bason C, Lorini R, Lunardi C, Dolcino M, Giannattasio A, d'Annunzio G, et al. In type 1 diabetes a subset of anti-coxsackievirus B4 antibodies recognize autoantigens and induce apoptosis of pancreatic beta cells. PLoS One 2013;8:e57729.
[17] Oldstone MB. Molecular mimicry, microbial infection, and autoimmune disease: evolution of the concept. Curr Top Microbiol Immunol 2005;296:1–17.
[18] Nair A, Wolter TR, Meyers AJ, Zipris D. Innate immune pathways in virus-induced autoimmune diabetes. Ann N Y Acad Sci 2008;1150:139–42.
[19] Van Speybroeck L. From epigenesis to epigenetics: the case of C. H. Waddington. Ann N Y Acad Sci 2002;981:61–81.
[20] Klutstein M, Nejman D, Greenfield R, Cedar H. DNA methylation in cancer and aging. Cancer Res 2016;76:3446–50.
[21] Du Q, Luu PL, Stirzaker C, Clark SJ. Methyl-CpG-binding domain proteins: readers of the epigenome. Epigenomics 2015;7:1051–73.

[22] Alam F, Islam MA, Gan SH, Mohamed M, Sasongko TH. DNA methylation: an epigenetic insight into type 2 diabetes mellitus. Curr Pharm Des 2016;22:4398–419.
[23] Bell CG, Teschendorff AE, Rakyan VK, Maxwell AP, Beck S, Savage DA. Genome-wide DNA methylation analysis for diabetic nephropathy in type 1 diabetes mellitus. BMC Med Genomics 2010;3:33.
[24] Miao F, Smith DD, Zhang L, Min A, Feng W, Natarajan R. Lymphocytes from patients with type 1 diabetes display a distinct profile of chromatin histone H3 lysine 9 dimethylation: an epigenetic study in diabetes. Diabetes 2008;57:3189–98.
[25] Rui J, Deng S, Lebastchi J, Clark PL, Usmani-Brown S, Herold KC. Methylation of insulin DNA in response to proinflammatory cytokines during the progression of autoimmune diabetes in NOD mice. Diabetologia 2016;59:1021–9.
[26] Felsenfeld G, Groudine M. Controlling the double helix. Nature 2003;421:448–53.
[27] Jenuwein T, Allis CD. Translating the histone code. Science 2001;293:1074–80.
[28] Brandl A, Heinzel T, Krämer OH. Histone deacetylases: salesmen and customers in the post-translational modification market. Biol Cell 2009;101:193–205.
[29] Hasan S, Hottiger MO. Histone acetyl transferases: a role in DNA repair and DNA replication. J Mol Med 2002;80:463–74.
[30] Roelfsema JH, Peters DJ. Rubinstein–Taybi syndrome: clinical and molecular overview. Expert Rev Mol Med 2007;9:1–16.
[31] Gregoretti IV, Lee YM, Goodson HV. Molecular evolution of the histone deacetylase family: functional implications of phylogenetic analysis. J Mol Biol 2004;338:17–31.
[32] Smith BC, Hallows WC, Denu JM. Mechanisms and molecular probes of sirtuins. Chem Biol 2008;15:1002–13.
[33] Minucci S, Pelicci PG. Histone deacetylase inhibitors and the promise of epigenetic (and more) treatments for cancer. Nat Rev Cancer 2006;6:38–51.
[34] Bolden JE, Peart MJ, Johnstone RW. Anticancer activities of histone deacetylase inhibitors. Nat Rev Drug Discov 2006;5:769–84.
[35] Marks PA, Richon VM, Breslow R, Rifkind RA. Histone deacetylase inhibitors as new cancer drugs. Curr Opin Oncol 2001;13:477–83.
[36] Sternson SM, Wong JC, Grozinger CM, Schreiber SL. Synthesis of 7200 small molecules based on a substructural analysis of the histone deacetylase inhibitors trichostatin and trapoxin. Org Lett 2001;3:4239–42.
[37] Takahashi I, Miyaji H, Yoshida T, Sato S, Mizukami T. Selective inhibition of IL-2 gene expression by trichostatin A, a potent inhibitor of mammalian histone deacetylase. J Antibiot 1996;49:453–7.
[38] de Ruijter AJ, van Gennip AH, Caron HN, Kemp S, van Kuilenburg AB. Histone deacetylases (HDACs): characterization of the classical HDAC family. Biochem J 2003;370:737–49.
[39] Bradner JE, West N, Grachan ML, Greenberg EF, Haggarty SJ, Warnow T, et al. Chemical phylogenetics of histone deacetylases. Nat Chem Biol 2010;6:238–43.
[40] Diana J, Simoni Y, Furio L, Beaudoin L, Agerberth B, Barrat F, et al. Crosstalk between neutrophils, B1a cells and plasmacytoid dendritic cells initiates autoimmune diabetes. Nat Med 2013;19:65–73.
[41] Parra M. Class IIa HDACs: new insights into their functions in physiology and pathology. FEBS J 2015;282:1736–44.
[42] Lundh M, Christensen DP, Rasmussen DN, Mascagni P, Dinarello CA, Billestrup N, et al. Lysine deacetylases are produced in pancreatic beta cells and are differentially regulated by proinflammatory cytokines. Diabetologia 2010;53:2569–78.
[43] Chou DH, Holson EB, Wagner FF, Tang AJ, Maglathlin RL, Lewis TA, et al. Inhibition of histone deacetylase 3 protects beta cells from cytokine-induced apoptosis. Chem Biol 2012;19:669–73.
[44] Skov S, Rieneck K, Bovin LF, Skak K, Tomra S, Michelsen BK, et al. Histone deacetylase inhibitors: a new class of immunosuppressors targeting a novel signal pathway essential for CD154 expression. Blood 2003;101:1430–8.
[45] Patel T, Patel V, Singh R, Jayaraman S. Chromatin remodeling resets the immune system to protect against autoimmune diabetes in mice. Immunol Cell Biol 2011;89:640–9.
[46] Jayaraman S, Patel A, Jayaraman A, Patel V, Holterman M, Prabhakar B. Transcriptome analysis of epigenetically modulated genome indicates signature genes in manifestation of type 1 diabetes and its prevention in NOD mice. PLoS One 2013;8:e55074.
[47] Jayaraman S. Novel methods of type 1 diabetes treatment. Discov Med 2014;17:347–55.
[48] Christensen DP, Gysemans C, Lundh M, Dahllöf MS, Noesgaard D, Schmidt SF, et al. Lysine deacetylase inhibition prevents diabetes by chromatin-independent immunoregulation and β-cell protection. Proc Natl Acad Sci U S A 2014;111:1055–9.

[49] Cabrera SM, Colvin SC, Tersey SA, Maier B, Nadler JL, Mirmira RG. Effects of combination therapy with dipeptidyl peptidase-IV and histone deacetylase inhibitors in the non-obese diabetic mouse model of type 1 diabetes. Clin Exp Immunol 2013;172:375–82.

[50] Hara N, Alkanani AK, Dinarello CA, Zipris D. Histone deacetylase inhibitor suppresses virus-induced proinflammatory responses and type 1 diabetes. J Mol Med 2014;92:93–102.

[51] Calderon B, Suri A, Unanue ER. In $CD4^+$ T-cell-induced diabetes, macrophages are the final effector cells that mediate islet beta-cell killing: studies from an acute model. Am J Pathol 2006;169:2137–47.

[52] Debray-Sachs M, Carnaud C, Boitard C, Cohen H, Gresser I, Bedossa P, et al. Prevention of diabetes in NOD mice treated with antibody to murine IFN gamma. J Autoimmun 1991;4:237–48.

[53] Bellemore SM, Nikoopour E, Schwartz JA, Krougly O, Lee-Chan E, Singh B. Preventative role of interleukin-17 producing regulatory T helper type 17 (Treg 17) cells in type 1 diabetes in non-obese diabetic mice. Clin Exp Immunol 2015;182:261–9.

[54] Emamaullee JA, Davis J, Merani S, Toso C, Elliott JF, Thiesen A, et al. Inhibition of Th17 cells regulates autoimmune diabetes in NOD mice. Diabetes 2009;58:1302–11.

[55] Cabrera SM, Wang X, Chen YG, Jia S, Kaldunski ML, Greenbaum CJ, et al. Interleukin-1 antagonism moderates the inflammatory state associated with type 1 diabetes during clinical trials conducted at disease onset. Eur J Immunol 2016;46:1030–46.

[56] Mosley AL, Ozcan S. Glucose regulates insulin gene transcription by hyperacetylation of histone H4. J Biol Chem 2003;278:19660–6.

[57] Choudhary C, Kumar C, Gnad F, Nielsen ML, Rehman M, Walther TC, et al. Lysine acetylation targets protein complexes and co-regulates major cellular functions. Science 2009;325:834–40.

[58] Glozak MA, Sengupta N, Zhang X, Seto E. Acetylation and deacetylation of non-histone proteins. Gene 2005;363:15–23.

[59] Mosley AL, Ozcan S. The pancreatic duodenal homeobox-1 protein (Pdx-1) interacts with histone deacetylases Hdac-1 and Hdac-2 on low levels of glucose. J Biol Chem 2004;279:54241–7.

[60] Haumaitre C, Lenoir O, Scharfmann R. Histone deacetylase inhibitors modify pancreatic cell fate determination and amplify endocrine progenitors. Mol Cell Biol 2008;28:6373–83.

[61] Thatava T, Ma B, Rohde M, Mayer H. Chromatin-remodeling factors allow differentiation of bone marrow cells into insulin-producing cells. Stem Cells 2006;24:2858–67.

[62] Montgomery RL, Potthoff MJ, Haberland M, Qi X, Matsuzaki S, Humphries KM, et al. Maintenance of cardiac energy metabolism by histone deacetylase 3 in mice. J Clin Invest 2008;118:3588–97.

[63] Raychaudhuri N, Raychaudhuri S, Thamotharan M, Devaskar SU. Histone code modifications repress glucose transporter 4 expression in the intrauterine growth-restricted offspring. J Biol Chem 2008;283:13611–26.

[64] McGee SL, van Denderen BJ, Howlett KF, Mollica J, Schertzer JD, Kemp BE, et al. AMP-activated protein kinase regulates GLUT4 transcription by phosphorylating histone deacetylase 5. Diabetes 2008;57:860–7.

[65] Beier UH, Wang L, Han R, Akimova T, Liu Y, Hancock WW. Histone deacetylases 6 and 9 and sirtuin-1 control $Foxp3^+$ regulatory T cell function through shared and isoform-specific mechanisms. Sci Signal 2012;5:45.

[66] Tao R, de Zoeten EF, Ozkaynak E, Chen C, Wang L, Porrett PM, et al. Deacetylase inhibition promotes the generation and function of regulatory T cells. Nat Med 2007;13:1299–307.

[67] Xiao H, Jiao J, Wang L, O'Brien S, Newick K, Wang LC, et al. HDAC5 controls the functions of Foxp3(+) T-regulatory and CD8(+) T cells. Int J Cancer 2016;138:2477–86.

[68] Peart MJ, Smyth GK, van Laar RK, Bowtell DD, Richon VM, Marks PA, et al. Identification and functional significance of genes regulated by structurally different histone deacetylase inhibitors. Proc Natl Acad Sci U S A 2005;102:3697–702.

[69] Mishra N, Reilly CM, Brown DR, Ruiz P, Gilkeson GS. Histone deacetylase inhibitors modulate renal disease in the MRL-lpr/lpr mouse. J Clin Invest 2003;11:539–52.

[70] Reilly CM, Thomas M, Gogal R Jr, Olgun S, Santo A, Sodhi R, et al. The histone deacetylase inhibitor trichostatin A upregulates regulatory T cells and modulates autoimmunity in NZB/W F1 mice. J Autoimmun 2008;31:123–30.

[71] Glauben R, Batra A, Fedke I, Zeitz M, Lehr HA, Leoni F, et al. Histone hyperacetylation is associated with amelioration of experimental colitis in mice. J Immunol 2006;176:5015–22.

[72] Choi JH, Oh SW, Kang MS, Kwon HJ, Oh GT, Kim DY. Trichostatin A attenuates airway inflammation in mouse asthma model. Clin Exp Allergy 2005;35:89–96.

[73] Camelo S, Iglesias AH, Hwang D, Due B, Ryu H, Smith K, et al. Transcriptional therapy with the histone deacetylase inhibitor trichostatin A ameliorates experimental autoimmune encephalomyelitis. J Neuroimmunol 2005;164:10–21.

[74] Jayaraman A, Soni A, Prabhakar B, Holterman M, Jayaraman S. The epigenetic drug trichostatin A ameliorates experimental autoimmune encephalomyelitis via T cell tolerance induction and impaired influx of T cells into the spinal cord. Neurobiol Dis 2017;108:1–12.
[75] André I, Gonzalez A, Wang B, Katz J, Benoist C, Mathis D. Checkpoints in the progression of autoimmune disease: lessons from diabetes models. Proc Natl Acad Sci U S A 1996;93:2260–3.
[76] Atkinson MA, Gianani R. The pancreas in human type 1 diabetes: providing new answers to age-old questions. Curr Opin Endocrinol Diabetes Obes 2009;16:279–85.
[77] Mishra N, Brown DR, Olorenshaw IM, Kammer GM. Trichostatin A reverses skewed expression of CD154, interleukin-10, and interferon-gamma gene and protein expression in lupus T cells. Proc Natl Acad Sci U S A 2001;98:2628–33.
[78] Moreira JM, Scheipers P, Sørensen P. The histone deacetylase inhibitor trichostatin A modulates $CD4^+$ T cell responses. BMC Cancer 2003;3:30.
[79] Shen L, Ciesielski M, Ramakrishnan S, Miles KM, Ellis L, Sotomayor P, et al. Class I histone deacetylase inhibitor entinostat suppresses regulatory T cells and enhances immunotherapies in renal and prostate cancer models. PLoS One 2012;7:e30815.
[80] You S, Slehoffer G, Barriot S, Bach JF, Chatenoud L. Unique role of $CD4^+CD62L^+$ regulatory T cells in the control of autoimmune diabetes in T cell receptor transgenic mice. Proc Natl Acad Sci U S A 2004;101:14580–5.
[81] Kukreja A, Cost G, Marker J, Zhang C, Sun Z, Lin-Su K, et al. Multiple immuno-regulatory defects in type-1 diabetes. J Clin Invest 2002;109:131–40.
[82] Jayaraman S, Balasingh S, Exposito J, Ponte G, Baidal D, Cure P, et al. Patients with Type 1 diabetes display selective defect in antigen receptor mediated T cell apoptosis. Open J Biol Biochem 2012;1:2.
[83] Putnam AL, Vendrame F, Dotta F, Gottlieb PA. $CD4^+$CD25high regulatory T cells in human autoimmune diabetes. J Autoimmun 2005;24:55–62.
[84] Brusko TM, Wasserfall CH, Clare-Salzler MJ, Schatz DA, Atkinson MA. Functional defects and the influence of age on the frequency of $CD4^+$ $CD25^+$ T- cells in type 1 diabetes. Diabetes 2005;54:1407–14.
[85] Curotto de Lafaille MA, Lino AC, Kutchukhidze N, Lafaille JJ. CD25- T cells generate $CD25^+Foxp3^+$ regulatory T cells by peripheral expansion. J Immunol 2004;173:7259–68.
[86] Koenen HJ, Smeets RL, Vink PM, van Rijssen E, Boots AM, Joosten I. Human CD25highFoxp3pos regulatory T cells differentiate into IL-17-producing cells. Blood 2008;112:2340–52.
[87] Gagliani N, Amezcua Vesely MC, Iseppon A, Brockmann L, Xu H, Palm NW, et al. Th17 cells transdifferentiate into regulatory T cells during resolution of inflammation. Nature 2015;523:221–5.
[88] Skyler JS, Krischer JP, Wolfsdorf J, Cowie C, Palmer JP, Greenbaum C, et al. Effects of oral insulin in relatives of patients with type 1 diabetes: The Diabetes Prevention Trial-Type 1. Diabetes Care 2005;28:1068–76.
[89] Wherrett DK, Bundy B, Becker DJ, DiMeglio LA, Gitelman SE, Goland R, et al. Antigen-based therapy with glutamic acid decarboxylase (GAD) vaccine in patients with recent-onset type 1 diabetes: a randomised double-blind trial. Lancet 2011;378: 319–27.
[90] Phillips B, Trucco M, Giannoukakis N. Current state of type 1 diabetes immunotherapy: incremental advances, huge leaps, or more of the same? Clin Dev Immunol 2011;2011:432016.
[91] Delong T, Baker RL, He J, Barbour G, Bradley B, Haskins K. Diabetogenic T-cell clones recognize an altered peptide of chromogranin A. Diabetes 2012;61:3239–46.
[92] Van Lint C, Emiliani S, Verdin E. The expression of a small fraction of cellular genes is changed in response to histone hyperacetylation. Gene Expr 1996;5:245–53.
[93] Bending D, De la Peña H, Veldhoen M, Phillips JM, Uyttenhove C, Stockinger B, et al. Highly purified Th17 cells from BDC2.5NOD mice convert into Th1-like cells in NOD/SCID recipient mice. J Clin Invest 2009;119:565–72.
[94] Villagra A, Sotomayor EM, Seto E. Histone deacetylases and the immunological network: implications in cancer and inflammation. Oncogene 2010;29:157–73.
[95] López JE, Sullivan ED, Fierke CA. Metal-dependent deacetylases: cancer and epigenetic regulators. ACS Chem Biol 2016;11:706–16.
[96] Rajput A, Singh B. Gene expression profiling in type 1 diabetes prone NOD mice immunized with a disease protective autoantigenic peptide. J Autoimmun 2004;23:311–21.
[97] Brenner D, Krammer PH, Arnold R. Concepts of activated T cell death. Crit Rev Oncol Hematol 2008;66:52–64.
[98] Yang W, Hussain S, Mi QS, Santamaria P, Delovitch TL. Perturbed homeostasis of peripheral T cells elicits decreased susceptibility to anti-CD3-induced apoptosis in prediabetic nonobese diabetic mice. J Immunol 2004;173:4407–16.
[99] Nikolic T, Welzen-Coppens JM, Leenen PJ, Drexhage HA, Versnel MA. Plasmacytoid dendritic cells in autoimmune diabetes-potential tools for immunotherapy. Immunobiology 2009;214:791–9.

[100] Baxter AG, Smyth MJ. The role of NK cells in autoimmune disease. Autoimmunity 2002;35:1–14.
[101] Fletcher MT, Baxter AG. Clinical application of NKT cell biology in type I (autoimmune) diabetes mellitus. Immunol Cell Biol 2009;87:315–23.
[102] Meyers AJ, Shah RR, Gottlieb PA, Zipris D. Altered toll-like receptor signaling pathways in human type 1 diabetes. J Mol Med 2010;88:1221–31.
[103] Eckenrode SE, Ruan Q, Yang P, Zheng W, McIndoe RA, She JX. Gene expression profiles define a key checkpoint for type 1 diabetes in NOD mice. Diabetes 2004;53:366–75.
[104] Irie J, Reck B, Wu Y, Wicker LS, Howlett S, Rainbow D, et al. Genome-wide microarray expression analysis of $CD4^+$ T Cells from nonobese diabetic congenic mice identifies Cd55 (Daf1) and Acadl as candidate genes for type 1 diabetes. J Immunol 2008;180:1071–9.
[105] Kodvawala A, Ghering AB, Davidson WS, Hui DY. Carboxyl ester lipase expression in macrophages increases cholesteryl ester accumulation and promotes atherosclerosis. J Biol Chem 2005;280:38592–8.
[106] Yoon JW, Jun HS, Santamaria P. Cellular and molecular mechanisms for the initiation and progression of beta cell destruction resulting from the collaboration between macrophages and T cells. Autoimmunity 1998;27:109–22.
[107] Lugrin J, Ding XC, Le Roy D, Chanson AL, Sweep FC, Calandra T, et al. Histone deacetylase inhibitors repress macrophage migration inhibitory factor (MIF) expression by targeting MIF gene transcription through a local chromatin deacetylation. Biochim Biophys Acta 2009;1793:1749–58.
[108] Bloom BR, Bennett B, Oettgen HF, McLean EP, Old LJ. Demonstration of delayed hypersensitivity to soluble antigens of chemically induced tumors by inhibition of macrophage migration. Proc Natl Acad Sci U S A 1969;64:1176–80.
[109] Jayaraman S, Muthukkaruppan VR. In vitro correlate of transplantation immunity: spleen cell migration inhibition in the lizard, *Calotes versicolor*. Dev Comp Immunol 1977;1:133–43.
[110] Xu H, Foltz L, Sha Y, Madlansacay MR, Cain C, Lindemann G, et al. Cloning and characterization of human erythroid membrane-associated protein, human ERMAP. Genomics 2001;76:2–4.
[111] Wang C, Faloon PW, Tan Z, Lv Y, Zhang P, Ge Y, et al. Mouse lysocardiolipin acyltransferase controls the development of hematopoietic and endothelial lineages during in vitro embryonic stem-cell differentiation. Blood 2007;110:3601–9.
[112] Marques M, Laflamme L, Gervais AL, Gaudreau L. Reconciling the positive and negative roles of histone H2A.Z in gene transcription. Epigenetics 2010;5:267–72.
[113] Patterson CC, Dahlquist GG, Gyürüs E, Green A, Soltész G. EURODIAB Study Group: incidence trends for childhood type 1 diabetes in Europe during 1989-2003 and predicted new cases 2005-20: a multicentre prospective registration study. Lancet 2009;373:2027–33.
[114] Shoda LK, Young DL, Ramanujan S, Whiting CC, Atkinson MA, Bluestone JA, et al. A comprehensive review of interventions in the NOD mouse and implications for translation. Immunity 2005;23:115–26.
[115] Jayaraman S. Epigenetic mechanisms of metabolic memory in diabetes. Circ Res 2012;110:1039–41.

CHAPTER

10

Systemic Sclerosis and Epigenetics

Steven O'Reilly

Northumbria University, Newcastle Upon Tyne, United Kingdom

OUTLINE

10.1 INTRODUCTION

Systemic sclerosis (SSc) is an autoimmune disease of unknown etiology, which is characterized by inflammation, vasculopathy, and fibrosis. Like all autoimmune diseases it has a female preponderance. Although the mechanisms that trigger autoimmunity remain under investigation,

The Epigenetics of Autoimmunity. http://dx.doi.org/10.1016/B978-0-12-809912-4.00010-6

it is clear that multiple are likely to exist. It has been suggested that the vasculature receives the initial insult, which then leads to the release of danger signals and apoptotic cells, activation of the immune system, and the release of proinflammatory cytokines [1]. Vascular abnormalities appear early on during disease development, with perivascular inflammation, autoimmunity, and inflammation leading to fibrosis.

The activation of fibroblasts culminates in their transformation into myofibroblasts. These cells are the abundant cell type in fibrosis and they are characterized by their increased levels of alpha smooth muscle actin in stress fibers and an excess of extracellular matrix (ECM) molecules (i.e., collagen, which is cross-linked and rigid once this scarring process has occurred). Fibroblasts are abundant throughout tissues and they secrete matrix components to keep the tissues together and respond accordingly to the local microenvironment.

Although some genetic associations have been discovered for SSc, most notably in the Major Histocompatibility Complex (MHC) locus, the relative effect of genetics appears to be modest at best; for example, the genetic heritability for SSc was estimated to be 0.008 [2]. The observation that Human Leukocyte antigen (HLA) genes were associated with susceptibility underpins the disease's autoimmune pathogenesis [3,4]; however, in comparison to the genetic component, epigenetics is likely to play a much larger role in the disease pathogenesis.

Recognition of the importance of epigenetics in a variety of diseases (i.e., cancer and autoinflammatory diseases) has produced new perspectives and enhanced our understanding of these diseases. Epigenetics was first described by Conrad Waddington in the 1940s and is now defined as "stably inherited from changes in a chromosome but without alterations to the underlying DNA sequence" [5]. Whereas an organisms genotype is relatively static throughout life, the epigenome is both plastic and malleable to the environment. Therefore epigenetic changes may explain the variability in the clinical course of the disease and the spontaneous regression that can occur. The clinical heterogeneity that is observed cannot be attributed to classical genetics alone [1]. Interestingly, epigenetic alteration may be passed not only through cellular generations but also through human generations, whereby the parental epimutation is transmitted to the subsequent generation [6]. Disruption and reduction of histone methylation in sperm is passed transgenerationally and results in impaired development of the offspring in association with aberrant RNA profiles [7]. A recent study also demonstrated the epigenetic inheritance of an acquired neuropathic phenotype, which was mediated by changes in multiple microRNAs (miRs) within the sperm [8].

Epigenetics is implicated in a plethora of diseases, including SSc, and all three epigenetic modifications have been found to be dysregulated in SSc [9]. This chapter will review all epigenetic aberrations in SSc and evaluate how these modifications may be targeted for therapeutic gain.

10.2 EPIGENETICS

Epigenetics literally means "above the gene." Every cell in the body contains the same DNA information; however, the cells ability to acquire a distinct phenotype is due to the epigenetic influence of the local environment [1]. Epigenetic modifications of DNA and histones leads to reorganization of the chromatin structure and therefore regulation of the interactions between transcription factors and DNA.

Of all the genes within the human genome, only 1%–2% encode for proteins, although approximately 95% are transcribed [10]. The observation that the majority of transcripts are noncoding challenges our perception of how genetic information is stored and expressed. It is known that the majority of these noncoding transcripts are longer than 200 nucleotides in length, and are therefore referred to as long noncoding RNAs (lncRNAs). It is generally accepted that these "junk RNA" are not nonfunctional but play active roles in regulating various processes. Smaller noncoding RNAs include miRs, PIWI-interacting RNAs, and endogenous interfering RNAs. Interest in this field has increased because the prevalence of noncoding RNAs is much higher than previously thought. Functional studies regarding the mechanisms of lncRNAs are ongoing and only some answers have emerged; however, it is known that lncRNAs are involved in genomic imprinting (i.e., *Air*), X-chromosome silencing, cardiac development, and cancer (i.e., *Hotair*) [11]. Like mRNAs, lncRNAs are generally transcribed by RNA polymerase II, spliced, capped at the 5′ end, and adenylated. They are also located proximally to the promoters of known coding genes. Many nuclear lncRNAs associate with chromatin remodeling complexes (i.e., the polycomb repressive complex, containing Ezh2) and guide them to specific genomic locations where they can block gene expression through a repressive heterochromatin state [12]. The lncRNA, *Hotair*, uses a similar mechanism to interact with the demethylase, LSD1, which subsequently removes the histone mark of active chromatin and reinforces the repression chromatin architecture on these loci. LncRNAs are also known to have enhancer functions, and a recent study demonstrated that they can increase gene expression by promoting looping of androgen receptor-bound enhancers at gene promoters [13].

The lncRNA, CHRF, regulates cardiac fibrosis by acting as a molecular sponge to delay the activity of miR-489. CHRF mops up miR-489, therefore leading to upregulation of its target mRNA (myeloid differentiation response gene), which is known to be involved in fibrosis [14]. The use of lncRNA tiling arrays in a mouse model of lung fibrosis showed that uc.77 lncRNA was highly upregulated after the initiation of fibrosis [15]. Overexpression of uc.77 caused fibrotic changes in lung cells and an upregulation of many fibrotic markers, including α-smooth muscle actin (α-Sma), which depends on ZEB2, as depletion of ZEB2 attenuated uc.77 induced changes [15].

Only one SSc study to date has examined a lncRNA (e.g., TSIX), in any detail [16]. TSIX is an lncRNA located on the X chromosome, which is interesting as X-chromosome inactivation is suspected to be involved in the female preponderance of SSc. The authors found that control skin contained barely detectable levels of TSIX; however, expression levels were elevated in the skin and fibroblasts from subjects with SSc [16]. Furthermore, the downregulation of TSIX in SSc fibroblasts led to a reduction of collagen in these cells [16]. The mechanism of how TSIX represses collagen expression is not clear; however, preliminary experiments suggest that it is mediated by an increased half-life of collagen mRNA. How the lncRNA, TSIX, influences mRNA stability remains to be elucidated, as do the identities of the factors that modulate the expression of TSIX in the dermal fibroblasts? Does TSIX alter the chromatin in association with the collagen promoter region thus allowing access?

10.3 MicroRNAs

miRs are a large family of highly conserved noncoding genes that operate in all eukaryotic cells. They have a much smaller length than lncRNAs but play a fundamental role in virtually all cellular processes through the negative regulation of gene expression [17].

miRs bind to the 3′ Untranslated Region (UTR) of their target mRNAs and trigger translational repression or mRNA destabilization [18]. One miR can target thousands of mRNAs and as such can have a huge effect on multiple pathways. Primary miRs are initially transcribed as much longer transcripts (between a few hundred to a few thousand base pairs long) before a microprocessor complex cleaves it into precursor miRs, which are generally 60–70 nucleotides in length. The endonuclease, Dicer1, cleaves the precursor miR to produce a mature miR. Profibrotic cytokines and chemokines can alter the expression of miRs (either positively or negatively) and may work in concert with each other to alter the levels of pro- or antifibrotic miRs. Other mechanisms that are likely to regulate miR expression include stabilization and lncRNA sponges.

One of the first miRs to be described in SSc (and fibrosis in general) was miR-29a [19]. miR-29a, sometimes described as the "fibromiR," is a critical antifibrotic miR that represses collagen [20]. Maurer et al. identified downregulated levels of miR-29a (and other miR-29 family members) in SSc skin biopsies and isolated dermal fibroblasts [19]. The authors used the 3′ UTR luciferase assay to demonstrate that collagen was a bona fide target of this miR. The 3′ UTR luciferase assay is known as the gold standard for confirming miR targets. Levels of miR-29a were also shown to be reduced in the bleomycin mouse model of fibrosis, and the use of the tyrosine-kinase inhibitor, imatinib, restored the reduced miR [19]. This suggests that miR-29a is under the control of the c-abl tyrosine kinase and some success has been achieved when treating SSc patients with imatinib; however, larger and more comprehensive trials are still needed [21].

Maurer et al. confirmed that collagen was the target of miR-29 and this has since been confirmed by other studies; however, it is now also known that other noncollagen genes (e.g., integrins) are also regulated by miR-29a [22]. Integrins are known to regulate the biological activity of transforming growth factor-β1 (TGF-β1). Reduced levels of miR-29a were detected in SSc dermal fibroblasts and TGF-β-activated kinase-1 (TAB-1) was identified as a novel target of miR-29a. TAB-1 alters the levels of TIMP-1, which inhibits the breakdown of collagen and increases its net deposition in the ECM. Sleeping beauty-mediated gene therapy of miR-29 attenuated fibrosis in a bleomycin mouse model [23] thus suggesting this is a primary target.

miR-155 has been shown to be associated with lung progression in SSc patients [24] and was reported to be elevated in fibroblasts from SSc skin biopsies [25]. miR-155 is highly upregulated in monocytes and macrophages following stimulation by Toll-Like Receptor (TLR) ligands. This occurs to restore the immune response so as not sustain an exaggerated and detrimental inflammatory response. Whether this is the reason for the elevation in SSc is unknown; however, our research group has also found elevated miR-155 levels in dermal fibroblasts from SSc patients. Knockout of miR-155 in a bleomycin mouse model of fibrosis protected the animals from skin fibrosis [26], and therefore it appears that miR-155 is also activating the inflammasome to mediate fibrosis [26].

miR-27b targets Gremlin-1 mRNA, and alterations in this miR alter the levels of collagen via Gremlin-1 expression [27]. Gremlin-1 is a Bone Mophogenetic Protein (BMP) antagonist that is hugely profibrotic and altered in SSc [28]. Small interfering (siRNA)-mediated silencing of Gremlin-1 in SSc dermal fibroblasts reduced collagen levels [28].

10.3.1 miR-21

miR-21 is an important miR in fibrosis in general [29]. miR-21 was found to be elevated in SSc tissue and isolated fibroblasts, and its putative target was hypothesized to be SMAD7. SMAD7 is an inhibitory protein that acts as a negative regulator of TGF-β signaling and thus its reduction may lead to unabated signaling.

10.3.2 miR-132

Downregulation of miR-132 in fibrosis leads to upregulation of methyl cap binding protein 2 (MeCP2), which is a critical step in the fibrotic cascade [30]. Indeed, reduction of MeCP2 in vitro leads to reduced collagen and α-Sma, and reduction of MeCP2 in vivo attenuates fibrosis in a carbon tetrachloride mouse model of liver fibrosis [30]. The reduction of miR-132 appears to be critical to the upregulation of MeCP2. Elevated MeCP2 alters the expression of Ezh2, deposits trimethylated lysines on histone 3, and leads to the repression of antifibrotic genes [30] (Fig. 10.1).

MeCP2 is also dependent on forming a repression complex with Histone deacetylases 1 and the corepressor Sin3A [31]. Elevated MeCP2 levels were observed in SSc tissues in association with increased collagen deposition. This prompted the examination of the levels of miR-132 in the matched tissues; however, no significant expression differences were observed [32]. Instead, reduced levels of miR-135b were reported, and the bone fide target of this miR is the transcription factor, signal transducer and activator of transcription 6 (STAT6) [32]. STAT6 was also found to be elevated in SSc and this was important in the IL-13-mediated increase in collagen and other fibrotic markers [32]. STAT6-mediated increases in fibrotic markers could be reduced by small-interfering RNA-mediated silencing of STAT6. Thus a complex interaction exists between a miR and a central transcription factor for enhanced collagen deposition through IL-13. Interestingly, DNA demethylation leads to reduced collagen, which suggests that miR-135b may also be regulated by methylation.

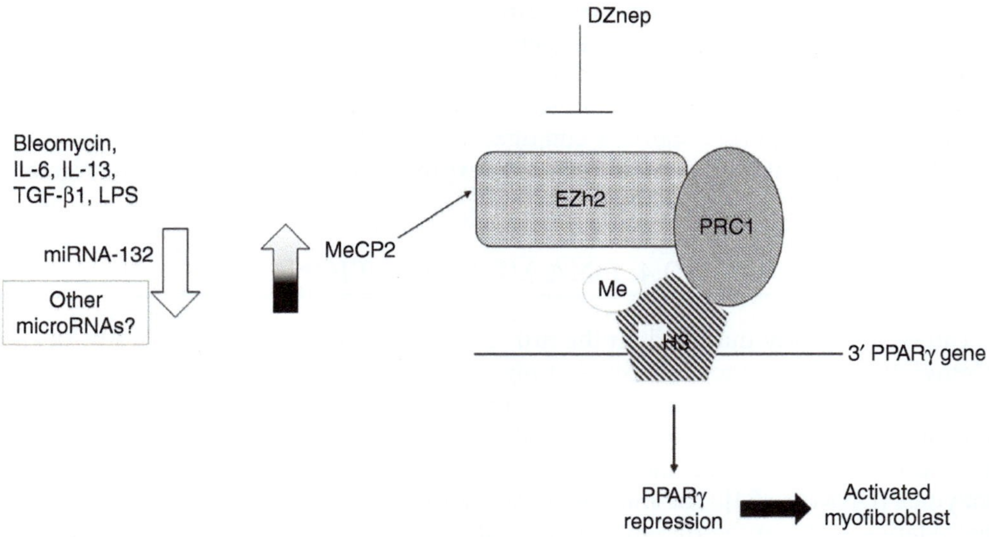

FIGURE 10.1 **Possible pathway of MeCP2 regulation in SSc.** The initial insult drives down the expression of miR-132 levels. It should be noted that other unidentified microRNAs may also be involved at this point in the pathway. The reduction of miR-132 leads to an increase of its target MeCP2 and this methyl binding protein regulates the expression of the histone methyltransferase Ezh2, which binds to PRC1 and catalyzes the addition of methyl groups to histone tails at the 3′ UTR of the PPARγ gene. This H3 methylation at the PPARγ promoter leads to repression of its expression, and PPARγ reduction derepresses extracellular matrix (ECM) genes (i.e., collagen, TGF-β, and α-Sma) leading to the activation of the myofibroblast. Points of intervention could be to mimic of miR-132, to restore the miR-132 levels, or to block Ezh2 through the use of DZnep. *3′UTR*, 3′ untranslated region; *Me*, methyl group.

10.3.3 miR-196

miR-196a is greatly downregulated in SSc dermal fibroblasts and skin tissue [25]. This downregulation is present in both the constitutive SSc cells and in normal fibroblasts stimulated with the profibrotic cytokine, TGF-β1. It has also been shown that this miR affects collagen 1 by binding to its 3′ UTR region, and miR protectors confirmed that this was a direct target of the miR and the effect was not mediated through another pathway. The authors also reported that serum levels of miR-196a were reduced in SSc patients with a higher modified Rodnan skin score and a higher frequency of digital ulcers.

Serum-derived miR-196, among others, was found to be dysregulated in SSc [33]. These dysregulated miRs in serum-derived exosomes, which are small lipid-coated vesicles that protect miRs, are suggested to lead to exaggerated wound healing responses and thus alter the levels of collagen and other ECM markers [33]. Although this was not shown directly by the authors, it is an interesting hypothesis.

Targeting miRs is now possible; for example, in cases where there are reduced levels of a miR, levels can be reconstituted using miR mimics. Restoration of the miR will lead to a reduction in the expression levels of its target, which may be involved in the fibrotic cascade or be an ECM molecule. In diseases where there are elevated miR concentrations, the aim would be to reduce these levels to derepress their target genes. This can be achieved using an antagomiR. Whilst this has been achieved in vitro and in some animal models in vivo, the data are still in their infancy as the compounds need to be chemically modified to increase their stability and to prevent their breakdown by endogenous RNase. This is now being investigated with the addition of a cholesterol moiety to the miRs to enhance stability. Targeting the miR mimic or antagomiR to the correct cell and tissue also remains an issue. The ideal scenario is to target the appropriate cell type where you want to restore the diminished miR or antagomiR; however, targeting these areas is still difficult but could be achieved experimentally by conjugating the miRs with a stabilizing moiety or using a monoclonal antibody that targets an antigen only expressed on the cell type to be targeted.

10.4 DNA METHYLATION

DNA can be covalently modified on the fifth carbon ring of cytosine, mainly in CpG dinucleotides. Methylation of this cytosine occurs at high densities within CpG islands, which are found in promoters, and this generally leads to transcriptional repression. Conversely, hypomethylation leads to increased gene expression. The methyl group is commonly donated by S-adenosyl-L-methionine (SAM) and its transfer is facilitated through the action of the DNA methyltransferase family of enzymes (DNMTs) [34]. The maintenance enzyme, DNMT1, is responsible for regenerating the bimethylation state on the daughter strand of DNA during cell division [35]. While DNMT1 is responsible for the maintenance of the methylation marks, DNMT3a and DNMT3b are responsible for de novo methylation and mammalian development [36] (Table 10.1). The transcriptional repression that occurs from hypermethylation is due to the recruitment of methyl-CpG binding proteins, which create a repressed chromatin environment through their methyl binding domains [37]. These may also recruit HDAC to further enhance repression.

DNA methylation plays a key role in a number of biological processes and it is now recognized that aberrant DNA methylation is associated with many diseases [38]. One of the first studies to demonstrate the role of DNA methylation, specifically hypermethylation, was in

TABLE 10.1 DNMTs and Function

DNMT1	Maintenance
DNMT3a	De novo
DNMT3b	De novo

kidney fibrosis [39]. In this study, hypermethylation of genes encoding RASAL1 repressed the gene expression and led to the activation of resident kidney myofibroblasts [39]. This was because RASAL1 encodes an inhibitor of the Ras oncogene, which is critical in kidney fibroblast activation [39].

The Wnt antagonists, dickkopf-related protein-1 (DKK1) and secreted frizzled related protein-1 (sFRP1), are downregulated in SSc, which results in amplification of the Wnt pathway [40]. The Wnt pathway is an evolutionary-conserved profibrotic signaling pathway in adult tissue. DKK1 and sFRP1 are hypermethylated at their respective promoter regions and are thus silenced. Treatment with 5′aza resulted in hypomethylation of these promoters, a derepression of gene and protein expression, and a reduction in the production of fibrotic markers [40]. Thus hypermethylation of the inhibitor reduces the protein levels and allows unrestrained Wnt signaling to occur, which is indeed profibrotic.

It has also been demonstrated that the maintenance enzyme, DNMT1, which maintains the methylation CpG mark over multiple generations, is elevated in SSc [41]. This may explain the persistent fibrosis seen in high-passage cells in in vitro cultures. It is interesting to note that betaine feeding into the methionine cycle can act as a precursor of SAM and help to provide the required methyl group. In joint fibroblasts, betaine supplementation can increase DNA methylation by altering the de novo methyltransferase, DNMT3A, and thus altering gene expression [42]. This could play a role in SSc. Chromosome 8 open reading frame 4 (cr8orf4) is a transcription regulator that was shown to be hypermethylated in lung fibroblasts. This hypermethylation leads to protein repression in these cells [43]; however, hypermethylation can be reversed following 5′aza incubation, which subsequently results in increased COX-2 levels coincident with restored prostaglandin E2 levels [43]. siRNA-mediated repression of cr8orf4 recapitulates the fibrotic phenotype [43]. Table 10.2 summarizes the differentially methylated genes in SSc.

As mentioned previously, the TET family of dioxygenase proteins are actively involved in DNA demethylation (Table 10.3). TET1 catalyzes the conversion of methylcytosine to 5-hydroxymethylcytosine, which can itself be further catalyzed. The current question is does 5-hydroymethylcytosine have a regulatory role in addition to being an intermediate in the demethylation of DNA? It has been suggested that this modification may have a functional role; however, this is currently an area of active debate. It was recently demonstrated that 5-hydroxymethylcytosine levels were elevated in SSc fibroblasts. Levels of TET1 mRNA were also elevated in SSc fibroblasts and this could be induced by hypoxia but not by TGF-β [44]. Although the levels of TET1 mRNA were elevated in the SSc fibroblasts [44]; no elevation in TET1 protein levels were observed [32]. The reason for this is not clear. The fact that there was a small global decrease in methylation in the SSc fibroblasts may not be overly important, and global analysis of methylation is a crude indicator of methylation; gene-specific methylation is a more informative readout. The authors also noted that hypoxia induced TET1 levels in this study and hypoxia is known to be important in SSc; however, the relative in vivo levels of hypoxia are not known and the levels used in vitro may be way beyond what is physiologically relevant. A genome-wide methylation study in dermal fibroblasts from

SSc patients (limited and diffuse) and controls identified significant differences in the methylation status between the two forms of the disease [45]. One of the most interesting genes that was differentially methylated between subjects and controls was the integrin, α-integrin 9, which was hypomethylated and thus should be overexpressed [45]. This is of interest because integrins activate TGF-β from its latent preprocessed form and thus promote fibrosis [46].

The transcription factor Fli-1 was shown to be hypermethylated in SSc dermal fibroblasts, which led to reduced expression levels. It has also been shown that this hypermethylation is maintained over time in culture [41]. The hypermethylated of Fli1 and the subsequent protein repression and overexpression of collagen 1 could be reversed with the addition of a demethylating agent in vitro [41].

It is well accepted that immune cells (i.e., monocytes and T cells) play a critical role in SSc. T cells are found in skin lesions in close proximity to myofibroblasts and can secrete a variety of factors that mediate fibrosis. DNA demethylation of CD40L in $CD4^+$ T cells in SSc has been found to cause an upregulation of CD40L [47]. This protein is important in the interaction and activation of B cells and fibrosis; indeed, blockade of the CD40/CD40L interaction reduced fibrosis in a tight skin mouse model [48]. ITGAL is an α chain subunit of the lymphocyte function associated antigen 1 system, whose interaction with ICAM regulates communication and interactions between T cells and other cells. Elevated levels of ITGAL have been found in the peripheral blood and T cells from SSc patients; for example, a recent study demonstrated higher levels of ITGAL on CD4 T cells in SSc, which was caused by hypomethylation of the ITGAL promoter [49]. It was also shown that

TABLE 10.2 Differential Methylation Genes in SSc

Gene	Methylation	Cell type
DKK	Hypermethylated	Skin fibroblast
SFRP1	Hypermethylated	Skin fibroblast
Cr8orf4	Hypermethylated	Lung fibroblast
α-Integrin9	Hypomethylated	Skin fibroblast
Fli1	Hypermethylated	Skin fibroblast
CD40L	Hypomethylated	$CD4^+$ T helper cell
ITGAL	Hypomethylated	T cells
FoxP3	Hypermethylated	T regulatory cells
CTGF	Hypomethylated	PBMCs
miR-135b	Hypermethylated	Skin fibroblast

TABLE 10.3 TETs and Their Products

TET1	5-Hydroxymethylcytosine
TET2	5-Formylcytosine
TET3	5-Carboxylcytosine

this change in expression level was correlated with disease activity [49]. Furthermore, incubation of healthy $CD4^+$ T cells with 5′aza resulted in hypomethylation of the ITGAL promoter and increased protein expression of ITGAL. There was also a strong correlation between the methylation status of the promoter and ITGAL surface expression [49].

Interestingly, this observation was only found in female SSc patients, which could be due to global hypomethylation and reactivation of the silent X chromosome in female patients with SSc [47]. This may explain the preponderance of females in the SSc population. A recent study reported that patients with SSc exhibited T cell hypermethylation of the transcription factor, FoxP3, when compared to controls, and that this hypermethylation led to reduced protein levels of FoxP3 [50]. FoxP3 is a transcription factor that is often used as a marker of T regulatory cells, which are immune modulating cells that secrete high levels of the antiinflammatory molecule IL-10. Hypermethylation of the FoxP3 promoter has been reported in other autoimmune diseases. In this instance, treatment of the hypermethylated T cells with 5′aza was able to demethylate the FoxP3 promoter and enhance FoxP3 expression and T regulatory cell numbers [50]. T regulatory cells are thought to be antifibrotic; however, some studies have shown different results and further studies are required.

MeCP2 is a methyl binding protein that mediates repression of gene expression by binding to methylated DNA in association with HDAC1. Elevated levels of MeCP2 have been observed in SSc [32], and this was shown to be induced by TGF-β1 and mediate collagen induction. Connective tissue growth factor (CTGF) has also been found to be hypomethylated in PBMCs from SSc subjects [51].

10.5 HISTONE MODIFICATIONS

The recognition that histones can be posttranslationally modified to alter the conformation of histones and gene expression has opened up a whole new area of research. In eukaryotic cells, DNA is condensed in chromatin and tightly packaged. Histones are globular structures except for their tail, where they possess multiple histone modifications including acetylation, methylation, phosphorylation, citrullination, ubiquitination, sumoylation, and ADP-ribosylation. The family of enzymes that facilitate the addition of these molecules to the histones are called histone transferases, and they include the histone acetyltransferases which mediate the addition of an acetyl group to the chromatin. Hyperacetylation of the histone tails results in an opening of the DNA structure, which makes it permissive to transcription. In contrast, trimethylation of lysine 9 of histone H3 and trimethylation of histone H3K27 lead to transcriptional silencing. HDACs are a family of enzymes that remove the acetyl mark from chromatin. HDACs can be classified into four groups and are predominantly found in the nucleus. Class I HDACS include HDAC1,2, 3, and 8 and are the most well studied. Inhibitors of HDACs are potent molecules that are used in cancer treatment. Trichostatin A (TSA), is a broad-spectrum HDAC inhibitor, which reduced the levels of collagen and fibronectin in SSc fibroblasts in the presence of TGF-β1 [52]. This collagen-diminishing effect was associated with a reduction in the DNA-binding abilities of Smad3 and Smad4 (as measured by EMSA) and a reduction in the nuclear translocation of Smad3 [52]. The bleomycin model of skin fibrosis is a well-used mimic of SSc, and treatment with TSA in this model abrogated the collagen increases with bleomycin and diminished the skin thickness [52]. This could also be mediated by an induction of cell cycle arrest. HDAC7, which is part of the class II family of HDACs, has also been found to be upregulated in SSc dermal fibroblasts [53]. SiRNA-mediated reduction of HDAC7 reduced ECM deposition in these cells [53] and, in cells isolated from idiopathic pulmonary fibrosis patients, targeting global HDACs resulted in a large suppression of HDAC7 and an increase in apoptosis [54]. The use of TSA in scleroderma fibroblasts has been shown to be antifibrotic by preventing the loss of Wnt-inhibitor factor-1 (WIF-1) and therefore leading to a depression of WIF-1 [55]. WIF-1 is a natural inhibitor of the Wnt pathway and is reduced in SSc [55], which subsequently leads to enhanced Wnt signaling through β-catenin and subsequent collagen deposition [56]. ATF and HDAC3 are recruited together to the WIF-1 promoter in SSc to silence WIF-1 expression [55]. A different study found that HDAC1 and HDAC6 are elevated in the SSc fibroblasts, and treatment with TSA reduced collagen expression, furthermore using Chromatin Immunoprecipitation (ChIP) it was found that H3 and H4 was much reduced in acetylation in the Fli1 gene promoter [41]. This suggests that the *Fli1* is epigenetically repressed, thus the deacetylation-mediated downregulation of Fli1 leads to an upregulation of collagen. Fli1 is a member of the Ets family of transcription factors and inhibits collagen via an Sp1-dependant pathway [57]. Fli1 deficiency due to epigenetic repression is also associated with the development of digital ulcers [58]. The promoter of KLF5 has also been reported to exhibit much reduced acetylated H3 and H4 levels in SSc fibroblasts than in controls [59]. Treatment with TSA restored these levels and subsequently reduced collagen levels. KLF5 is a transcription factor that is hugely reduced in SSc skin cells and is thought to repress CTGF expression. Indeed, KLF5-Fli1 double knockout mice spontaneously developed skin fibrosis and vascular abnormalities [59]. This is interesting as although there are animal models of SSc, none really demonstrate both the fibrotic and vascular abnormalities found in human SSc (Table 10.4).

Sirtuins are members of the class III family of HDACs, and are unique as they require NAD^+ as a cofactor. Sirtuins are mammalian orthologues of the yeast silent information regulator 2, and are linked to the regulation of lifespan and metabolism. Sirtuin-1 has been found to be reduced in SSc [60]. This enzyme is important in mediating fibrosis, and sirtuin-1 induction using either resveratrol or overexpression vectors was able to reduce collagen and α-Sma levels in dermal fibroblasts from SSc patients [60]. In the bleomycin model of skin fibrosis, resveratrol totally attenuated bleomycin-induced fibrosis and appeared to do this by the induction of sirtuin1 and independently of inflammation. It has been suggested that the reduction of sirtuin1 in SSc mediates an increase in the levels of the acetyltransferase, P300, at the collagen promoter and the subsequent hyperacetylation that is driving the increase in ECM expression [60]. Chronic hypoxia (24 hours) in dermal fibroblasts was also able to massively repress sirtuin1 levels, thus suggesting a link between the hypoxic environment, epigenetic changes, and collagen deposition.

Histone K27 trimethylation is a methyl mark on the histone at lysine 27, which is mediated by Ezh2 enzymes that normally produce a repressive chromatin state. SSc monocytes stimulated with a single-stranded RNA TLR8 agonist were able to produce elevated levels of TIMP-1 [61]. TIMP-1 is a profibrotic molecule and thus of interest in SSc. The induction of TIMP-1 via TLR was investi-

TABLE 10.4 Upregulated HDACs in SSc

HDAC1
HDAC6
HDAC7

gated further and the role of the transcription factor, Fra-2, was examined in detail. Fra2 appears to be upregulated after TLR8 agonist stimulation, and ChIP analysis revealed that Fra2 binding to the TIMP-1 promoter mediates its induction and that this interaction is enhanced following DZnep incubation [61]. DZnep is a validated inhibitor of Ezh2 and actively demethylates histone K27. Thus histone K27 methylation normally represses Fra-2 expression; upon demethylation this enhances Fra2 and promotes TIMP-1 elevation. Furthermore, conditioned media from the TLR8 and DZnep incubated monocytes was able to upregulate collages levels when it was transferring to isolated in vitro-cultured fibroblasts [61]. Thus histone methylation is important in this mechanism although the precise molecular mechanism of how it mediates its effects on fibrosis remain unclear.

Lysine methyltransferases catalyze the addition of a methyl group onto the lysine residues in histone tails. Although none have been described in SSc, the lysine methyltransferase ASH1 has been demonstrated to be upregulated in hepatic stellate cells. ASK1 deposits lysine specific marks on the histone tails of genes that mediate fibrosis, such as collagen1, TGF-β1, and TIMP-1 [62]. Genetic reduction of ASH1 results in decreased expression of collagen and TIMP-1 in stellate cells. Whilst this has not been shown in SSc, it is speculated that this is also occurring as there are clear alterations in SSc cells at the level of histone methylation. Furthermore, it is known that MeCP2 itself regulates ASH1 expression [62], and elevated MeCP2 expression levels have already been identified in SSc fibroblasts [32]. This suggests that MeCP2 could possibly be a therapeutic target in SSc.

10.6 CONCLUSIONS

There are now multiple lines of evidence that suggest that epigenetic modifications are critical in SSc pathogenesis. miRs are particularly perturbed in SSc and may represent not just clinical therapeutic targets but also biomarkers in the disease. The fact that they are stable in body fluids is also of major interest. The lncRNA species have only recently been described in fibrotic conditions, with only one described in the context of SSc. These species will likely emerge as major regulators of gene expression with multiple different mechanisms of action. Their modes of action are not at all well described and are largely unknown, but may include the alteration of histone accessibility and their ability to act as "sponges" for sequestering miRs. In terms of measuring epigenetic marks such as histone lysine methylation and miRs, the use of whole tissue is now of very little value due to cell-specific alterations and the dynamic nature of epigenetic changes. One should use laser capture microdissection to isolate single cells for analysis as the extra labor is worthwhile for true meaningful data.

10.7 FUTURE DIRECTIONS

Epigenetics are now appreciated as playing a key role in multiple diseases including SSc. Various epigenetic modifications are present in multiple cell types, including immune cells and stromal cells. miRs are cell-type specific and can have a combined effect on their targets. Thus the measurement of miRs in whole skin tissue is of limited value and single-cell isolation is the way to derive meaningful information. Single-cell sequencing is now a reality and can be performed with ease and for reduced costs. Methylation analysis has been performed on dermal fibroblasts in SSc; however, relatively little analysis has been performed on immune cells from patients with this disease. A small amount of analysis has been performed using whole PBMCs; however, this tissue contains multiple immune cell subtypes and important methylation information could be being missed. It would be more useful to isolate the immune cell subtypes (i.e., $CD4^+$ T cells) and then analyze the methylation status. Overall the net methylation appears to increase in the disease and the use of demethylating agents such as 5′aza could be useful therapeutic strategies. This is a global demethylating agent that is used in the cancer setting; however, in vitro it has been shown to reduce collagen expression in SSc fibroblasts [32] and in the bleomycin model of skin fibrosis [40].

10.8 SUMMARY

Epigenetics are key to the pathogenesis of SSc, and many dysregulated miRs lead to the profibrotic activation of fibroblasts. Hypo- and hypermethylation of various genes are involved in the disease; however, overall there is a tendency towards global hypermethylation. Epigenetic drugs such as 5-aza could be useful therapeutic strategies. However, much more research is required in this novel and important area.

References

[1] Ciechomska M, van Laar J, O'Reilly S. Current frontiers in systemic sclerosis pathogenesis. Exp Dermatol 2015;24(6):401–6.
[2] Feghali-Bostwick C, Medsger TA, Wright TM. Analysis of systemic sclerosis in twins reveals low concordance for disease and high concordance for the presence of antinuclear antibodies. Arthritis Rheum 2003;48(7):1956–63.
[3] Gladman DD, Kung TN, Siannis F, Pellett F, Farewell VT, Lee P. HLA markers for susceptibility and expression in scleroderma. J Rheumatol 2005;32(8):1481–7.
[4] Vlachoyiannopoulos PG, Dafni UG, Pakas I, Spyropoulou-Vlachou M, Stavropoulos-Giokas C, Moutsopoulos HM. Systemic scleroderma in Greece: low mortality and strong linkage with HLA-DRB1*1104 allele. Ann Rheum Dis 2000;59(5):359–67.
[5] Waddington CH. Canalization of development and the inheritance of acquired characters. Nature 1942;150(3811):563–5.
[6] Stringer JM, Barrand S, Western P. Fine-tuning evolution: germ-line epigenetics and inheritance. Reproduction 2013;146(1):R37–48.
[7] Siklenka K, Erkek S, Godmann M, Lambrot R, McGraw S, Lafleur C, et al. Disruption of histone methylation in developing sperm impairs offspring health transgenerationally. Science 2015;350:6261.
[8] Gapp K, Jawaid A, Sarkies P, Bohacek J, Pelczar P, Prados J, et al. Implication of sperm RNAs in transgenerational inheritance of the effects of early trauma in mice. Nat Neurosci 2014;17(5):667–9.
[9] O'Reilly S. MicroRNAs in fibrosis: opportunities and challenges. Arthritis Res Ther 2016;18(1):1–10.
[10] Dinger ME, Amaral PP, Mercer TR, Mattick JS. Pervasive transcription of the eukaryotic genome: functional indices and conceptual implications. Brief Funct Genomic Proteomic 2009;8(6):407–23.
[11] Yoshimizu T, Miroglio A, Ripoche M-A, Gabory A, Vernucci M, Riccio A, et al. The H19 locus acts in vivo as a tumor suppressor. Proc Natl Acad Sci USA 2008;105(34):12417–22.
[12] Gupta RA, Shah N, Wang KC, Kim J, Horlings HM, Wong DJ, et al. Long non-coding RNA HOTAIR reprograms chromatin state to promote cancer metastasis. Nature 2010;464(7291):1071–6.
[13] Yang L, Lin C, Jin C, Yang JC, Tanasa B, Li W, et al. LncRNA-dependent mechanisms of androgen-receptor-regulated gene activation programs. Nature 2013;500(7464):598–602.
[14] Wang K, Liu F, Zhou L-Y, Long B, Yuan S-M, Wang Y, et al. The long noncoding RNA CHRF regulates cardiac hypertrophy by targeting miR-489. Circ Res 2014;114(9):1377–88.
[15] Sun H, Chen J, Qian W, Kang J, Wang J, Jiang L, et al. Integrated long non-coding RNA analyses identify novel regulators of epithelial-mesenchymal transition in the mouse model of pulmonary fibrosis. J Cell Mol Med 2016;20(7):1234–46.

[16] Wang Z, Jinnin M, Nakamura K, Harada M, Kudo H, Nakayama W, et al. Long non-coding RNA TSIX is upregulated in scleroderma dermal fibroblasts and controls collagen mRNA stabilization. Exp Dermatol 2016;25(2):131–6.
[17] Horsburgh S, Fullard N, Roger M, Degnan A, Todryk S, Przyborski S, et al. MicroRNAs in the skin: role in development, homoeostasis and regeneration. Clin Sci 2017;131(15):1923–40.
[18] Béthune J, Artus-Revel CG, Filipowicz W. Kinetic analysis reveals successive steps leading to miRNA-mediated silencing in mammalian cells. EMBO Rep 2012;13(8):716–23.
[19] Maurer B, Stanczyk J, Jüngel A, Akhmetshina A, Trenkmann M, Brock M, et al. MicroRNA-29, a key regulator of collagen expression in systemic sclerosis. Arthritis Rheum 2010;62(6):1733–43.
[20] Roderburg C, Urban G-w, Bettermann K, Vucur M, Zimmermann H, Schmidt S, et al. Micro-RNA profiling reveals a role for miR-29 in human and murine liver fibrosis. Hepatology 2011;53(1):209–18.
[21] Fraticelli P, Gabrielli B, Pomponio G, Valentini G, Bosello S, Riboldi P, et al. Low-dose oral imatinib in the treatment of systemic sclerosis interstitial lung disease unresponsive to cyclophosphamide: a phase II pilot study. Arthritis Res Ther 2014;16(4):1–10.
[22] Cushing L, Kuang PP, Qian J, Shao F, Wu J, Little F, et al. miR-29 is a major regulator of genes associated with pulmonary fibrosis. Am J Respir Cell Mol Biol 2011;45(2):287–94.
[23] Xiao J, Meng X-M, Huang XR, Chung ACK, Feng Y-L, Hui DSC, et al. miR-29 inhibits bleomycin-induced pulmonary fibrosis in mice. Mol Ther 2012;20(6):1251–60.
[24] Christmann RB, Wooten A, Sampaio-Barros P, Borges CL, Carvalho CRR, Kairalla RA, et al. miR-155 in the progression of lung fibrosis in systemic sclerosis. Arthritis Res Ther 2016;18(1):1–13.
[25] Honda N, Jinnin M, Kajihara I, Makino T, Makino K, Masuguchi S, et al. TGF-β–mediated downregulation of microRNA-196a contributes to the constitutive upregulated type I collagen expression in scleroderma dermal fibroblasts. J Immunol 2012;188(7):3323–31.
[26] Artlett CM, Sassi-Gaha S, Hope JL, Feghali-Bostwick CA, Katsikis PD. Mir-155 is overexpressed in systemic sclerosis fibroblasts and is required for NLRP3 inflammasome-mediated collagen synthesis during fibrosis. Arthritis Res Ther 2017;19(1):144.
[27] Graham JR, Williams CMM, Yang Z. MicroRNA-27b targets gremlin 1 to modulate fibrotic responses in pulmonary cells. J Cell Biochem 2014;115(9):1539–48.
[28] O'Reilly S, Ciechomska M, Cant R, van Laar JM. Interleukin-6 (IL-6) trans signaling drives a STAT3-dependent pathway that leads to hyperactive transforming growth factor-β (TGF-β) signaling promoting SMAD3 activation and fibrosis via gremlin protein. J Biol Chem 2014;289(14):9952–60.
[29] Zhu H, Li Y, Qu S, Luo H, Zhou Y, Wang Y, et al. MicroRNA expression abnormalities in limited cutaneous scleroderma and diffuse cutaneous scleroderma. J Clin Immunol 2012;32(3):514–22.
[30] Mann J, Chu DCK, Maxwell A, Oakley F, Zhu NL, Tsukamoto H, et al. MeCP2 controls an epigenetic pathway that promotes myofibroblast transdifferentiation and fibrosis. Gastroenterology 2010;138(2). 705-14.e4.
[31] Nan X, Ng HH, Johnson CA, Laherty CD, Turner BM, Eisenman RN, et al. Transcriptional repression by the methyl-CpG-binding protein MeCP2 involves a histone deacetylase complex. Nature 1998;393(6683):386–9.
[32] O'Reilly S, Ciechomska M, Fullard N, Przyborski S, van Laar JM. IL-13 mediates collagen deposition via STAT6 and microRNA-135b: a role for epigenetics. Sci Rep 2016;6:25066.
[33] Nakamura K, Jinnin M, Harada M, Kudo H, Nakayama W, Inoue K, et al. Altered expression of CD63 and exosomes in scleroderma dermal fibroblasts. J Dermatol Sci 2016;84(1):30–9.
[34] Cheng X, Roberts RJ. AdoMet-dependent methylation: DNA methyltransferases and base flipping. Nucleic Acids Res 2001;29(18):3784–95.
[35] Goyal R, Reinhardt R, Jeltsch A. Accuracy of DNA methylation pattern preservation by the Dnmt1 methyltransferase. Nucleic Acids Res 2006;34(4):1182–8.
[36] Okano M, Bell DW, Haber DA, Li E. DNA methyltransferases Dnmt3a and Dnmt3b are essential for de novo methylation and mammalian development. Cell 1999;99(3):247–57.
[37] Bird A. DNA methylation patterns and epigenetic memory. Genes Dev 2002;16(1):6–21.
[38] Baylin SB, Jones PA. A decade of exploring the cancer epigenome: biological and translational implications. Nat Rev Cancer 2011;11(10):726–34.
[39] Bechtel W, McGoohan S, Zeisberg EM, Muller GA, Kalbacher H, Salant DJ, et al. Methylation determines fibroblast activation and fibrogenesis in the kidney. Nat Med 2010;16(5):544–50.
[40] Dees C, Schlottmann I, Funke R, Distler A, Palumbo-Zerr K, Zerr P, et al. The Wnt antagonists DKK1 and SFRP1 are downregulated by promoter hypermethylation in systemic sclerosis. Ann Rheum Dis 2014;73(6):1232–9.

[41] Wang Y, Fan P-S, Kahaleh B. Association between enhanced type I collagen expression and epigenetic repression of the FLI1 gene in scleroderma fibroblasts. Arthritis Rheum 2006;54(7):2271–9.
[42] Gaur N, Karouzakis E, Glück S, Bagdonas E, Jüngel A, Michel BA, et al. MicroRNAs interfere with DNA methylation in rheumatoid arthritis synovial fibroblasts. RMD Open 2016;2(2).
[43] Evans Iona C, Barnes Josephine L, Garner Ian M, Pearce David R, Maher Toby M, Shiwen X, et al. Epigenetic regulation of cyclooxygenase-2 by methylation of c8orf4 in pulmonary fibrosis. Clin Sci 2016;130(8):575–86.
[44] Hattori M, Yokoyama Y, Hattori T, Motegi S-i, Amano H, Hatada I, et al. Global DNA hypomethylation and hypoxia-induced expression of the ten eleven translocation (TET) family, TET1, in scleroderma fibroblasts. Exp Dermatol 2015;24(11):841–6.
[45] Altorok N, Tsou P-S, Coit P, Khanna D, Sawalha AH. Genome-wide DNA methylation analysis in dermal fibroblasts from patients with diffuse and limited systemic sclerosis reveals common and subset-specific DNA methylation aberrancies. Ann Rheum Dis 2015;74(8):1612–20.
[46] Margadant C, Sonnenberg A. Integrin–TGF-β crosstalk in fibrosis, cancer and wound healing. EMBO Rep 2010;11(2):97–105.
[47] Lian X, Xiao R, Hu X, Kanekura T, Jiang H, Li Y, et al. DNA demethylation of CD40L in $CD4^+$ T cells from women with systemic sclerosis: a possible explanation for female susceptibility. Arthritis Rheum 2012;64(7):2338–45.
[48] Komura K, Fujimoto M, Yanaba K, Matsushita T, Matsushita Y, Horikawa M, et al. Blockade of CD40/CD40 ligand interactions attenuates skin fibrosis and autoimmunity in the tight-skin mouse. Ann Rheum Dis 2008;67(6):867–72.
[49] Wang Y, Shu Y, Xiao Y, Wang Q, Kanekura T, Li Y, et al. Hypomethylation and overexpression of ITGAL (CD11a) in $CD4^+$ T cells in systemic sclerosis. Clin Epigenet 2014;6(1):25.
[50] Wang YY, Wang Q, Sun XH, Liu RZ, Shu Y, Kanekura T, et al. DNA hypermethylation of the forkhead box protein 3 (FOXP3) promoter in $CD4^+$ T cells of patients with systemic sclerosis. Br J Dermatol 2014;171(1):39–47.
[51] Matatiele P, Tikly M, Tarr G, Gulumian M. DNA methylation similarities in genes of black South Africans with systemic lupus erythematosus and systemic sclerosis. J Biomed Sci 2015;22(1):34.
[52] Huber LC, Distler JHW, Moritz F, Hemmatazad H, Hauser T, Michel BA, et al. Trichostatin A prevents the accumulation of extracellular matrix in a mouse model of bleomycin-induced skin fibrosis. Arthritis Rheum 2007;56(8):2755–64.
[53] Hemmatazad H, Rodrigues HM, Maurer B, Brentano F, Pileckyte M, Distler JHW, et al. Histone deacetylase 7, a potential target for the antifibrotic treatment of systemic sclerosis. Arthritis Rheum 2009;60(5):1519–29.
[54] Korfei M, Skwarna S, Henneke I, MacKenzie B, Klymenko O, Saito S, et al. Aberrant expression and activity of histone deacetylases in sporadic idiopathic pulmonary fibrosis. Thorax 2015;70(11):1022–32.
[55] Svegliati S, Marrone G, Pezone A, Spadoni T, Grieco A, Moroncini G, et al. Oxidative DNA damage induces the ATM-mediated transcriptional suppression of the Wnt inhibitor WIF-1 in systemic sclerosis and fibrosis. Sci Signal 2014;7(341). ra84-ra.
[56] Wei J, Fang F, Lam AP, Sargent JL, Hamburg E, Hinchcliff ME, et al. Wnt/β-catenin signaling is hyperactivated in systemic sclerosis and induces Smad-dependent fibrotic responses in mesenchymal cells. Arthritis Rheum 2012;64(8):2734–45.
[57] Czuwara-Ladykowska J, Shirasaki F, Jackers P, Watson DK, Trojanowska M. Fli-1 inhibits collagen type I production in dermal fibroblasts via an Sp1-dependent pathway. J Biol Chem 2001;276(24):20839–48.
[58] Saigusa R, Asano Y, Taniguchi T, Yamashita T, Takahashi T, Ichimura Y, et al. A possible contribution of endothelial CCN1 downregulation due to Fli1 deficiency to the development of digital ulcers in systemic sclerosis. Exp Dermatol 2015;24(2):127–32.
[59] Noda S, Asano Y, Nishimura S, Taniguchi T, Fujiu K, Manabe I, et al. Simultaneous downregulation of KLF5 and Fli1 is a key feature underlying systemic sclerosis. Nat Commun 2014;5:5797.
[60] Wei J, Ghosh AK, Chu H, Fang F, Hinchcliff ME, Wang J, et al. The histone deacetylase sirtuin 1 is reduced in systemic sclerosis and abrogates fibrotic responses by targeting transforming growth factor β signaling. Arthritis Rheumatol 2015;67(5):1323–34.
[61] Ciechomska M, O'Reilly S, Przyborski S, Oakley F, Bogunia-Kubik K, van Laar JM. Histone demethylation and toll-like receptor 8–dependent cross-talk in monocytes promotes transdifferentiation of fibroblasts in systemic sclerosis via Fra-2. Arthritis Rheumatol 2016;68(6):1493–504.
[62] Perugorria MJ, Wilson CL, Zeybel M, Walsh M, Amin S, Robinson S, et al. Histone methyltransferase ASH1 orchestrates fibrogenic gene transcription during myofibroblast transdifferentiation. Hepatology 2012;56(3):1129–39.

CHAPTER

11

Primary Sjögren's Syndrome and Epigenetics

Amandine Charras, Christelle Le Dantec*, Orsia D. Konsta*,**, Wesley H. Brooks†, Anne Bordron*, Yves Renaudineau*,‡*

*European University of Brittany, Brest, France; **National University of Athens, Greece; †University of South Florida, Tampa, FL, United States; ‡Brest University Medical School, CHU Morvan, Brest, France

OUTLINE

The Epigenetics of Autoimmunity. http://dx.doi.org/10.1016/B978-0-12-809912-4.00011-8

11.1 INTRODUCTION

Primary Sjögren's syndrome (pSS) is a chronic autoimmune disorder referred to as an autoimmune exocrinopathy or autoimmune epithelitis [1]. The clinical manifestations can be subdivided into those that are related to exocrine dysfunctions, which mainly lead to xerostomia (dry mouth) and keratoconjunctivitis sicca (dry eyes), and those that affect parenchymal organs, which lead to systemic features and are characterized by periepithelial mononuclear cell infiltrates. Systemic manifestations could also result from immune complex depositions due to B-cell hyperactivation. Most patients will experience an indolent and benign evolution; however, approximately 5% of patients will develop a B-cell lymphoma [2–4]. An altered health-related quality of life and psychological features are also frequently reported in patients with pSS in addition to the classical sicca features. The treatment provided is based on the disease expression (i.e., the symptoms and the extraglandular manifestations), and a systemic disease activity index (ESSDAI) has been developed for follow-ups [5,6].

T cells predominate in exocrine glands during the initiation and moderate stages of the disease [7]; however, B cells and dendritic cells are recruited and activated as the disease evolves. B-cell hyperactivation is a dominant feature of pSS, and is characterized by: (1) the presence of circulating autoantibodies (Ab) against the sicca syndrome ribonucleoprotein particles (SSA/Ro and SSB/La), usually in association with rheumatoid factors and hypergammaglobulinemia [8]; (2) salivary gland echostructure abnormalities, which reflect inflammation and fibrosis [9,10]; (3) lymphocyte organization with germinal center-like structures; and (4) an abnormal peripheral blood B-cell profile [11–13]. Lymphocyte infiltration correlates poorly with decrease in lacrimal and salivary secretions. Epithelial cells are not only the target of the immune system in this physiopathological model, but they are also implicated in its genesis, as they act as nonclassical antigen presenting cells and produce adhesion molecules, costimulatory molecules, chemokines, and cytokines (further details are provided in [14,15]).

The etiology and pathogenesis of pSS remain unclear; however, environmental factors (e.g., viruses and drugs), genetic predisposition, and epigenetic factors have all been implicated [16–19]. Genetic risk factors associated with pSS include *HLA-DR* allele subtypes and up to 40 polymorphisms associated with key pathological pathways, including type I interferon (IFN)-regulated or induced genes (e.g., *IRF5* and *STAT4*), the NF-κB pathway (e.g., *TNIP1* and *TNFAIP3*), and B-cell activation (e.g., *CXCR5* and *BLK*) [20,21]. It is important to note that most of these genetic mutations are present within regulatory sequences, which are far from DNA-coding regions and within cell-specific epigenetic areas [22,23].

Epigenetics was initially introduced to explain how cells defined by a common genotype can differentiate into alternative phenotypes. Such a definition has evolved, and epigenetics now refers to the heritable changes in gene expression that do not entail modifications in the DNA sequence. Recent data also suggest that epigenetic mechanisms are not only important to regulate development, control the cell cycle, and respond to external changes, but that they are also able to drive complex diseases [24–26].

11.2 INVOLVEMENT OF EPIGENETICS IN pSS

In the 1960s, the oral administration of hydralazine or isoniazid to mice for several weeks was shown to increase the incidence of antinuclear Ab. The authors also established a link with the relative frequency of systemic lupus erythematosus (SLE)-like syndromes in patients treated with the same drugs [27]. It is important to note that variations were observed depending on the animal strain, age, and sex; however, this effect disappeared after discontinuation of the drug. Both hydralazine and procainamide have the capacity to inhibit DNA methyltransferase (DNMT) activity and remove methyl groups from cytosines within phosphorylated cytosine–guanine (CpG) sites. Although hydralazine was used to cure hypertension, it was also reported to induce SS in humans on occasion [28,29].

Two research groups have demonstrated that autoreactivity in SLE is characterized by DNA hypomethylation in both $CD4^+$ T cells and $CD19^+$ B cells [30,31]. Conversely, experiments involving the passive transfer of DNMT inhibitor-pretreated $CD4^+$ T cells into mice induced the production of anti-dsDNA Ab and the generation of severe immune complex glomerulonephritis [32]. Similar results were also obtained when B cells, instead of $CD4^+$ T cells, were treated with demethylating drugs [33]. Further studies characterized the defective pathways that led to decreased DNMT1 expression in $CD4^+$ T cells, identifying the oxidative stress, PKCδ, and MAPK/ERK pathways [34,35].

11.3 ALU, LONG INTERSPERSED NUCLEAR ELEMENTS, AND HUMAN ENDOGENOUS RETROVIRUSES AS SENSORS OF DNA DEMETHYLATION

Transposons, frequently referred to as "junk" DNA, are subdivided into long interspersed nuclear elements (LINE, which occupy 16.9% of the genome), short interspersed nuclear elements [i.e., short stretch of DNA characterized by the action of the *Arthrobacter luteus (Alu)* restriction endonuclease, which occupy 10.6% of the genome], and long terminal repeat elements, which include the 400,000 copies of human endogenous retroviruses (HERVs). The latter have colonized up to 7% of the human genome by reversing the normal flow of genetic information from DNA to RNA [36]. HERV genes are mostly nonfunctional due to the presence of deletions, stop codons, or frame shifts; however, a few copies have the ability to generate functional proteins, although they are normally epigenetically suppressed. Tugnet et al. reported that the HERV-K family is the only group that can produce intact viral particles [37]. Moreover, members of the HERV-K family, together with HERV-E, are the most transcriptionally active as they retain intact open reading frames (ORFs), which therefore increases the possibility that HERVs may play a role in human diseases [38–40]. HERV-K elements are strongly suspected to be involved in amyotrophic lateral sclerosis, multiple sclerosis, and cancer [41], and an aberrant accumulation of retroelements is associated with two genetic diseases with SLE features (familial chilblain lupus and Aicardi–Goutieres syndrome) [42].

Alu, LINE-1, and HERV elements are overexpressed in the minor salivary glands (MSGs) of pSS patients, and this deregulation often results in defective control of DNA methylation (Table 11.1) [19,31,37,38,40,43–45]. The hypothesis that retroelements contribute to human autoimmunity and

TABLE 11.1 Cell-Specific Epigenetic Modifications and Associated Environmental Factors in Patients with Primary Sjögren's Syndrome

	MSG epithelial cells	B cells	T cells	Monocytes	References
Environmental factors	Drugs, viruses, and stress	Drugs and EBV	Drugs	Viruses	[29,32,33]
Transposons	Altered Alu, LINE1, HERV-E, and HRES1	HRES1	Unknown	Unknown	[19,43,45,48]
DNA methylation	—	Altered	Altered	Unknown	See Table 11.3 [54]
DNA methylation pathways	Reduced PKC/Erk, DNMT1, and GADD45A	Unknown	Unknown	Unknown	[55]
Methylation-sensitive genes	Increased *SSB/LA*, *KRT19*, and *AQP5*	Unknown	*cd70*, *lta*, and *foxP3*	Unknown	See Table 11.2
Histone modifications	H4 acetylation	Unknown	Unknown	Unknown	[56]
MiRNAs	Increased miR-146a, miR-574, and miR-768	Unknown	MiR-146a	MiR-34b	See Table 11.4
Reversibility	Anti-CD20 (DNA methylation)	Unknown	Unknown	Unknown	[55]

EBV, Epstein–Barr virus; *MSG*, minor salivary gland.

pSS is supported by different observations. First, numerous retroelements have been shown to be overexpressed in MSG since the discovery of the human intracisternal A-type retroviral particle [46–48], the human T-cell lymphotropic virus (HTLV) related endogenous sequence (HRES1) [48], and HERV-E elements (i.e., HERV-K113 and HRV-5) [19,49,50]. HRES1 was demonstrated to be regulated by DNA methylation. Another way to explore HERV overexpression (and DNA demethylation) in pSS is to test auto-Ab production against HERV antigens (i.e., *GAG* or *POLI*). Brookes et al. used synthetic peptides derived from HERV *HRES1* and highlighted anti-HRES1 Ab in 32% of pSS patients compared to <5% of controls [48]. Hishikawa et al. also detected anti-HERV clone 4-1 p30 gag Abs in 35% of pSS patients compared to 0% of controls [51]. Cross-reactivity was also tested for this group of auto-Abs, which illustrated that the anti-HRES1 Ab cross-reacted with the HTLV1 p19 capsid antigen, and that the anti-HERV clone 4-1 p30 gag Ab cross-reacted with the Smith and U1 small nuclear ribonucleoprotein particle autoantigens [52]. Second, after observing that a LINE-1 promoter was demethylated and its transcript overexpressed in MSG tissue from pSS patients, Mavragani et al. showed that transfection of plasmacytoid dendritic cells with a LINE-1-containing plasmid had the capacity to induce type 1 IFN [43]. Third, the overexpression of Alu in peripheral blood can induce an inflammatory response (e.g., IFNα, IL-6, and TNFα) that is reversible in the presence of chloroquine (a molecular agent used in SLE and pSS). Alu elements purified from SLE patients can bind the autoantigen, SSA/Ro60, and form immune complexes with anti-SSA/Ro Abs. This mechanism is also suspected to occur in pSS [53]. Further experiments are needed to establish the contribution of retroelements in the pathophysiology of pSS and to characterize the main cellular subset that produces retroelements, as almost all experiments performed to date have used whole MSG tissues instead of isolated cell subsets.

11.4 X CHROMOSOME IN pSS

Several theories support a role for the X chromosome in the development of pSS: (1) pSS occurs predominantly in women with a ratio of 9:1 (females:males); (2) there are reports showing that trisomy X (47,XXX), a super female phenotype (mosaic of XXXXX/XXXX/XXX/XX/XO), or disomy X (47, XXY), which is referred to as Klinefelter's syndrome in males, may increase the risk of pSS occurrence [57–59]; and (3) females inherit both maternal and paternal X chromosomes whereas men receive the maternal X and the paternal Y chromosome, therefore a mechanism of dosage compensation is required to equalize X-linked gene expression that is not gender specific. Consequently, X-chromosome inactivation is a major epigenetic event in mammals, so that females only contain one transcriptionally active X chromosome [60,61]. The active chromosome is referred to as Xa and the inactive chromosome as Xi. The choice between the maternally- or paternally-derived Xi is random and made in somatic cells in early embryo development [62]. After this, each daughter cell from that somatic cell will inherit the decision and maintain the same parentally derived X.

Overexpression of the X-linked CD40 ligand (CD40L) in $CD4^+$ T cells from female subjects with SLE was explained by increased DNA demethylation of the Xi [63]. Belkhir et al. observed that CD40L was overexpressed in ex vivo activated $CD4^+$ T cells from female patients with pSS. The authors tested its control by DNA methylation but failed to demonstrate a link with the cytosine methylation status at CpG sites within the regulatory regions of CD40L. This suggests that CD40L overexpression in pSS $CD4^+$ T cells may occur through nonepigenetic regulatory mechanisms [64].

11.5 DNA METHYLATION IS AFFECTED IN pSS

11.5.1 Global DNA Methylation

A recent study reported that pSS patients showed global DNA demethylation in MSG epithelial cells when compared to biopsies from controls. These defects were heritable, as observed during long-term primary culture of the salivary gland epithelial cells (SGEC) [55]. Global DNA demethylation in MSG cells was observed in patients with lymphocyte infiltration and for those patients that were positive for anti-SSB/La Ab (Fig. 11.1) [54,65]. SSB/La overexpression in SGEC from anti-SSB/La auto-Ab positive pSS patients was also shown to be associated with *SSB/LA* promoter demethylation (Figs. 11.2 and 11.3 adapted from [54,66]). However, no differences were observed in relation to age, sex, disease duration, clinical and biological parameters, drug intake, or disease activity (ESSDAI score).

Treatment with the demethylating drug, 5-aza-2′-deoxycytidine (5-Aza), increases the expression of the autoantigen, SSB/LA, and cytokeratin-19 (*KRT19*) in a human salivary gland (HSG) cell line (Table 11.2) [54,65]. Incubation with 5-Aza also upregulated aquaporin 5 (*AQP5*) gene expression and increased fluid secretion in the HSG ductal NS-SV-DC cell line [67]. Abnormal subcellular localization of the AQP5 protein was also reported, thus providing an explanation for the abnormal salivary flux that characterizes pSS patients [68]. However, treatment of CD4$^+$ T cells from pSS patients and controls with 5-Aza was ineffective at inducing IRF5 and CD40L upregulation [64,69].

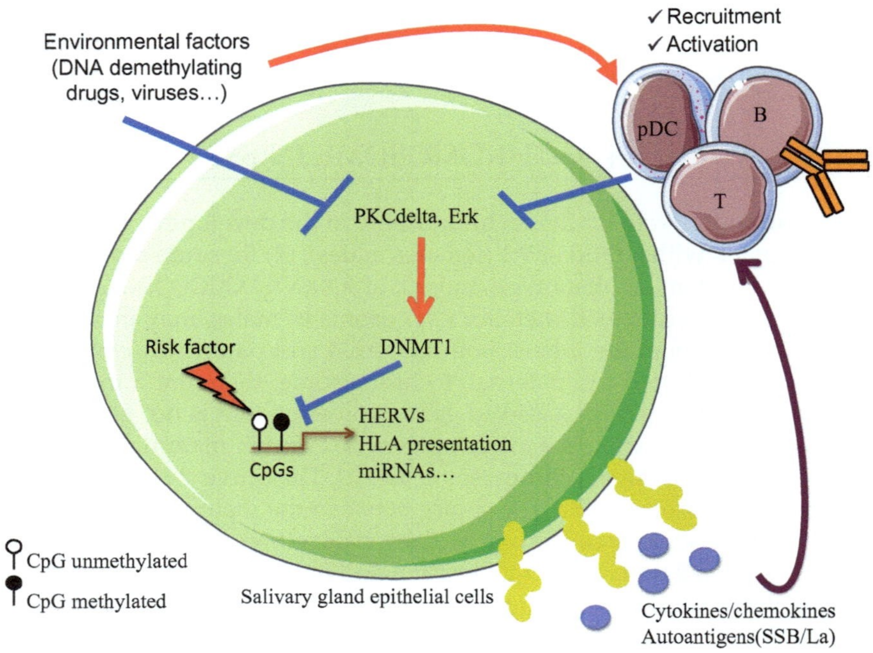

FIGURE 11.1 **Involvement of epigenetic mechanisms in minor salivary glands from patients with Sjögren's syndrome.** *Ab*, antibody; *B*, B cell; *DNMT1*, DNA methyltransferase; *HERV*, human endogeneous retrovirus; *HLA*, human leukocyte antigen; *IFN*, interferon; *pDC*, plasmacytoid dendritic cell; *SGEC*, salivary gland epithelial cell; *SSB/La*, anti-sicca syndrome B autoantibodies; *T*, T cell.

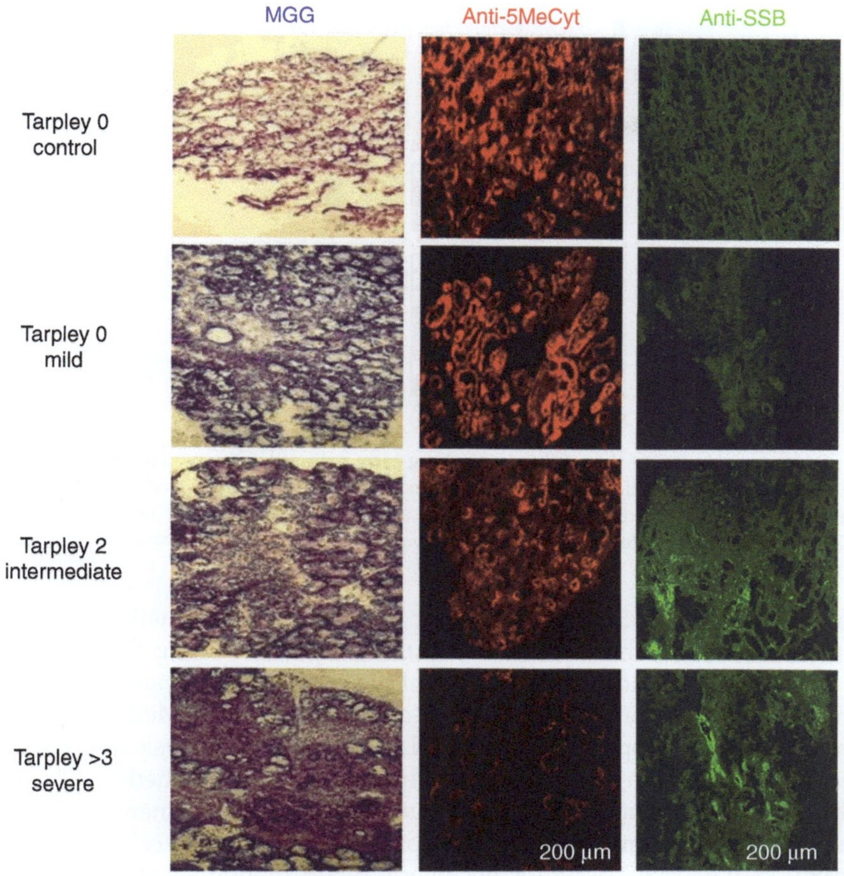

FIGURE 11.2 **Defective DNA methylation in salivary gland epithelial acini from patients with primary Sjögren's syndrome (pSS) is associated with lymphocyte infiltration and *ssb/la* gene expression.** Photomicrographs of MSG biopsy sections stained with May-Günwald-Giemsa (MGG, *blue*), an anti-5mC antibody (clone 33D3, *red*), and an anti-SSB/La Ab (*green*) in a representative control and pSS patients (for more information see [54]). PSS patients were classified as mild, intermediate, or severe according to the grade of the inflammatory Tarpley's score.

Gonzalez et al. observed hypermethylation of the dystonin (*DST*) gene promoter and overexpression of an epithelial splice variant of *DST* (the bullous pemphigoid antigen 1) in acinar cells from pSS patients [73]. Hypotheses have since arisen related to the use of an alternative promoter [77] or to posttranslational modifications.

11.5.2 Genome-Wide Methylome Analysis

11.5.2.1 Infinium Technology

The most widely used strategy for studying DNA methylation in human pSS tissues is the Infinium Human Methylation 450K (HM450K) BeadChips (Illumina) (Table 11.3) [78]. This technology interrogates human genomic DNA after bisulfite conversion (which converts unmethylated

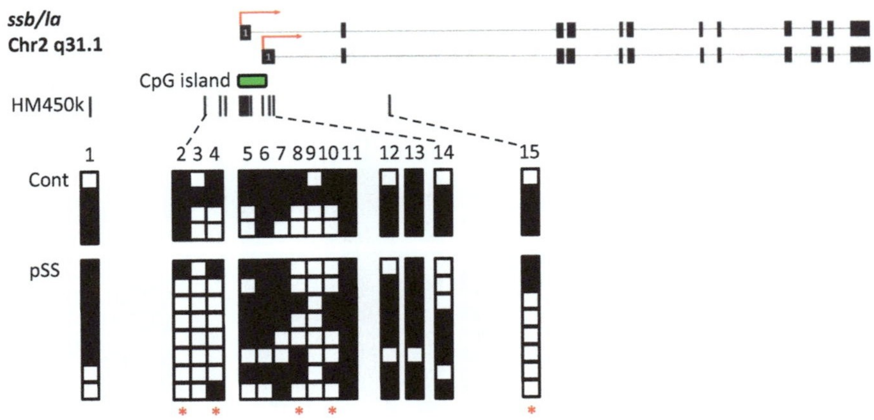

FIGURE 11.3 **DNA methylation of the *ssb/la* gene in long-term cultured salivary gland epithelial cells (SGEC) from primary Sjögren's syndrome (pSS) patients and controls (Cont) (from [66]).** The *ssb/la* locus depicts the location of the CpG probes tested using the HM450K (Illumina). *White* boxes indicate CpG sites with a value below the reference and *red* stars (*) represent CpG probes that were differentially methylated between controls and pSS patients.

cytosines to uracils) using two 50-mer probes per CpG locus: one unmethylated and one methylated. The methylation state at each probe is calculated by determining the fluorescence ratio between beads. The HM450K BeadChips cover 2% of the 28 million CpG sites in the human genome and 99% of the genes, with an average of 17 CpG sites per gene. The Methylation EPIC BeadChip (Illumina) was recently developed to further DNA methylation analysis and cover 850,000 CpG sites; however, pSS studies using this technology have yet to be published.

Genome-wide DNA methylation arrays need an appropriate experimental design to avoid misinterpretations [81]. First, and most important, epigenetic profiles are cell-specific. Therefore heterogeneous mixtures [i.e., MSG tissue or peripheral blood mononuclear cells (PBMCs)] must be adjusted to analyze and decipher specific changes in the epigenetic profile as opposed to changes in the cellular composition. Second, the integration of quality controls for samples (i.e., probes, background, and adjustments for batch effects such as laboratory conditions and experiment time) can help to remove nonspecific signals. Third, probes residing in the X chromosome or including single nucleotide polymorphisms could or could not be filtered out. It should also be noted that methylation levels tend to be tightly correlated between nearby CpG loci, that somatic rather than germinal mutations affect methylation levels, that genes and gene regions (e.g., promoters) are not equally covered, and that gene pathway analysis can be biased by the selection of the CpG array. Therefore the fourth and final essential step in the experimental design is the validation of significant hits with independent data resources and other experimental approaches.

11.5.2.2 Whole Blood, T Cells, and B Cells

Genome-wide DNA methylation profiling was conducted in PBMCs, B cells, and T cells (whole and naive) using the HM450K beads [71,72,79,82] (Table 11.3). Although the methodological analyses used were distinct, several important points should be highlighted from these studies. First, although global DNA was unaffected, there was a trend for DNA hypermethylation in B cells and DNA demethylation in T cells when comparing differentially methylated beads. Second, B cells demonstrated approximately 50-times more DNA methylation changes than T cells, and such an

TABLE 11.2 DNA Methylation Sensitive and Insensitive Genes in Patients With Primary Sjögren's Syndrome

Promoter					
Hypomethylated	Normal or hypermethylated	Cell type	Expression	Function	References
AQP5		NS-SV-DC	Up	Salivary flux	[67]
CD70		CD4+	Up	Costimulatory cytokine	[70]
GTF2I		SGEC and HSG	NT	Regulator of transcription	[66]
HRES-1		MSG, CD19+	Up	HTLV-1-related endogenous retrovirus	[45,48]
ICA1		SGEC and HSG	NT	Autoantigen	[66]
IFI44L		CD19+	NT	IFN pathway	[71]
IFITM, IFITM3, AND		CD19+	NT	IFN pathway	[71]
KRT19		SGEC and HSG	Up	Intermediate filament for epithelial cells	[65]
LINE1		MSG	Up	Induction of IFN-alpha	[43]
LTA		CD4+	Up	T-cell cytokine	[72]
NLRP3		SGEC and HSG	NT	NF-kB pathway	[66]
RUNX3		CD19+	NT	Transcription factor (Th17)	[71]
SLC25A40		SGEC and HSG	NT	Mitochondrial carrier	[66]
SSB/LA		SGEC and HSG	Up	Autoantigen and RNA metabolism	[54]
TNFAIP8		CD19+	NT	Antiapoptotic	[71]
	CD40L	CD4+	Unaffected	T-cell cosignal	[64]
	DST	MSG	Up BP320	Anchoring protein	[73]
	FOXP3	CD4+	Down	Regulatory T cells	[74,75]
	IFI44L	CD4+	NT	IFN pathway	[76]
	IRF5	CD19+ and CD4+	Unaffected	Transcription factor (IFN)	[69,71]

CD4+, CD4+T cells; *CD19+*, CD19+ B cells; *down*, downregulated; *HSG cell line*, human salivary gland epithelial cells; *IFN*, interferon; *MSG*, minor salivary gland; *NS-SV-DC cell line*, human salivary gland ductal; *NT*, not tested; *SGEC*, long-term cultured salivary gland epithelial cells; *up*, upregulated.

effect was increased when considering patients with anti-SSA/SSB Abs. Third, common identified pathways were related to antigen presentation and type I IFN, while specific pathways were related to receptor signaling and development genes in B cells, and solute carrier proteins and the runt-related transcription factor 1 (*RUNX1*) in T cells. Fourth, mapping of differentially methylated CpGs (DMCs) revealed that they were predominantly present within open-sea or intergenic

TABLE 11.3 Genome-Wide Methylome Analysis Using the Human Methylation 450K BeadChip (Illumina) and Signaling Pathways in Cell Subtypes of Patients with Primary Sjögren's Syndrome

DNA methylation status according to control					
DMCs (n) (selection criteria)	**Hypomethylated (%)**	**Hypermethylated (%)**	**Cell type**	**Pathways**	**References**
45 ($P < 1.3 \times 10^{-7}$)	73.3	26.7	MSG	—	[79]
11,785 ($P < 1.3 \times 10^{-7}$)	52.4	47.6	PBMCs	Antigen presentation, IFN signaling, and Graft-versus-host disease signaling	[79]
453 ($P < 1.3 \times 10^{-7}$)	22	78	$CD19^+$	IFN signaling	[79]
6707 (Median >7%, $P < 0.01$)	44	56	$CD19^+$	IL4 signaling, IL8 signaling, CXCR4 signaling, PTEN signaling, B-cell receptor signaling, B-cell development, and altered T- and B-cell signaling in RA and SLE signaling	[71]
753 (Mean >1.2, $P \leq 0.01$)	73.4	26.6	$CD4^+$	Apoptosis and inflammatory pathways	[72]
7820	73	27	MSG	Immune response and signal transduction	[80]
4662 (Median >7%, $P < 0.01$)	21	79	SGEC	Calcium and Wnt pathways	[66]

CD4+, CD4+ naive T cells; *CD19+*, CD19+ B cells; *DMCs*, differentially methylated CpGs; *MSG*, minor salivary glands; *PBMCs*, peripheral blood mononuclear cells; *SGEC*, long-term cultured salivary gland epithelial cells.

genomic areas that were far from promoters and CpG islands. Fifth, overlaps between genes with DMCs and pSS-associated genetic risk factors were observed.

Analysis of promoter DNA methylation in CD4$^+$ T cells revealed demethylation and overexpression of *CD70* (TNF receptor-associated factor 7) [70]. Several genes involved in the IFN pathway were demethylated in naive CD4$^+$ T cells, including lymphotoxin alpha (*LTA*), signal transducer and activator of transcription (*STAT*)1, IFN-induced protein 44 like (*IFI44L*), IFN-induced transmembrane protein 1 (*IFITM1*), and ubiquitin specific peptidase 18 (*USP18*); *RUNX1* was hypermethylated [72]. However, promoter demethylation at the *IFI44L* gene locus in pSS patients was not confirmed in a separate study [76]. Yu et al. identified DNA hypermethylation in the forkhead box P3 (*FOXP3*) promoter in pSS patients, which corresponded to a decrease in the expression of FOXP3 at both the mRNA and protein levels in peripheral CD4$^+$ T cells of patients compared to controls [75]. Of note, FOXP3+ regulatory T cells (T_{reg}) frequency in the MSG correlated positively with the inflammation grade, C3 and C4 hypocomplementemia, and cryoglobulinemia [83,84].

11.5.2.3 *Minor Salivary Glands (MSG)*

Two studies used the HM450K beads to analyze MSG samples from pSS patients [79,80]; however, these studies were limited because of the small number of samples tested (n = 13–15) and because of the heterogeneity of the MSG tissue, as it consist of infiltrating mononuclear cells (e.g., T cells, monocytes, B cells, and dendritic cells), acinar cells, ductal epithelial cells, fibroblasts, and myoepithelial cells [78]. Cole et al. [80] reported inverse correlations between the DNA methylation levels of the top five DMCs and the focus score (a measure of lymphocytic infiltration) and the ocular staining score (a measure of tear film reduction). An enrichment of DMCs in the promoter regions of genes involved in the stress response, the immune response, and signal transduction was reported. Moreover, analysis of transcription factor binding motifs within promoter regions around pSS-associated DMCs revealed antioxidant response elements, IFN-stimulated response elements, and PU-1 binding motifs.

However, results obtained from genome-wide analysis were different when long-term cultured SGEC were analyzed with HM450K instead of MSG samples [66]. PSS patients were discriminated from controls using a supervised hierarchical clustering of DNA methylation, and two pathways were shown to be enriched: (1) the calcium pathway (DNA hypomethylation), which is important for the control of salivary flux; and (2) the Wnt pathway (DNA hypermethylation), which is important for epithelial cell development and proliferation. An IFN signature was also highlighted for differentially methylated genes in pSS SGEC. Patients were further subdivided according to treatment, particularly hydroxychloroquine intake, which led to characterization of the phosphatidyl inositol pathway. It should be noted that several DMCs were also present within pSS-associated risk factors (i.e., *ICA1*, NRLP3, SLC25A40, and GTF2I), which were sensitive to the actions of 5-Aza in the HSG cell line.

11.5.3 DNA Methylation Enzymes

Global DNA demethylation in long-term cultured SGEC is associated with a decrease in the methylating enzyme, DNMT1, and an increase in the demethylating partner, GADD45α [55]. The contribution of B cells in the demethylation process was further highlighted following the treatment of pSS patients with rituximab, a chimeric anti-CD20 monoclonal Ab, in the TEARS open clinical trial [55,85]. Global DNA methylation levels were higher at 4 months after B-cell depletion in comparison with the MSG biopsy obtained at the initiation of rituximab therapy, thus suggesting that DNA demethylation in SGEC may be partly attributed to the presence of infiltrating B cells. This is supported by the inverse correlation between global DNA methylation and lymphocyte infiltration [54,65]. This hypothesis was tested using coculture experiments, and B cells were confirmed to mediate DNA demethylation in the HSG cell line through altered ERK/DNMT1 signaling, similar to that reported in SLE [86].

11.6 HISTONES IN pSS

Global histone H3 and H4 hyperacetylation, and histone H3 demethylation, at lysine 4 characterize CD4$^+$ T cells from SLE patients [87]. In addition to positively affecting transcription, these modifications can also lead to the development of specific auto-Ab. This was demonstrated with the lupus-derived monoclonal Ab BT-164 and KM-2, which recognize histone H3 trimethylated lysine 27 and histone H4 acetylated lysine 8, 12, and 16, respectively [88–90]. Blocking histone hyperacetylation represents an interesting therapeutic alternative as two histone deacetylase inhibitors (Trichostatin A and suberoylanilide hydroxamic acid) can revert histone modifications and improve SLE symptoms in mice without affecting auto-Ab titers [91,92]. These results provide arguments to suggest that the DNA methylation modifications that characterize pSS patients are associated with histone modifications, and that some of the antihistone auto-Abs detected in pSS target post-translational histone modifications [93]. TNFα inhibition of *AQP5* expression in the NS-SV-AC cell line was tested, and demonstrated that such inhibition was independent of DNA methylation but controlled by histone H4 acetylation [56].

11.7 miRNA IN pSS

11.7.1 Description

MicroRNAs (miRNAs) are small noncoding single-strand RNA transcripts of 19–22 nucleotides that regulate gene expression at the post-transcriptional level and are important in a wide range of physiological and pathological processes [94]. Primary miRNA transcripts are generated by RNA polymerase II in the nucleus and cleaved by an RNase III enzyme, Drosha. After cleavage, miRNAs are transported to the cytoplasm by exportin 5, where they are further processed by Dicer into mature miRNA duplexes. They are finally separated into single strands at the core of the multiprotein RNA-induced silencing complex (RISC) by Argonaute proteins. Most miRNAs bind to the 3′ untranslated region of their target mRNAs and cause translation repression or mRNA degradation. Dysregulated miRNA expression is associated with DNA methylation and histone modifications in numerous autoimmune diseases.

11.7.2 MiRNA in MSG and Saliva from pSS Patients

Investigation of miRNAs in MSG samples revealed important differences between pSS patients and controls (Table 11.4) [95–97]. A predictive analysis of biologic pathways under miRNA control suggested that they regulate salivation through the regulation of neural pathways [95]. It was also claimed that downregulation of miRNAs targeting *TRIM21* and *TROVE2* in the MSG of pSS patients may partially explain the lack of transcriptional regulation of the sicca-associated

TABLE 11.4 Deregulated Expression of miRNAs in Patients With Primary Sjögren's Syndrome

MiRNAs					
Upregulated	**Downregulated**	**Cell type**	**mRNA Target**	**Function**	**References**
miR-146a		PBMCs	*IRAK1/TRAF6*	NF-κB and IFN regulator	[98,99]
	miR-155	PBMCs	*SOCS1*, *TAB2*, and *cMAF*	B-cell development	[99]
miR-181a		PBMCs and MSG		B-cell development and salivary flux	[97,100]
miR-16		PBMCs and MSG	*TRIM21*	Autoantigen and salivary flux	[97,101]
miR-223		PBMCs	*TROVE2*	Autoantigen	[97]
miR-200b		SGEC	*TRIM21* and *TROVE2*	Autoantigen	[97]
Let-7b		SGEC SSA+*	*TRIM21* and *TROVE2*	Autoantigen	[97]
miR-34b-3p		Monocytes	*GRB2*, *P38*, and *MEKK1*	MAPK pathway	[102]
miR-877-3p		Monocytes	*SMAD2*, *SOS1*, *ERK1*, *RAC1*, *HGK*, and *STAT6*	TGFβ, MAPK, and JAK-STAT pathways	[102]
miR-3162-3p		Monocytes	Unknown	Unknown	[102]
miR-4701-5p		Monocytes	*SHP2*, *HIPK2*, *MKK7*, and *IFNAR1*	MAPK and JAK-STAT pathways	[102]
	miR-1207-5p	MSG	*TRIM21*	Autoantigen	[103]
	miR-1695-3p	MSG	*TRIM21*	Autoantigen	[103]
miR-203		Saliva	Unknown	Unknown	[104]
miR-768-3p		MSG and saliva	Unknown	Unknown	[95,104]
	miR-574-3p	MSG and saliva	Unknown	Unknown	[95,104]
	miR-17-92		Unknown	Unknown	[105]
miR-4524		MSG and HSG	Unknown	Unknown	[96]
miR-5571		MSG and HSG	Unknown	Unknown	[96]
miR-5100		MSG and HSG	Unknown	Unknown	[96]
miR-5572		MSG and HSG	Unknown	Unknown	[96]
miR-BART13-3p		B cells and saliva	*STIM1*	Ca^{2+} pathway in HSG	[106]

HSG, Human salivary gland cell line; *MSG*, minor salivary gland tissues; *PBMCs*, peripheral blood mononuclear cells; *SGEC*, long-term cultured salivary gland epithelial cells; *SSA+*, SGEC from patients that had autoantibodies to SSA/Ro and SSB/La; *TRIM21*, SSA/Ro 52 kDa; *TROVE 2*, SSA/Ro 60 kDa.

autoantigens, SSA/Ro 52 kDa and 60 kDa [97], while the SSB/LA autoantigen was proposed to be controlled by DNA methylation [54].

11.7.3 MiRNA-146a in PBMCs and Exosomes from pSS Patients

Pauley et al. reported that miR-146a was upregulated in PBMCs from pSS patients and SS-prone mouse models [107]. This observation has since been confirmed by other groups. MiR146a levels increase before pSS onset in PBMCs and during disease progression in the MSG, thus suggesting that miR-146a acts early in the disease pathogenesis. MiR-146a is also present within saliva exosomes, which are microvesicles secreted by cells including lymphocytes and epithelial cells [108]. Mir146a plays a critical role in increasing phagocytic activity and repressing inflammatory cytokine production in monocytes. The transcription factor, NF-κB, controls miR-146a and the downstream TLR/INF pathway through TNF-associated factor 6 (TRAF6), IL-1 receptor-associated kinase (IRAK1), STAT1, and IRF5 [98]. Zilahi et al. tested PBMCs from pSS patients and healthy controls for expression levels of miR146a and miR146b and their target genes *IRAK1, IRAK4,* and *TRAF6*. The authors established that miR146a/b and *TRAF6* were overexpressed in pSS patients, while *IRAK1* expression was significantly decreased. It was therefore proposed that *TRAF6* contributes to the increased activation of the NF-κB pathway through the involvement of PKCξ, as is typically seen in the disease [109].

11.8 CROSS TALK WITH GENETIC RISK FACTORS

The primary role of the epigenome is to regulate transcription, and recent research from the Encyclopedia of DNA Elements (ENCODE) and the Roadmap Epigenomic Program have highlighted that epigenetics can affect: (1) the transcriptional machinery, when present within promoters, enhancers, and insulators (e.g., gene regulatory elements that control interactions between enhancers and promoters); (2) the control of intron excision, when present in areas controlling spliceosome complex formation; and (3) the control of mRNA, through the control of nonsense mediated decay and stabilization via miRNAs.

It has also been shown that epigenetic signatures fall within pSS-associated genetic risk factors. This was demonstrated in B cells when analyzing both the active histone mark H3K27Ac [21] and DMCs [71]. This cross talk was shown to be cell-specific due to the overrepresentation of the association in cell-specific regulatory elements. Analysis of transcription factor binding sites within pSS risk variants revealed immune-related transcription factors (i.e., NF-kB, STATs, and PU-1), which are downstream of the type I and NF-kB pathways that are known to be affected in pSS [20].

11.9 CONCLUSIONS

PSS is a complex and heterogenous disease involving several cell types (i.e., B lymphocytes, T lymphocytes, and epithelial cells). Evidence suggests that epigenetic changes (i.e., DNA methylation, histone modifications, and miRNAs) confer a risk for pSS in genetically predisposed individuals. Another important point is cellular specificity, as there are important differences between epithelial cells and lymphocytes. However, these processes have been shown to be reversible and involve specific pathways, therefore better understanding of the mechanisms controlling these pathways would undoubtedly benefit pSS diagnosis, prognosis, prevention, and treatment.

Given the large volume of experimental evidence for the transgenerational inheritance of epigenetic trails [110], the next step will be to test the contribution of parental epigenetic alterations in determining the risk for chronic diseases (i.e., pSS). If true, identification of such mechanisms may hold great promise for predicting susceptibility traits and for developing epigenetic drugs. These new drugs may lead to personalized treatment according to a patient's epigenetic status in their lymphocytes or SGECs.

Acknowledgments

We are grateful to the "Association Française du Gougerot–Sjögren et des Syndromes Secs," the region Bretagne for their support, and to Genevieve Michel and Simone Forest for their help writing the chapter.

References

[1] Mavragani CP, Moutsopoulos HM. Sjogren's syndrome. Annu Rev Pathol 2014;9:273–85.
[2] Voulgarelis M, Dafni UG, Isenberg DA, Moutsopoulos HM. Malignant lymphoma in primary Sjogren's syndrome: a multicenter, retrospective, clinical study by the European Concerted Action on Sjogren's Syndrome. Arthritis Rheum 1999;42(8):1765–72.
[3] Baimpa E, Dahabreh IJ, Voulgarelis M, Moutsopoulos HM. Hematologic manifestations and predictors of lymphoma development in primary Sjogren syndrome: clinical and pathophysiologic aspects. Medicine 2009;88(5):284–93.
[4] Tobon GJ, Renaudineau Y, Hillion S, Cornec D, Devauchelle-Pensec V, Youinou P, et al. The Fms-like tyrosine kinase 3 ligand, a mediator of B cell survival, is also a marker of lymphoma in primary Sjogren's syndrome. Arthritis Rheum 2010;62(11):3447–56.
[5] Seror R, Bowman SJ, Brito-Zeron P, Theander E, Bootsma H, Tzioufas A, et al. EULAR Sjogren's syndrome disease activity index (ESSDAI): a user guide. RMD Open 2015;1(1):e000022.
[6] Seror R, Ravaud P, Bowman SJ, Baron G, Tzioufas A, Theander E, et al. EULAR Sjogren's syndrome disease activity index: development of a consensus systemic disease activity index for primary Sjogren's syndrome. Ann Rheum Dis 2010;69(6):1103–9.
[7] Christodoulou MI, Kapsogeorgou EK, Moutsopoulos HM. Characteristics of the minor salivary gland infiltrates in Sjogren's syndrome. J Autoimmun 2010;34(4):400–7.
[8] Capaldo C, Carvajal Alegria G, Cornec D, Jousse-Joulin S, Devauchelle-Pensec V, Renaudineau Y. The active immunological profile in patients with primary Sjogren's syndrome is restricted to typically encountered autoantibodies. Clin Exp Rheumatol 2016;34(4):722.
[9] Jousse-Joulin S, Devauchelle-Pensec V, Morvan J, Guias B, Pennec Y, Pers JO, et al. Ultrasound assessment of salivary glands in patients with primary Sjogren's syndrome treated with rituximab: quantitative and Doppler waveform analysis. Biologics 2007;1(3):311–9.
[10] Cornec D, Jousse-Joulin S, Marhadour T, Pers JO, Boisrame-Gastrin S, Renaudineau Y, et al. Salivary gland ultrasonography improves the diagnostic performance of the 2012 American College of Rheumatology classification criteria for Sjogren's syndrome. Rheumatology 2014;53(9):1604–7.
[11] Cornec D, Saraux A, Cochener B, Pers JO, Jousse-Joulin S, Renaudineau Y, et al. Level of agreement between 2002 American-European Consensus Group and 2012 American College of Rheumatology classification criteria for Sjogren's syndrome and reasons for discrepancies. Arthritis Res Ther 2014;16(2):R74.
[12] Simonin L, Pasquier E, Leroyer C, Cornec D, Lemerle J, Bendaoud B, et al. Lymphocyte disturbances in primary antiphospholipid syndrome and application to venous thromboembolism follow-up. Clin Rev Allergy Immunol 2017;53(1):14–27.
[13] Alonso R, Buors C, Le Dantec C, Hillion S, Pers JO, Saraux A, et al. Aberrant expression of CD6 on B-cell subsets from patients with Sjogren's syndrome. J Autoimmun 2010;35(4):336–41.
[14] Voulgarelis M, Tzioufas AG. Pathogenetic mechanisms in the initiation and perpetuation of Sjogren's syndrome. Nat Rev Rheumatol 2010;6(9):529–37.

[15] Brito-Zeron P, Baldini C, Bootsma H, Bowman SJ, Jonsson R, Mariette X, et al. Sjogren syndrome. Nature Rev Dis Primers 2016;2:16047.
[16] Liu A, La Cava A. Epigenetic dysregulation in systemic lupus erythematosus. Autoimmunity 2014;47(4):215–9.
[17] Mavragani CP, Moutsopoulos HM. The geoepidemiology of Sjogren's syndrome. Autoimmun Rev 2010;9(5):A305–10.
[18] Cobb BL, Lessard CJ, Harley JB, Moser KL. Genes and Sjogren's syndrome. Rheum Dis Clin North Am 2008;34(4):847–68. (vii).
[19] Le Dantec C, Varin MM, Brooks WH, Pers JO, Youinou P, Renaudineau Y. Epigenetics and Sjogren's syndrome. Curr Pharm Biotechnol 2012;13(10):2046–53.
[20] Lessard CJ, Li H, Adrianto I, Ice JA, Rasmussen A, Grundahl KM, et al. Variants at multiple loci implicated in both innate and adaptive immune responses are associated with Sjogren's syndrome. Nat Genet 2013;45(11):1284–92.
[21] Konsta OD, Le Dantec C, Charras A, Brooks WH, Arleevskaya MI, Bordron A, et al. An in silico approach reveals associations between genetic and epigenetic factors within regulatory elements in B cells from primary Sjogren's syndrome patients. Front Immunol 2015;6:437.
[22] Dozmorov MG, Wren JD, Alarcon-Riquelme ME. Epigenomic elements enriched in the promoters of autoimmunity susceptibility genes. Epigenetics 2014;9(2):276–85.
[23] Farh KK, Marson A, Zhu J, Kleinewietfeld M, Housley WJ, Beik S, et al. Genetic and epigenetic fine mapping of causal autoimmune disease variants. Nature 2015;518(7539):337–43.
[24] Brooks WH, Le Dantec C, Pers JO, Youinou P, Renaudineau Y. Epigenetics and autoimmunity. J Autoimmun 2010;34(3):J207–19.
[25] Renaudineau Y. The revolution of epigenetics in the field of autoimmunity. Clin Rev Allergy Immunol 2010;39(1):1–2.
[26] Delgado-Vega AM, Alarcon-Riquelme ME, Kozyrev SV. Genetic associations in type I interferon related pathways with autoimmunity. Arthritis Res Ther 2010;12(Suppl 1):S2.
[27] Cannat A, Seligmann M. Induction by isoniazid and hydrallazine of antinuclear factors in mice. Clin Exp Immunol 1968;3(1):99–105.
[28] Brooks WH. X chromosome inactivation and autoimmunity. Clin Rev Allergy Immunol 2010;39(1):20–9.
[29] Darwaza A, Lamey PJ, Connell JM. Hydrallazine-induced Sjogren's syndrome. Int J Oral Maxillofac Surg 1988;17(2):92–3.
[30] Absher DM, Li X, Waite LL, Gibson A, Roberts K, Edberg J, et al. Genome-wide DNA methylation analysis of systemic lupus erythematosus reveals persistent hypomethylation of interferon genes and compositional changes to CD4+T-cell populations. PLoS Genet 2013;9(8):e1003678.
[31] Garaud S, Le Dantec C, Jousse-Joulin S, Hanrotel-Saliou C, Saraux A, Mageed RA, et al. IL-6 modulates CD5 expression in B cells from patients with lupus by regulating DNA methylation. J Immunol 2009;182(9):5623–32.
[32] Yung R, Powers D, Johnson K, Amento E, Carr D, Laing T, et al. Mechanisms of drug-induced lupus. II. T cells overexpressing lymphocyte function-associated antigen 1 become autoreactive and cause a lupus-like disease in syngeneic mice. J Clin Invest 1996;97(12):2866–71.
[33] Mazari L, Ouarzane M, Zouali M. Subversion of B lymphocyte tolerance by hydralazine, a potential mechanism for drug-induced lupus. Proc Natl Acad Sci U S A 2007;104(15):6317–22.
[34] Strickland FM, Li Y, Johnson K, Sun Z, Richardson BC. CD4(+) T cells epigenetically modified by oxidative stress cause lupus-like autoimmunity in mice. J Autoimmun 2015;62:75–80.
[35] Gorelik G, Sawalha AH, Patel D, Johnson K, Richardson B. T cell PKCδ kinase inactivation induces lupus-like autoimmunity in mice. Clin Immunol 2015;158(2):193–203.
[36] Balada E, Ordi-Ros J, Vilardell-Tarres M. Molecular mechanisms mediated by human endogenous retroviruses (HERVs) in autoimmunity. Rev Med Virol 2009;19(5):273–86.
[37] Tugnet N, Rylance P, Roden D, Trela M, Nelson P. Human endogenous retroviruses (HERVs) and autoimmune rheumatic disease: is there a link? Open Rheumatol J 2013;7:13–21.
[38] Renaudineau Y, Vallet S, Le Dantec C, Hillion S, Saraux A, Youinou P. Characterization of the human CD5 endogenous retrovirus-E in B lymphocytes. Genes Immun 2005;6(8):663–71.
[39] Renaudineau Y, Hillion S, Saraux A, Mageed RA, Youinou P. An alternative exon 1 of the CD5 gene regulates CD5 expression in human B lymphocytes. Blood 2005;106(8):2781–9.
[40] Le Dantec C, Vallet S, Brooks WH, Renaudineau Y. Human endogenous retrovirus group E and its involvement in diseases. Viruses 2015;7(3):1238–57.
[41] Li W, Lee MH, Henderson L, Tyagi R, Bachani M, Steiner J, et al. Human endogenous retrovirus-K contributes to motor neuron disease. Sci Transl Med 2015;7(307):307ra153.
[42] Volkman HE, Stetson DB. The enemy within: endogenous retroelements and autoimmune disease. Nat Immunol 2014;15(5):415–22.

[43] Mavragani CP, Sagalovskiy I, Guo Q, Nezos A, Kapsogeorgou EK, Lu P, et al. Expression of long interspersed nuclear element 1 retroelements and induction of type I interferon in patients with systemic autoimmune disease. Arthritis Rheumatol 2016;68(11):2686–96.
[44] Garaud S, Le Dantec C, Berthou C, Lydyard PM, Youinou P, Renaudineau Y. Selection of the alternative exon 1 from the CD5 gene down-regulates membrane level of the protein in B lymphocytes. J Immunol 2008;181(3):2010–8.
[45] Fali T, Le Dantec C, Thabet Y, Jousse S, Hanrotel C, Youinou P, et al. DNA methylation modulates HRES1/p28 expression in B cells from patients with lupus. Autoimmunity 2014;47(4):265–71.
[46] Garry RF, Fermin CD, Hart DJ, Alexander SS, Donehower LA, Luo-Zhang H. Detection of a human intracisternal A-type retroviral particle antigenically related to HIV. Science 1990;250(4984):1127–9.
[47] La Placa M, Vitone F, Bianchi T, Vincenzi C, Gibellini D, Re MC, et al. Serum antibodies against human intracisternal A-type particle (HIAP) endogenous retrovirus in alopecia areata patients: a hallmark of autoimmune disease? J Invest Dermatol 2004;123(2):407–9.
[48] Brookes SM, Pandolfino YA, Mitchell TJ, Venables PJ, Shattles WG, Clark DA, et al. The immune response to and expression of cross-reactive retroviral gag sequences in autoimmune disease. Br J Rheumatol 1992;31(11):735–42.
[49] Moyes DL, Martin A, Sawcer S, Temperton N, Worthington J, Griffiths DJ, et al. The distribution of the endogenous retroviruses HERV-K113 and HERV-K115 in health and disease. Genomics 2005;86(3):337–41.
[50] Murovska M, Lejniece S, Kozireva S, Koulikovska M, Yin H, Blomberg J. Human retrovirus 5 sequences in peripheral blood cells of patients with B-cell non-Hodgkin's lymphoma. Int J Cancer 2000;85(6):762–70.
[51] Hishikawa T, Ogasawara H, Kaneko H, Shirasawa T, Matsuura Y, Sekigawa I, et al. Detection of antibodies to a recombinant GAG protein derived from human endogenous retrovirus clone 4-1 in autoimmune diseases. Viral Immunol 1997;10(3):137–47.
[52] Piotrowski PC, Duriagin S, Jagodzinski PP. Expression of human endogenous retrovirus clone 4-1 may correlate with blood plasma concentration of anti-U1 RNP and anti-Sm nuclear antibodies. Clin Rheumatol 2005;24(6):620–4.
[53] Hung T, Pratt GA, Sundararaman B, Townsend MJ, Chaivorapol C, Bhangale T, et al. The Ro60 autoantigen binds endogenous retroelements and regulates inflammatory gene expression. Science 2015;350(6259):455–9.
[54] Konsta OD, Le Dantec C, Charras A, Cornec D, Kapsogeorgou EK, Tzioufas AG, et al. Defective DNA methylation in salivary gland epithelial acini from patients with Sjogren's syndrome is associated with SSB gene expression, anti-SSB/LA detection, and lymphocyte infiltration. J Autoimmun 2016;68:30–8.
[55] Thabet Y, Le Dantec C, Ghedira I, Devauchelle V, Cornec D, Pers JO, et al. Epigenetic dysregulation in salivary glands from patients with primary Sjogren's syndrome may be ascribed to infiltrating B cells. J Autoimmun 2013;41:175–81.
[56] Yamamura Y, Motegi K, Kani K, Takano H, Momota Y, Aota K, et al. TNF-alpha inhibits aquaporin 5 expression in human salivary gland acinar cells via suppression of histone H4 acetylation. J Cell Mol Med 2012;16(8):1766–75.
[57] Ozcelik T. X chromosome inactivation and female predisposition to autoimmunity. Clin Rev Allergy Immunol 2008;34(3):348–51.
[58] Harris VM, Sharma R, Cavett J, Kurien BT, Liu K, Koelsch KA, et al. Klinefelter's syndrome (47,XXY) is in excess among men with Sjogren's syndrome. Clin Immunol 2016;168:25–9.
[59] Liu K, Kurien BT, Zimmerman SL, Kaufman KM, Taft DH, Kottyan LC, et al. X chromosome dose and sex bias in autoimmune diseases: increased prevalence of 47,XXX in systemic lupus erythematosus and Sjogren's syndrome. Arthritis Rheumatol 2016;68(5):1290–300.
[60] Brooks WH, Renaudineau Y. Epigenetics and autoimmune diseases: the X chromosome-nucleolus nexus. Front Genet 2015;6:22.
[61] Brooks WH, Renaudineau Y. The 'inactive X-chromosome-nucleolus nexus' hypothesis of autoimmune diseases and its implications. Eur Med J 2016;2(2):82–9.
[62] Lessing D, Anguera MC, Lee JT. X chromosome inactivation and epigenetic responses to cellular reprogramming. Ann Rev Genomics Hum Genet 2013;14:85–110.
[63] Lu Q, Wu A, Tesmer L, Ray D, Yousif N, Richardson B. Demethylation of CD40LG on the inactive X in T cells from women with lupus. J Immunol 2007;179(9):6352–8.
[64] Belkhir R, Gestermann N, Koutero M, Seror R, Tost J, Mariette X, et al. Upregulation of membrane-bound CD40L on CD4+ T cells in women with primary Sjogren's syndrome. Scand J Immunol 2014;79(1):37–42.
[65] Konsta OD, Charras A, Le Dantec C, Kapsogeorgeou E, Bordron A, Brooks WH, et al. Epigenetic modifications in salivary glands from patients with Sjogren's syndrome affect cytokeratin 19 expression. Bull Group Int Rech Sci Stomatol Odontol 2016;53(1):e01.
[66] Charras A, Konsta OD, Le Dantec C, Bagacean C, Kapsogeorgou EK, Tzioufas AG, et al. Cell-specific epigenome-wide DNA methylation profile in long-term cultured minor salivary gland epithelial cells from patients with Sjogren's syndrome. Ann Rheum Dis 2017;76(3):625–8.

[67] Motegi K, Azuma M, Tamatani T, Ashida Y, Sato M. Expression of aquaporin-5 in and fluid secretion from immortalized human salivary gland ductal cells by treatment with 5-aza-2′-deoxycytidine: a possibility for improvement of xerostomia in patients with Sjogren's syndrome. Lab Invest 2005;85(3):342–53.
[68] Lee BH, Gauna AE, Perez G, Park YJ, Pauley KM, Kawai T, et al. Autoantibodies against muscarinic type 3 receptor in Sjogren's syndrome inhibit aquaporin 5 trafficking. PLoS One 2013;8(1):e53113.
[69] Gestermann N, Koutero M, Belkhir R, Tost J, Mariette X, Miceli-Richard C. Methylation profile of the promoter region of IRF5 in primary Sjogren's syndrome. Eur Cytokine Netw 2012;23(4):166–72.
[70] Yin H, Zhao M, Wu X, Gao F, Luo Y, Ma L, et al. Hypomethylation and overexpression of CD70 (TNFSF7) in CD4+ T cells of patients with primary Sjogren's syndrome. J Dermatol Sci 2010;59(3):198–203.
[71] Miceli-Richard C, Wang-Renault SF, Boudaoud S, Busato F, Lallemand C, Bethune K, et al. Overlap between differentially methylated DNA regions in blood B lymphocytes and genetic at-risk loci in primary Sjogren's syndrome. Ann Rheum Dis 2016;75(5):933–40.
[72] Altorok N, Coit P, Hughes T, Koelsch KA, Stone DU, Rasmussen A, et al. Genome-wide DNA methylation patterns in naive CD4+ T cells from patients with primary Sjogren's syndrome. Arthritis Rheumatol 2014;66(3):731–9.
[73] Gonzalez S, Aguilera S, Alliende C, Urzua U, Quest AF, Herrera L, et al. Alterations in type I hemidesmosome components suggestive of epigenetic control in the salivary glands of patients with Sjogren's syndrome. Arthritis Rheum 2011;63(4):1106–15.
[74] Sharma R, Deshmukh US, Zheng L, Fu SM, Ju ST. X-linked Foxp3 (Scurfy) mutation dominantly inhibits submandibular gland development and inflammation respectively through adaptive and innate immune mechanisms. J Immunol 2009;183(5):3212–8.
[75] Yu X, Liang G, Yin H, Ngalamika O, Li F, Zhao M, et al. DNA hypermethylation leads to lower FOXP3 expression in CD4+ T cells of patients with primary Sjogren's syndrome. Clin Immunol 2013;148(2):254–7.
[76] Zhao M, Zhou Y, Zhu B, Wan M, Jiang T, Tan Q, et al. IFI44L promoter methylation as a blood biomarker for systemic lupus erythematosus. Ann Rheum Dis 2016;75(11):1998–2006.
[77] Elliott DA, Kim WS, Gorissen S, Halliday GM, Kwok JB. Leucine-rich repeat kinase 2 and alternative splicing in Parkinson's disease. Mov Disord 2012;27(8):1004–11.
[78] Renaudineau Y, Ballestar E. Epigenetics: DNA methylation signatures in Sjogren syndrome. Nat Rev Rheumatol 2016;12(10):565–6.
[79] Imgenberg-Kreuz J, Sandling JK, Almlof JC, Nordlund J, Signer L, Norheim KB, et al. Genome-wide DNA methylation analysis in multiple tissues in primary Sjogren's syndrome reveals regulatory effects at interferon-induced genes. Ann Rheum Dis 2016;75(11):2029–36.
[80] Cole MB, Quach H, Quach D, Baker A, Taylor KE, Barcellos LF, et al. Epigenetic signatures of salivary gland inflammation in Sjogren's syndrome. Arthritis Rheumatol 2016;68(12):2936–44.
[81] Wilhelm-Benartzi CS, Koestler DC, Karagas MR, Flanagan JM, Christensen BC, Kelsey KT, et al. Review of processing and analysis methods for DNA methylation array data. Br J Cancer 2013;109(6):1394–402.
[82] Le Dantec C, Charras A, Bordron A, Brooks WH, Renaudineau Y. Epigenetic mechanisms in Sjögren's syndrome. Eur Med J 2016;1(5):21–8.
[83] Christodoulou MI, Kapsogeorgou EK, Moutsopoulos NM, Moutsopoulos HM. Foxp3+ T-regulatory cells in Sjogren's syndrome: correlation with the grade of the autoimmune lesion and certain adverse prognostic factors. Am J Pathol 2008;173(5):1389–96.
[84] Sarigul M, Yazisiz V, Bassorgun CI, Ulker M, Avci AB, Erbasan F, et al. The numbers of Foxp3 + T_{reg} cells are positively correlated with higher grade of infiltration at the salivary glands in primary Sjogren's syndrome. Lupus 2010;19(2):138–45.
[85] Devauchelle-Pensec V, Pennec Y, Morvan J, Pers JO, Daridon C, Jousse-Joulin S, et al. Improvement of Sjogren's syndrome after two infusions of rituximab (anti-CD20). Arthritis Rheum 2007;57(2):310–7.
[86] Lu Q, Renaudineau Y, Cha S, Ilei G, Brooks WH, Selmi C, et al. Epigenetics in autoimmune disorders: highlights of the 10th Sjogren's syndrome symposium. Autoimmun Rev 2010;9(9):627–30.
[87] Zhou Y, Qiu X, Luo Y, Yuan J, Li Y, Zhong Q, et al. Histone modifications and methyl-CpG-binding domain protein levels at the TNFSF7 (CD70) promoter in SLE CD4+ T cells. Lupus 2011;20(13):1365–71.
[88] van Bavel CC, Dieker JW, Kroeze Y, Tamboer WP, Voll R, Muller S, et al. Apoptosis-induced histone H3 methylation is targeted by autoantibodies in systemic lupus erythematosus. Ann Rheum Dis 2011;70(1):201–7.
[89] Dieker JW, Fransen JH, van Bavel CC, Briand JP, Jacobs CW, Muller S, et al. Apoptosis-induced acetylation of histones is pathogenic in systemic lupus erythematosus. Arthritis Rheum 2007;56(6):1921–33.
[90] Thabet Y, Canas F, Ghedira I, Youinou P, Mageed RA, Renaudineau Y. Altered patterns of epigenetic changes in systemic lupus erythematosus and auto-antibody production: is there a link? J Autoimmun 2012;39(3):154–60.

[91] Reilly CM, Mishra N, Miller JM, Joshi D, Ruiz P, Richon VM, et al. Modulation of renal disease in MRL/lpr mice by suberoylanilide hydroxamic acid. J Immunol 2004;173(6):4171–8.
[92] Mishra N, Reilly CM, Brown DR, Ruiz P, Gilkeson GS. Histone deacetylase inhibitors modulate renal disease in the MRL-lpr/lpr mouse. J Clin Invest 2003;111(4):539–52.
[93] Dantec L, Charras A, Brooks W, Renaudineau Y. Similarities and differences of epigenetic mechanisms in lupus and Sjögren's syndrome. Lupus 2015;1:e101.
[94] Zare-Shahabadi A, Renaudineau Y, Rezaei N. MicroRNAs and multiple sclerosis: from physiopathology toward therapy. Expert Opin Ther Targets 2013;17(12):1497–507.
[95] Alevizos I, Alexander S, Turner RJ, Illei GG. MicroRNA expression profiles as biomarkers of minor salivary gland inflammation and dysfunction in Sjogren's syndrome. Arthritis Rheum 2011;63(2):535–44.
[96] Tandon M, Gallo A, Jang SI, Illei GG, Alevizos I. Deep sequencing of short RNAs reveals novel microRNAs in minor salivary glands of patients with Sjogren's syndrome. Oral Dis 2012;18(2):127–31.
[97] Kapsogeorgou EK, Gourzi VC, Manoussakis MN, Moutsopoulos HM, Tzioufas AG. Cellular microRNAs (miRNAs) and Sjogren's syndrome: candidate regulators of autoimmune response and autoantigen expression. J Autoimmun 2011;37(2):129–35.
[98] Pauley KM, Stewart CM, Gauna AE, Dupre LC, Kuklani R, Chan AL, et al. Altered miR-146a expression in Sjogren's syndrome and its functional role in innate immunity. Eur J Immunol 2011;41(7):2029–39.
[99] Shi H, Zheng LY, Zhang P, Yu CQ. MiR-146a and miR-155 expression in PBMCs from patients with Sjogren's syndrome. J Oral Pathol Med 2014;43(10):792–7.
[100] Peng L, Ma W, Yi F, Yang YJ, Lin W, Chen H, et al. MicroRNA profiling in Chinese patients with primary Sjogren syndrome reveals elevated miRNA-181a in peripheral blood mononuclear cells. J Rheumatol 2014;41(11):2208–13.
[101] Gourzi VC, Kapsogeorgou EK, Kyriakidis NC, Tzioufas AG. Study of microRNAs (miRNAs) that are predicted to target the autoantigens Ro/SSA and La/SSB in primary Sjogren's syndrome. Clin Exp Immunol 2015;182(1):14–22.
[102] Williams AE, Choi K, Chan AL, Lee YJ, Reeves WH, Bubb MR, et al. Sjogren's syndrome-associated microRNAs in CD14(+) monocytes unveils targeted TGFbeta signaling. Arthritis Res Ther 2016;18(1):95.
[103] Yang Y, Peng L, Ma W, Yi F, Zhang Z, Chen H, et al. Autoantigen-targeting microRNAs in Sjogren's syndrome. Clin Rheumatol 2016;35(4):911–7.
[104] Michael A, Bajracharya SD, Yuen PS, Zhou H, Star RA, Illei GG, et al. Exosomes from human saliva as a source of microRNA biomarkers. Oral Dis 2010;16(1):34–8.
[105] Iborra M, Bernuzzi F, Invernizzi P, Danese S. MicroRNAs in autoimmunity and inflammatory bowel disease: crucial regulators in immune response. Autoimmun Rev 2012;11(5):305–14.
[106] Gallo A, Jang SI, Ong HL, Perez P, Tandon M, Ambudkar I, et al. Targeting the Ca2+ sensor STIM1 by exosomal transfer of EBV-miR-BART13-3p is associated with Sjogren's syndrome. EBioMedicine 2016;216–26.
[107] Pauley KM, Cha S, Chan EK. MicroRNA in autoimmunity and autoimmune diseases. J Autoimmun 2009;32(3–4): 189–94.
[108] Gallo A, Tandon M, Alevizos I, Illei GG. The majority of microRNAs detectable in serum and saliva is concentrated in exosomes. PLoS One 2012;7(3):e30679.
[109] Zilahi E, Tarr T, Papp G, Griger Z, Sipka S, Zeher M. Increased microRNA-146a/b, TRAF6 gene and decreased IRAK1 gene expressions in the peripheral mononuclear cells of patients with Sjogren's syndrome. Immunol Lett 2012;141(2):165–8.
[110] Haque MM, Nilsson EE, Holder LB, Skinner MK. Genomic clustering of differential DNA methylated regions (epimutations) associated with the epigenetic transgenerational inheritance of disease and phenotypic variation. BMC Genomics 2016;17:418.

CHAPTER

12

Autoimmune Thyroid Diseases and Epigenetics

Silvia Martina Ferrari, Fabio Coppedè*, Giusy Elia*, Francesca Ragusa*, Salvatore Benvenga**,†, Roberto Vita**, Alessandro Antonelli*, Poupak Fallahi**

***University of Pisa, Pisa, Italy; **University of Messina School of Medicine, Messina, Italy; †University Hospital, Messina, Italy**

OUTLINE

12.1 INTRODUCTION

Epigenetics literally means above genetics and refers to the molecular mechanisms that regulate gene expression levels without altering the nucleotide sequence [1]. Epigenetic mechanisms include DNA methylation, histone tail modifications, nucleosome positioning, and gene silencing mediated by noncoding RNAs (ncRNAs). These mechanisms are involved in several physiological processes, including cell differentiation and development, maintenance of cellular identity, silencing of repetitive elements in differentiated cells, genomic imprinting, X-chromosome

The Epigenetics of Autoimmunity. http://dx.doi.org/10.1016/B978-0-12-809912-4.00012-X

inactivation (XCI), and learning and memory processes [2]. However, epigenetic changes also occur with aging and contribute to age-related diseases [1–4]. External and environmental factors (i.e., tobacco smoking); in utero and perinatal exposures; maternal psychosocial stress during pregnancy; malnutrition or overfeeding in early life; exposure to metals, bisphenol A, or ionizing radiations; and dietary and lifestyle habits are also able to switch genes on and off by altering the epigenetic marks or impairing the machineries that establish these marks [3]. Most epigenetic alterations are heritable, can contribute to cancer development if transmitted during mitosis in somatic cells, or can be passed through gametes [3]. Indeed, increasing evidence suggests that epigenetic modifications contribute to several human disorders, including autoimmune diseases (ADs) [4], and might partially explain the "missing heritability" in these conditions (i.e., the missing link of gene–environment interactions) [3].

DNA methylation is one of the most studied epigenetic marks and consists of the addition of a methyl group to cytosine, forming 5-methylcytosine (5-mC). Methylation of CpG rich regions in the promoter, called CpG islands, results in gene silencing [5]. De novo DNA methyltransferases (DNMTs) (namely, DNMT3a and DNMT3b) establish DNA methylation patterns during early development, and DNMT1 is then required to maintain these epigenetic signatures during cell divisions. Ten-eleven translocation family proteins catalyze the transformation from 5-mC to 5-hydroxymethylcytosine, a key intermediate during the active demethylation of cytosine. Methyl-CpG-binding proteins (i.e., MeCP2) recognize and bind to the methylated CpG islands, and enhance DNA methylation-mediated transcriptional repression [5].

Eukaryotic genomes contain numerous nucleosomes, which are the structural and functional units of chromatin that consist of approximately 147 base pairs of DNA wrapped twice around a histone octamer consisting of two copies each of the core histones (H2A, H2B, H3, and H4). Nucleosomes are connected by stretches of "linker DNA" and linker histones and are involved in chromatin compaction [6]. Nucleosome positioning determines the accessibility of DNA to regulatory proteins, gene expression levels, and cell differentiation, and is to some degree inheritable. Connections between nucleosome positioning and other epigenetic factors, such as DNA methylation and hydroxymethylation, have also been shown [7].

Several posttranslational modifications occur on histone tails, including lysine acetylation, ubiquitination, and sumoylation; lysine and arginine methylation; and serine and threonine phosphorylation [8]. The combinations of these modifications ("the histone code") regulate the chromatin structure in a dynamic fashion. For example, lysine acetylation on histone tails decreases their affinity for DNA, leading to an open and transcriptionally-active chromatin structure. In contrast, methylation on lysine or arginine tail residues is associated with condensation or relaxation of the chromatin structure; for example, di- and trimethylation of H3K4 (H3K4me2 or H3K4me3, respectively) and acetylation of H3 or H4 produces an open chromatin structure that allows gene transcription, while histone deacetylation and trimethylation at H3K9 and H3K27 (H3K9me3 and H3K27me3, respectively) condenses the chromatin and leads to gene silencing [8].

NcRNAs are small or long molecules that act as epigenetic elements to regulate gene expression levels. Among them, microRNAs (miRNAs) are small molecules of approximately 22 nucleotides that bind to the 3′ untranslated region of target mRNAs and mediate their posttranscriptional regulation by degrading the transcript or inhibiting translation. MiRNAs target approximately 60% of all genes and interact with other epigenetic mechanisms (i.e., DNA methylation and histone tail modifications) to regulate the global gene expression profile [9].

Many long ncRNAs (lncRNAs), which are greater than 200 nucleotides in length, have also been identified. LncRNAs are divided into five groups based on their genomic proximity to protein-coding genes: sense, antisense, intronic, intergenic, and bidirectional [10]. Unlike miRNAs, lncRNAs can regulate gene expression both negatively and positively, and may function by forming lncRNA–RNA, lncRNA–protein, or lncRNA–chromatin interactions [11]. LncRNAs contribute to a range of human diseases [12]; however, only a few lncRNAs regulate immune responses [13,14], such as T-cell differentiation, dendritic-cell function, and cytokine production. Therefore lncRNAs play a role in autoimmune disorders in a cell-type specific manner [15]; for example, upregulation of a lncRNA could be implicated in the pathogenesis of Sjögren syndrome [16].

Epigenetics has many potential medical applications [17], and there is increasing interest in the use of epigenetic biomarkers as diagnostic or prognostic tools in human diseases [18]. Furthermore, many epigenetic marks are reversible, and compounds that are able to interfere with DNA methylation or histone tail modifications are already approved for the treatment of myelodysplastic syndromes or cutaneous T-cell lymphoma, and are undergoing clinical trials for many other diseases [19]. Natural compounds such as dietary folates, curcumin, resveratrol, quercetin, genistein, epigallocatechin 3 gallate, and vitamin B12 have epigenetic properties and may be able to counteract environmentally-inducible epigenetic changes [20]. This chapter provides an overview of the epigenetics of autoimmune thyroid diseases (AITD).

12.2 AUTOIMMUNE DISEASES

Autoimmune disorders constitute a group of diseases characterized by an immune attack, mediated by autoreactive T cells and autoantibodies, on components of an individual's own body. The immune attack is caused by a breakdown of immune tolerance [21] and results in inflammatory changes in the target organ. Cytokines and chemokines are very important in the initiation and perpetuation of this autoimmune inflammation [22].

More than 80 ADs have been identified; they affect approximately 5% of the global population, and are more frequent in women of reproductive age. This observation draws attention to the role of sex hormones and X-chromosome genes in autoimmune disorders [23].

ADs can be classified into two canonical groups: organ specific or systemic. Organ-specific autoimmune disorders are characterized by the immune response specifically reacting against autoantigens within a single organ (e.g., Hashimoto's thyroiditis (HT), Graves' disease (GD), diabetes mellitus type 1 (T1D), and primary biliary cirrhosis) [24]. Systemic autoimmune disorders are characterized by a systemic distribution of the autoantigens leading to a multiorgan attack (e.g., systemic lupus erythematosus (SLE), rheumatoid arthritis (RA), Sjögren's syndrome, and psoriasis) [25].

Individuals with an AD frequently suffer from more than one autoimmune disorder, either at the same time or during different stages of their lives [26].

Furthermore, individuals with ADs are at higher risk of developing some types of cancers; for example, patients with SLE or RA have a higher risk of developing specific types of lymphoma compared to the general population [27,28]. Solid tumors are also more frequent in individuals with ADs [29,30].

The treatment strategy depends on the type of AD. Nonsteroidal antiinflammatory drugs, immunosuppressants [31], or intravenous immunoglobulins may be used [32]. While treatment usually improves the symptoms, it does not typically cure the disease [31]. Autoimmune processes arise from a combination of genetic susceptibility and environmental factors, therefore it is critical to understand their interactions to define risk levels and to develop therapeutic strategies.

12.3 GENETICS AND EPIGENETICS OF AUTOIMMUNE DISEASES

Genetic studies in patients with SLE, coupled with evidence of discordance in disease inheritance in twins, have revealed that genetics alone cannot fully explain disease heritability. Increasing evidence has also supported a contribution of epigenetics in RA and many other ADs [33,34].

12.4 MONOZYGOTIC TWIN STUDIES IN AUTOIMMUNE DISEASES

Evidence from monozygotic (MZ) twin studies shows that the concordance of autoimmunity is 20%–30%, suggesting that genetics alone is not responsible for the development of ADs [35]. The concordance rates for some ADs changed extensively, and factors such as the latitude of the country where the subjects lived and the age of disease onset in the proband, influenced the concordance rate. This led researchers to hypothesize that the environment and aging are major modifiers of the genetic effects. Moreover, the time after diagnosis in the proband also seems to be a determining factor in the concordance rate [35].

Several studies have evaluated DNA methylation in MZ twins discordant for ADs. DNA methylation changes were identified in SLE [36], multiple sclerosis [37–39], psoriasis [40], and T1D [41]. DNA methylation is not the only epigenetic change that could explain the lack of concordance in MZ twins; however, there are only a few studies investigating other potential mechanisms [42] in a small number of twin pairs. Most of those studies have used peripheral blood as the source of DNA, which contains multiple cell types and therefore multiple epigenomes, thus skewing the methylation differences and making it difficult to generate conclusions [35].

12.4.1 Autoimmune Thyroid Disorders

Each AD involves the immune system, which is the effector of the autoimmune damage, and a target organ, which can variably resist to the autoimmune damage. Epigenetic changes to the immune system or the target organ can play a role in the disease development [35].

HT is a Th1 AD in which the thyroid gland is gradually destroyed, with some affected individuals eventually developing hypothyroidism. HT affects approximately 5% of the population, begins between the ages of 30 and 50, and is more common in women than men (4–10:1, female:male) [24]. HT is hypothesized to be due caused by the interaction of genetic predisposition and environmental factors. The strong genetic component was identified in MZ twin studies, with a disease concordance rate of 38%–55%, with an even higher concordance of circulating thyroid antibodies (up to 80%) [43]. In addition to the MHC class II genes (*HLA-DR3*, *HLA-DR5*), there are several other genetic loci associated with AITD, including immune-regulatory (*CD40*, *CTLA-4*, *PTPN22*, *FOXP3*, and *CD25*) and thyroid-specific [thyroglobulin and thyroid stimulating hormone receptor (*TSHR*)] genes [44]. The AITDs predominate in females, and an epigenetic phenomenon, XCI, has also been shown to be skewed in female AITD patients. One study analyzed genomic DNA from 87 female patients with GD, 47 female patients with HT, and 69 healthy female controls [45]. Skewed heterozygous XCI (≥80%) was identified in 29% of GD patients and 26% of HT patients, in contrast to only 9% of healthy controls. This suggests that the degree of XCI is an important contributor to the increased risk of AITD in females [45]. These results have been confirmed by an additional study [46]. The presence of sexual dimorphism in ADs suggests that epigenetic factors are more frequently found in one sex than the other, or that members of one sex are more vulnerable than the other. The increase in some ADs among women, and the societal changes that have affected women over recent years, has led some to hypothesize that habits associated with these societal and cultural changes might be the cause of the increased AD incidence in females [35].

GD, or toxic diffuse goiter, is a Th1 AITD and is the most common cause of hyperthyroidism. A combination of genetic, epigenetic, and environmental factors causes an autoimmune response to the thyroid gland in GD, which is characterized by lymphocytic infiltrations and autoantibodies targeting the TSHR [24,47,48]. A genome-wide analysis of DNA methylation and enrichment of H3K4me3 and H3K27ac histone marks in sorted $CD4^+$ and $CD8^+$ T cells was performed to identify the epigenetic changes involved in GD [49]. The authors identified 365 and 3322 differentially methylated CpG sites in $CD4^+$ and $CD8^+$ T cells, respectively. Among the hypermethylated CpG sites, there was an enrichment for genes involved in T-cell signaling (*CD247, LCK, ZAP70, CD3D, CD3E, CD3G, CTLA4,* and *CD8A*), and decreased expression of the *CD3* gene family was also identified. In addition, hypermethylation of the first intron of the *TSHR* gene was also observed, which is where several GD-associated polymorphisms are located. These results demonstrated the involvement of dysregulated DNA methylation and histone modifications in T-cell signaling genes in patients with GD [49].

Liu et al. [50] characterized the miRNA expression profile in peripheral blood mononuclear cells (PBMC) from GD patients (41 who met the initial criteria for GD, and 13 who were remission) and 35 healthy subjects, and investigated the expression changes following T(3) treatment. The authors found that 16 miRNAs were differentially expressed in initial GD patients compared to control subjects; moreover, the expression levels of miR-154*, miR-376b, and miR-431* were suppressed in PBMC from initial GD patients but recovered in those in remission. T(3) treatment directly inhibited the expression of these miRNAs in cultured PBMC from healthy controls [50], and miR-146a1 was downregulated in GD patients [51]. More recently, microarray analysis has revealed that the expression levels of miR-22 and miR-183 were increased in thyroid tissue from GD patients, while expression levels of their potential target mRNAs were decreased. Conversely, expression levels of miR-101, miR-197, and miR-660 were decreased while the expression levels of their potential target mRNAs were increased [52]. An additional study also identified moderate expression levels of specific miRNAs (146b, 181b, 21, 221, and 222) in thyroid tissue from patients with GD, which fell between the expression levels found in normal thyroid glands and papillary thyroid carcinomas [53]. Furthermore, circulating miR-23b-5p and miR-92a-39 levels were shown to be increased in GD patients in remission compared to those with intractable GD, whilst circulating levels of let-7g-3p and miR-339-5p were decreased [54]. It has been suggested that increased miR-155 and decreased miR-146a expression levels promote ocular inflammation and proliferation in Graves' ophthalmopathy [55], and that circulating levels of miR-146a and interleukin-17 are significantly correlated with the clinical activity of Graves' ophthalmopathy [56]. Moreover, it was also shown that miR-346 regulates $CD4^+CXCR5^+$ T cells by targeting Bcl6, a positive regulator of follicular helper T cells, and has a determinant role in the pathogenesis of GD [57,58].

Downregulation of miR-155_2 and upregulation of miR-200a1 were observed in thyroid tissue from patients with HT [51]. Tissue samples from patients with HT were also found to display significantly reduced miR-141 levels compared to both papillary thyroid carcinomas and control thyrocytes biopsies, thus suggesting that miR-141 expression could differentiate HT from papillary thyroid carcinomas and benign thyrocytes [59]. Furthermore, decreased expression of miR-125a-3p was shown to upregulate the interleukin-23 receptor in thyroid tissue from HT patients [60], and single nucleotide polymorphisms within miR-499a and miR-125a have also been associated with AITDs [61]. Expression levels of miR-142-5p, miR-142-3p, and miR-146a were also found to be elevated in thyroid tissue from patients with HT [62]. Moreover, miR-142-5p was also detected in the serum of patients with HT and positively correlated with thyroglobulin antibody levels [62]. Overexpression of miR-142-5p in HT thyrocytes resulted in reduced claudin-1 mRNA and protein levels [62]. Claudin proteins are important constituents of the tight junction complexes regulating epithelia permeability, and the reduced expression of claudin-1 mediated by miR-142-5p was hypothesized to cause increased permeability of the thyrocyte monolayer [62].

Studies regarding the relationship between miRNA gene polymorphisms and increased AITD risk have revealed a possible role for miRNAs in the pathogenesis of AITD. Therefore there is increasing interest in clarifying the variability of miRNA expression to discriminate those miRNAs that are deregulated in a disease from those that could account for several ADs [58]. A better understanding of miRNA-mediated networks in ADs, and their crosstalk with other epigenetic mechanisms that regulate gene expression levels, is essential to clarify the translational implications of these biomarkers.

12.5 CONCLUSIONS

There is evidence to suggest that both genetic and epigenetic factors contribute to AITD. Genetic investigations in autoimmune conditions revealed that they could not be explained by genetic susceptibility alone, and low concordance for most ADs in MZ twins further suggests that epigenetics and the environment may play significant roles [63]. This chapter summarizes the relevant epigenetic changes observed in cells and tissues from individuals with different thyroid autoimmune disorders, highlights their involvement in disease pathogenesis, and opens the way to potential targets for therapeutic strategies [64].

Indeed, the reversible nature of epigenetic mechanisms may offer a new class of drugs for modulating the immune system and inflammation. Epigenetic drugs are already used to treat some types of cancer, and many clinical trials are ongoing in other human diseases. Therefore epigenetic-based therapeutics that control autoimmunity and chronic inflammatory processes could have broad implications for the management of ADs [65]. However, further studies are needed to better understand the epigenetically-regulated networks involved in the pathogenesis of ADs to identify the optimal therapeutic targets.

References

[1] Handel AE, Ebers GC, Ramagopalan SV. Epigenetics: molecular mechanisms and implications for disease. Trends Mol Med 2010;16:7–16.

[2] Martín-Subero JI. How epigenomics brings phenotype into being. Pediatr Endocrinol Rev 2011;9(Suppl. 1):506–10.

[3] Hoffmann A, Spengler D. Environmental factors and epigenetics of neuropsychiatric disorders. In: Yasui D, Peedicayil J, Grayson DR, editors. Neuropsychiatric disorders and epigenetics. Academic Press; 2016. p. 9–34.
[4] Coppedè F, Migliore L. Epigenetics of autoimmune diseases. In: Maulik N, Karagiannis T, editors. Molecular mechanisms and physiology of disease: implications for epigenetics and health. Springer; 2014. p. 151–73.
[5] Jones PA, Liang G. Rethinking how DNA methylation patterns are maintained. Nat Rev Genet 2009;10:805–11.
[6] Luger K, Mäder AW, Richmond RK, Sargent DF, Richmond TJ. Crystal structure of the nucleosome core particle at 2.8 Å resolution. Nature 1997;389:251–60.
[7] Teif VB, Beshnova DA, Vainshtein Y, Marth C, Mallm JP, Höfer T, et al. Nucleosome repositioning links DNA (de) methylation and differential CTCF binding during stem cell development. Genome Res 2014;24:1285–95.
[8] Martin C, Zhang Y. The diverse functions of histone lysine methylation. Nat Rev Mol Cell Biol 2005;6:838–49.
[9] Sato F, Tsuchiya S, Meltzer SJ, Shimizu K. MicroRNAs and epigenetics. FEBS J 2011;278:1598–609.
[10] Rinn JL, Chang HY. Genome regulation by long noncoding RNAs. Annu Rev Biochem 2012;81:145–66.
[11] Johnsson P, Ackley A, Vidarsdottir L, Lui WO, Corcoran M, Grander D, et al. A pseudogene long-noncoding-RNA network regulates PTEN transcription and translation in human cells. Nat Struct Mol Biol 2013;20:440–6.
[12] Li Z, Chao TC, Chang KY, Lin N, Patil VS, Shimizu C, et al. The long noncoding RNA THRIL regulates TNFα expression through its interaction with hnRNPL. Proc Natl Acad Sci USA 2014;111:1002–7.
[13] Hu G, Tang Q, Sharma S, Yu F, Escobar TM, Muljo SA, et al. Expression and regulation of intergenic long noncoding RNAs during T cell development and differentiation. Nat Immunol 2013;14:1190–8.
[14] Heward JA, Lindsay MA. Long non-coding RNAs in the regulation of the immune response. Trends Immunol 2014;35:408–19.
[15] Hrdlickova B, Kumar V, Kanduri K, Zhernakova DV, Tripathi S, Karjalainen J, et al. Expression profiles of long noncoding RNAs located in autoimmune disease-associated regions reveal immune cell-type specificity. Genome Med 2014;6:88.
[16] Wang J, Peng H, Tian J, Ma J, Tang X, Rui K, et al. Upregulation of long noncoding RNA TMEVPG1 enhances T helper type 1 cell response in patients with Sjögren syndrome. Immunol Res 2016;64:489–96.
[17] Chahwan R, Wontakal SN, Roa S. The multidimensional nature of epigenetic information and its role in disease. Discov Med 2011;11:233–43.
[18] Costa-Pinheiro P, Montezuma D, Henrique R, Jerónimo C. Diagnostic and prognostic epigenetic biomarkers in cancer. Epigenomics 2015;7:1003–15.
[19] Ramachandran S, Ient J, Göttgens EL, Krieg AJ, Hammond EM. Epigenetic therapy for solid tumors: highlighting the impact of tumor hypoxia. Genes (Basel) 2015;6:935–56.
[20] Shankar E, Kanwal R, Candamo M, Gupta S. Dietary phytochemicals as epigenetic modifiers in cancer: promise and challenges. Semin Cancer Biol 2016;40–41:82–99.
[21] Cho JH, Gregersen PK. Genomics and the multifactorial nature of human autoimmune disease. N Engl J Med 2011;365:1612–23.
[22] Antonelli A, Ferrari SM, Giuggioli D, Ferrannini E, Ferri C, Fallahi P. Chemokine (C-X-C motif) ligand (CXCL)10 in autoimmune diseases. Autoimmun Rev 2014;13:272–80.
[23] Javierre BM, Hernando H, Ballestar E. Environmental triggers and epigenetic deregulation in autoimmune disease. Discov Med 2011;12:535–45.
[24] Antonelli A, Ferrari SM, Corrado A, Di Domenicantonio A, Fallahi P. Autoimmune thyroid disorders. Autoimmun Rev 2015;14:174–80.
[25] Javierre BM, Esteller M, Ballestar E. Epigenetic connections between autoimmune disorders and haematological malignancies. Trends Immunol 2008;29:616–23.
[26] Fallahi P, Ferrari SM, Ruffilli I, Elia G, Biricotti M, Vita R, et al. The association of other autoimmune diseases in patients with autoimmune thyroiditis: review of the literature and report of a large series of patients. Autoimmun Rev 2016;15:1125–8.
[27] Hansen A, Lipsky PE, Dorner T. B-cell lymphoproliferation in chronic inflammatory rheumatic diseases. Nat Clin Pract Rheumatol 2007;3:561–9.
[28] Bernatsky S, Ramsey-Goldman R, Labrecque J, Joseph L, Boivin JF, Petri M, et al. Cancer risk in systemic lupus: an updated international multi-centre cohort study. J Autoimmun 2013;42:130–5.
[29] Antonelli A, Mosca M, Fallahi P, Neri R, Ferrari SM, D'Ascanio A, et al. Thyroid cancer in systemic lupus erythematosus: a case-control study. J Clin Endocrinol Metab 2010;95:314–8.
[30] Antonelli A, Ferri C, Ferrari SM, Di Domenicantonio A, Giuggioli D, Galleri D, et al. Increased risk of papillary thyroid cancer in systemic sclerosis associated with autoimmune thyroiditis. Rheumatology (Oxford) 2016;55:480–4.

[31] Goldmuntz E, Penn AS. Autoimmune diseases fact sheet. OWH, https://www.womenshealth.gov/publications/our-publications/fact-sheet/autoimmune-diseases.html; July 16, 2012.
[32] Katz U, Shoenfeld Y, Zandman-Goddard G. Update on intravenous immunoglobulins (IVIg) mechanisms of action and off-label use in autoimmune diseases. Curr Pharm Des 2011;17:3166–75.
[33] Bird A. Perceptions of epigenetics. Nature 2007;447:396–8.
[34] Coppedè F, Migliore L. DNA damage in neurodegenerative diseases. Mutat Res 2015;776:84–97.
[35] Greer JM, McCombe PA. The role of epigenetic mechanisms and processes in autoimmune disorders. Biologics 2012;6:307–27.
[36] Javierre BM, Fernandez AF, Richter J, Al-Shahrour F, Martin-Subero JI, Rodriguez-Ubreva J, et al. Changes in the pattern of DNA methylation associate with twin discordance in systemic lupus erythematosus. Genome Res 2010;20:170–9.
[37] Baranzini SE, Mudge J, van Velkinburgh JC, Khankhanian P, Khrebtukova I, Miller NA, et al. Genome, epigenome and RNA sequences of monozygotic twins discordant for multiple sclerosis. Nature 2010;464:1351–6.
[38] Breitling LP, Yang R, Korn B, Burwinkel B, Brenner H. Tobacco-smoking-related differential DNA methylation: 27K discovery and replication. Am J Hum Genet 2011;88:450–7.
[39] Rakyan VK, Down TA, Balding DJ, Beck S. Epigenome-wide association studies for common human diseases. Nat Rev Genet 2011;12:529–41.
[40] Gervin K, Vigeland MD, Mattingsdal M, Hammerø M, Nygård H, Olsen AO, et al. DNA methylation and gene expression changes in monozygotic twins discordant for psoriasis: identification of epigenetically dysregulated genes. PLoS Genet 2012;8:e1002454.
[41] Rakyan VK, Beyan H, Down TA, et al. Identification of type 1 diabetes-associated DNA methylation variable positions that precede disease diagnosis. PLoS Genet 2011;7:e1002300.
[42] Heijmans BT, Mill J. Commentary: the seven plagues of epigenetic epidemiology. Int J Epidemiol 2012;41:74–8.
[43] Chistiakov DA. Immunogenetics of Hashimoto's thyroiditis. J Autoimmune Dis 2005;2:1.
[44] Hasham A, Tomer Y. Genetic and epigenetic mechanisms in thyroid autoimmunity. Immunol Res 2012;54:204–13.
[45] Yin X, Latif R, Tomer Y, Davies TF. Thyroid epigenetics: X chromosome inactivation in patients with autoimmune thyroid disease. Ann N Y Acad Sci 2007;1110:193–200.
[46] Brix TH, Hegedüs L. Twins as a tool for evaluating the influence of genetic susceptibility in thyroid autoimmunity. Ann Endocrinol (Paris) 2011;72:103–7.
[47] Brent GA. Clinical practice: Graves' disease. N Engl J Med 2008;358:2594–605.
[48] Burch HB, Cooper DS. Management of Graves' disease: a review. JAMA 2015;314:2544–54.
[49] Limbach M, Saare M, Tserel L, Kisand K, Eglit T, Sauer S, et al. Epigenetic profiling in $CD4^+$ and $CD8^+$ T cells from Graves' disease patients reveals changes in genes associated with T cell receptor signaling. J Autoimmun 2016;67:46–56.
[50] Liu R, Ma X, Xu L, Wang D, Jiang X, Zhu W, et al. Differential microRNA expression in peripheral blood mononuclear cells from Graves' disease patients. J Clin Endocrinol Metab 2012;97:E968–72.
[51] Bernecker C, Lenz L, Ostapczuk MS, Schinner S, Willenberg H, Ehlers M, et al. MicroRNAs miR-146a1, miR-155_2, and miR-200a1 are regulated in autoimmune thyroid diseases. Thyroid 2012;22:1294–5.
[52] Qin Q, Wang X, Yan N, Song RH, Cai TT, Zhang W, et al. Aberrant expression of miRNA and mRNAs in lesioned tissues of Graves' disease. Cell Physiol Biochem 2015;35:1934–42.
[53] Pohl M, Grabellus F, Worm K, Arnold G, Walz M, Schmid KW, et al. Intermediate microRNA expression profile in Graves' disease falls between that of normal thyroid tissue and papillary thyroid carcinoma. J Clin Pathol 2017;70:33–9.
[54] Hiratsuka I, Yamada H, Munetsuna E, Hashimoto S, Itoh M. Circulating microRNAs in Graves' disease in relation to clinical activity. Thyroid 2016;26:1431–40.
[55] Li K, Du Y, Jiang BL, He JF. Increased microRNA-155 and decreased microRNA-146a may promote ocular inflammation and proliferation in Graves' ophthalmopathy. Med Sci Monit 2014;20:639–43.
[56] Wei H, Guan M, Qin Y, Xie C, Fu X, Gao F, et al. Circulating levels of miR-146a and IL-17 are significantly correlated with the clinical activity of Graves' ophthalmopathy. Endocr J 2014;61:1087–92.
[57] Chen J, Tian J, Tang X, Rui K, Ma J, Mao C, et al. MiR-346 regulates $CD4^+CXCR5^+$ T cells in the pathogenesis of Graves' disease. Endocrine 2015;49:752–60.
[58] Coppedè F. Epigenetics and autoimmune thyroid diseases. Front Endocrinol (Lausanne) 2017;8:149. doi: 10.3389/fendo.2017.00149.
[59] Dorris ER, Smyth P, O'Leary JJ, Sheils O. *MIR141* expression differentiates Hashimoto thyroiditis from PTC and benign thyrocytes in Irish archival thyroid tissues. Front Endocrinol (Lausanne) 2012;3:102.

[60] Peng H, Liu Y, Tian J, Ma J, Tang X, Yang J, et al. Decreased expression of microRNA-125a-3p upregulates interleukin-23 receptor in patients with Hashimoto's thyroiditis. Immunol Res 2015;62:129–36.

[61] Cai T, Li J, An X, Yan N, Li D, Jiang Y, et al. Polymorphisms in *MIR499A* and *MIR125A* gene are associated with autoimmune thyroid diseases. Mol Cell Endocrinol 2016;440:106–15.

[62] Zhu J, Zhang Y, Zhang W, Zhang W, Fan L, Wang L, et al. MicroRNA-142-5p contributes to Hashimoto's thyroiditis by targeting CLDN1. J Transl Med 2016;14:166.

[63] Perricone C, Versini M, Ben-Ami D, Gertel S, Watad A, Segel MJ, et al. Smoke and autoimmunity: the fire behind the disease. Autoimmun Rev 2016;15:354–74.

[64] Wu H, Zhao M, Yoshimura A, Chang C, Lu Q. Critical link between epigenetics and transcription factors in the induction of autoimmunity: a comprehensive review. Clin Rev Allergy Immunol 2016;50:333–44.

[65] Ciechomska M, O'Reilly S. Epigenetic modulation as a therapeutic prospect for treatment of autoimmune rheumatic diseases. Mediators Inflamm 2016;2016. Article ID 9607946.

CHAPTER

13

The Epigenetics of Primary Biliary Cholangitis

Simona Marzorati, Ana Lleo*,†, Pietro Invernizzi***

***Liver Unit and Center for Autoimmune Liver Diseases, Humanitas Clinical and Research Center, Rozzano, Milan, Italy; **University of Milan-Bicocca, Milan, Italy; †Humanitas University, Rozzano, Milan, Italy**

OUTLINE

The Epigenetics of Autoimmunity. http://dx.doi.org/10.1016/B978-0-12-809912-4.00013-1

13.1 INTRODUCTION

Primary biliary cirrhosis, recently redefined as primary biliary cholangitis (PBC) [1], is a complex and chronic idiopathic autoimmune liver disease in which an immune-driven biliary injury leads to secondary cholestasis [2,3]. As with other autoimmune inflammatory diseases (AIDs), the decisive role of genetic factors has been widely demonstrated in PBC, as has the contribution of certain environmental factors that could lead to the breakdown of immune tolerance [4–8].

Recent progress in high-throughput genetic technologies has helped to identify numerous genetic risk loci that are associated with susceptibility or predisposition to PBC [9–11]. Genome-wide association studies (GWAS) produced an accumulation of disease-associated genetic variants [12,13]; however, as previously reported with other AIDs [14–25], one disappointment from these studies is that few have manage to localize the actual causal variants. Therefore the role of these genetic variants in the development and progression of the disease, and their causal mechanisms (i.e., causal nucleotide changes, rare genetic variants or the missing heritability sequencing) [26–28], remain largely unknown. Less than 20% of the heritability of autoimmune cholestatic liver diseases has been uncovered by GWAS. Moreover, progress toward understanding disease mechanisms has been limited by difficulties in assigning a molecular function to those GWAS hits that do not affect protein-coding sequences. Current models suggest that many environmental factors (e.g., viruses, hormones, nutrition, and chemicals) may also trigger PBC [9,29,30]; however, these factors alone are not sufficient to explain the deficiency of the genetic basis of PBC (i.e., the female predominance) [10]. In conclusion, just how genetic and environmental factors cause PBC remains to be determined, but it seems highly likely that other mechanisms are also implicated in the pathogenesis of this disease. It is increasingly apparent that environmental insults exert their effects via epigenetic mechanisms; for this reason, epigenetic studies could help to elucidate and provide explanations as to how genetic and environmental factors influence the onset and evolution of PBC (Fig. 13.1). Epigenetics refers to both heritable and acquired changes in gene function that occur without a change in the DNA sequence [11]. These changes, including nucleosome dynamics and histone modifications, cause structural alterations in the chromatin structure and greatly affect gene expression.

Since epigenetic modifications alter the chromatin structures and consequently affect transcription and cellular functions, epigenetic mechanisms ultimately influence life processes (e.g., cellular growth or differentiation). Indeed, epigenetic dysregulation has been associated with the pathogenesis of numerous complex diseases, including AIDs.

The first epigenetics-based article related to AIDs was published in the mid-1990s, and interest in this field began to grow precipitously by the beginning of 2007. Moreover, increased attention has been paid to the possible involvement of epigenetic defects in AIDs, including PBC, since 2011. Epigenetic mechanisms act as a lens through which one can understand the possible mechanisms involved in the pathogenesis of PBC. Since epigenetic abnormalities are reversible (to a certain degree), the identification of cell-specific targets of epigenetic deregulation could also serve as clinical markers for diagnosis, disease progression, and therapeutic approaches. This chapter presents an up-to-date analysis of the recent findings concerning the epigenetic mechanisms involved in PBC, either related to the immune system or to the target organ. Since many risk loci for PBC are also risk loci for other AIDs, it is possible to hypothesize that a shared genetic basis and similar molecular pathways may exist. Therefore epigenetic phenomenon to be explored in PBC in the future will also be discussed, using other autoimmune disorders as

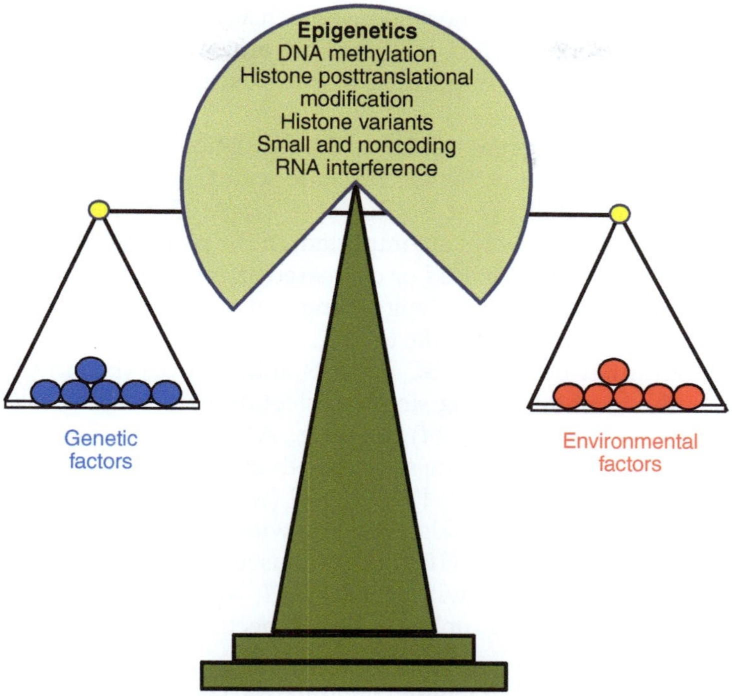

FIGURE 13.1 **Schematic illustration of the relationship between epigenetic, genetic, and environment factors in the development of primary biliary cholangitis (PBC).**

a reference. Some epigenetic defects that are already known to be involved in X-chromosome inactivation, or in immune cells in PBC patients, will also be introduced and discussed.

13.2 GENETICS AND THE ENVIRONMENT IN PBC

Knowledge of the genetic factors associated with PBC has expanded in the last decade due to the use of high-throughput whole-genome array technologies, which have permitted the genotyping of large numbers of individual single-nucleotide variants and GWAS [31]. Several susceptibility loci have been determined to date; however, the main proportion of PBC risk has still not been explained by GWAS. Indeed, studies on disease clusters of nonrelated individuals suggested that environmental factors may contribute to the breaking of immunological tolerance and lead to the onset of the disease [29,30]. Environmental components reported previously include infectious agents, chemical xenobiotics, lifestyle variables, and an exposure to pollutants.

13.2.1 Genetics

Twin and familial aggregation studies suggest that PBC is a heritable condition [32]. Data from a large cohort of patients, obtained using GWAS and iCHIP, suggested that most of these genetic associations were derived from human leukocyte antigen (HLA) loci [28,33]. HLA loci contain many genes related to the adaptive immune response, such as antigens originating extracellularly from T lymphocytes; however, only a few HLA loci have provided prognostic information for PBC. The HLA genes most commonly implicated in PBC susceptibility are the Class II *DRB1*08* allele family [28].

In addition to HLA associations, genes related to the regulation of the immune system also appear to be involved in PBC susceptibility. To date, six large-scale GWAS have identified 27 non-HLA loci associated with PBC susceptibility ($P < 5 \times 10^{-8}$) [14] (Table 13.1). The identified risk variants indicate that immunoregulatory pathways are important in the pathogenesis of PBC, although the results do not identify if there is an increased or decreased effect. Importantly, such perturbations are shared among a diverse number of AIDs, suggesting that the risk may determine a generalized propensity to autoimmunity that is not specific to PBC.

Some variants did not differ between PBC patients and controls (i.e., in CD40 [15] or CD14 [16] ligands); however, the most interesting single-nucleotide polymorphisms (SNP) occurred in the cytotoxic T lymphocyte antigen 4 (*CTLA4*) gene (e.g., A49G) [17,18]. Some studies also identified non-HLA susceptibility genes that were shared with other AIDs, these included variants in Ikaros family zinc finger protein 3 (*IKZF3*) [19], *SH2B3* (which encodes a member of the SH2B adaptor proteins) [20], and *IL12* [21], and also variants within the JAK-STAT signaling pathway [27]. Ethnic differences in HLA and non-HLA genetic susceptibility loci in PBC have also been reported; for example, a variant located within *POU2AF1*, which encodes a transcription factor (TF) involved in the transcription of a number of B-cell specific proteins, has only been reported in a Japanese population [22].

Although GWAS have been critical for understanding the common genetic variation involved in PBC susceptibility, their limitations pertaining to sample availability, power, and strong linkage disequilibrium across genes has meant that they have failed to completely explain the genetic basis of PBC.

13.2.2 Environmental Factors

Despite numerous studies underlining the importance of genetics in the etiology of PBC, only a few have managed to define a weak association between genetic loci and PBC susceptibility. This has led many to suggest that there must be a role for a widespread environmental agent in the initiation and perpetuation of the disease, and the importance of environmental factors in PBC has been supported by numerous studies. Reports on disease clusters of nonrelated individuals suggested that environmental factors helped to break immunological tolerance and lead to the onset of the disease [23,30]. Environmental components include infectious agents, chemical xenobiotics, lifestyle variables, and an exposure to pollutants.

Molecular mimicry, a mechanism explaining how chemical or infectious agents can break tolerance through posttranslational modifications [24], is likely to explain why environmental factors are associated with several AIDs [25,36,37]. This paradigm suggests that pathogens present epitopes (i.e., peptides, carbohydrates, or DNA sequences) that share some degree of homology with self-epitopes, thus resulting in a promiscuous antibody- and cell-mediated immune response that can react with both the pathogen and the host [38]. Agents, such as bacteria, viruses, parasites, and fungi have all been implicated in the pathogenesis of PBC. Direct infection with mycobacterial *hsp65* [39], *Lactobacillus delbrueckii* (the target of IgG3), or the intestinal commensal proteobacterium *Novosphingobium aromaticivorans* [40–42] have also been suggested to play a role in the disease pathogenesis. Other

TABLE 13.1 Non-HLA Risk Loci Associated With PBC Susceptibility at Genome-Wide Significance

Study	Cases (n)	Controls (n)	Loci	Associated genes	OR (95% confidence interval)	*P* value
Hirschfield et al.[21]	1031	2713	1p31	*IL12RB2*	1.51 (1.33–1.70)	2.76×10^{-11}
			3q25	*IL12A*	1.54 (1.38–1.72)	2.42×10^{-14}
Liu et al. [12]	945	4651	7q32	*IRF5-TNPO3*	1.57 (1.38–1.77)	8.66×10^{13}
			17q12	*IKZF3* and *ORMDL3*	1.38	1.69×10^{-9}
			19q13	*SPIB*	1.46	7.97×10^{-11}
Mells et al. [35]	1840	5163	1q31	*DENND1B*	1.34 (1.25–1.45)	2.06×10^{14}
			2q32	*STAT4* and *STAT1*	1.50 (1.37–1.64)	2.35×10^{-19}
			3q13	*CD80*	1.35 (1.23–1.47)	2.53×10^{-11}
			3p24	*PLCL2*	1.20 (1.12–1.27)	2.28×10^{-8}
			4q24	*NFKB1*	1.26 (1.18–1.34)	4.06×10^{-12}
			5p13	*IL7R*	1.30 (1.21–1.40)	1.02×10^{-11}
			7p14	*ELMO1*	1.25 (1.16–1.36)	4.44×10^{-8}
			11q13	*RPS6KA4*	1.23 (1.15–1.31)	2.06×10^{-10}
			11q23	*CXCR5* and *DDX6*	1.37 (1.25–1.50)	2.69×10^{-12}
			12p13	*TNFRSF1A*	1.22 (1.14–1.30)	1.80×10^{-9}
			14q24	*RAD51B*	1.29 (1.20–1.39)	1.76×10^{-11}
			14q32	*TNFAIP2*	1.22 (1.16–1.27)	2.61×10^{-13}
			16p13	*CLEC16A* and *SOCS1*	1.29 (1.20–1.38)	2.95×10^{-12}
			16q24	*IRF8*	1.31 (1.21–1.43)	4.66×10^{-11}
			22q13	*TAB1* and *SYNGR1*	1.27 (1.18–1.38)	1.08×10^{-9}
Nakamura et al. [22]	1274	1091	9q32	*TNFSF15*	1.56 (1.39–1.76)	2.84×10^{14}
			11q23	*POU2AF1*	1.39 (1.24–1.56)	2.38×10^{-8}

(*Continued*)

TABLE 13.1 Non-HLA Risk Loci Associated With PBC Susceptibility at Genome-Wide Significance (*cont.*)

Study	Cases (n)	Controls (n)	Loci	Associated genes	OR (95% confidence interval)	*P* value
Juran et al. [8]	2426	5731	13q14	*TNFSF11*	1.33 (1.20–1.47)	2.18×10^{-8}
Liu et al. [14]	2861	8514	12q24	*SH2B3*	1.27 (1.19–1.34)	1.18×10^{-14}
			17q21	*MAPT*	1.25 (1.16–1.35)	2.15×10^{-9}
			19p12	*TYK2*	1.29 (1.21–1.38)	1.29×10^{-13}

Modified from Gulamhusein AF, Juran BD, Lazaridis KN. Genome-wide association studies in primary biliary cirrhosis. Semin Liver Dis 2015;35:392–401 [34].

infective pathogens, such as lipopolysaccharide, lipoteichoic acid, *Helicobacter pylori*, *Escherichia coli*, and *Chlamydia* are also associated with PBC [43]; however, unlike the previous pathogens, they are not supported by strong evidence. Several viral infections have also been proposed to be involved in the pathogenesis of PBC; for example, a human retrovirus was identified in the perihepatic lymph nodes in PBC patients, although the findings were not reproducible in subsequent studies [44]. Increased titers of the Epstein–Barr virus antigen were also found in the sera of patients with early PBC, suggesting its potential involvement in PBC. Similar peptide sequences between cytomegalovirus and human mitochondrial antigens suggest that cytomegalovirus may be involved in the initiation of PBC.

Xenobiotics are environmental low molecular weight chemicals that alter and complex to defined self- or nonself-proteins. They induce modifications in the molecular structure of native proteins, thereby triggering a well-organized immune response and perpetuating chronic autoimmune responses. Several studies have suggested that xenobiotics may also be implicated in the pathogenesis of PBC, and four main etiological mechanisms have been proposed: (1) liver cells (hepatocytes and biliary epithelial) are the main victims when the primary detoxifying organ is exposed to detrimental environmental chemicals; (2) immune repertoire neoantigen specific T and B cells, once primed, could cross-react with native autoantigens; (3) chemical modification of the native cellular protein could alter its processing in antigen-presenting cells and thereby result in the presentation of cryptic peptides. Large-scale epidemiological studies evaluating environmental risk factors for PBC demonstrated that smoking, cosmetics, and previous pregnancies were associated with an increased risk of PBC [45]; and (4) the vulnerability of the main mitochondrial autoantigen, the E2 subunit of pyruvate dehydrogenase (PDC-E2), to environmental xenobiotic modification could be explained by molecular mimicry. PDC-E2 has been shown to be susceptible to aberrant chemical modifications [46], which is perpetuated by compounds that are widely used in perfumes, lipsticks, and food flavorings [47,48].

Other environmental factors have also been proposed to increase the risk of developing PBC; however, their role remains unclear. For examples, significant clusters of PBC were found surrounding Superfund [23] and radiation sites [49], thus suggesting that toxin exposure might be a risk factor for PBC. However, further evidence is required to validate these findings.

13.3 EPIGENETIC MECHANISMS AND PBC

Nongenetic alternations are tightly regulated by two major epigenetic modifications: chemical modification to the cytosine residues of DNA (DNA methylation) and modification to the histone proteins associated with DNA (histone modifications). It is now recognized that these two major epigenetic mechanisms, together with noncoding RNAs (ncRNAs), affect AIDs and PBC in different ways.

The basic unit of chromatin (the nucleosome) is formed from 146 base pairs of DNA wrapped around histone octamers (containing two copies each of H2A, H2B, H3, and H4). The main functions of chromatin include DNA packaging, gene expression, and DNA replication. Posttranslational modifications (i.e., acetylation, methylation, phosphorylation, citrullination, ADP-ribosylation, sumoylation, and ubiquitination) of histones regulate gene expression by allowing DNA to condense into heterochromatin or open into euchromatin [50].

Alterations in the heterochromatin region result in repressed genes that are associated with morphogenesis or differentiation (i.e., imprinting or X-chromosome inactivation) [51,52]. Epigenetic factors can deposit, interpret, or eliminate epigenetic information and can be divided into three groups: (1) "writers," which are enzymes that modify DNA and histones; (2) "readers," which contain specific protein domains that recognize DNA or histone marks; and, (3) "erasers," which can delete existing signals to create space for new modifications.

Frequent epigenetic mechanism errors have been identified in inflammatory diseases and AIDs, which have resulted in the miswriting, misreading, or the faulty removal of epigenetic signals. Four groups of epigenetic mechanisms (DNA methylation, histone posttranslational modification, histone variants, and small and noncoding RNA interference) that regulate gene expression (activation or repression) are discussed below in relation to PBC [53] (Table 13.2). Due to the striking female predominance [54], the high prevalence of Y-chromosome loss in male PBC patients [55], and the fact that the X chromosome contains several genes associated with immunological tolerance (i.e., *IL2RG* and *FOXP3*) [56,57], particular attention will be paid to the numerous epigenetic mechanisms that can affect the X chromosome [58].

13.3.1 DNA Methylation Changes in PBC

DNA methylation is a postreplication modification that involves the addition of a methyl group to cytosine residues by DNA methyltransferases (DNMTs) [74]. At least three DNMTs (DNMT1,

TABLE 13.2 Epigenetic Modifications Described in AIDs and Cholestatic Diseases, Including PBC

Epigenetic dysregulation	Cell targets	Disease or model	Gene or pathway involved	References
DNA methylation	CD4 and CD8	PBC	Methylation of the *CD40L* promoter	[59]
	Liver cells	Animal model	Hypomethylation of *FOXA2*	[60,61]
		Cholestasis	TGFβ	[62]
	T cells	Immune system	INFα	[63]
		Tregs	*FOXP3*	[64]
		Thymic epithelial cells, thymomas, and normal tissues	*AIRE*	[65]
Histone posttranslational modification	T cells	SLE	Acetylation of histone 4	[66]
		PBC	Dysregulation of *PIN4*	[67]
Small and noncoding RNA interference	CD8 T cells	Immune system	*lncRAK053349*	[68,69]
	PBMCs	PBC	miR26-a	[70]
	T cells	PBC	miR21	[71]
	Liver cells	PBC	miR506	[72]
X-chromosome related	T and B cells	PBC	Monosomy	[73]

DNMT3A, and DNMT3B) have been described to date. Biochemical studies suggest that DNA methylation is just one component of a wider epigenetic program that includes postsynthesis chromatin modifications [75]. DNA methylation is a dynamic process that varies among tissues [76,77] and can be either passive or active [53]. It typically occurs at cytosine–guanine dinucleotide (CpG) sites located within the 5′ upstream promoter regions of genes. Methylation of a promoter CpG site generally blocks the accessibility of transcriptional activators and thereby inhibits gene transcription. There are four possible DNA methylation patterns: (1) methylation of CpG islands in gene promoter regions; (2) methylation of CpG island shores; (3) methylation of gene bodies, which are methylated in their basal state to facilitate correct transcription and prevent spurious initiations (typical of stem cells); and (4) hypermethylation of repetitive sequences, which protects chromosomal integrity by preventing reactivation of endoparasitic sequences that cause chromosomal instability, translocations, and gene disruption.

Methylation represses transcription where the hypermethylation of promoter regions is associated with gene silencing, whereas intragenic hypermethylation is linked to gene activation [78,79]. Conversely, when the promoter site is unmethylated, transcription is permitted. DNA methylation is the most described epigenetic mechanism in AIDs; for example, global hypomethylation, as well as hypo- and hypermethylation at specific loci [24,80], has already been described in systemic lupus erythematosus (SLE), rheumatoid arthritis (RA), and ulcerative colitis.

FoxA2 and Sox17 are two key transcription factors that regulate cell differentiation and tissue regeneration [81,82]. McDaniel et al. identified several CpG islands approximately 2000 bp upstream of the 5′ coding regions in the promoter regions of these two genes [62]. They demonstrated the FoxA2 promoter resulted hypomethylated, and therefore FoxA2 notably enhanced, in murine liver progenitor cells and small mouse cholangiocytes; on the contrary, it was silenced in human PSC and PBC liver tissues. Bile duct growth and proliferation are tightly regulated by factors, such as bile acids, hormones, and cytokines [83]. Of the cytokines involved in this process, there is ample evidence for increased TGF-β activation [60,61]. DNA methylation regulated by TGF-β activation has since been demonstrated during chronic liver injury and malignant transformation.

The importance of methylation in immune system cells has also been underlined in numerous studies. In fact, DNA methylation plays an essential role in normal transcriptional control of key factors in T-cell differentiation and function, including the interferon (IFN) γ locus [63] and the TFs FoxP3 [64] and AIRE [65]. Global and single-gene methylation irregularities have been identified in PBC patients; however, these alterations of methylation status have not been correlated with gene expression changes in PBC.

In 2014 Selmi et al. performed a genome-wide investigation to determine the differences in DNA methylation and copy number variation (CNV) between PBC affected and nonaffected monozygotic twins [84]. Promoter methylation influences gene expression levels; however, CNV, which is the result of de novo duplications and other rearrangements [85] that occur at much higher rates than single nucleotide variants, may also regulate gene expression [86]. DNA hypermethylation was reported in 60 differentially-methylated regions of the X chromosome, which corresponded to 51 genes. DNA hypermethylation was also observed in an additional nine genes on autosomal chromosomes in the affected twin compared to the nonaffected twin.

Numerous studies have also reported gene-specific DNA methylation changes in PBC. The CD40 ligand (*CD40L*) plays an important role in the pathogenesis of several AIDs [87,88], including PBC. When a genetic defect in this gene is reported in a patient, an elevation of serum immunoglobulin M (IgM) levels and immunodeficiency are also observed. PBC patients often present with elevated IgM levels. Based on the hypothesis that elevated serum IgM levels may be indicative of a *CD40L* defect, CD4+ and CD8+ T cells were isolated from circulating mononuclear cells of PBC patients and analyzed for genetic and epigenetic modifications of this gene [59]. A significantly lower level of *CD40L* promoter methylation was observed in the CD4+ T cells of PBC-affected patients. Importantly, this epigenetic dysregulation was also inversely correlated with serum IgM levels. This observation led to the hypothesis that abnormal methylation of *CD40L* may be involved in the pathogenesis of PBC [59].

13.3.2 Histone Posttranslational Modifications

The proteins that are deputized to package and organize DNA are histones, which are categorized as either core (H2A, H2B, H3, and H4) or linker (H1 and H5) histones. Acetylation, methylation, phosphorylation, citrullination, ADP-ribosylation, sumoylation, and ubiquitination are some of the most common posttranslational histone modifications [89]. Recent studies suggest that the accumulation of toxic bile acids induces epigenetic alterations, particularly the acetylation and methylation of histones [90].

These modifications control the accessibility of chromatin and transcriptional activities inside a cell. Histone methyltransferases and demethylases, and histone acetyltransferases and deacetylases (HDAC), regulate these epigenetic modifications and control the balance of histone modifications [91].

The correlation between histone modifications and gene expression in autoimmune diseases (i.e., RA and SLE) has been well investigated [66]; however, only a few studies have focused on dysregulated histone modifications in PBC. One study in affected PBC subjects reported an upregulation of acetylation of histone 4 in the promoter regions of *CD40L, LIGHT, IL17,* and *IFN-γ* in autoreactive T cells, and a downregulation in the promoter regions of *TRAIL, Apo2,* and *HDAC7A* [92]

One X-linked gene involved in the pathogenesis of PBC is *PIN4* [67], which is part of the parvulin family of peptidyl-prolyl *cis/trans* isomerases. This family of proteins actively participate in chromatin remodeling through a mechanism involving ATP-dependent changes in histone modifications. *PIN4* dysregulation could therefore affect chromatin remodeling, alter transcription, and regulate multiple susceptibility or protective PBC genes [67,93].

13.3.3 Histone Variants

All conventional histones, except H4, have histone variants [94], even if they are only expressed at low levels compared to their conventional counterparts. Variants may differ by a few amino acids or may incorporate large nonhistone domains, thus leading to the production of proteins that differ from their canonical counterparts in both their genomic and proteomic features. So far no studies have reported correlation between histone variants and PBC pathogenesis. In addition, a plethora of studies have shown that histone variants contribute to many DNA-based cellular processes [95], and their ability to function as damage-associated molecular pattern molecules (DAMPs) means that it is reasonable to assume that this epigenetic mechanism may be relevant to the pathogenesis of PBC. Besides acting as direct autoantigens in autoimmune disorders, extracellular histones can also prevent DNA degradation through the formation of histone–DNA complexes, which enhances the autoimmune response [96]. The histone H2A family contains numerous variants [97], of which MacroH2A is localized on the inactive X-chromosome [98]. It is still under debate whether this variant initiates X inactivation [99], enzymatically represses transcription [100], or sterically blocks access to TFs and coactivators [101–103]. Relatively little is known about histone variants in immune cells [104], probably due to the requirement of millions of cells for analysis using the established protocol. As mentioned earlier, histones act as DAMPs when they are released into the extracellular space. When activated immune cells [105,106] or injured cells release histones into the extracellular space, they bind to receptors (e.g., TLRs) and trigger activation of multiple signaling pathways (e.g., MAPK, NF-κB, and AKT) and the inflammasome in a single or combined manner. The ability of histones to translocate from the nucleus to the extranuclear space may be used as a biomarker to improve the diagnosis, prognosis, and management of human diseases. For example, serum levels of histones have already been correlated with the severity of SLE [107]. Antihistone-based therapeutic strategies may also be useful for treating several diseases; however, understanding the structure, modification, and function of histones inside and outside the cell remains a challenge [108].

13.3.4 Small and Noncoding RNA Interference

One epigenetic control tool that has been widely studied is the RNA-based mechanism, which occurs through ncRNAs. One limitation in our understanding of GWAS data is based on our incomplete knowledge of noncoding regulatory elements, their mechanisms of action, and the cellular states and processes in which they function. ncRNAs are functional RNA molecules that are transcribed from DNA but not translated into proteins. They regulate gene expression posttranscriptionally and are key players in a wide range of biological process [109]. They can be further subdivided into two main groups: short (< 30 nucleotides) and long (> 200 nucleotides) ncRNAs.

Short ncRNA, also known as small interfering RNAs (siRNAs), are RNA molecules of 21–25 nucleotides that convey RNAi sequence specificity [110]. MicroRNAs (miRNA) are small endogenous RNA molecules that regulate major cellular processes, such as apoptosis, differentiation, the cell cycle, and immune function. More than 1000 miRNAs have been identified in the human genome and they target approximately 60% of mammalian genes. It is evident that miRNAs are the most abundant class of ncRNA regulators.

Alterations in miRNA expression and function lead to dysfunction of the innate and adaptive immune system and susceptibility to autoimmune diseases. Thus it is not surprising that miRNAs control the expression of genes that promote or suppress autoreactivity in AIDs. Indeed, the involvement of miRNAs in the development of autoreactive lymphocytes in SLE, synoviocyte proliferation in RA, and neural demyelination in MS has already been shown [111]. While some miRNAs are common across AIDs, some are specific to certain diseases or are directly linked to immune cells or target organs for a specific disease. Numerous studies have shown distinct spatial miRNA profiles for different tissues. A recent study described over 200 differentially-expressed miRNAs in peripheral blood mononuclear cells (PBMCs) from PBC patients [70,72], among which miR26-a was found to be downregulated [72]. The predicted target processes of this miRNA include cell proliferation, apoptosis, inflammation, oxidative stress, and metabolism associated with the development of PBC; however, the exact role of miR-26 in the pathogenesis of PBC remains to be clarified. Sander et al. suggested that miR26-a contributes to posttranscriptional regulation by allowing the overexpression of a protein (polycomb group protein EZH2) that modulates biliary epithelial cell (BEC) senescence in PBC [71].

Another miRNA involved in the pathogenesis of PBC is miR-506 [112], which functions to regulate the pH of cholangiocytes. The authors demonstrated that miR-506 expression was increased in the livers of subjects with PBC. Increased levels of this miRNA decreased the level of InsP3R3 [113], an intracellular calcium channel that regulates bicarbonate flow along the subapical portion of the endoplasmic reticulum.

In a mouse model of PBC, Ando et al. [113a] demonstrated that hepatic CD44+CD62L− effector CD8+ T cells had a significantly higher expression of miR-21 compared to the same cells from the spleen or mesenteric lymph nodes. The authors demonstrated that miR-21 transfection into CD8+ T cells in C57BL/6 mice produced an upregulation of proinflammatory cytokines (i.e., INF-γ and IL17) and an inhibition of autoreactive CD8+ T-cell apoptosis. Although this study was performed in transgenic mice, it suggested that miR-21 could be a contributing factor to liver-specific autoimmune cholangitis.

How miRNAs are regulated remains under debate. The number of studies documenting the epigenetic regulation of miRNAs has increased dramatically over recent years. The methylation and histone modification status of miRNA promoter regions has already been documented for a number of disorders, but not for PBC. It is also recognized that SNPs, one of the most common types of genetic variation, can affect the transcribed region of mRNA and thus result in altered mRNA processing and turnover [114].

GWAS studies have already been performed on PBC samples and underline how hundreds of SNPs seem to influence the phenotype through regulatory changes in mRNA expression. Among the SNPs identified in PBC cases, rs10923 was found to locate within an experimentally-validated RISC binding site [115] and within the seed complementary region of a miRNA that is upregulated in PBC patients [72].

The majority of noncoding RNA transcripts belong to the group called the "long noncoding RNAs (lncRNAs)." LncRNAs are important regulatory molecules in tumor suppression and oncogenic pathways. They can complex with chromatin to modify its state (i.e., by transcriptional regulation, posttranscriptional regulation, and intracellular trafficking) and work as precursors for siRNAs to influence gene expression. One subgroup of lncRNAs, the large intergenic noncoding RNAs, has been associated with chromatin modifying complexes, which target specific genomic loci to promote epigenetic states. Although miRNAs are the most comprehensive and well-studied ncRNA, many reports suggest that lncRNAs could mediate host responses and immune function [116]. In fact, numerous lncRNAs have recently been linked to diverse human diseases that are related to the regulation of the immune system [117–119], and hence they could theoretically play an active role in many AIDs (i.e., SLE) [120,121]. However, knowledge of PBC-related lncRNAs remains limited.

Amaral et al. demonstrated that more than 200 lncRNAs were implicated in imprinting; however, only a small proportion of these have been functionally identified [122]. Different lncRNAs participate in the regulation of X-chromosome inactivation, a mechanism that is linked to the female preponderance of AIDs. Circumstantially, X-inactive specific transcript (*Xist*) was one of the first lncRNAs discovered in mammals [68,123]. As reported earlier for mRNA, an up-to-date genome screen has shown that approximately 10% of all AID-associated SNPs localize to lncRNA genes [69]. Specifically, 38 of the SNPs localized to lncRNA seem to be strongly associated with PBC. One interesting SNP is AK053349, which is located within a lncRNA that is highly upregulated in activated CD8+ T lymphocytes in PBC, and shows a positive correlation with the Mayo Risk Score [124].

13.4 TRANSCRIPTION FACTORS

The accessibility of DNA depends on nonhistone chromatin architectural proteins such as TFs. TFs typically bind to *cis*-acting elements within promoters, enhancers, and other gene regulatory regions, and function within regulatory complexes to control gene activity. Knowledge about how TFs and chromatin control gene activity has expanded dramatically in recent years.

The interactions between TFs and chromatin are dynamic. TFs must gain access to their binding sites within the chromatin complex; however, once TFs are bound to DNA they can modify the chromatin landscape. Unsurprisingly, mutations in TFs and epigenetic enzymes are frequently observed in and associated with diseases. Further analysis has also revealed that regions bound by specific TFs can vary from individual to individual; for example, nuclear factor kappa-B (NF-κB) is an important TF that regulates inflammation and AIDs; however, approximately 7%–10% of NF-κB binding sites differ between individuals, and many contain polymorphisms [125]. A similar study showed that approximately 11% of CCCTC-binding factor binding sites were allele-specific in cell lines derived from parents and children from two lineages [126]. Two TFs [transcription initiation factor 250 kDa subunit (TAFII-250) and PAX3/forkhead transcription factor] have been shown to be unregulated in PBC-affected patients [127,128]. TAF(II)250 is a master regulator in the initial step of RNA polymerase II transcription [129]. A transcriptional regulator (SMARC1), which modulates the chromatin structure and is involved in apoptosis, DNA damage, and differentiation, was also hypermethylated in PBC patients [84].

13.5 X CHROMOSOME

PBC shows a striking female predominance of approximately 10:1 [130], thus suggesting a potential role of the X chromosome in female PBC patients. Furthermore, genes involved in immunological tolerance are also located on the X chromosome, which may also partially explain the general predominance of autoimmune disease in females [131]. Interestingly, a higher frequency of monosomy X was observed in peripheral leukocytes of women with PBC [73].

In healthy women, one X chromosome homologue is usually inactivated by epigenetic mechanisms (e.g., heterochromatin packaging, DNA methylation, and histone modifications, such as H3K27me3, H2K119 ubiquitination, H4K20me1, H3K9me2, and H4 hypoacetylation) to silence the genes. Studies have shown that epigenetics plays a crucial role in X-chromosome inactivation (XCI) and escape [132,133],therefore epigenetic dysregulation could explain the female PBC predominance. While X inactivation is random in somatic cells, PBC-affected patients show preferential loss in cells with X monosomy [131,134,135].

Therefore it could be hypothesized that PBC susceptibility is either influenced by epigenetic alteration in the silencing of protective genes or by susceptibility X-chromosome genes variably escaping XCI. Finally, patients with PBC are also characterized by an enhanced X monosomy rate in PBMCs, particularly T and B lymphocytes [54].

The striking female predominance of PBC suggests that this predisposition may be secondary to epigenetic alterations on the X chromosome. To address this hypothesis, X-chromosome methylation profiles were determined in CD4, CD8, and CD14 cells from PBC patients and controls. A total of 20, 15, and 19 gene promoters showed a significant difference in DNA methylation in CD4+, CD8+, and CD14+ T cells, respectively, in patients with PBC. Interestingly, *FUNDC2* was hypermethylation in CD8+ T cells and *CXCR3* was strikingly demethylated in CD4+ T cells, which inversely correlated with *CXCR3* expression levels in CD4+ T cells from early-stage PBC patients [136].

The X chromosome is enriched with miRNAs, which is exemplified by the fact that 7% of all miRNAs identified to date are encoded on it. MiRNAs can regulate XCI in female subjects by epigenetic processes, and some studies suggest that this is because miRNAs are able to escape X-linked inactivation [137,138]. This results in the suppression of many miRNA-regulated target genes that are involved in immunosuppressive pathways, which leads to a heightened autoimmune response in females and their predisposition to AIDs.

Two lncRNAs (Xist and its antisense transcript Tsix, which is a negative regulator of Xist) are also deregulated during XCI. Prior to differentiation, XCI is a random event and Xist and Tsix are actively transcribed due to H3K4 dimethylation of the *Xist* gene. *Xist* expression is elevated upon differentiation, which results in *Xist* RNA coating the future inactive X chromosome and triggering extensive histone methylation and chromosome inactivation [139]. Numerous studies have shown that women affected by AIDs overexpress X-linked miRNA; however, little is known about how female-biased X-linked miRNA expression contributes to PBC.

13.6 EPIGENETIC CONTROL OF IMMUNE SYSTEM CELLS AND CHOLANGIOCYTES IN PBC

The autoimmune basis of PBC is supported by the presence of highly-specific antimitochondrial antibodies (AMAs) and autoreactive T cells [140]. It was suggested that PBC ensues from a multilineage T- and B-cell response against the immunodominant mitochondrial autoantigen, PDC-E2. One of question that remain about PBC is the specificity of the immune attack on the BECs. During apoptosis, BECs translocate mitochondrial antigens into apoptotic bodies, and in the presence of apoptotic BECs and AMAs, PBC macrophages polarize to a proinflammatory phenotype. These observations suggest that different immune cells (circulating or resident in the liver) are involved in the pathogenesis of PBC. Circulating immune cells are constantly exposed to environmental factors, and therefore how epigenetic modifications contribute to the proper acquisition of cell fate in the immune system has already been documented.

H3K4 methylation is one of the mechanisms through which adult somatic cells activate or maintain a specific differentiation program, hence building the case for "memory" processes, both in the brain and in the immune system [141,142].

Histone modifications correlate with gene transcription levels in T cells [143,144]. Among the immune cells residing in the liver, macrophages are able to acquire a specific epigenetic plasticity in response to external stimuli. Evidence indicates that HDACs are crucial regulators of inflammation and immunity. The DNA methylation pattern is also impaired in PBC, thus affecting immune-related genes and immune responses. Indeed, global DNA hypomethylation or hypermethylation is observed in PBMCs from PBC patients.

It is clear that all autoimmune diseases contain epigenetic aberrations, each one specific for the disease in question. In PBC patients, cholangiocytes show specific immunobiological features that are responsible for their selective targeting by the immune system. Cholangiocytes are active players in both innate and adaptive immune responses through various immunological pathways, and they have unique biological properties that can trigger the development of autoimmune cholangitis in the right setting [145]. Heterogeneity in gene and protein expression along the biliary epithelium is well described [146,147]; for example, miRNAs (i.e., miR506) have been shown to be upregulated in PBC cholangiocytes [112], thus demonstrating how the epigenetic regulation of gene expression is important to understand. The reasons why PBC selectively targets small bile ducts are still unclear.

13.7 NOVEL THERAPIES

International human epigenome projects are currently working to catalog genome-wide epigenetic markers in all major tissues. As previously discussed, numerous studies have suggested the involvement of deregulated epigenetic processes in AIDs, therefore mapping epigenetic patterns in the immune system or target organs is critical. Epigenetic biomarkers can be used for clinical prediction, diagnosis, or therapeutic development. Moreover, epigenetic states are dynamically regulated and potentially reversible, and thereby provide an ideal opportunity for pharmacological manipulation [13,148]. Drugs that target the epigenetic regulatory machinery are currently undergoing clinical trials in a variety of autoimmune diseases. Of the therapeutic outcomes, the ability to manipulate mRNA expression is especially appealing, and miRNA mimics, miRNA inhibitors, miRNA sponges, small-molecule inhibitors, and antisense oligonucleotides have already been trialed. No study is currently under development for PBC; however, the encouraging results in other autoimmune diseases could be used as a starting point for other similar diseases. Recent proof-of-concept studies have shown that mRNA agonists targeted to the liver can modulate the expression of downstream genes. Among the epigenetic regulators, histone acetylation contributes to epigenetic dysfunction. Some drugs have been developed that use control chromatin as the target proteins (i.e., HDAC inhibitors (HDACIs) [149] and DNMTs [150]). These drugs have stimulated a tremendous shift in the medical treatment of most autoimmune conditions. One inhibitor that is well-studied in autoimmune models (but not in PBC) is the HDACi, which has been extensively explored as an antiinflammatory and immunosuppressive agent.

Finally, since broad implications have already been demonstrated for epigenetic-based therapeutics in the control of autoimmunity and chronic inflammatory processes in rheumatic diseases, this approach also seems to be appealing for PBC [151].

13.8 CONCLUSIONS

Epigenetic alterations have been linked to disease phenotypes in humans. More specifically, the role of epigenetic dysregulation in numerous AIDs has been extensively studied. Hedrich et al. reported that cross-talk between various ncRNAs and between ncRNAs and other epigenetic mechanisms played a major role in the pathogenesis of SLE [152]. However, data regarding the involvement of epigenetics in the induction of PBC are still under debate. It is almost universally accepted that alterations in the DNA methylation pattern and histone modifications may be partially responsible for the low concordance rate of PBC between monozygotic twins. Unfortunately, epigenetic dysregulations are not sufficient to explain the deficiencies in our understandings of the genetic basis of PBC: one example is the lack of an explanation for female predominance. Thanks to high-throughput sequencing technologies, it is now evident that epigenetic mechanisms determine gene function and activities, but they are highly sensitive and could be reversed by aging and environmental influences. Therefore these mechanisms could be reversed in a rapid way, particularly during the cell cycle.

The ability to integrate multiple layers of epigenetic data with transcriptomic and proteomic data will be necessary to increase our understanding of how the interaction of genetic and epigenetic factors increases susceptibility or protects against PBC. One example of this approach is seen in the elegant work of Farh et al. [153], who developed a fine-mapping algorithm to identify causal variants for 21 autoimmune diseases starting from genotyping data. They integrated these predictions with transcription and *cis*-regulatory element annotations, and observed that most causal variants mapped to enhancers, and more specifically that causal SNPs in PBC preferentially mapped to enhancers and promoters of B-cell elements [153]. In addition, concentrating on the common epigenetic changes or pathogenic pathways in autoimmune system cells that are known to be involved in the pathogenesis of PBC will be pivotal to accelerate the identification of therapeutic targets for this disease. Epigenetic changes are particularly notable in macrophages, where numerous mechanisms (i.e., miRNAs, histone methylation, and histone acetylation) seem to be involved in regulating phagocyte activation and function [154].

This chapter has sought to highlight the most important epigenetic dysfunctions involved in the development of PBC. All the studies performed to date regarding epigenetic dysfunction in PBC, have raised the question as to whether the reported epigenetic alterations play a causative role or are the consequences of other pathologic processes that occur during PBC development. Further epigenome-wide studies on all cell types involved in the pathology, exploration of a larger repertoire of epigenetic signals, and investigation of the epigenetic landscape at different phases of the disease will increase our knowledge of this phenomenon.

LIST OF ABBREVIATIONS

AID	Autoimmune diseases
CNV	Copy number variations
CTCF	CCCTC-binding factor
DMR	Differentially methylated regions
Dnmts	DNA methyl transferases
eRNA	Enhancer RNAs
EWAS	Epigenome-wide association studies
Forkhead box P3	Transcription factors FoxP3
GWAS	Genome-wide association studies
HAT	Histone acetyltransferases
HDAC	Deacetylases
HDACi	HDAC inhibitors
HDACIs	Deacetylase inhibitors
HDM	Histone demethylases
HMT	Histone methyltransferases
lincRNAs	Large intergenic non-coding RNAs
lncRNAs	Long noncoding RNAs
miRNA	microRNAs
ncRNAs	Noncoding RNAs
NF-κB	Nuclear factor kappa-B

PBC	Primary biliary cirrhosis
RA	Rheumatoid Arthritis
siRNAs	Small interfering RNAs
SLE	Systemic lupus erythematosus
SNPs	Single-nucleotide polymorphisms
TFs	Transcription factors
TSSs	Transcription start sites
UC	Ulcerative colitis
Xist	X-inactive-specific transcript

References

[1] Beuers U, Gershwin ME, Gish RG, Invernizzi P, Jones DE, Lindor K, Ma X, et al. Changing nomenclature for PBC: from 'cirrhosis' to 'cholangitis'. Hepatology 2015;62:1620–2.

[2] Invernizzi P. Primary biliary cirrhosis. Foreword. Semin Liver Dis 2014;34:253–4.

[3] Invernizzi P. Human leukocyte antigen in primary biliary cirrhosis: an old story now reviving. Hepatology 2011;54:714–23.

[4] Vyse TJ, Kotzin BL. Genetic basis of systemic lupus erythematosus. Curr Opin Immunol 1996;8:843–51.

[5] Invernizzi P, Mackay IR. Autoimmune liver diseases. World J Gastroenterol 2008;14:3290–1.

[6] Invernizzi P, Gershwin ME. The genetics of human autoimmune disease. J Autoimmun 2009;33:290–9.

[7] Lleo A, Moroni L, Caliari L, Invernizzi P. Autoimmunity and Turner's syndrome. Autoimmun Rev 2012;11:A538–43.

[8] Juran BD, Hirschfield GM, Invernizzi P, Atkinson EJ, Li Y, Xie G, Kosoy R, et al. Immunochip analyses identify a novel risk locus for primary biliary cirrhosis at 13q14, multiple independent associations at four established risk loci and epistasis between 1p31 and 7q32 risk variants. Hum Mol Genet 2012;21:5209–21.

[9] Smyk D, Rigopoulou EI, Baum H, Burroughs AK, Vergani D, Bogdanos DP. Autoimmunity and environment: am I at risk? Clin Rev Allergy Immunol 2012;42:199–212.

[10] Kaplan MM, Gershwin ME. Primary biliary cirrhosis. N Engl J Med 2005;353:1261–73.

[11] Pepke S, Wold B, Mortazavi A. Computation for ChIP-seq and RNA-seq studies. Nat Methods 2009;6:S22–32.

[12] Liu X, Invernizzi P, Lu Y, Kosoy R, Bianchi I, Podda M, Xu C, et al. Genome-wide meta-analyses identify three loci associated with primary biliary cirrhosis. Nat Genet 2010;42:658–60.

[13] Carbone M, Lleo A, Sandford RN, Invernizzi P. Implications of genome-wide association studies in novel therapeutics in primary biliary cirrhosis. Eur J Immunol 2014;44:945–54.

[14] Liu JZ, Almarri MA, Gaffney DJ, Mells GF, Jostins L, Cordell HJ, Ducker SJ, et al. Dense fine-mapping study identifies new susceptibility loci for primary biliary cirrhosis. Nat Genet 2012;44:1137–41.

[15] Higuchi M, Horiuchi T, Kojima T, Nishizaka H, Ishibashi H, Hayashi K, Niho Y, et al. Analysis of CD40 ligand gene mutations in patients with primary biliary cirrhosis. Scand J Clin Lab Invest 1998;58:429–32.

[16] Corpechot C, Poupon R. Promoter polymorphism of the CD14 endotoxin receptor gene and primary biliary cirrhosis. Hepatology 2002;35:242–3.

[17] Agarwal K, Jones DE, Daly AK, James OF, Vaidya B, Pearce S, Bassendine MF. CTLA-4 gene polymorphism confers susceptibility to primary biliary cirrhosis. J Hepatol 2000;32:538–41.

[18] Fan LY, Tu XQ, Cheng QB, Zhu Y, Feltens R, Pfeiffer T, Zhong RQ. Cytotoxic T lymphocyte associated antigen-4 gene polymorphisms confer susceptibility to primary biliary cirrhosis and autoimmune hepatitis in Chinese population. World J Gastroenterol 2004;10:3056–9.

[19] Quintana FJ, Jin H, Burns EJ, Nadeau M, Yeste A, Kumar D, Rangachari M, et al. Aiolos promotes TH17 differentiation by directly silencing Il2 expression. Nat Immunol 2012;13:770–7.

[20] Katayama H, Mori T, Seki Y, Anraku M, Iseki M, Ikutani M, Iwasaki Y, et al. Lnk prevents inflammatory CD8(+) T-cell proliferation and contributes to intestinal homeostasis. Eur J Immunol 2014;44:1622–32.

[21] Hirschfield GM, Liu X, Xu C, Lu Y, Xie G, Gu X, Walker EJ, et al. Primary biliary cirrhosis associated with HLA, IL12A, and IL12RB2 variants. N Engl J Med 2009;360:2544–55.

[22] Nakamura M, Nishida N, Kawashima M, Aiba Y, Tanaka A, Yasunami M, Nakamura H, et al. Genome-wide association study identifies TNFSF15 and POU2AF1 as susceptibility loci for primary biliary cirrhosis in the Japanese population. Am J Hum Genet 2012;91:721–8.

[23] Ala A, Stanca CM, Bu-Ghanim M, Ahmado I, Branch AD, Schiano TD, Odin JA, et al. Increased prevalence of primary biliary cirrhosis near Superfund toxic waste sites. Hepatology 2006;43:525–31.

[24] Selmi C, Leung PS, Sherr DH, Diaz M, Nyland JF, Monestier M, Rose NR, et al. Mechanisms of environmental influence on human autoimmunity: a National Institute of Environmental Health Sciences expert panel workshop. J Autoimmun 2012;39:272–84.

[25] James JA, Robertson JM. Lupus and Epstein-Barr. Curr Opin Rheumatol 2012;24:383–8.

[26] Kar SP, Seldin MF, Chen W, Lu E, Hirschfield GM, Invernizzi P, Heathcote J, et al. Pathway-based analysis of primary biliary cirrhosis genome-wide association studies. Genes Immun 2013;14:179–86.

[27] Cordell HJ, Han Y, Mells GF, Li Y, Hirschfield GM, Greene CS, Xie G, et al. International genome-wide meta-analysis identifies new primary biliary cirrhosis risk loci and targetable pathogenic pathways. Nat Commun 2015;6:8019.

[28] Invernizzi P, Ransom M, Raychaudhuri S, Kosoy R, Lleo A, Shigeta R, Franke A, et al. Classical HLA-DRB1 and DPB1 alleles account for HLA associations with primary biliary cirrhosis. Genes Immun 2012;13:461–8.

[29] Selmi C, De Santis M, Cavaciocchi F, Gershwin ME. Infectious agents and xenobiotics in the etiology of primary biliary cirrhosis. Dis Markers 2010;29:287–99.

[30] Selmi C, Gershwin ME. The role of environmental factors in primary biliary cirrhosis. Trends Immunol 2009;30:415–20.

[31] Bianchi I, Carbone M, Lleo A, Invernizzi P. Genetics and epigenetics of primary biliary cirrhosis. Semin Liver Dis 2014;34:255–64.

[32] Selmi C, Mayo MJ, Bach N, Ishibashi H, Invernizzi P, Gish RG, Gordon SC, et al. Primary biliary cirrhosis in monozygotic and dizygotic twins: genetics, epigenetics, and environment. Gastroenterology 2004;127:485–92.

[33] Dilthey AT, Moutsianas L, Leslie S, McVean G. HLA*IMP: an integrated framework for imputing classical HLA alleles from SNP genotypes. Bioinformatics 2011;27:968–72.

[34] Gulamhusein AF, Juran BD, Lazaridis KN. Genome-wide association studies in primary biliary cirrhosis. Semin Liver Dis 2015;35:392–401.

[35] Mells GF, Floyd JA, Morley KI, Cordell HJ, Franklin CS, Shin SY, Heneghan MA, et al. Genome-wide association study identifies 12 new susceptibility loci for primary biliary cirrhosis. Nat Genet 2011;43:329–32.

[36] Rashid T, Ebringer A. Autoimmunity in rheumatic diseases is induced by microbial infections via crossreactivity or molecular mimicry. Autoimmune Dis 2012;2012:539282.

[37] Mark LP, Gill MC, Mahut M, Derrick PJ. Dual nano-electrospray for probing solution interactions and fast reactions of complex biomolecules. Eur J Mass Spectrom 2012;18:439–46.

[38] Van de Water J, Ishibashi H, Coppel RL, Gershwin ME. Molecular mimicry and primary biliary cirrhosis: premises not promises. Hepatology 2001;33:771–5.

[39] Bogdanos DP, Gilbert D, Bianchi I, Leoni S, Mitry RR, Ma Y, Mieli-Vergani G, et al. Antibodies to soluble liver antigen and alpha-enolase in patients with autoimmune hepatitis. J Autoimmune Dis 2004;1:4.

[40] Selmi C, Balkwill DL, Invernizzi P, Ansari AA, Coppel RL, Podda M, Leung PS, et al. Patients with primary biliary cirrhosis react against a ubiquitous xenobiotic-metabolizing bacterium. Hepatology 2003;38:1250–7.

[41] Kaplan MM. *Novosphingobium aromaticivorans*: a potential initiator of primary biliary cirrhosis. Am J Gastroenterol 2004;99:2147–9.

[42] Mattner J, Savage PB, Leung P, Oertelt SS, Wang V, Trivedi O, Scanlon ST, et al. Liver autoimmunity triggered by microbial activation of natural killer T cells. Cell Host Microbe 2008;3:304–15.

[43] Harada K, Nakanuma Y. Biliary innate immunity: function and modulation. Mediat Inflamm 2010;2010:1–9.

[44] Mason AL. The evidence supports a viral aetiology for primary biliary cirrhosis. J Hepatol 2011;54:1312–4.

[45] Gershwin ME, Selmi C, Worman HJ, Gold EB, Watnik M, Utts J, Lindor KD, et al. Risk factors and comorbidities in primary biliary cirrhosis: a controlled interview-based study of 1032 patients. Hepatology 2005;42:1194–202.

[46] Leung PS, Wang J, Naiyanetr P, Kenny TP, Lam KS, Kurth MJ, Gershwin ME. Environment and primary biliary cirrhosis: electrophilic drugs and the induction of AMA. J Autoimmun 2013;41:79–86.

[47] Amano K, Leung PS, Rieger R, Quan C, Wang X, Marik J, Suen YF, et al. Chemical xenobiotics and mitochondrial autoantigens in primary biliary cirrhosis: identification of antibodies against a common environmental, cosmetic, and food additive 2-octynoic acid. J Immunol 2005;174:5874–83.

[48] Rieger R, Leung PS, Jeddeloh MR, Kurth MJ, Nantz MH, Lam KS, Barsky D, et al. Identification of 2-nonynoic acid, a cosmetic component, as a potential trigger of primary biliary cirrhosis. J Autoimmun 2006;27:7–16.

[49] Ohba K, Omagari K, Kinoshita H, Soda H, Masuda J, Hazama H, Tagawa M, et al. Primary biliary cirrhosis among atomic bomb survivors in Nagasaki Japan. J Clin Epidemiol 2001;54:845–50.

[50] Siddiqi S, Mills J, Matushansky I. Epigenetic remodeling of chromatin architecture: exploring tumor differentiation therapies in mesenchymal stem cells and sarcomas. Curr Stem Cell Res Ther 2010;5:63–73.

[51] Reik W. Stability and flexibility of epigenetic gene regulation in mammalian development. Nature 2007;447:425–32.

[52] Feinberg AP, Tycko B. The history of cancer epigenetics. Nat Rev Cancer 2004;4:143–53.
[53] Allis TJ CD, Reinberg D, Caparros ML. Epigenetics. New York: Cold Spring Harbor; 2007.
[54] Invernizzi P. The X chromosome in female-predominant autoimmune diseases. Ann N Y Acad Sci 2007;1110:57–64.
[55] Lleo A, Oertelt-Prigione S, Bianchi I, Caliari L, Finelli P, Miozzo M, Lazzari R, et al. Y chromosome loss in male patients with primary biliary cirrhosis. J Autoimmun 2013;41:87–91.
[56] Selmi C, Invernizzi P, Miozzo M, Podda M, Gershwin ME. Primary biliary cirrhosis: does X mark the spot? Autoimmun Rev 2004;3:493–9.
[57] Kumar A, Teuber SS, Gershwin ME. Current perspectives on primary immunodeficiency diseases. Clin Dev Immunol 2006;13:223–59.
[58] Marzorati S, Lleo A, Carbone M, Gershwin ME, Invernizzi P. The epigenetics of PBC: the link between genetic susceptibility and environment. Clin Res Hepatol Gastroenterol 2016;40(6):650–9.
[59] Lleo A, Liao J, Invernizzi P, Zhao M, Bernuzzi F, Ma L, Lanzi G, et al. Immunoglobulin M levels inversely correlate with CD40 ligand promoter methylation in patients with primary biliary cirrhosis. Hepatology 2012;55:153–60.
[60] Li Z, White P, Tuteja G, Rubins N, Sackett S, Kaestner KH. Foxa1 and Foxa2 regulate bile duct development in mice. J Clin Invest 2009;119:1537–45.
[61] Fu S, Fei Q, Jiang H, Chuai S, Shi S, Xiong W, Jiang L, et al. Involvement of histone acetylation of Sox17 and Foxa2 promoters during mouse definitive endoderm differentiation revealed by microRNA profiling. PLoS One 2011;6:e27965.
[62] McDaniel K, Meng F, Wu N, Sato K, Venter J, Bernuzzi F, Invernizzi P, et al. Forkhead box A2 regulated biliary heterogeneity and senescence during cholestatic liver injury. Hepatology; 2017. 65, 544-596.
[63] Sawalha AH. Epigenetics and T-cell immunity. Autoimmunity 2008;41:245–52.
[64] Lal G, Zhang N, van der Touw W, Ding Y, Ju W, Bottinger EP, Reid SP, et al. Epigenetic regulation of Foxp3 expression in regulatory T cells by DNA methylation. J Immunol 2009;182:259–73.
[65] Kont V, Murumagi A, Tykocinski LO, Kinkel SA, Webster KE, Kisand K, Tserel L, et al. DNA methylation signatures of the AIRE promoter in thymic epithelial cells, thymomas and normal tissues. Mol Immunol 2011;49:518–26.
[66] Wu H, Zhao M, Chang C, Lu Q. The real culprit in systemic lupus erythematosus: abnormal epigenetic regulation. Int J Mol Sci 2015;16:11013–33.
[67] Mitchell MM, Lleo A, Zammataro L, Mayo MJ, Invernizzi P, Bach N, Shimoda S, et al. Epigenetic investigation of variably X chromosome inactivated genes in monozygotic female twins discordant for primary biliary cirrhosis. Epigenetics 2011;6:95–102.
[68] Lee JT. Epigenetic regulation by long noncoding RNAs. Science 2012;338:1435–9.
[69] Ricano-Ponce I, Wijmenga C. Mapping of immune-mediated disease genes. Annu Rev Genomics Hum Genet 2013;14:325–53.
[70] Ninomiya M, Kondo Y, Funayama R, Nagashima T, Kogure T, Kakazu E, Kimura O, et al. Distinct microRNAs expression profile in primary biliary cirrhosis and evaluation of miR 505-3p and miR197-3p as novel biomarkers. PLoS One 2013;8:e66086.
[71] Sander S, Bullinger L, Klapproth K, Fiedler K, Kestler HA, Barth TF, Moller P, et al. MYC stimulates EZH2 expression by repression of its negative regulator miR-26a. Blood 2008;112:4202–12.
[72] Padgett KA, Lan RY, Leung PC, Lleo A, Dawson K, Pfeiff J, Mao TK, et al. Primary biliary cirrhosis is associated with altered hepatic microRNA expression. J Autoimmun 2009;32:246–53.
[73] Invernizzi P, Miozzo M, Battezzati PM, Bianchi I, Grati FR, Simoni G, Selmi C, et al. Frequency of monosomy X in women with primary biliary cirrhosis. Lancet 2004;363:533–5.
[74] Bjanesoy TE, Andreassen BK, Bratland E, Reiner A, Islam S, Husebye ES, Bakke M. Altered DNA methylation profile in Norwegian patients with autoimmune Addison's disease. Mol Immunol 2014;59:208–16.
[75] Grewal SI, Elgin SC. Heterochromatin: new possibilities for the inheritance of structure. Curr Opin Genet Dev 2002;12:178–87.
[76] Niles KM, Chan D, La Salle S, Oakes CC, Trasler JM. Critical period of nonpromoter DNA methylation acquisition during prenatal male germ cell development. PLoS One 2011;6:e24156.
[77] Liang P, Song F, Ghosh S, Morien E, Qin M, Mahmood S, Fujiwara K, et al. Genome-wide survey reveals dynamic widespread tissue-specific changes in DNA methylation during development. BMC Genomics 2011;12:231.
[78] Aran D, Toperoff G, Rosenberg M, Hellman A. Replication timing-related and gene body-specific methylation of active human genes. Hum Mol Genet 2011;20:670–80.
[79] Hon GC, Hawkins RD, Caballero OL, Lo C, Lister R, Pelizzola M, Valsesia A, et al. Global DNA hypomethylation coupled to repressive chromatin domain formation and gene silencing in breast cancer. Genome Res 2012;22:246–58.
[80] Lu Q. The critical importance of epigenetics in autoimmunity. J Autoimmun 2013;41:1–5.

[81] Yao S, Chen S, Clark J, Hao E, Beattie GM, Hayek A, Ding S. Long-term self-renewal and directed differentiation of human embryonic stem cells in chemically defined conditions. Proc Natl Acad Sci USA 2006;103:6907–12.
[82] Christodoulou C, Longmire TA, Shen SS, Bourdon A, Sommer CA, Gadue P, Spira A, et al. Mouse ES and iPS cells can form similar definitive endoderm despite differences in imprinted genes. J Clin Invest 2011;121:2313–25.
[83] Liu Z, Sakamoto T, Ezure T, Yokomuro S, Murase N, Michalopoulos G, Demetris AJ. Interleukin-6, hepatocyte growth factor, and their receptors in biliary epithelial cells during a type I ductular reaction in mice: interactions between the periductal inflammatory and stromal cells and the biliary epithelium. Hepatology 1998;28:1260–8.
[84] Selmi C, Cavaciocchi F, Lleo A, Cheroni C, De Francesco R, Lombardi SA, De Santis M, et al. Genome-wide analysis of DNA methylation, copy number variation, and gene expression in monozygotic twins discordant for primary biliary cirrhosis. Front Immunol 2014;5:128.
[85] Wong KK, deLeeuw RJ, Dosanjh NS, Kimm LR, Cheng Z, Horsman DE, MacAulay C, et al. A comprehensive analysis of common copy-number variations in the human genome. Am J Hum Genet 2007;80:91–104.
[86] Stranger BE, Forrest MS, Dunning M, Ingle CE, Beazley C, Thorne N, Redon R, et al. Relative impact of nucleotide and copy number variation on gene expression phenotypes. Science 2007;315:848–53.
[87] Mehling A, Loser K, Varga G, Metze D, Luger TA, Schwarz T, Grabbe S, et al. Overexpression of CD40 ligand in murine epidermis results in chronic skin inflammation and systemic autoimmunity. J Exp Med 2001;194:615–28.
[88] Kato K, Santana-Sahagun E, Rassenti LZ, Weisman MH, Tamura N, Kobayashi S, Hashimoto H, et al. The soluble CD40 ligand sCD154 in systemic lupus erythematosus. J Clin Invest 1999;104:947–55.
[89] Freitas MA, Sklenar AR, Parthun MR. Application of mass spectrometry to the identification and quantification of histone post-translational modifications. J Cell Biochem 2004;92:691–700.
[90] Hamilton JP. Epigenetic mechanisms involved in the pathogenesis of hepatobiliary malignancies. Epigenomics 2010;2:233–43.
[91] Kouzarides T. Chromatin modifications and their function. Cell 2007;128:693–705.
[92] Hu Z, Huang Y, Liu Y, Sun Y, Zhou Y, Gu M, Chen Y, et al. Beta-arrestin 1 modulates functions of autoimmune T cells from primary biliary cirrhosis patients. J Clin Immunol 2011;31:346–55.
[93] Surmacz TA, Bayer E, Rahfeld JU, Fischer G, Bayer P. The N-terminal basic domain of human parvulin hPar14 is responsible for the entry to the nucleus and high-affinity DNA-binding. J Mol Biol 2002;321:235–47.
[94] Boulard M, Bouvet P, Kundu TK, Dimitrov S. Histone variant nucleosomes: structure, function and implication in disease. Subcell Biochem 2007;41:71–89.
[95] Bonisch C, Nieratschker SM, Orfanos NK, Hake SB. Chromatin proteomics and epigenetic regulatory circuits. Expert Rev Proteom 2008;5:105–19.
[96] Xu J, Zhang X, Pelayo R, Monestier M, Ammollo CT, Semeraro F, Taylor FB, et al. Extracellular histones are major mediators of death in sepsis. Nat Med 2009;15:1318–21.
[97] Zink LM, Hake SB. Histone variants: nuclear function and disease. Curr Opin Genet Dev 2016;37:82–9.
[98] Costanzi C, Pehrson JR. Histone macroH2A1 is concentrated in the inactive X chromosome of female mammals. Nature 1998;393:599–601.
[99] Mermoud JE, Costanzi C, Pehrson JR, Brockdorff N. Histone macroH2A12 relocates to the inactive X chromosome after initiation and propagation of X-inactivation. J Cell Biol 1999;147:1399–408.
[100] Ladurner AG. Inactivating chromosomes: a macro domain that minimizes transcription. Mol Cell 2003;12:1–3.
[101] Perche PY, Vourc'h C, Konecny L, Souchier C, Robert-Nicoud M, Dimitrov S, Khochbin S. Higher concentrations of histone macroH2A in the Barr body are correlated with higher nucleosome density. Curr Biol 2000;10:1531–4.
[102] Angelov D, Molla A, Perche PY, Hans F, Cote J, Khochbin S, Bouvet P, et al. The histone variant macroH2A interferes with transcription factor binding and SWI/SNF nucleosome remodeling. Mol Cell 2003;11:1033–41.
[103] Abbott DW, Laszczak M, Lewis JD, Su H, Moore SC, Hills M, Dimitrov S, et al. Structural characterization of macroH2A containing chromatin. Biochemistry 2004;43:1352–9.
[104] Naranbhai V, Fairfax BP, Makino S, Humburg P, Wong D, Ng E, Hill AV, et al. Genomic modulators of gene expression in human neutrophils. Nat Commun 2015;6:7545.
[105] Brinkmann V, Reichard U, Goosmann C, Fauler B, Uhlemann Y, Weiss DS, Weinrauch Y, et al. Neutrophil extracellular traps kill bacteria. Science 2004;303:1532–5.
[106] von Kockritz-Blickwede M, Goldmann O, Thulin P, Heinemann K, Norrby-Teglund A, Rohde M, Medina E. Phagocytosis-independent antimicrobial activity of mast cells by means of extracellular trap formation. Blood 2008;111:3070–80.
[107] Schmiedeke TM, Stockl FW, Weber R, Sugisaki Y, Batsford SR, Vogt A. Histones have high affinity for the glomerular basement membrane: relevance for immune complex formation in lupus nephritis. J Exp Med 1989;169:1879–94.

[108] Chen R, Kang R, Fan XG, Tang D. Release and activity of histone in diseases. Cell Death Dis 2014;5:e1370.

[109] Lee RC, Feinbaum RL, Ambros V. The *C. elegans* heterochronic gene lin-4 encodes small RNAs with antisense complementarity to lin-14. Cell 1993;75:843–54.

[110] Mello CC, Conte D Jr. Revealing the world of RNA interference. Nature 2004;16:338–42.

[111] Dai R, Ahmed SA. MicroRNA, a new paradigm for understanding immunoregulation, inflammation, and autoimmune diseases. Transl Res 2011;157:163–79.

[112] Banales JM, Saez E, Uriz M, Sarvide S, Urribarri AD, Splinter P, Tietz Bogert PS, et al. . Up-regulation of microRNA 506 leads to decreased Cl-/HCO3- anion exchanger 2 expression in biliary epithelium of patients with primary biliary cirrhosis. Hepatology; 2012. 56, 687-697.

[113] Ananthanarayanan M, Banales JM, Guerra MT, Spirli C, Munoz-Garrido P, Mitchell-Richards K, Tafur D, et al. Post-translational regulation of the type III inositol 1,4,5-trisphosphate receptor by miRNA-506. J Biol Chem 2015;290:184–96.

[113a] Ando Y, Yang GX, Kenny TP, Kawata K, Zhang W, Huang W, et al. Overexpression of microRNA-21 is associated with elevated pro-inflammatory cytokines in dominant-negative TGF-β receptor type II mouse. J Autoimmun 2013;41:111–9.

[114] Johnson AD, Zhang Y, Papp AC, Pinsonneault JK, Lim JE, Saffen D, Dai Z, et al. Polymorphisms affecting gene transcription and mRNA processing in pharmacogenetic candidate genes: detection through allelic expression imbalance in human target tissues. Pharmacogenet Genom 2008;18:781–91.

[115] Hafner M, Landthaler M, Burger L, Khorshid M, Hausser J, Berninger P, Rothballer A, et al. Transcriptome-wide identification of RNA-binding protein and microRNA target sites by PAR-CLIP. Cell 2010;141:129–41.

[116] Wilusz JE, Freier SM, Spector DL. 3′ end processing of a long nuclear-retained noncoding RNA yields a tRNA-like cytoplasmic RNA. Cell 2008;135:919–32.

[117] Guttman M, Amit I, Garber M, French C, Lin MF, Feldser D, Huarte M, et al. Chromatin signature reveals over a thousand highly conserved large non-coding RNAs in mammals. Nature 2009;458:223–7.

[118] Collier SP, Collins PL, Williams CL, Boothby MR, Aune TM. Cutting edge: influence of Tmevpg1, a long intergenic noncoding RNA, on the expression of IFNg by Th1 cells. J Immunol 2012;189:2084–8.

[119] Carpenter S, Aiello D, Atianand MK, Ricci EP, Gandhi P, Hall LL, Byron M, et al. A long noncoding RNA mediates both activation and repression of immune response genes. Science 2013;341:789–92.

[120] Wu GC, Pan HF, Leng RX, Wang DG, Li XP, Li XM, Ye DQ. Emerging role of long noncoding RNAs in autoimmune diseases. Autoimmun Rev 2015;14:798–805.

[121] Sigdel KR, Cheng A, Wang Y, Duan L, Zhang Y. The emerging functions of long noncoding RNA in immune cells: autoimmune diseases. J Immunol Res 2015;2015:848790.

[122] Amaral PP, Clark MB, Gascoigne DK, Dinger ME, Mattick JS. LncRNAdb: a reference database for long noncoding RNAs. Nucleic Acids Res 2011;39:D146–51.

[123] Engreitz JM, Pandya-Jones A, McDonel P, Shishkin A, Sirokman K, Surka C, Kadri S, et al. The Xist lncRNA exploits three-dimensional genome architecture to spread across the X chromosome. Science 2013;341:1237973.

[124] Zhang L, Huang Y, Wang H, Kong W, Ye X, Chen Y, Liu T, Qin Q, Deng. A. Increased expression of lncRNA AK053349 in peripheral blood mononuclear cells from patients with primary biliary cirrhosis and its clinical significance. Int J Lab Med 2013;34:2656–9.

[125] Kasowski M, Grubert F, Heffelfinger C, Hariharan M, Asabere A, Waszak SM, Habegger L, et al. Variation in transcription factor binding among humans. Science 2010;328:232–5.

[126] McDaniell R, Lee BK, Song L, Liu Z, Boyle AP, Erdos MR, Scott LJ, et al. Heritable individual-specific and allele-specific chromatin signatures in humans. Science 2010;328:235–9.

[127] Shackel NA, Seth D, Haber PS, Gorrell MD, McCaughan GW. The hepatic transcriptome in human liver disease. Comp Hepatol 2006;5:6.

[128] Kopycinska J, Kempinska-Podhorodecka A, Haas T, Elias E, DePinho RA, Paik J, Milkiewicz P, et al. Activation of FoxO3a/Bim axis in patients with primary biliary cirrhosis. Liver Int 2013;33:231–8.

[129] Wassarman DA, Sauer F. TAF(II)250: a transcription toolbox. J Cell Sci 2001;114:2895–902.

[130] Lleo A, Battezzati PM, Selmi C, Gershwin ME, Podda M. Is autoimmunity a matter of sex? Autoimmun Rev 2008;7:626–30.

[131] Invernizzi P, Pasini S, Selmi C, Gershwin ME, Podda M. Female predominance and X chromosome defects in autoimmune diseases. J Autoimmun 2009;33:12–6.

[132] Heard E, Kress C, Mongelard F, Courtier B, Rougeulle C, Ashworth A, Vourc'h C, et al. Transgenic mice carrying an Xist-containing YAC. Hum Mol Genet 1996;5:441–50.

[133] Prothero KE, Stahl JM, Carrel L. Dosage compensation and gene expression on the mammalian X chromosome: one plus one does not always equal two. Chromosome Res 2009;17:637–48.
[134] Miozzo M, Selmi C, Gentilin B, Grati FR, Sirchia S, Oertelt S, Zuin M, et al. Preferential X chromosome loss but random inactivation characterize primary biliary cirrhosis. Hepatology 2007;46:456–62.
[135] Bianchi I, Lleo A, Gershwin ME, Invernizzi P. The X chromosome and immune associated genes. J Autoimmun 2012;38:J187–92.
[136] Lleo A, Zhang W, Zhao M, Tan Y, Bernuzzi F, Zhu B, Liu Q, et al. DNA methylation profiling of the X chromosome reveals an aberrant demethylation on CXCR3 promoter in primary biliary cirrhosis. Clin Epigenet 2015;7:61.
[137] Sharma S, Eghbali M. Influence of sex differences on microRNA gene regulation in disease. Biol Sex Differ 2014;5:3.
[138] Hewagama A, Eghbali M. Role of X-chromosome encoded miRNAs in autoimmunity: suppressing the suppressor and female predisposition. Rheumatol Curr Res. 2013;3:118.
[139] Kaikkonen MU, Lam MT, Glass CK. Non-coding RNAs as regulators of gene expression and epigenetics. Cardiovasc Res 2011;90:430–40.
[140] Marzorati S, Invernizzi P, Lleo A. Making sense of autoantibodies in cholestatic liver diseases. Clin Liver Dis 2016;20:33–46.
[141] Eissenberg JC, Shilatifard A. Histone H3 lysine 4 (H3K4) methylation in development and differentiation. Dev Biol 2010;339:240–9.
[142] Wei G, Wei L, Zhu J, Zang C, Hu-Li J, Yao Z, Cui K, et al. Global mapping of H3K4me3 and H3K27me3 reveals specificity and plasticity in lineage fate determination of differentiating CD4+ T cells. Immunity 2009;30:155–67.
[143] Barski A, Cuddapah S, Cui K, Roh TY, Schones DE, Wang Z, Wei G, et al. High-resolution profiling of histone methylations in the human genome. Cell 2007;129:823–37.
[144] Wang Z, Zang C, Rosenfeld JA, Schones DE, Barski A, Cuddapah S, Cui K, et al. Combinatorial patterns of histone acetylations and methylations in the human genome. Nat Genet 2008;40:897–903.
[145] Lleo A, Maroni L, Glaser S, Alpini G, Marzioni M. Role of cholangiocytes in primary biliary cirrhosis. Semin Liver Dis 2014;34:273–84.
[146] Marzioni M, Glaser SS, Francis H, Phinizy JL, LeSage G, Alpini G. Functional heterogeneity of cholangiocytes. Semin Liver Dis 2002;22:227–40.
[147] Glaser S, Francis H, Demorrow S, Lesage G, Fava G, Marzioni M, Venter J, et al. Heterogeneity of the intrahepatic biliary epithelium. World J Gastroenterol 2006;12:3523–36.
[148] Szyf M. Epigenetic therapeutics in autoimmune disease. Clin Rev Allergy Immunol 2010;39:62–77.
[149] Regna NL, Christopher MR. Isoform-selective HDAC inhibition in autoimmune disease. J Clin Cell Immunol 2014;5:207.
[150] de Andres MC, Perez-Pampin E, Calaza M, Santaclara FJ, Ortea I, Gomez-Reino JJ, Gonzalez A. Assessment of global DNA methylation in peripheral blood cell subpopulations of early rheumatoid arthritis before and after methotrexate. Arthritis Res Ther 2015;17:233.
[151] Ciechomska M, O'Reilly S. Epigenetic modulation as a therapeutic prospect for treatment of autoimmune rheumatic diseases. Mediators Inflamm 2016;2016:9607946.
[152] Hedrich CM, Rauen T, Apostolidis SA, Grammatikos AP, Rodriguez N, Ioannidis C, et al. Stat3 promotes IL-10 expression in lupus T cells through transactivation and chromatin remodeling. Proc Natl Acad Sci USA 2014;111:13457–62.
[153] Farh KK, Marson A, Zhu J, Kleinewietfeld M, Housley WJ, Beik S, Shoresh N, et al. Genetic and epigenetic fine mapping of causal autoimmune disease variants. Nature 2015;518:337–43.
[154] Ostuni R, Piccolo V, Barozzi I, Polletti S, Termanini A, Bonifacio S, Curina A, et al. Latent enhancers activated by stimulation in differentiated cells. Cell 2013;152:157–71.

CHAPTER

14

Immune-Mediated Pulmonary Disease and Epigenetics

Sabrina Ramelli, William T. Gerthoffer***

*University of South Alabama, Mobile, AL, United States; **The Center for Lung Biology, University of South Alabama, Mobile, AL, United States

OUTLINE

The Epigenetics of Autoimmunity. http://dx.doi.org/10.1016/B978-0-12-809912-4.00014-3

14.1 INFLAMMATION AND TISSUE REMODELING IN ASTHMA AND PULMONARY ARTERIAL HYPERTENSION

Inflammation of the lung in asthma and pulmonary arterial hypertension (PAH) triggers maladaptation of the tissue phenotype. Inflammation invokes multiple epigenetic mechanisms to alter transcription, RNA processing, and translation. Therefore, the abundance of proteins expressed varies depending on the types, intensity, and duration of the external cues. In airway and vascular remodeling, these initiating signals lead to a vicious positive-feedback loop where inflammatory mediators beget more mediators of inflammation. Inhaled corticosteroids are the current drugs of choice for treating asthma, and they usually stop such positive-feedback signaling loops; however, there is no comparable antiinflammatory therapy for PAH (producing novel antiinflammatory therapies is one of the goals of PAH drug development) [1,2]. This chapter reviews recent studies of the epigenetic processes involved in lung inflammation. Focus will be placed on asthma and PAH as pathologies that illustrate the profound effect that inflammatory gene expression has on lung structure and function. Prior reviews by Liu et al. [3] and Comer et al. [4] provide more comprehensive descriptions of epigenetics in other inflammatory lung diseases.

Many studies have identified altered DNA methylation patterns in inflammatory genes, pathological changes in histone modifications, and patterns of expression of noncoding RNAs (miRNAs and long noncoding RNAs). These studies are rapidly expanding our understanding of the molecular pathology of inflammatory lung diseases. Deeper knowledge of the molecular basis of epigenetics and the epigenome will aid in drug development and help in the management of difficult diseases. New drug targets and new therapeutics should evolve from a better understanding of the epigenetic determinants of both immune-cell and structural-cell phenotypes in the lung.

Asthma is a syndrome identified clinically by wheezing, shortness of breath, coughing, and air trapping due to obstruction of the airways. The airways are obstructed because of mucous plugging, reversible airway smooth muscle contraction, and irreversible airway lumen narrowing due to airway wall remodeling (Fig. 14.1). In allergic asthma, the characteristic symptoms are triggered by allergens (e.g., house dust mite (HDM), cockroach, cat dander, or molds). Asthma attacks may also be exacerbated by infections, exercise, or cold air. Most people with mild-to-moderate asthma find that their acute wheezing and shortness of breath is readily reversible with inhaled bronchodilators (i.e., β2-adrenergic agonists). Long-term control in these patients is usually achieved by administering inhaled corticosteroids, which control inflammation and prevent subsequent acute attacks. However, 5%–10% of asthmatics suffer from irreversible airway obstruction [5,6]. Failure to respond to β-agonists and corticosteroids can be fatal [7], therefore this subset of asthmatics with uncontrolled disease supports the case for identifying mechanistically-new bronchodilators [8], new antiremodeling therapies [9,10], and new immune-based therapies. Recent progress in developing therapeutic antibodies against IgE, IL-5, and TSLP has been significant, and interested readers should consult several excellent reviews of progress in asthma immunotherapy [11–14]. However, this review will focus on novel ways to modify the immunological component of asthma by specifically targeting epigenetic mechanisms. The guiding idea is that antagonizing causative epigenetic events and adaptive epigenetic responses will induce lung repair [15]. Fig. 14.2 illustrates these principles and introduces the idea that new approaches are evolving to effect "therapeutic reprogramming" of epigenetic processes. Epigenetic features of the immune response are rational and achievable drug targets in contrast to single-nucleotide polymorphisms in genomic DNA. DNA methylation and histone modifications are dynamic, reversible, and are mediated by enzymes that are druggable targets. MicroRNA (miRNA) expression is dynamic and is amenable to disruption by RNA interference (RNAi)-based approaches. The first steps in hitting epigenetic targets includes describing the patterns of DNA methylation, histone marks, and noncoding RNAs that are expressed in normal and disease states. This review will summarize the recent studies regarding the epigenetics of immune and structural cells in adaptive and innate immunity in the lung.

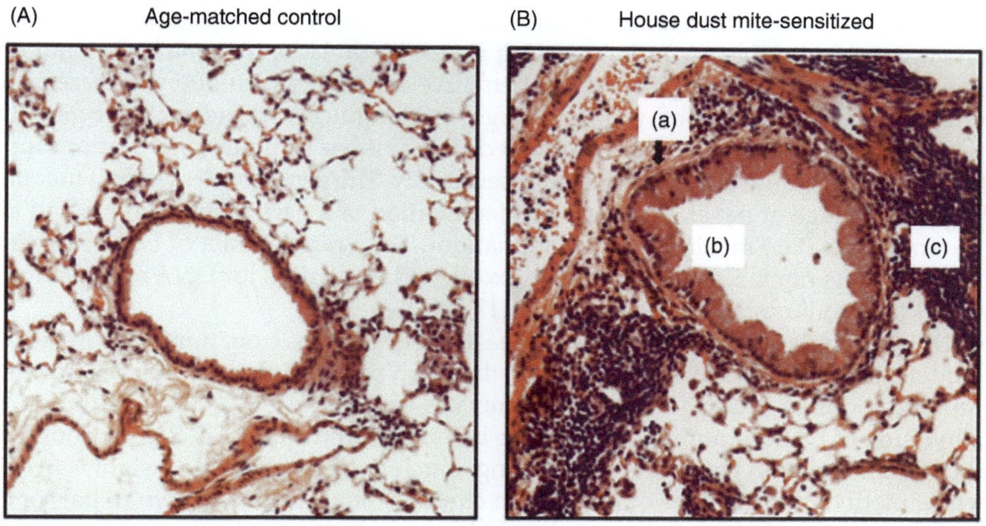

FIGURE 14.1 **Airway-wall remodeling in the house dust mite mouse model of asthma.** (A) Normal airway structure in a nonsensitized BALB/c female mouse. (B) A similar size airway from a mouse sensitized and challenged with house dust mite extract for 18 days, showing (a) airway smooth muscle hypertrophy, (b) mucosal metaplasia, and (c) immune-cell infiltration.

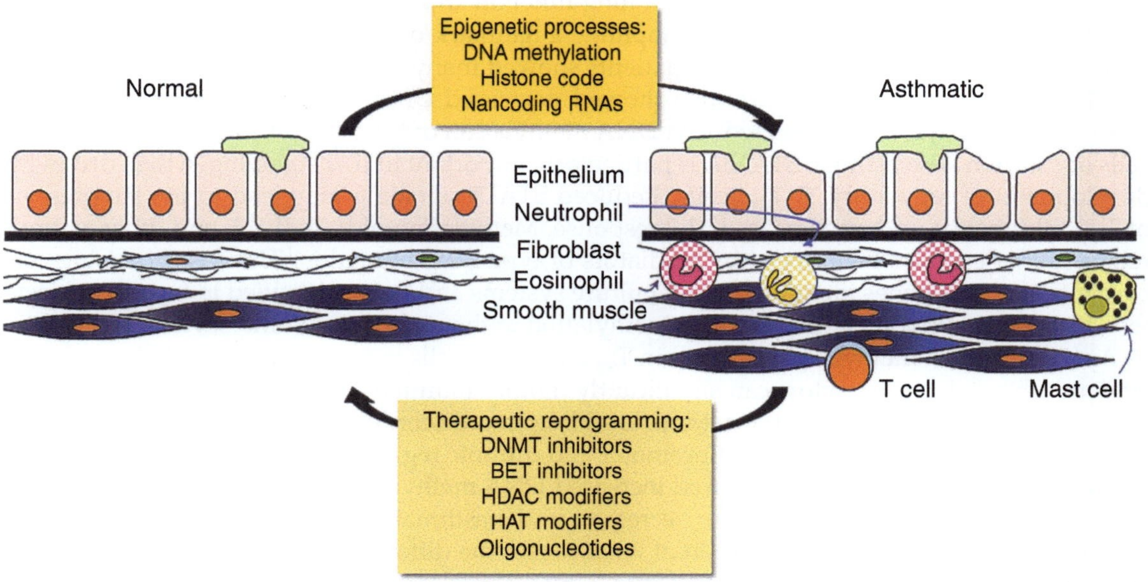

FIGURE 14.2 **Therapeutic reprogramming of pathological epigenetic processes is a goal of new therapeutic approaches in pulmonary diseases.**

14.2 EPIGENETIC CONTROL OF INFLAMMATORY GENE EXPRESSION IN ASTHMA

14.2.1 DNA Methylation and Inhibitors

Methylation of cytosine (5mC) residues at cytosine–guanine (CpG) dinucleotides occurs in both promoters and gene bodies. Promoter methylation acts in concert with histone modifications to control the transcription of many protein-coding genes. Methylation in gene bodies can influence the rate of transcription and the patterns of alternative splicing, which often increase protein expression. For example, subtle methylation changes in the glucocorticoid receptor gene (*NR3C1*) alters the expression levels and isoforms expressed [16]. This has profound implications for asthma therapy and may contribute to glucocorticoid insensitivity. The genes expressed will therefore vary depending on the stage of development, the differentiation of a particular cell type, and the presence of external cues (i.e., mediators of inflammation). The Encyclopedia of DNA Elements (ENCODE) and the Epigenome Road Map are two excellent data sources for DNA methylation sites in immune cells and nonimmune cells of the lung [17,18].

Global DNA methylation levels have been measured in samples from humans with asthma and mouse models of asthma. Variations in DNA methylation patterns in asthma have been described in human lung biopsies, leukocytes, and oral mucosa samples [19–22]. Several studies describe hypomethylation of DNA in immune cells in the blood and in the lung [22,23]. Although changes in global DNA methylation levels are enlightening, much more mechanistic insight is provided by gene-specific methylation changes or changes in chromosomal regions known to harbor disease-specific polymorphisms. In a study of peripheral blood monocytes in children with atopic asthma, Yang et al. [22] found decreased methylation of genes required for type 2 inflammation (i.e., *IL-4* and *IL-13*) and T-cell maturation (i.e., *RUNX3*). A study in IL-13-stimulated airway epithelial cells showed two distinct clusters of genes that underwent changes in DNA methylation [24]:one cluster was enriched for genes associated with lung function and asthma severity, and the other was enriched for genes associated with eosinophilia, a hallmark of type 2 inflammation in the lung. Epithelial cells isolated from asthmatic patients showed many of the same methylation events as the IL-13-stimulated cells. *STAT5A* also showed decreased DNA methylation in atopic asthma. Stefanowicz et al. [25] described reduced DNA methylation of the STAT5A promoter in epithelial cells of children with asthma. STAT5A is particularly important in IL-2 signaling, which drives the development of regulatory T (T_{reg}) cells. Reduced STAT5 signaling may lead to the suppression of T_{reg} cell function and an increased Th2 response. Methylation of IL-4, IL-13, and other asthma-related genes can also be induced by mediators of type 2 inflammation. There is promising concordance between genes identified in cell-culture systems and those identified in human subjects, which strongly supports a role for DNA methylation in the pathogenesis and severity of asthma.

A study of DNA methylation patterns in T_{reg} and $T_{effector}$ cells in twins also provided convincing evidence that epigenetic factors can significantly modify immune cell phenotype [21,26]. Runyon et al. reported on cases where the asthmatic sibling in a discordant twin pair showed enhanced CpG methylation of both the *FOXP3* promoter and intronic regions in T_{reg} cells [26]. $T_{effector}$ cells in the same asthmatic siblings also showed increased DNA methylation in the *IFNG* gene. The functions of both T_{reg} and $T_{effector}$ cells were decreased in the asthmatic sibling, therefore studies using discordant twin pairs begs the question of what causes the difference in epigenetic marks and the asthma phenotype?

Air pollution and exposure to inhaled particulates is thought to have an important effect on lung immune-cell phenotypes in the general population [27]. Inhaled pollutants might increase asthma attacks by modifying the methylation status of genes that regulate T-cell and mast-cell differentiation (i.e., *IL-13, IL-9, CD40L, FCER1A*) and eosinophilia (i.e., eotaxin and major basic protein) [20]. Reduced methylation at the *IFNG* and *IL-4* gene promoters can be elicited by the stimulation of $CD4^+$ T cells [23], which shows that DNA methylation of inflammatory genes is not static. This is consistent with the idea that genetic predisposition and the immune-cell environment in the lung collaborate to determine the asthma phenotype [28–30].

Changes in DNA methylation patterns in lung immune and structural cells in asthma might predispose a subject to allergic asthma [10]; however, they might also arise secondarily. Allergen challenge unleashes dramatic signaling events that are instigated by pro- and antiinflammatory mediators. Downstream signaling pathways are very likely to modulate the activity of DNA methyltransferases (DNMT) and demethylases which results in both local and global changes in DNA methylation. Data from mouse models of asthma support this notion; for example, global DNA methylation levels in lung tissues decreased in a chronic HDM mouse model [31], which corresponded with reduced DNMT3A activity. Interestingly, hydroxymethylation of cytosine increased in this model, although the function of this modified base is not fully understood. In the classical ovalbumin (OVA)-sensitized mouse, DNMT1 expression decreased, which is consistent with reduced DNA methylation [32].

Changes in DNA methylation levels in whole lung tissue are interesting but do not directly address the question of whether the immune response is altered. More direct support comes from methylation studies that have been performed in immune cells that are known to modify lung function in asthma. For example, Brand et al. investigated methylation patterns in genes that modify the Th1–Th2 balance using an OVA-sensitized mouse model [33]. There was increased methylation of the *IFNG* promoter, which correlated with reduced IFNγ production, but no change in methylation of the CNS1 noncoding locus. *IFNG* promoter methylation was inhibited by 5-azacytidine, suggesting that methylation at this locus favors a shift to the Th2 phenotype in atopic asthma.

De novo DNMTs are necessary for proper T-cell differentiation; however, their function in asthma is not entirely clear. Several mouse model studies suggest that DNMT activity is required for the Th2 high-asthma phenotype. In DNMT3A-floxed Cd4-Cre knockout mice, OVA challenge resulted in profound inflammation, greater eosinophilia, and enhanced IL-13 secretion from cultured Th2 T cells [34]. These results suggest that DNMT activity may normally limit asthma severity by decreasing the Th2 response following allergen challenge.

The timing of a therapeutic intervention that reduces DNA methylation might be critical. If DNA methylation changes occur in childhood in response to secondhand smoke or air pollution, then the changes may increase the incidence of asthma in those children who are genetically predisposed [35]. Intervening prior to the initial onset of asthma in children is unrealistic without specific predictive biomarkers. Similarly, asthma in adults may present well after the methylation events have occurred. However, intervening between asthma attacks in both children and adults is a realistic goal and is the standard of care in patients controlled with inhaled corticosteroids. Whether DNA methylation can be selectively modulated to prevent attacks using new classes of drugs is unclear; however, there is reason for optimism because de novo DNA methylation events are dynamic and can be reversed by DNA demethylases. Therefore blocking DNMT activity might be effective in preventing recurring attacks and attenuating asthma exacerbations. Proof of this principle comes from a prevention study in OVA-challenged mice. Wu et al. found that inhibiting DNMT activity with 5-azacytidine reduced inflammation and stopped the development of airway hyperreactivity [36]. The antiinflammatory effect was ascribed to the upregulation of T_{reg} cell maturation and the reduced Th2 response. One important limitation of this study is that 5-azacytidine has additional pharmacological effects. Therefore more selective agents with lung-restricted distribution must be developed to minimize the off-target effects. One alternate approach might be to target methyl-CpG-binding proteins in the nucleosome remodeling and histone deacetylation (NuRD) complex, which reads DNA methylation patterns in gene silencing [37,38].

14.2.2 Histone Acetylation and Inhibitors

There has been significant progress in describing the histone marks associated with asthma. Patterns of histone modifications vary during disease development and with disease severity. The patterns and functional outcomes also vary with the type of cell (i.e., airway epithelium, immune cell, or structural cell). Many of the disease-related histone modifications are caused by changes in the expression and activity of histone modifying enzymes, especially increased histone acetyltransferase (HAT) and decreased histone deacetylase (HDAC) levels [39–41]. This results in an imbalance between permissive and restrictive histone modifications regulating proinflammatory gene expression. It is also important to note that the protein deacetylases and acetyltransferases that catalyze histone modifications are not selective for histones [42,43], but are general modifiers of protein acetylation in both the nucleus and the cytoplasm. Histones, transcription factors (i.e., NFκB, STAT3, Myc, and p53), glucocorticoid receptors, and heat shock proteins are all substrates that regulate immune-cell functions. Therefore studies linking changes in HAT and HDAC activity to inflammation and asthma should be interpreted cautiously, as should studies using broad spectrum nonselective HDAC inhibitors (i.e., trichostatin and butyrate). Not all changes in cell differentiation, cell function, and lung function will be caused by changes in histone-mediated chromatin remodeling. Establishing the role of histone modifications in asthma will require careful studies that correlate biochemical and physiological findings. Loss-of-function and gain-of-function molecular approaches will also be necessary to define cause and effect.

Some of the earliest surveys of histone modifying enzymes in asthma reported increased HAT and decreased HDAC activity, which resulted in increased histone acetylation [39,41]. Lower HDAC activity levels may be caused in part by reactive oxygen species, which are known to increase in the asthmatic lung [44]. An important question being addressed is exactly where in the genome, and in which cells, histone acetylation is increased. A summary of some of the key genes and cells that are relevant to the immune response in the lung are described below. The reader is also advised to refer to earlier reviews of airways inflammation and histone modifications [45,46].

Histone modifications vary with asthma severity and can be modified by antiinflammatory therapies. Subjects with severe steroid-insensitive asthma have very low levels of HDAC1 and HDAC2 protein expression and deacetylase activity [47]. Corticosteroid treatment increases HDAC activity and reduces HAT activity [39]. HDAC2 expression is repressed in animal models of severe asthma by miR-21, which directly downregulates phosphatase and tensin homolog (PTEN) expression [48]. The proposed signaling pathway is illustrated in Fig. 14.3. Downregulation of PTEN results in increased PI3K phosphorylation and increased AKT phosphorylation, which in turn reduces HDAC2 expression (Fig.14.3A). A miR-21 antagonist reverses the effects of PI3K phosphorylation and

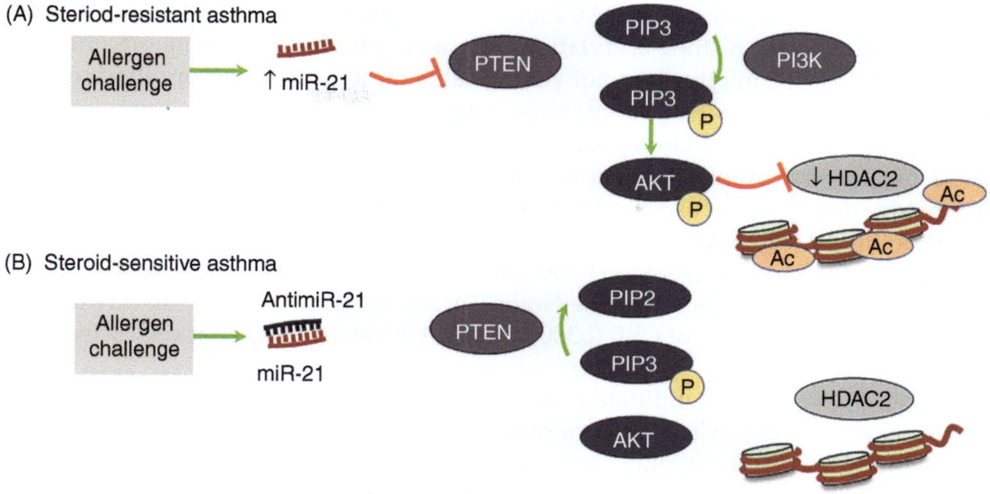

FIGURE 14.3 **MicroRNA, protein kinases, and HDACs all collaborate in the pathogenesis of severe steroid-resistant asthma. Kim et al. demonstrated that increased miR-21 is part of the mechanism of steroid-resistance in a mouse model of severe asthma [48].** (A) Elevated miR-21 reduces PTEN expression, which enhances AKT activity. Increased AKT represses HDAC2 expression leading to steroid resistance. (B) A miR-21 antagonist (antimiR) antagonizes this pathway and reestablishes steroid sensitivity.

HDAC2 protein levels, and restores corticosteroid sensitivity (Fig.14.3B). The authors hypothesized that a shift in the HAT to HDAC activity ratio would increase histone and NFκB acetylation, with the net effect being increased production of proinflammatory cytokines and growth factors. This appears to be the case in immune cells such as T cells [49], macrophages, and monocytes [41,47]. In addition to regulating immune-cell cytokine production, the ratio of HAT to HDAC activity also controls the production of inflammatory mediators in airway smooth muscle cells [50]. It is not yet clear whether people who develop asthma are genetically predisposed to an imbalance of HAT to HDAC activity or if the imbalance is induced during the pathogenesis of asthma. In either case, reestablishing a normal HAT to HDAC activity ratio could be an effective therapeutic approach.

14.2.3 Drugs Acting on HDACs and HATs as Asthma Therapies

Early studies using the nonselective HDAC inhibitor, trichostatin, showed that it increased the expression of IL-8 and granulocyte-macrophage colony stimulating factor (GM-CSF) in airway epithelial cells and macrophages [41,51]. Therefore inflammatory gene expression in lung cells may normally be repressed by hypoacetylation and increased by protein acetylation. As histone hyperacetylation dominates in asthma, drugs that inhibit HAT activity or increase HDAC activity are theoretically appealing. Fig.14.3 summarizes one example of this principle. As described above, the increased miR-21 expression during lung inflammation would represses PTEN expression, and this event would occur in a variety of cell types in the asthmatic lung, including vascular and airway smooth muscle cells and fibroblasts. Reduced PTEN levels would resulted in increased PI3K activity, increased AKT activity, and reduced HDAC2 expression. A miR-21 antagomir would upregulate HDAC2 and restore steroid sensitivity [48], although the relative significance of this pathway in immune cells versus structural cells in asthma is not clear yet.

However, the miR-21–PTEN–PI3K–HDAC2 axis is an excellent example of two different epigenetic processes (miRNA and histone acetylation) interacting to modulate lung inflammation.

Clinically useful HAT inhibitors are not yet available for antiinflammatory therapy as their development has lagged behind HDAC inhibitors somewhat; however, there are novel bisubstrate inhibitors in development, and library screening for novel classes of small-molecule inhibitors is ongoing [52–54]. Several proof-of-principle studies suggest that HAT inhibitors may be effective antiinflammatory drugs. Cui et al. found that garcinol inhibited HAT activity in $CD4^+$ T cells from an asthmatic rat model [55]. Several histone acetylation marks associated with the Notch1 promoter were significantly decreased as were the levels of several Th2 cytokines (IL-4, IL-5, and IL-13), thus suggesting that histone methylation at the Notch1 promoter contributed to airway inflammation. Garcinol is an effective inhibitor of CBP (p300) and PCAF; however, it is not very selective for HAT family members.

C646 is another interesting candidate that competitively inhibits p300 with modest selectivity over other acetyltransferases. C646 reduces histone acetylation and causes cell cycle arrest in leukemia cells [56–58]. The HAT-inhibitory effects of C646 were tested in RAW264 cells and found to reduce NFκB activity. This is promising as peripheral blood monocytes from patients with severe asthma have stubbornly high NFκB-mediated gene expression levels even during steroid therapy [59]. Unfortunately, NFκB activity and proinflammatory gene expression levels in RAW264 cells were only modestly inhibited by concentrations of C646 that were well above the K_i for HAT inhibition [60]. Mass spectroscopy analysis of histone H3 showed increased rather than the predicted decrease in acetylation, which occurred because of the high drug concentrations used to inhibit HDAC activity. Therefore the structure of C646 and its related homologs need to be optimized to develop more potent and selective HAT inhibitors. Any new agents that pass early molecular and cell screening would then need to be tested in cells from asthma patients and in animal models of asthma.

14.2.4 Histone Methylation and Inhibitors

Acetylation of lysine residues in histones is frequently permissive for transcription. Methylation of histones can also be permissive; however, it can also be restrictive depending upon the lysine residue and the particular histone family member; for example, H3K4me2 and H3K4me3 marks are permissive but H3K9me2 and H3K9me3 marks are restrictive. Methylation reactions are catalyzed by histone lysine methyltransferases (HMT) or protein arginine methyltransferases (PRMT); histone demethylases remove the methyl groups. Both lysine and arginine methylation events are altered in pulmonary inflammation and both modify the expression of proinflammatory genes; for example, PRMT1 is upregulated by allergic inflammation in a rat model, and a PRMT inhibitor (AMI-1) reduces eosinophilia and IL-4 production, which are two hallmarks of type 2 inflammation, in asthma [61]. Epithelial cytokines may be necessary signals for the upregulation of PRMT1; for example, the stimulation of lung fibroblasts and epithelial cells with IL-1β and IL-4 upregulated *PRMT1* expression [62]. Fibroblasts were also transactivated by IL-1β released from IL-4 stimulated epithelial cells. Prior work from the same group showed that this mechanism also increased *COX2* expression in lung fibroblasts [63] and eotaxin-1 in airway epithelial cells [61]. This suggests that cytokines produced by immune cells and structural cells during an asthma attack can modify the histone arginine methylation capacity and alter gene expression. There is also clear evidence for histone lysine modifications in both immune and structural cells. VEGF165a production is enhanced in human airway smooth muscle cells from asthma patients compared to healthy controls [64]. VEGF165a is normally repressed by trimethylation of H3K4 in these cells; however, this repression is lost in asthma and results in elevated VEGF expression, which may contribute to the increased airway vascularization seen in asthma.

The role of histone acetylation and methylation in Th1 and Th2 differentiation is well defined and described in several excellent reviews [3,65,66]. The histone marks that drive T cell differentiation are being reported in increasing detail, and they are deciphered by protein complexes such as the NuRD and REST corepressor 1 (CoREST). T cell development depends on proper NuRD function; for example, a T-cell knockout of the NuRD complex component, Mta2, increased T-cell proliferation and increased IL-4 and IFNγ production under Th2 (IL-4) conditions but not under Th1 (IL-12) conditions in mice [67]. The roles of MTA2 and other NuRD complex proteins in asthma pathogenesis are not as well defined; however, the NuRD complex may be a useful drug target. Lysine-specific demethylase 1 (LSD1) is also a component of NuRD and CoREST and may prove to be a high-value drug target [68]. LSD1 interacts with MTA2 in the human NuRD complex and demethylates H3K4. In breast cancer cells, upregulation of LSD1 correlates with downregulation of TGF-β1, increased epithelial-mesenchymal transition (EMT), and increased metastasis [69]. TGF-β1 promotes epithelial cell EMT, myofibroblast differentiation, and smooth muscle cell hypertrophy during airway remodeling in asthma. The roles of LSD1 and other NuRD and CoREST complex proteins have not been fully defined in asthma and airway inflammation; however, this is a worthwhile future direction given the central role that TGF-β1 plays in both immune cell and structural cell differentiation in the lung.

14.2.5 Bromodomain Proteins and Inhibitors

Bromodomain proteins regulate chromatin structure and gene expression by "reading" the acetylation patterns of histones. Members of the BET subfamily (BRD3 and BRD4) are thought to promote airway inflammation and airway hyperreactivity, and BET inhibitors are antiinflammatory in mouse models of asthma. Both neutrophil and macrophage numbers were reduced by the BET inhibitor, I-BET-762 [70], and the decreased immune-cell infiltration was associated with the decreased production of inflammatory cytokines (i.e., IL-27, CCL2, and IFNγ). In human bronchial epithelial cells, BET inhibitors (JQ1 and PFI-1) reduced the expression of IL-6 and CXCL8 due to the reduced recruitment of BRD4 to their respective promoters [71]. Similar findings were also reported for human airway smooth muscle cells [72]. BRD4 inhibition reduced the secretion of CXCL8 by disrupting the histone reader functions of BET proteins in both epithelium and smooth muscle cells. Multiple cell types in the lung appear to require members of the BET family for cytokine release, therefore small molecule inhibitors of BET protein function are obvious candidates as new antiinflammatory drugs.

14.2.6 MicroRNAs

Corticosteroid sensitivity can be improved in patients with chronic obstructive pulmonary (COPD) by inhibiting the translation of cytokine genes using rapamycin [73], and low doses of the mechanistic target of Rapamycin (mTOR) inhibitor, FK506, reversed arterial remodeling in PAH [74]. These studies suggest that inhibiting the translation of multiple proteins is a rational goal in the treatment of inflammatory lung diseases. However, the strategy is risky because both maladaptive and protective changes in protein synthesis will be affected. This could increase the risk of off-target effects, which has been seen when rapamycin (Sirolimus) is used chronically as an immunosuppressant to prevent transplant rejection. An alternative approach would be to inhibit the synthesis of a smaller set of proteins specifically in the lung using miRNA mimics and antagonists. Multiple proteins will be affected by miRNAs either by reducing the mRNA stability and half-life or by blocking translation (Fig. 14.4); however, reducing the synthesis of a modest number of coregulated proteins in a restricted region (i.e., the lung) might be tolerable. Appropriate miRNAs, the target protein networks, and the optimum drug delivery vehicles are still being defined; however, several miRNAs are emerging from preclinical studies of allergic inflammation as high value targets (Fig. 14.4B).

Previous reviews have already listed the miRNAs that are altered in humans and in animal models of asthma [4,75,76]. The interested reader is referred to these more comprehensive reviews for details of which miRNAs are altered in disease. The focus here is on three miRNAs that were identified as being important for lung inflammation and airway structure: miR-21, miR-155, and miR-221 (Fig.14.4B). All three miRNAs have been described in humans and animal models of asthma and show high concordance. All three have also been targeted with oligonucleotide antagonists in cell systems and in preclinical efficacy trials in animals. These miRNAs are farthest along the drug-development

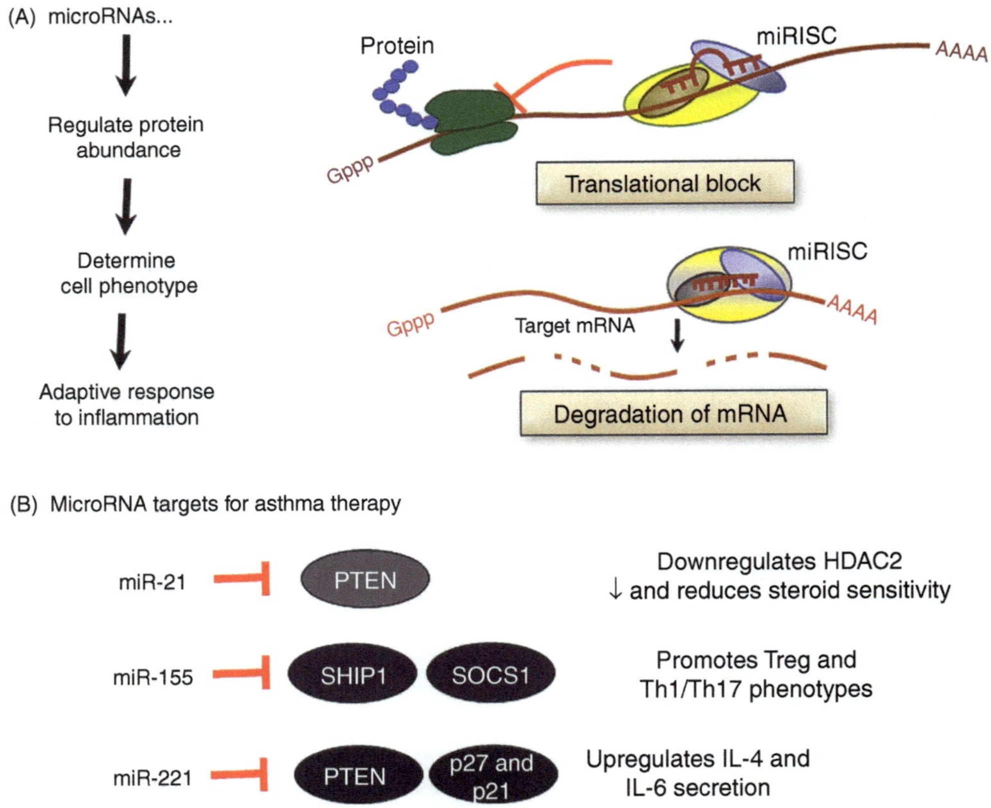

FIGURE 14.4 **MicroRNAs are valid therapeutic targets in inflammatory pulmonary diseases.** (A) miRNAs repress synthesis of numerous protein targets by two major mechanisms: mRNA degradation when base pairing is perfect, or near perfect; and inhibition of translation due to slight sequence mismatches with the target mRNA. (B) Several proinflammatory microRNAs are potential targets for antisense oligonucleotide therapy of asthma because they regulate important signal transduction events in inflammation. A selected subset of target proteins is shown that modify the inflammatory response in asthma and in animal models of allergic inflammation. Delivery of miRNA antagonists against each of these miRNAs are being investigated as novel antiinflammatory treatments for asthma.

pipeline and their antagonists have shown important antiinflammatory effects. Lung-directed or lung-restricted delivery should mean that the off-target effects are tolerable; however, little preclinical animal data are published and no trials in humans have yet begun.

14.2.6.1 MiR-21

MiRNA expression studies in humans and mouse models of asthma identified miR-21 as being upregulated [77,78]. MiR-21 has previously been suggested as a serum biomarker in asthma, as asthmatic children demonstrated elevated serum miR-21 levels, which positively correlated with serum eosinophilia [79]. Forced expiratory volume, 1 sec (FEV1) and serum IL-12 were also both reduced in subjects with elevated serum miR-21. Previous studies of IL-13-overexpressing and OVA-challenged mice reported that miR-21 was able to directly target IL-12p35 mRNA, thus promoting Th2 T-cell polarization [80]. Knockout of miR-21 increased IL-12 production by LPS-stimulated dendritic cells and increased IFNγ production by $CD4^+$ T cells [81]. These data suggest that miR-21 antagonists may be effective antiinflammatory treatments; however, the initial study using a miR-21 antagonist in a HDM model failed to show efficacy [82]. A subsequent report from the same group did show significant efficacy of a miR-21 antagonist in several pathogen-mediated mouse models of severe asthma [48]. In these models, miR-21 appears to repress PTEN expression, which enhances PI3K-mediated downregulation of HDAC2. A miR-21 antagonist enhanced HDAC2 expression, which is highly desirable for restoring normal epigenetic regulation in humans with severe asthma (Fig. 14.3).

14.2.6.2 MiR-155

MiR-155 is a widely expressed multifunctional miRNA that is present in human lungs at moderate levels [75]. It regulates T-cell differentiation by modulating the development of Th2, Th17, and T_{reg} subtypes [83,84]. The effects of miR-155 on immune cells are complex, context-dependent, and vary with the disease model; however, there is evidence from animal studies to suggest that miR-155 is necessary for chronic inflammatory responses [83,85]. It also regulates B-cell development and promotes the production of antibodies. Many direct protein targets of miR-155 have been identified in immune cells, including Pu.1, SHIP1, SOCS1, and c-Maf. Expression of miR-155 was not elevated in bronchial biopsies from humans with mild asthma [75], but was significantly elevated in epithelial cell brushings [86]. Although miR-155 is an obvious target for developing an antiinflammatory antagomir, the problem may be more difficult than first thought. When a 2′-O-methyl anti-miR-155 oligonucleotide was delivered intranasally to OVA-sensitized mice, it reached the airway wall cells and some, but not all, immune cells [87]. Myeloid cells took up the antagomir, but lung and spleen lymphocytes did not. The antagomir had no significant effects on airway resistance, compliance, or immune cells in bronchoalveolar lavage (BALF). These results emphasize the importance of developing novel methods of oligonucleotide delivery to enable the relevant target cells to be reached. Furthermore, the miRNA antagomir must be delivered to subcellular compartments in these target cells at sufficiently high concentration to hybridize and neutralize the native miRNA target.

14.2.6.3 MiR-221

mir-221 is modestly expressed in human lungs [75], but has significant effects on lung inflammation. Peripheral blood monocytes (PBMCs) from children with asthma express increased levels of miR-221 [88], and inhibition of this miRNA reduced airway inflammation in a mouse model [48]. miR-221 regulates the actin cytoskeleton in multiple cell types, and following mast cell activation, miR-221 modulates the expression of proteins that are necessary for adherence and degranulation.

Sensitized mice treated with a miR-221 inhibitor showed reduced inflammation, presumably due to decreased mast cell degranulation [89]. The ability of IgE to trigger mast cell degranulation is enhanced by miR-221 via PI3K-mediated regulation of calcium signaling [90]. Mast cells also secrete IL-4, which is a major stimulus for lung inflammation in asthma. Zhou et al. described the mechanisms regulating IL-4 secretion in P815 mastocytoma cells [91], and identified that miR-221 targets PTEN, which blocks IL-4 secretion. Like miR-21, miR-221 also promotes inflammation by enhancing IL-4 secretion by a pathway requiring PTEN and PI3K. Whether miR-221 also increases the expression of HDACs remains to be determined, but upregulating HDAC expression would be very useful in the treatment of asthma.

In addition to influencing the PI3K–PTEN–AKT pathway and mast cell secretion, miR-221 can also modify the functions of cell cycle control proteins. MiR-221 expression is increased in airway smooth muscle cells from subjects with severe asthma, and delivering exogenous miR-221 to cultured airway smooth muscle cells increased their proliferation and IL-6 expression [92]. Studies using mimics and inhibitors showed that these miR-221 effects were correlated with downregulation of the checkpoint control proteins, $p21^{CIP1}$ and $p27^{KIP1}$. These are promising effects, but it is not clear yet how they can be translated to reverse obstructive airway remodeling in vivo.

These three miRNAs emerged from analyses of the cell physiology of their target transcripts and from understanding the role of the targeted proteins in asthma. In the case of miR-21 and miR-155, there is strong in silico and in vitro evidence that has not yet translated into positive therapeutic responses in animal models [82,87]; however, the use of novel oligonucleotide chemistry and better drug delivery methods may solve this problem. Delivering therapeutic oligonucleotides to the proper cell type and the correct subcellular site of action is critical for taking advantage of novel miRNA-based therapeutic targets. Another known limitation of using oligonucleotides as drugs is their "off-target" effects on other organ systems. Lung-directed and cell-type directed oligonucleotide delivery should pave the way for testing many new miRNA antagonists, siRNAs, and antisense oligonucleotides that target a number of protein, mRNA, and noncoding RNA targets.

14.3 EPIGENETIC CONTROL OF INFLAMMATORY GENE EXPRESSION IN PULMONARY ARTERIAL HYPERTENSION

Patients with severe PAH (group I PH) suffer from right heart failure due to persistent abnormally-high right ventricular pressure. Elevated pulmonary arterial pressure is caused by a combination of arterial wall thickening and the formation of occlusive vascular lesions, which are thought to have an inflammatory component [93–95]. Neointimal arterial lesions and complex plexiform lesions are hallmarks of severe disease and are associated with a poor prognosis [96]. The neointimal and plexiform lesions are thought to arise from the increased proliferation and decreased apoptosis of endothelial cells, smooth muscle cells, and adventitial fibroblasts. Inflammation is a coincident finding that may contribute to the extent of lesion formation [95,97,98]. Immune cell migration from the blood and differentiation of immune cells residing in the artery adventitia has been observed, and inflammatory cells in or near the arterial walls are thought to promote the formation of occlusive arterial lesions [99,100]. Circulating cytokines and chemokines are also elevated in PAH patients, and there is evidence of increased NFκB levels in macrophages and endothelial cells [101]. Circulating autoantibodies in PAH patients are also thought to alter endothelial and fibroblast function [102,103]. The biochemical triggers for the immune cell infiltration, mediator production, and autoantibody production are not fully defined, but they may be due to genetic predisposition plus a "second hit," such as a redox condition or a toxin [98,104–106].

Hypoxia initiates multiple signaling events in vascular wall cells and in immune cells [107,108]; for example, the subcellular localization and DNA binding of transcription factors that regulate homeostatic gene expression are sensitive to redox conditions [109]. Reactive oxygen and reactive nitrogen species influence the transcription of genes that are relevant to pulmonary vascular inflammation, including Ref-1 and HIF-1α. The Ref-1 apurinic/apyrimidinic (AP) endonuclease is a particularly interesting protein as it is a redox-sensor protein that can modulate lung inflammation. Inhibiting Ref-1 function in OVA-specific T-cell mice leads to an increased Th1 response [110]. Whether this occurs in PAH is not clear, but it seems plausible that hypoxic conditions could change Ref-1-mediated regulation of redox-sensitive transcription factors and modify perivascular inflammation. There is currently great interest in redox-dependent epigenetic factors that influence vascular inflammation. Significant progress has been made on defining the miRNAs that are necessary for vessel remodeling and vascular inflammation [111]; however, progress has been slower in establishing the causes and consequences of DNA methylation and histone modifications in immune cells in PAH. Both DNA methylation and histone modifications are potential targets for new drug development [2,112,113].

14.3.1 DNA Methylation

There is one report of altered DNA methylation patterns in the superoxide dismutase gene in the vascular smooth muscle cells in PAH [114]. Aside from this study there is no direct evidence for 5mC or 5-hydroxymethylcytosine (5hmC) modifications that correlate with inflammatory gene expression in PAH. However, there are studies of related conditions that are instructive. For example, one study described increased methylation of the granulysin gene (*GLYN*) in the peripheral blood monocytes of patients with pulmonary venoocclusive disease, which was absent in patients with PAH [115]. The authors concluded that the *GLYN* genes in cytotoxic, natural killer, and natural killer-like T cells were unaffected by PAH. Markers of other relevant circulating leukocytes (e.g., monocytes) were not reported in this study. Another potentially relevant study appeared in the scleroderma literature. The bone morphogenetic protein receptor type II (BMPR2) promoter was methylated at CpG sites in subjects with scleroderma [116]. Some patients with scleroderma also develop secondary pulmonary hypertension, and BMPR2 dysfunction is a primary cause of inheritable PAH. It is therefore possible that subjects with scleroderma are predisposed to PAH when the BMPR2 promoter is methylated and BMPR2 expression is reduced. However, BMPR2 silencing by DNA methylation has not yet been reported in subjects with PAH. There is also a remarkable lack of information on the DNA 5mC and 5hmC status of proinflammatory genes in PAH. Given that immune cells infiltrate occlusive vascular lesions in humans and animal models of PAH, this is a surprising gap in the knowledge. A genome-wide association (GWAS) study of systemic hypertension revealed several loci where DNA methylation patterns associated with the disease [117]. Comparable genome-wide DNA methylation studies in PAH models and in human lung tissues need to be conducted to establish the significance of altered 5mC and 5hmC patterns in the initiation and progression of PAH; however, this is a challenging task given the rarity of PAH. Loci identified in humans with systemic hypertension should serve as a good starting point, and epigenetic studies in appropriate animal models should be useful for identifying key DNA methylation patterns that precede or parallel perivascular inflammation and arteriopathy.

14.3.2 Histone Modifications

Relatively few studies have described the role of histone acetylation in genes that are responsible for vascular remodeling in PAH, and progress is limited by the modest understanding of exactly which genes to focus on. The most direct evidence for HDAC influence in lung inflammation comes from a bovine model of hypoxic pulmonary hypertension, which is somewhat different from PAH in humans [118]. Class I HDAC inhibitors markedly reduced the ability of fibroblasts from the pulmonary artery of chronically hypoxic calves to synthesize proinflammatory cytokines and adhesion molecules. The fibroblasts promoted monocyte migration, which was inhibited by the HDAC inhibitors. Humans with hereditary PAH demonstrate higher expression levels of HDAC1, 4, and 5 and elevated global histone H3 and H4 acetylation [119]. Heart and lung levels of HDAC1 and 5 correlated positively with increased pulmonary artery pressure in a chronic hypoxia rat model, and HDAC inhibitors (i.e., valproic acid and SAHA) prevented pathological vascular remodeling in the rat model [119]. These are provocative results that must be tempered by the fact that HDACs and HATs regulate the acetylation of many proteins, not just histones.

Part of the therapeutic effect of HDAC inhibitors in animal models of PAH may also be due to altered ROS production by immune cells. For example, HDAC inhibitors reduced NOX1, 2, and 4 expression by reducing Pol II and p300 HAT loading on the promoters of NOX genes in macrophages and monocytic leukemia (THP-1) cells [120]. HDAC inhibition also decreased H3K4me3 and H3K9ac histone marks at these promoter sites in lung fibroblasts. Unfortunately, the effects of valproic acid on perivascular inflammation were not reported. In summary, several studies of valproic acid in rat models of PAH report beneficial preventative and reversal effects [119,121]. Studies of HDAC inhibitors in PAH suggest that the regulation of proinflammatory genes, progrowth genes, and promigratory signaling genes are highly relevant to the initiation of pulmonary inflammation and ultimately to PAH severity. Further mechanistic studies in immune cells and pulmonary artery cells in multiple animal models of PAH are required to define which HDAC classes, which isoforms, and which chemical inhibitors will be most effective as a PAH therapy. Careful biochemical analysis will also be required to ascertain if the acetylation events are modified primarily or solely in histones versus other known HAT and HDAC substrates.

14.3.3 HATs

Compared to studies of HDACs and HDAC inhibitor drugs, there are even fewer studies of HATs in PAH. HAT activity is upregulated in the lung tissue of PAH patients [122], which may influence cytokine gene expression by altering histone acetylation. Part of the beneficial effect of the prostacyclin analog, iloprost, may be due to its antiinflammatory histone modifications. Iloprost reduced the secretion of IL-8, CCL2, RANTES, and TNFα in activated monocytes [123], and the inhibition of cytokine production was mediated by the inhibition of STAT1 phosphorylation. This correlated with reduced localization of the HAT, p300, at the STAT1 promoter. These results suggest that HAT inhibition may evolve into an adjunctive therapy in PAH; however, significant information is needed to define the proinflammatory genes that are sensitive to HAT inhibitors. At this point, there are no preclinical animal studies supporting the efficacy of HAT inhibitors in PAH models.

14.3.4 Histone Methylation

Several inflammatory conditions can modify both permissive (H4K4me3 and H3K4me1) and repressive (H3K27me3) histone marks. Histone methylation and DNA de novo methylation can also interact [124]. SET HMT proteins catalyze the formation of H3K9me3, which favors de novo DNA methylation by DNMT3a and DNMT3b during embryonic development [124]. Therefore H3K9 methylation may exert similar effects in perivascular immune cells and the structural cells of inflamed arteries in PAH; however, only indirect evidence currently supports this hypothesis. The most convincing evidence comes from studies of the proinflammatory cytokine, IL-33. This cytokine is a soluble signaling protein that acts on cell membrane ST2 receptors and nuclear proteins to regulate gene expression through helix-loop-helix motifs. The contribution of IL-33 to perivascular inflammation in PAH is not entirely clear, but IL-33 mRNA is reduced in lung tissue and nuclear IL-33 is downregulated in the arteries of PAH patients [125]. Knockdown of IL-33 in cultured endothelial cells increased IL-6 and increased soluble ST2 expression. Chromatin immunoprecipitation (ChIP) assays revealed that IL-33 binds to the homeodomains of several inflammatory genes and is included in a complex with SUV38H1. SUV38H1 is an HMT that promotes the recruitment of DNMTs and thus permits de novo DNA methylation. These data collectively suggest that the innate immune response in PAH is promoted by derepression of cytokine genes by the IL-33 interaction with SUV38H1. However, it is not known if H3K9me3 or DNA methylation also changes in cytokine gene promoters or cytokine gene bodies in PAH. It is also not clear whether the IL-33–ST2–SUV38H1 axis is a cause or consequence of PAH. As IL-33 favors the Th2 phenotype and is atheroprotective [126], it would be useful to know what effect IL-33 has on lymphocyte gene expression programs and the arteriopathy of PAH.

14.3.5 Bromodomain Proteins

Vascular remodeling in PAH depends on gene expression changes in multiple cell types via a variety of transcriptional regulation pathways. NFκB-regulated genes are often upregulated in inflammation, including in vascular inflammation. A recent study of NFκB-regulated genes in normal pulmonary vascular endothelial cells showed that the BET inhibitor, JQ1, caused cell cycle arrest [122] by increasing the expression of cell cycle arrest proteins, $p19^{INK4D}$ and $p21^{CIP1}$, and by reducing the expression of several cyclin-dependent kinases. These effects correlated with increased HAT activity, presumably due to the inhibition of normal BET and HAT binding in reader–writer complexes. BET inhibition had similar antiproliferative effects on pulmonary artery vascular smooth muscle cells [105], and BRD4 was the proposed target of JQ1 in both endothelium and smooth muscle cells. It is possible that BET inhibitors may have beneficial antiinflammatory effects in PAH, but the perivascular lymphocyte infiltration effects have not been reported.

14.3.6 MiRNAs

Several correlative studies in human PAH samples show that miRNA expression is altered by the disease. MiR-21, miR-145, miR-204, and the miR-17~92 cluster were among the first miRNAs to be described in human PAH samples [127–131]. Several labs have used a variety of rat and mouse models to describe miRNAs that are altered in PAH and some that may be useful therapeutic targets. More comprehensive lists of miRNAs that are associated with PAH in humans and animal models are available in previous reviews [4,132]. Early reports of dysregulated miRNA expression in PAH lungs and hearts quickly led to translational studies that described the therapeutic approaches to normalize dysregulated miRNA expression. From the initial reports, it seemed likely that miRNA-based agents may be developed to halt or reverse the progression of PAH in humans. Subsequent studies of miR-204 [129], miR-17 [133], miR-21 [134], miR-20a [135], miR-145 [130], miR-223 [136], miR-424, and miR-503 [137] provided ample proof that RNAi-based targeting of miRNAs can attenuate experimental pulmonary hypertension. However, none of these studies employed lung-directed intravenous therapy, nor did they show reversal of occlusive vascular remodeling in an animal model that mimicked the complex occlusive lesions seen in severe PAH in humans. A lung-directed lipid nanoparticle was recently administered intravenously to deliver a miR-145 antagonist to treat severe PAH in a rat model. The miR-145 antagomir partially reversed vascular wall thickening and cardiac dysfunction [138], and a modest reduction of perivascular inflammation was also observed (unpublished), although the targeted immune cells were not identified. Similar antiinflammatory effects were reported by Gubrij et al. using a miR-223 antagonist in the monocrotaline rat model of PAH [136].

Most of the mechanistic work cited above has naturally focused on endothelial cell and vascular smooth muscle cell biology; however, several of the miRNAs associated with PAH in humans (i.e., miR-145, miR-155, miR-21, and the miR-17~92 cluster) are also important regulators of immune cell function. One of the earliest reports of miRNAs in PAH showed that IL-6 upregulated the expression of the miR-17~92 cluster, which targets BMPR2. Reduced BMPR2 signaling is thought to be the "first hit" in subjects with hereditary forms of PAH. Inflammation may also exacerbate PAH pathology by inhibiting BMPR2 signaling. Inflammation may in turn be regulated by miR-145, which is master regulator of the smooth muscle contractile phenotype that also promotes M1 to M2 macrophage polarization [139] and Th2 development [82]. A mir-145 antagonist inhibited type II inflammation in an asthma model by decreasing the expression of IL-5 and IL-13 in the lung [82]. miR-21 also has antiinflammatory effects in mouse models of severe asthma [48]. miR-124 has an indirect effect on vascular inflammation by silencing MCP1 expression in adventitial fibroblasts [140]. However, most of this evidence is circumstantial and only a few studies have clearly defined the effect of miRNA mimics and antagonists on perivascular inflammation in PAH. One promising future direction is to develop miRNA antagonists that target the expression of multiple protein networks that are required for both vascular wall remodeling and perivascular inflammation. Such drugs would constitute a novel class of therapeutics that may complement, and even improve, the efficacy of current vasodilator therapies by reversing arterial wall thickening.

14.4 SUMMARY AND FUTURE DIRECTIONS

Epigenetic processes that regulate immune cell differentiation and function are now under intense study in asthma and PAH; however, this work is still in the discovery stage. With a few notable exceptions, most investigations are cataloging the types of DNA and histone modifications during development and the changes induced by epigenetic modifying drugs. Mouse models of asthma, rat models of PAH, and numerous surveys of human lung samples in both diseases have been used to aid progress towards novel antiinflammatory therapies. Gain-of-function and loss-of-function studies in cells and mouse models point out some clear trends, such as the increased HDAC levels in PAH but reduced HDAC levels in asthma. Pharmacological inhibition studies have also revealed that HDAC inhibition may be therapeutic for PAH and that HAT inhibition could be therapeutic for asthma. However, none of the therapies have yet moved into human trials. This will not be true for long as there is ample evidence that HDAC inhibition with vorinostat is effective in cancer chemotherapy and is well tolerated.

One important future direction is to define which approaches have the greatest efficacy and the least toxicity. It is also important to establish whether epigenetic-targeted monotherapy is more effective than combinations of agents in different mechanistic classes. More information is also needed regarding the gene-specific histone modifications and the gene-specific DNA methylation patterns for the key immune cells in asthma and PAH. New data are needed that describe the miRNAs and long noncoding RNAs that are expressed in health and disease. It will be important to establish which miRNAs cause disease and which mediate adaptive responses. In both cases, human studies of serial changes in key epigenetic regulators will be necessary to differentiate cause from consequence. A snapshot taken at the end stage of a disease, which is often all that can be done in humans, is a very restricted image. Endpoint observations are often a poor reflection of the cause of the disease. They are a function of both causative phenomena and adaptation to disease conditions. They may or may not reveal valid drug targets for disease-modifying therapy. Therefore it is also instructive to investigate parallel pathways and cell functions that can "rescue" a disease phenotype even if the pathway does not cause the disease. In addition, future research should establish the efficacy of the lung-restricted delivery of epigenetic modifiers in vivo. If effective and safe agents can be delivered to the lungs, they could become important adjuncts to previously-established agents used in asthma and PAH.

CONFLICT OF INTEREST

SR and WTG declare they have no conflicts of interest.

Acknowledgments

This manuscript was support in part by NIH grants to WTG (AI116985) and SR (T32 HL076125).

ABBREVIATIONS

5hmC 5-hydroxymethylcytosine
5mC 5-methylcytosine
BMPR2 bone morphogenetic protein receptor type II
DNMT DNA methyltransferase
EMT epithelial-mesenchymal transition
HAT histone acetyltransferase
HDAC histone deacetylase
HDM house dust mite
HIF1α hypoxia inducible factor-1α
HMT histone lysine methyltransferase
IL interleukin

IL-β interleukin-1β
IFNγ interferon-γ
miRNA microRNA
OVA ovalbumin
PAH pulmonary arterial hypertension
PCAF lysine acetyltransferase 2B
PI-3K phosphatidyl inositol 3 kinase
PRMT protein arginine methyltransferase
PTEN phosphatase and tensin homolog
SAHA suberoylanilide hydroxamic acid
TGF-β1 transforming growth factor β1
TNFα tumor necrosis factor-α
TSLP thymic stromal lymphopoietin

References

[1] Montani D, Chaumais MC, Guignabert C, Gunther S, Girerd B, Jais X, et al. Targeted therapies in pulmonary arterial hypertension. Pharmacol Ther 2014;141(2):172–91.
[2] Cohen-Kaminsky S, Hautefort A, Price L, Humbert M, Perros F. Inflammation in pulmonary hypertension: what we know and what we could logically and safely target first. Drug Discov Today 2014;19(8):1251–6.
[3] Liu Y, Li H, Xiao T, Lu Q. Epigenetics in immune-mediated pulmonary diseases. Clin Rev Allergy Immunol 2013;45(3):314–30.
[4] Comer BS, Ba M, Singer CA, Gerthoffer WT. Epigenetic targets for novel therapies of lung diseases. Pharmacol Ther 2015;147:91–110.
[5] Chipps BE, Zeiger RS, Borish L, Wenzel SE, Yegin A, Hayden ML, et al. Key findings and clinical implications from the epidemiology and natural history of asthma: Outcomes and Treatment Regimens (TENOR) study. J Allergy Clin Immunol 2012;130(2):332–42.
[6] Sullivan SD, Wenzel SE, Bresnahan BW, Zheng B, Lee JH, Pritchard M, et al. Association of control and risk of severe asthma-related events in severe or difficult-to-treat asthma patients. Allergy 2007;62(6):655–60.
[7] Wenzel S. Severe asthma in adults. Am J Respir Crit Care Med 2005;172(2):149–60.
[8] Gerthoffer WT, Solway J, Camoretti-Mercado B. Emerging targets for novel therapy of asthma. Curr Opin Pharmacol 2013;13(3):324–30.
[9] Comer BS, Camoretti-Mercado B, Kogut PC, Halayko AJ, Solway J, Gerthoffer WT. Cyclooxygenase-2 and microRNA-155 expression are elevated in asthmatic airway smooth muscle cells. Am J Respir Cell Mol Biol 2015;52(4):438–47.
[10] Brook PO, Perry MM, Adcock IM, Durham AL. Epigenome-modifying tools in asthma. Epigenomics 2015;7(6):1017–32.
[11] Bel EH, Wenzel SE, Thompson PJ, Prazma CM, Keene ON, Yancey SW, et al. Oral glucocorticoid-sparing effect of mepolizumab in eosinophilic asthma. N Engl J Med 2014;371(13):1189–97.
[12] Humbert M, Busse W, Hanania NA, Lowe PJ, Canvin J, Erpenbeck VJ, et al. Omalizumab in asthma: an update on recent developments. J Allergy Clin Immunol Pract 2014;2(5):525–536e1.
[13] Gauvreau GM, O'Byrne PM, Boulet LP, Wang Y, Cockcroft D, Bigler J, et al. Effects of an anti-TSLP antibody on allergen-induced asthmatic responses. N Engl J Med 2014;370(22):2102–10.
[14] Ortega HG, Liu MC, Pavord ID, Brusselle GG, FitzGerald JM, Chetta A, et al. Mepolizumab treatment in patients with severe eosinophilic asthma. N Engl J Med 2014;371(13):1198–207.
[15] Barnes PJ. Targeting the epigenome in the treatment of asthma and chronic obstructive pulmonary disease. Proc Am Thorac Soc 2009;6(8):693–6.
[16] Leenen FA, Muller CP, Turner JD. DNA methylation: conducting the orchestra from exposure to phenotype? Clin Epigenetics 2016;8:92.
[17] Hoffman MM, Ernst J, Wilder SP, Kundaje A, Harris RS, Libbrecht M, et al. Integrative annotation of chromatin elements from ENCODE data. Nucleic Acids Res 2013;41(2):827–41.
[18] Romanoski CE, Glass CK, Stunnenberg HG, Wilson L, Almouzni G. Epigenomics: roadmap for regulation. Nature 2015;518(7539):314–6.

[19] White GP, Hollams EM, Yerkovich ST, Bosco A, Holt BJ, Bassami MR, et al. CpG methylation patterns in the IFNgamma promoter in naive T cells: variations during Th1 and Th2 differentiation and between atopics and non-atopics. Pediatr Allergy Immunol 2006;17(8):557–64.

[20] Sofer T, Baccarelli A, Cantone L, Coull B, Maity A, Lin X, et al. Exposure to airborne particulate matter is associated with methylation pattern in the asthma pathway. Epigenomics 2013;5(2):147–54.

[21] Murphy TM, Wong CC, Arseneault L, Burrage J, Macdonald R, Hannon E, et al. Methylomic markers of persistent childhood asthma: a longitudinal study of asthma-discordant monozygotic twins. Clin Epigenetics 2015;7:130.

[22] Yang IV, Pedersen BS, Liu A, O'Connor GT, Teach SJ, Kattan M, et al. DNA methylation and childhood asthma in the inner city. J Allergy Clin Immunol 2015;136(1):69–80.

[23] Kwon NH, Kim JS, Lee JY, Oh MJ, Choi DC. DNA methylation and the expression of IL-4 and IFN-gamma promoter genes in patients with bronchial asthma. J Clin Immunol 2008;28(2):139–46.

[24] Nicodemus-Johnson J, Naughton KA, Sudi J, Hogarth K, Naurekas ET, Nicolae DL, et al. Genome-wide methylation study identifies an IL-13-induced epigenetic signature in asthmatic airways. Am J Respir Crit Care Med 2016;193(4):376–85.

[25] Stefanowicz D, Hackett T-l, Garmaroudi FS, Günther OP, Neumann S, Sutanto EN, et al. DNA methylation profiles of airway epithelial cells and PBMCs from healthy, atopic and asthmatic children. PLoS One 2012;7(9).

[26] Runyon RS, Cachola LM, Rajeshuni N, Hunter T, Garcia M, Ahn R, et al. Asthma discordance in twins is linked to epigenetic modifications of T cells. PLoS One 2012;7(11):e48796.

[27] Nadeau K, McDonald-Hyman C, Noth EM, Pratt B, Hammond SK, Balmes J, et al. Ambient air pollution impairs regulatory T-cell function in asthma. J Allergy Clin Immunol 2010;126(4):845–852.e10.

[28] Ober C, Vercelli D. Gene-environment interactions in human disease: nuisance or opportunity? Trends Genet 2011;27(3):107–15.

[29] Myers RA, Scott NM, Gauderman WJ, Qiu W, Mathias RA, Romieu I, et al. Genome-wide interaction studies reveal sex-specific asthma risk alleles. Hum Mol Genet 2014;23(19):5251–9.

[30] Sharma S, Litonjua A. Asthma, allergy, and responses to methyl donor supplements and nutrients. J Allergy Clin Immunol 2014;133(5):1246–54.

[31] Cheng RY, Shang Y, Limjunyawong N, Dao T, Das S, Rabold R, et al. Alterations of the lung methylome in allergic airway hyper-responsiveness. Environ Mol Mutagen 2014;55(3):244–55.

[32] Verma M, Chattopadhyay BD, Paul BN. Epigenetic regulation of DNMT1 gene in mouse model of asthma disease. Mol Biol Rep 2013;40(3):2357–68.

[33] Brand S, Kesper DA, Teich R, Kilic-Niebergall E, Pinkenburg O, Bothur E, et al. DNA methylation of TH1/TH2 cytokine genes affects sensitization and progress of experimental asthma. J Allergy Clin Immunol 2012;129(6):1602–1610.e6.

[34] Yu Q, Zhou B, Zhang Y, Nguyen ET, Du J, Glosson NL, et al. DNA methyltransferase 3a limits the expression of interleukin-13 in T helper 2 cells and allergic airway inflammation. Proc Natl Acad Sci U S A 2012;109(2):541–56.

[35] Breton CV, Siegmund KD, Joubert BR, Wang X, Qui W, Carey V, et al. Prenatal tobacco smoke exposure is associated with childhood DNA CpG methylation. PLoS One 2014;9(6):e99716.

[36] Wu CJ, Yang CY, Chen YH, Chen CM, Chen LC, Kuo ML. The DNA methylation inhibitor 5-azacytidine increases regulatory T cells and alleviates airway inflammation in ovalbumin-sensitized mice. Int Arch Allergy Immunol 2013;160(4):356–64.

[37] Schubeler D. Function and information content of DNA methylation. Nature 2015;517(7534):321–6.

[38] Wood KH, Zhou Z. Emerging molecular and biological functions of MBD2, a reader of DNA methylation. Front Genet 2016;7:93.

[39] Ito K, Caramori G, Lim S, Oates T, Chung KF, Barnes PJ, et al. Expression and activity of histone deacetylases in human asthmatic airways. Am J Respir Crit Care Med 2002;166(3):392–6.

[40] Ito K, Lim S, Caramori G, Cosio B, Chung KF, Adcock IM, et al. A molecular mechanism of action of theophylline: induction of histone deacetylase activity to decrease inflammatory gene expression. Proc Natl Acad Sci U S A 2002;99(13):8921–6.

[41] Cosio BG, Mann B, Ito K, Jazrawi E, Barnes PJ, Chung KF, et al. Histone acetylase and deacetylase activity in alveolar macrophages and blood mononocytes in asthma. Am J Respir Crit Care Med 2004;170(2):141–7.

[42] Kouzarides T. Acetylation: a regulatory modification to rival phosphorylation? EMBO J 2000;19(6):1176–9.

[43] Witt O, Lindemann R. HDAC inhibitors: magic bullets, dirty drugs or just another targeted therapy. Cancer Lett 2009;280(2):123–4.

[44] Marwick JA, Ito K, Adcock IM, Kirkham PA. Oxidative stress and steroid resistance in asthma and COPD: pharmacological manipulation of HDAC-2 as a therapeutic strategy. Expert Opin Ther Targets 2007;11(6):745–55.
[45] Adcock IM, Tsaprouni L, Bhavsar P, Ito K. Epigenetic regulation of airway inflammation. Curr Opin Immunol 2007;19(6):694–700.
[46] Adcock IM. HDAC inhibitors as anti-inflammatory agents. Br J Pharmacol 2007;150(7):829–31.
[47] Hew M, Bhavsar P, Torrego A, Meah S, Khorasani N, Barnes PJ, et al. Relative corticosteroid insensitivity of peripheral blood mononuclear cells in severe asthma. Am J Respir Crit Care Med 2006;174(2):134–41.
[48] Kim RY, Horvat JC, Pinkerton JW, Starkey MR, Essilfie AT, Mayall JR, et al. MicroRNA-21 drives severe, steroid-insensitive experimental asthma by amplifying phosphoinositide 3-kinase-mediated suppression of histone deacetylase 2. J Allergy Clin Immunol 2017;139:519–32.
[49] Grausenburger R, Bilic I, Boucheron N, Zupkovitz G, El-Housseiny L, Tschismarov R, et al. Conditional deletion of histone deacetylase 1 in T cells leads to enhanced airway inflammation and increased Th2 cytokine production. J Immunol 2010;185(6):3489–97.
[50] Keslacy S, Tliba O, Baidouri H, Amrani Y. Inhibition of tumor necrosis factor-alpha-inducible inflammatory genes by interferon-gamma is associated with altered nuclear factor-kappaB transactivation and enhanced histone deacetylase activity. Mol Pharmacol 2007;71(2):609–18.
[51] Ito K, Barnes PJ, Adcock IM. Glucocorticoid receptor recruitment of histone deacetylase 2 inhibits interleukin-1beta-induced histone H4 acetylation on lysines 8 and 12. Mol Cell Biol 2000;20(18):6891–903.
[52] Lau OD, Kundu TK, Soccio RE, Ait-Si-Ali S, Khalil EM, Vassilev A, et al. HATs off: selective synthetic inhibitors of the histone acetyltransferases p300 and PCAF. Mol Cell 2000;5(3):589–95.
[53] Yang C, Ngo L, Zheng YG. Rational design of substrate-based multivalent inhibitors of the histone acetyltransferase Tip60. ChemMedChem 2014;9(3):537–41.
[54] Wapenaar H, Dekker FJ. Histone acetyltransferases: challenges in targeting bi-substrate enzymes. Clin Epigenetics 2016;8:59.
[55] Cui ZL, Gu W, Ding T, Peng XH, Chen X, Luan CY, et al. Histone modifications of Notch1 promoter affect lung $CD4^+$ T cell differentiation in asthmatic rats. Int J Immunopathol Pharmacol 2013;26(2):371–81.
[56] Bowers EM, Yan G, Mukherjee C, Orry A, Wang L, Holbert MA, et al. Virtual ligand screening of the p300/CBP histone acetyltransferase: identification of a selective small molecule inhibitor. Chem Biol 2010;17(5):471–82.
[57] Gao XN, Lin J, Ning QY, Gao L, Yao YS, Zhou JH, et al. A histone acetyltransferase p300 inhibitor C646 induces cell cycle arrest and apoptosis selectively in AML1-ETO-positive AML cells. PLoS One 2013;8(2):e55481.
[58] Oike T, Komachi M, Ogiwara H, Amornwichet N, Saitoh Y, Torikai K, et al. C646, a selective small molecule inhibitor of histone acetyltransferase p300, radiosensitizes lung cancer cells by enhancing mitotic catastrophe. Radiother Oncol 2014;111(2):222–7.
[59] Gagliardo R, Chanez P, Mathieu M, Bruno A, Costanzo G, Gougat C, et al. Persistent activation of nuclear factor-kappaB signaling pathway in severe uncontrolled asthma. Am J Respir Crit Care Med 2003;168(10):1190–8.
[60] van den Bosch T, Boichenko A, Leus NG, Ourailidou ME, Wapenaar H, Rotili D, et al. The histone acetyltransferase p300 inhibitor C646 reduces pro-inflammatory gene expression and inhibits histone deacetylases. Biochem Pharmacol 2016;102: 130–40.
[61] Sun Q, Yang X, Zhong B, Jiao F, Li C, Li D, et al. Upregulated protein arginine methyltransferase 1 by IL-4 increases eotaxin-1 expression in airway epithelial cells and participates in antigen-induced pulmonary inflammation in rats. J Immunol 2012;188(7):3506–12.
[62] Liu L, Sun Q, Bao R, Roth M, Zhong B, Lan X, et al. Specific regulation of PRMT1 expression by PIAS1 and RKIP in BEAS-2B epithelia cells and HFL-1 fibroblasts in lung inflammation. Sci Rep 2016;6:21810.
[63] Sun Q, Liu L, Roth M, Tian J, He Q, Zhong B, et al. PRMT1 upregulated by epithelial proinflammatory cytokines participates in COX2 expression in fibroblasts and chronic antigen-induced pulmonary inflammation. J Immunol 2015;195(1):298–306.
[64] Clifford RL, John AE, Brightling CE, Knox AJ. Abnormal histone methylation is responsible for increased vascular endothelial growth factor 165a secretion from airway smooth muscle cells in asthma. J Immunol 2012;189(2):819–31.
[65] Bowen H, Kelly A, Lee T, Lavender P. Control of cytokine gene transcription in Th1 and Th2 cells. Clin Exp Allergy 2008;38(9):1422–31.
[66] Miller RL, Ho SM. Environmental epigenetics and asthma: current concepts and call for studies. Am J Respir Crit Care Med 2008;177(6):567–73.
[67] Lu X, Kovalev GI, Chang H, Kallin E, Knudsen G, Xia L, et al. Inactivation of NuRD component Mta2 causes abnormal T cell activation and lupus-like autoimmune disease in mice. J Biol Chem 2008;283(20):13825–33.

[68] Helin K, Dhanak D. Chromatin proteins and modifications as drug targets. Nature 2013;502(7472):480–8.
[69] Wang Y, Zhang H, Chen Y, Sun Y, Yang F, Yu W, et al. LSD1 is a subunit of the NuRD complex and targets the metastasis programs in breast cancer. Cell 2009;138(4):660–72.
[70] Nguyen TH, Maltby S, Eyers F, Foster PS, Yang M. Bromodomain and extra terminal (BET) inhibitor suppresses macrophage-driven steroid-resistant exacerbations of airway hyper-responsiveness and inflammation. PLoS One 2016;11(9):e0163392.
[71] Khan YM, Kirkham P, Barnes PJ, Adcock IM. Brd4 is essential for IL-1beta-induced inflammation in human airway epithelial cells. PLoS One 2014;9(4):e95051.
[72] Clifford RL, Patel JK, John AE, Tatler AL, Mazengarb L, Brightling CE, et al. CXCL8 histone H3 acetylation is dysfunctional in airway smooth muscle in asthma: regulation by BET. Am J Physiol Lung Cell Mol Physiol 2015;308(9):L962–72.
[73] Mitani A, Ito K, Vuppusetty C, Barnes PJ, Mercado N. Restoration of corticosteroid sensitivity in chronic obstructive pulmonary disease by inhibition of mammalian target of rapamycin. Am J Respir Crit Care Med 2016;193(2):143–53.
[74] Spiekerkoetter E, Sung YK, Sudheendra D, Bill M, Aldred MA, van de Veerdonk MC, et al. Low-dose FK506 (Tacrolimus) in end-stage pulmonary arterial hypertension. Am J Respir Crit Care Med 2015;192(2):254–7.
[75] Williams AE, Larner-Svensson H, Perry MM, Campbell GA, Herrick SE, Adcock IM, et al. MicroRNA expression profiling in mild asthmatic human airways and effect of corticosteroid therapy. PLoS One 2009;4(6):e5889.
[76] Perry MM, Tsitsiou E, Austin PJ, Lindsay MA, Gibeon DS, Adcock IM, et al. Role of non-coding RNAs in maintaining primary airway smooth muscle cells. Respir Res 2014;15(1):58.
[77] Garbacki N, Di Valentin E, Huynh-Thu VA, Geurts P, Irrthum A, Crahay C, et al. MicroRNAs profiling in murine models of acute and chronic asthma: a relationship with mRNAs targets. PLoS One 2011;6(1):e16509.
[78] Wu XB, Wang MY, Zhu HY, Tang SQ, You YD, Xie YQ. Overexpression of microRNA-21 and microRNA-126 in the patients of bronchial asthma. Int J Clin Exp Med 2014;7(5):1307–12.
[79] Elbehidy RM, Youssef DM, El-Shal AS, Shalaby SM, Sherbiny HS, Sherief LM, et al. MicroRNA-21 as a novel biomarker in diagnosis and response to therapy in asthmatic children. Mol Immunol 2016;71:107–14.
[80] Lu TX, Munitz A, Rothenberg ME. MicroRNA-21 is up-regulated in allergic airway inflammation and regulates IL-12p35 expression. J Immunol 2009;182(8):4994–5002.
[81] Lu TX, Hartner J, Lim EJ, Fabry V, Mingler MK, Cole ET, et al. MicroRNA-21 limits in vivo immune response-mediated activation of the IL-12/IFN-gamma pathway, Th1 polarization, and the severity of delayed-type hypersensitivity. J Immunol 2011;187(6):3362–73.
[82] Collison A, Mattes J, Plank M, Foster PS. Inhibition of house dust mite-induced allergic airways disease by antagonism of microRNA-145 is comparable to glucocorticoid treatment. J Allergy Clin Immunol 2011;128(1):160–7.
[83] Seddiki N, Brezar V, Ruffin N, Levy Y, Swaminathan S. Role of miR-155 in the regulation of lymphocyte immune function and disease. Immunology 2013;142(1):32–8.
[84] Rodriguez A, Vigorito E, Clare S, Warren MV, Couttet P, Soond DR, et al. Requirement of bic/microRNA-155 for normal immune function. Science 2007;316(5824):608–11.
[85] Malmhall C, Alawieh S, Lu Y, Sjostrand M, Bossios A, Eldh M, et al. MicroRNA-155 is essential for T2-mediated allergen-induced eosinophilic inflammation in the lung. J Allergy Clin Immunol 2013;133:1429–38.
[86] Jardim MJ, Dailey L, Silbajoris R, Diaz-Sanchez D. Distinct microRNA expression in human airway cells of asthmatic donors identifies a novel asthma-associated gene. Am J Respir Cell Mol Biol 2012;47(4):536–42.
[87] Plank MW, Maltby S, Tay HL, Stewart J, Eyers F, Hansbro PM, et al. MicroRNA expression is altered in an ovalbumin-induced asthma model and targeting miR-155 with antagomirs reveals cellular specificity. PLoS One 2015;10(12):e0144810.
[88] Liu F, Qin H-b, Xu B, Zhou H, Zhao D-y. Profiling of miRNAs in pediatric asthma: upregulation of miRNA-221 and miRNA-485-3p. Mol Med Report 2012;6(5):1178–82.
[89] Mayoral RJ, Deho L, Rusca N, Bartonicek N, Saini HK, Enright AJ, et al. MiR-221 influences effector functions and actin cytoskeleton in mast cells. PLoS One 2011;6(10).
[90] Xu H, Gu LN, Yang QY, Zhao DY, Liu F. MiR-221 promotes IgE-mediated activation of mast cells degranulation by PI3K/Akt/PLCgamma/Ca(2^+) pathway. J Bioenerg Biomembr 2016;48(3):293–9.
[91] Zhou Y, Yang Q, Xu H, Zhang J, Deng H, Gao H, et al. miRNA-221-3p enhances the secretion of interleukin-4 in mast cells through the phosphatase and tensin homolog/p38/nuclear factor-kappaB pathway. PLoS One 2016;11(2):e0148821.
[92] Perry MM, Baker JE, Gibeon DS, Adcock IM, Chung KF. Airway smooth muscle hyperproliferation is regulated by microRNA-221 in severe asthma. Am J Respir Cell Mol Biol 2014;50(1):7–17.
[93] Chan SY, Loscalzo J. Pathogenic mechanisms of pulmonary arterial hypertension. J Mol Cell Cardiol 2008;44(1):14–30.

[94] Humbert M, Morrell NW, Archer SL, Stenmark KR, MacLean MR, Lang IM, et al. Cellular and molecular pathobiology of pulmonary arterial hypertension. J Am Coll Cardiol 2004;43(Suppl 12):13S–24S.
[95] Rabinovitch M. Molecular pathogenesis of pulmonary arterial hypertension. J Clin Invest 2012;122(12):4306–13.
[96] Tuder RM, Marecki JC, Richter A, Fijalkowska I, Flores S. Pathology of pulmonary hypertension. Clin Chest Med 2007;28(1):23–42.
[97] Hassoun PM. Pulmonary arterial hypertension complicating connective tissue diseases. Semin Respir Crit Care Med 2009;30(4):429–39.
[98] Savai R, Pullamsetti SS, Kolbe J, Bieniek E, Voswinckel R, Fink L, et al. Immune and inflammatory cell involvement in the pathology of idiopathic pulmonary arterial hypertension. Am J Respir Crit Care Med 2012;186(9):897–908.
[99] Kherbeck N, Tamby MC, Bussone G, Dib H, Perros F, Humbert M, et al. The role of inflammation and autoimmunity in the pathophysiology of pulmonary arterial hypertension. Clin Rev Allergy Immunol 2013;44(1):31–8.
[100] Vaillancourt M, Ruffenach G, Meloche J, Bonnet S. Adaptation and remodelling of the pulmonary circulation in pulmonary hypertension. Can J Cardiol 2015;31(4):407–15.
[101] Price LC, Caramori G, Perros F, Meng C, Gambaryan N, Dorfmuller P, et al. Nuclear factor kappa-B is activated in the pulmonary vessels of patients with end-stage idiopathic pulmonary arterial hypertension. PLoS One 2013;8(10):e75415.
[102] Tamby MC, Chanseaud Y, Humbert M, Fermanian J, Guilpain P, Garcia-de-la-Pena-Lefebvre P, et al. Anti-endothelial cell antibodies in idiopathic and systemic sclerosis associated pulmonary arterial hypertension. Thorax 2005;60(9):765–72.
[103] Terrier B, Tamby MC, Camoin L, Guilpain P, Broussard C, Bussone G, et al. Identification of target antigens of antifibroblast antibodies in pulmonary arterial hypertension. Am J Respir Crit Care Med 2008;177(10):1128–34.
[104] Voelkel NF, Tamosiuniene R, Nicolls MR. Challenges and opportunities in treating inflammation associated with pulmonary hypertension. Expert Rev Cardiovasc Ther 2016;14(8):939–51.
[105] Meloche J, Renard S, Provencher S, Bonnet S. Anti-inflammatory and immunosuppressive agents in PAH. Handb Exp Pharmacol 2013;218:437–76.
[106] Dunham-Snary KJ, Danchen W, Sykes EA, Thakrar A, Parlow LR, Mewburn JD, et al. Hypoxic pulmonary vasoconstriction: from molecular mechanisms to medicine. Chest 2016;151:181–92.
[107] Liu H, Fisher SA. Hypoxia-inducible transcription factor-1alpha triggers an autocrine survival pathway during embryonic cardiac outflow tract remodeling. Circ Res 2008;102(11):1331–9.
[108] Bosco MC, Puppo M, Blengio F, Fraone T, Cappello P, Giovarelli M, et al. Monocytes and dendritic cells in a hypoxic environment: spotlights on chemotaxis and migration. Immunobiology 2008;213(9–10):733–49.
[109] Semenza GL. Oxygen sensing, homeostasis, and disease. N Engl J Med 2011;365(6):537–47.
[110] Akhter N, Takeda Y, Nara H, Araki A, Ishii N, Asao N, et al. Apurinic/apyrimidinic endonuclease 1/redox factor-1 (Ape1/Ref-1) modulates antigen presenting cell-mediated T helper cell type 1 responses. J Biol Chem 2016;291(45):23672–80.
[111] Courboulin A, Ranchoux B, Cohen-Kaminsky S, Perros F, Bonnet S. MicroRNA networks in pulmonary arterial hypertension: share mechanisms with cancer? Curr Opin Oncol 2016;28(1):72–82.
[112] Saco TV, Parthasarathy PT, Cho Y, Lockey RF, Kolliputi N. Role of epigenetics in pulmonary hypertension. Am J Physiol Cell Physiol 2014;306(12):C1101–5.
[113] Loscalzo J, Handy DE. Epigenetic modifications: basic mechanisms and role in cardiovascular disease (2013 Grover Conference series). Pulm Circ 2014;4(2):169–74.
[114] Archer SL, Marsboom G, Kim GH, Zhang HJ, Toth PT, Svensson EC, et al. Epigenetic attenuation of mitochondrial superoxide dismutase 2 in pulmonary arterial hypertension: a basis for excessive cell proliferation and a new therapeutic target. Circulation 2010;121(24):2661–71.
[115] Perros F, Cohen-Kaminsky S, Gambaryan N, Girerd B, Raymond N, Klingelschmitt I, et al. Cytotoxic cells and granulysin in pulmonary arterial hypertension and pulmonary veno-occlusive disease. Am J Respir Crit Care Med 2013;187(2):189–96.
[116] Wang Y, Kahaleh B. Epigenetic repression of bone morphogenetic protein receptor II expression in scleroderma. J Cell Mol Med 2013;17(10):1291–9.
[117] Kato N, Loh M, Takeuchi F, Verweij N, Wang X, Zhang W, et al. Trans-ancestry genome-wide association study identifies 12 genetic loci influencing blood pressure and implicates a role for DNA methylation. Nat Genet 2015;47(11):1282–93.

[118] Li M, Riddle SR, Frid MG, El Kasmi KC, McKinsey TA, Sokol RJ, et al. Emergence of fibroblasts with a proinflammatory epigenetically altered phenotype in severe hypoxic pulmonary hypertension. J Immunol 2011;187(5):2711–22.
[119] Zhao L, Chen CN, Hajji N, Oliver E, Cotroneo E, Wharton J, et al. Histone deacetylation inhibition in pulmonary hypertension: therapeutic potential of valproic acid and suberoylanilide hydroxamic acid. Circulation 2012;126(4):455–67.
[120] Chen F, Li X, Aquadro E, Haigh S, Zhou J, Stepp DW, et al. Inhibition of histone deacetylase reduces transcription of NADPH oxidases and ROS production and ameliorates pulmonary arterial hypertension. Free Radic Biol Med 2016;99:167–78.
[121] Lan B, Hayama E, Kawaguchi N, Furutani Y, Nakanishi T. Therapeutic efficacy of valproic acid in a combined monocrotaline and chronic hypoxia rat model of severe pulmonary hypertension. PLoS One 2015;10(1):e0117211.
[122] Mumby S, Gambaryan N, Meng C, Perros F, Humbert M, Wort SJ, et al. Bromodomain and extra-terminal protein mimic JQ1 decreases inflammation in human vascular endothelial cells: implications for pulmonary arterial hypertension. Respirology 2016;22:157–64.
[123] Strassheim D, Riddle SR, Burke DL, Geraci MW, Stenmark KR. Prostacyclin inhibits IFN-gamma-stimulated cytokine expression by reduced recruitment of CBP/p300 to STAT1 in a SOCS-1-independent manner. J Immunol 2009;183(11):6981–8.
[124] Cedar H, Bergman Y. Linking DNA methylation and histone modification: patterns and paradigms. Nat Rev Genet 2009;10(5):295–304.
[125] Shao D, Perros F, Caramori G, Meng C, Dormuller P, Chou PC, et al. Nuclear IL-33 regulates soluble ST2 receptor and IL-6 expression in primary human arterial endothelial cells and is decreased in idiopathic pulmonary arterial hypertension. Biochem Biophys Res Commun 2014;451(1):8–14.
[126] Miller AM, Xu D, Asquith DL, Denby L, Li Y, Sattar N, et al. IL-33 reduces the development of atherosclerosis. J Exp Med 2008;205(2):339–46.
[127] Brock M, Trenkmann M, Gay RE, Michel BA, Gay S, Fischler M, et al. Interleukin-6 modulates the expression of the bone morphogenic protein receptor type II through a novel STAT3-microRNA cluster 17/92 pathway. Circ Res 2009;104(10):1184–91.
[128] Caruso P, MacLean MR, Khanin R, McClure J, Soon E, Southgate M, et al. Dynamic changes in lung microRNA profiles during the development of pulmonary hypertension due to chronic hypoxia and monocrotaline. Arterioscler Thromb Vasc Biol 2010;30(4):716–23.
[129] Courboulin A, Paulin R, Giguere NJ, Saksouk N, Perreault T, Meloche J, et al. Role for miR-204 in human pulmonary arterial hypertension. J Exp Med 2011;208(3):535–48.
[130] Caruso P, Dempsie Y, Stevens HC, McDonald RA, Long L, Lu R, et al. A role for miR-145 in pulmonary arterial hypertension: evidence from mouse models and patient samples. Circ Res 2012;111(3):290–300.
[131] Bockmeyer CL, Maegel L, Janciauskiene S, Rische J, Lehmann U, Maus UA, et al. Plexiform vasculopathy of severe pulmonary arterial hypertension and microRNA expression. J Heart Lung Transplant 2012;31(7):764–72.
[132] Boucherat O, Potus F, Bonnet S. MicroRNA and pulmonary hypertension. Adv Exp Med Biol 2015;888:237–52.
[133] Pullamsetti SS, Doebele C, Fischer A, Savai R, Kojonazarov B, Dahal BK, et al. Inhibition of microRNA-17 improves lung and heart function in experimental pulmonary hypertension. Am J Respir Crit Care Med 2012;185(4):409–19.
[134] Parikh VN, Jin RC, Rabello S, Gulbahce N, White K, Hale A, et al. MicroRNA-21 integrates pathogenic signaling to control pulmonary hypertension: results of a network bioinformatics approach. Circulation 2012;125(12):1520–32.
[135] Brock M, Samillan VJ, Trenkmann M, Schwarzwald C, Ulrich S, Gay RE, et al. AntagomiR directed against miR-20a restores functional BMPR2 signalling and prevents vascular remodelling in hypoxia-induced pulmonary hypertension. Eur Heart J 2012;35:3203–11.
[136] Gubrij IB, Pangle AK, Pang L, Johnson LG. Reversal of microRNA dysregulation in an animal model of pulmonary hypertension. PLoS One 2016;11(1):e0147827.
[137] Kim J, Kang Y, Kojima Y, Lighthouse JK, Hu X, Aldred MA, et al. An endothelial apelin-FGF link mediated by miR-424 and miR-503 is disrupted in pulmonary arterial hypertension. Nat Med 2013;19(1):74–82.
[138] McLendon JM, Joshi SR, Sparks J, Matar M, Fewell JG, Abe K, et al. Lipid nanoparticle delivery of a microRNA-145 inhibitor improves experimental pulmonary hypertension. J Control Release 2015;210:67–75.
[139] Freilich RW, Woodbury ME, Ikezu T. Integrated expression profiles of mRNA and miRNA in polarized primary murine microglia. PLoS One 2013;8(11):e79416.
[140] Wang D, Zhang H, Li M, Frid MG, Flockton AR, McKeon BA, et al. MicroRNA-124 controls the proliferative, migratory, and inflammatory phenotype of pulmonary vascular fibroblasts. Circ Res 2014;114(1):67–78.

CHAPTER

15

The Epigenetics of Autoimmunity and Epigenetic Drug Discovery

David F. Tough, Inmaculada Rioja, Paul-Peter Tak, Rab K. Prinjha

GlaxoSmithKline, Medicines Research Centre, Stevenage, United Kingdom

OUTLINE

15.1 INTRODUCTION

15.1.1 Immunity, Memory, and Epigenetics

The cells of the immune system are typically classified into innate and adaptive subtypes based on the receptors they employ to recognize pathogens. Innate cells utilize germ line encoded receptors to sense infectious agents; encoding of these receptors in the germ line is possible because they are directed towards conserved features that are essential for the viability of the microbe or the coevolution of the organism with its host. The ligands recognized by such receptors may be

The Epigenetics of Autoimmunity. http://dx.doi.org/10.1016/B978-0-12-809912-4.00015-5

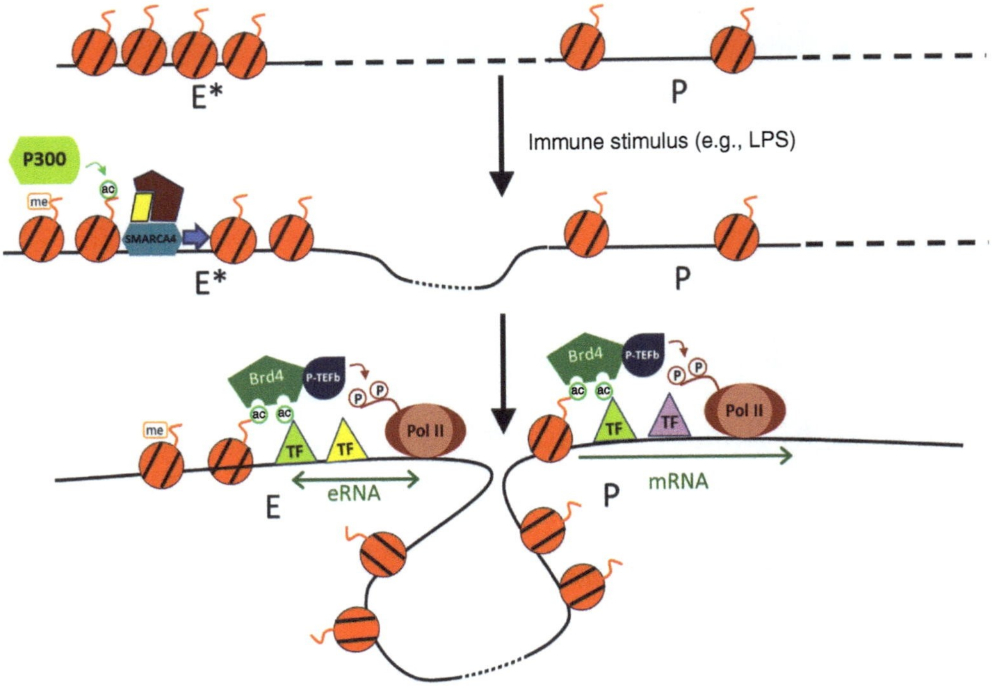

FIGURE 15.1 **Epigenetic regulation of activation-induced inflammatory gene expression in innate immune cells.** The Fig. 15.1 represents a highly simplified depiction of some of the epigenetic events involved in the regulation of lipopolysaccharide (LPS)-induced secondary response genes in innate immune cells. Signaling events triggered by LPS lead to proteins that impart posttranslational modifications on histones [including acetylation (ac) and methylation (me)] and remodel chromatin being recruited to potential enhancer regions (E*). Recruited proteins also mediate sliding or eviction of nucleosomes to generate accessible regions of DNA and promote long range interactions between the de novo enhancers (E) and relevant gene promoters (P). BRD4 localizes in a signal-induced manner to both enhancers and promoters, and functions to enhance transcriptional elongation at least in part through the recruitment of the P-TEFb complex, with the generation of eRNA contributing to the induction of gene mRNA expression. De novo enhancers can persist and lead to the continued gene expression or rapid reexpression upon restimulation of the cell.

characteristic of particular organisms (e.g., lipopolysaccharide (LPS), which is a conserved component of the outer wall of Gram-negative bacteria). Based on these properties, such features of infectious agents have been termed pathogen-associated molecular patterns (PAMPs) and the receptors that recognize them pattern recognition receptors (PRRs) [1]. In addition to recognizing microbes directly, PRRs can function by sensing host molecules that are released or modified in response to cellular stress or damage, these are so-called damage-associated molecular patterns (DAMPs) [2]. Since germ line encoded PRRs can be expressed broadly in innate immune cells, they serve as an effective surveillance system for the presence of pathogens. PRR-derived signals can also provide information about the broad class of infectious agents encountered (since specific components of Gram-negative bacteria, Gram-positive bacteria, DNA viruses, RNA viruses, and fungi trigger distinct PRRs) as well as an indication of their potential threat based on the location in the body in which the pathogen is sensed and its association with DAMPs. However, the ability of PRRs to distinguish different pathogens and hence mediate specific immunity is limited (Fig. 15.1).

In contrast to the germ line encoded receptors utilized by innate immune cells, adaptive immune cells (i.e., T and B lymphocytes) recognize pathogens through receptors that have been generated through the somatic recombination of variable gene segments. This process occurs during T cell and B cell development in the thymus and bone marrow, respectively, and generates a vast array of ligand specificities. The diversity of receptor specificity is so great that the adaptive immune system is prepared to recognize virtually any antigenic epitope present on infectious agents, and can adapt to antigenic variation as pathogens evolve. However, T cell and B cell antigen receptors are distributed in a clonal manner such that individual adaptive immune cells express receptors of only a single specificity. Therefore there are very few T and B cells able to recognize any particular pathogen, at least in an immunologically naive state (i.e., in an individual who has not previously encountered that specific pathogen). Hence naive T and B cells do not function effectively as surveillance systems for the detection of pathogens, and the development of adaptive immunity requires a lag period in which rare antigen-specific T and B cells are expanded in number. However, once expanded after the primary encounter with a pathogen, adaptive immune cells can mediate a range of effector mechanisms to efficiently clear or contain the pathogen and provide selective and effective protection from reinfection with the same organism (the phenomenon of immunological memory, which is described further below).

15.1.2 Innate Immune Responses

After recognizing infectious agents, innate immune cells respond by rapidly producing large quantities of inflammatory mediators that are involved in: (1) the direct killing or containment of the pathogen, (2) the recruitment of additional innate cells to the site of infection, and (3) the activation of adaptive immune cells. Innate cells that are of importance for this latter purpose include dendritic cells, which are essential for the initial activation of naive T cells, and macrophages, which interact with and stimulate the function of preactivated T cells. The triggering of PRRs is associated with dramatic changes in gene expression in innate cells; for example, stimulation of macrophages with LPS results in the altered expression of hundreds of genes within a few hours [3].

PRR-induced changes in gene transcription are linked with extensive alterations at the level of chromatin (the form in which DNA is packaged in the nucleus in close association with histone proteins). The structure of chromatin is based on the nucleosome, which is comprised of 146 bp of DNA wrapped around core histone proteins (two copies each of histones H3, H4, H2A, and H2B), with the further organization of nucleosomes into higher-order structures through interactions with additional histone and nonhistone proteins [4,5]. Chromatin can vary in the extent to which DNA is contained in nucleosomes and in the degree of nucleosome "packing." DNA in tightly packed "heterochromatin" is inaccessible to DNA-binding proteins such as transcription factors (TFs), while DNA regions involved in active transcription possess a more open chromatin structure. Chromatin remodeling that leads to the generation of accessible or nonaccessible DNA regions is mediated by a range of "epigenetic" mechanisms and is discussed further below.

Chromatin changes that occur in response to PRR signaling in innate immune cells appear to be particularly prominent in regions of the genome involved in the regulation of gene expression. For instance, it has been shown that the genome-wide distribution of gene enhancers is markedly altered following LPS stimulation of macrophages [6,7]. Enhancers are regulatory elements that function alongside promoters to control gene expression, but may be separated from their target gene by long stretches of DNA (in some cases more than 1 Mb) [8,9]. These long-range effects are mediated through "looping" of the intervening DNA, which brings enhancers into close proximity with promoters and provides a focal point for the binding of TFs and proteins that execute RNA transcription [10]. Enhancers are typically a few hundred base pairs in size, but can be present in clusters that span large expanses of DNA (up to 50 kb). These clusters have variously been termed locus control regions (LCR), stretch enhancers, or super-enhancers (SEs), and are associated with high levels of gene expression [11–15]. In addition to SEs, enhancers located in distinct locations can also converge on the same gene promoter to drive high levels of gene expression.

The ability of a DNA region to act as an enhancer is intimately linked to its chromatin state. Thus active enhancers require an open chromatin structure to allow the DNA to access TFs and the transcriptional machinery. They can be identified experimentally on this basis as well as by characteristic histone modifications present on the neighboring nucleosomes. Enhancer DNA is itself transcribed bidirectionally, generating enhancer RNAs (eRNAs) that are thought to play an important role in driving the expression of enhancer-associated genes [6]. Moreover, functional enhancer activity is dependent on chromatin remodeling events that bring these distal regulatory regions near to gene promoters.

LPS stimulation of macrophages forms thousands of new enhancers (de novo enhancers), while many of the enhancers that were present before stimulation are lost [6,7]; a large proportion of the de novo enhancers have the characteristics of SEs [16]. Since the remodeling of the enhancer landscape is closely linked to the marked changes in gene expression in these cells, these findings indicate that chromatin modifications play a fundamental role in shaping the nature of the innate immune response [17]. Thus PRR-induced remodeling of the enhancer landscape enables the development of a new proinflammatory gene expression pattern in innate immune cells.

As the production of proinflammatory mediators is potentially highly damaging to the host, it is important that the PRR-induced gene expression pattern is transient and shuts down when the pathogen is contained or cleared. However, it is evident that some PRR-stimulated epigenetic modifications can persist in innate immune cells [18]. These enduring chromatin changes do not necessarily result in continued expression of proinflammatory genes, but rather place these genes in a primed state, allowing for a more rapid response to a subsequent stimulation. This has been proposed to provide a form of innate cell "memory" and contribute to an increased nonspecific resistance to reinfection.

15.1.3 Adaptive Immune Responses

T and B cells also undergo marked changes in gene expression and functional properties during the response to infection. This includes the induction of genes linked to the cell cycle; T and B cells are largely noncycling in the absence of antigenic stimulation but proliferate extensively during a pathogen-specific response, which allows for a dramatic expansion of the pathogen-specific cell population from the very low numbers found amongst naive T and B cells. Cells also acquire the ability to express a variety of molecules that are capable of combating infection during this process. For B cells, this involves their differentiation into plasma cells, which are capable of producing and secreting large quantities of antibodies. For T cells, the response to pathogens is associated with the expression of a range of cell-associated and soluble mediators that can directly kill the pathogen or pathogen-infected cells or stimulate other cell types to do so. This contrasts with the early response of naive T cells (which have not previously responded to antigen) to T-cell antigen receptor (TCR)-mediated activation. Naive T cells initially produce only a very limited range of cytokines (predominantly IL-2) that contribute to T-cell proliferation and expansion.

T cells can develop distinct patterns of gene expression depending on the signals present in the microenvironment during T-cell activation (in addition to the triggering of the TCR). Such modulatory (costimulatory) signals derive from activated innate immune cells and differ based on the type of pathogen encountered (which is linked to the type of PAMP expressed by the pathogen and the PRR through which this is recognized). The nature of the immune response can be tailored to that which is most effective at containing a particular pathogen; for instance, $CD4^+$ T cells typically adopt a "Th1" gene expression pattern when responding to facultative intracellular bacteria, and the "signature" cytokine produced by Th1 cells, IFN-γ, is a potent inducer of intracellular killing pathways in infected cells.

T cell proliferation and differentiation into cells with specialized effector functions is accompanied by extensive epigenetic modifications at gene regulatory sites [19]. Key drivers in the chromatin remodeling events are TFs, which are induced by TCR-triggering and costimulatory signals and direct T cells towards distinct differentiation pathways through the generation of new enhancers and SEs [20]. Consequently, many of the genes encoding effector molecules that are characteristic of different T-cell subpopulations (i.e., cytokines and cytokine receptors) become associated with SEs in T cells [21]. For instance, the *Ifng*, *Il13*, and *Il17*a genes were shown to be linked to SEs in Th1, Th2, and Th17 cells, respectively [22].

As for the innate immune response, the T cell response is usually highly inflammatory and must be turned off once the infection has been cleared. Accordingly, most effector T cells disappear through a process of programmed cell death after the antigenic stimulus has been removed; however, not all the expanded T cells die. A proportion of these cells return to a "resting" state and survive as a heterogeneous population of memory T cells [23]. These antigen-specific cells can persist at a substantially increased frequency (10,000-fold or more) relative to that preinfection. In addition, while memory T cells do not constitutively produce most effector molecules in a resting state, they retain the epigenetic modifications generated during their initial activation and differentiation [24–27]. These epigenetic modifications include those that poise genes for rapid reexpression upon T-cell restimulation; this is the case for effector genes (i.e., proinflammatory cytokines). In addition, some epigenetic modifications that persist into memory can alter the baseline gene expression, for example, changing the expression of chemokine receptors and adhesion molecules so that some memory T cells are able to localize within tissues at the sites of pathogen entry (particularly "resident" memory T cells). Combined with a greatly increased precursor frequency, these properties of memory T cells allow them to provide a much more rapid and effective response upon subsequent encounters with the same pathogen. In contrast, effector B cells (i.e., plasma cells) can persist and continue to express their effector function (antibody secretion) for long periods of time (>80 years for some infections) [28,29]. Since these antibodies are typically highly specific for the pathogen that elicited their production, their continued production is not intrinsically inflammatory. Circulating antibodies likely provide the most effective mechanism for protecting against reinfection [30].

15.1.4 Dysregulation of Immunity in Immune-Mediated Disease

Although the transient generation of inflammatory immune responses plays a vital role in the protection against infection, inappropriate activation of the immune system causes a wide range of human diseases. Some diseases result from the generation of immune responses against self antigens. These autoimmune diseases include multiple sclerosis (MS), which is a disorder of the central nervous system (CNS) in which T cells respond to antigens present in the myelin sheath surrounding nerve fibers, and myasthenia gravis, which is characterized by autoantibodies blocking or destroying the nicotinic acetylcholine receptor at the junction between the nerve and the muscle. Other immune-mediated diseases are caused by the generation of inflammatory immune responses against otherwise harmless foreign antigens. For example, T-cell responses against the gliadin protein component of cereal gluten results in celiac disease, whereas immune responses to pollen antigens cause allergies. Similarly, inflammatory responses to "commensal" bacteria are thought to be linked to Crohn's disease, psoriasis, and other inflammatory diseases [31].

The development of disease and the specific pathology observed is linked to the type of immune response generated; for example, the symptoms of allergy are caused by Th2-associated mechanisms whereas IL-17 is a key mediator in psoriasis. An imbalance in the ratio of proinflammatory versus antiinflammatory mediators can also convert a benign immune response to one that is pathogenic. This is illustrated in the case of Th17 cells, where a noninflammatory immune response is associated with Th17 cells that produce the antiinflammatory cytokine IL-10, whereas lack of IL-10 expression and the presence of a distinct gene expression signature is linked to pathogenic autoimmune Th17 cells [32,33]. Therefore the precise gene expression pattern is an important determinant in immune-mediated disease.

15.1.5 Growing Understanding of Immune-Mediated Diseases From Genetic Studies

Insight into the genetic basis of immune-mediated disease has been aided by two types of findings: (1) the identification of rare mutations that cause profound autoimmunity, and (2) the demonstration of a statistical association between single-nucleotide polymorphisms (SNPs) and susceptibility to common immune-mediated diseases, as realized through genome-wide association studies (GWAS).

Rare mutations that cause autoimmunity have demonstrated the fundamental requirement for certain proteins in establishing and maintaining a state of immune tolerance to self-tissues. Two examples in this respect are the aptly named autoimmune regulator (AIRE) and forkhead box P3 (FOXP3). AIRE is a transcriptional regulator principally expressed within a subset of epithelial cells in the medulla of the thymus (mTECs) and plays a key role in shaping the repertoire of TCRs expressed by T cells [34]. As discussed previously, the generation of TCRs through somatic recombination of variable gene segments generates receptors with a vast range of antigen specificities, which are distributed in a clonal manner on T cells. However, as the random generation of TCRs can also produce reactivity to self-antigens, the T-cell pool must be purged of such cells during their development in the thymus.

AIRE enables this process by driving the expression of genes in mTECs that are normally only expressed in peripheral tissues [35,36]. Interestingly, AIRE stimulates the transcription of these tissue-specific antigens at least in part through binding to SEs [37]. Consequently, T cells bearing self-reactive TCRs that encounter these antigens during development in the thymus are largely deleted from the repertoire through negative selection [38]. A proportion of self-reactive T cells also survive but differentiate into regulatory T cells (T_{regs}), which function to actively suppress immune responses to self-antigens in the periphery [39]. The essential role of AIRE in this process is underlined by the fact that loss-of-function mutations in AIRE lead to autoimmune polyendocrinopathy candidiasis ectodermal dystrophy, a disease characterized by autoimmune destruction of numerous tissues and organs, mainly of endocrine origin [40,41].

FOXP3 is an X-chromosome encoded member of the forkhead TF family, which has been shown to be essential for the differentiation and function of T_{regs} [42]. After recognizing self-antigens, T_{regs}

inhibit the proinflammatory function of multiple immune cell types (i.e., effector T cells and innate immune cells) through a range of mechanisms [43]. The critical role that T_{regs} play in suppressing immune self-reactivity is evident from the consequences of rare loss-of-function mutations in the human *FOXP3* gene, which cause the immune dysregulation, polyendocrinopathy, enteropathy and X-linked inheritance (IPEX) autoimmune disease [44,45].

While rare loss-of-function mutations can lead to devastating autoimmune diseases, the genetic basis of most immune-mediated diseases is more complex. Findings from GWAS typically indicate that the genetic component of disease susceptibility is comprised of contributions from multiple genes. However, an unexpected finding from GWAS is that the majority of disease-associated SNPs do not lead to changes in the coding sequence of genes (i.e., mutations in proteins) but are found within noncoding regions of the genome [46,47].

The significance of these findings has begun to emerge through the use of novel genome-wide approaches to elucidate the function of nonprotein-coding DNA. For example, several high-throughput methods for identifying putative enhancers on a genome-wide level have been developed and employed. Some of these approaches are based on the observation that enhancer regions should be accessible to TFs and hence possess a relatively open chromatin structure. For instance, exposure of nuclear DNA to deoxyribonuclease I (DNase I, which cuts preferentially at sites of nucleosome depletion) followed by genome-wide DNA sequencing identified an enriched number of DNase I hypersensitive sites (DHSs) in gene regulatory regions. When combined with other complementary genome-wide measurements [i.e., chromatin immunoprecipitation (ChIP)] followed by massively parallel DNA sequencing (ChIP-seq), which detects the presence of specific histone modifications, this approach can be used to discriminate between different types of regulatory elements (e.g., enhancers or promoters). A large number of DHSs (~200,000 in each cell type) and enhancers have been detected using this methodology [48]. This number greatly exceeds the number of protein coding genes (~20,000), which is in accordance with the view that individual genes are controlled by multiple regulatory elements, including clustered SEs or LCRs.

These genome-wide approaches have made it clear that enhancers are highly cell-type specific, which is reflected in the cell-specific patterns of gene expression, and that SEs appear to play a particularly important role in driving cell-specific gene expression [14]. The enhancer landscape has also been shown to be highly flexible within a given cell type. As discussed previously, this flexibility is prominent in the immune system where large numbers of new enhancers appear while others disappear during processes such as PRR-mediated activation of innate immune cells or T-cell differentiation into specialized effector cell subsets. This also applies to other cells in the context of inflammation; for example, large-scale remodeling of the enhancer landscape was observed endothelial cells were exposed to the proinflammatory cytokine, TNF [16]. It has also become clear that GWAS SNPs preferentially localize within regions of the genome that are involved in gene regulation, particularly enhancers. The implication of these findings is that disease-associated SNPs may be commonly linked to quantitative changes in gene expression and that dysregulation of gene expression is a key driver of human disease [49].

A key observation gleaned from the integration of GWAS data and genome-wide chromatin studies is that disease-associated SNPs are enriched within cell-type specific enhancers and SEs, including those that are generated in response to cellular differentiation, activation, and stress [11,13,14]. The manner in which these genes are regulated is distinct from constitutively expressed "housekeeping" genes, which generally either do not require enhancer activity or contain an integrated enhancer [50]. Moreover, it is evident that disease-associated SNPs are particularly enriched within cell-type specific enhancers in a manner that fits with known or predicted disease mechanisms. This includes increasing evidence that GWAS SNPs associated with immune diseases localize preferentially within gene regulatory regions in immune cells. For example, one study of 349 cell and tissue samples demonstrated selective enrichment of GWAS SNPs for Crohn's disease within DHSs present in Th17 and Th1 cells and SNPs associated with MS within DHSs in T and B cells [48]. Another group found that 60% of causal nucleotide variants linked to 21 autoimmune diseases mapped to immune cell enhancers [51]. In addition, other studies focused on specific diseases have reported similar associations, such as the enrichment of asthma-associated SNPs in TH2-specific enhancers [52]; the localization of type 1 diabetes-associated SNPs within enhancer sequences in thymus T cells, B cells, and $CD34^+$ stem cells [53]; and the presence of an ankylosing spondylitis-associated SNP within a Th1-specific enhancer [54].

There is also evidence that GWAS SNPs are particularly prominent within SEs. SNPs associated with rheumatoid arthritis (RA), type 1 diabetes, MS, celiac disease, and ulcerative colitis (UC) were all located in SEs in Tfh cells, which are a subset of $CD4^+$ T cells that specialize in promoting B-cell antibody production [55]. Similarly, a SNP associated with the autoimmune disease, vitiligo, was shown to localize within an SE and to increase the expression of major histocompatibility (MHC) genes [56]. MHC proteins present peptides derived from pathogens and self-proteins on the cell surface for recognition by peptide-specific T cells, therefore dysregulation of MHC expression may contribute to this disease. Although the mechanism by which this vitiligo-associated SNP alters MHC gene expression was not reported, it was recently shown that SNPs associated with the mucosal inflammatory diseases Crohn's disease, UC, and celiac disease altered the binding of TBX21, a key TF involved in Th1 differentiation [57].

The localization of immune disease-associated SNPs in the enhancers of immune cells implicates the dysregulation of gene expression in these diseases; however, much remains to be learned about the genes that are regulated by such SNPs. As enhancers can be separated from their targets genes by long stretches of DNA, it is generally not possible to assign an enhancer to its target gene based on knowledge of the linear sequence of DNA alone. However, techniques have been developed that can detect interactions between distinct regions of chromatin, such as those that occur between enhancers and gene promoters. These approaches are now utilized to detect long-range DNA interactions, including disease-associated SNPs. For example, promoter capture Hi-C was used to identify interacting regions for over 30,000 gene promoters in 17 human primary hematopoietic cell types [58]. The authors found that promoter interactions were highly cell-type specific and were enriched for links between active promoters and enhancers. Most notably, the genes interacting with GWAS autoimmune disease-associated variants were found to be enriched in lymphoid cells and specifically in inflammation and immune response-related pathways (i.e., interleukin and TCR signaling).

15.1.6 Link to Enhancers Suggests the Potential for Epigenetic Mechanisms to Treat Autoimmunity

The evidence that dysregulated gene expression is a key driver of immune-mediated diseases suggests that approaches aimed at normalizing gene expression levels could have therapeutic benefit. However, such approaches were deemed to be unfeasible as it was unclear how to selectively modify the expression of genes contributing to disease pathogenesis while sparing those genes that were essential for cell and tissue homeostasis. The discovery that GWAS immune disease-associated SNPs frequently localize in de novo enhancers or SEs under conditions of immune activation suggest that distinct mechanisms are involved in the regulation of disease-associated versus homeostatic genes. There has also been a growing understanding of the epigenetic mechanisms involved in regulating enhancer function and gene expression, which has led to the identification of tractable drug targets for the selective modification of gene expression.

As discussed previously, the ability of a DNA region to act as an enhancer is highly dependent on the local structure of chromatin. Chromatin is subject to a range of modifications, both on DNA and histones, which can change its structure and its accessibility to TFs and other proteins involved in gene expression regulation. DNA can be modified through the activity of DNA methyltransferases to convert cytosine residues to 5-methylcytosine (5mC); this process is reversible, with 5mC being converted back to cytosine through the sequential activity of the ten-eleven translocation family of dioxygenases and thymine DNA glycosylase [59–63]. Histones can be modified through post-translational modifications, which can be added to different amino acid residues, primarily within the extended "tails" that extend from the main globular domain of the histones [64]. Commonly described histone modifications include acetylation; methylation and ubiquitination of lysine residues; methylation of arginine residues; and phosphorylation of serine, threonine, or histidine residues. The generation of these modifications is mediated by specific families of enzymatic proteins termed epigenetic "writers." Families of proteins that are able to remove these modifications are called epigenetic "erasers," while proteins that are capable of recognizing and binding to specific modifications and thereby assembling gene-regulating macromolecular complexes are known as epigenetic "readers."

DNA accessibility can be altered by chromatin modifications in three ways: (1) the acetylation of lysine residues within histones neutralizes their positive charge thus weakening the strength of histone binding to negatively charged DNA [65]; (2) the conversion of cytosine to 5mC within TF recognition motifs can disrupt the binding of TFs [66,67], and (3) chromatin modifications generate recognition signals for epigenetic reader proteins, including 5mC binding proteins and protein families that recognize specific histone modifications. Reader proteins may either block or enhance the recruitment of TFs upon binding to specific chromatin marks. In addition, readers can bring with them a range of enzymatic activities that are capable of remodeling chromatin. This can lead to the local removal of histones to create a nucleosome-free region of DNA, and hence an active regulatory region. Conversely, some readers bring catalytic activities that promote the generation of a closed chromatin structure, or transcriptional repressor complexes that shutdown gene transcription.

Insight into the mechanisms involved in the generation of active enhancers can be gained from the characteristic epigenetic features associated with these regions. For instance, the nucleosomes within enhancers possess high levels of histone H3 lysine 27 acetylation (H3K27ac) and H3 lysine 4 mono- or dimethylation (H3K4me1/2), and SEs are characterized by extended stretches of these marks [19,68,69]. The histones flanking enhancers also show a high rate of turnover, indicating that the local chromatin structure is maintained through a dynamic and active process [70,71]. In accordance with this, multiple chromatin regulating proteins have been shown to be highly enriched within SEs [13,72]. These include writer proteins (i.e., the histone acetyltransferases, P300 and CBP), reader proteins (i.e., CHD7, BRD4, and SMARCA4), and eraser proteins (i.e., the histone deacetylases, HDAC1 and HDAC2, and the histone demethylase, KDM1a) [13]. Several of these proteins also contain intrinsic chromatin remodeling activity (e.g., CHD7 and SMARCA4), or are components of multiprotein complexes with this activity (e.g., KDM1 and HDAC1/2), and therefore could be involved the removal or sliding of nucleosomes [73–75]. The reader protein, BRD4, binds to acetyl lysine residues via its bromodomains and interacts with key regulators of gene transcription (i.e., the mediator and P-TEFb protein) [76].

The observation that proteins involved in epigenetic mechanisms are found at high levels within SEs suggests that these proteins could be targets for the selective modulation of SE-driven genes, and hence for the treatment of immune-mediated diseases. The ability to explore this hypothesis has been aided by the development of small-molecule inhibitors of several epigenetic proteins. Some of the more advanced classes of epigenetic inhibitors (Table 15.1) are described below and insights gained from the use of these tools into the fundamental mechanisms regulating gene expression in immune cells as well as the progress towards therapeutic application of these approaches are summarized.

15.1.7 Lysine Deacetylase Inhibitors

Lysine deacetylases (KDACs) catalyze the removal of acetyl groups from the ε-N-acetylated lysine residues of protein substrates, including histone and nonhistone proteins [77]. This family of proteins include 18 enzymes that can be divided into four subclasses based on their catalytic-core homologies. Class I, II, and IV enzymes are termed histone deacetylases (HDACs), while the class III deacetylases are named sirtuins (SIRTs) [78]. In accordance with the key role of acetylation in

TABLE 15.1 Examples of Drugs Targeting Epigenetic Mechanisms That Have Progressed Into Clinical Trials

Compound Name	Target(s)	Phase	Disease indications
Vorinostat (SAHA)	Class I and II HDACs	Launched	Cancer (T-cell lymphoma)
Romidepsin (FK228, Istodax)	Class I HDACs	Launched	Cancer (T-cell lymphoma)
Belinostat (PXD-101)	Class I, II, and IV HDACs	Launched	Cancer (T-cell lymphoma)
Panobinostat (Farydak)	Class I and II HDACs	Launched	Cancer (myeloma)
Entinostat (MS-275)	Class I HDACs	Phase III	Cancer (breast)
Mocetinostat (MGCD0103)	Class I HDACs	Phase II	Cancer
Tazemetostat (EPZ 5687)	EZH2	II	Cancer
GSK126	EZH2	I	Cancer
CPI-1205	EZH2	I	Cancer
Pinometostat (EPZ 5676)	DOT1L	I	Cancer
Apabetalone (RVX-208)	BET family bromodomains	III	Atherosclerosis and type 2 diabetes
GSK-525762 (IBET-762)	BET family bromodomains	II	Cancer
Birabresib (OTX-015)	BET family bromodomains	II	Cancer
ABBV-075	BET family bromodomains	I	Cancer
BI-894999	BET family bromodomains	I	Cancer
AZD-5153	BET family bromodomains	I	Cancer
BI-894999	BET family bromodomains	I	Cancer
PLX-51107	BET family bromodomains	I	Cancer

BET, Bromodomain and extra terminal domain.

regulating chromatin structure, protein networks, and cellular signaling pathways, HDACs have well-recognized roles in the regulation of gene expression and in the control of cellular processes (i.e., cytoskeletal assembly, proliferation, DNA repair, and apoptosis) [79–81].

HDACs can be recruited to chromatin as components of nuclear repressor complexes, which promote histone deacetylation and lead to DNA compaction and gene repression [82]. However, certain HDACs (i.e., HDAC1 and HDAC2) are also known to be located in enhancers and SEs [13]. HDACs can also affect gene expression through deacetylation of other proteins, including TFs [83]. Their diverse functions mean that HDACs can be involved in both the promotion and inhibition of gene expression, and this has been shown to be the case within the immune system [82,84]. Therefore alterations in HDACs expression or activity have been linked to different pathogenic states, including cancers, tissue fibrosis, metabolic disorders, autoimmune diseases, and inflammatory diseases [85]. HDACs represent attractive targets for drug development, and considerable effort has gone into the development of increasingly selective HDAC inhibitors (HDACi).

Many of the HDACi generated to date (i.e., trichostatin A and vorinostat) are active against nearly all HDACs, while others have a more restricted selectivity profile and show preferential activities against HDAC subfamilies. For example, butyric acid, valproic acid, and givinostat (ITF2357) inhibit class I and IIa HDACs; romidepsin, mocetinostat, and entinostat (MS-275) show preferential activity towards class I HDACs; and trifluoromethyloxadiazole compounds are active against class IIa HDACs [86].

Although HDACi are mainly developed as anticancer drugs, many of them are also explored for their efficacy in treating CNS, metabolic, immunologic, and inflammatory disorders [85]. Support for their use in immune-mediated diseases has come from gene-targeting approaches, predominantly in mouse models, which have shown a role for specific HDACs in the regulation of immune responses. An important caveat to these studies is that removal of the protein through gene targeting would be expected to disrupt multiprotein complexes associated with HDAC proteins, and hence might not reflect the effects of HDAC enzymatic inhibition [87]. However, a substantial body of evidence from preclinical and clinical studies utilizing structurally diverse HDACi supports the potential for this family of proteins to be a therapeutic target in immune-mediated diseases [85,88]. An extensive review of HDACi-mediated immune-cell function has been published elsewhere [89]. This chapter focuses on the most advanced approaches to HDACi that are now entering the clinic for the treatment of immune-mediated disease.

Givinostat (ITF2357, Italfarmaco, Italy) is an orally active class I and II HDACi [90] and is the only HDACi that has progressed to phase II clinical trials for both oncology (relapsed leukemias and myelomas) and chronic autoinflammatory diseases (systemic juvenile idiopathic arthritis patients) [91]. Progression of givinostat into clinical trials for immune-mediated diseases is supported by a wealth of studies showing the broad antiinflammatory effects of this compound in immune cells and in a range of autoimmune or inflammatory models of diabetes [92,93], arthritis [94], colitis [95,96], chronic asthma [97], and acute liver injury [90]. ITF2357 was effective at reducing the production of a range of proinflammatory cytokines (i.e., TNF, IL-1α, IL-1β, and IFN-γ) in LPS-stimulated cultured human peripheral blood mononuclear cells, and oral administration of the compound to mice led to reduced LPS-induced serum TNF and IFN-γ [90]. Givinostat provided protection in a mouse model of diabetes, apparently due to the effects on both immune cells and pancreatic islet cells. Treatment of splenocytes and macrophages with the compound led to a reduced production of inflammatory mediators (i.e., nitrites, TNF, and IFN-γ), while the treatment of isolated mouse islet cells enhanced their viability and insulin secretion [93].

Givinostat was also shown to ameliorate inflammation in experimental colitis by reducing the production of IFN-γ and enhancing the production of IL-10, effectively inhibiting the inflammation-associated tumor growth in mice [95]. Separate studies demonstrated that givinostat exerts antiinflammatory effects in experimental colitis by modifying the polarization of $CD4^+$ T cells via the IL-6 pathway. The HDACi suppressed T-cell polarization toward the proinflammatory Th17 phenotype, and enhanced the generation of $FOXP3^+$ T_{regs} at the site of inflammation in the lamina propria of ITF2357-treated animals, which resulted in the amelioration of the disease [96].

Givinostat has also been shown to prevent joint inflammation and cartilage and bone destruction in experimental arthritis. In an acute arthritis model (*Streptococcus pyogenes* cell wall arthritis), ITF2357 treatment led to reduced joint swelling, reduced cell influx into the joint cavity, and improved metabolic function of chondrocytes. The production of proinflammatory cytokines (i.e., TNF and IL-1β) in the synovial tissue was also reduced after ITF2357 treatment. Moreover, ITF2357 treatment reduced the severity scores and prevented bone destruction in the chronic models of rat adjuvant arthritis and mouse collagen type II arthritis. This was in agreement with the inhibition of bone resorption observed in vitro, where ITF2357 suppressed calcium release induced by IL-1β, TNF, or other stimuli. ITF2357 was also shown to inhibit RA fibroblast-like synoviocyte (FLS) IL-6 production induced by IL-1β, TNF, and toll-like receptor ligands by reducing the stability of IL-6 mRNA, hence providing mechanistic evidence of how HDACi may suppress inflammatory cytokine production in RA [98].

Some of the effects observed preclinically with HDACi have also been observed in the clinic. In a phase I (safety) trial of givinostat, peripheral blood from drug-treated patients stimulated with LPS ex vivo showed reduced production of inflammatory cytokines (i.e., TNF, IL-1β, IL-6, and IFN-γ) but not antiinflammatory cytokines (i.e., IL-10 and IL-1Ra) compared to controls [99]. Clinical benefits have also been observed in some HDACi trials; for example, the effects of givinostat were assessed in a small number of patients with the autoinflammatory syndromes, TNF receptor-associated periodic syndrome (TRAPS), hyper-IgD and periodic fever syndrome (HIDS), or Schnitzler syndrome, which are all characterized by symptomatic episodes associated with increased serum concentrations of both proinflammatory cytokines and antiinflammatory mediators [100]. Partial remission was observed in patients with Schnitzler syndrome and TRAPS, but no beneficial effects were seen in patients with HIDS. Encouraging results have also been obtained for givinostat in a small trial (17 patients) of patients with systemic-onset juvenile idiopathic arthritis (SOJIA) [91]. The drug was well tolerated after 12 weeks of treatment in this study and showed significant benefits, particularly in reducing the arthritis and pain components of the disease, which manifested in a reduction of the number of joints with active arthritis and limited range of motion, a significant reduction in the neutrophilia, and an improvement of the Childhood Health Assessment Questionnaire (CHAQ). The clinical benefit appeared after a short period of treatment (4–6 weeks) and the durability of the response was documented for at least 3 months after the discontinuation of the drug. Levels of CD40L, IL-1α, and IFN-γ in whole blood lysates were decreased at weeks 2 and 4 compared with baseline levels, thus indicating that givinostat therapy reduced cytokine-mediated markers of disease activity in SOJIA. The clinical data are consistent with those from animal models of RA and suggest that trials with HDACi are promising as safe oral alternatives to anticytokine biologicals and methotrexate.

Thus HDACi can target several mechanisms that are thought to be relevant in human autoimmune and inflammatory diseases, and there is some evidence that the beneficial effects of HDAC inhibition could translate to the clinic. However, despite having shown a much-improved toxicity profile compared to conventional cytotoxic chemotherapies, the dose-limiting toxic side effects of the currently approved HDACi (romidepsin and vorinostat, which have been approved for use in oncology) limit their use for the treatment of chronic immune-mediated diseases. Further progress in this area may require the identification of isoform or class-selective HDACi to reduce the toxicity and side-effects. Some progress has been made in this regard and compounds with reported selectivity for individual HDACs (i.e., HDAC1, HDAC3, HDAC6, and HDAC8) have recently been described [86]. Although information is limited regarding the individual contribution of HDACs to the antiinflammatory effects mediated by pan-HDACi, several studies have reported an important role of HDAC3 in regulating inflammatory responses through the NF-κB pathway [101]. For example, inhibition of HDAC3 in RA FLS largely recapitulated the effects of pan-HDACi in suppressing inflammatory gene expression, including type I IFN [102], thus highlighting HDAC3 as a potential therapeutic target in the treatment of RA and type I IFN-driven autoimmune diseases. Information on the adverse events associated with more selective HDAC inhibition may soon be available from oncology clinical trials initiated with HDAC1- and HDAC6-selective inhibitors.

An alternative strategy that is currently being evaluated to reduce the side effects associated with pan-HDAC inhibition is the development of compounds that preferentially target cell types involved in disease pathology. For example, tefinostat (CHR-2845) is a monocyte- and macrophage-targeted HDACi that was designed as a drug–ester conjugate. This compound is specifically hydrolyzed into a pharmacologically active acid by the intracellular esterase, human carboxylesterase-1 (hCE-1), which is selectively expressed within myeloid cells [103]. The acidic form of the molecule is retained within the targeted hCE-1^{+} cells due to its low permeability, thus leading to increased intracellular drug concentrations specifically within these cells. This approach is expected to provide potential benefits in terms of drug potency, longevity of action and, directing the pharmacological effects to cells expressing hCE-1 (i.e., monocytes, macrophages, and dendritic cells). Tefinostat's tolerability has been evaluated in a phase I clinical trial (NCT00820508) in patients with advanced hematological malignancies, and shown to be safe when given once daily for up to 15 months [104]. This study showed pharmacologically targeted relevant monocyte-specific protein acetylation in the absence of dose-limiting toxicities that are commonly seen with nontargeted HDACi (e.g., fatigue, diarrhea, and thrombocytopenia). The early signs of efficacy warrant further evaluation of tefinostat in larger patient populations in which monocytes or macrophages are important drivers of disease. A phase I/II dose-escalation trial of tefinostat for cancer-associated inflammation in patients with advanced hepatocellular carcinoma is currently being conducted (EudraCT number: 2012-000326-22). Recently published data also provides a strong rationale for the further clinical evaluation of tefinostat in monocytoid-lineage hematological neoplasms, including chronic myelomonocytic leukemias and monocyte-lineage acute myeloid leukemias (AMLs) [105]. Tefinostat, under development by Chroma Therapeutics and GlaxoSmithKline (GSK), is currently being progressed to phase II clinical trials for the treatment of RA (NCT02965599).

In addition to the progress being made on HDACi, efforts have also been focused on the identification of pharmacological modulators of the SIRT family of KDACs. Mounting evidence supports a role for SIRT1 activation in suppressing inflammation in vitro and in vivo [106]. Taken together with the more recent demonstration that SIRT1 regulates immune, metabolic, and mitochondrial bioenergy networks [107], this provides a rationale for the identification of SIRT1 activators for the treatment of autoimmune and inflammatory diseases. Small molecules that are capable of SIRT1 activation have been identified, including a range of natural products, derivatives, and synthetic small molecules [108]. A series of compounds that act as allosteric activators of SIRT1 have also been described [109].

Resveratrol was the first reported small molecule SIRT1 activator. It is a low potency polyphenol found in grapes and red wine and has been associated with a range of health benefits through many different mechanisms of action [110]. In vitro investigations have shown that resveratrol has immunomodulatory properties on T and B lymphocytes and exerts antiinflammatory, anticatabolic, antiapoptotic, and antioxidative effects in a range of cell types, including chondrocytes and synoviocytes. Resveratrol has shown protective effects in the joints in preclinical models of RA and osteoarthritis through modulation of inflammation, chondrolysis, and angiogenesis [111,112]; however, despite abundant preclinical studies in animal models, clinical investigations regarding the safety and beneficial effects of resveratrol in humans are scarce and have mainly focused on pharmacokinetics and metabolism [113].

Early phase clinical trials investigating the safety of higher potency SIRT1 activators (i.e., SRT2104) in metabolic diseases and inflammation indicate that SRT2104 is safe and well tolerated [114,115]. Despite the lack of significant clinical activity achieved in patients with mild-to-moderate UC [116], encouraging results were obtained when SRT2104 was used to treat moderate-to-severe psoriasis patients. Histological improvements were observed in association with modulation of the IL-17 and TNF signaling pathways and keratinocyte differentiation target genes [117]. Given the interesting clinical activity and the generally favorable safety profile seen in this study, further investigations of SIRT1 activators for the treatment of psoriasis are warranted.

15.1.8 Lysine Methyltransferase Inhibitors

Methylation of lysine residues within N-terminal histone tails regulates transcriptional activity and is mediated by histone methyltransferases (HMTs) [117a]. The catalytic activity of the lysine-specific methyltransferases resides within the evolutionarily conserved SET domain found in many chromatin regulators. Over 20 human lysine methyltransferases have been described to date (reviewed in Ref. [118]) and efforts to identify inhibitors of HMT catalytic activity have led to the development of small molecule inhibitors of a range of HMTs, including CARM1, DOT1L, G9a/GLP, SMYD2, and EZH2 [119–122].

One of the most studied HMTs is the enhancer of zeste homolog-2 (EZH2), the SET domain-containing catalytic subunit of the polycomb complex 2 (PRC2), which induces gene silencing by transferring mono-, di-, or trimethyl groups onto lysine 27 of histone3 (H3K27) [123,124]. The histone-modifying function of EZH2 is consistent with its predominant nuclear localization; however, while H3K27 appears to be its predominant substrate, EZH2 can also be found in the cytosol of several cell types and additional substrates (i.e., the TF STAT3) [125] have also been identified. This suggests that alternative functions of EZH2 exist beyond chromatin remodeling, which could be relevant in immune cells [126].

The selective deletion of EZH2 in multiple mouse tissues revealed critical roles for EZH2-mediated H3K27m3 in the regulation of cell differentiation, cell proliferation, and stem cell identity and pluripotency. Aberrant activation of EZH2 has also been associated with the development of various malignances [127], and EZH2 also has an important role in the differentiation and function of T cells. Several studies have evaluated the role of EZH2 in the regulation of T-cell activation, differentiation, and plasticity, as well as in the development of autoimmune diseases and graft-versus-host disease (GVHD) [128].

Early reports suggested that cytosolic EZH2 was essential in actin polymerization-dependent processes such as antigen receptor signaling in T cells. T-cell knockout of *EZH2* in these studies led to a substantial decrease in TCR-mediated T-cell activation in vitro [126], which is an inhibitory effect associated with the silencing of proapoptotic genes [128,129], and a disruption of the association between cytosolic EZH2, VAX, and NCK, which is a complex that contributes to the TCR-mediated activation of ERK [126]. It was also shown that EZH2-deficient T cells fail to respond to antigen stimulation in vivo, thus contributing to severe T cell-mediated immune-deficiency in mice [130]. Moreover, mice harboring EZH2-deficient T cells were shown to succumb to *Toxoplasma gondii* infection due to the suboptimal Th1 response that failed to limit the parasite burden [130]. In addition to affecting conventional T cells, recent studies performed using EZH2 conditional knockout mice ($EZH2^{cKO}$) demonstrated that T-cell-specific depletion of EZH2 induced a pronounced expansion of innate-like natural killer T (NKT) cells. This appeared to occur through a chromatin-modifying independent mechanism, since levels of H3K27me3 in $EZH2^{cKO}$ NKT cells were unaltered [131].

EZH2 has also been implicated in T_{reg} cell function, including a demonstration that EZH2 can directly control FOXP3 expression [132]. Following an inflammatory insult, EZH2 was also shown to be recruited by FOXP3 to repress key genes in T_{regs} [133]. Genetic ablation of EZH2 does not disrupt induced T_{reg} cell generation in vitro [134,135]; however, a critical role for EZH2 in the maintenance of T_{reg} cell identity during cellular activation has been reported in studies demonstrating that EZH2 was required to support a FOXP3-dependent gene expression pattern following CD28-mediated cellular activation [136]. Thus mice with EZH2-deficient T_{reg} cells failed to maintain immune tolerance, developed spontaneous autoimmunity associated with reduced numbers of $FOXP3^+$ cells in nonlymphoid tissues, and showed impaired resolution of inflammation in the CNS tissues upon acute induction of autoimmunity. Separate genetic studies in mice showed that ablation of EZH2 in the T-cell compartment led to more activated $CD4^+$ T cells and a reduced percentage of $FOXP3^+CD4^+$ cells in the secondary lymphoid organs [137]. EZH2-deficient T_{reg} cells from these mice failed to prevent T-cell-mediated colitis in mice using an adoptive transfer model, thus suggesting that EZH2-deficient T_{regs} were functionally defective. Consistent with this observation, mice with targeted deficiency of EZH2 in T_{regs} were shown to spontaneously develop inflammatory bowel disease [138].

These findings in genetically modified mice indicate an important role for EZH2 in the immune system, particularly in T cells; however, limited studies have been performed in which the therapeutic potential of EZH2 targeting has been investigated using inhibitors of EZH2. GSK126 has been shown to be a potent highly selective S-adenosyl-methionine-competitive small-molecule inhibitor of EZH2 methyltransferase activity. It functions by decreasing global H3K27me3 levels and reactivating silenced PRC2 target genes without degrading the PRC2 complex [139]. As genetic deletion of T-cell EZH2 can prevent GVHD following allogeneic hematopoietic stem cell transplantation in mice [128], the effects of GSK126 were evaluated in a mouse GVHD model [140]; however, enzymatic inhibition of EZH2 by GSK126 unexpectedly failed to prevent the disease. Hence the role of EZH2 in immune function remains to be fully understood. Nevertheless, GSK126 has shown potent antiproliferative effects in a range of cancer cell lines and effectively inhibited tumor growth in mouse xenograft models [139,141,142]. These studies support EZH2 inhibition as a promising therapeutic target for cancer, and GSK126 is currently undergoing phase I trials for hypermethylation-related malignancies [solid tumors and B-cell lymphomas (NCT01897571)] to test this hypothesis. Work continues towards the development of alternative EZH2 enzyme inhibitors, including derivatives of GSK126 and structurally distinct small molecules [143].

15.1.9 Bromodomain and Extra Terminal Domain (BET) Bromodomain Inhibitors

The BET family of proteins is comprised of four members: BRD2, BRD3, BRD4, and BRDT. BRD2, BRD3, and BRD4 are ubiquitously expressed while BRDT is expressed exclusively in germ cells [144]. BET proteins function as epigenetic readers as they contain bromodomains, which are protein modules that aid in the recognition of acetylated lysine residues [145]. BET bromodomains can bind to acetylated histones and TFs, with both activities thought to contribute to the recruitment of BET proteins to specific genomic sites. BRD4 has been shown to be enriched within enhancers and SEs, and a similar localization pattern has been reported for BRD2 and BRD3 [146,147].

Given the caveats of gene deletions and its effects on protein complex formation, the true understanding of BET protein function has been significantly aided by the development of potent and selective small molecule inhibitors of BET bromodomains. The acetyl lysine binding sites present in these bromodomains are highly homologous, therefore most of the inhibitors are active against all eight bromodomains in the four BET proteins. For this reason, the data generated with these compounds do not typically distinguish between the specific BET family members. In addition, the function of BRDT has not been extensively studied due to its restricted expression pattern. Nevertheless, great insights into the role of BET proteins in the regulation of gene expression have been obtained using BET bromodomain inhibitors. The focus here is on the role of the BET proteins in the regulation of gene expression in the immune system, and on the therapeutic potential of BET inhibitors in immune-mediated diseases.

The first demonstration that BET inhibition could selectively modulate inflammation-associated gene expression in innate immune cells came from studies using the pan-BET inhibitor, I-BET762 [148]. Treatment with I-BET762 selectively reduced the expression of a range of cytokines and chemokines by LPS-stimulated mouse macrophages. However, only a subset of LPS-induced genes were affected, which were the so-called secondary response genes that are expressed with relatively slow kinetics after LPS stimulation. Conversely, the genes that are induced most rapidly by LPS, the primary response genes, were not affected. Primary response genes have promoters and enhancers that are constitutively permissive for transcription, whereas secondary response genes require chromatin remodeling events including the generation of de novo enhancers and SEs before transcription can occur. These findings suggest a specific role for BET proteins in genes that are regulated through such epigenetic processes [149]. No effects were observed on the expression of housekeeping genes, which do not require this type of chromatin remodeling, therefore further supporting this hypothesis [50].

Subsequent studies have directly linked the selective modification of gene expression by BET inhibitors in macrophages to BET protein function at enhancers, specifically SEs. BET bromodomain inhibitors were shown to preferentially reduce the expression of genes driven by de novo generated SEs following LPS treatment of macrophages [16], and to block eRNA expression from de novo enhancers [6]. The selective inhibition of LPS-induced inflammatory gene expression in macrophages by I-BET762 has been confirmed, and has been extended to incorporate several chemically distinct pan-BET inhibitor compounds, which produced similar effects in various innate immune cells (i.e., macrophages, monocytes, and dendritic cells) in response to a variety of innate-cell activating stimuli [150–157]. Such observations in macrophages have also been extend to inflammatory responses in other cells; for example, BET bromodomain inhibition was shown to block TNF-induced de novo SE-driven gene expression in endothelial cells [16] and to inhibit the induction of inflammatory gene eRNA and mRNA in a gastric adenocarcinoma cell line by *Helicobactor pylori* (the causative agent of stomach ulcers) [158]. SE-associated oncogene expression in cancer cell lines, including viral oncogene expression driven by a SE generated upon genomic integration of human papillomavirus, is also highly sensitive to BET inhibitors [12,159].

BET inhibitors also affect the responses of adaptive immune cells, which show similarly highly selective modulation of gene expression [160]. I-BET762 and JQ1 were shown to inhibit the differentiation of CD4$^+$ cells into polarized effector subsets and the ability of predifferentiated Th cells to produce effector cytokines [160,161]. SE-driven genes appear to be particularly sensitive to BET inhibitors in T cells; for example, JQ1 was reported to preferentially inhibit the expression of SE-regulated disease-associated genes in CD4$^+$ memory and effector T cells isolated from the inflamed joint synovial fluid of patients with juvenile idiopathic arthritis [162]. JQ1 also provided pathological protection in a mouse model of autoimmune uveitis, which was associated with the selective inhibition of Th1-associated SE-regulated genes [20]. BET inhibitor-sensitive genes coincided with those driven by the Th1 lineage-defining TF, TBX21 (T-bet). Interestingly, SNPs associated with multiple immune-mediated diseases have been shown to alter the binding of TBX21 [57]. In addition to T-cell effects, JQ1 was also reported to inhibit multiple aspects of B-cell function, including proliferation, expansion, and antibody isotype switching [163,164].

In accordance with their ability to selectively modulate immune cell gene expression in vitro, BET inhibitors have shown striking antiinflammatory effects in vivo. Protective effects have been observed in a wide range of animal models of inflammation and autoimmunity, including sepsis, MS, atherosclerosis, periodontitis, uveitis, HIV-induced kidney injury, psoriasis, type I diabetes, and arthritis [16,20,148,150,151,153,160, 161,165–170]. In these in vivo studies it is important to consider that, in addition to the immune cell effects, selective modulation of gene expression in other cells that contribute to the disease (i.e., synovial fibroblasts in RA and pancreatic islet β-cells in type 1 diabetes) will contribute to the BET inhibitor-mediated protection from pathology [168,171]. De novo enhancers and SEs also play an important role in driving inflammation-associated genes in nonimmune cells, which may potentially explain their sensitivity to modulation by BET inhibitors.

15.2 CONCLUSION

Human genetic data have highlighted the prevalent role that dysregulated gene expression plays in immune-mediated diseases. Therapeutic approaches to treat these diseases will therefore depend on an understanding of the fundamental mechanisms that regulate gene expression and the identification of tractable drug targets that can correct defective gene regulation. Progress in this respect has been greatly aided by the expanding knowledge of enhancer biology, including an understanding of the epigenetic mechanisms that control their function, a growing realization that inflammation-associated gene expression is linked to particular types of enhancers, and evidence that genetic polymorphisms linked to immune-mediated diseases localize within such regions. Epigenetic proteins in multiple target classes have also proven to be tractable to inhibition with small molecules. The development of potent and selective compounds has enabled investigations into the function of numerous epigenetic proteins, and has revealed marked and selective effects of epigenetic inhibitors on the expression of inflammation-associated genes. While much remains to be learned about the epigenetic mechanisms that control gene expression in the immune system, continued progress in rapidly advancing these inhibitors into the clinic is expected, which will provide new opportunities for the treatment of human diseases.

References

[1] Janeway CA Jr. Approaching the asymptote? Evolution and revolution in immunology. Cold Spring Harb Symp Quant Biol 1989;1–13.

[2] Matzinger P. Tolerance, danger, and the extended family. Ann Rev Immunol 1994;991–1045. 4/1/1994, Annual Reviews.

[3] Ramsey SA, Klemm SL, Zak DE, Kennedy KA, Thorsson V, Li B, et al. Uncovering a macrophage transcriptional program by integrating evidence from motif scanning and expression dynamics. PLoS Comput Biol 2008;4(3):e1000021.

[4] Davey CA, Sargent DF, Luger K, Maeder AW, Richmond TJ. Solvent mediated interactions in the structure of the nucleosome core particle at 1.9Å resolution. J Mol Biol 2002;319(5):1097–113.

[5] Luger K, Mader AW, Richmond RK, Sargent DF, Richmond TJ. Crystal structure of the nucleosome core particle at 2.8Å resolution. Nature 1997;389(6648):251–60.

[6] Kaikkonen M-U, Spann N-J, Heinz S, Romanoski C-E, Allison K-A, Stender J-D, et al. Remodeling of the enhancer landscape during macrophage activation is coupled to enhancer transcription. Mol Cell 2013;51(3):310–25.

[7] Ostuni R, Piccolo V, Barozzi I, Polletti S, Termanini A, Bonifacio S, et al. Latent enhancers activated by stimulation in differentiated cells. Cell 2013;152(1–2):157–71.

[8] Bulger M, Groudine M. Functional and mechanistic diversity of distal transcription enhancers. Cell 2011;144(3): 327–39.

[9] Symmons O, Uslu VV, Tsujimura T, Ruf S, Nassari S, Schwarzer W, et al. Functional and topological characteristics of mammalian regulatory domains. Genome Res 2014;24(3):390–400.

[10] de Laat W, Duboule D. Topology of mammalian developmental enhancers and their regulatory landscapes. Nature 2013;502(7472):499–506.

[11] Whyte W-A, Orlando D-A, Hnisz D, Abraham B-J, Lin C-Y, Kagey M-H, et al. Master transcription factors and mediator establish super-enhancers at key cell identity genes. Cell 2013;153(2):307–19.

[12] Loven J, Hoke H-A, Lin C-Y, Lau A, Orlando D-A, Vakoc C-R, et al. Selective inhibition of tumor oncogenes by disruption of super-enhancers. Cell 2013;153(2):320–34.

[13] Hnisz D, Abraham B-J, Lee T-I, Lau A, Saint-Andre V, Sigova A-A, et al. Super-enhancers in the control of cell identity and disease. Cell 2013;155(4):934–47.

[14] Parker SCJ, Stitzel ML, Taylor DL, Orozco JM, Erdos MR, Akiyama JA, et al. Chromatin stretch enhancer states drive cell-specific gene regulation and harbor human disease risk variants. Proc Natl Acad Sci USA 2013;110(44):17921–6.

[15] Hay D, Hughes JR, Babbs C, Davies JOJ, Graham BJ, Hanssen LLP, et al. Genetic dissection of the α-globin super-enhancer in vivo. Nat Genet 2016;48(8):895–903.

[16] Brown J-D, Lin C-Y, Duan Q, Griffin G, Federation AJ, Paranal R-M, et al. NF-κB directs dynamic super enhancer formation in inflammation and atherogenesis. Mol Cell 2014;56:1–13.

[17] Hah N, Benner C, Chong L-W, Yu RT, Downes M, Evans RM. Inflammation-sensitive super enhancers form domains of coordinately regulated enhancer RNAs. Proc Natl Acad Sci USA 2015;112(3):E297–302.

[18] Netea MG, Quintin J, van der Meer JW. Trained immunity: a memory for innate host defense. Cell Host Microbe 2011;9(5):355–61.

[19] Roh T-Y, Cuddapah S, Zhao K. Active chromatin domains are defined by acetylation islands revealed by genome-wide mapping. Genes Dev 2005;19(5):542–52.

[20] Hertweck A, Evans CM, Eskandarpour M, Lau JC, Oleinika K, Jackson I, et al. T-bet activates Th1 genes through mediator and the super elongation complex. Cell Rep 2016;15(12):2756–70.

[21] Vahedi G, Kanno Y, Furumoto Y, Jiang K, Parker SCJ, Erdos MR, et al. Super-enhancers delineate disease-associated regulatory nodes in T cells. Nature 2015;520(7548):558–62.

[22] Fang Z, Hecklau K, Gross F, Bachmann I, Venzke M, Karl M, et al. Transcription factor co-occupied regions in the murine genome constitute T-helper-cell subtype-specific enhancers. Eur J Immunol 2015;45(11):3150–7.

[23] Sprent J, Tough DF. T cell death and memory. Science 2001;293(5528):245–8.

[24] Northrop JK, Wells AD, Shen H. Cutting edge: chromatin remodeling as a molecular basis for the enhanced functionality of memory CD8 T cells. J Immunol 2008;181(2):865–8.

[25] Zediak VP, Johnnidis JB, Wherry EJ, Berger SL. Cutting edge: persistently open chromatin at effector gene loci in resting memory $CD8^+$ T cells independent of transcriptional status. J Immunol 2011;186(5):2705–9.

[26] Messi M, Giacchetto I, Nagata K, Lanzavecchia A, Natoli G, Sallusto F. Memory and flexibility of cytokine gene expression as separable properties of human Th1 and Th2 lymphocytes. Nat Immunol 2003;4(1):78–86.

[27] Youngblood B, Hale JS, Ahmed R. T-cell memory differentiation: insights from transcriptional signatures and epigenetics. Immunology 2013;139(3):277–84.

[28] Halliley LJ, Tipton MC, Liesveld J, Rosenberg FA, Darce J, Gregoretti VI, et al. Long-lived plasma cells are contained within the $CD19^{-}CD38hiCD138^{+}$ subset in human bone marrow. Immunity 2015;43(1):132–45.

[29] Yu X, Tsibane T, McGraw PA, House FS, Keefer CJ, Hicar MD, et al. Neutralizing antibodies derived from the B cells of 1918 influenza pandemic survivors. Nature 2008;455(7212):532–6.

[30] Zinkernagel RM. Immunological memory ≠ protective immunity. Cell Mol Life Sci 2012;69(10):1635–40.

[31] Manichanh C, Borruel N, Casellas F, Guarner F. The gut microbiota in IBD. Nat Rev Gastroenterol Hepatol 2012;9(10):599–608.

[32] McGeachy MJ, Bak-Jensen KS, Chen Y, Tato CM, Blumenschein W, McClanahan T, et al. TGF-β and IL-6 drive the production of IL-17 and IL-10 by T cells and restrain Th-17 cell-mediated pathology. Nat Immunol 2007;8(12):1390–7.

[33] Lee Y, Awasthi A, Yosef N, Quintana FJ, Xiao S, Peters A, et al. Induction and molecular signature of pathogenic Th17 cells. Nat Immunol 2012;13(10):991–9.

[34] Danso-Abeam D, Humblet-Baron S, Dooley J, Liston A. Models of Aire-dependent gene regulation for thymic negative selection. Front Immunol 2011;2(14).

[35] Anderson MS, Venanzi ES, Klein L, Chen Z, Berzins SP, Turley SJ, et al. Projection of an immunological self shadow within the thymus by the Aire protein. Science 2002;298(5597):1395–401.

[36] Metzger TC, Anderson MS. Control of central and peripheral tolerance by Aire. Immunol Rev 2011;241(1):89–103.

[37] Bansal K, Yoshida H, Benoist C, Mathis D. The transcriptional regulator Aire binds to and activates super-enhancers. Nat Immunol 2017;18(3):263–73.

[38] Klein L, Kyewski B, Allen PM, Hogquist KA. Positive and negative selection of the T cell repertoire: what thymocytes see (and don't see). Nat Rev Immunol 2014;14(6):377–91.

[39] Yang S, Fujikado N, Kolodin D, Benoist C, Mathis D. Regulatory T cells generated early in life play a distinct role in maintaining self-tolerance. Science 2015;348(6234):589–94.

[40] Nagamine K, Peterson P, Scott HS, Kudoh J, Minoshima S, Heino M, et al. Positional cloning of the APECED gene. Nat Genet 1997;17(4):393–8.

[41] Finnish-German AC. An autoimmune disease, APECED, caused by mutations in a novel gene featuring two PHD-type zinc-finger domains. Nat Genet 1997;17(4):399–403.

[42] Plitas G, Rudensky AY. Regulatory T cells: differentiation and function. Cancer Immunol Res 2016;4(9):721–5.

[43] Sojka DK, Huang Y-H, Fowell DJ. Mechanisms of regulatory T-cell suppression: a diverse arsenal for a moving target. Immunology 2008;124(1):13–22.

[44] Bennett CL, Christie J, Ramsdell F, Brunkow ME, Ferguson PJ, Whitesell L, et al. The immune dysregulation, polyendocrinopathy, enteropathy, X-linked syndrome (IPEX) is caused by mutations of FOXP3. Nat Genet 2001;27(1):20–1.

[45] Wildin RS, Ramsdell F, Peake J, Faravelli F, Casanova JL, Buist N, et al. X-linked neonatal diabetes mellitus, enteropathy and endocrinopathy syndrome is the human equivalent of mouse scurfy. Nat Genet 2001;27(1):18–20.

[46] Hindorff LA, Sethupathy P, Junkins HA, Ramos EM, Mehta JP, Collins FS, et al. Potential etiologic and functional implications of genome-wide association loci for human diseases and traits. Proc Natl Acad Sci USA 2009;106(23):9362–7.

[47] Manolio TA. Genome-wide association studies and assessment of the risk of disease. New Engl J Med 2010;363(2):166–76.

[48] Maurano MT, Humbert R, Rynes E, Thurman RE, Haugen E, Wang H, et al. Systematic localization of common disease-associated variation in regulatory DNA. Science 2012;337(6099):1190–5.

[49] Nicolae DL, Gamazon E, Zhang W, Duan S, Dolan ME, Cox NJ. Trait-associated SNPs are more likely to be eQTLs: annotation to enhance discovery from GWAS. PLoS Genet 2010;6(4):e1000888.

[50] Zabidi MA, Arnold CD, Schernhuber K, Pagani M, Rath M, Frank O, et al. Enhancer-core-promoter specificity separates developmental and housekeeping gene regulation. Nature 2015;518(7540):556–9.

[51] Farh KK-H, Marson A, Zhu J, Kleinewietfeld M, Housley WJ, Beik S, et al. Genetic and epigenetic fine mapping of causal autoimmune disease variants. Nature 2015;518(7539):337–43.

[52] Seumois G, Chavez L, Gerasimova A, Lienhard M, Omran N, Kalinke L, et al. Epigenomic analysis of primary human T cells reveals enhancers associated with Th2 memory cell differentiation and asthma susceptibility. Nat Immunol 2014;15(8):777–88.

[53] Onengut-Gumuscu S, Chen WM, Burren O, Cooper NJ, Quinlan AR, Mychaleckyj JC, et al. Fine mapping of type 1 diabetes susceptibility loci and evidence for colocalization of causal variants with lymphoid gene enhancers. Nat Genet 2015;47(4):381–6.

[54] Roberts AR, Vecellio M, Chen L, Ridley A, Cortes A, Knight JC, et al. An ankylosing spondylitis-associated genetic variant in the IL23R-IL12RB2 intergenic region modulates enhancer activity and is associated with increased Th1-cell differentiation. Ann Rheum Dis 2016;75(12):2150–6.

[55] Weinstein JS, Lezon-Geyda K, Maksimova Y, Craft S, Zhang Y, Su M, et al. Global transcriptome analysis and enhancer landscape of human primary T follicular helper and T effector lymphocytes. Blood 2014;124(25):3719–29.

[56] Cavalli G, Hayashi M, Jin Y, Yorgov D, Santorico SA, Holcomb C, et al. MHC class II super-enhancer increases surface expression of HLA-DR and HLA-DQ and affects cytokine production in autoimmune vitiligo. Proc Natl Acad Sci USA 2016;113(5):1363–8.
[57] Soderquest K, Hertweck A, Giambartolomei C, Henderson S, Mohamed R, Goldberg R, et al. Genetic variants alter T-bet binding and gene expression in mucosal inflammatory disease. PLoS Genet 2017;13(2):e1006587.
[58] Javierre BM, Burren OS, Wilder SP, Kreuzhuber R, Hill SM, Sewitz S, et al. Lineage-specific genome architecture links enhancers and non-coding disease variants to target gene promoters. Cell 2016;167(5). 1369.e19-1384.e19.
[59] Tahiliani M, Koh KP, Shen Y, Pastor WA, Bandukwala H, Brudno Y, et al. Conversion of 5-methylcytosine to 5-hydroxymethylcytosine in mammalian DNA by MLL partner TET1. Science 2009;324(5929):930–5.
[60] He YF, Li BZ, Li Z, Liu P, Wang Y, Tang Q, et al. Tet-mediated formation of 5-carboxylcytosine and its excision by TDG in mammalian DNA. Science 2011;333(6047):1303–7.
[61] Ito S, Shen L, Dai Q, Wu SC, Collins LB, Swenberg JA, et al. Tet proteins can convert 5-methylcytosine to 5-formylcytosine and 5-carboxylcytosine. Science 2011;333(6047):1300–3.
[62] Cortellino S, Xu J, Sannai M, Moore R, Caretti E, Cigliano A, et al. Thymine DNA glycosylase is essential for active DNA demethylation by linked deamination-base excision repair. Cell 2011;146(1):67–79.
[63] Maiti A, Drohat AC. Thymine DNA glycosylase can rapidly excise 5-formylcytosine and 5-carboxylcytosine: potential implications for active demethylation of CpG sites. J Biol Chem 2011;286(41):35334–8.
[64] Kouzarides T. Chromatin modifications and their function. Cell 2007;128(4):693–705.
[65] Hong L, Schroth GP, Matthews HR, Yau P, Bradbury EM. Studies of the DNA binding properties of histone H4 amino terminus. Thermal denaturation studies reveal that acetylation markedly reduces the binding constant of the H4 "tail" to DNA. J Biol Chem 1993;268(1):305–14.
[66] Campanero MR, Armstrong MI, Flemington EK. CpG methylation as a mechanism for the regulation of E2F activity. Proc Natl Acad Sci USA 2000;97(12):6481–6.
[67] Iguchi-Ariga SM, Schaffner W. CpG methylation of the cAMP-responsive enhancer/promoter sequence TGACGTCA abolishes specific factor binding as well as transcriptional activation. Genes Dev 1989;3(5):612–9.
[68] Heintzman ND, Hon GC, Hawkins RD, Kheradpour P, Stark A, Harp LF, et al. Histone modifications at human enhancers reflect global cell-type-specific gene expression. Nature 2009;459(7243):108–12.
[69] Creyghton MP, Cheng AW, Welstead GG, Kooistra T, Carey BW, Steine EJ, et al. Histone H3K27ac separates active from poised enhancers and predicts developmental state. Proc Natl Acad Sci 2010;107(50):21931–6.
[70] Kraushaar DC, Jin W, Maunakea A, Abraham B, Ha M, Zhao K. Genome-wide incorporation dynamics reveal distinct categories of turnover for the histone variant H3.3. Genome Biol 2013;14(10):1–15.
[71] Deaton AM, Gómez-Rodríguez M, Mieczkowski J, Tolstorukov MY, Kundu S, Sadreyev RI, et al. Enhancer regions show high histone H3.3 turnover that changes during differentiation. elife 2016;5:e15316.
[72] Rada-Iglesias A, Bajpai R, Swigut T, Brugmann SA, Flynn RA, Wysocka J. A unique chromatin signature uncovers early developmental enhancers in humans. Nature 2011;470(7333):279–83.
[73] Schnetz MP, Handoko L, Akhtar-Zaidi B, Bartels CF, Pereira CF, Fisher AG, et al. CHD7 targets active gene enhancer elements to modulate ES cell-specific gene expression. PLoS Genet 2010;6(7):e1001023.
[74] Kadoch C, Copeland RA, Keilhack H. PRC2 and SWI/SNF chromatin remodeling complexes in health and disease. Biochemistry 2016;55(11):1600–14.
[75] Torchy MP, Hamiche A, Klaholz BP. Structure and function insights into the NuRD chromatin remodeling complex. Cell Mol Life Sci 2015;72(13):2491–507.
[76] Jang MK, Mochizuki K, Zhou M, Jeong HS, Brady JN, Ozato K. The bromodomain protein Brd4 is a positive regulatory component of P-TEFb and stimulates RNA polymerase II-dependent transcription. Mol Cell 2005;19(4):523–34.
[77] Choudhary C, Kumar C, Gnad F, Nielsen ML, Rehman M, Walther TC, et al. Lysine acetylation targets protein complexes and co-regulates major cellular functions. Science 2009;325(5942):834–40.
[78] Gregoretti IV, Lee YM, Goodson HV. Molecular evolution of the histone deacetylase family: functional implications of phylogenetic analysis. J Mol Biol 2004;338(1):17–31.
[79] Roos WP, Krumm A. The multifaceted influence of histone deacetylases on DNA damage signalling and DNA repair. Nucleic Acids Res 2016;44(21):10017–30.
[80] Matthews GM, Newbold A, Johnstone RW. Intrinsic and extrinsic apoptotic pathway signaling as determinants of histone deacetylase inhibitor antitumor activity. Adv Cancer Res 2012;116:165–97.
[81] Xu WS, Parmigiani RB, Marks PA. Histone deacetylase inhibitors: molecular mechanisms of action. Oncogene 2007;26(37):5541–52.
[82] Yang X-J, Seto E. The Rpd3/Hda1 family of lysine deacetylases: from bacteria and yeast to mice and men. Nat Rev Mol Cell Biol 2008;9(3):206–18.

[83] Zhang H, Shang Y-P, Chen H-Y, Li J. Histone deacetylases function as novel potential therapeutic targets for cancer. Hepatology Res 2017;47(2):149–59.
[84] Cheng F, Lienlaf M, Perez-Villarroel P, Wang HW, Lee C, Woan K, et al. Divergent roles of histone deacetylase 6 (HDAC6) and histone deacetylase 11 (HDAC11) on the transcriptional regulation of IL10 in antigen presenting cells. Mol Immunol 2014;60(1):44–53.
[85] Tang J, Yan H, Zhuang S. Histone deacetylases as targets for treatment of multiple diseases. Clin Sci (Lond) 2013;124(11):651–62.
[86] Qin HT, Li HQ, Liu F. Selective histone deacetylase small molecule inhibitors: recent progress and perspectives. Expert Opin Ther Pat 2016;1–15.
[87] Bantscheff M, Hopf C, Savitski MM, Dittmann A, Grandi P, Michon AM, et al. Chemoproteomics profiling of HDAC inhibitors reveals selective targeting of HDAC complexes. Nat Biotechnol 2011;29(3):255–65.
[88] Hull EE, Montgomery MR, Leyva KJ. HDAC inhibitors as epigenetic regulators of the immune system: impacts on cancer therapy and inflammatory diseases. Biomed Res Int 2016;2016:8797206.
[89] Tough DF, Tak PP, Tarakhovsky A, Prinjha RK. Epigenetic drug discovery: breaking through the immune barrier. Nat Rev Drug Discov 2016;15(12):835–53.
[90] Leoni F, Fossati G, Lewis EC, Lee JK, Porro G, Pagani P, et al. The histone deacetylase inhibitor ITF2357 reduces production of pro-inflammatory cytokines in vitro and systemic inflammation in vivo. Mol Med 2005;11(1–12):1–15.
[91] Vojinovic J, Damjanov N. HDAC inhibition in rheumatoid arthritis and juvenile idiopathic arthritis. Mol Med 2011;17(5–6):397–403.
[92] Hara N, Alkanani AK, Dinarello CA, Zipris D. Histone deacetylase inhibitor suppresses virus-induced proinflammatory responses and type 1 diabetes. J Mol Med (Berl) 2014;92(1):93–102.
[93] Lewis EC, Blaabjerg L, Storling J, Ronn SG, Mascagni P, Dinarello CA, et al. The oral histone deacetylase inhibitor ITF2357 reduces cytokines and protects islet beta cells in vivo and in vitro. Mol Med 2011;17(5–6):369–77.
[94] Joosten LA, Leoni F, Meghji S, Mascagni P. Inhibition of HDAC activity by ITF2357 ameliorates joint inflammation and prevents cartilage and bone destruction in experimental arthritis. Mol Med 2011;17(5–6):391–6.
[95] Glauben R, Batra A, Stroh T, Erben U, Fedke I, Lehr HA, et al. Histone deacetylases: novel targets for prevention of colitis-associated cancer in mice. Gut 2008;57(5):613–22.
[96] Glauben R, Sonnenberg E, Wetzel M, Mascagni P, Siegmund B. Histone deacetylase inhibitors modulate interleukin 6-dependent CD4+ T cell polarization in vitro and in vivo. J Biol Chem 2014;289(9):6142–51.
[97] Su XM, Ren Y, Kong LF, Kang J. Therapeutic effects of histone deacetylase inhibitor givinostat on air inflammation and high airway resistance in a murine asthma model. Zhonghua Nei Ke Za Zhi 2017;56(2):121–6.
[98] Grabiec AM, Korchynskyi O, Tak PP, Reedquist KA. Histone deacetylase inhibitors suppress rheumatoid arthritis fibroblast-like synoviocyte and macrophage IL-6 production by accelerating mRNA decay. Ann Rheum Dis 2012;71(3):424–31.
[99] Furlan A, Monzani V, Reznikov LL, Leoni F, Fossati G, Modena D, et al. Pharmacokinetics, safety and inducible cytokine responses during a phase 1 trial of the oral histone deacetylase inhibitor ITF2357 (givinostat). Mol Med 2011;17(5–6):353–62.
[100] Bodar EJ, Simon A, van der Meer JW. Effects of the histone deacetylase inhibitor ITF2357 in autoinflammatory syndromes. Mol Med 2011;17(5–6):363–8.
[101] Leus NG, Zwinderman MR, Dekker FJ. Histone deacetylase 3 (HDAC 3) as emerging drug target in NF-kappaB-mediated inflammation. Curr Opin Chem Biol 2016;33:160–8.
[102] Angiolilli C, Kabala PA, Grabiec AM, Van Baarsen IM, Ferguson BS, Garcia S, et al. Histone deacetylase 3 regulates the inflammatory gene expression programme of rheumatoid arthritis fibroblast-like synoviocytes. Ann Rheum Dis 2017;76(1):277–85.
[103] Needham LA, Davidson AH, Bawden LJ, Belfield A, Bone EA, Brotherton DH, et al. Drug targeting to monocytes and macrophages using esterase-sensitive chemical motifs. J Pharmacol Exp Ther 2011;339(1):132–42.
[104] Ossenkoppele GJ, Lowenberg B, Zachee P, Vey N, Breems D, Van de Loosdrecht AA, et al. A phase I first-in-human study with tefinostat – a monocyte/macrophage targeted histone deacetylase inhibitor – in patients with advanced haematological malignancies. Br J Haematol 2013;162(2):191–201.
[105] Zabkiewicz J, Gilmour M, Hills R, Vyas P, Bone E, Davidson A, et al. The targeted histone deacetylase inhibitor tefinostat (CHR-2845) shows selective in vitro efficacy in monocytoid-lineage leukaemias. Oncotarget 2016;7(13):16650–62.
[106] Xie J, Zhang X, Zhang L. Negative regulation of inflammation by SIRT1. Pharmacol Res 2013;67(1):60–7.
[107] Vachharajani VT, Liu T, Wang X, Hoth JJ, Yoza BK, McCall CE. Sirtuins link inflammation and metabolism. J Immunol Res 2016;2016:8167273.

[108] Kumar A, Chauhan S. How much successful are the medicinal chemists in modulation of SIRT1: a critical review. Eur J Med Chem 2016;119:45–69.
[109] Hubbard BP, Gomes AP, Dai H, Li J, Case AW, Considine T, et al. Evidence for a common mechanism of SIRT1 regulation by allosteric activators. Science 2013;339(6124):1216–9.
[110] Smoliga JM, Baur JA, Hausenblas HA. Resveratrol and health: a comprehensive review of human clinical trials. Mol Nutr Food Res 2011;55(8):1129–41.
[111] Elmali N, Baysal O, Harma A, Esenkaya I, Mizrak B. Effects of resveratrol in inflammatory arthritis. Inflammation 2007;30(1–2):1–6.
[112] Nguyen C, Savouret JF, Widerak M, Corvol MT, Rannou F. Resveratrol: potential therapeutic interest in joint disorders: a critical narrative review. Nutrients 2017;9(1).
[113] Tome-Carneiro J, Larrosa M, Gonzalez-Sarrias A, Tomas-Barberan FA, Garcia-Conesa MT, Espin JC. Resveratrol and clinical trials: the crossroad from in vitro studies to human evidence. Curr Pharm Des 2013;19(34):6064–93.
[114] Hoffmann E, Wald J, Lavu S, Roberts J, Beaumont C, Haddad J, et al. Pharmacokinetics and tolerability of SRT2104, a first-in-class small molecule activator of SIRT1, after single and repeated oral administration in man. Br J Clin Pharmacol 2013;75(1):186–96.
[115] Libri V, Brown AP, Gambarota G, Haddad J, Shields GS, Dawes H, et al. A pilot randomized, placebo controlled, double blind phase I trial of the novel SIRT1 activator SRT2104 in elderly volunteers. PLoS One 2012;7(12):e51395.
[116] Sands BE, Joshi S, Haddad J, Freudenberg JM, Oommen DE, Hoffmann E, et al. Assessing colonic exposure, safety, and clinical activity of SRT2104, a novel oral SIRT1 activator, in patients with mild to moderate ulcerative colitis. Inflamm Bowel Dis 2016;22(3):607–14.
[117] Krueger JG, Suarez-Farinas M, Cueto I, Khacherian A, Matheson R, Parish LC, et al. A randomized, placebo-controlled study of SRT2104, a SIRT1 activator, in patients with moderate to severe psoriasis. PLoS One 2015;10(11):e0142081.
[117a] Kouzarides T. Curr Opin Genet Dev. 2002;12(2):198-209. Review. Erratum in: Curr Opin Genet Dev 2002;12(3):371. PMID:11893494.
[118] Tough DF, Lewis HD, Rioja I, Lindon MJ, Prinjha RK. Epigenetic pathway targets for the treatment of disease: accelerating progress in the development of pharmacological tools: IUPHAR review 11. Br J Pharmacol 2014;171(22): 4981–5010.
[119] Verma SK, Tian X, LaFrance LV, Duquenne C, Suarez DP, Newlander KA, et al. Identification of potent, selective, cell-active inhibitors of the histone lysine methyltransferase EZH2. ACS Med Chem Lett 2012;3(12):1091–6.
[120] Allan M, Manku S, Therrien E, Nguyen N, Styhler S, Robert MF, et al. N-Benzyl-1-heteroaryl-3-(trifluoromethyl)-1H-pyrazole-5-carboxamides as inhibitors of co-activator associated arginine methyltransferase 1 (CARM1). Bioorg Med Chem Lett 2009;19(4):1218–23.
[121] Liu F, Chen X, Allali-Hassani A, Quinn AM, Wasney GA, Dong A, et al. Discovery of a 2,4-diamino-7-aminoalkoxyquinazoline as a potent and selective inhibitor of histone lysine methyltransferase G9a. J Med Chem 2009;52(24):7950–3.
[122] Yao Y, Chen P, Diao J, Cheng G, Deng L, Anglin JL, et al. Selective inhibitors of histone methyltransferase DOT1L: design synthesis, and crystallographic studies. J Am Chem Soc 2011;133(42):16746–9.
[123] Su IH, Basavaraj A, Krutchinsky AN, Hobert O, Ullrich A, Chait BT, et al. Ezh2 controls B cell development through histone H3 methylation and Igh rearrangement. Nat Immunol 2003;4(2):124–31.
[124] Kadoch C, Copeland RA, Keilhack H. PRC2 and SWI/SNF chromatin remodeling complexes in health and disease. Biochemistry 2016;55(11):1600–14.
[125] Dasgupta M, Dermawan JK, Willard B, Stark GR. STAT3-driven transcription depends upon the dimethylation of K49 by EZH2. Proc Natl Acad Sci USA 2015;112(13):3985–90.
[126] Su IH, Dobenecker MW, Dickinson E, Oser M, Basavaraj A, Marqueron R, et al. Polycomb group protein ezh2 controls actin polymerization and cell signaling. Cell 2005;121(3):425–36.
[127] Christofides A, Karantanos T, Bardhan K, Boussiotis VA. Epigenetic regulation of cancer biology and anti-tumor immunity by EZH2. Oncotarget 2016;7(51):85624–40.
[128] He S, Xie F, Liu Y, Tong Q, Mochizuki K, Lapinski PE, et al. The histone methyltransferase Ezh2 is a crucial epigenetic regulator of allogeneic T-cell responses mediating graft-versus-host disease. Blood 2013;122(25):4119–28.
[129] Zhang Y, Kinkel S, Maksimovic J, Bandala-Sanchez E, Tanzer MC, Naselli G, et al. The polycomb repressive complex 2 governs life and death of peripheral T cells. Blood 2014;124(5):737–49.
[130] Yang XP, Jiang K, Hirahara K, Vahedi G, Afzali B, Sciume G, et al. EZH2 is crucial for both differentiation of regulatory T cells and T effector cell expansion. Sci Rep 2015;5:10643.
[131] Vasanthakumar A, Xu D, Lun AT, Kueh AJ, van Gisbergen KP, Iannarella N, et al. A non-canonical function of Ezh2 preserves immune homeostasis. EMBO Rep 2017;18(4):619–31.

[132] Xiong Y, Khanna S, Grzenda AL, Sarmento OF, Svingen PA, Lomberk GA, et al. Polycomb antagonizes p300/CREB-binding protein-associated factor to silence FOXP3 in a Kruppel-like factor-dependent manner. J Biol Chem 2012;287(41):34372–85.
[133] Arvey A, van der Veeken J, Samstein RM, Feng Y, Stamatoyannopoulos JA, Rudensky AY. Inflammation-induced repression of chromatin bound by the transcription factor Foxp3 in regulatory T cells. Nat Immunol 2014;15(6):580–7.
[134] Tumes Damon J, Onodera A, Suzuki A, Shinoda K, Endo Y, Iwamura C, et al. The polycomb protein Ezh2 regulates differentiation and plasticity of CD4+ T helper type 1 and type 2 Cells. Immunity 2013;39(5):819–32.
[135] Zhang Y, Kinkel S, Maksimovic J, Bandala-Sanchez E, Tanzer MC, Naselli G, et al. The polycomb repressive complex 2 governs life and death of peripheral T cells. Blood 2014;124(5):737–49.
[136] DuPage M, Chopra G, Quiros J, Rosenthal WL, Morar MM, Holohan D, et al. The chromatin-modifying enzyme Ezh2 is critical for the maintenance of regulatory T cell identity after activation. Immunity 2015;42(2):227–38.
[137] Yang X-p, Jiang K, Hirahara K, Vahedi G, Afzali B, Sciume G, et al. EZH2 is crucial for both differentiation of regulatory T cells and T effector cell expansion. Sci Rep 2015;5:10643.
[138] Sarmento OF, Svingen PA, Xiong Y, Sun Z, Bamidele AO, Mathison AJ, et al. The role of the histone methyltransferase enhancer of zeste homolog 2 (EZH2) in the pathobiological mechanisms underlying inflammatory bowel disease (IBD). J Biol Chem 2017;292(2):706–22.
[139] McCabe MT, Ott HM, Ganji G, Korenchuk S, Thompson C, Van Aller GS, et al. EZH2 inhibition as a therapeutic strategy for lymphoma with EZH2-activating mutations. Nature 2012;492(7427):108–12.
[140] Huang Q, He S, Tian Y, Gu Y, Chen P, Li C, et al. Hsp90 inhibition destabilizes Ezh2 protein in alloreactive T cells and reduces graft-versus-host disease in mice. Blood 2017;129(20):2737–48.
[141] Yeh CM, Shay J, Zeng TC, Chou JL, Huang TH, Lai HC, et al. Epigenetic silencing of ARNTL, a circadian gene and potential tumor suppressor in ovarian cancer. Int J Oncol 2014;45(5):2101–7.
[142] Tiffen JC, Gunatilake D, Gallagher SJ, Gowrishankar K, Heinemann A, Cullinane C, et al. Targeting activating mutations of EZH2 leads to potent cell growth inhibition in human melanoma by derepression of tumor suppressor genes. Oncotarget 2015;6(29):27023–36.
[143] Wu Y, Hu J, Ding H, Chen L, Zhang Y, Liu R, et al. Identification of novel EZH2 inhibitors through pharmacophore-based virtual screening and biological assays. Bioorg Med Chem Lett 2016;26(15):3813–7.
[144] Barda S, Paz G, Yogev L, Yavetz H, Lehavi O, Hauser R, et al. Expression of BET genes in testis of men with different spermatogenic impairments. Fertil Steril 2012;97(1):46–52.
[145] Zeng L, Zhou MM. Bromodomain: an acetyl-lysine binding domain. FEBS Lett February 20, 2002;124–8.
[146] Pinz S, Unser S, Rascle A. Signal transducer and activator of transcription STAT5 is recruited to c-Myc super-enhancer. BMC Mol Biol 2016;17(1):1–11.
[147] Cheung KL, Zhang F, Jaganathan A, Sharma R, Zhang Q, Konuma T, et al. Distinct roles of Brd2 and Brd4 in potentiating the transcriptional program for Th17 cell differentiation. Mol Cell 2017;65(6):1068–80.
[148] Nicodeme E, Jeffrey KL, Schaefer U, Beinke S, Dewell S, Chung CW, et al. Suppression of inflammation by a synthetic histone mimic. Nature 2010;468(7327):1119–23.
[149] Ramirez-Carrozzi VR, Nazarian AA, Li CC, Gore SL, Sridharan R, Imbalzano AN, et al. Selective and antagonistic functions of SWI/SNF and Mi-2β nucleosome remodeling complexes during an inflammatory response. Genes Dev 2006;20(3):282–96.
[150] Belkina AC, Nikolajczyk BS, Denis GV. BET protein function is required for inflammation: Brd2 genetic disruption and BET inhibitor JQ1 impair mouse macrophage inflammatory responses. J Immunol 2013;190(7):3670–8.
[151] Barrett E, Brothers S, Wahlestedt C, Beurel EO. I-BET151 selectively regulates IL-6 production. Biochim Biophys Acta (BBA) – Mol Basis Dis 2014;1842(9):1549–55.
[152] Wienerroither S, Rauch I, Rosebrock F, Jamieson AM, Bradner J, Muhar M, et al. Regulation of NO synthesis, local inflammation, and innate immunity to pathogens by BET family proteins. Mol Cell Biol 2014;34(3):415–27.
[153] Meng S, Zhang L, Tang Y, Tu Q, Zheng L, Yu L, et al. BET inhibitor JQ1 blocks inflammation and bone destruction. J Dental Res 2014;93(7):657–62.
[154] Qiao Y, Giannopoulou E-G, Chan C-H, Park SH, Gong S, Chen J, et al. Synergistic activation of inflammatory cytokine genes by interferon-γ-induced chromatin remodeling and toll-like receptor signaling. Immunity 2013;39(3):454–69.
[155] Chan CH, Fang C, Qiao Y, Yarilina A, Prinjha RK, Ivashkiv LB. BET bromodomain inhibition suppresses transcriptional responses to cytokine-Jak-STAT signaling in a gene-specific manner in human monocytes. Eur J Immunol 2015;45(1):287–97.
[156] Toniolo PA, Liu S, Yeh JE, Moraes-Vieira PM, Walker SR, Vafaizadeh V, et al. Inhibiting STAT5 by the BET bromodomain inhibitor JQ1 disrupts human dendritic cell maturation. J Immunol 2015;194(7):3180–90.

[157] Schilderink R, Bell M, Reginato E, Patten C, Rioja I, Hilbers FW, et al. BET bromodomain inhibition reduces maturation and enhances tolerogenic properties of human and mouse dendritic cells. Mol Immunol 2016;79:66–76.
[158] Chen J, Wang Z, Hu X, Chen R, Romero-Gallo J, Peek RM, et al. BET inhibition attenuates *Helicobacter pylori*–induced inflammatory response by suppressing inflammatory gene transcription and enhancer activation. J Immunol 2016;196(10):4132–42.
[159] Dooley KE, Warburton A, McBride AA. Tandemly integrated HPV16 can form a Brd4-dependent super-enhancer-like element that drives transcription of viral oncogenes. mBio 2016;7(5).
[160] Bandukwala HS, Gagnon J, Togher S, Greenbaum JA, Lamperti ED, Parr NJ, et al. Selective inhibition of $CD4^+$ T-cell cytokine production and autoimmunity by BET protein and c-Myc inhibitors. Proc Natl Acad Sci USA 2012;109(36):14532–7.
[161] Mele DA, Salmeron A, Ghosh S, Huang HR, Bryant BM, Lora JM. BET bromodomain inhibition suppresses TH17-mediated pathology. J Exp Med 2013;210(11):2181–90.
[162] Peeters JGC, Vervoort SJ, Tan SC, Mijnheer G, de Roock S, Vastert SJ, et al. Inhibition of super-enhancer activity in autoinflammatory site-derived T cells reduces disease-associated gene expression. Cell Rep 2015;12(12):1986–96.
[163] Belkina AC, Blanton WP, Nikolajczyk BS, Denis GV. The double bromodomain protein Brd2 promotes B cell expansion and mitogenesis. J Leukoc Biol 2014;95(3):451–60.
[164] Stanlie A, Yousif Ashraf S, Akiyama H, Honjo T, Begum Nasim A. Chromatin reader Brd4 functions in Ig class switching as a repair complex adaptor of nonhomologous end-joining. Mol Cell 2014;55(1):97–110.
[165] Michaeloudes C, Mercado N, Clarke C, Bhavsar PK, Adcock IM, Barnes PJ, et al. Bromodomain and extraterminal proteins suppress NF-E2-related factor 2-mediated antioxidant gene expression. J Immunol 2014;192(10):4913–20.
[166] Zhang G, Liu R, Zhong Y, Plotnikov AN, Zhang W, Zeng L, et al. Down-regulation of NF-κB transcriptional activity in HIV-associated kidney disease by BRD4 inhibition. J Biol Chem 2012;287(34):28840–51.
[167] Nadeem A, Al-Harbi NO, Al-Harbi MM, El-Sherbeeny AM, Ahmad SF, Siddiqui N, et al. Imiquimod-induced psoriasis-like skin inflammation is suppressed by BET bromodomain inhibitor in mice through RORC/IL-17A pathway modulation. Pharmacol Res 2015;99(0):248–57.
[168] Fu W, Farache J, Clardy SM, Hattori K, Mander P, Lee K, et al. Epigenetic modulation of type-1 diabetes via a dual effect on pancreatic macrophages and β cells. eLife 2014;3:e04631.
[169] Park-Min KH, Lim E, Lee MJ, Park SH, Giannopoulou E, Yarilina A, et al. Inhibition of osteoclastogenesis and inflammatory bone resorption by targeting BET proteins and epigenetic regulation. Nat Commun 2014;5.
[170] Zhang QG, Qian J, Zhu YC. Targeting bromodomain-containing protein 4 (BRD4) benefits rheumatoid arthritis. Immunol Lett 2015;166(2):103–8.
[171] Klein K, Kabala PA, Grabiec AM, Gay RE, Kolling C, Lin LL, et al. The bromodomain protein inhibitor I-BET151 suppresses expression of inflammatory genes and matrix degrading enzymes in rheumatoid arthritis synovial fibroblasts. Ann Rheum Dis 2014;75(2):422–9.

CHAPTER

16

Treatment of Autoimmune Diseases and Prevention of Transplant Rejection and Graft-Versus-Host Disease by Regulatory T Cells: The State of the Art and Perspectives

Giuseppe Nocentini, Luigi Cari, Graziella Migliorati, Carlo Riccardi
University of Perugia, Perugia, Italy

OUTLINE

The Epigenetics of Autoimmunity. http://dx.doi.org/10.1016/B978-0-12-809912-4.00016-7

16.1 INTRODUCTION

Disruption of immune homeostasis not only causes autoimmune diseases, but it also leads to inflammatory diseases, chronic infection, cardiovascular diseases, obesity, and cancer. Immune homeostasis can be defined as the dynamic balance between the activating and inhibitory effects of immune cells on the immune system. Cells of the innate immune system [i.e., neutrophils, monocytes, macrophages, dendritic cells (DCs), eosinophils, mast cells, basophils, and natural killer cells], cells of the adaptive immune system (i.e., T and B lymphocytes), and endothelial cells participate in the activation of immune or inflammatory responses. Furthermore, subsets of cells belonging to the innate and adaptive immune systems also participate in the negative regulation of the immune response. Several subsets of cells participate in immune homeostasis, including non-αβ $CD4^+$ T lymphocytes (e.g., γδ $CD4^+$ regulatory T cells (T_{regs}), $CD4^-CD8^-$ T cells, natural killer T cells, and $CD8^+$ T_{regs}), regulatory B cells, macrophages (M2 macrophages), and DCs (tolerogenic DCs). However, the most studied and possibly the most active subsets involved in the regulation of the immune response are αβ $CD4^+$ T_{regs}. Several studies have reported that the development and maintenance of autoimmune diseases, such as rheumatoid arthritis, systemic lupus erythematosus (SLE), type 1 diabetes, and relapsing–remitting multiple sclerosis, is associated with a decrease in the T_{reg} population or a defect in the suppressive activity of T_{regs} [1–6].

Partial resistance of conventional T cells to T_{regs} sometimes associated with an increase in the number of fully active T_{reg} subsets is observed in certain patients with autoimmune diseases [4,6–8]. The increase in the number of T_{reg} subsets appears to play a role via partial control of autoreactive conventional T cells, as patients with an increased number of T_{regs} had lower disease activity [7,9].

Thus in cases where there is a decrease in T_{reg} number or activity or where there is resistance to T_{reg} suppression by autoreactive conventional T cells, an increase in the number of fully active T_{regs} may prove beneficial. The number of T_{regs} can be increased by treatment with purified or ex vivo-differentiated T_{regs}, provided that the T_{regs} are infused in sufficient quantities. Moreover, the T_{regs} administered should not impede the normal activity of the immune system (e.g., by inducing generalized immunosuppression or tumor growth). Therefore it may be more appropriate to use antigen-specific T_{regs} compared to polyclonal T_{regs}. Moreover, a stable phenotype of T_{regs} is very important as the differentiation of T_{regs} into conventional T cells over time may worsen the disease prognosis.

This chapter reviews the methods that have been used to increase the number of T_{reg} subsets in patients with autoimmune diseases and patients who have undergone organ or bone marrow transplantation. The benefits and limitations of these methods are discussed and the research questions that still need to be answered are presented. The findings may be beneficial as a guide for future research into T_{regs} and their treatment benefits.

16.2 T_{REG} SUBSETS

"Suppressor T cells" were first reported in the late 1960s and were defined as a specialized subset of lymphocytes that limited the immune response [10]; however, some contradictory reports were published in the early 1980s and studies on "suppressor T cells" were discontinued within a few years [11]. The subsequent identification and characterization of $CD4^+CD25^+$ T cells in healthy mice, and the discovery of the transcription factor forkhead box P3 (Foxp3) as a marker of $CD4^+CD25^+$ T cells [12–14], opened up this field of study again and T_{regs} now have a prominent position in immune research. T_{regs} isolated in vivo are classified as thymus-derived T_{regs} (tT_{regs}) or peripherally derived T_{regs} (pT_{regs}) and are collectively referred to as natural T_{regs} (nT_{regs}). T_{regs} differentiated ex vivo are called induced T_{regs} (iT_{regs}) and are generated in the laboratory. This group of T_{regs} is discussed in detail in Section 16.4.2.

16.2.1 Thymus-Derived T_{regs}

tT_{regs} develop in the thymus as a distinct lineage via a two-step process [15,16]. The first signaling event is driven by the T cell receptor (TCR), through relatively strong stimulation, and its coreceptor CD28, which leads to expression of the IL-2 receptor (CD25) and chromatin remodeling at the Foxp3 locus (T_{reg} progenitors). The genome organizer Satb1 favors activation of the tT_{reg}-specific super-enhancer in this phase, which represents a crucial epigenetic event in the development of tT_{regs} [17]. The second signaling event is TCR independent and leads to Foxp3 expression following triggering of glucocorticoid-induced TNF receptor-related (GITR, also called TNFRSF18), OX40 (TNFRSF4, CD134), and TNFR2 [18]. The phenotype and activity of tT_{regs} is attributable to a specific hypomethylation pattern (see Section 16.5.1) and the expression of the transcription factors Foxp3 and IKAROS family zinc finger 2 (Helios) [19,20].

Human tT_{regs} were originally described as $CD25^+$ or $CD25^{high}$, but are currently described as $Foxp3^{low}CD45RA^+CD25^{low}$ in the naive or resting form and $Foxp3^{high}CD45RA^-CD25^{high}$ [conventional T_{regs} or effector T_{regs} (eT_{regs})] after antigenic stimulation. Moreover, tT_{regs} are characterized by many other surface markers including GITR (expressed in most murine $CD4^+CD25^+Foxp3^+$ cells and in a smaller percentage of human $CD4^+CD25^+Foxp3^+$ cells), OX40, CTLA-4, CD103, LAG-3, CD39, LAP, PD-1, and PD-L1 [4,21]. It is yet to be established whether differences in the expression of these markers define different tT_{reg} subsets or tT_{reg} cells that have a specific activity at a specific time. For example, murine tT_{regs} have recently been divided into $triple^{high}$ ($CD4^+Foxp3^+GITR^{high}PD\text{-}1^{high}CD25^{high}$) and $triple^{low}$ ($CD4^+Foxp3^+GITR^{low}PD\text{-}1^{low}CD25^{low}$) cells [22]. $Triple^{high}$ T_{regs} are highly self-reactive and control lymphoid proliferation in peripheral lymph nodes, while triple^{low} are less self-reactive and promote the conversion of CD4$^+$ conventional T cells into pT$_{regs}$. High levels of CD127 expression are typically observed in activated CD4$^+$CD25$^+$ T cells but not tT$_{regs}$ [23], therefore tT$_{regs}$ are defined as CD4$^+$CD25$^{+/high}$ CD127$^{-/low}$.

tT$_{regs}$ are required to control immune homeostasis but they are not capable of suppressing chronic inflammation or autoimmunity on their own. They are specific for self-antigens expressed within the thymus, are capable of expansion, and exert regulatory effects via cell-to-cell contact, but they also produce cytokines with immunosuppressive effects [24]. tT$_{regs}$ have recently been shown to play a crucial role in tissue repair and maintenance, at least in muscles and lungs [25–27]. This mechanism is distinct from their role in the suppression of immune responses and inflammation, is elicited by IL-18 and IL-33, and mainly consists of the production of amphiregulin.

16.2.2 Peripherally Derived T$_{regs}$

pT$_{regs}$ are derived from activated CD4$^+$ conventional T cells in the presence of the appropriate microenvironmental signals. The cytokines involved in the induction of pT$_{regs}$ are transforming growth factor (TGF)-β and interleukin (IL)-10 [28–31]. Unlike tT$_{regs}$, pT$_{regs}$ may not express CD25 and FoxP3. Moreover, they are defined as memory T cells (CD45RA$^-$) and are specific to the self-antigens expressed outside the thymus and to foreign antigens. They do not have a homogeneous phenotype or identical functions, and the best characterized pT$_{regs}$ are T helper (Th)3 cells, T regulatory type 1 (Tr1) cells, and CD25$^{-/low}$Foxp3$^{-/low}$GITR$^+$ (GITR single-positive, GITRsp) cells.

Th3 cells were originally described in mice fed low doses of myelin basic protein in which tolerance to the antigen was induced [32]. Th3 cells express Foxp3, and their regulatory activity is mainly due to the production of TGF-β, which is in turn responsible for Th3 differentiation [32,33]. Tr1 cells were first described in severe combined immunodeficient (SCID) patients who had developed long-term tolerance to stem cell allografts [34]. Tr1 cells do not express Foxp3 and secrete high levels of IL-10, which is the main cytokine responsible for Tr1 differentiation [35]. CD49b and LAG-3 were recently identified as specific markers of Tr1 cells [36], but another studies have since indicated the existence of other Tr1 markers [37]. Finally, GITRsp cells express low levels of Foxp3 and produce both TGF-β and IL-10 [38]. The number of GITRsp cells is increased in patients with autoimmune diseases and low disease activity, which may represent an attempt to maintain homeostasis [7,9].

In addition to LAG-3 and GITR, many other surface markers (i.e., CTLA-4, CD39, CD73, and ICOS) may characterize pT$_{regs}$ [4,24]. The only clear difference between pT$_{reg}$ subsets is the expression of Foxp3, which is expressed by Th3 cells but expressed at much lower levels by Tr1 and GITRsp cells. However, most of the other markers overlap and the cytokines involved in suppression are the same. Therefore the distinction between subsets is only interesting from a methodological viewpoint (ex vivo differentiation and isolation) and does not necessarily indicate distinguishable in vivo subsets with distinct functions.

Overall, pT$_{regs}$ are required for the control of immune homeostasis and are crucial in the development of tumors [39,40]. They exert regulatory activity by secreting immunosuppressive cytokines (e.g., IL-10 and TGF-β) and express molecules that are involved in the inhibition of cell-to-cell contact [24].

16.2.3 Distinguishing Between Different Types of T_{regs}

T_{reg} markers (i.e., CD25 and GITR) are also expressed by activated $CD4^+$ conventional T cells, which complicates the purification of T_{regs} from patients or "healthy donors" in whom the immune system has been activated for asymptomatic infections or autoimmune diseases. In this instance, there is a risk of isolating more activated conventional T cells than T_{regs}, particularly when CD25 is used as the only marker; for example, one study showed that in approximately 50% of the "healthy donors," $CD4^+CD25^+$ cells did not exhibit regulatory activity [41]. Even Foxp3 can be expressed by human (but not murine) conventional T cells following activation [14,42,43], but these cells are not suppressive and easily lose Foxp3 expression.

Many T_{reg} surface markers are shared by tT_{regs} and pT_{regs}, and there is currently no clear evidence to show that pT_{regs} are physiologically produced by the human immune system [20]. Therefore from a therapeutic perspective it would be more feasible to focus on the stability, activity, and function of T_{reg} subsets rather than on their derivation. For example, Smigiel et al. have divided murine T_{regs} into $CD4^+Foxp3^+CD44^{lo}CD62L^+CCR7^{hi}$ central T_{regs} (cT_{regs}) and $CD4^+Foxp3^+CD44^{hi}CD62L^{lo/-}CCR7^{lo}$ $ICOS^+$ effector T_{regs} (eT_{regs}) [44]. cT_{regs} reside in secondary lymphoid tissues and are IL-2 dependent, while eT_{regs} reside in nonlymphoid tissues, are IL-2 independent, and need TCR and costimulatory signals. Interestingly, tissue-resident T_{regs} comprise 30%–60% of the total number of $CD4^+$ lymphocytes infiltrating a tissue and display a tissue-specific TCR repertoire and a unique phenotype [45]. Despite the difficulties in their investigation, they appear to be crucial for the control of immune responses.

Miyara et al. recently demonstrated that CD15s (Sialyl Lewis x) is highly specific for activated, terminally differentiated, and highly suppressive $Foxp3^{high}$ human eT_{regs} [46]. Moreover, T_{regs} can express transcription factors that are normally expressed by specific Th subsets at sites of inflammation and therefore produce specific cytokines and receptors [47].

Therefore there are no highly specific markers for each T_{reg} subset. Nonetheless, certain markers can be used to isolate T_{regs} with regulatory activities, in which case it is necessary to understand the degree of stability and regulatory activity of the sorted T_{regs} (Section 16.5 and Tables 16.1–16.3).

16.3 IN VIVO EXPANSION OR DIFFERENTIATION OF T_{REGS}

Administration of T_{regs} to patients with autoimmune disease following their isolation (or induction) and expansion (presented and discussed in Section 16.4) is expensive and associated with a certain degree of risk, such as the possibility of contamination. An alternative approach is to treat patients with drugs that promote the in vivo expansion and differentiation of T_{reg} populations. The cost would be lower than that of T_{reg} therapy, and the treatment could be repeated if required. Moreover, this approach may be interesting not only as a monotherapy but also in association with T_{reg} therapy to stabilize or induce the expansion of infused cells.

16.3.1 Effects of Old or Nonbiotechnological Drugs

Drugs such as glucocorticoids, rapamycin and its analogues, retinoic acid (vitamin A), and vitamin D exert immunosuppressive effects through several mechanisms, including the expansion of T_{regs} [93–96]. Moreover, when IL-10 or IL-10-inducing agents are used in combination with standard

TABLE 16.1 Effect of tT_{reg} Infusion in Representative Models of graft versus host disease (GVHD), Graft Rejection, and Autoimmune Disease

Model	Drug	Phenotype of sorted cells	Treatment for in vitro expansion	Phenotype of infused cells	Effect	References
GVHD in mice	Expanded polyclonal tT_{regs}	$CD4^+$ $CD25^+$	Anti-CD3 IL-2	$CD4^+$ $CD25^+$	Significant inhibition of rapidly lethal GVHD	Taylor et al. [70]
Heart graft in mice	Expanded donor alloantigen-specific tT_{regs} (4C TCR transgenic)	4C TCR transgenic $CD4^+$ $CD25^+$ $CD62L^{high}$	Anti-CD3 Anti-CD28 IL-2	4C TCR transgenic $CD4^+$ $CD25^+$ $CD62L^{high}$	Expanded allo-tT_{regs} were able to significantly prolong the survival of cardiac grafts	Brennan et al. [71]
Heart graft in mice	Expanded polyclonal tT_{regs}	$CD4^+$ $CD25^+$ $CD62L^{high}$	Anti-CD3 Anti-CD28 IL-2	$CD4^+$ $CD25^+$ $CD62L^{high}$	Expanded polyclonal tT_{regs} had no effect on graft survival	Brennan et al. [71]
Heart graft in cynomolgus monkeys	Expanded donor alloantigen-specific tT_{regs}	$CD4^+$ $CD25^+$ $CD127^-$	Artificial APC	$CD4^+$ $CD25^+$ $CD127^-$ $Foxp3^+$	tT_{regs} failed to prolong heart graft survival	Ezzelarab et al. [72]
Kidney graft in cynomolgus monkeys	Expanded donor alloantigen-specific tT_{reg}	$CD4^+$ $CD25^+$	DC Anti-CD3 IL-2	$CD4^+$ $CD25^+$	Expanded donor alloantigen specific tT_{regs} combined with sirolimus and ATG prolongs renal allograft survival	Ma et al. [73]
Skin graft in humanized mice	Expanded polyclonal tT_{regs}	$CD4^+$ $CD25^+$ $CD127^{low}$	Anti-CD3 Anti-CD28 IL-2	$CD4^+$ $CD25^+$ $CD127^{low}$	tT_{regs} favored a long-term human skin transplant survival along with a reduction in the $CD8^+$ human cellular graft infiltrate	Issa et al. [74]
Skin graft in humanized mice	Expanded donor alloantigen-specific tT_{regs}	$CD4^+$ $CD25^+$ $Foxp3^+$	APC Anti-CD3 Anti-CD28 IL-2	$CD4^+$ $CD25^+$ $Foxp3^+$	tT_{regs} given at a 1:5 T_{reg}:T effector ratio induced 100% long-term graft survival	Landwehr-Kenzel et al. [75]
Skin graft in humanized mice	Expanded polyclonal tT_{regs}	$CD4^+$ $CD25^+$ $Foxp3^+$	Anti-CD3 Anti-CD28 IL-2	$CD4^+$ $CD25^+$ $Foxp3^+$	tT_{regs} given at 1:5 T_{reg}:T effector ratio delayed rejection by >1 week. However, at day 37 all grafts were fully rejected. tT_{regs} given at 1:1 T_{reg}:T effector ratio promoted graft survival >100 days	Landwehr-Kenzel et al. [75]

Arteriosclerosis of human arteries in humanized mice	Expanded polyclonal tT_{regs}	$CD4^+$ $CD25^+$ or $CD4^+$ $CD25^+$ $CD127^{low}$	Anti-CD3 Anti-CD28 IL-2	$CD4^+$ $CD25^+$ $CD127^{low}$ and $CD4^+$ $CD25^{bright}$	Development of arteriosclerosis in human arteries was prevented with the infusion of tT_{regs}. $CD4^+CD25^+CD127^{high}$ tT_{regs} provide a more potent therapy than $CD4^+CD25^+$ tT_{regs}	Nadig et al. [76]
2,4,6-trinitro-benzene sulfonic acid (TNBS)-mediated colitis in mice	Antigen-specific tT_{reg} (transgenic mice expressing TNP-specific CAR)	TNP-specific $CD4^+$ $CD25^+$	Not expanded	TNP-specific $CD4^+$ $CD25^+$	TNP-specific CAR tT_{regs} was associated with significant amelioration of colitis and improved survival	Elinav et al. [77]
Myelin basic protein (MBP)-induced EAE in mice	Expanded or not expanded antigen-specific tT_{reg} (MBP)	$CD4^+$ $CD25^+$	Anti-CD3 Anti-CD28 IL-2	$CD4^+$ $CD25^+$	Antigen-specific tT_{regs} from TCR transgenic mice, but not polyclonal tT_{regs}, protect against EAE	Stephens et al. [78]
Spontaneous diabetes in NOD mice	Expanded donor alloantigen-specific from BDC2.5 TCR transgenic mice tT_{regs}	$CD4^+$ $CD25^+$	DC Anti-CD3 IL-2	$CD4^+$ $CD25^+$	5000 antigen-specific tT_{regs} inhibited diabetes development	Tarbell et al. [79]
Spontaneous diabetes in NOD mice	Expanded polyclonal tT_{regs}	$CD4^+$ $CD25^+$	Anti-CD3 IL-2	$CD4^+$ $CD25^+$	Polyclonal tT_{regs} failed to inhibit diabetes development even when used at high doses (10^5 cells)	Tarbell et al. [79]
Induced diabetes in $NOD.RAG^{-/-}$ mice	Expanded donor alloantigen-specific from BDC2.5 TCR transgenic mice tT_{regs}	$CD4^+$ $CD25^+$ $CD62L^+$	DC Anti-CD3 Anti-CD28 IL-2	$CD4^+$ $CD25^+$ $CD62L^+$	Antigen-specific tT_{regs} can cure diabetes after disease onset	Tang et al. [80]

TABLE 16.2 Effects of iT_{reg} Infusion in Representative Models of GVHD, Graft Rejection, and Autoimmune Disease

Murine model	Drug	Phenotype of sorted cells	In vitro treatment (induction)	Phenotype of infused cells	Effect	References
GVHD	Alloantigen-specific iTh3 cells	$CD4^+$	APC TGF-β	—	iTh3 cells failed to prevent lethal GVHD (2 of 15 mice survived)	Zeller et al. [48]
GVHD	Alloantigen-specific iTr1 cells	$CD4^+$	APC IL-10	—	iTr1 cells failed to prevent lethal GVHD (no survivors)	Zeller et al. [48]
GVHD	Alloantigen-specific iT_{regs}	$CD4^+$	APC IL-10 TGF-β	—	iT_{reg} inhibited lethal GVHD (11 out of 15 mice survived). Survivors were not entirely GVHD free by histological criteria	Zeller et al. [48]
GVHD	Donor-alloantigen-specific iTh3 cells	$CD4^+$ $Foxp3^-$	DC IL-2 TGF Retinoic acid	$CD4^+$ $Foxp3^+$	iTh3 cells failed to prevent lethal GVHD	Koenecke et al. [49]
GVHD	Donor-alloantigen-specific iTh3 cells	$CD4^+$ $CD25^-$ $Foxp3^-$	$CD11c^+$ DC TGF-β Retinoic acid	$CD4^+$ $CD25^+$ $Foxp3^+$	iTh3 cells prevented GVHD caused by cotransferred $CD45RB^{high}$ T cells. Similar findings were made when iTh3 cells were transferred after onset of GVHD	Sela et al. [50]
GVHD	Antigen-specific iTh3 cells [ovalbumin, (OVA) peptide]	$CD4^+$ $CD25^-$	OVA peptide APC IL-2 TGF-β	$CD4^+$ $CD25^+$	OVA-specific iT_{regs} were able to prevent lethal GVHD induced by polyclonal T cells in OVA-transgenic, but not in WT recipient mice	Semple et al. [51]
GVHD	iTh3 cells	$CD4^+$ $Foxp3^-$	Anti-CD3 Anti-CD28 IL-2 TGF-β	$CD4^+$ $Foxp3^+$	iTh3 cells failed to protect mice from lethal GVHD even when administered at high T_{reg}-to-conventional T cell ratios	Beres et al. [52]
GVHD	iTh3 cells	$CD4^+$ splenocytes	Anti-CD3 Anti-CD28 IL-2 TGF-β Rapamycin	79% was $CD4^+$ $CD25^+$	iTh3 cells exacerbated GVHD and downregulated Foxp3 expression in iTh3. When iTh3 transfused mice were treated with rapamycin, GVHD was prevented in almost all mice	Zhang et al. [53]

GVHD	Antigen-specific iTh3 cells (HY peptide)	$CD4^+$ $CD25^-$	HY peptide APC IL-2 TGF-β Retinoic acid	$CD4^+$ $CD25^+$	iTh3 cells prevented GVHD rejection of male (HY^+) but not female (HY^-) mice	Li et al. [54]
GVHD	Donor-alloantigen-specific iTh3 cells	$CD4^+$ $CD25^-$	$CD11c^+$ DC IL-2 TGF-β Retinoic acid	85% was $CD4^+$ $CD25^+$	iTh3 cells were effective in preventing GVHD, but abrogated the graft-versus-leukemia effect against aggressive leukemia	Heinrichs et al. [55]
Pancreatic islet transplant	Polyclonal iTr1 cells	$CD4^+$	Anti-CD3 Anti-CD28 IL-2 IL-10	20%–25% was iTr1	iTr1 cells prevented transplant rejection in 5 out of 8 mice until 25 days after transplant. Long-term graft survival (100 days after transplant) was observed in 50% of mice receiving iTr1 cells	Gagliani et al. [56]
Skin graft	Antigen-specific iTh3 cells (HY peptide)	$RAG1^{-/-}$ Marilyn. $Foxp3^{hCD2}$ splenocytes	HY peptide DC TGF-β	—	iTh3 only prevented skin-graft rejection in recipients expressing the HY peptide	Hilbrands et al. [57]
Lupus-like syndrome	iTh3 cells	$CD4^+$ $CD25^-$	Antigen presenting cells (APC) IL-2 TGF-β	$CD4^+$ $CD25^+$	iTh3 cells prevented from induction of lymphoid hyperplasia, B-cell activation, and immune complex glomerulonephritis	Zheng et al. [58]
Lupus-like syndrome	Polyclonal iTh3 cells	Naive $CD4^+$ $CD62L^+$ $CD25^-$ $CD44^{low}$	Anti-CD3 Anti-CD28 IL-2 TGF-β	—	Polyclonal iTh3 cells suppressed a lupus-like chronic GVHD	Lan et al. [59]
$CD4^+CD45RB^{high}$ induced colitis in SCID mice	Antigen-specific iTr1 cells (OVA peptide)	OVA-specific $CD4^+$	OVA peptide APC Anti-CD3 Il-2 IL10	—	iTr1 prevented colitis	Groux et al. [60]
Inflammatory bowel disease	Antigen-specific iTr1 cells (OVA peptide)	$CD4^+$	OVA peptide APC IL-10	—	Late injection of iTr1 cells efficiently cured ongoing bowel inflammation	Foussat et al. [61]

(*Continued*)

TABLE 16.2 Effects of iT_{reg} Infusion in Representative Models of GVHD, Graft Rejection, and Autoimmune Disease (*Cont.*)

Murine model	Drug	Phenotype of sorted cells	In vitro treatment (induction)	Phenotype of infused cells	Effect	References
Oxazolone-induced skin inflammation	Antigen-specific iTr1 cells (OVA peptide)	$CD4^+$	OVA peptide APC IL-10	—	Late injection of iTr1 cells efficiently cured ongoing skin inflammatory responses	Foussat et al. [61]
Autoimmune gastritis	Antigen-specific iTh3 cells (TxA23 antigen)	TxA23 transgenic mice $CD4^+$ $CD8^-$ $CD25^-$	Anti-CD3 Anti-CD28 IL-2 TGF-β	$CD4^+$ $CD8^-$ $CD25^+$	Cotransfer of iTh3 cells and TxA23 antigen prevented the development of autoimmune gastritis	Huter et al. [62]
Spontaneous diabetes in nonobese diabetic (NOD) mice	Antigen-specific iTh3 cells (islet-specific)	$CD4^+$	Anti-CD3 TGF-β Anti-IFN-γ	—	iTh3 cells prevented spontaneous development of type 1 diabetes in NOD mice and inhibited the development of pancreatic infiltrates and disease onset orchestrated by Th1 effectors	Weber et al. [63]
Experimental autoimmune encephalomyelitis (EAE)	Antigen-specific iTr1 cells (OVA peptide)	Naive $CD4^+$ $CD62L^{high}$	OVA peptide APC Anti-CD3 Anti-CD28 Dexamethasone Vitamin D3 Anti–IL-4 Anti–IFN γ Anti–IL-12	—	One day before cell transfer, OVA adsorbed to alum was injected intracranially to provide a reservoir of antigen. iTr1 cells were able to prevent EAE with absolute abrogation of disease onset and the absence of clinical signs	Barrat et al. [64]
EAE	iTh3 cells	$CD4^+$ $Foxp3^-$	Anti-CD3 Anti-CD28 IL-2 TGF-β	$CD4^+$ $Foxp3^+$	iTh3 cells suppressed EAE by an IL-10-dependent manner through the induction of anergy in effector T cells and the induction of protective T lymphocytes	Selvaraj and Geiger [65]

EAE	Antigen-specific iTr1 cells (MOG peptide)	Splenocytes	MOG peptide Anti-CD3 Anti-CD28 CXCL12-Ig IL-12 Anti-IL-4	23% was $CD4^+$ $CD25^-$ $Foxp3^-$ $IL\text{-}10^{high}$	iTr1 treated mice went into remission within 4–5 days	Meiron et al. [66]
EAE	iTh3 cells	$CD4^+$ $Foxp3^-$	Anti-CD3 Anti-CD28 IL-2 TGF-β	$CD4^+$ $Foxp3^+$	iTh3 cells protected against the development of EAE	O'Connor et al. [67]
Autoimmune arthritis	Antigen-specific iTh3 cells (type II collagen)	Naive $CD4^+$ $CD62L^+$ $CD25^-$ $CD44^{low}$ splenocytes	Type II collagen IL-2 TGF-β	—	iTh3 cells transiently decreased arthritis severity and decreased the number of $IL\text{-}17^+$ T cells	Kong et al. [68]
Autoimmune arthritis	Antigen-specific iTr1 cells (type II collagen)	Splenocytes	Type II collagen IL-10 Anti-IL-4	—	iTr1 reduced the incidence and clinical symptoms of arthritis in both preventive and curative settings (with a significant impact on type II collagen antibodies). iTr1 cells also decreased the proliferation of antigen-specific effector T cells	Asnagli et al. [69]

TABLE 16.3 Effect of T_{reg} Infusion in Patients (Published and Ongoing Studies)

PI and institution	Indication	Drug	Phenotype of sorted cells	Phenotype of infused cells	Effect	References/study IDs
Martelli, University of Perugia	GVHD prophylaxis	Polyclonal tT_{regs}	$CD4^+$ $CD25^+$	93% was $CD4^+$ $CD25^+$ 69% was $CD4^+$ $CD25^+$ $Foxp3^+$ 17% was $CD4^+$ $CD25^+$ $CD127^+$	No GVHD was observed in 17 out of 20 patients. Two patients developed grade I cutaneous self-limited untreated GVHD and one patient developed grade III GVHD. Absence of posttransplantation immunosuppression with improved immunity against opportunistic pathogens. A full graft-versus-leukemia effect was observed	Di Ianni et al. [81]
Brunstein, University of Minnesota	GVHD prophylaxis	Polyclonal expanded tT_{regs}	$CD4^+$ $CD25^+$	86% was $CD4^+$ $CD25^+$ 64% was $CD4^+$ $CD127^-$ $Foxp3^+$	The incidence of grade II–IV acute GVHD was reduced (43% vs. 61%). No deleterious effect on the risks of infection, relapse, or early mortality was observed	Brunstein et al. [82]/ID: NCT00602693
Martelli, University of Perugia	GVHD prophylaxis	Polyclonal tT_{regs}	$CD4^+$ $CD25^+$	81% was $Foxp3^+$ 11% was $CD127^+$	95% of patients achieved full donor-type engraftment. No posttransplant immunosuppression was given. 15% developed grade 2–4 acute GVHD. The very low cumulative incidence of relapse (0.05) was significantly better than that in the historical controls	Martelli et al. [83]
Roncarolo, San Raffaele University	GVHD prophylaxis	iTr1 (IL10-DLI)	PBMC	—	Fast immune reconstitution was observed in 5 out of 12 patients. In 4 out of the 5 patients who are alive, complete remission was observed and they were free of immunosuppression at 7.2 years (median follow-up time) after haplohematopoietic stem cell transplantation. Transient GVHD was observed in the patients	Bacchetta et al. [84]/ID: ALT-TEN

Meyer, Stanford University Hospitals and Clinics	Acute GVHD treatment	Polyclonal tT_{regs}	$CD4^+$ $CD25^+$ $Foxp3^+$	—	Estimated primary completion date: October 2017	ID: NCT02526329
Trzonkowski, Medical University of Gdańsk	Chronic GVHD treatment	Polyclonal expanded tT_{regs}	$CD4^+$ $CD25^{high}$ $CD127^-$	—	In the 2 treated patients, significant alleviation of the symptoms and reduction of pharmacologic immunosuppression were observed	Trzonkowski et al. [85]/ID: NKEBN/458-310/2008
Bornhäuser, DFG Center for Regenerative Therapies	Chronic GVHD treatment	Polyclonal expanded tT_{regs}	$CD4^+$ $CD25^{high}$	84% was $CD4^+$ $CD25^{high}$ $cd127^{low}$ $Foxp3^+$	Two of the five patients showed improvement of chronic GVHD. The other 3 patients showed stable chronic GVHD symptoms for up to 21 months. Immunosuppressive treatment could be reduced. One patient developed a malignant melanoma and another patient had Bowen skin cancer at 4 months and 11 months after T_{reg} transfusion, respectively	Theil et al. [86]/ID: EK206082008
Baron, University Hospital of Liege	Chronic GVHD treatment	Polyclonal tT_{regs}	$CD4^+$ $CD25^+$ $Foxp3^+$	$CD4^+$ $CD25^+$ $Foxp3^+$	Estimated study completion date: December 2016	ID: NCT01903473
Lacerda, Instituto de Medicina Molecular	Steroid-refractory chronic GVHD	Polyclonal tT_{regs}	$CD4^+$ $CD25^+$	—	Estimated primary completion date: March 2019	ID: NCT02385019
Koreth, Dana-Farber Cancer Institute	Steroid-refractory chronic GVHD	Polyclonal tT_{regs}	$CD4^+$ $CD25^+$	—	Estimated study completion date: July 2018	ID: NCT01937468
Johnston, Stanford University	Steroid-refractory chronic GVHD	Polyclonal tT_{regs}	$CD4^+$ $CD25^+$	—	Estimated primary completion date: January 2016	ID: NCT01911039

(Continued)

TABLE 16.3 Effect of T_{reg} Infusion in Patients (Published and Ongoing Studies) (*Cont.*)

PI and institution	Indication	Drug	Phenotype of sorted cells	Phenotype of infused cells	Effect	References/study IDs
Lombardi, King's College	Liver transplantation	Polyclonal tT_{regs}	—	—	Estimated primary completion date: June 2019	Trzonkowski et al. [87]/ID: ThRIL NCT02166177
Feng, Bluestone, Kang, and Tang, University of California	Liver transplantation	Donor-alloantigen reactive T_{regs}	—	—	Estimated primary completion date: January 2022	Trzonkowski et al. [87]/ID: NCT02188719
Feng, University of California	Liver transplantation	Donor-alloantigen reactive T_{regs}	—	—	Estimated primary completion date: April 2017	Trzonkowski et al. [87]/ID: NCT02474199
Todo, Hokkaido University Graduate School of Medicine	Liver transplantation	iT_{regs}	Enriched lymphocytes	28% was $CD4^+$ $CD25^+$ $Foxp3^+$ 26% was $CD4^+$ $CD127^{low}$ $Foxp3^+$	No significant adverse events were observed; 7 out of 10 patients were successfully weaned off treatment and immunosuppressive therapy was discontinued. The other 3 patients developed mild rejection during weaning, after which conventional low-dose immunotherapy was resumed	Todo et al. [88]
Wang, Nanjing Medical University	Liver transplantation	Donor-alloantigen reactive tT_{regs}	$CD4^+$ $CD25^+$ $CD127^-$	$CD4^+$ $CD25^+$ $CD127^-$	Estimated primary completion date: December 2015	ID: NCT01624077
Lombardi, King's College Hospital NHS Trust	Kidney transplantation	Polyclonal expanded tT_{regs}	$CD4^+$ $CD25^+$	—	Estimated primary completion date: December 2016	Geissler [89]/ID: NCT02129881
Reinke, Charite University	Kidney transplantation	Polyclonal expanded tT_{regs}	$CD4^+$ $CD25^+$ $Foxp3^+$	—	Estimated primary completion date: April 2017	Geissler [89]/ID: NCT02371434
Kang, University of California	Kidney transplantation	Donor-alloantigen reactive T_{regs}	—	—	Estimated primary completion date: December 2017	Geissler [89]/ID: NCT02244801

Markmann, Massachusetts General Hospital	Kidney transplantation	Drug-conditioned iT_{regs}	PBMC	—	Estimated primary completion date: May 2017	Geissler [89]/ID: NCT02091232
Vincenti, University of California	Subclinical rejection in kidney transplantation	Polyclonal expanded tT_{regs}	$CD4^+$ $CD25^+$ $CD127^{low/-}$	—	Estimated primary completion date: December 2016	Trzonkowski et al. [87]/ID: NCT02088931
Skaro, Northwestern University	Kidney transplantation	Polyclonal expanded tT_{regs}	$CD4^+$ $CD25^+$	—	Estimated primary completion date: December 2020	ID: NCT02145325
Trzonkowski, Medical University of Gdansk	Recent type 1 diabetes	Polyclonal expanded tT_{regs}	$CD3^+$ $CD4^+$ $CD25^{high}$ $CD127^-$	>90% was $CD3^+$ $CD4^+$ $CD25^{high}$ $CD127^{low}$ $Foxp3^+$	At 6 months after diabetes onset (4–5 months after T_{reg} infusion), 8 out of 10 patients still required 0.5 UI/kg per day of insulin (2 patients did not need insulin)	Marek-Trzonkowska et al. [90]/ID: ISRCTN06128462
Tang, University of California	Type 1 diabetes	Polyclonal expanded tT_{regs}	$CD4^+$ $CD25^+$ $CD127^{low/-}$	92% was $Foxp3^+$	No infusion reactions or cell therapy-related high-grade adverse events were observed. C-peptide levels persisted up to 2+ years after transfer in more than 50% of patients	Bluestone et al. [91]/ID: NCT01210664
Colombel, Huriez Hospital	Refractory Crohn's disease	Antigen-specific iTr1	PBMC	OVA-Tr1	After treatment (5 weeks), clinically significant improvements were observed in 6 of the 8 patients who received 10^6 cells. The clinical effect was maximal at 5 weeks posttreatment and subsequently reduced progressively	Desreumaux et al. [92]/ID: CATS1
Chwojnicki, Medical University of Gdansk	Multiple sclerosis	Polyclonal expanded tT_{regs}	$CD4^+$ $CD25^+$ $CD127^-$	—	—	Trzonkowski et al. [87]/ID: VAC2.0EudraCT: 2014-004320-22

(Continued)

TABLE 16.3 Effect of T_{reg} Infusion in Patients (Published and Ongoing Studies) (*Cont.*)

PI and institution	Indication	Drug	Phenotype of sorted cells	Phenotype of infused cells	Effect	References/study IDs
Dall'Era, Haemel, Bluestone, Rosenblum, and Wofsy, University of California	Systemic lupus Erythematosus	Polyclonal expanded tT_{regs}	$CD4^+$ $CD25^+$ $CD127^{low/-}$	—	Estimated primary completion date: December 2018	ID: NCT02428309
Bodaghi, Assitance publique - Hopitaux de Paris	Uveitis	Polyclonal expanded tT_{regs}	$CD4^+$ $CD25^+$ $Foxp3^+$	—	Estimated study completion date: December 2017	ID: NCT02494492

immunosuppressive treatments, Tr1 cells can be generated in murine models of autoimmunity or allogeneic transplantation [35].

The main mechanism of action of fingolimod, an immunosuppressive and antiinflammatory drug, is the trapping of conventional T cells within lymph nodes. Its effects on T_{regs} differ between T_{reg} subtypes; for example, it appears to inhibit tT_{reg} proliferation but increases their suppressive activity, while it promotes the differentiation of pT_{regs} [97–100]. In a murine model of arthritis, the combined treatment of fingolimod and an arthritis-inducing peptide was able to decrease the arthritis clinical scores and prevent relapse upon retreatment with an arthritis-inducing peptide 1 month later [101,102]. Only the GITRsp pT_{reg} subset showed expansion with this treatment, therefore it is likely that GITRsp expansion is another mechanism of action of fingolimod.

16.3.2 Effects of Epigenetic Drugs

Given the role of epigenetic mechanisms in the development and maintenance of T_{regs}, (see Section 16.5), drugs that can modulate DNA accessibility have been investigated with regard to the promotion of pT_{reg} development or stabilization of the T_{reg} phenotype.

Histone acetylation generally increases DNA accessibility and promotes gene transcription. Therefore the higher the histone deacetylase (HDAC) activity, the lower the gene transcriptional activity. HDAC inhibitors (e.g., trichostatin A, suberoylanilide hydroxamic acid, and panobinostat) favor acetylation of histones and gene transcription, including that of Foxp3 [103]. Several studies have demonstrated the usefulness of these drugs in the potentiation of murine and human T_{reg} activity in in vitro experiments [103–107] and their beneficial effects in transplantation [103] and murine models of autoimmune disease [104]. HDAC inhibitors stimulate the thymic production of tT_{regs}, promote the peripheral conversion of conventional T cells into pT_{regs}, and enhance T_{regs} suppressive function. The effects of HDAC inhibitors on the inflammatory response and cell polarization are also attributable to modulation of the expression of other proteins, such as the IL-6 receptor, which suppresses the polarization of T_{regs} towards the proinflammatory Th17 cell type [108,109]. However, the use of HDAC inhibitors is associated with adverse effects, including profound and debilitating fatigue [110].

HDACs are subdivided into four classes of enzymes, with the first two being the most studied. Class I HDAC enzymes are expressed in all cell types and their inhibition does not have T_{reg}-specific effects, whereas class II HDAC enzymes show tissue-specific expression and T_{reg}-specific effects. Therefore the effects of inhibitors that are specific to certain HDAC enzymes are currently being evaluated with regard to the prevention of rejection and the treatment of collagen-induced arthritis (CIA) and other autoimmune diseases [105,111,112].

Studies published between 2007 and 2009 described the presence of CpG-rich islands in the Foxp3 locus, which is the main point of epigenetic regulation in T_{regs} [113,114]. For example, Floess et al. demonstrated that an evolutionarily conserved region upstream of Foxp3 exon 1 showed complete demethylation of CpG motifs in ex vivo isolated tT_{regs} but not in naïve $CD4^+$ conventional T cells. Moreover, Lal et al. identified another CpG motif in the Foxp3 nonintronic upstream enhancer with the same methylation pattern. Demethylation of these CpG motifs using the DNA methyltransferase inhibitors, azacytidine or decitabine, stabilized Foxp3 expression in naïve $CD4^+$ conventional T cells treated with TGFβ (Th3 pT_{reg}) [114,115]. The effects of hypomethylating agents are not confined to Foxp3 modulation; for example, Stamou et al. demonstrated that human activated T cells treated with hypomethylating agents remain negative for Foxp3 but express human leukocyte antigen G (HLA-G) and acquire regulatory properties [116]. It has been reported that treatment with hypomethylating agents prevents diabetes in cyclophosphamide-treated NOD mice and mitigates graft-versus-host disease (GVHD) without compromising on the graft-versus-leukemia effects in a murine model [117,118].

16.3.3 Induction of T_{reg} Expansion by Biotechnological Drugs

The main mechanism of action of some monoclonal antibodies is T_{reg} expansion, and the most interesting of these is anti-CD3 antibodies, with the first report about its curative effect in autoimmune disease published more than 20 years ago [119,120]. Several studies have since demonstrated the effect of anti-CD3 treatment in autoimmune disease models, including experimental autoimmune encephalomyelitis (EAE), multiple sclerosis, colitis, and rheumatoid arthritis [119]. It is hypothesized that CD3 triggering in activated conventional T cells leads to their anergy and apoptosis and favors TGF-β production by macrophages [121]. TGF-β creates a tolerogenic microenvironment and promotes Th3 differentiation. In the intestinal environment and in patients with diabetes, anti-CD3 antibodies favor IL-10 production and differentiation or expansion of $CD4^{+}CCR4^{+}IL\text{-}10^{+}$ T_{regs} [122,123]. In summary, the aim of anti-CD3 antibody treatment should be to induce anergy and apoptosis of conventional cells (T_{regs} exhibit greater resistance to anti-CD3-induced cell death) and increase the number of pT_{regs} [124]. Anti-CD3 antibodies can also be orally and intranasally administered, and these routes have been shown to have fewer adverse effects and stable efficacies in some disease models [119]. Four antihuman-CD3 antibodies are currently being studied (otelixizumab, teplizumab, visilizumab, and foralumab), and the phase I and II results are promising, except for visilizumab, which exhibits low tolerability [119].

A daily low dose of recombinant IL-2 increased the number of tT_{regs} eightfold in patients with chronic GVHD refractory to glucocorticoid therapy, with few effects on $CD4^{+}$ conventional T cells [125]. The T_{reg}-to-conventional T cell ratio increased by more than fivefold and was still twice the baseline level at 4 weeks after the IL-2 treatment was discontinued. Even though IL-2 treatment resulted in expansion of the T_{reg} population, only 50% of patients experienced any clinical benefits and 30% experienced grade three (out of four) adverse events. In contrast, Saadoun et al. found that treatment with low-dose IL-2 led to clinical improvement in patients with vasculitis [126]. Interestingly, antibodies against IL-2 have a selective IL-2-dependent effect on certain T cell subsets [127,128]. For example, anti-IL-2 JES6-1 antibodies induce T_{reg} growth and inhibit autoimmune disease progression in mice, which indicates that cotreatment with IL-2 and JES6-1 may increase the specific effects of IL-2 on T_{regs}.

The main effect of anti-CD28 superagonist antibodies is the expansion of T_{reg} populations. These antibodies are therefore effective in the treatment of autoimmunity models (i.e., diabetes, arthritis, and glomerulonephritis) and the prevention of organ rejection following transplantation [129]. The CD28-targeting superagonist antibody, TGN1412 (now called TAB08), resulted in a life-threatening cytokine release syndrome in a phase I study. The antibody is currently being evaluated in vitro at lower doses and seems to induce the expansion of T_{reg} populations without favoring the release of proinflammatory cytokines. This finding may mean that it can be used in humans [129].

GITR is a good marker of both tT_{regs} and pT_{regs} [130,131]. Triggering of GITR delivers a powerful costimulatory signal in T_{regs}, as observed in in vitro studies and in mice that are transgenic for the expression of the GITR ligand (GITRL), which show expansion of tT_{regs} and Tr1/GITRsp cells [132]. However, GITR is also a costimulatory molecule of conventional T cells, and GITRL transgenic mice also demonstrate expansion of conventional memory $CD4^+$ cells, albeit to a smaller extent. Preferential expansion of T_{regs} is possibly related to the higher expression levels of GITR in T_{regs} than in conventional T cells [4], and is also observed in healthy mice after treatment with nondepleting Fc-GITRL fusion proteins [133]. Agonist antibodies against murine GITR favor the expansion of both conventional T cells and tT_{regs}; however, the agonist antibody, DTA-1, has a more pronounced effect on $CD4^+$ conventional T cells, while the agonist antibody, G3C, has a more prominent effect on tT_{regs}[134]. These findings indicate that it is possible to preferentially target GITR expressed on T_{regs} or conventional T cells. Studies on the effect of GITR triggering on the treatment of autoimmune diseases are ongoing and the results look promising (Cari et al., unpublished).

In each of the treatments summarized above, the main limitation is that the target receptor is expressed in both T_{regs} and activated conventional T cells; however, antibodies that differ in terms of their affinity and epitope specificity may obtain the desired effects.

16.4 T_{REGS} USED IN T_{REG} THERAPY

16.4.1 Ex Vivo Sorted, Expanded, or Conditioned $Foxp3^+$ nT_{regs}

T_{regs} can be isolated using different techniques such as magnetic sorting, flow cytometry, and microfluidic switch technology, as discussed in Trzonkowski et al. [87] Magnetic sorting is a rapid and almost closed system, which means that it is associated with few regulatory challenges; however, optimal purification is difficult using this technique as only one positive marker can be used for sorting. Flow cytometry has a higher purification rate as specific T_{reg} subsets can be isolated; however, it is more complex and time consuming and presents considerable regulatory challenges.

As briefly discussed, there are no highly specific markers that can be used to isolate T_{regs} and there is no consensus about which of the known markers is best. Presently, CD25 is the most commonly used marker. Several studies have suggested that GITR is the best marker of activated murine T_{regs} [4,130] and that it is transiently expressed in conventional T cells soon after activation. Therefore GITR may be a good candidate for T_{reg} purification, even if T_{regs} are transiently inhibited after purification by anti-GITR antibodies.

Direct infusion of isolated T_{regs} can be used to prevent organ rejection or GVHD following organ or bone marrow transplantation, respectively, and only a low number of T_{regs} are required [135]. In contrast, a high number of polyclonal T_{regs} need to be infused for the successful treatment of autoimmune diseases [135]. Therefore it may be necessary to induce the expansion of ex vivo purified T_{regs}. Some studies have reported that tT_{regs} lose their Foxp3 expression and regulatory properties during expansion. This highlights the need for methods that stabilize the T_{reg} phenotype during expansion, such as retinoic acid treatment or a combination of IL-2, rapamycin, HDAC inhibitors, and hypomethylating agents [136–138].

Other studies have reported that stability does not seem to be a problem. For example, combined stimulation by CD3 and CD28, in the presence of endogenous CD32 (FcγRII) from the cell line, induced the proliferation of human tT_{regs}, which was further enhanced by IL-2 [139]. A 40,000-fold expansion of fully active T_{regs} was achieved within 1 month. In a phase I study on diabetic patients, tT_{regs} ($CD4^+CD25^+CD127^-$) were selected by flow cytometry and induced to expand using anti-CD3 and anti-CD28 coated Dynabeads and recombinant IL-2 [91]. Recipient analysis demonstrated that a subset of the adoptively transferred tT_{regs} were present and long-lived. Therefore the expansion of antigen-specific T_{regs} instead of polyclonal T_{regs} may be a viable option that has better efficacy and selectivity [75,140].

The effects of polyclonal and antigen-specific tT_{reg} infusion following isolation or isolation, conditioning, and stabilization in preclinical studies are reported in Table 16.1. The data suggest that expanded tT_{regs} inhibit GVHD and transplant rejection, especially when they are specific to a particular antigen. The low number of studies that have used tT_{regs} to treat or prevent autoimmune diseases may suggest that tT_{regs} are not very active in this setting. Moreover, as indicated by the data, only antigen-specific tT_{regs} can be successfully used for treatment.

16.4.2 Ex Vivo Induced T_{regs}

In vivo, Th3 and Tr1 cells (T_{regs}) differentiate from $CD4^+$ conventional T cells in microenvironments that are rich in TGF-β and IL-10, respectively. Ex vivo-induced Th3 and Tr1 cells will henceforth be referred to as iTh3 and iTr1 cells, respectively.

Murine iTh3 cells are obtained from freshly isolated $CD4^+CD25^-$ naïve T cells that are cultured for 1 week with anti-CD3, antigen-presenting cells (APCs), and TGF-β [30]. T cells are either $CD25^+$ or $CD25^-$ after differentiation, but both cell types appear anergic, which is indicative of a T_{reg} phenotype. However, only $CD4^+CD25^+$ iTh3 cells have been observed to exhibit suppressive activity. Another protocol involves the culture of T-enriched cells in serum-free synthetic medium and stimulation with an equal number of irradiated non-T cells in the presence of TGF-β and low-dose IL-2 for 6 days [58]. Human iTh3 cells can be obtained by culturing $CD4^+$ T cells and irradiated B cells with Staphylococcal enterotoxin B and TGF-β in serum-free synthetic medium for 6 days [29].

Production and purification of human iTr1 cells can be performed using three methods, which have been reported in detail elsewhere [87,141]. IL-10 DLI cells are Tr1 cells that are derived from donor peripheral blood mononuclear cells (PBMCs) in the presence of IL-10 and irradiated T-cell-depleted PBMCs from patients scheduled for transplantation. DC-10 DLI iTr1 cells are derived from mononuclear cells cultured with DCs (DC-10) generated from monocytes cultured with granulocyte-macrophage colony-stimulating factor, IL-4, and IL-10. Both types of iTr1 cells were used in a clinical trial on patients undergoing hematopoietic stem cell transplantation; however, the regulatory capacity of DC-10 was not sufficient in the presence of PBMCs from uremic patients. Therefore a third protocol produces iTr1 cells (called T_{10} Tr1 cells) from the coculture of DC-10 cells, which are generated from living kidney donors, and $CD4^+$ T cells, which are purified from uremic recipients. One problem with iTr1 production is that the resulting population includes a significant fraction of contaminating non-Tr1 cells. Andolfi et al. generated a homogeneous IL-10-producing cell population by transducing human $CD4^+$ T cells with a bidirectional lentiviral vector (LV) encoding human IL-10 and the marker gene, green fluorescent protein. Transduced human $CD4^+$ T cells displayed typical Tr1 features [142].

The effects of iT_{reg} infusion in representative preclinical studies are presented in Table 16.2. Data show that iT_{regs} (both polyclonal and antigen specific) are effective in preventing and treating several autoimmune diseases in murine models. Conversely, alloantigen-specific iT_{regs} fail to prevent GVHD in several murine models.

16.4.3 Chimeric Antigen Receptor and TCR-Transferred T_{regs}

Chimeric antigen receptor (CAR) technology redirects T cells towards a specific antigen and CAR-expressing conventional T cells are used to treat cancer in mice and humans [143]. CARs are formed from the fusion of an extracellular single-chain antigen-recognition domain (commonly a single-chain antibody fragment) with a membrane-spanning element and an intracellular T-cell-signaling domain. In "second-generation" CARs, the signaling domain comprises that of a co-stimulatory molecule (e.g., CD28 or 4-1BB) and that of the ζ chain of the CD3 complex. CARs are introduced into normal T cells via gene transfer techniques, which redirect the cells towards a specific antigen in a manner that is independent of the major histocompatibility complexes [143].

Murine and human CAR T_{regs} were successfully prepared by transferring CARs to ex vivo sorted tT_{regs} [144–146]. Moreover, antigen-specific T_{regs} were obtained by transferring genes of a TCR to ex vivo sorted tT_{regs} (TCR-transferred T_{regs}) [147,148]. Finally, some studies evaluated the possibility of producing CAR T_{regs} from human $CD4^+$ conventional T cells that were engineered to express Foxp3 and CAR via infection with a LV [146,149].

Elinav et al. demonstrated that nT_{regs} expressing an anti-2,4,6-trinitrophenol (TNP) CAR, which were expanded following stimulation with TNP, maintained their Foxp3 expression level and specifically suppressed conventional T cells [144]. Furthermore, adoptive transfer of CAR nT_{regs} with TNP-specific chimeric receptor (TNP-TPCR) expression in a transgenic mouse model of colitis was associated with antigen-specific, dose-dependent amelioration of the disease. Interestingly, inhibition of conventional T cells by T_{regs} was induced by the antigen but was not specific to antigen-specific conventional T cells (known as the bystander effect). Thus CAR T_{regs} can be recruited and activated by a specific antigen; however, they also exert regulatory effects in the microenvironment. This mechanism was clearly demonstrated in another study in which tT_{regs} were transduced with OTII-TCR, which is specific for an OVA-derived peptide presented by MHC class II molecules [150]. These T_{regs} were used to treat arthritis that was induced by coinjection of the arthritis-inducing antigen and OVA, and swelling was found to be substantially decreased. However, the treatment was not successful in cases where arthritis was induced by injection of the arthritis-inducing antigen alone. This not only demonstrates the bystander effect of T_{regs} but also that antigen-specific T_{regs} are concentrated in the area where the antigen is expressed.

Foxp3-expressing CAR T cells exhibited several phenotypic features that are characteristic of T_{regs}, including upregulation of CD25 and GITR and loss of CD127; however, they did not express high CTLA-4 levels. The efficacy of Foxp3-transduced CAR conventional cells was tested in murine models, but the results were contradictory, with some studies indicating that the effect was better than that of tT_{regs} and others reporting otherwise [146].

Other studies describe the preparation of antigen-specific CAR T_{regs} for some applications. For example, Yoon et al. described the potential of CAR T_{regs} to inhibit a patient's immune response to factor VIII when it was administered as a replacement therapy to hemophilia A patients [151]. Hull et al. prepared human islet-specific TCR-transferred T_{regs}, which had the potential to prevent the development of diabetes [152]. Other studies also described the potential of CAR T_{regs} to prevent human skin transplant rejection and allograft rejection [153,154].

The risk of using CAR or TCR-transferred T_{regs} for the treatment of autoimmune diseases appears to be less than the risk associated with the use of CAR conventional T cells for treating tumors. For example, cytokine storms are unlikely with CAR T_{regs}; however, the lack of purity or an unstable T_{reg} phenotype may result in autoimmune disease. Moreover, vector-specific toxicity, which is related to the replication potential of vectors, may also be a limitation.

Choosing an appropriate target of CARs or TCRs is crucial in the attempt to translate antigen-specific T_{reg} technologies to humans. The major problem is that the antigens associated with autoimmune diseases are largely unknown. In a computational analysis of inflammatory bowel disease

(IBD) patients, Yuan et al. identified 41 genes that could effectively distinguish IBD patient samples from those of healthy donors [155]. However, the only transmembrane proteins that were found to be overexpressed in IBD samples were those expressed by activated cells of the immune system, such as CD3ζ, Fas ligand, and toll-like receptor 4. These findings indicate the presence of an inflammatory infiltrate, but they do not provide information about potential antigens that could be targeted by engineered T_{regs}. A way to circumvent needing to know the antigens involved in the pathogenesis of autoimmune diseases would be to direct the CAR T_{regs} and TCR-transferred T_{regs} towards antigens that are specifically expressed by the target tissue (i.e., in which autoreactive conventional T cells are present). There is substantial evidence to indicate a bystander effect of T_{regs} [144,150] that is produced as a consequence of the T_{reg}-dependent increase in suppressive cytokines in the microenvironment. In this context, tissue-resident T cells [45] may be useful for choosing appropriate TCRs.

Unpredicted antigenic specificity (e.g., cross-reactivity) of engineered T_{regs} may be detrimental if it impedes the immune response against a pathogen or tumor cell. Therefore it may be prudent to use CAR T_{regs} that express a suicide gene, which can be activated if needed [156]. Finally, when using Foxp3-transduced conventional cells to produce CAR T_{regs}, it is not clear whether they will have a stable regulatory phenotype (see Section 16.5). Moreover, the same problem could also occur when engineered tT_{regs} are used.

16.5 THE EPIGENOME, STABILITY, AND PLASTICITY OF T_{REGS}

16.5.1 $Foxp3^+$ nT_{regs} (tT_{regs} and Th3 pT_{regs})

The range of antigen reactivity required to trigger Foxp3 expression in the thymus is broad; however, several lines of evidence suggest that tT_{reg} differentiation is preferentially induced under conditions of heightened TCR reactivity [157]. Exposure to antigens under subimmunogenic conditions, as well as chronic antigen exposure, triggers pT_{reg} cell differentiation [158]. This means that stable pT_{regs} can be generated by low doses of a strong agonist [159]. If high-affinity antigens drive tT_{reg} and pT_{reg} differentiation, it would be risky to infuse patients or perform in vivo expansion of T_{regs} with an unstable phenotype (i.e., differentiation into conventional T cells). Moreover, T_{reg} stability may also influence the long-term benefits of a treatment.

Floess et al. demonstrated that an evolutionarily conserved region upstream of Foxp3 exon 1 showed complete demethylation of CpG motifs in ex vivo isolated $CD4^+Foxp3^+CD25^+$ T_{regs} but not in naive conventional cells [113]. More recently, Ohkura et al. showed that two independent processes were required for the establishment of lineage commitment [160]: the induction of Foxp3 expression and the establishment of T_{reg}-specific demethylated regions (TSDRs). The T_{reg} epigenome and Foxp3 expression are independent factors, and the former is more closely associated with T_{regs} regulatory activity and phenotype stability [20]. As stated in Section 16.2.1, Satb1 appears to be crucial in the development of the tT_{reg} epigenome; in fact, its ablation in thymocytes before Foxp3 expression was shown to impair tT_{regs} development and promote the development of autoimmune diseases [17].

The genes that contain TSDRs include transcription factor-encoding genes [i.e., *Foxp3*, *Ikzf2* (encoding Helios), and *Ikzf4* (encoding Eos)] and T_{reg} marker-encoding genes [i.e., *Il2ra* (encoding CD25), *Tnfrsf18* (encoding GITR), and *Ctla4* (encoding CTLA-4)]. Each region contains

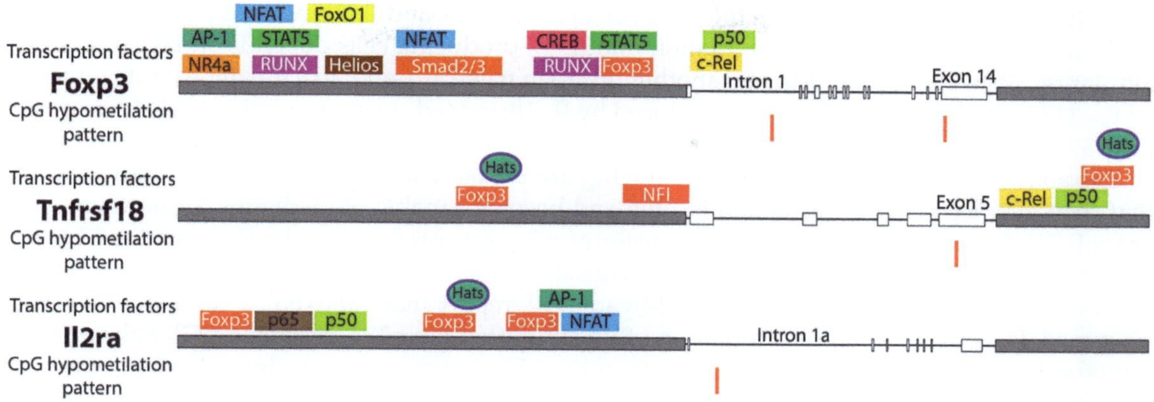

FIGURE 16.1 **Transcription factors modulating the expression of Foxp3, Tnfrsf18 (coding GITR), and Il2ra (coding CD25) and the main hypomethylation regions found in tTregs.** Transcription factors modulate *Foxp3* expression by binding to its promoter region and intron 1. They include activator protein 1 (AP-1), nuclear receptor subfamily 4 group A (NR4a), nuclear factor of activated T-cells (NFAT), signal transducer and activator of transcription 5 (STAT5), runt related transcription factor (RUNX), forkhead box O1 (FoxO1), IKAROS family zinc finger 2 (Helios), SMAD family members (SMAD), CAMP responsive element binding protein (CREB), REL proto-oncogene, NF-κB subunit (c-Rel), nuclear factor κB subunit 1 (p50), and Foxp3 itself. Transcription factors including nuclear factor I C (NFI) modulate *Tnfrsf18* expression by binding to its promoter region and to an enhancer located downstream of the *Tnfrsf18* coding region. Foxp3-dependent modulation of *Il2ra* expression is due to transcription factors including RELA proto-oncogene, NF-KB subunit (p65) binding its promoter region. Histone acetylation in the regulatory regions of *Tnfrs18* and *Il2ra* by recruited histone acetyltransferases (Hats) is described. tT_{reg}-specific hypomethylated regions are found in intron 1 and exon 14 of *Foxp3*, exon 5 of *Tnfrsf18*, and intron1a of *Il2ra*. (only the best regions are shown)

approximately 10 CpG residues, which are not CpG islands [160,161]. T_{reg}-specific downregulated genes are mostly Foxp3-dependent and TSDR-independent, while T_{reg}-specific upregulated genes are largely TSDR-dependent and Foxp3-independent [20]. However, Foxp3 is also able to upregulate gene expression, as has recently been demonstrated for GITR [162]. $Foxp3^{high}$ $CD45RA^-CD25^{high}$ nT_{regs} would be functionally stable as they possess TSDRs ($epigenome^+$), even though they are $Foxp3^-$ cells [20]. The published findings regarding TSDRs and promoter regulation of the TSDR-related genes (*Foxp3*, *Il2ra*, and *Tnfrsf18*) are summarized in Fig. 16.1.

Some studies suggest that the only condition required for maintaining the T_{reg} phenotype is the continuous triggering of the TCR [163], and several studies conducted before the tT_{reg} epigenome was discovered demonstrated that $Foxp3^+$ T_{regs} (tT_{regs} and Th3) represented a stable cell lineage. For example, in an EAE model, antigen-specific $Foxp3^+$ T_{regs} and $Foxp3^+$ conventional cells displayed distinct TCR CDR3 sequences, meaning that $Foxp3^+$ T_{regs} were tT_{reg} and not pT_{reg} [164]. Rudensky et al. showed that $Foxp3^+$ T_{regs} were remarkably stable, not only under steady state conditions but also under various inflammatory conditions [165].

However, other studies have demonstrated that $epigenome^+$ T_{regs} are unstable. For example, Foxp3 expression in $Foxp3^+$ T_{regs} is lost during prolonged CD3 and CD28 triggering [166], repetitive TCR stimulation [167], or costimulation through Tim-1 and OX40 [168,169]. Moreover, when treated in vitro with IL-6, $CD4^+CD25^+Foxp3^+$ and $CD4^+CD25^+CD62L^+$ T_{regs} produce IL-17, exhibit lower regulatory activities, and induce the transformation of T_{regs} into Th17 cells [170,171]. In vitro human studies have also documented a close relationship between nT_{regs} and Th17 cells. For example, the stimulation of memory nT_{regs} ($CD4^+CD25^+Foxp3^+CD127^-CD27^+$) under inflammatory conditions (coculture with allogeneic APCs in the presence of IL-2, IL-15, and IL-1) induced them to differentiate into Th17 cells [108]. T_{reg} differentiation into IL-17-producing cells depends on HDAC activity. Similar results were also reported by another group in naïve T_{regs} and, to a greater extent, in memory T_{regs} [172]. Conversion of T_{regs} is also induced by IL-1β; indeed, a subpopulation of IL-17-secreting Foxp3$^+$ T_{regs} was identified in human circulating CD4$^+$ T cells, which express the IL-1 receptor [173].

The abovementioned studies imply that nT_{regs} are unstable in proinflammatory conditions and that they may lose Foxp3 expression in vivo and become capable of mediating autoimmune diseases [174]. Interestingly, the adoptive transfer of nT_{regs} can prevent the development of CIA, but its therapeutic effect on preestablished CIA is unsatisfactory [136]. Unsatisfactory results were also observed in other disease models [175]. Miyao et al. report that exFoxp3$^+$ T cells are not reprogrammed T_{regs} but are a minor population of uncommitted Foxp3$^+$ T cells [176,177], although this hypothesis needs to be confirmed by further studies. However, even if the hypothesis is proven to be true, it does not resolve the issue of how these cells can be discriminated from true nT_{regs}.

In conclusion, tT_{regs} appear to be active and stable in physiological conditions. When infused, they can prevent transplant rejection and GVHD (Table 16.1), but they are unstable in the inflammatory environment.

Miyara et al. demonstrated that IL-2, rapamycin, HDAC inhibitors, and DNA methyltransferase inhibitors act in synergy to permit the expansion of human T_{regs}. This was accompanied by high and sustained expression of Foxp3 and CD15s and potent in vitro suppressive capabilities [137]. However, this in vivo effect was demonstrated in a murine xeno-GVHD reaction, which does not demonstrate the stability of treated T_{regs} in an inflammatory environment.

16.5.2 iTh3 iT_{regs}

TSDRs are highly specific to tT_{regs} and Th3 pT_{regs}; however, they are not found in iT_{regs} produced by in vitro antigenic stimulation with TGF-β and IL-2 (iTh3) or in conventional T cells that are retrovirally transduced with Foxp3 [20,113,160]. Therefore nT_{regs} maintain high Foxp3 levels and regulatory activity after 6 days of culture, while iTh3 cells lose Foxp3 expression and regulatory activity when cultured without TGF-β for the same period of time [113]. Thus it seems that iTh3 cells may not be useful for treating autoimmune diseases as their phenotype is much more dependent on the microenvironment than that observed in nT_{regs}.

However, some studies have demonstrated that iTh3 cells have a stable phenotype. For example, iTh3 cells do not differentiate into Th17 cells in vitro when stimulated with IL-6 or TGF-β plus IL-6. This stability is due to their very low expression of the IL-6 receptor [170]. Hilbrands et al. demonstrated that iTh3 cells that were specific for the male *Dby* peptide were able to suppress graft rejection in female mice with cells (from a male skin graft) expressing the *Dby* peptide [57]. The iTh3 cells survived for a long time in these mice and maintained their suppressive phenotype; furthermore, the iTh3 cells appeared to have a T_{reg}-specific epigenome 1 month after treatment. These data demonstrate that at least some of the T cells induced in vitro to express Foxp3 can be stabilized in vivo if they are reexposed to the antigen, and that they can acquire epigenetic maturation that is comparable to wildtype pT_{regs}. Murine models in which autoimmune diseases were treated with iTh3 have also confirmed their stability. For example, adoptive transfer of iTh3 cells was shown to attenuate murine airway allergic inflammation, and the number of Foxp3$^+$ cells was observed to approximately double at 3–6 days after the initial iTh3 cell infusion [178]. iTh3 cells were also reported to suppress lupus-like chronic GVHD by inducing protective DCs that generated additional T_{regs} [59]. The protective effects of infused iT_{regs} involve the action of both IL-10 and TGF-β, but the tolerogenic effects of iTh3 cells on DCs and the immunosuppressive effects of these DCs are TGF-β dependent. Adoptive transfer of antigen-specific iTh3 cells before CIA induction decreased the incidence of CIA; moreover, the CIA score was ameliorated when the infusion was performed 14 days after the initial CIA induction (no improvement was observed after nT_{reg} infusion in these animals) [68]. Finally, iTh3 cells demonstrated a phenotype that was more stable than that of nT_{regs}, as T_{reg} cells were predominant in the draining lymph nodes of iTh3-treated mice while Th17 cells were predominant in nT_{reg}-treated mice.

In conclusion, iTh3 cells maintain (or even potentiate) their regulatory activity in an inflammatory microenvironment even though they lack the T_{reg} epigenome (or acquire the T_{reg} epigenome in vivo). On the contrary, iTh3 cells are not always effective at attenuating GVHD, even when used in combination with in vivo treatments that stabilize their phenotype [49,53,54].

16.5.3 iTr1 iT_{regs}

In vitro-induced Tr1 (iTr1) cells do not express Foxp3. Moreover, their epigenetic phenotype has not been studied, therefore their stability can only be evaluated by studying their cytokine production and regulatory activity over time.

Infusion of OVA-specific Tr1 clones was used to successfully treat colitis in a SCID mouse model in which the mice were fed OVA [60]. The effects of the treatment lasted for at least 2 months. In an encephalitis model, the effect of the treatment lasted for at least 2 weeks [66]. Conversely, Tr1 cells were unable to prevent GVHD in a murine model [48]. However, $CD4^+$ T cells cultured with IL-10 and TGF-β (iT_{regs}) demonstrated a 73% success rate for the prevention of GVHD in a mouse model. Interestingly, the mice in which GVHD was prevented by iT_{reg} infusion survived for at least 2 months even though they showed some histological signs of GVHD. In summary, iTr1 cells appear to be more active in inflammatory and autoimmune diseases and not as involved in their prevention, at least in murine models.

In a proof-of-concept study, 12 patients were treated with Tr1 cells to prevent GVHD; however, the treatment was only successful in four of these patients. Despite this, the remission in these individuals was long lasting (median follow-up, 7.2 years), which could indicate that Tr1 cells have a stable phenotype [84]. In another human study, OVA-specific T_{regs} were generated from cultured PBMCs from patients with refractory Crohn's disease. Cells were generated in the presence of OVA, IL-2, and IL-4, and were expanded on a layer of transfected feeder cells (Schneider cells) [92]. The cells expressed high levels of IL-10 postculture (indicating that the cells were iTr1 cells); however, 50% of the cells also expressed Foxp3 and 40% expressed CD25, which implies that they were not iTr1 cells. After the infusion of 1 million iT_{regs}, patients were kept on an OVA-enriched diet to ensure activation of T_{regs}, and three of the eight patients underwent remission that lasted for 4 weeks.

16.5.4 $GITR^+$ T_{regs} and GITRsp pT_{regs}

GITR is expressed in both tT_{regs} and pT_{regs}, particularly when activated. Therefore studies on GITR-expressing T_{regs} can concern the whole $GITR^+$ T_{reg} population or, more frequently, $CD25^+GITR^+Foxp3^+$ T_{regs}. However, some studies have evaluated the properties of GITRsp pT_{regs} and reported interesting results.

The first studies to demonstrate that GITR was a murine T_{reg} marker reported that GITR binding by anti-GITR antibodies inhibited T_{reg} regulatory activity in in vitro experiments [179,180]. This inhibitory activity was also confirmed in human studies and was found to be independent of the agonistic function of the antibodies [181,182]. The in vivo effects of GITR triggering on T_{reg} stability have been studied in tumors. GITR triggering by an anti-GITR antibody decreased the number of tumor-infiltrating T_{regs}, promoted the differentiation of T_{regs} into Th9 cells [183,184], increased the posttranscriptional modifications that reduced Foxp3 protein expression, decreased the Helios and IL-10 expression, and increased IFNγ expression [185,186]. The obvious conclusion from the above mentioned data is that GITR triggering destabilizes the T_{reg} phenotype.

However, there is much evidence to suggest that these effects are limited to the tumor environment in which GITR is triggered with pharmacological doses of the anti-GITR antibody, and that the effect of GITR triggering in the context of immune homeostasis and autoimmune diseases is completely different. First, mice that are transgenic for GITRL have functionally active T_{regs} [132,187], which implies that inhibition of T_{reg} activity by the anti-GITR antibody is transient or is observed only at pharmacological doses. Moreover, tT_{reg} and GITRsp pT_{regs} are expanded in GITRL transgenic mice [132,187]. The decrease in the number of tumor-infiltrating T_{regs} following treatment with anti-GITR is mainly due to antibody-dependent cell-mediated cytotoxicity, which affects $GITR^+$ T_{regs} and depends on FcγR-expressing cells infiltrating the tumor [188–190]. Interestingly, following anti-GITR antibody treatment, Foxp3 expression is decreased in tumor-infiltrating lymphocytes but not in lymphocytes from lymph nodes of the same mice. Moreover, adoptively transferred $Foxp3^+$ T_{regs} from tumor-bearing mice lose Foxp3 expression in the host when treated with an anti-GITR antibody, whereas T_{regs} from naive mice maintain Foxp3 expression [185]. If the anti-GITR antibody effects are truly tumor dependent (as suggested by these studies), the effect on tT_{reg} stability in other contexts may be irrelevant. The crucial role of physiological GITR triggering in the control of T_{reg} number is summarized elsewhere [102,182]. In a murine model of arthritis, the expansion of GITRsp pT_{regs} due to combined treatment with fingolimod and the arthritis-inducing peptide decreased the arthritis clinical scores and prevented relapse on retreatment with the arthritis-inducing peptide 1 month later [101,102]. This indicates that GITRsp cells participate in the recovery of homeostasis of the immune system and have a rather stable phenotype. The number of GITRsp cells is also increased in some patients affected by SLE and Sjogren syndrome, where they appear to play a role in the control of autoreactive conventional T cells, as patients with increased numbers of T_{regs} had lower disease activity [7,9]. Moreover, preliminary data from our group demonstrated that GITRsp cells can be expanded by GITR triggering and that their activity prevents the development of autoimmune diseases in murine models (Cari et al., unpublished).

In conclusion, the small amount of data about $GITR^+$ T_{regs} and GITRsp pT_{regs} indicate that these T_{regs} participate in the control of immune response and have a stable phenotype.

16.6 CONCLUSIONS AND PERSPECTIVES

T_{reg} therapy has the potential to become a revolutionary immunosuppressive treatment method (Fig. 16.2); however, the information currently available regarding T_{reg} activity and plasticity are not enough to plan T_{reg} therapy on a rational basis. The scientific community is still searching for clear insight into these issues.

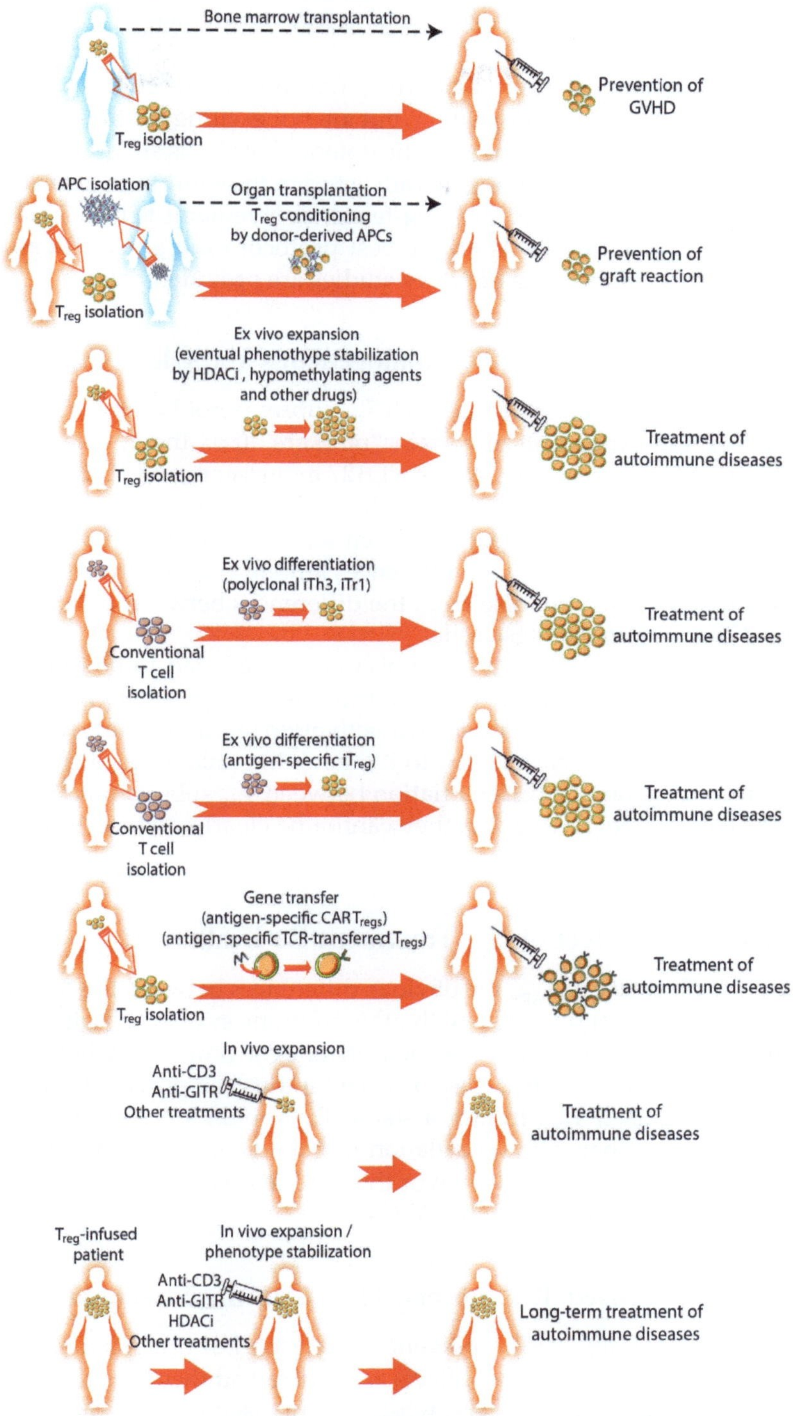

FIGURE 16.2 **Different techniques for increasing the number of Tregs for the prevention of GVHD and graft rejection and the treatment of autoimmune diseases.** The *red figures* represent patients, and the *blue figures* represent healthy donors. Additional treatment of the patients administered with T_{reg} infusions may promote the expansion of T_{regs} or stabilization of the T_{reg} phenotype, and may be repeated over time. *APCs*, Antigen presenting cells; *CAR*, chimeric antigen receptor; HDACi, histone deacetylase inhibitors; *iTh3*, in vitro induced Th3 cells; *iTr1*, in vitro induced Tr1 cells.

The few results from clinical trials (Table 16.3) indicate that the effects of T_{regs} are satisfactory for the prevention of GVHD and are interesting, but are not exciting for the treatment of autoimmune diseases. However, research on the therapeutic potential of T_{regs} is still in its initial stages. We know that the safety profile of T_{reg} infusion is generally good in the short term, even though some adverse events have been reported; however, the long-term safety remains to be evaluated. Therefore much more research into T_{regs}, with a focus on the best protocol for a given disease, is required. Some perspectives that can form the basis of future studies are provided below.

16.6.1 Are T_{reg} Subsets Formed by Homogeneous and Clear-Cut Cell Populations?

There is clear evidence to indicate that each T_{reg} subset is not homogeneous; in fact, each subset is characterized by the expression of certain "markers" (i.e., the epigenome phenotype; Foxp3, CD25, GITR, and LAG-3 expression; lack of CD127 expression; and IL-10 and TGF-β production). However, when the cells of one subset are analyzed for the expression of other T_{reg} markers, they do not exhibit the same expression pattern. Even when induced in vitro, iT_{regs} may not represent a homogeneous population. Moreover, differences among iT_{regs} are derived not only from the different protocols that are used but also from the differences between patients and the effect of the disease on the differentiation of these cells.

The lack of homogeneity within a T_{reg} subset is not believed to be due to technical issues or interlaboratories differences, but it is likely to represent the plasticity of T_{regs}, which acquire new markers and lose others in order to alter their activation status, homing, and mechanism of regulation. The plasticity of T_{regs} is clearly tied to the potential risk of T_{regs} differentiating into conventional T cells. In conclusion, the differentiation between T_{reg} subsets (and their markers) is useful from a methodological perspective, but they cannot be clearly defined in vivo as distinguishable populations.

16.6.2 How Can the Effect of T_{reg} Treatment Be Evaluated?

Considering the plasticity of T_{regs}, the obvious differences between the in vitro and in vivo properties of T_{regs}, and the interaction of T_{regs} with other immune system cells, it would be more interesting to evaluate the in vivo persistence of their effect over time (i.e., whether they can be used to successfully treat the autoimmune disease or whether they can prevent transplant rejection) rather than the survival of infused T_{regs}. In fact, a stable T_{reg} population with low activity may prove to have less curative effects than a T_{reg} population that tends to disappear but efficiently shapes the immune system. For example, T_{regs} may favor the generation of a new T_{reg} population ($CD4^+$, $CD8^+$, or $B220^+$) or the maturation of tolerogenic APCs.

16.6.3 Stability of Infused T_{reg} Subsets: How Can the Effects be Prolonged?

In the case of T_{regs} that are used to prevent the rejection of transplanted organs or GVHD, T_{reg} activity may be unnecessary once their effects are realized ad therefore only the timing of the infusion is crucial for their effect. Conversely, when T_{regs} are used to cure autoimmune diseases, a new homeostatic balance is required, which appears to be crucial to the persistence of the treatment effect (derived from T_{regs} or other T_{reg}-promoted cells). However, the effects of T_{reg} infusion are transient in some experimental settings. This may be due to T_{reg} instability, lack of T_{reg} proliferation and activation, or T_{reg} apoptosis. Therefore the association of T_{reg} therapies with treatments that stabilize the T_{reg} phenotype or promote T_{reg} survival and expansion would be interesting (Fig. 16.2). In vitro stabilization of T_{regs} is possible with epigenetic drugs (e.g., HDAC inhibitors and hypomethylating agents), TGF-β, transretinoic acid, and vitamin D. Moreover, in vivo treatment can also be performed after infusion using epigenetic drugs, immunosuppressive small molecules (e.g., glucocorticoids, rapamycin, and fingolimod), or biotechnological drugs that favor T_{reg} expansion (e.g., the anti-CD3 antibody or the anti-GITR antibody), as discussed in Section 16.3.

16.6.4 Polyclonal Versus Antigen-Specific T_{regs}

Most clinical trials have been performed using polyclonal T_{regs}; however, studies on murine models indicate that antigen-specific T_{regs} are much more active due to activation and homing. Antigen-specific T_{regs} may also be safer because they do not cause generalized immunosuppression. However, these results of transgenic studies in animals cannot be extrapolated to human disease as the antigens associated with autoimmune diseases in humans are almost unknown. A relevant exception is the use of alloantigen-specific T_{regs} in the prevention of GVHD and graft rejection.

In the treatment of autoimmune diseases, T_{regs} in the T_{reg} therapy (i.e., those purified from the patients or induced by T cells purified from the patients) may be enriched with T_{regs} that are specific for self-antigens (i.e., those involved in the disease development). In some diseases, some T_{reg} subsets (e.g., GITRsp cells) are expanded. Appropriate sorting strategies can favor the isolation of T_{regs} enriched with antigen-specific T_{regs}.

Autoreactive conventional T cells are expanded in several patients with autoimmune diseases. If iT_{regs} are prepared from memory conventional T cells from the patient, consistent enrichment of T cells that are specific to the antigens responsible for the disease can be achieved and consequently most of the iT_{regs} would be antigen-specific.

16.6.5 Which T_{reg} Subset is Better?

The immune homeostasis in mammals indicates that tT_{regs} are stable throughout the lifetime of the organism and that their plasticity is not pathogenetic. The nT_{reg} epigenome may be responsible for the stability of the phenotype. However, overall analysis of the data indicate that tT_{regs} do not have a stable phenotype within inflamed tissues. Therefore it is possible that the T_{reg} epigenome is modulated by microenvironmental signals in favor of activation of the immune system and that tT_{regs} lose their phenotype in this context.

Conversely, most iT_{reg} subsets (and possibly pT_{reg} subsets) lack a T_{reg} epigenome, sometimes also lack Foxp3 expression, and appear to have an unstable phenotype. However, their phenotype appears to be stable in inflammatory or autoimmune diseases where they appear to be more active than tT_{regs}.

Future studies should evaluate the preliminary evidence concerning the superiority of tT_{regs} in the prevention of immune reactions and the superiority of pT_{regs} and iT_{regs} in the treatment of inflammatory and autoimmune diseases. They should also compare the efficacy and safety of the different T_{reg} subsets and CAR or TCR-transferred T_{regs} in the treatment of different autoimmune diseases. T_{reg} therapy in association with in vivo treatments used to expand or stabilize T_{regs} should be carefully evaluated. The long-term safety of polyclonal and antigen-specific T_{regs} should be investigated in depth before considering T_{reg} therapy, particularly when life expectancy is good.

Acknowledgments

This work was supported by an Associazione Italiana per la Ricerca sul Cancro (AIRC) grant to Carlo Riccardi (IG-14291), and a Fondo di Ateneo per la Ricerca di Base 2015 grant to Giuseppe Nocentini.

References

[1] Lawson CA, Brown AK, Bejarano V, Douglas SH, Burgoyne CH, Greenstein AS, et al. Early rheumatoid arthritis is associated with a deficit in the $CD4^+CD25^{high}$ regulatory T cell population in peripheral blood. Rheumatology (Oxford) 2006;45:1210–7.
[2] Gerli R, Nocentini G, Alunno A, Bocci EB, Bianchini R, Bistoni O, et al. Identification of regulatory T cells in systemic lupus erythematosus. Autoimmun Rev 2009;8:426–30.
[3] Miyara M, Gorochov G, Ehrenstein M, Musset L, Sakaguchi S, Amoura Z. Human $FoxP3^+$ regulatory T cells in systemic autoimmune diseases. Autoimmun Rev 2011;10:744–55.
[4] Petrillo MG, Ronchetti S, Ricci E, Alunno A, Gerli R, Nocentini G, et al. $GITR^+$ regulatory T cells in the treatment of autoimmune diseases. Autoimmun Rev 2015;14:117–26.
[5] Brusko TM, Putnam AL, Bluestone JA. Human regulatory T cells: role in autoimmune disease and therapeutic opportunities. Immunol Rev 2008;223:371–90.
[6] Long SA, Buckner JH. $CD4^+FOXP3^+$ T regulatory cells in human autoimmunity: more than a numbers game. J Immunol 2011;187:2061–5.
[7] Nocentini G, Alunno A, Petrillo MG, Bistoni O, Bartoloni E, Caterbi S, et al. Expansion of regulatory $GITR^+CD25^{low/-}CD4^+$ T cells in systemic lupus erythematosus patients. Arthritis Res Ther 2014;16:444–5.
[8] Bonelli M, Savitskaya A, Steiner CW, Rath E, Smolen JS, Scheinecker C. Phenotypic and functional analysis of $CD4^+CD25^-Foxp3^+$ T cells in patients with systemic lupus erythematosus. J Immunol 2009;182:1689–95.
[9] Alunno A, Petrillo MG, Nocentini G, Bistoni O, Bartoloni E, Caterbi S, et al. Characterization of a new regulatory $CD4^+$ T cell subset in primary Sjogren's syndrome. Rheumatology (Oxford) 2013;52:1387–96.
[10] Gershon RK, Kondo K. Infectious immunological tolerance. Immunology 1971;21:903–14.
[11] Germain RN. Special regulatory T-cell review: a rose by any other name: from suppressor T cells to Tregs, approbation to unbridled enthusiasm. Immunology 2008;123:20–7.
[12] Sakaguchi S, Sakaguchi N, Asano M, Itoh M, Toda M. Immunologic self-tolerance maintained by activated T cells expressing IL-2 receptor alpha-chains (CD25) breakdown of a single mechanism of self-tolerance causes various autoimmune diseases. J Immunol 1995;155:1151–64.
[13] Wildin RS, Ramsdell F, Peake J, Faravelli F, Casanova JL, Buist N, et al. X-linked neonatal diabetes mellitus, enteropathy and endocrinopathy syndrome is the human equivalent of mouse scurfy. Nat Genet 2001;27:18–20.
[14] Hori S, Nomura T, Sakaguchi S. Control of regulatory T cell development by the transcription factor Foxp3. Science 2003;299:1057–61.
[15] Lio CW, Hsieh CS. A two-step process for thymic regulatory T cell development. Immunity 2008;28:100–11.
[16] Burchill MA, Yang J, Vang KB, Moon JJ, Chu HH, Lio CW, et al. Linked T cell receptor and cytokine signaling govern the development of the regulatory T cell repertoire. Immunity 2008;28:112–21.
[17] Kitagawa Y, Ohkura N, Kidani Y, Vandenbon A, Hirota K, Kawakami R, et al. Guidance of regulatory T cell development by Satb1-dependent super-enhancer establishment. Nat Immunol 2017;18:173–83.
[18] Mahmud SA, Manlove LS, Schmitz HM, Xing Y, Wang Y, Owen DL, et al. Costimulation via the tumor-necrosis factor receptor superfamily couples TCR signal strength to the thymic differentiation of regulatory T cells. Nat Immunol 2014;15:473–81.
[19] Thornton AM, Korty PE, Tran DQ, Wohlfert EA, Murray PE, Belkaid Y, et al. Expression of Helios, an Ikaros transcription factor family member, differentiates thymic-derived from peripherally induced $Foxp3^+$ T regulatory cells. J Immunol 2010;184:3433–47.
[20] Morikawa H, Sakaguchi S. Genetic and epigenetic basis of Treg cell development and function: from a FoxP3-centered view to an epigenome-defined view of natural Treg cells. Immunol Rev 2014;259:192–205.
[21] Rodriguez-Perea AL, Arcia ED, Rueda CM, Velilla PA. Phenotypical characterization of regulatory T cells in humans and rodents. Clin Exp Immunol 2016;185:281–91.

[22] Wyss L, Stadinski BD, King CG, Schallenberg S, McCarthy NI, Lee JY, et al. Affinity for self-antigen selects Treg cells with distinct functional properties. Nat Immunol 2016;17:1093–4.
[23] Seddiki N, Santner-Nanan B, Martinson J, Zaunders J, Sasson S, Landay A, et al. Expression of interleukin (IL)-2 and IL-7 receptors discriminates between human regulatory and activated T cells. J Exp Med 2006;203:1693–700.
[24] Vignali DA, Collison LW, Workman CJ. How regulatory T cells work. Nat Rev Immunol 2008;8:523–32.
[25] Arpaia N, Green JA, Moltedo B, Arvey A, Hemmers S, Yuan S, et al. A distinct function of regulatory T cells in tissue protection. Cell 2015;162:1078–89.
[26] Burzyn D, Kuswanto W, Kolodin D, Shadrach JL, Cerletti M, Jang Y, et al. A special population of regulatory T cells potentiates muscle repair. Cell 2013;155:1282–95.
[27] Mock JR, Garibaldi BT, Aggarwal NR, Jenkins J, Limjunyawong N, Singer BD, et al. Foxp3$^+$ regulatory T cells promote lung epithelial proliferation. Mucosal Immunol 2014;7:1440–51.
[28] Seo N, Hayakawa S, Takigawa M, Tokura Y. Interleukin-10 expressed at early tumour sites induces subsequent generation of CD4$^+$ T-regulatory cells and systemic collapse of antitumour immunity. Immunology 2001;103:449–57.
[29] Zheng SG, Gray JD, Ohtsuka K, Yamagiwa S, Horwitz DA. Generation ex vivo of TGF-β-producing regulatory T cells from CD4$^+$CD25$^-$ precursors. J Immunol 2002;169:4183–93.
[30] Chen W, Jin W, Hardegen N, Lei KJ, Li L, Marinos N, et al. Conversion of peripheral CD4$^+$CD25$^-$ naive T cells to CD4$^+$CD25$^+$ regulatory T cells by TGF-β induction of transcription factor Foxp3. J Exp Med 2003;198:1875–86.
[31] Fantini MC, Becker C, Monteleone G, Pallone F, Galle PR, Neurath MF. Cutting edge: TGF-β induces a regulatory phenotype in CD4$^+$CD25$^-$ T cells through Foxp3 induction and down-regulation of Smad7. J Immunol 2004;172:5149–53.
[32] Chen Y, Kuchroo VK, Inobe J, Hafler DA, Weiner HL. Regulatory T cell clones induced by oral tolerance: suppression of autoimmune encephalomyelitis. Science 1994;265:1237–40.
[33] Carrier Y, Yuan J, Kuchroo VK, Weiner HL. Th3 cells in peripheral tolerance I. Induction of Foxp3-positive regulatory T cells by Th3 cells derived from TGF-β T cell-transgenic mice. J Immunol 2007;178:179–85.
[34] Bacchetta R, Bigler M, Touraine JL, Parkman R, Tovo PA, Abrams J, et al. High levels of interleukin 10 production in vivo are associated with tolerance in SCID patients transplanted with HLA mismatched hematopoietic stem cells. J Exp Med 1994;179:493–502.
[35] Roncarolo MG, Gregori S, Bacchetta R, Battaglia M. Tr1 cells and the counter-regulation of immunity: natural mechanisms and therapeutic applications. Curr Top Microbiol Immunol 2014;380:39–68.
[36] Gagliani N, Magnani CF, Huber S, Gianolini ME, Pala M, Licona-Limon P, et al. Coexpression of CD49b and LAG-3 identifies human and mouse T regulatory type 1 cells. Nat Med 2013;19:739–46.
[37] White AM, Wraith DC, Tr1-Like T. Cells: an enigmatic regulatory T cell lineage. Front Immunol 2016;7:355.
[38] Bianchini R, Bistoni O, Alunno A, Petrillo MG, Ronchetti S, Sportoletti P, et al. CD4$^+$ CD25^{low} GITR$^+$) cells: a novel human CD4$^+$ T-cell population with regulatory activity. Eur J Immunol 2011;41:2269–78.
[39] Tanoue T, Atarashi K, Honda K. Development and maintenance of intestinal regulatory T cells. Nat Rev Immunol 2016;16:295–309.
[40] Whiteside TL. What are regulatory T cells (Treg) regulating in cancer and why? Semin Cancer Biol 2012;22:327–34.
[41] Tsaknaridis L, Spencer L, Culbertson N, Hicks K, LaTocha D, Chou YK, et al. Functional assay for human CD4$^+$CD25$^+$ Treg cells reveals an age-dependent loss of suppressive activity. J Neurosci Res 2003;74:296–308.
[42] Allan SE, Crome SQ, Crellin NK, Passerini L, Steiner TS, Bacchetta R, et al. Activation-induced FOXP3 in human T effector cells does not suppress proliferation or cytokine production. Int Immunol 2007;19:345–54.
[43] Gavin MA, Torgerson TR, Houston E, DeRoos P, Ho WY, Stray-Pedersen A, et al. Single-cell analysis of normal and FOXP3-mutant human T cells: FOXP3 expression without regulatory T cell development. Proc Natl Acad Sci USA 2006;103:6659–64.
[44] Smigiel KS, Richards E, Srivastava S, Thomas KR, Dudda JC, Klonowski KD, et al. CCR7 provides localized access to IL-2 and defines homeostatically distinct regulatory T cell subsets. J Exp Med 2014;211:121–36.
[45] Burzyn D, Benoist C, Mathis D. Regulatory T cells in nonlymphoid tissues. Nat Immunol 2013;14:1007–13.
[46] Miyara M, Chader D, Sage E, Sugiyama D, Nishikawa H, Bouvry D, et al. Sialyl Lewis x (CD15s) identifies highly differentiated and most suppressive FOXP3^{high} regulatory T cells in humans. Proc Natl Acad Sci USA 2015;112:7225–30.
[47] Hall AO, Beiting DP, Tato C, John B, Oldenhove G, Lombana CG, et al. The cytokines interleukin 27 and interferon-gamma promote distinct Treg cell populations required to limit infection-induced pathology. Immunity 2012;37:511–23.
[48] Zeller JC, Panoskaltsis-Mortari A, Murphy WJ, Ruscetti FW, Narula S, Roncarolo MG, et al. Induction of CD4$^+$ T cell alloantigen-specific hyporesponsiveness by IL-10 and TGF-β. J Immunol 1999;163(7):3684–91.

[49] Koenecke C, Czeloth N, Bubke A, Schmitz S, Kissenpfennig A, Malissen B, et al. Alloantigen-specific de novo-induced Foxp3+ Treg revert in vivo and do not protect from experimental GVHD. Eur J Immunol 2009;39(11):3091–6.

[50] Sela U, Olds P, Park A, Schlesinger SJ, Steinman RM. Dendritic cells induce antigen-specific regulatory T cells that prevent graft versus host disease and persist in mice. J Exp Med 2011;208(12):2489–96.

[51] Semple K, Yu Y, Wang D, Anasetti C, Yu XZ. Efficient and selective prevention of GVHD by antigen-specific induced Tregs via linked-suppression in mice. Biol Blood Marrow Transplant 2011;17(3):309–18.

[52] Beres A, Komorowski R, Mihara M, Drobyski WR. Instability of Foxp3 expression limits the ability of induced regulatory T cells to mitigate graft versus host disease. Clin Cancer Res 2011;17(12):3969–83.

[53] Zhang P, Tey SK, Koyama M, Kuns RD, Olver SD, Lineburg KE, et al. Induced regulatory T cells promote tolerance when stabilized by rapamycin and IL-2 in vivo. J Immunol 2013;191(10):5291–303.

[54] Li J, Heinrichs J, Haarberg K, Semple K, Veerapathran A, Liu C, et al. HY-specific induced regulatory T cells display high specificity and efficacy in the prevention of acute graft-versus-host disease. J Immunol 2015;195(2):717–25.

[55] Heinrichs J, Li J, Nguyen H, Wu Y, Bastian D, Daethanasanmak A, et al. CD8+ Tregs promote GVHD prevention and overcome the impaired GVL effect mediated by CD4+ Tregs in mice. Oncoimmunology 2016;5(6):e1146842.

[56] Gagliani N, Jofra T, Stabilini A, Valle A, Atkinson M, Roncarolo MG, et al. Antigen-specific dependence of Tr1-cell therapy in preclinical models of islet transplant. Diabetes 2010;59(2):433–9.

[57] Hilbrands R, Chen Y, Kendal AR, Adams E, Cobbold SP, Waldmann H, et al. Induced Foxp3+ T cells colonizing tolerated allografts exhibit the hypomethylation pattern typical of mature regulatory T cells. Front Immunol 2016;7:124.

[58] Zheng SG, Wang JH, Koss MN, Quismorio F Jr, Gray JD, Horwitz DA. CD4+ and CD8+ regulatory T cells generated ex vivo with IL-2 and TGF-β suppress a stimulatory graft-versus-host disease with a lupus-like syndrome. J Immunol 2004;172(3):1531–9.

[59] Lan Q, Zhou X, Fan H, Chen M, Wang J, Ryffel B, et al. Polyclonal CD4+Foxp3+ Treg cells induce TGFβ-dependent tolerogenic dendritic cells that suppress the murine lupus-like syndrome. J Mol Cell Biol. Oxford University Press December 2012;4(6):409–19.

[60] Groux H, O'Garra A, Bigler M, Rouleau M, Antonenko S, de Vries JE, et al. A CD4+ T-cell subset inhibits antigen-specific T-cell responses and prevents colitis. Nature 1997;389(October):737–42.

[61] Foussat A, Cottrez F, Brun V, Fournier N, Breittmayer JP, Groux H. A comparative study between T regulatory Type 1 and CD4+CD25+ T cells in the control of inflammation. J Immunol 2003;171(10):5018–26.

[62] Huter EN, Stummvoll GH, DiPaolo RJ, Glass DD, Shevach EM. Cutting edge: antigen-specific TGF β-induced regulatory T cells suppress Th17-mediated autoimmune disease. J Immunol 2008;181(12):8209–13.

[63] Weber SE, Harbertson J, Godebu E, Mros GA, Padrick RC, Carson BD, et al. Adaptive islet-specific regulatory CD4 T cells control autoimmune diabetes and mediate the disappearance of pathogenic Th1 cells in vivo. J. Immunol 2006;176(8):4730–9.

[64] Barrat FJ, Cua DJ, Boonstra A, Richards DF, Crain C, Savelkoul HF, et al. In vitro generation of interleukin 10-producing regulatory CD4+ T cells is induced by immunosuppressive drugs and inhibited by T helper type 1 (Th1)- and Th2-inducing cytokines. J Exp Med 2002;195(5):603–16.

[65] Selvaraj RK, Geiger TL. Mitigation of experimental allergic encephalomyelitis by TGF-induced Foxp3+ regulatory T lymphocytes through the induction of anergy and infectious tolerance. J Immunol 2008;180(5):2830–8.

[66] Meiron M, Zohar Y, Anunu R, Wildbaum G, Karin N. CXCL12 (SDF-1alpha) suppresses ongoing experimental autoimmune encephalomyelitis by selecting antigen-specific regulatory T cells. J Exp Med October 27, 2008;205(11):2643–55.

[67] O'Connor RA, Leech MD, Suffner J, Hammerling GJ, Anderton SM. Myelin-reactive TGF-β-induced regulatory T cells can be programmed to develop Th1-like effector function but remain less proinflammatory than myelin-reactive Th1 effectors and can suppress pathogenic T cell clonal expansion in vivo. J. Immunol. 2010;185(12).

[68] Kong N, Lan Q, Chen M, Wang J, Shi W, Horwitz DA, et al. Antigen-specific transforming growth factor β-induced Treg cells, but not natural Treg cells, ameliorate autoimmune arthritis in mice by shifting the Th17/Treg cell balance from Th17 predominance to Treg cell predominance. Arthritis Rheum August 2012;64(8):2548–58.

[69] Asnagli H, Martire D, Belmonte N, Quentin J, Bastian H, Boucard-Jourdin M, et al. Type 1 regulatory T cells specific for collagen-type II as an efficient cell-based therapy in arthritis. Arthritis Res Ther 2014;16(3):R115.

[70] Taylor PA, Lees CJ, Blazar BR. The infusion of ex vivo activated and expanded CD4+CD25+ immune regulatory cells inhibits graft-versus-host disease lethality. Blood 2002;99(10):3493–9.

[71] Brennan TV, Tang Q, Liu F-C, Hong V, Bi M, Bluestone JA, et al. Requirements for prolongation of allograft survival with regulatory T cell infusion in lymphosufficient hosts. J Surg Res July 5, 2011;169(1):e69–75.

[72] Ezzelarab MB, Zhang H, Guo H, Lu L, Zahorchak AF, Wiseman RW, et al. Regulatory T cell infusion can enhance memory T cell and alloantibody responses in lymphodepleted nonhuman primate heart allograft recipients. Am J Transplant 2016;16(7):1999–2015.
[73] Ma A, Qi S, Song L, Hu Y, Dun H, Massicotte E, et al. Adoptive transfer of $CD4^+CD25^+$ regulatory cells combined with low-dose sirolimus and anti-thymocyte globulin delays acute rejection of renal allografts in Cynomolgus monkeys. Int Immunopharmacol 2011;11(5):618–29.
[74] Issa F, Hester J, Goto R, Nadig S, Goodacre TE, Wood K. Ex vivo-expanded human regulatory T cells prevent the rejection of skin allografts in a humanised mouse model. Transplantation December 27, 2010;90(12):1321–7.
[75] Landwehr-Kenzel S, Issa F, Luu SH, Schmuck M, Lei H, Zobel A, et al. Novel GMP-compatible protocol employing an allogeneic B cell bank for clonal expansion of allospecific natural regulatory T cells. Am J Transplant 2014;14(3):594–606.
[76] Nadig SN, Wieckiewicz J, Wu DC, Warnecke G, Zhang W, Luo S, et al. In vivo prevention of transplant arteriosclerosis by ex vivo-expanded human regulatory T cells. Nat Med 2010;16(7):809–13.
[77] Elinav E, Waks T, Eshhar Z. Redirection of regulatory T cells with predetermined specificity for the treatment of experimental colitis in mice. Gastroenterology 2008;134(7).
[78] Stephens LA, Malpass KH, Anderton SM. Curing CNS autoimmune disease with myelin-reactive $Foxp3^+$ Treg. Eur J Immunol 2009;39(4):1108–17.
[79] Tarbell KV, Yamazaki S, Olson K, Toy P, Steinman RM. $CD25^+$ $CD4^+$ T Cells, expanded with dendritic cells presenting a single autoantigenic peptide, suppress autoimmune DIABETES. J Exp Med June 7, 2004;199(11):1467–77.
[80] Tang Q, Henriksen KJ, Bi M, Finger EB, Szot G, Ye J, et al. In vitro-expanded antigen-specific regulatory T cells suppress autoimmune diabetes. J Exp Med 2004;199(11):1455–65.
[81] Di Ianni M, Falzetti F, Carotti A, et al. Adoptive immunotherapy with T-regs prevents GVHD and favours immune reconstitution after HLA haploidentical transplants for hematological malignancies. Blood 2009;4.
[82] Brunstein CG, Miller JS, Cao Q, McKenna DH, Hippen KL, Curtsinger J, et al. Infusion of ex vivo expanded T regulatory cells in adults transplanted with umbilical cord blood: safety profile and detection kinetics. Blood 2011;117(3):1061–70.
[83] Martelli MF, Di Ianni M, Ruggeri L, Falzetti F, Carotti A, Terenzi A, et al. HLA-haploidentical transplantation with regulatory and conventional T-cell adoptive immunotherapy prevents acute leukemia relapse. Blood 2014;124(4):638–44.
[84] Bacchetta R, Lucarelli B, Sartirana C, Gregori S, Lupo Stanghellini MT, Miqueu P, et al. Immunological outcome in haploidentical-HSC transplanted patients treated with IL-10-anergized donor T cells. Front Immunol 2014;5:16.
[85] Trzonkowski P, Bieniaszewska M, Juścińska J, Dobyszuk A, Krzystyniak A, Marek N, et al. First-in-man clinical results of the treatment of patients with graft versus host disease with human ex vivo expanded $CD4^+CD25^+CD127^-$ T regulatory cells. Clin Immunol 2009;133(1):22–6.
[86] Theil A, Tuve S, Oelschlägel U, Maiwald A, Döhler D, Oßmann D, et al. Adoptive transfer of allogeneic regulatory T cells into patients with chronic graft-versus-host disease. Cytotherapy 2015;17(4):473–86.
[87] Trzonkowski P, Bacchetta R, Battaglia M, Berglund D, Bohnenkamp HR, ten Brinke A, et al. Hurdles in therapy with regulatory T cells. Sci Transl Med September 9, 2015;7(304). 304ps18.
[88] Todo S, Yamashita K, Goto R, Zaitsu M, Nagatsu A, Oura T, et al. A pilot study of operational tolerance with a regulatory T cell-based Cell therapy in living donor liver transplantation. Hepatology 2016;552.
[89] Geissler EK. The ONE Study compares cell therapy products in organ transplantation: introduction to a review series on suppressive monocyte-derived cells. Transplant. Res. BioMed Central 2012;1:11. September 28.
[90] Marek-Trzonkowska N, Mysliwiec M, Dobyszuk A, Grabowska M, Techmanska I, Juscinska J, et al. Administration of $CD4^+$ $CD25^{high}$ $CD127^-$ regulatory T cells preserves B-cell function in type 1 diabetes in children. Diabetes Care 2012;35:3–6.
[91] Bluestone JA, Buckner JH, Fitch M, Gitelman SE, Gupta S, Hellerstein MK, et al. Type 1 diabetes immunotherapy using polyclonal regulatory T cells. Sci Transl Med 2015;7(315). 315ra189.
[92] Desreumaux P, Foussat A, Allez M, Beaugerie L, Hebuterne X, Bouhnik Y, et al. Safety and efficacy of antigen-specific regulatory T-cell therapy for patients with refractory Crohn's disease. Gastroenterology 2012;143(5). 1207–17.e1–2.
[93] Nocentini G, Migliorati G, Riccardi C. The molecular and cellular mechanisms responsible for the anti-inflammatory and immunosuppressive effects of glucocorticoids. In: Cimaz R, editor. Systemic corticosteroids for inflammatory disorders in pediatrics. Switzerland: Springer International Publishing AG; 2015. p. 25–42.
[94] Perl A. Activation of mTOR (mechanistic target of rapamycin) in rheumatic diseases. Nat Rev Rheumatol 2016;12:169–82.

[95] Mucida D, Park Y, Kim G, Turovskaya O, Scott I, Kronenberg M, et al. Reciprocal TH17 and regulatory T cell differentiation mediated by retinoic acid. Science 2007;317:256–60.
[96] Hayes CE, Hubler SL, Moore JR, Barta LE, Praska CE, Nashold FE. Vitamin D actions on CD4+ T cells in autoimmune disease. Front Immunol 2015;6:100–1.
[97] Wolf AM, Eller K, Zeiser R, Durr C, Gerlach UV, Sixt M, et al. The sphingosine 1-phosphate receptor agonist FTY720 potently inhibits regulatory T cell proliferation in vitro and in vivo. J Immunol 2009;183:3751–60.
[98] Sawicka E, Dubois G, Jarai G, Edwards M, Thomas M, Nicholls A, et al. The sphingosine 1-phosphate receptor agonist FTY720 differentially affects the sequestration of CD4+/CD25+ T-regulatory cells and enhances their functional activity. J Immunol 2005;175:7973–80.
[99] Zhou PJ, Wang H, Shi GH, Wang XH, Shen ZJ, Xu D. Immunomodulatory drug FTY720 induces regulatory CD4+CD25+ T cells in vitro. Clin Exp Immunol 2009;157:40–7.
[100] Sun Y, Wang W, Shan B, Di J, Chen L, Ren L, et al. FTY720-induced conversion of conventional Foxp3− CD4+ T cells to Foxp3+ regulatory T cells in NOD mice. Am J Reprod Immunol 2011;66:349–62.
[101] Yoshida Y, Mikami N, Matsushima Y, Miyawaki M, Endo H, Banno R, et al. Combination treatment with fingolimod and a pathogenic antigen prevents relapse of glucose-6-phosphate isomerase peptide-induced arthritis. Immun Inflamm Dis 2016;4:263–73.
[102] Nocentini G, Cari L, Migliorati G, Riccardi C. The role of GITR single-positive cells in immune homeostasis. Immun Inflamm Dis 2017;5:4–6.
[103] Tao R, de Zoeten EF, Ozkaynak E, Chen C, Wang L, Porrett PM, et al. Deacetylase inhibition promotes the generation and function of regulatory T cells. Nat Med 2007;13:1299–307.
[104] Saouaf SJ, Li B, Zhang G, Shen Y, Furuuchi N, Hancock WW, et al. Deacetylase inhibition increases regulatory T cell function and decreases incidence and severity of collagen-induced arthritis. Exp Mol Pathol 2009;87:99–104.
[105] de Zoeten EF, Wang L, Sai H, Dillmann WH, Hancock WW. Inhibition of HDAC9 increases T regulatory cell function and prevents colitis in mice. Gastroenterology 2010;138:583–94.
[106] Akimova T, Ge G, Golovina T, Mikheeva T, Wang L, Riley JL, et al. Histone/protein deacetylase inhibitors increase suppressive functions of human FOXP3+ Tregs. Clin Immunol 2010;136:348–63.
[107] Lucas JL, Mirshahpanah P, Haas-Stapleton E, Asadullah K, Zollner TM, Numerof RP. Induction of Foxp3+ regulatory T cells with histone deacetylase inhibitors. Cell Immunol 2009;257:97–104.
[108] Koenen HJ, Smeets RL, Vink PM, van Rijssen E, Boots AM, Joosten I. Human CD25highFoxp3pos regulatory T cells differentiate into IL-17-producing cells. Blood 2008;112:2340–52.
[109] Glauben R, Sonnenberg E, Wetzel M, Mascagni P, Siegmund B. Histone deacetylase inhibitors modulate interleukin 6-dependent CD4+ T cell polarization in vitro and in vivo. J Biol Chem 2014;289:6142–51.
[110] Akimova T, Beier UH, Liu Y, Wang L, Hancock WW. Histone/protein deacetylases and T-cell immune responses. Blood 2012;119:2443–51.
[111] Ellis JD, Neil DA, Inston NG, Jenkinson E, Drayson MT, Hampson P, et al. Inhibition of histone deacetylase 6 reveals a potent immunosuppressant effect in models of transplantation. Transplantation 2016;100:1667–74.
[112] Oh BR, Suh DH, Bae D, Ha N, Choi YI, Yoo HJ, et al. Therapeutic effect of a novel histone deacetylase 6 inhibitor, CKD-L, on collagen-induced arthritis in vivo and regulatory T cells in rheumatoid arthritis in vitro. Arthritis Res Ther 2017;19:154–73.
[113] Floess S, Freyer J, Siewert C, Baron U, Olek S, Polansky J, et al. Epigenetic control of the foxp3 locus in regulatory T cells. PLoS Biol 2007;5. e38.
[114] Lal G, Zhang N, van der Touw W, Ding Y, Ju W, Bottinger EP, et al. Epigenetic regulation of Foxp3 expression in regulatory T cells by DNA methylation. J Immunol 2009;182:259–73.
[115] Polansky JK, Kretschmer K, Freyer J, Floess S, Garbe A, Baron U, et al. DNA methylation controls Foxp3 gene expression. Immunol 2008;38:1654–63.
[116] Stamou P, Marioli D, Patmanidi AL, Sgourou A, Vittoraki A, Theofani E, et al. Simple in vitro generation of human leukocyte antigen-G-expressing T-regulatory cells through pharmacological hypomethylation for adoptive cellular immunotherapy against graft-versus-host disease. Cytotherapy 2017;19:521–30.
[117] Choi J, Ritchey J, Prior JL, Holt M, Shannon WD, Deych E, et al. In vivo administration of hypomethylating agents mitigate graft-versus-host disease without sacrificing graft-versus-leukemia. Blood 2010;116:129–39.
[118] Zheng Q, Xu Y, Liu Y, Zhang B, Li X, Guo F, et al. Induction of Foxp3 demethylation increases regulatory CD4+CD25+ T cells and prevents the occurrence of diabetes in mice. J Mol Med (Berl) 2009;87:1191–205.
[119] Kuhn C, Weiner HL. Therapeutic anti-CD3 monoclonal antibodies: from bench to bedside. Immunotherapy 2016;8:889–906.

[120] Chatenoud L, Thervet E, Primo J, Bach JF. Anti-CD3 antibody induces long-term remission of overt autoimmunity in nonobese diabetic mice. Proc Natl Acad Sci USA 1994;91:123–7.
[121] Perruche S, Zhang P, Liu Y, Saas P, Bluestone JA, Chen W. CD3-specific antibody-induced immune tolerance involves transforming growth factor-β from phagocytes digesting apoptotic T cells. Nat Med 2008;14:528–35.
[122] Herold KC, Taylor L. Treatment of type 1 diabetes with anti-CD3 monoclonal antibody: induction of immune regulation? Immunol Res 2003;28:141–50.
[123] Waldron-Lynch F, Henegariu O, Deng S, Preston-Hurlburt P, Tooley J, Flavell R, et al. Teplizumab induces human gut-tropic regulatory cells in humanized mice and patients. Sci Transl Med 2012;4(11). 8ra12.
[124] Spence A, Klementowicz JE, Bluestone JA, Tang Q. Targeting Treg signaling for the treatment of autoimmune diseases. Curr Opin Immunol 2015;37:11–20.
[125] Koreth J, Matsuoka K, Kim HT, McDonough SM, Bindra B, Alyea EP 3rd, et al. Interleukin-2 and regulatory T cells in graft-versus-host disease. N Engl J Med 2011;365:2055–66.
[126] Saadoun D, Rosenzwajg M, Joly F, Six A, Carrat F, Thibault V, et al. Regulatory T-cell responses to low-dose interleukin-2 in HCV-induced vasculitis. N Engl J Med 2011;365:2067–77.
[127] Nocentini G, Migliorati G, Riccardi C. Are we able to harness the immunomodulatory power of cytokines for novel autoimmune disease treatments? Am J Pharmacol Toxicol 2015;10:37–9.
[128] Spangler JB, Tomala J, Luca VC, Jude KM, Dong S, Ring AM, et al. Antibodies to interleukin-2 elicit selective T cell subset potentiation through distinct conformational mechanisms. Immunity 2015;42:815–25.
[129] Tabares P, Berr S, Romer PS, Chuvpilo S, Matskevich AA, Tyrsin D, et al. Human regulatory T cells are selectively activated by low-dose application of the CD28 superagonist TGN1412/TAB08. Eur J Immunol 2014;44:1225–336.
[130] Ronchetti S, Ricci E, Petrillo MG, Cari L, Migliorati G, Nocentini G, et al. Glucocorticoid-induced tumour necrosis factor receptor-related protein: a key marker of functional regulatory T cells. J Immunol Res 2015;2015. 171520.
[131] Nocentini G, Riccardi C. GITR a modulator of immune response and inflammation. Adv Exp Med Biol 2009;647:156–73.
[132] Carrier Y, Whitters MJ, Miyashiro JS, LaBranche TP, Ramon HE, Benoit SE, et al. Enhanced GITR/GITRL interactions augment IL-27 expression and induce IL-10-producing Tr-1 like cells. Eur J Immunol 2012;42:1393–404.
[133] Ephrem A, Epstein AL, Stephens GL, Thornton AM, Glass D, Shevach EM. Modulation of Treg cells/T effector function by GITR signaling is context-dependent. Eur J Immunol 2013;43:2421–9.
[134] Nishioka T, Nishida E, Iida R, Morita A, Shimizu J. In vivo expansion of $CD4^+Foxp3^+$ regulatory T cells mediated by GITR molecules. Immunol Lett 2008;121:97–104.
[135] Tang Q, Bluestone JA. Regulatory T-cell therapy in transplantation: moving to the clinic. Cold Spring Harb Perspect Med 2013;3.
[136] Zhou X, Kong N, Wang J, Fan H, Zou H, Horwitz D, et al. Cutting edge: all-trans retinoic acid sustains the stability and function of natural regulatory T cells in an inflammatory milieu. J Immunol 2010;185:2675–9.
[137] Miyara M, Chader D, Burlion A, Goldstein J, Sterlin D, Norol F, et al. Combination of IL-2, rapamycin, DNA methyltransferase and histone deacetylase inhibitors for the expansion of human regulatory T cells. Oncotarget 2016.
[138] Putnam AL, Brusko TM, Lee MR, Liu W, Szot GL, Ghosh T, et al. Expansion of human regulatory T-cells from patients with type 1 diabetes. Diabetes 2009;58:652–62.
[139] Hoffmann P, Eder R, Kunz-Schughart LA, Andreesen R, Edinger M. Large-scale in vitro expansion of polyclonal human $CD4^+CD25^{high}$ regulatory T cells. Blood 2004;104:895–903.
[140] Putnam AL, Safinia N, Medvec A, Laszkowska M, Wray M, Mintz MA, et al. Clinical grade manufacturing of human alloantigen-reactive regulatory T cells for use in transplantation. Am J Transplant 2013;13:3010–20.
[141] Chihara N, Madi A, Karwacz K, Awasthi A, Kuchroo VK. Differentiation and characterization of Tr1 Cells. Curr Protoc Immunol 2016;113(3):271–310.
[142] Andolfi G, Fousteri G, Rossetti M, Magnani CF, Jofra T, Locafaro G, et al. Enforced IL-10 expression confers type 1 regulatory T cell (Tr1) phenotype and function to human $CD4^+$ T cells. Mol Ther 2012;20:1778–90.
[143] Frey NV, Porter DL. The promise of chimeric antigen receptor T-cell therapy. Oncology (Williston Park) 2016;30.
[144] Elinav E, Adam N, Waks T, Eshhar Z. Amelioration of colitis by genetically engineered murine regulatory T cells redirected by antigen-specific chimeric receptor. Gastroenterology 2009;136:1721–31.
[145] Hombach AA, Kofler D, Rappl G, Abken H. Redirecting human $CD4^+CD25^+$ regulatory T cells from the peripheral blood with pre-defined target specificity. Gene Ther 2009;16:1088–96.
[146] Jethwa H, Adami AA, Maher J. Use of gene-modified regulatory T-cells to control autoimmune and alloimmune pathology: is now the right time? Clin Immunol 2014;150:51–63.
[147] Brusko TM, Koya RC, Zhu S, Lee MR, Putnam AL, McClymont SA, et al. Human antigen-specific regulatory T cells generated by T cell receptor gene transfer. PLoS One 2010;5. e11726.

[148] Tsang JY, Tanriver Y, Jiang S, Xue SA, Ratnasothy K, Chen D, et al. Conferring indirect allospecificity on CD4⁺CD25⁺ Tregs by TCR gene transfer favors transplantation tolerance in mice. J Clin Invest 2008;118:3619–28.
[149] Allan SE, Alstad AN, Merindol N, Crellin NK, Amendola M, Bacchetta R, et al. Generation of potent and stable human CD4⁺ T regulatory cells by activation-independent expression of FOXP3. Mol Ther 2008;16:194–202.
[150] Wright GP, Notley CA, Xue SA, Bendle GM, Holler A, Schumacher TN, et al. Adoptive therapy with redirected primary regulatory T cells results in antigen-specific suppression of arthritis. Proc Natl Acad Sci USA 2009;106:19078–83.
[151] Yoon J, Schmidt A, Zhang A-h, Königs C, Kim YC, Scott DW. FVIII-specific human chimeric antigen receptor T-regulatory cells suppress T- and B-cell responses to FVIII. Blood 2017;129:238–45.
[152] Hull CM, Nickolay LE, Estorninho M, Richardson MW, Riley JL, Peakman M, et al. Generation of human islet-specific regulatory T cells by TCR gene transfer. J Autoimmun 2017;79:63–73.
[153] Noyan F, Zimmermann K, Hardtke-Wolenski M, Knoefel A, Schulde E, Geffers R, et al. Prevention of allograft rejection by use of regulatory T cells with an MHC-specific chimeric antigen receptor. Am J Transplant 2017;17:917–30.
[154] Boardman DA, Philippeos C, Fruhwirth GO, Ibrahim MA, Hannen RF, Cooper D, et al. Expression of a chimeric antigen receptor specific for donor HLA class I enhances the potency of human regulatory T cells in preventing human skin transplant rejection. Am J Transplant 2017;17:931–43.
[155] Yuan F, Zhang YH, Kong XY, Cai YD. Identification of candidate genes related to inflammatory bowel disease using minimum redundancy maximum relevance, incremental feature selection, and the shortest-path approach. Biomed Res Int 2017;2017:5741948.
[156] Minagawa K, Zhou X, Mineishi S, Di Stasi A. Seatbelts in CAR therapy: how safe are CARS? Pharmaceuticals (Basel) 2015;8:230–49.
[157] Li MO, Rudensky AY. T cell receptor signalling in the control of regulatory T cell differentiation and function. Nat Rev Immunol 2016;16:220–33.
[158] Apostolou I, von Boehmer H. In vivo instruction of suppressor commitment in naive T cells. J Exp Med 2004;199:1401–8.
[159] Gottschalk RA, Corse E, Allison JP. TCR ligand density and affinity determine peripheral induction of Foxp3 in vivo. J Exp Med 2010;207:1701–11.
[160] Ohkura N, Hamaguchi M, Morikawa H, Sugimura K, Tanaka A, Ito Y, et al. T cell receptor stimulation-induced epigenetic changes and Foxp3 expression are independent and complementary events required for Treg cell development. Immunity 2012;37:785–99.
[161] Morikawa H, Ohkura N, Vandenbon A, Itoh M, Nagao-Sato S, Kawaji H, et al. Differential roles of epigenetic changes and Foxp3 expression in regulatory T cell-specific transcriptional regulation. Proc Natl Acad Sci USA 2014;111:5289–94.
[162] Tone Y, Kidani Y, Ogawa C, Yamamoto K, Tsuda M, Peter C, et al. Gene expression in the Gitr locus is regulated by NF-kappaB and Foxp3 through an enhancer. J Immunol 2014;192:3915–24.
[163] Vahl JC, Drees C, Heger K, Heink S, Fischer JC, Nedjic J, et al. Continuous T cell receptor signals maintain a functional regulatory T cell pool. Immunity 2014;41:722–36.
[164] Liu X, Nguyen P, Liu W, Cheng C, Steeves M, Obenauer JC, et al. T cell receptor CDR3 sequence but not recognition characteristics distinguish autoreactive effector and Foxp3⁺ regulatory T cells. Immunity 2009;31:909–20.
[165] Rubtsov YP, Niec RE, Josefowicz S, Li L, Darce J, Mathis D, et al. Stability of the regulatory T cell lineage in vivo. Science 2010;329:1667–71.
[166] Gabrysova L, Christensen JR, Wu X, Kissenpfennig A, Malissen B, O'Garra A. Integrated T-cell receptor and costimulatory signals determine TGF-β-dependent differentiation and maintenance of Foxp3⁺ regulatory T cells. Eur J Immunol 2011;41:1242–8.
[167] Hoffmann P, Boeld TJ, Eder R, Huehn J, Floess S, Wieczorek G, et al. Loss of FOXP3 expression in natural human CD4⁺CD25⁺ regulatory T cells upon repetitive in vitro stimulation. Eur J Immunol 2009;39:1088–97.
[168] Degauque N, Mariat C, Kenny J, Zhang D, Gao W, Vu MD, et al. Immunostimulatory Tim-1-specific antibody deprograms Tregs and prevents transplant tolerance in mice. J Clin Invest 2008;118:735–41.
[169] Vu MD, Xiao X, Gao W, Degauque N, Chen M, Kroemer A, et al. OX40 costimulation turns off Foxp3⁺ Tregs. Blood 2007;110:2501–10.
[170] Zheng SG, Wang J, Horwitz DA. Cutting edge: Foxp3⁺CD4⁺CD25⁺ regulatory T cells induced by IL-2 and TGF-β are resistant to Th17 conversion by IL-6. J Immunol 2008;180:7112–6.
[171] Xu L, Kitani A, Fuss I, Strober W. Cutting edge: regulatory T cells induce CD4⁺CD25-Foxp3⁻ T cells or are self-induced to become Th17 cells in the absence of exogenous TGF-β. J Immunol 2007;178:6725–9.
[172] Deknuydt F, Bioley G, Valmori D, Ayyoub M. IL-1β and IL-2 convert human Treg into T(H)17 cells. Clin Immunol 2009;131:298–307.

[173] Raffin C, Raimbaud I, Valmori D, Ayyoub M. Ex vivo IL-1 receptor type I expression in human $CD4^+$ T cells identifies an early intermediate in the differentiation of Th17 from $FOXP3^+$ naive regulatory T cells. J Immunol 2011;187:5196–202.
[174] Zhou X, Bailey-Bucktrout SL, Jeker LT, Penaranda C, Martinez-Llordella M, Ashby M, et al. Instability of the transcription factor Foxp3 leads to the generation of pathogenic memory T cells in vivo. Nat Immunol 2009;10:1000–7.
[175] Bailey-Bucktrout SL, Bluestone JA. Regulatory T cells: stability revisited. Trends Immunol 2011;32:301–6.
[176] Miyao T, Floess S, Setoguchi R, Luche H, Fehling HJ, Waldmann H, et al. Plasticity of $Foxp3^+$ T cells reflects promiscuous Foxp3 expression in conventional T cells but not reprogramming of regulatory T cells. Immunity 2012;36:262–75.
[177] Komatsu N, Mariotti-Ferrandiz ME, Wang Y, Malissen B, Waldmann H, Hori S. Heterogeneity of natural $Foxp3^+$ T cells: a committed regulatory T-cell lineage and an uncommitted minor population retaining plasticity. Proc Natl Acad Sci USA 2009;106:1903–8.
[178] Xu W, Lan Q, Chen M, Chen H, Zhu N, Zhou X, et al. Adoptive transfer of induced-Treg cells effectively attenuates murine airway allergic inflammation. PLoS One 2012;7. e40314.
[179] Shimizu J, Yamazaki S, Takahashi T, Ishida Y, Sakaguchi S. Stimulation of $CD25^+CD4^+$ regulatory T cells through GITR breaks immunological self-tolerance. Nat Immunol 2002;3:135–42.
[180] McHugh RS, Whitters MJ, Piccirillo CA, Young DA, Shevach EM, Collins M, et al. $CD4^+CD25^+$ immunoregulatory T cells: gene expression analysis reveals a functional role for the glucocorticoid-induced TNF receptor. Immunity 2002;16:311–23.
[181] Nocentini G, Ronchetti S, Petrillo MG, Riccardi C. Pharmacological modulation of GITRL/GITR system: therapeutic perspectives. Br J Pharmacol 2012;165:2089–99.
[182] Nocentini G, Cari L, Riccardi C. GITR: a modulator of regulatory and effector T-cell activity crucial in tumour rejection and autoimmune diseases. eLS. Chichester: John Wiley & Sons Ltd; 2017.
[183] Xiao X, Shi X, Fan Y, Zhang X, Wu M, Lan P, et al. GITR subverts $Foxp3^+$ Tregs to boost Th9 immunity through regulation of histone acetylation. Nat Commun 2015;6:8266.
[184] Kim IK, Kim BS, Koh CH, Seok JW, Park JS, Shin KS, et al. Glucocorticoid-induced tumor necrosis factor receptor-related protein co-stimulation facilitates tumor regression by inducing IL-9-producing helper T cells. Nat Med 2015;21:1010–7.
[185] Schaer DA, Budhu S, Liu C, Bryson C, Malandro N, Cohen A, et al. GITR pathway activation abrogates tumor immune suppression through loss of regulatory T cell lineage stability. Cancer Immunol Res 2013;1.
[186] Cohen AD, Schaer DA, Liu C, Li Y, Hirschhorn-Cymmerman D, Kim SC, et al. Agonist anti-GITR monoclonal antibody induces melanoma tumor immunity in mice by altering regulatory T cell stability and intra-tumor accumulation. PLoS One 2010;5. e10436.
[187] van Olffen RW, Koning N, van Gisbergen KP, Wensveen FM, Hoek RM, Boon L, et al. GITR triggering induces expansion of both effector and regulatory CD4+ T cells in vivo. J Immunol 2009;182:7490–500.
[188] Bulliard Y, Jolicoeur R, Windman M, Rue SM, Ettenberg S, Knee DA, et al. Activating Fc gamma receptors contribute to the antitumor activities of immunoregulatory receptor-targeting antibodies. J Exp Med 2013;210:1685–93.
[189] Nocentini G, Cari L, Ronchetti S, Riccardi C. Modulation of tumor immunity: a patent evaluation of WO2015026684A1. Exp Opin Ther Pat 2016;26(3):417–25.
[190] Cari L, Nocentini G, Migliorati G, Riccardi C. Potential effect of tumor-specific Treg-targeted antibodies in the treatment of human cancers: a bioinformatics analysis. Oncoimmunology 2017;e1387705. [Epub ahead of print].

CHAPTER

17

Noncoding RNA-Targeted Therapeutics in Autoimmune Diseases: From Bench to Bedside

*Zimu Zhang**, *Zhi Yao**, *Rongxin Zhang**,**

*School of Basic Medical Sciences, Tianjin Medical University, Tianjin, China; **School of Life Sciences and Biopharmaceutics, Guangdong Pharmaceutical University, Guangzhou, China

OUTLINE

The Epigenetics of Autoimmunity. http://dx.doi.org/10.1016/B978-0-12-809912-4.00017-9

17.1 INTRODUCTION

The mammalian genome is transcribed into multiple types of RNAs; however, only approximately 2% of transcripts can be translated into functional proteins. The remaining transcripts serve as epigenetic modifiers and play critical roles in gene regulation. These transcripts are defined as noncoding RNAs (ncRNAs) and can be grouped into several categories, including microRNAs (miRNAs) and long noncoding RNAs (lncRNAs) [1].

miRNAs are short (approximately 22 nucleotides) single-stranded ncRNAs that play a crucial role in regulating gene expression in a posttranscriptional manner. LncRNAs are larger than 200 nucleotides and lack the ability to code for proteins. They can also be classified into long intergenic ncRNA (lincRNA), intronic lncRNA, antisense lncRNA, transcribed pseudogene lncRNA, and enhancer RNA according to their proximity to protein-coding mRNAs. With respect to their function, lncRNAs can be classified as nonfunctional lncRNA, lncRNAs that act alone without a necessary transcript, and functional lncRNAs that can act in cis or trans. miRNAs and lncRNAs can participate in almost every genetic pathway including the immune response, which is indispensable for maintaining homeostasis and defending the host against antigens [2–4].

Autoimmune diseases such as systemic lupus erythematosus (SLE), multiple sclerosis (MS), rheumatoid arthritis (RA), and type 1 diabetes (T1DM) are characterized by the development of unexpected autoantibodies or the presence of deviant autoreactive immune cells [5]. Functional studies have revealed that a spectrum of aberrantly expressed miRNAs and lncRNAs exert significant roles in the pathogenesis of most autoimmune diseases by modulating autoimmunity-related genes and epigenetic processes [3]. Increasing evidence suggests that the pattern of circulating cell-free miRNAs and lncRNAs in the serum can serve as diagnostic and prognostic biomarkers, which allows molecular dynamic calculations to be performed using noninvasive liquid biopsies. Moreover, various mimics and inhibitors have been shown to efficiently modulate ncRNA function in vitro and have partly shown efficacy in preclinical studies, thus accelerating the development of novel drugs and fueling hope for the success of clinical trials for autoimmune diseases.

This chapter aims to briefly summarize the knowledge of ncRNA-targeted inhibition or replacement strategies. Special focus will be placed on small interfering RNAs (siRNAs) and antisense oligonucleotide (ASO)-based therapeutics as well as genome editing approaches, with a focus on in vivo delivery strategies and the ongoing challenge to improve the efficacy and specificity of ncRNA-targeted therapeutics. Finally, progress on the preclinical and clinical development of ncRNA-based autoimmune disease therapeutics will be reviewed and potential future steps to be taken in this field will be outlined.

17.2 TARGETING ncRNA

miRNA genes are typically transcribed by RNA polymerase II and then processed into mature miRNAs. The miRNA-induced silencing complex (miRISC) can lead to translational repression, mRNA degradation, or mRNA destabilization by binding to complementary sequences in target mRNAs [6]. LncRNAs are very much like mRNAs in terms of their biogenesis and form. They are found to be transcribed pervasively but are highly diverse in their sequence, structure, and function [3,4].

Two major approaches are applied for ncRNA-targeted therapeutics: the first approach uses ncRNA inhibition, which suppresses the action of endogenous ncRNA using siRNAs, ASOs, or other strategies; and the second approach uses ncRNA replacement, which introduces synthetic ncRNAs (mimics) or activates gene expression to restore the function of the endogenous ncRNA.

17.2.1 ncRNA Inhibition Strategies

Therapeutic strategies for miRNA and lncRNA inhibition can be categorized according to their mechanism of action. Small RNA sponges, siRNAs, ASOs, genome editing, and small-molecule inhibitors are the main approaches that have been used to date.

17.2.1.1 Small RNA Sponges

Small RNA sponges are vector-based strategies, which are constructed using mRNA sequences that contain multiple artificial small RNA binding sites. These binding sites can serve as a decoy or a "sponge". When the sponge-encoded vectors are transfected into cultured cells, they can selectively deplete miRNAs and other endogenous small RNAs, consequently allowing the translation of target mRNAs [7] (Fig. 17.1A). Du et al. showed that in vivo silencing of miR-326 by miRNA sponges triggered a decrease in the number of Th17 cells and a relief in disease severity, and vice versa. This study suggests that miR-326 plays important roles in Th17 polarization and MS pathogenesis [8]. Several lncRNAs and other ncRNAs could act as natural sponges to abrogate the availability of miRNAs and alter the expression of downstream genes; for instance, lncRNA-ATB was reported to compete with endogenous RNA for binding to the miR-200 family [9]. Similarly, the circular RNA, ciRS-7, was reported to function as an efficient miRNA sponge and strongly suppress miR-7 activity [10].

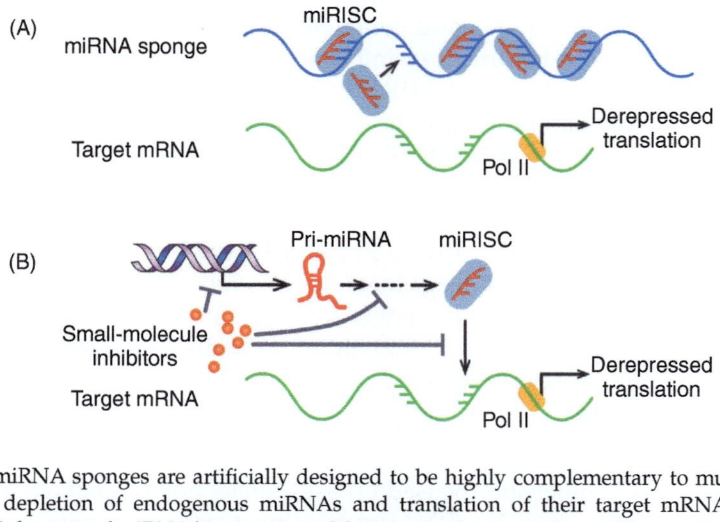

FIGURE 17.1 (A) miRNA sponges are artificially designed to be highly complementary to multiple miRNAs, which leads to the selective depletion of endogenous miRNAs and translation of their target mRNAs. (B) Small-molecule inhibitors control multiple steps of miRNA biogenesis and function and prevent the inhibitory effect of miRNA on mRNA translation.

17.2.1.2 siRNAs

RNA interference (RNAi) was first reported by Fire et al. in 1998 when they identified that double-stranded RNA (dsRNA) in *Caenorhabditis elegans* could induce potent gene silencing in a sequence-specific manner [11]. In *Drosophila* cells, duplexes of 21- and 22-nucleotide RNAs were identified to be the main effectors of RNAi activity and were termed siRNAs [12]. RNAi has since become an attractive field of study and the mechanism has been shown to be well-conserved in several species. In human cells, a dsRNA-specific ribonuclease (Dicer) initiates RNAi in the nucleus by cleaving long dsRNA into 20–24-nucleotide double-strand siRNAs with 3′ OH and 5′ phosphate (PO_4) groups. These siRNA duplexes assemble an RNA-induced silencing complex (RISC) with argonaute 2 (AGO2) and other proteins. siRNA is unwound in the RISC and its guide strand is released to bind to mRNA where endonucleolytic cleavage by AGO2 occurs at approximately 10–12 nucleotides downstream of the 5′ end of the binding site [13].

In 2001, Elbashir et al. reported that exogenous 21-nucleotide siRNA duplexes could enter the RISC and effectively suppress gene expression through perfect sequence complementarity to target mRNA transcripts. This provided a promising tool for investigating gene function and gene-specific therapeutics. [14]. In 2003, Song et al. successfully performed an in vivo siRNA study in a mouse model of autoimmune hepatitis. Treatment with naked siRNA caused a significant reduction in *Fas* mRNA and Fas protein levels, which protected the mice from hepatocyte necrosis and inflammatory infiltration [15]. SiRNA was more recently used to knockdown the lncRNA, *HULC*, to determine its role in the miR-200a-3p/ZEB1-mediated epithelial-to-mesenchymal transition in hepatocellular carcinoma [16].

It is noteworthy that the half-life of unmodified siRNA is too short to reach target organs and cells in systemic delivery due to degradation by endogenous nucleases in the bloodstream. Therefore most of the initial clinical trials used siRNAs administered locally via a pulmonary route, eye drops, or intravitreal injections. For instance, with the ability of specifically targeting beta-2 adrenergic receptor mRNA, the siRNA drug, SYL040012 (bamosiran), completed a phase II clinical trial (ClinicalTrials.gov identifier NCT02250612) for the treatment of elevated intraocular pressure in patients with open-angle glaucoma or ocular hypertension via eye drop administration [17]. Another siRNA drug, ALN-RSV01, has also completed a phase II clinical trial (ClinicalTrials.gov identifier NCT01065935) for the treatment of infection with respiratory syncytial virus via nebulization. The systemic delivery of modified siRNA complexes is currently being tested by several companies; for example, Alnylam Pharmaceuticals has developed a novel GalNAc-siRNA delivery platform through which the siRNA drug, ALN-TTR02 (patisiran), can be administered by intravenous infusion. Clinical trials for this drug have now entered phase III (ClinicalTrials.gov identifier NCT02510261) [18].

As of January 2017, no siRNA drugs were approved by the Food and Drug Administration (FDA); however, the number of siRNA drugs that have entered clinical trials has increased rapidly. RNAi research has focused on the development of siRNA as a promising therapeutic platform and provided new insights into the therapeutics of autoimmune diseases such as autoimmune uveitis and immune-mediated pulmonary diseases.

17.2.1.3 ncRNA-Targeted ASOs

In 1978, Zamecnik and Stephenson focused on a 13-nucleotide oligodeoxynucleotide that was able to target Rous sarcoma virus 35S RNA by binding to complementary sequences through

Watson–Crick base pairing. The oligodeoxynucleotide disrupted the viral replication and protein translation in vitro and led to the creation of the field of ASOs [19].

ASOs inhibit gene expression or interfere with functional RNAs through multiple general mechanisms, including RNase H-mediated cleavage, steric blocking of translation, and splice switching [20,21]. The most common mechanism of ASO activity is RNase H-mediated cleavage. ASOs bind to the target RNA sequence and recruit RNase H1, which initiates the RNA cleavage. Without RNase H, the ASO–RNA complexes do not lead to enzyme recruitment or RNA degradation but suppress gene expression and protein translation via sterically blocking ribosomal access to RNA (Fig. 17.2A). Splice-switching activity is another mechanism that occurs without RNase H, whereby ASOs can modulate pre-mRNA splicing and the presence of splice variant proteins. Synthetic oligonucleotides targeting miRNAs usually function through the first two mechanisms, whereas those targeting lncRNAs induce inhibitory activities via multiple mechanisms, such as mRNA-targeted ASOs.

Compared to siRNAs, which are double-stranded duplexes with an average molecular weight of 13.3 kDa, ASOs are single-stranded molecules with an average molecular weight of 5–8 kDa. They are designed to be highly complementary to the target sequences, which allows for more efficient and flexible binding to the target RNA [22]. Boutla et al. were the first to report that miRNA-specific ASOs could be used to inactivate miRNA function in Drosophila melanogaster embryos [23]. Numerous miRNA-targeted ASOs were subsequently transfected into human cells

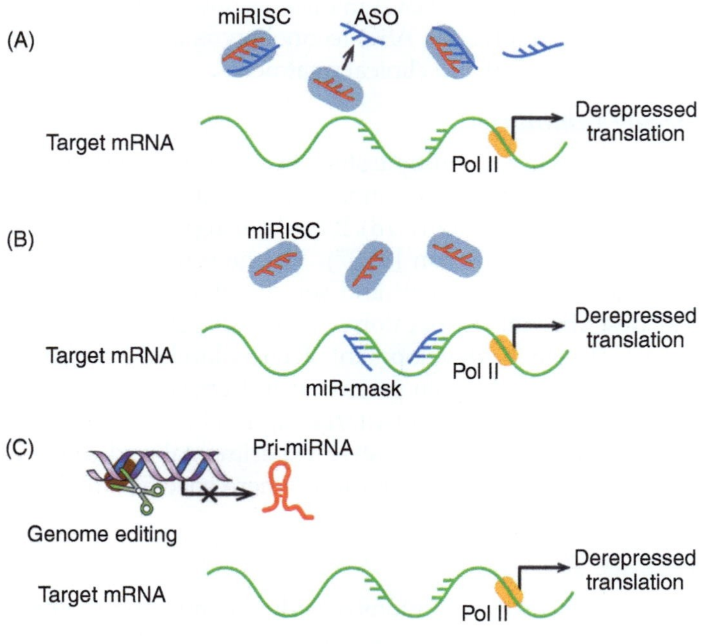

FIGURE 17.2 (A) miRNA-targeted ASOs can bind to and specifically restrain the function of target miRNAs based on their highly complementary sequences to the target miRNA. (B) miR-masks cover the target sites on mRNAs where they could be bound by specific miRNA, therefore derepressing the inhibitory effect on mRNA translation. (C) Genome-editing approaches cut the genome at desired locations and remove the target genes, therefore offering precise modification of the genome.

and were used as novel biological tools to investigate miRNA function [24,25]. miRNA-masking (miR-mask) technology is another method through which to realize the ASO approach. A miR-mask covers up the miRNA-complementary sites on mRNAs and derepresses their translation [26] (Fig.17.2B). A recent study disrupted nonsmall cell lung cancer (NSCLC) proliferation and metastasis by designing a miR-mask to block the effects of miR-522 targeting *DENND2D* [27].

Clinical trials of miRNA-targeted ASOs started in 2008. The first miRNA-targeted drug, miravirsen (also known as SPC3649), was developed by Santaris Pharma for the treatment of chronic hepatitis C virus infection by targeting miR-122. Miravirsen has been reported to be safe and effective in the long term [28,29] and is currently being tested in phase II clinical trials (ClinicalTrials.gov identifier NCT02452814). Although miRNA-targeted drugs are all still in clinical trials, four approved mRNA- or pre-mRNA-targeting drugs (fomivirsen, mipomersen, eteplirsen, and nusinersen) have shown promise for use as ASO drugs (Table 17.1).

It is noteworthy that the RISC-mediated activity of siRNA is mainly localized in the cytoplasm, which is thought to prevent it from targeting nuclear-retained RNAs. However, RNase H is found in the cell nucleus, mitochondria, and cytoplasm, which endows ASOs with the ability to efficiently reduce the levels of transcripts localized in the nucleus. This is especially relevant for those lncRNAs that are present exclusively in the nucleus or in both the nucleus and the cytoplasm [34]. A recent study used an ASO targeting taurine upregulated gene 1 (*TUG1*), which is a cancer-related lncRNA. Coupled with a drug delivery system, this intravenously administered ASO was able to effectively inhibit cytoplasmic- and nuclear-localized TUG1 in mice, as determined by RNA-fluorescence in situ hybridization (FISH) analysis [35]. Overall, ASO technology is a powerful experimental approach that opens new avenues for preclinical and clinical treatments.

17.2.1.4 Small-Molecule Inhibitors

Small-molecule inhibitors can target multiple steps of miRNA biogenesis and function, including primary miRNA transcription, miRNA maturation, the formation of the miRISC, and interactions between miRISC and the target mRNA (Fig.17.1B). It has been reported that azobenzene or aryl amide derivatives could inhibit miR-21 expression [36,37]. Via the screening of small molecule libraries, several compounds were identified as potent and selective inhibitors of miR-122 or miR-221/222 [38,39]. Velagapudi et al. described the use of Inforna, which is a computation-based rational design of small RNA-targeting molecules, in the development of a new kind of small molecule, targaprimir-96. This molecule was able to target the hairpin precursor and Drosha processing site of pri-miR-96 in breast cancer cells [40]. Small-molecule miRNA-targeting inhibitors have yet to be studied in autoimmune diseases but they represent an alternative experimental and therapeutic approach. Such studies may advance a disease–gene–drug paradigm where specific small molecules are designed with the help of data from genome sequencing.

17.2.1.5 Genome Editing

Most ncRNA genes lie within intronic regions of coding genes, and their expression is generally subject to the same transcriptional processes as other genes. Although transcriptional regulation is proposed to be an important step in controlling the expression of ncRNAs, siRNA- and ASO-mediated gene silencing could not maintain long-term control and would diminish with time and cell division. Recently developed genome-editing approaches, such as zinc finger nucleases (ZFNs), transcription activator-like effector nucleases (TALENs), and clustered regularly interspaced short palindromic repeat (CRISPR)-associated (Cas) systems, offer precise modification of the genome and the potential to permanently correct the underlying cause of a disease.

TABLE 17.1 ASO Drugs Approved by the FDA

Drug name	Target	Use	Administration route	Sponsors	Approval date	References
Fomivirsen (Vitravene)	*UL123* (a key CMV gene)	CMV infection in patients with AIDS	Intraocular injection	Isis Pharmaceuticals	August 1998	[30]
Mipomersen (Kynamro)	Apolipoprotein B-100	HoFH	Subcutaneous injection	Ionis Pharmaceuticals	January 2013	[31]
Eteplirsen (Exondys 51 or AVI-4658)	Dystrophin gene	DMD	Intravenous infusion	Sarepta Therapeutics	September 2016	[32]
Nusinersen (Spinraza)	*SMN1* gene	Spinal muscular atrophy	Intrathecal injection	Biogen	December 2016	[33]

ASO, Antisense oligonucleotide; *CMV*, cytomegalovirus retinitis; *DMD*, Duchenne muscular dystrophy; *FDA*, Food and Drug Administration; *HoFH*, homozygous familial hypercholesterolemia.

ZFNs are artificial restriction enzymes generated by fusing a custom-designed DNA-binding zinc finger domain to a DNA-cleavage domain derived from the class-IIS restriction endonuclease, FokI [41]. Engrafted with engineered ZFNs that disrupted endogenous *CCR5*, Perez et al. generated a HIV-resistant genotype in HIV-1-infected mice, which displayed lower viral loads and higher $CD4^+$ T-cell counts [42]. In 2014, a remarkable clinical trial was conducted to investigate the efficacy and safety of the adoptive transfer of ex vivo expanded *CCR5*-targeted ZFN-modified autologous $CD4^+$ T cells into HIV patients, which eventually produced partially satisfying results [43]. TALENs are another class of artificial restriction enzymes that consist of a FokI-derived nonspecific DNA cleavage domain and transcription activator-like effectors, which are proteins secreted by *Xanthomonas* bacteria that can be engineered to bind desired DNA sequences [44]. TALENs have a similar efficiency and precision to ZFNs and are also prominent tools in the field of genome editing.

CRISPR/Cas9, as a representative version of the CRISPR/Cas system which is originally a prokaryotic immune system that confers resistance to invading phage, revolutionized molecular biology and has been extensively used to edit genomes in recent years. After direction by a synthetic single-guide RNA (sgRNA), the Cas9 nuclease can cut the genome at a desired location and remove the target genes [45]. A comparative study revealed that the efficiency of CRISPR/Cas9-induced silencing activity was significantly higher (10- to 100-fold) than that of TALENs in human induced pluripotent stem cells (iPSCs) [46]. Given its advantages in terms of simplicity and scale, the CRISPR/Cas9 system was applied for the inhibition of miRNA and lncRNA activity and showed robust repression in vitro and in vivo [47,48] (Fig.17.2C). It is noteworthy that CRISPR/Cas9-based therapeutics reached human testing for the first time in October 2016. Researchers from the University of Sichuan, China, modified autogenous immune cells using CRISPR/Cas9 and increased the cell number in vitro before injecting the cells back into lung cancer patients to promote the immune response to attack and defeat the tumor cells [49]. It is expected that CRISPR/Cas9-mediated genome editing will be a therapeutic strategy with great efficiency and stability.

17.2.2 ncRNA Replacement Strategies

NcRNAs such as miRNAs and lncRNAs are indispensable for the development and homeostasis of the immune system [2,3]. In cases where an autoimmune disease results from the lack of expression of a specific ncRNA, replacement strategies are warranted to restore the lost function.

Synthetic miRNA mimics are generally short double-stranded oligonucleotides and are among the most widely used approaches for restoring miRNA levels. After transfer into cells, miRNA mimics are processed into their single-strand form and regulate the target mRNAs in a miRNA-like manner [50]. For example, miR-125a was found to be downregulated in peripheral $CD4^+$ T cells of patients and mouse models with SLE, MS, and Crohn's disease. After intravenous injection of chemically synthesized miR-125a mimics, the expression of miR-125a in mouse $CD4^+$ T cells was elevated and the pathogenesis of experimental autoimmune encephalomyelitis (EAE) was ameliorated [51]. Several miRNA mimic-based drugs have been developed by pharmaceutical companies and have entered clinical trials (Table 17.2).

Besides oligonucleotide-based mimics, some small molecules are also capable of restoring miRNA levels. Decitabine and 5-azacytidine are epigenetic drugs that can modify methylation profiles

TABLE 17.2 Selected miRNA-targeted Therapeutics Currently in Clinical Development

MicroRNA	Drug name	Drug format	Sponsors	Use	Clinical trial stage	ClinicalTrials.gov identifier
miR-21	RG-012	2′-F- and 2′-MOE-modified antisense inhibitor	Regulus Therapeutics and Sanofi Genzyme	Alport syndrome	Phase II	NCT02855268
miR-29b	MRG-201	miRNA mimic	miRagen Therapeutics	Pathologic fibrosis	Phase I	NCT02603224
miR-155	MRG-106	LNA-modified antisense inhibitor	miRagen Therapeutics	Hematological malignancies	Phase I	NCT02580552
miR-155	MRG-107	LNA-modified antisense inhibitor	miRagen Therapeutics	Amyotrophic lateral sclerosis	Preclinical	ND
miR-34	MRX34	miRNA mimic	Mirna Therapeutics	Melanoma	Phase II	NCT02862145
miR-122	SPC3649 (Miravirsen)	LNA-modified antisense inhibitor	Santaris Pharma[a]	HCV infection	Phase II	NCT02508090
miR-122	RG-101	GalNAc-conjugated antisense inhibitor	Regulus Therapeutics	HCV infection	Phase II	ND
miR-103/107	RG-125 (AZD4076)	GalNAc-conjugated antisense inhibitor	Regulus Therapeutics and AstraZeneca	Nonalcoholic steatohepatitis in patients with type 2 diabetes	Phase II	NCT02826525

(Continued)

TABLE 17.2 Selected miRNA-targeted Therapeutics Currently in Clinical Development (*Cont.*)

MicroRNA	Drug name	Drug format	Sponsors	Use	Clinical trial stage	ClinicalTrials.gov identifier
miR-17	ND	Antisense inhibitor	Regulus Therapeutics	Autosomal dominant polycystic kidney disease	Preclinical	ND
miR-27	ND	Antisense inhibitor	Regulus Therapeutics	Cholestatic disease	Preclinical	ND
miR-10b	ND	Antisense inhibitor	Regulus Therapeutics	Glioblastoma multiforme	Preclinical	ND
miR-92a	MRG-110	Antisense inhibitor	miRagen Therapeutics	Chronic ischemic disorders	Preclinical	ND
miR-215	ND	miRNA mimic	Mirna Therapeutics	Cancer (details undisclosed)	Preclinical	ND
miR-101	ND	miRNA mimic	Mirna Therapeutics	Cancer (details undisclosed)	Preclinical	ND
miR-16	ND	miRNA mimic	Mirna Therapeutics	Cancer (details undisclosed)	Preclinical	ND
let-7	ND	miRNA mimic	Mirna Therapeutics	Cancer (details undisclosed)	Preclinical	ND

2′-F, 2′-fluoro; *2′-MOE*, 2′-O-methyoxyethyl; *GalNAc*, N-acetylgalactosamine; *HCV*, hepatitis C virus; *LNA*, locked nucleic acid; *ND*, not determined.
a Santaris Pharma was acquired by F. Hoffmann-La Roche in 2014.

and simultaneously induce the nonspecific expression of several miRNAs [52]. Enoxacin, a fluoroquinolone compound, was similarly reported to bind to the miRNA biosynthesis protein, TAR RNA-binding protein 2, to enhance the production of a spectrum of miRNAs [53].

Compared with restoring miRNA levels through synthetic miRNA mimics, restoring biological activities of lncRNAs seems to be more difficult to achieve [54]. Unlike miRNAs, lncRNAs can fold into secondary and higher-order structures whose sequences are hard to predict and difficult to replace. In addition, many lncRNAs are localized in the cell nucleus, which requires an additional step to deliver the mimic to the correct location. LncRNAs are also more easily degraded than miRNAs and their underlying mechanisms are not well defined. If a lncRNA loses its function, it is impossible to restore it without expressing it at the specific genomic locus; however, a recent study by Konermann et al. may shed light on the restoration of lncRNA levels. This study demonstrated that the structure-guided engineered CRISPR/Cas9 complex could induce efficient transcriptional activation instead of its previous inhibitory activities. This new system demonstrated potent activation of endogenous genes as well as upregulation of lincRNA transcripts, providing a powerful genetic perturbation technology [55].

17.3 CHALLENGES FOR ncRNA-TARGETED THERAPEUTICS

Unmodified RNA and DNA oligonucleotides are usually unstable in vivo and are extremely susceptible to serum nuclease activity. Therefore chemical modifications are required to increase the resistance to nucleolytic degradation and simultaneously enhance the binding affinity to targeted RNA. Naked oligonucleotides also display low cellular penetration and inefficient delivery to specific organs, tissues, and cell types. Thus appropriate in vivo delivery strategies are necessary to direct the oligonucleotides to the target location and permit their entry into the cell. Off-target effects, immunostimulatory effects, toxicity, and efficacy-related concerns are also inevitable outcomes when bringing RNA-targeted therapeutics to the clinic. Therefore a variety of approaches have been developed to overcome these obstacles and enhance the drug-like properties of oligonucleotides.

17.3.1 Chemical Modification of Oligonucleotides

The mechanism through which siRNAs interact with the cellular RISC machinery may only tolerate limited chemical modifications; however, ASOs may bind to their protein-free target mRNA and allow more extensive and complex modifications, which would permit ASOs to remain functional in the cells. Numerous modifications to improve the stability and efficacy of siRNAs and ASOs have been developed over recent years; however, considerably more attention has been paid to ASO technology, and a spectrum of clinical trials are underway with these therapeutics.

17.3.1.1 Backbone Modification

The first chemical modification was the improvement of the oligonucleotide backbone to increase its nuclease resistance. In 1987, Matsukura et al. developed a phosphorothioate backbone to replace the inherent phosphodiester linkage in oligonucleotides. This novel backbone exhibited significant stability in plasma, tissues, and cells and assisted the oligonucleotides in reaching their target RNA [56]. A later study demonstrated that phosphorothioate oligonucleotides were able to strongly bind to plasma proteins in vivo, conferring effective cellular uptake and delayed renal excretion [57]. Fomivirsen (Vitravene) is a synthetic 21-oligonucleotide ASO with phosphorothioate linkages. It was the first antisense drug to be approved by the FDA in August 1998, and it is now administered via intraocular injection to treat immunocompromised patients with cytomegalovirus retinitis [30].

17.3.1.2 2′ Sugar Modifications

Modifications to the 2′ position of the sugar moiety (i.e., 2′-O-alkylribonucleotides and 2′-deoxy-2′-fluoro) were developed to block exonuclease attack and enhance the binding affinity to targeted RNA [58,59]; however, these modifications may impair RNase H-mediated cleavage of the RNA strand to some extent. To resolve this limitation, chimeric oligonucleotides were designed that contained a central 10-nucleotide phosphorothioate gap flanked by 2′ sugar modifications ("gapmers") to allow cleavage of the targeted RNA strand by RNase H [60]. It also been reported that oligonucleotides containing 2′-O-methylphosphonate (2′-OMe) or 2′-O-methoxyethyl (2′-MOE) exhibit high affinity and sufficiency to elicit RNase H cleavage [61–63]. Additional studies suggested that 2′-OMe-modified phosphorothioate siRNAs improved in vivo stabilization and effective gene silencing [64]. Furthermore, 2′-OMe-modified ASOs were found to effectively inactive miRISC and block miRNA function in human cell extracts and cultured human cells [65]. "Antagomirs" were developed in 2005 to incorporate the 2′-OMe modifications, a 3′ cholesterol tail, and a phosphorothioate backbone. These antagomirs showed good tissue distribution and were effective at silencing target miRNAs following intravenous administration to mice [66]. 2′-MOE-modified phosphorothioate ASOs were also administered to mice by intraperitoneal or subcutaneous injection, and were able to significantly reduce the level of target miRNA in the liver. Luciferase sensor reporter assays have demonstrated that 2′-MOE-modified ASOs are more effective than 2′-OMe-modified ASOs at inhibiting miRNA activity in HeLa cells and primary hepatocytes [67].

Several leading drug companies have performed clinical trials using the gapmers with 2′ sugar modifications. Drisapersen (PRO051) is a 2′-OMe phosphorothioate oligonucleotide drug that alters the splicing of the dystrophin RNA transcript to eliminate exon 51 from the mature dystrophin mRNA [68]. The drug was developed by BioMarin Pharmaceutical, and it has been used in phase III clinical trials for the treatment of Duchenne muscular dystrophy (ClinicalTrials.gov identifier NCT02636686). Mipomersen (Kynamro) from Ionis pharmaceuticals is a 2′-MOE phosphorothioate oligonucleotide drug that targets apolipoprotein B100 mRNA in the liver. The drug can be systematically delivered and received FDA approval for the treatment of patients with homozygous familial hypercholesterolemia in January 2013 [31].

A locked nucleic acid (LNA) is a novel oligonucleotide analogue that contains a conformationally restricted nucleotide with a 2′-O, 4′-C-methylene bridge, which locks the sugar ring into a 3′ endo conformation. The LNA modification was first developed in 1998 and offered dramatically enhanced stability towards cellular nucleases and efficient hybridization to the target RNA [69,70]. Subsequent studies systematically administered LNA-modified oligonucleotides that targeted miRNA or lncRNA to explore their in vivo function in nonhuman primates and mice [71,72]. It has also been reported that the effective doses of LNA-modified oligonucleotides are much lower than that of other modified oligonucleotides, and that their efficacy could be further enhanced by combining them with other modifications (i.e., 2′-fluoro or 2′-OMe) [73].

The excellent drug-like properties of LNA-modified oligonucleotides attracted the attention of many companies who wanted to bring these compounds to the clinic. SPC2996 (a novel LNA-modified phosphorothioate antisense molecule targeting *Bcl-2*) was developed by Santaris Pharma for the treatment of chronic lymphocytic leukemia. It has completed phase II clinical trials where it exhibited rapid leukemic cell clearance and immune activation (ClinicalTrials.gov identifier NCT00285103) [74]. Miravirsen (SPC3649), another LNA-modified ASO drug developed by Santaris Pharma, can bind with high affinity to miR-122 and showed promise in a phase II clinical trial for the treatment of hepatitis C [28,63]. Mixmer oligos, consisting of LNA interspersed with 2′-OMe nucleotides for high-affinity binding to lncRNAs, were generated by RaNA Therapeutics, and preclinical trials have been performed to test their treatment of spinal muscular atrophy. The lncRNA, SMN-antisense 1 (SMN-AS1), was found to be a novel regulator and repressed *SMN2* expression by recruiting the polycomb repressive complex 2 to its locus [75].

17.3.1.3 *Additional Modifications*

N-diethyl-4-(4-nitronaphthalen-1-ylazo)-phenylamine (ZEN) and 2′,4′-constrained ethyl (cEt) are novel modifications that have exhibited promise and advanced to human clinical trials [76,77]. The ZEN modification combined with the 2′-OMe RNA backbone showed enhanced miRNA inhibition in vitro and high specificity and low toxicity in cell culture [77]. AZD9150 (STAT3-Rx) was the first Gen 2.5 cEt-modified ASO to enter clinical trials [78]. It functions as an inhibitor of the previously undruggable transcription factor, STAT3, and has entered phase II clinical trials for the treatment of advanced tumors (ClinicalTrials.gov identifier NCT02983578). These studies reveal the considerable progress that has been made over recent decades. Strategies that combine multiple chemical modifications may provide a critical step forward in therapeutic applications.

17.3.2 In Vivo Delivery Strategies

Various patterns of chemical modifications have attempted to enhance the drug-like properties of oligonucleotides; however, how to direct these oligonucleotides to target organs and deliver them into cells is another major barrier to be overcome. An appropriate in vivo delivery vector or carrier is necessary to increase the tissue distribution and delay the rapid excretion by the liver and kidneys. The 5′ end of the antisense strand performs the key inhibitory activity, therefore the 3′ end of the siRNAs or ASOs may tolerate being conjugated to ligands that can interact with membrane acceptor sites and thus facilitate cellular uptake [64,66]. As both siRNAs and ASOs display a negatively charged surface, it is reasonable to suppose that the factors influencing the in vivo delivery of one molecule may be similar to the other.

17.3.2.1 *Lipid-Based Delivery*

Lipid-based vectors are among the most widely used for systemic nonviral gene delivery as they have similar structures to cell and organelle membranes. A cholesterol tail (a major helper lipid) connected to the 3′ end of miRNA-targeting ASOs resulted in specific, efficient, and long-lasting effects in multiple organs and tissues [66]. Similarly, covalently linked cholesterol derivatives (i.e., lithocholic acid or lauric acid) on siRNAs enhanced the cellular uptake and gene silencing in liver cells [79]. It has also been reported that high-density lipoprotein can direct siRNA delivery into the liver, gut, kidney, and steroidogenic organs, while low-density lipoprotein (LDL)-incorporated siRNAs are primarily targeted to the liver after systemic administration [80].

Subsequent studies have shown that cholesterol-modified siRNAs can be taken up by dendritic cells (DCs) in vitro and in vivo, thus allowing the development of targeted therapies for a larger spectrum of organ-specific inflammatory autoimmune diseases [81].

Liposomes are spherical vesicles with a typical lipid bilayer that can incorporate nucleic acids to form a lipoplex [82]. In vivo liposome-mediated siRNA delivery was first reported by Sørensen et al. These authors incorporated siRNAs into a cationic liposome, DOTAP (1,2-dioleoyl-3-trimethylammonium-propane), to silence tumor necrosis factor α (TNF-α) in mice. DOTAP-based liposomes were able to deliver siRNAs into various cell types following intravenous or intraperitoneal injection of fluorescein-labeled siRNAs [83]. In addition to DOTAP, a number of other liposome-based transfection systems are commercially available, including Lipofectamine, Oligofectamine, DharmaFECT, and TransIT-TKO. Lipofectamine RNAiMAX is the mostly commonly used transfection agent and is efficient at delivering siRNA [84], while Lipofectamine CRISPRMAX is an ideal carrier for the delivery of Cas9 protein and gRNA complexes (Cas9 ribonucleoprotein) [85].

Polyethylene glycol (PEG)ylated liposomes are effective tools for enhancing the pharmacokinetics of therapeutic molecules in vivo, and PEGylated liposomal doxorubicin (Doxil or Caelyx) is FDA approved for clinical use [86]. An siRNA targeting the hepatitis B virus (HBV) mRNA was incorporated into a PEGylated liposome to form a stable nucleic acid–lipid particle (SNALP). Intravenous injection of the compound was able to significantly silence HBV replication in mice [87]. Similar studies were conducted in nonhuman primates and the animals treated with SNALP-formulated *APOB*-specific siRNAs had a maximal silencing of >90% and significant reductions in ApoB protein, serum cholesterol, and LDL levels [88]. In a recent study, PEGylated liposomes containing echo-contrast gas, "bubble liposomes," were shown to be useful tools for diagnostic ultrasound imaging and miRNA delivery [89].

pH-responsive liposomes are lipid-based vectors with amphoteric properties, which are capable of deciding whether to facilitate desired encapsulation or to release their cargo in different pH environments. Liposomes made of phosphatidylethanolamine remain stable in plasma but are able to transit their structures in response to a low pH environment within the endosomal system, thus realizing the "smart" delivery of oligonucleotides into the cytoplasm [90]. A novel amphoteric liposome, Nov038, was generated from a specific ratio of 1-palmitoyl-2-oleoyl-sn-glycero-3-phosphocholine, 1,2-dioleoyl-sn-glycero-3-phosphoethanolamine, cholesteryl hemisuccinate, and α-3′-O-cholesteryloxycarbonyl)-δ-(N-ethylmorpholine)-succinamide. Nov038-incorporated CD40-specific ASOs were efficiently delivered into DCs and macrophages and achieved potent gene knockdown effects in vivo. Furthermore, intravenous administration of this compound into the arthritic joints of mice with experimental arthritis was shown to be effective [91]. pH-responsive liposomes were successful both in vitro and in vivo, and their safety profile and tolerability led to them being used in clinical practice. MRX34 is a pH-responsive liposomal formulation complexed with miR-34a mimics. It was developed by Mirna Therapeutics and entered phase II clinical trials (ClinicalTrials.gov identifier NCT02862145) for the treatment of advanced melanoma [92].

17.3.2.2 Polymer Nanoparticle-Mediated Delivery

The rapid development of novel materials and nanoengineering has seen polymer-based nanoparticles emerge as powerful vehicles for the in vivo delivery of oligonucleotides. Polyethylenimine

(PEI) is one of the most extensively used cationic polymers for oligonucleotide delivery due to its strong compaction capacity and its intrinsic endosomolytic activity, and PEI-derived transfection reagents (i.e., JETPEI) are now commercially available. Low molecular weight branched PEI was introduced as a miRNA delivery platform. Intraperitoneal and intratumoral injections of PEI-complexed miR-145 or miR-33a were effectively delivered and inhibiting tumor growth in mice with xenograft colon carcinoma [93]. CD98-specific siRNA–PEI-loaded nanoparticles were recently generated and encapsulated in a hydrogel. Oral administration of the nanoparticles to mice with induced colitis resulted in a significant downregulation of CD98 expression in intestinal epithelial cells and macrophages and decreased intestinal inflammation. Therefore this therapy represents a promising strategy for the treatment and prevention of inflammatory bowel disease (IBD) and colitis-associated cancer [94]. To achieve more specific transfection efficacy, a transferrin-PEI delivery system was developed to direct siRNAs to activated T cells due to the increased expression of transferrin receptors on activated T cells. This targeting system may be useful for treating autoimmune diseases with more efficacies and less side effects [95].

Dendrimers are also artificial polymers with well-defined highly branched structures, of which the most widely used is polyamidoamine (PAMAM). Similar to PEI, PAMAM-carrying nucleic acids could be endocytosed by cells and deliver nucleic acids into the cytoplasm. Ren et al. designed a sequential codelivery system using a PAMAM-modified hollow gold nanoparticle with a near-infrared-radiation (NIR) response [96]. The sequential miRNA–drug delivery system successfully released miRNA-specific ASOs to sensitize the cancer cells, followed by the NIR-triggered release of the antitumor drug. This novel PAMAM-based delivery system achieved synergistic efficacy and provided a promising technology for advanced combination therapy.

Nanoparticles with a diameter <10 nm may be rapidly eliminated by the kidneys, while nanoparticles with diameters >200 nm may activate the complement system and be quickly removed from the blood stream. While liposomes are usually heterogeneous in size, polymer-based nanoparticles could be produced in relatively homogeneous sizes. Although the optimal nanoparticle size depends on the specific location and type of tissue in vivo, a diameter of approximately 50 nm performs excellently in most studies [97]. Due to the finely controlled assembly process and efficient cellular uptake, polymer nanoparticle-mediated delivery can act as an effective therapeutic and diagnostic tool in a variety of applications.

17.3.2.3 *Exosome-Mediated Delivery*

Exosomes are naturally occurring biological transport nanovesicles that are released by immune, tumor, and brain cells. They are now considered as a prominent and universal form of cell communication. Exosomes can contain, protect, and deliver RNA and appear to be well-tolerated for transferring cargo between different cells. They have attracted enormous research interest and have been used as delivery vectors for gene therapies [98].

Exosomes are capable of traversing complex biological barriers, especially the blood-brain barrier. Alvarez-Erviti et al. used self-derived DCs for exosome production and loaded the exosomes with exogenous siRNAs using electroporation. Following intravenous injection, these engineered exosomes effectively delivered siRNAs into the brain, resulting in a specific knockdown without attack from the complement system, coagulation factors, or circulating antibodies [99]. Exosomes can also selectively target monocyte-derived myeloid cells, as activated myeloid cells usually act as scavengers and have a high vesicle-uptake capacity [100]. Intranasally administered exosomes were shown to successfully deliver loaded antiinflammatory drugs to the brain where they were taken up by microglial cells and alleviated autoimmune encephalomyelitis and Parkinson's disease in mouse models [101,102].

A novel artificial exosome mimetic was recently constructed using grapefruit-derived lipids and shared the same spherical phospholipid bilayer structure as natural exosomes. The grapefruit-derived nanovectors (GNVs) efficiently delivered their therapeutic cargo (i.e., chemotherapeutic drugs, DNA expression vectors, siRNAs, and antibodies) in in vitro cell culture and mouse tumor models. It is noteworthy that the GNVs were unable to pass the placental barrier following systemic injection into pregnant mice, reflecting an advantage that most other artificial nanoparticles may not confer [103]. Zhuang et al. further improved the GNV hybrid using PEI and folic acid (FA-pGNVs) to carry therapeutic miRNAs for the treatment of brain tumors in mice. Following intranasal delivery, FA-pGNV/miR-17 nanoparticles were effectively transported to the brain and inhibited the tumor growth in a tumor-bearing mouse model. This provides a promising noninvasive approach for the treatment of inflammatory-related brain diseases such as autoimmune encephalomyelitis and malignant gliomas [104].

Naturally secreted exosomes containing lncRNAs have also been reported in a series of studies [105,106]; however, loading exosomes with exogenous lncRNAs warrants further exploration.

17.3.2.4 *Peptide- and Protein-Mediated Delivery*

Peptide- and protein-based carriers have been widely explored for oligonucleotide delivery, either through their electrostatic interactions or chemical conjugations between nucleotides and amino acids. Advantages of this method include their efficient cellular delivery, their increased stability against exonucleases, and their improved binding to complementary sequences.

Cell-penetrating peptides (CPPs) are a popular and distinct class of short peptides that facilitate the cellular uptake of nanoparticles and large fragments of nucleic acids. Most CPP–cargo conjugates enter cells by direct penetration or endocytosis and subsequently release their cargo into the cytoplasm or nucleus [107]. Crombez et al. reported on an improved variant of the amphipathic CPP, MPG-8, which could assemble nanoparticles with siRNAs and promote their efficient delivery into cells. Following systemic administration, cholesterol-modified nanoparticles of MPG-8 and siRNA targeting cyclin B1 displayed extensive tissue distribution and long-term stability and prevented tumor growth in a xenografted tumor mouse model [108].

Cell-targeting peptides are another class of short peptides that specifically interact with a cell-surface receptor on a particular cell type [107]. For example, the $\alpha v\beta 3/5$ integrins were shown to be overexpressed by cancer cells and specifically recognized by the short peptide motif Arg–Gly–Asp (RGD). Kim et al. developed novel tumor-targeted nanoparticles by coupling cyclic RGD peptides so that they could specifically direct siRNAs into cancer cells without affecting normal cells [109]. CD40L is also known as a protein that is primarily expressed on activated T cells and it serves as an important target for the treatment of autoimmune diseases. Using the crystal structure and molecular docking properties of CD40L, Ding et al. designed a CD40L-specific peptide ligand (A25) and conjugated it to the surface of liposomes. The A25-modified liposomes containing antiinflammatory drugs significantly reduced the $CD40L^+$ cell ratios and greatly improved the clinical scores in mice with EAE [110]. It is feasible that the A25-modified vectors could specifically deliver siRNAs or miRNAs to CD40L-expressing T cells for the treatment of autoimmune diseases.

Similar to the interaction between a ligand and a receptor, engineered vectors could be modified using an antibody fragment to bind to a specific antigen. This technique has been extensively applied for directing cargoes to target tissues or cells with high specificity and strong affinity.

Kumar et al. generated a novel siRNA vector conjugated with a single-chain antibody targeting CD7, which is a surface antigen present on most human T cells. After antibody binding, the CD7 receptor was rapidly internalized and the siRNAs delivered into the target T cells. This approach demonstrated the feasibility of suppressing HIV infection in humanized mice and created opportunities for highly specific treatments of many inflammation-related diseases without significant side effects. [111]

17.3.2.5 Carbohydrate- and Oligonucleotide-Mediated Delivery

Carbohydrate and oligonucleotide have been investigated as guides to deliver cargoes to specific tissues or cells. N-Acetylgalactosamine (GalNAc) is an amino sugar derivative of galactose that serves as a target ligand for asialoglycoprotein receptors (ASGPR), which are overexpressed on hepatocytes. A synthetic trivalent GalNAc ligand conjugated to siRNA was designed by Alnylam Pharmaceuticals to bind to ASGPR and facilitate liver-specific gene silencing in vivo [112]. Alnylam Pharmaceuticals have conducted several clinical trials using the GalNAc-siRNA delivery platform for the treatment of diseases such as hemophilia, paroxysmal nocturnal hemoglobinuria, familial amyloidotic cardiomyopathy, and polyneuropathy (ClinicalTrials.gov identifiers NCT02554773, NCT02352493, NCT02319005, and NCT01960348, respectively).

CpG oligonucleotides are short single-stranded synthetic DNA molecules that contain cytosine–phosphate–guanine motifs and are recognized by toll-like receptor 9 (TLR9) in immune system cells (i.e., B cells and plasmacytoid DCs). *Stat3*-targeting siRNAs were conjugated with CpG-containing oligonucleotides and injected into mice either locally at the tumor site or intravenously. CpG–*Stat3* siRNAs targeted and silenced the expression of the immunoregulatory transcription factor, STAT3, in $TLR9^+$ myeloid cells and B cells, triggering potent antitumor immune responses [113].

Aptamers are oligonucleotide molecules created from a large random sequence pool. They can form stable three-dimensional shapes by specifically binding to target-cell membrane receptors and be used to mediate in vivo delivery. Aptamer–siRNA chimeras were first reported in 2006 by McNamara et al. The authors screened the aptamer for binding to prostate-specific membrane antigens through systematic evolution of ligands by an exponential enrichment procedure. The chimeras specifically inhibited the growth of prostate cancer cells and mediated tumor regression in a xenograft mouse model following intratumoral injection [114]. These chimeras were modified in subsequent studies and exhibited enhanced silencing potency and in vivo stability following intravenous injection [115]. An aptamer–CPP complex was recently generated and produced a more efficient siRNA delivery and potent suppression of prostate tumor growth in vivo [116]. Aptamer–miRNA chimeras are also being investigated as replacement therapeutics; for example, an aptamer–miRNA-212 conjugate was designed to target the tyrosine kinase receptor, Axl, and was able to specifically restore TNF-related apoptosis-inducing ligand-mediated cytotoxicity in human nonsmall cell lung cancer [117].

17.3.2.6 Viral-Mediated Delivery

The viral vectors used in most in vivo studies mainly use retroviruses, adenoviruses, and adeno-associated viruses (AAV) to express plasmid DNA encoding short hairpin RNA (shRNA) for inhibitory functions or miRNA for replacement treatment. ShRNA is an artificial RNA molecule that contains an intramolecular stem-loop structure, which can be packed into viral vectors and continually synthesized in the host cells [118]. Viral vectors efficiently transport their genome into the nucleus, inducing long-term gene expression compared to the aforementioned nonviral vectors; however, the DNA packaging capacity of viral vectors is generally less than that of nonviral vectors. The clinical application of retroviral vectors is mainly limited by its genotoxicity and insertional mutagenesis, while adenoviral vectors exhibit a high risk of immunogenicity and have other safety concerns. Conversely, recombinant AAVs (rAAV) have a much smaller risk of genome insertion, deliver to nondividing cells, have poor pathogenicity, and a good safety profile [119]. Based on rAAV vectors, alipogene tiparvovec (Glybera) was developed to deliver an intact copy of the human lipoprotein lipase (LPL) gene to muscle cells to treat LPL deficiency. This was the first gene therapy drug approved in Europe [120]. RAAV vectors provide powerful tools for gene editing techniques such as ZFN, TALEN, and CRISPR/Cas9 [121]. The use of vector-engineering efforts may bring about the development of next-generation viral-mediated delivery with minimal adverse effects.

17.3.3 Off-Target Effects

Off-target effects are another major challenge associated with in vivo ncRNA-targeted therapeutics. Positions 2–7 at the 5′ end of miRNA, the so-called "seed sequence," are essential for target recognition and so that miRNA can bind to the 3′ untranslated region (UTR) of mRNAs containing only a partial miRNA complementary site [6]. Similarly, siRNAs could lead to miRNA-like effects by affecting partial complementary sequences, and induce unanticipated biological consequences (off-target effects), which underscores the importance of stringent target selectivity [122,123]. Later studies reported that siRNAs modified with 2′-OMe, an unlocked nucleic acid modification at position 7, or an abasic pivot substitution at position 6 could reduce most off-target effects without sacrificing silencing of the intended target gene [124–126]. Furthermore, owing to their dose-dependent properties, off-target effects could be diluted by using an enzymatic approach in which highly complex siRNA pools containing up to 60 accurately defined siRNAs were generated to synergistically silence one target gene and each individual siRNA was present at very low concentrations [127].

In addition, many miRNAs share a similar seed sequence and belong to the same miRNA family [6]. MiRNA-targeting ASOs are generally unable to distinguish between members from the same miRNA family under physiological conditions, thereby cross-inhibiting miRNAs with an identical seed sequence. An alternative approach to eliminate these off-target effects could be to disrupt the generation of miRNA precursors. As each individual pri-miRNA sequence is usually unique among different miRNAs, ASOs could be designed to bind specific pri-miRNA to inhibit the miRNA expression with minimum off-target effects.

A high frequency of off-target activity was also observed in the CRISPR/Cas9 system, which can induce mutations at sites other than the intended on-target site and hinder its therapeutic applications. The binding sites of sgRNA, the sequence of the protospacer adjacent motif (PAM), the delivery pattern of the Cas9 protein, and the cell type could all be involved in the mechanism of off-target effects [128–130]. Multiple efforts have been made to improve the CRISPR/Cas9 system; for example, truncation of the 3′ end of the sgRNA improved target specificity and led to a 5000-fold decrease in undesired mutagenesis at some off-target sites without affecting the on-target genome editing efficiency [131]. Moreover, the D10 mutant nickase version of Cas9 was able to replace the wildtype Cas9 nuclease, and a CRISPR/Cas9 system improved by a paired-nicking strategy with two sgRNAs exhibited enhanced genome editing specificity [132].

17.3.4 Immunostimulatory and Toxic Effects

Exogenous nucleotide sequences and delivery vectors may be detected by both the innate and adaptive immune system, therefore immunostimulatory effects are serious toxicological concerns for in vivo ncRNA-targeted therapeutics. Toll-like receptors (TLRs) are proteins that play an essential role in the innate immune system, which is responsible for recognizing pathogen-derived molecules (i.e., viral RNAs). SiRNAs and ASOs have similar properties to viral RNAs and are recognized by TLR3, TLR7, and TLR8 [133,134]. Although reduced immunostimulatory effects were triggered by nucleotide sequences lacking GU-rich motifs, siRNAs could still evoke GU content-independent IFN-α production in plasmacytoid DCs [135]. Subsequent studies reported that the associated IFN-α production may be associated with a UCA motif in oligonucleotides and the presence of a UGGC motif may induce toxic effects [136,137]. A series of chemical modifications of siRNA and ASOs have since been developed to abrogate immune activation. For example, modification at the 2′ position of the sugar ring (i.e., 2′-fluoro, 2′-OMe, and 2′-LNA) eliminated immune recognition by TLRs without reducing silencing potency [135,138]; however, LNA-induced toxic effects such as hepatotoxicity and body-weight loss should not be ignored in further investigations [139]. Inhibition of blood-coagulation and activation of the complement system were also observed in preclinical animal tests of phosphorothioate oligonucleotides [140,141].

Regarding the delivery vectors, the clinical use of PEI was limited by its high cytotoxicity (i.e., membrane disruption and cell apoptosis), and efforts to modify PEI nanoparticles have been made to overcome this problem. Conjugation of hydrophilic PEG onto PEI was proposed as a solution to reduce the toxicity [142]. Furthermore, the incorporation of poly(D,L-lactide-co-glycolide) permitted PEI nanoparticles to deliver siRNAs into cells and achieve efficient cellular uptake and targeted gene inhibition in vitro without significant toxicity [143]. Although rAAV vectors provide promising tools for delivery of the CRISPR/Cas9 system, the risk of immunogenicity is still the major concern for viral vector-based gene therapeutics. In contrast, it is noteworthy that exosomes, especially those naturally derived from endogenous cells, are well-tolerated by the immune system and exhibit a low toxicity, therefore demonstrating the potential to revolutionize and open up new avenues for widespread and efficient treatments [98].

17.4 DEVELOPMENT OF ncRNA-BASED AUTOIMMUNE DISEASE THERAPEUTICS

Non-coding RNA-targeted therapeutics offer several advantages over traditional small molecule-based drugs, most notably that oligonucleotides can be designed to target almost any gene, including the "undruggable" ones. Owing to intense interest from biomedical research institutions and pharmaceutical companies, a large body of work has been devoted to the preclinical and clinical development of ncRNA-based autoimmune diseases therapeutics.

17.4.1 miR-21

The link between miR-21 and inflammation was first reported in 2007 by Moschos et al., who found that miR-21 was significantly upregulated in a time-dependent fashion in the lung tissue of mice exposed to aerosolized lipopolysaccharides (LPS) [144]. MiR-21 was subsequently shown to regulate IL-12p35, a key cytokine associated with balancing T helper cell polarization, and contribute to allergic lung inflammation in mice with experimental asthma [145]. Aberrant expression of miR-21 was later confirmed in multiple autoimmune diseases, including SLE, MS, T1DM, and psoriasis. Overexpression of miR-21 suppresses of a series of genes, including *PDCD4*, *SMAD7*, and *BCL2*. Thus miR-21 serves as a proinflammatory miRNAs whose inhibition ameliorated autoimmune diseases in mouse models and could be a promising candidate for clinic treatment [146–149]. A single-stranded chemically modified oligonucleotide drug targeting miR-21, RG-012, has been developed by Regulus Therapeutics and has entered phase II clinical trials (ClinicalTrials.gov identifier NCT02855268).

17.4.2 miR-29b

MiR-29b was found to be aberrantly upregulated in many autoimmune diseases. Mechanistic studies indicated that miR-29b could regulate inflammatory cytokine expression by directly targeting T-bet and Eomes or via the Sp1–NFκB–HDAC–miR-29b regulatory network, thus suggesting an important role for miR-29b in inflammation and the immune response [150,151]. A subsequent study on miR-29-deficient mice showed that miR-29b inhibited the activation of the proinflammatory T cell phenotype in vivo [152]. In addition, systemic delivery of miR-29b conferred protection against T1DM in mice after the adoptive transfer of the disease by $CD8^+$ T cells [153]. MiR-29b therapy was also capable of inhibiting the Th1 immune response and attenuating renal inflammation and fibrosis, thus playing a protective role in mice with diabetic nephropathy [154]. A novel drug, MRG-201, was designed by miRagen Therapeutics to mimic the activities of miR-29b. This drug is undergoing a phase I clinical trial (ClinicalTrials.gov identifier: NCT02603224) to investigate its safety, tolerability, pharmacokinetic activity, and pharmacodynamic activity in healthy volunteers.

17.4.3 miR-155

MiR-155 was one of the first miRNAs shown to be induced during the inflammatory response [155]. The close association between miR-155 expression and autoimmune diseases was first reported by Pauley et al., who showed that peripheral blood mononuclear cells from RA patients displayed significantly increased miR-155 expression compared to those from healthy individuals [156]. This correlation was later proposed to be due to the miR-155-mediated inhibition of Jarid2, which is a DNA-binding protein that activates the polycomb-mediated repression of proinflammatory genes and contributes to defective Th17 cell function [157]. Reduced miR-155 also increased the expression of PU.1 and inhibited B-cell receptor crosslinking-induced antibody production, suggesting that miR-155 is an important regulator of B-cell activation in RA [158]. MiR-155-deficient mice also showed significantly reduced local bone destruction when challenged with the induction of collagen-induced arthritis [159]. In addition, miR-155 was also reported to be proinflammatory in MS and SLE [160,161]. A synthetic miRNA antagonist (LNA antimiR) of miR-155 has been developed by miRagen Therapeutics and a phase I clinical trial is being conducted (ClinicalTrials.gov identifier NCT02580552).

17.4.4 lncRNAs

Although the association between lncRNAs and specific autoimmune diseases has not yet been reported, the lncRNA-mediated regulation of the immune response has been extensively investigated. Wang et al. demonstrated that lnc-DC regulated the differentiation of human and mouse monocytes into DCs. Knockdown of lnc-DC altered the expression of 664 protein-coding genes involved in antigen uptake, cytokine release, and allogenic $CD4^+$ T cell production. These genes may disrupt monocyte and DC function and show potential relevance to autoimmune diseases [162]. In addition, lincRNA-Cox2 mediates the stimulation or repression of immune response-related genes (*Il6*, *Stat1*, and *Irf7*) in mouse macrophages [163]. NRON is a lncRNA that is transcribed from intronic regions and serves as a noncoding repressor of nuclear factor of activated T cells [164]. Whether these lncRNAs contribute to the pathogenesis of autoimmune diseases remains to be determined.

Several additional miRNA-targeted drugs have entered into clinical trials (Table 17.2). Most of the ncRNA-based therapeutics are still at the preclinical stage but have shown positive activities in animal models of autoimmune diseases. With the rapid advances in chemical modifications and in vivo delivery strategies for oligonucleotides, we are optimistic that more ncRNA-based autoimmune disease trials will be conducted and that more novel drugs will be designed by pharmaceutical companies.

17.5 CONCLUSIONS AND FUTURE PERSPECTIVES

Remarkable progress has been made in elucidating the functions of ncRNAs in the immune response and autoimmune diseases, and hundreds of miRNAs have been identified as potential diagnostic biomarkers and therapeutic targets.

Suppressive and replacement therapies are major approaches applied for ncRNA-targeted therapeutics. In the context of suppressive therapy, siRNAs and ASOs are the most extensively used tools to inhibit the abnormal overexpression of ncRNAs in diseases. The siRNA- and ASO-mediated suppression of cytoplasm-localized mature miRNAs could be enhanced by improving chemical modifications and delivery strategies. In contrast, lncRNAs reside predominantly in the nucleus, in the cytoplasm, or in both compartments, therefore the wide variety of subcellular distributions may greatly affect the siRNA- and ASO-mediated silencing.

The efficacy of siRNA- and ASO-mediated lncRNA-knockdown methods was recently evaluated in different cellular compartments, and provided valuable insights for choosing an optimal inhibition strategy in lncRNA-targeted therapeutics [165]. The results demonstrated that nuclear lncRNAs (i.e., *MALAT1* and *NEAT1*) were more effectively suppressed by ASOs, while cytoplasmic lncRNAs (i.e., *DANCR* and *OIP5-AS1*) were altered more significantly by siRNA-mediated knockdown. Dual-localized lncRNAs (i.e., *TUG1* and *HOTAIR*) were equally suppressed by both methods and the knockdown efficacy could be improved by a mixed-modality approach combining siRNA and ASO reagents.

Genome editing techniques have been rapidly developed in recent years and have received a tremendous amount of attention. As a representative technique, the CRISPR/Cas9 system is capable of not only exhibiting inhibitory activities for suppressive therapy, but also of activating gene transcription for replacement therapy. The launch of CRISPR/Cas9-based human clinical trials has provided new insights into targeted therapies and raised enormous expectations among researchers and clinicians.

Despite the remarkable progress, several challenges (i.e., off-target, immunostimulatory, and toxic effects) remain to be overcome before further practical applications are made. In addition, it is now known that the same miRNA can have different functions in different immune cells; for example, silencing miR-155 expression in B and Th17 cells could reduce proinflammatory activity to alleviate RA, but miR-155 also maintains competitive fitness in regulatory T cells, which are required for immune homeostasis [157,158,166]. Therefore the ncRNA-targeted drugs should be designed with high specificity and an optimal volume of distribution and drug clearance to achieve the therapeutic efficacy without evoking additional immune disorders. Since many autoimmune diseases are chronic disorders, it is indispensable to evaluate the long-term effects of ncRNA-targeted therapeutics. Moreover, noninvasive routes of administration (i.e., pulmonary, eye drops, intranasal, and oral) are worthy of further investigations as these are easily accepted and lead to better patient compliance.

As the field continues to evolve, the endeavor to completely understand ncRNA biogenesis and function will provide more research orientations. The desired performance of the approved ASO drugs (i.e., eteplirsen and nusinersen) and the positive results from clinical trials of miRNA-targeted drugs (i.e., Miravirsen and MRX34) show that ncRNA-targeted therapeutics are close to achieving their clinical potential. We are optimistic that personalized ncRNA-targeted autoimmune disease treatments based on an individual's immune profile will succeed and prosper as a therapeutically viable technology in the coming years.

References

[1] Mattick JS. Non-coding RNAs: the architects of eukaryotic complexity. EMBO Rep 2001;2(11):986–91.
[2] Chen CZ, Schaffert S, Fragoso R, Loh C. Regulation of immune responses and tolerance: the microRNA perspective. Immunol Rev 2013;253(1):112–28.
[3] Heward JA, Lindsay MA. Long non-coding RNAs in the regulation of the immune response. Trends Immunol 2014;35(9):408–19.
[4] Quinn JJ, Chang HY. Unique features of long non-coding RNA biogenesis and function. Nat Rev Genet 2016; 17(1):47–62.
[5] Zhernakova A, Withoff S, Wijmenga C. Clinical implications of shared genetics and pathogenesis in autoimmune diseases. Nat Rev Endocrinol 2013;9(11):646–59.
[6] Krol J, Loedige I, Filipowicz W. The widespread regulation of microRNA biogenesis, function and decay. Nat Rev Genet 2010;11(9):597–610.
[7] Ebert MS, Neilson JR, Sharp PA. MicroRNA sponges: competitive inhibitors of small RNAs in mammalian cells. Nat Methods 2007;4(9):721–6.
[8] Du C, Liu C, Kang J, Zhao G, Ye Z, Huang S, et al. MicroRNA miR-326 regulates TH-17 differentiation and is associated with the pathogenesis of multiple sclerosis. Nat Immunol 2009;10(12):1252–9.
[9] Yuan JH, Yang F, Wang F, Ma JZ, Guo YJ, Tao QF, et al. A long noncoding RNA activated by TGF-beta promotes the invasion-metastasis cascade in hepatocellular carcinoma. Cancer Cell 2014;25(5):666–81.
[10] Hansen TB, Jensen TI, Clausen BH, Bramsen JB, Finsen B, Damgaard CK, et al. Natural RNA circles function as efficient microRNA sponges. Nature 2013;495(7441):384–8.
[11] Fire A, Xu S, Montgomery MK, Kostas SA, Driver SE, Mello CC. Potent and specific genetic interference by double-stranded RNA in *Caenorhabditis elegans*. Nature 1998;391(6669):806–11.
[12] Elbashir SM, Lendeckel W, Tuschl T. RNA interference is mediated by 21- and 22-nucleotide RNAs. Genes Dev 2001;15(2):188–200.
[13] Davidson BL, McCray PB Jr. Current prospects for RNA interference-based therapies. Nat Rev Genet 2011; 12(5):329–40.
[14] Elbashir SM, Harborth J, Lendeckel W, Yalcin A, Weber K, Tuschl T. Duplexes of 21-nucleotide RNAs mediate RNA interference in cultured mammalian cells. Nature 2001;411(6836):494–8.
[15] Song E, Lee SK, Wang J, Ince N, Ouyang N, Min J, et al. RNA interference targeting Fas protects mice from fulminant hepatitis. Nat Med 2003;9(3):347–51.

[16] Li SP, Xu HX, Yu Y, He JD, Wang Z, Xu YJ, et al. LncRNA HULC enhances epithelial-mesenchymal transition to promote tumorigenesis and metastasis of hepatocellular carcinoma via the miR-200a-3p/ZEB1 signaling pathway. Oncotarget 2016;7(27):42431–46.
[17] Martinez T, Gonzalez MV, Roehl I, Wright N, Paneda C, Jimenez AI. In vitro and in vivo efficacy of SYL040012, a novel siRNA compound for treatment of glaucoma. Mol Ther 2014;22(1):81–91.
[18] Zimmermann TS, Karsten V, Chan A, Chiesa J, Boyce M, Bettencourt BR, et al. Clinical proof of concept for a novel hepatocyte-targeting GalNAc-siRNA conjugate. Mol Ther 2017;25(1):71–8.
[19] Zamecnik PC, Stephenson ML. Inhibition of Rous sarcoma virus replication and cell transformation by a specific oligodeoxynucleotide. Proc Natl Acad Sci U S A 1978;75(1):280–4.
[20] Crooke ST. Molecular mechanisms of antisense oligonucleotides. Nucleic Acid Ther 2017;27:70–7.
[21] Bennett CF, Baker BF, Pham N, Swayze E, Geary RS. Pharmacology of antisense drugs. Annu Rev Pharmacol Toxicol 2017;57:81–105.
[22] Zhou T, Kim Y, MacLeod AR. Targeting long noncoding RNA with antisense oligonucleotide technology as cancer therapeutics. Methods Mol Biol 2016;1402:199–213.
[23] Boutla A, Delidakis C, Tabler M. Developmental defects by antisense-mediated inactivation of micro-RNAs 2 and 13 in Drosophila and the identification of putative target genes. Nucleic Acids Res 2003;31(17):4973–80.
[24] Esau C, Kang X, Peralta E, Hanson E, Marcusson EG, Ravichandran LV, et al. MicroRNA-143 regulates adipocyte differentiation. J Biol Chem 2004;279(50):52361–5.
[25] Cheng AM, Byrom MW, Shelton J, Ford LP. Antisense inhibition of human miRNAs and indications for an involvement of miRNA in cell growth and apoptosis. Nucleic Acids Res 2005;33(4):1290–7.
[26] Wang Z. The principles of MiRNA-masking antisense oligonucleotides technology. Methods Mol Biol 2011;676:43–9.
[27] Zhang T, Hu Y, Ju J, Hou L, Li Z, Xiao D, et al. Downregulation of miR-522 suppresses proliferation and metastasis of non-small cell lung cancer cells by directly targeting DENN/MADD domain containing 2D. Sci Rep 2016;6:19346.
[28] Janssen HL, Reesink HW, Lawitz EJ, Zeuzem S, Rodriguez-Torres M, Patel K, et al. Treatment of HCV infection by targeting microRNA. N Engl J Med 2013;368(18):1685–94.
[29] van der Ree MH, van der Meer AJ, de Bruijne J, Maan R, van Vliet A, Welzel TM, et al. Long-term safety and efficacy of microRNA-targeted therapy in chronic hepatitis C patients. Antiviral Res 2014;111C:53–9.
[30] de Smet MD, Meenken CJ, van den Horn GJ. Fomivirsen: a phosphorothioate oligonucleotide for the treatment of CMV retinitis. Ocul Immunol Inflamm 1999;7(3–4):189–98.
[31] Hair P, Cameron F, McKeage K. Mipomersen sodium: first global approval. Drugs 2013;73(5):487–93.
[32] Railroading at the FDA, Nat Biotech. 2016 34 (11) 1078I 11. doi: 10.1038/nbt.3733.
[33] Khorkova O, Wahlestedt C. Oligonucleotide therapies for disorders of the nervous system. Nat Biotechnol 2017;35(3):249–63.
[34] Zong X, Huang L, Tripathi V, Peralta R, Freier SM, Guo S, et al. Knockdown of nuclear-retained long noncoding RNAs using modified DNA antisense oligonucleotides. Methods Mol Biol 2015;1262:321–31.
[35] Katsushima K, Natsume A, Ohka F, Shinjo K, Hatanaka A, Ichimura N, et al. Targeting the Notch-regulated non-coding RNA TUG1 for glioma treatment. Nat Commun 2016;7:13616.
[36] Gumireddy K, Young DD, Xiong X, Hogenesch JB, Huang Q, Deiters A. Small-molecule inhibitors of microRNA miR-21 function. Angew Chem Int Ed Engl 2008;47(39):7482–4.
[37] Naro Y, Thomas M, Stephens MD, Connelly CM, Deiters A. Aryl amide small-molecule inhibitors of microRNA miR-21 function. Bioorg Med Chem Lett 2015;25(21):4793–6.
[38] Connelly CM, Deiters A. Identification of inhibitors of microRNA function from small molecule screens. Methods Mol Biol 2014;1095:147–56.
[39] Tan SB, Li J, Chen X, Zhang W, Zhang D, Zhang C, et al. Small molecule inhibitor of myogenic microRNAs leads to a discovery of miR-221/222-myoD-myomiRs regulatory pathway. Chem Biol 2014;21(10):1265–70.
[40] Velagapudi SP, Cameron MD, Haga CL, Rosenberg LH, Lafitte M, Duckett DR, et al. Design of a small molecule against an oncogenic noncoding RNA. Proc Natl Acad Sci U S A 2016;113(21):5898–903.
[41] Lombardo A, Genovese P, Beausejour CM, Colleoni S, Lee YL, Kim KA, et al. Gene editing in human stem cells using zinc finger nucleases and integrase-defective lentiviral vector delivery. Nat Biotechnol 2007;25(11):1298–306.
[42] Perez EE, Wang J, Miller JC, Jouvenot Y, Kim KA, Liu O, et al. Establishment of HIV-1 resistance in CD4+ T cells by genome editing using zinc-finger nucleases. Nat Biotechnol 2008;26(7):808–16.
[43] Tebas P, Stein D, Tang WW, Frank I, Wang SQ, Lee G, et al. Gene editing of CCR5 in autologous CD4 T cells of persons infected with HIV. N Engl J Med 2014;370(10):901–10.

[44] Hockemeyer D, Wang H, Kiani S, Lai CS, Gao Q, Cassady JP, et al. Genetic engineering of human pluripotent cells using TALE nucleases. Nat Biotechnol 2011;29(8):731–4.

[45] Jinek M, Chylinski K, Fonfara I, Hauer M, Doudna JA, Charpentier E. A programmable dual-RNA-guided DNA endonuclease in adaptive bacterial immunity. Science 2012;337(6096):816–21.

[46] Smith C, Abalde-Atristain L, He C, Brodsky BR, Braunstein EM, Chaudhari P, et al. Efficient and allele-specific genome editing of disease loci in human iPSCs. Mol Ther 2015;23(3):570–7.

[47] Han J, Zhang J, Chen L, Shen B, Zhou J, Hu B, et al. Efficient in vivo deletion of a large imprinted lncRNA by CRISPR/Cas9. RNA Biol 2014;11(7):829–35.

[48] Chang H, Yi B, Ma R, Zhang X, Zhao H, Xi Y. CRISPR/cas9, a novel genomic tool to knock down microRNA in vitro and in vivo. Sci Rep 2016;6:22312.

[49] Cyranoski D. CRISPR gene-editing tested in a person for the first time. Nature 2016;539(7630):479.

[50] Bader AG, Brown D, Stoudemire J, Lammers P. Developing therapeutic microRNAs for cancer. Gene Ther 2011;18(12):1121–6.

[51] Pan W, Zhu S, Dai D, Liu Z, Li D, Li B, et al. MiR-125a targets effector programs to stabilize Treg-mediated immune homeostasis. Nat Commun 2015;6:7096.

[52] Lujambio A, Ropero S, Ballestar E, Fraga MF, Cerrato C, Setien F, et al. Genetic unmasking of an epigenetically silenced microRNA in human cancer cells. Cancer Res 2007;67(4):1424–9.

[53] Melo S, Villanueva A, Moutinho C, Davalos V, Spizzo R, Ivan C, et al. Small molecule enoxacin is a cancer-specific growth inhibitor that acts by enhancing TAR RNA-binding protein 2-mediated microRNA processing. Proc Natl Acad Sci U S A 2011;108(11):4394–9.

[54] Ling H. Non-coding RNAs: therapeutic strategies and delivery systems. Adv Exp Med Biol 2016;937:229–37.

[55] Konermann S, Brigham MD, Trevino AE, Joung J, Abudayyeh OO, Barcena C, et al. Genome-scale transcriptional activation by an engineered CRISPR-Cas9 complex. Nature 2015;517(7536):583–8.

[56] Matsukura M, Shinozuka K, Zon G, Mitsuya H, Reitz M, Cohen JS, et al. Phosphorothioate analogs of oligodeoxynucleotides: inhibitors of replication and cytopathic effects of human immunodeficiency virus. Proc Natl Acad Sci U S A 1987;84(21):7706–10.

[57] Bigelow JC, Chrin LR, Mathews LA, McCormick JJ. High-performance liquid chromatographic analysis of phosphorothioate analogues of oligodeoxynucleotides in biological fluids. J Chromatogr 1990;533:133–40.

[58] Lamond AI, Sproat BS. Antisense oligonucleotides made of 2′-O-alkylRNA: their properties and applications in RNA biochemistry. FEBS Lett 1993;325(1–2):123–7.

[59] Kawasaki AM, Casper MD, Freier SM, Lesnik EA, Zounes MC, Cummins LL, et al. Uniformly modified 2′-deoxy-2′-fluoro phosphorothioate oligonucleotides as nuclease-resistant antisense compounds with high affinity and specificity for RNA targets. J Med Chem 1993;36(7):831–41.

[60] Monia BP, Lesnik EA, Gonzalez C, Lima WF, McGee D, Guinosso CJ, et al. Evaluation of 2′-modified oligonucleotides containing 2′-deoxy gaps as antisense inhibitors of gene expression. J Biol Chem 1993;268(19):14514–22.

[61] Agrawal S, Mayrand SH, Zamecnik PC, Pederson T. Site-specific excision from RNA by RNase H and mixed-phosphate-backbone oligodeoxynucleotides. Proc Natl Acad Sci U S A 1990;87(4):1401–5.

[62] Cummins LL, Owens SR, Risen LM, Lesnik EA, Freier SM, McGee D, et al. Characterization of fully 2′-modified oligoribonucleotide hetero- and homoduplex hybridization and nuclease sensitivity. Nucleic Acids Res 1995;23(11):2019–24.

[63] Baker BF, Lot SS, Condon TP, Cheng-Flournoy S, Lesnik EA, Sasmor HM, et al. 2′-O-(2-methoxy)ethyl-modified anti-intercellular adhesion molecule 1 (ICAM-1) oligonucleotides selectively increase the ICAM-1 mRNA level and inhibit formation of the ICAM-1 translation initiation complex in human umbilical vein endothelial cells. J Biol Chem 1997;272(18):11994–2000.

[64] Soutschek J, Akinc A, Bramlage B, Charisse K, Constien R, Donoghue M, et al. Therapeutic silencing of an endogenous gene by systemic administration of modified siRNAs. Nature 2004;432(7014):173–8.

[65] Meister G, Landthaler M, Dorsett Y, Tuschl T. Sequence-specific inhibition of microRNA- and siRNA-induced RNA silencing. RNA 2004;10(3):544–50.

[66] Krutzfeldt J, Rajewsky N, Braich R, Rajeev KG, Tuschl T, Manoharan M, et al. Silencing of microRNAs in vivo with 'antagomirs'. Nature 2005;438(7068):685–9.

[67] Esau C, Davis S, Murray SF, Yu XX, Pandey SK, Pear M, et al. miR-122 regulation of lipid metabolism revealed by in vivo antisense targeting. Cell Metab 2006;3(2):87–98.

[68] van Deutekom JC, Janson AA, Ginjaar IB, Frankhuizen WS, Aartsma-Rus A, Bremmer-Bout M, et al. Local dystrophin restoration with antisense oligonucleotide PRO051. N Engl J Med 2007;357(26):2677–86.

[69] Koshkin AA, Singh SK, Nielsen P, Rajwanshi VK, Kumar R, Meldgaard M, et al. LNA (locked nucleic acids): synthesis of the adenine, cytosine, guanine, 5-methylcytosine, thymine and uracil bicyclonucleoside monomers, oligomerisation, and unprecedented nucleic acid recognition. Tetrahedron 1998;54(14):3607–30.
[70] Nielsen CB, Singh SK, Wengel J, Jacobsen JP. The solution structure of a locked nucleic acid (LNA) hybridized to DNA. J Biomol Struct Dyn 1999;17(2):175–91.
[71] Shaked I, Meerson A, Wolf Y, Avni R, Greenberg D, Gilboa-Geffen A, et al. MicroRNA-132 potentiates cholinergic anti-inflammatory signaling by targeting acetylcholinesterase. Immunity 2009;31(6):965–73.
[72] Xing Z, Lin A, Li C, Liang K, Wang S, Liu Y, et al. lncRNA directs cooperative epigenetic regulation downstream of chemokine signals. Cell 2014;159(5):1110–25.
[73] Lennox KA, Behlke MA. A direct comparison of anti-microRNA oligonucleotide potency. Pharm Res 2010;27(9):1788–99.
[74] Durig J, Duhrsen U, Klein-Hitpass L, Worm J, Hansen JB, Orum H, et al. The novel antisense Bcl-2 inhibitor SPC2996 causes rapid leukemic cell clearance and immune activation in chronic lymphocytic leukemia. Leukemia 2011;25(4):638–47.
[75] Woo CJ, Maier VK, Davey R, Brennan J, Li G, Brothers J II, et al. Gene activation of SMN by selective disruption of lncRNA-mediated recruitment of PRC2 for the treatment of spinal muscular atrophy. Proc Natl Acad Sci U S A 2017;114(8):E1509–18.
[76] Seth PP, Vasquez G, Allerson CA, Berdeja A, Gaus H, Kinberger GA, et al. Synthesis and biophysical evaluation of 2′,4′-constrained 2′O-methoxyethyl and 2′,4′-constrained 2′O-ethyl nucleic acid analogues. J Org Chem 2010;75(5):1569–81.
[77] Lennox KA, Owczarzy R, Thomas DM, Walder JA, Behlke MA. Improved performance of anti-miRNA oligonucleotides using a novel non-nucleotide modifier. Mol Ther Nucleic Acids 2013;2:e117.
[78] Hong D, Kurzrock R, Kim Y, Woessner R, Younes A, Nemunaitis J, et al. AZD9150, a next-generation antisense oligonucleotide inhibitor of STAT3 with early evidence of clinical activity in lymphoma and lung cancer. Sci Transl Med 2015;7(314):314ra185.
[79] Lorenz C, Hadwiger P, John M, Vornlocher H-p, Unverzagt C. Steroid and lipid conjugates of siRNAs to enhance cellular uptake and gene silencing in liver cells. Bioorg Med Chem Lett 2004;14(19):4975–7.
[80] Wolfrum C, Shi S, Jayaprakash KN, Jayaraman M, Wang G, Pandey RK, et al. Mechanisms and optimization of in vivo delivery of lipophilic siRNAs. Nat Biotechnol 2007;25(10):1149–57.
[81] Bruck J, Pascolo S, Fuchs K, Kellerer C, Glocova I, Geisel J, et al. Cholesterol modification of p40-specific small interfering RNA enables therapeutic targeting of dendritic cells. J Immunol 2015;195(5):2216–23.
[82] Torchilin VP. Recent advances with liposomes as pharmaceutical carriers. Nat Rev Drug Discov 2005;4(2):145–60.
[83] Sorensen DR, Leirdal M, Sioud M. Gene silencing by systemic delivery of synthetic siRNAs in adult mice. J Mol Biol 2003;327(4):761–6.
[84] Zhao M, Yang H, Jiang X, Zhou W, Zhu B, Zeng Y, et al. Lipofectamine RNAiMAX: an efficient siRNA transfection reagent in human embryonic stem cells. Mol Biotechnol 2008;40(1):19–26.
[85] Yu X, Liang X, Xie H, Kumar S, Ravinder N, Potter J, et al. Improved delivery of Cas9 protein/gRNA complexes using lipofectamine CRISPRMAX. Biotechnol Lett 2016;38(6):919–29.
[86] Barenholz YC. Doxil: the first FDA-approved nano-drug: lessons learned. J Control Release 2012;160(2):117–34.
[87] Morrissey DV, Lockridge JA, Shaw L, Blanchard K, Jensen K, Breen W, et al. Potent and persistent in vivo anti-HBV activity of chemically modified siRNAs. Nat Biotechnol 2005;23(8):1002–7.
[88] Zimmermann TS, Lee AC, Akinc A, Bramlage B, Bumcrot D, Fedoruk MN, et al. RNAi-mediated gene silencing in non-human primates. Nature 2006;441(7089):111–4.
[89] Endo-Takahashi Y, Negishi Y, Suzuki R, Maruyama K, Aramaki Y. MicroRNA imaging in combination with diagnostic ultrasound and bubble liposomes for microRNA delivery. Methods Mol Biol 2016;1372:209–13.
[90] Fattal E, Couvreur P, Dubernet C. "Smart" delivery of antisense oligonucleotides by anionic pH-sensitive liposomes. Adv Drug Deliv Rev 2004;56(7):931–46.
[91] Andreakos E, Rauchhaus U, Stavropoulos A, Endert G, Wendisch V, Benahmed AS, et al. Amphoteric liposomes enable systemic antigen-presenting cell-directed delivery of CD40 antisense and are therapeutically effective in experimental arthritis. Arthritis Rheum 2009;60(4):994–1005.
[92] Bouchie A. First microRNA mimic enters clinic. Nat Biotechnol 2013;31(7):577.
[93] Ibrahim AF, Weirauch U, Thomas M, Grunweller A, Hartmann RK, Aigner A. MicroRNA replacement therapy for miR-145 and miR-33a is efficacious in a model of colon carcinoma. Cancer Res 2011;71(15):5214–24.
[94] Laroui H, Geem D, Xiao B, Viennois E, Rakhya P, Denning T, et al. Targeting intestinal inflammation with CD98 siRNA/PEI-loaded nanoparticles. Mol Ther 2014;22(1):69–80.

[95] Xie Y, Kim NH, Nadithe V, Schalk D, Thakur A, Kilic A, et al. Targeted delivery of siRNA to activated T cells via transferrin-polyethylenimine (Tf-PEI) as a potential therapy of asthma. J Control Release 2016;229:120–9.
[96] Ren Y, Wang R, Gao L, Li K, Zhou X, Guo H, et al. Sequential co-delivery of miR-21 inhibitor followed by burst release doxorubicin using NIR-responsive hollow gold nanoparticle to enhance anticancer efficacy. J Control Release 2016;228:74–86.
[97] Hoshyar N, Gray S, Han H, Bao G. The effect of nanoparticle size on in vivo pharmacokinetics and cellular interaction. Nanomedicine 2016;11(6):673–92.
[98] Zhou Y, Zhou G, Tian C, Jiang W, Jin L, Zhang C, et al. Exosome-mediated small RNA delivery for gene therapy. Wiley Interdiscip Rev RNA 2016;7(6):758–71.
[99] Alvarez-Erviti L, Seow Y, Yin H, Betts C, Lakhal S, Wood MJ. Delivery of siRNA to the mouse brain by systemic injection of targeted exosomes. Nat Biotechnol 2011;29(4):341–5.
[100] Sun D, Zhuang X, Xiang X, Liu Y, Zhang S, Liu C, et al. A novel nanoparticle drug delivery system: the anti-inflammatory activity of curcumin is enhanced when encapsulated in exosomes. Mol Ther 2010;18(9):1606–14.
[101] Zhuang X, Xiang X, Grizzle W, Sun D, Zhang S, Axtell RC, et al. Treatment of brain inflammatory diseases by delivering exosome encapsulated anti-inflammatory drugs from the nasal region to the brain. Mol Ther 2011;19(10):1769–79.
[102] Haney MJ, Klyachko NL, Zhao Y, Gupta R, Plotnikova EG, He Z, et al. Exosomes as drug delivery vehicles for Parkinson's disease therapy. J Control Release 2015;207:18–30.
[103] Wang Q, Zhuang X, Mu J, Deng ZB, Jiang H, Zhang L, et al. Delivery of therapeutic agents by nanoparticles made of grapefruit-derived lipids. Nat Commun 2013;4:1867.
[104] Zhuang X, Teng Y, Samykutty A, Mu J, Deng Z, Zhang L, et al. Grapefruit-derived nanovectors delivering therapeutic miR17 through an intranasal route inhibit brain tumor progression. Mol Ther 2016;24(1):96–105.
[105] Qu L, Ding J, Chen C, Wu ZJ, Liu B, Gao Y, et al. Exosome-transmitted lncARSR promotes sunitinib resistance in renal cancer by acting as a competing endogenous RNA. Cancer Cell 2016;29(5):653–68.
[106] Yang N, Li S, Li G, Zhang S, Tang X, Ni S, et al. The role of extracellular vesicles in mediating progression, metastasis and potential treatment of hepatocellular carcinoma. Oncotarget 2017;8(2):3683–95.
[107] Vives E, Schmidt J, Pelegrin A. Cell-penetrating and cell-targeting peptides in drug delivery. Biochim Biophys Acta 2008;1786(2):126–38.
[108] Crombez L, Morris MC, Dufort S, Aldrian-Herrada G, Nguyen Q, Mc Master G, et al. Targeting cyclin B1 through peptide-based delivery of siRNA prevents tumour growth. Nucleic Acids Res 2009;37(14):4559–69.
[109] Kim HA, Nam K, Kim SW. Tumor targeting RGD conjugated bio-reducible polymer for VEGF siRNA expressing plasmid delivery. Biomaterials 2014;35(26):7543–52.
[110] Ding Q, Si X, Liu D, Peng J, Tang H, Sun W, et al. Targeting and liposomal drug delivery to CD40L expressing T cells for treatment of autoimmune diseases. J Control Release 2015;207:86–92.
[111] Kumar P, Ban HS, Kim SS, Wu H, Pearson T, Greiner DL, et al. T cell-specific siRNA delivery suppresses HIV-1 infection in humanized mice. Cell 2008;134(4):577–86.
[112] Matsuda S, Keiser K, Nair JK, Charisse K, Manoharan RM, Kretschmer P, et al. siRNA conjugates carrying sequentially assembled trivalent N-acetylgalactosamine linked through nucleosides elicit robust gene silencing in vivo in hepatocytes. ACS Chem Biol 2015;10(5):1181–7.
[113] Kortylewski M, Swiderski P, Herrmann A, Wang L, Kowolik C, Kujawski M, et al. In vivo delivery of siRNA to immune cells by conjugation to a TLR9 agonist enhances antitumor immune responses. Nat Biotechnol 2009;27(10):925–32.
[114] McNamara JO II, Andrechek ER, Wang Y, Viles KD, Rempel RE, Gilboa E, et al. Cell type-specific delivery of siRNAs with aptamer-siRNA chimeras. Nat Biotechnol 2006;24(8):1005–15.
[115] Dassie JP, Liu XY, Thomas GS, Whitaker RM, Thiel KW, Stockdale KR, et al. Systemic administration of optimized aptamer-siRNA chimeras promotes regression of PSMA-expressing tumors. Nat Biotechnol 2009;27(9):839–49.
[116] Diao Y, Liu J, Ma Y, Su M, Zhang H, Hao X. A specific aptamer-cell penetrating peptides complex delivered siRNA efficiently and suppressed prostate tumor growth in vivo. Cancer Biol Ther 2016;17(5):498–506.
[117] Iaboni M, Russo V, Fontanella R, Roscigno G, Fiore D, Donnarumma E, et al. Aptamer-miRNA-212 conjugate sensitizes NSCLC cells to TRAIL. Mol Ther Nucleic Acids 2016;5:e289.
[118] Paddison PJ, Caudy AA, Bernstein E, Hannon GJ, Conklin DS. Short hairpin RNAs (shRNAs) induce sequence-specific silencing in mammalian cells. Genes Dev 2002;16(8):948–58.
[119] Valdmanis PN, Lisowski L, Kay MA. rAAV-mediated tumorigenesis: still unresolved after an AAV assault. Mol Ther 2012;20(11):2014–7.
[120] Yla-Herttuala S. Endgame: glybera finally recommended for approval as the first gene therapy drug in the European Union. Mol Ther 2012;20(10):1831–2.

[121] Valdmanis PN, Kay MA. Future of RAAV gene therapy: platform for RNAi, gene editing, and beyond. Hum Gene Ther 2017;28:361–72.
[122] Jackson AL, Bartz SR, Schelter J, Kobayashi SV, Burchard J, Mao M, et al. Expression profiling reveals off-target gene regulation by RNAi. Nat Biotechnol 2003;21(6):635–7.
[123] Lin X, Ruan X, Anderson MG, McDowell JA, Kroeger PE, Fesik SW, et al. siRNA-mediated off-target gene silencing triggered by a 7 nt complementation. Nucleic Acids Res 2005;33(14):4527–35.
[124] Jackson AL, Burchard J, Leake D, Reynolds A, Schelter J, Guo J, et al. Position-specific chemical modification of siRNAs reduces "off-target" transcript silencing. RNA 2006;12(7):1197–205.
[125] Bramsen JB, Pakula MM, Hansen TB, Bus C, Langkjaer N, Odadzic D, et al. A screen of chemical modifications identifies position-specific modification by UNA to most potently reduce siRNA off-target effects. Nucleic Acids Res 2010;38(17):5761–73.
[126] Lee HS, Seok H, Lee DH, Ham J, Lee W, Youm EM, et al. Abasic pivot substitution harnesses target specificity of RNA interference. Nat Commun 2015;6:10154.
[127] Hannus M, Beitzinger M, Engelmann JC, Weickert MT, Spang R, Hannus S, et al. siPools: highly complex but accurately defined siRNA pools eliminate off-target effects. Nucleic Acids Res 2014;42(12):8049–61.
[128] Pattanayak V, Lin S, Guilinger JP, Ma E, Doudna JA, Liu DR. High-throughput profiling of off-target DNA cleavage reveals RNA-programmed Cas9 nuclease specificity. Nat Biotechnol 2013;31(9):839–43.
[129] Kim S, Kim D, Cho SW, Kim J, Kim JS. Highly efficient RNA-guided genome editing in human cells via delivery of purified Cas9 ribonucleoproteins. Genome Res 2014;24(6):1012–9.
[130] Duan J, Lu G, Xie Z, Lou M, Luo J, Guo L, et al. Genome-wide identification of CRISPR/Cas9 off-targets in human genome. Cell Res 2014;24(8):1009–12.
[131] Fu Y, Sander JD, Reyon D, Cascio VM, Joung JK. Improving CRISPR-Cas nuclease specificity using truncated guide RNAs. Nat Biotechnol 2014;32(3):279–84.
[132] Ran FA, Hsu PD, Lin CY, Gootenberg JS, Konermann S, Trevino AE, et al. Double nicking by RNA-guided CRISPR Cas9 for enhanced genome editing specificity. Cell 2013;154(6):1380–9.
[133] Kariko K, Bhuyan P, Capodici J, Weissman D. Small interfering RNAs mediate sequence-independent gene suppression and induce immune activation by signaling through toll-like receptor 3. J Immunol 2004;172(11):6545–9.
[134] Heil F, Hemmi H, Hochrein H, Ampenberger F, Kirschning C, Akira S, et al. Species-specific recognition of single-stranded RNA via toll-like receptor 7 and 8. Science 2004;303(5663):1526–9.
[135] Hornung V, Guenthner-Biller M, Bourquin C, Ablasser A, Schlee M, Uematsu S, et al. Sequence-specific potent induction of IFN-alpha by short interfering RNA in plasmacytoid dendritic cells through TLR7. Nat Med 2005;11(3):263–70.
[136] Jurk M, Chikh G, Schulte B, Kritzler A, Richardt-Pargmann D, Lampron C, et al. Immunostimulatory potential of silencing RNAs can be mediated by a non-uridine-rich toll-like receptor 7 motif. Nucleic Acid Ther 2011;21(3):201–14.
[137] Fedorov Y, Anderson EM, Birmingham A, Reynolds A, Karpilow J, Robinson K, et al. Off-target effects by siRNA can induce toxic phenotype. RNA 2006;12(7):1188–96.
[138] Sioud M. Single-stranded small interfering RNA are more immunostimulatory than their double-stranded counterparts: a central role for 2′-hydroxyl uridines in immune responses. Eur J Immunol 2006;36(5):1222–30.
[139] Swayze EE, Siwkowski AM, Wancewicz EV, Migawa MT, Wyrzykiewicz TK, Hung G, et al. Antisense oligonucleotides containing locked nucleic acid improve potency but cause significant hepatotoxicity in animals. Nucleic Acids Res 2007;35(2):687–700.
[140] Henry SP, Novotny W, Leeds J, Auletta C, Kornbrust DJ. Inhibition of coagulation by a phosphorothioate oligonucleotide. Antisense Nucleic Acid Drug Dev 1997;7(5):503–10.
[141] Galbraith WM, Hobson WC, Giclas PC, Schechter PJ, Agrawal S. Complement activation and hemodynamic changes following intravenous administration of phosphorothioate oligonucleotides in the monkey. Antisense Res Dev 1994;4(3):201–6.
[142] Ogris M, Brunner S, Schuller S, Kircheis R, Wagner E. PEGylated DNA/transferrin-PEI complexes: reduced interaction with blood components, extended circulation in blood and potential for systemic gene delivery. Gene Ther 1999;6(4):595–605.
[143] Patil Y, Panyam J. Polymeric nanoparticles for siRNA delivery and gene silencing. Int J Pharm 2009;367(1-2):195–203.
[144] Moschos SA, Williams AE, Perry MM, Birrell MA, Belvisi MG, Lindsay MA. Expression profiling in vivo demonstrates rapid changes in lung microRNA levels following lipopolysaccharide-induced inflammation but not in the anti-inflammatory action of glucocorticoids. BMC Genomics 2007;8:240.
[145] Lu TX, Munitz A, Rothenberg ME. MicroRNA-21 is up-regulated in allergic airway inflammation and regulates IL-12p35 expression. J Immunol 2009;182(8):4994–5002.

[146] Stagakis E, Bertsias G, Verginis P, Nakou M, Hatziapostolou M, Kritikos H, et al. Identification of novel microRNA signatures linked to human lupus disease activity and pathogenesis: miR-21 regulates aberrant T cell responses through regulation of PDCD4 expression. Ann Rheum Dis 2011;70(8):1496–506.
[147] Murugaiyan G, da Cunha AP, Ajay AK, Joller N, Garo LP, Kumaradevan S, et al. MicroRNA-21 promotes Th17 differentiation and mediates experimental autoimmune encephalomyelitis. J Clin Invest 2015;125(3):1069–80.
[148] Sims EK, Lakhter AJ, Anderson-Baucum E, Kono T, Tong X, Evans-Molina C. MicroRNA 21 targets BCL2 mRNA to increase apoptosis in rat and human beta cells. Diabetologia 2017;60:1057–65.
[149] Guinea-Viniegra J, Jimenez M, Schonthaler HB, Navarro R, Delgado Y, Concha-Garzon MJ, et al. Targeting miR-21 to treat psoriasis. Sci Transl Med 2014;6(225):225re1.
[150] Steiner DF, Thomas MF, Hu JK, Yang Z, Babiarz JE, Allen CD, et al. MicroRNA-29 regulates T-box transcription factors and interferon-gamma production in helper T cells. Immunity 2011;35(2):169–81.
[151] Liu S, Wu LC, Pang J, Santhanam R, Schwind S, Wu YZ, et al. Sp1/NFkappaB/HDAC/miR-29b regulatory network in KIT-driven myeloid leukemia. Cancer Cell 2010;17(4):333–47.
[152] Smith KM, Guerau-de-Arellano M, Costinean S, Williams JL, Bottoni A, Mavrikis Cox G, et al. miR-29ab1 deficiency identifies a negative feedback loop controlling Th1 bias that is dysregulated in multiple sclerosis. J Immunol 2012;189(4):1567–76.
[153] Salama A, Fichou N, Allard M, Dubreil L, De Beaurepaire L, Viel A, et al. MicroRNA-29b modulates innate and antigen-specific immune responses in mouse models of autoimmunity. PLoS One 2014;9(9):e106153.
[154] Chen HY, Zhong X, Huang XR, Meng XM, You Y, Chung AC, et al. MicroRNA-29b inhibits diabetic nephropathy in db/db mice. Mol Ther 2014;22(4):842–53.
[155] O'Connell RM, Taganov KD, Boldin MP, Cheng G, Baltimore D. MicroRNA-155 is induced during the macrophage inflammatory response. Proc Natl Acad Sci U S A 2007;104(5):1604–9.
[156] Pauley KM, Satoh M, Chan AL, Bubb MR, Reeves WH, Chan EK. Upregulated miR-146a expression in peripheral blood mononuclear cells from rheumatoid arthritis patients. Arthritis Res Ther 2008;10(4):101.
[157] Escobar TM, Kanellopoulou C, Kugler DG, Kilaru G, Nguyen CK, Nagarajan V, et al. miR-155 activates cytokine gene expression in Th17 cells by regulating the DNA-binding protein Jarid2 to relieve polycomb-mediated repression. Immunity 2014;40(6):865–79.
[158] Alivernini S, Kurowska-Stolarska M, Tolusso B, Benvenuto R, Elmesmari A, Canestri S, et al. MicroRNA-155 influences B-cell function through PU1 in rheumatoid arthritis. Nat Commun 2016;7:12970.
[159] Bluml S, Bonelli M, Niederreiter B, Puchner A, Mayr G, Hayer S, et al. Essential role of microRNA-155 in the pathogenesis of autoimmune arthritis in mice. Arthritis Rheum 2011;63(5):1281–8.
[160] Moore CS, Rao VT, Durafourt BA, Bedell BJ, Ludwin SK, Bar-Or A, et al. miR-155 as a multiple sclerosis-relevant regulator of myeloid cell polarization. Ann Neurol 2013;74(5):709–20.
[161] Zhou S, Wang Y, Meng Y, Xiao C, Liu Z, Brohawn P, et al. In vivo therapeutic success of microRNA-155 antagomir in a mouse model of lupus alveolar hemorrhage. Arthritis Rheumatol 2016;68(4):953–64.
[162] Wang P, Xue Y, Han Y, Lin L, Wu C, Xu S, et al. The STAT3-binding long noncoding RNA lnc-DC controls human dendritic cell differentiation. Science 2014;344(6181):310–3.
[163] Carpenter S, Aiello D, Atianand MK, Ricci EP, Gandhi P, Hall LL, et al. A long noncoding RNA mediates both activation and repression of immune response genes. Science 2013;341(6147):789–92.
[164] Sharma S, Findlay GM, Bandukwala HS, Oberdoerffer S, Baust B, Li Z, et al. Dephosphorylation of the nuclear factor of activated T cells (NFAT) transcription factor is regulated by an RNA-protein scaffold complex. Proc Natl Acad Sci U S A 2011;108(28):11381–6.
[165] Lennox KA, Behlke MA. Cellular localization of long non-coding RNAs affects silencing by RNAi more than by antisense oligonucleotides. Nucleic Acids Res 2016;44(2):863–77.
[166] Lu LF, Thai TH, Calado DP, Chaudhry A, Kubo M, Tanaka K, et al. Foxp3-dependent microRNA155 confers competitive fitness to regulatory T cells by targeting SOCS1 protein. Immunity 2009;30(1):80–91.

CHAPTER

18

Future Challenges and Prospects for the Epigenetics of Autoimmunity

Mahdi Mahmoudi, Saeed Aslani

Rheumatology Research Center, Tehran University of Medical Sciences, Tehran, Iran

OUTLINE

The Epigenetics of Autoimmunity. http://dx.doi.org/10.1016/B978-0-12-809912-4.00018-0

18.1 ONGOING ATTEMPTS AND UPCOMING CHALLENGES

Personalized medicine, or individualized medicine, was first introduced in the field of genetics. It is defined as a type of medical modeling (commonly known as customization of individual health care) that considers the unique structure of the patient by using molecular technology in the process of medical practice [1]. The specific genetic background of each individual is regarded as an essential part of personalized medicine; however, this has now been extended to include all types of personalized measurements (i.e., epigenetics, which is a heritable form of gene expression regulation that does not involve DNA sequence changes). The final goal of personalized medicine is to treat each patient based on their background (e.g., genetic variations, epigenetic modifications, and other environmental circumstances). Therefore the use of epigenetics could be a new prospect for offering a personalized therapeutic approach. However, novel precise methods need to be devised to measure these epigenomic changes in individuals with diseases (i.e., autoimmune diseases). It will also be beneficial to develop new analysis models to interpret these changes before they can be applied in practice.

18.2 PERSONALIZED MEDICINE

The development of personalized epigenetics for the management of different disorders and diseases has led to a rapid expansion in the types of tools that are available to help clinicians prevent and control medical conditions with an epigenetic basis in their etiology or pathogenesis. For example, chronic pain is a component of numerous medical conditions, and several studies have established an important role for epigenetics in the development and maintenance of chronic pain [2]. Furthermore, chronic pain management is largely dependent on individual's makeup, therefore controlling the epigenomic markers in chronic pain sufferers could be of significant importance when selecting suitable antiinflammatory drugs.

Multiple studies have shown that autoimmune disorders have a remarkable epigenetic basis, which can be largely attributed to the effects of the environmental contributing factors on the immune system [3]. Being armed with an understanding of the precise epigenetic modifications in a patient with an autoimmune disorder may be beneficial when determining which gene may need to be regulated and through which epigenetic mechanism [4]. Autoimmune disorders are often associated with the age of patients, and the increasing epigenetic diverge that occurs during aging demonstrates the important role of personalized epigenetics in managing and monitoring these aging-related diseases.

Nowadays, there is consensus that personalized epigenetics has the potential to revolutionize the diagnosis, prognosis, and management of autoimmune diseases [5]. The core of personalized epigenetics is considered to be the interindividual variations in the epigenetic signatures, including DNA methylation, histone modifications, and microRNAs (miRNAs). Moreover, computational epigenetics has provided new opportunities to interpret and understand the complexities of epigenomic data, and show how it can be used in practice to determine the diagnosis, prognosis, and therapy choice in patients. Being armed with comprehensive understanding of epigenetic biomarkers as well as personalized epigenetic responses to drugs and toxic environmental agents seems to be appropriate tools in suiting the exertion of epigenetics in personalized epigenetic medicine [6]. It seems that the future of personalized epigenetics is promising and there is no doubt that this area will develop and ultimately revolutionize clinical practice, especially for patients with autoimmune disorders.

18.3 MEASURING THE EPIGENETIC PROFILE

DNA methylation can be measured using small amounts of genomic DNA, extracted from either tissue samples [7] or peripheral blood [8]. Sodium bisulfite treatment converts unmethylated cytosine residues to uracil residues while methylated cytosine residues are left unchanged. This has been considered as the gold standard in DNA methylation analysis (Fig. 18.1) [9]. This specific DNA base change can be measured using polymerase chain reaction amplification, DNA sequencing, or mass spectroscopy. DNA methylation can also be analyzed on a genome-wide scale using a variety of enzyme-based or antibody-affinity techniques [10–15]. Low-resolution array methods can be used to rapidly recognize differentially methylated genes in case–control studies [16]. Moreover, the global methylation level in a tissue can be analyzed by directly assessing the cytosine and 5′-methylcytosine content [17–20] or by using a combination of methylation-dependent DNA digestion and fluorescent tagging [21]. Ultra-high-throughput DNA sequencing technologies have been developed for the direct sequencing of methylated DNA fragments and bisulfite-converted genomic DNA [22,23], which will facilitate the evaluation of comprehensive methylation profiles. The new techniques and technologies have been developed to provide a better understanding of epigenetic marks in human diseases and will hopefully open doors to future studies in the field of epigenetics (Table 18.1).

18.4 EPIGENOME-WIDE ASSOCIATION STUDIES

Although genome-wide association studies (GWAS) have been useful for discovering loci associated with common autoimmune diseases, several unanswered questions regarding causality still exist. Recent genomic technological breakthroughs have provided the opportunity to conduct large-scale investigations of human disease-associated epigenetic modifications, specifically those affecting DNA methylation states. These studies, which are figuratively called epigenome-wide association studies (EWAS), have produced novel data in the context of the epigenome; however, they have also produced new challenges that were not significant in GWAS. Nevertheless, it is hoped that integrating EWAS and GWAS will help to resolve complex GWAS haplotypes regarding functional analysis.

Understanding the genetic and nongenetic factors that are implicated in human disorders is one of the major challenges in biomedical research. GWAS have identified more than 800 single nucleotide polymorphisms (SNPs) that are associated with more than 150 human diseases [24]. Although comprehensive genetic understanding of most human disorders is not yet known, resequencing of whole genomes seems to be a promising approach for identifying disease-causing genetic variations. However, attention is also being drawn towards exploring nongenetic variations, mainly epigenetic factors, which influence the etiopathology of complex human disease [25–27].

The epigenome status of a cell is highly dynamic, with a complex collaboration and interaction of genetic and environmental factors eventually determining the epigenotype signature [11]. Maintaining normal epigenomic homeostasis governs the normal cellular function, and turbulence

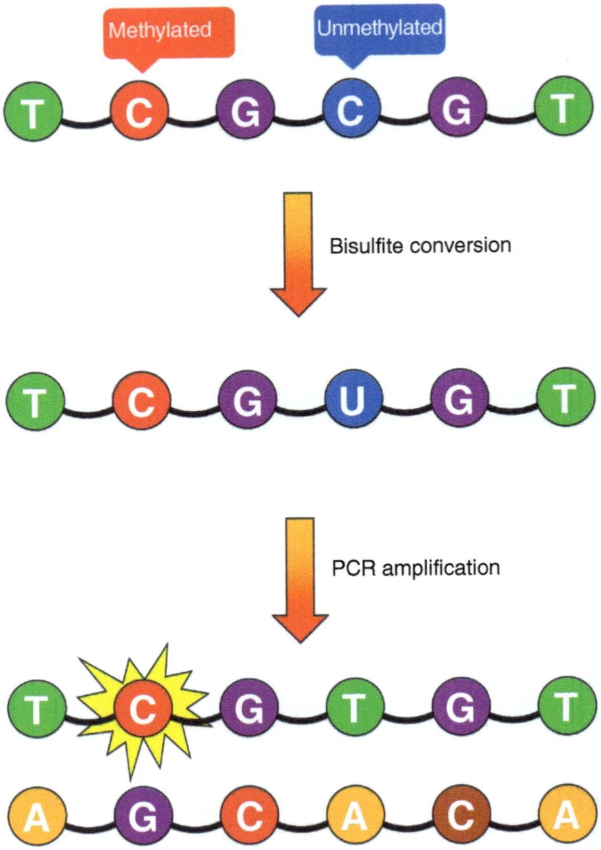

FIGURE 18.1 Schematic diagram of the sodium bisulfite conversion of genomic DNA to discriminate and detect unmethylated and methylated cytosine residues. During the conversion, unmethylated cytosine residues are deaminated to uracil while methylated cytosine residues remain methylated. After polymerase chain reaction (PCR) amplification, unmethylated cytosine will be detected as thymidine, while methylated cytosine will be identified as cytosine.

TABLE 18.1 An Outline of the Specifications of the Methods for DNA Methylation Evaluation

Technology or approach	Characteristics	References
Sodium bisulfite conversion	Gold standard for methylome analysis	[9]
Ultra-high-throughput DNA sequencing technologies	Direct sequencing of methylated DNA fragments or of bisulfite-converted genomic DNA	[22,23]
Enzyme-based or antibody-affinity techniques	Genome-wide analysis	[10–15]
Low-resolution array method	Rapid identification	[16]
Combination of methylation-dependent DNA digestion and fluorescent tagging	Measuring the overall methylation level	[17–21]

in this balance has been reported to be associated with human diseases, especially cancer [27]. The drawbacks of most associated studies performed to date have been insufficient coverage of the genome and inadequate sample sizes. Therefore the rate of phenotypic change that is attributable to interindividual epigenomic variation has yet to be determined. This issue can only be rectified by using large-scale EWAS, both in terms of genome coverage and sample size. Experimental technologies for the evaluation of DNA methylation are now comparable, in terms of their resolution and throughput, with GWAS chips, and should be able to measure approximately 500,000 SNPs.

Other than the common association-study concerns (i.e., sufficient sample size and sophisticated assay technology), there are additional points that must be considered when designing an EWAS, namely, the tissue type. Patterns of DNA methylation are tissue specific and developmental-stage specific, both of which are variables that change with time. Moreover, the associations resulting from EWAS can be either causal or consequential for the assessed phenotype, hence providing additional analytical challenges. These considerations must be discussed in respect to the design and analysis of EWAS data. In a nutshell, we should be mindful that EWAS techniques and approaches, just like GWAS, will be subject to development as we become armed with more experience.

18.4.1 Epigenetic Information in EWAS

In mammalian cells, information in epigenetic marks is commonly transmitted through stable DNA methylation, posttranslational regulation of histone proteins, or noncoding RNAs (ncRNAs) such as miRNAs. The most common type of DNA methylation modification is mediated by methylation of cytosine residues in cytosine–guanine dinucleotides (CpG); however, some studies suggest that CpH methylation (where H can be C, A, or T nucleotides) may be more important than previously thought [28,29]. The 5′-hydroxymethylation of cytosine residues (hmC) has been suggested as another type of DNA methylation regulation [30,31]. Despite limiting details, mounting evidence supports a role for hmC in gene regulation and differentiation [32]. Conversely, histone modifications include, but are not limited to, biochemical changes such as methylation, acetylation, and citrullination of amino acids in histone tails [11]. NcRNAs have the ability to self-propagate and can be transmitted regardless of the related DNA, which means that they can transmit regulatory information "epigenetically" [33,34]. Common ncRNAs include short miRNAs, PIWI-interacting RNAs, and large intergenic noncoding RNAs [33].

18.4.2 Profiling the Epigenetic Variation

One of the main advancements that facilitated large-scale GWAS was the establishment of powerful but low-cost genetic profiling techniques, especially SNP arrays. The technologies for epigenomic profiling have recently reached a level where the conductance of large-scale EWAS is now practical. To perform such studies, the target mark or molecule should be stable, the technique should carry out a high-throughput analysis, and the sampling should be easy to perform in routine clinical procedures. Moreover, automatable methods for whole-genome profiling should be available to speed up such studies. The most favorable epigenetic mark for EWAS is the evaluation of DNA methylation, particularly methylation of CpG sites. Although other epigenetic marks may be significantly important, they may not be easily available in routine clinical samples or not affordable for high-throughput processing. Mounting numbers of studies have also demonstrated well-established correlations between various epigenetic marks; therefore DNA methylation profiling can provide data about histone modification states, at least through computational epigenetics [11].

Theoretically, EWAS can benefit from both sequencing and array-based profiling approaches. The most common platforms have been extensively evaluated for these techniques [35] and benchmark scores have been measured [36,37]. The well-suited choice of EWAS should have harmony for some items, including epigenome coverage, specificity, accuracy, resolution, throughput, and cost [38]. Finally, although sequencing-based approaches are just becoming common place, the array-based methods (i.e., the SNP arrays used in GWAS) are the most demanding and favorable technologies for EWAS. The Illumina 450K Infinium Methylation BeadChip is believed to be promising for the initial steps of EWAS as it offers a satisfactory balance of epigenome coverage, resolution, and throughput. It should be noted that the Illumina MethylationEPIC BeadChip microarray for evaluation of whole-genome DNA methylation is a significant improvement compared with the Illumina HumanMethylation450 (HM450) array, which was unable to cover distal regulatory elements. This recent technological advance has surpassed the previous level of coverage of regulatory regions and shows increased reliability and reproducibility, conferring a favorable approach for high-throughput genomic methylation analyses [39]. However, the development of such a technology is in its infancy and will require further improvements in the future.

18.4.3 Integration of EWAS and GWAS

Genetic variants affecting the methylation status are called methylation quantitative trait loci (methQTLs) [39a], and some studies have identified correlations between epigenotypes and genotypes, raising hopes for future integrated analyses. One study analyzed SNPs, gene expression data, and DNA methylation data in HapMap cell lines. The authors discovered that both gene expression and DNA methylation were under the control of specific SNPs, hence providing valuable evidence of interactions between genetic and epigenetic mechanisms that result in synergistic consequences [40]. From this point of view, EWAS can be performed to evaluate the genetic susceptibility that exploits their function. Another study successfully integrated GWAS and EWAS information to determine the haplotype-specific methylation in a type 2 diabetes and obesity susceptibility locus [41]. In the future, it may be possible to perform integrated evaluations for multiple genetic and epigenetic marks to obtain disease-causing chromatin states, which is not feasible or easy when using DNA methylation data only. One additional study applied multivariate hidden Markov analysis for recurrent and spatially coherent interactions of epigenetic signs. The authors identified distinguished chromatin regions in human T cells that appeared to be promising for integrating with GWAS data to provide new insights into blood-based disorders [42]. The identification of disease-associated genetic variations through GWAS will contribute to the advancement of complementary methods for identifying additional variants that can potentially be integrated with GWAS.

GWAS had the advantage of already having a predetermined map of the human genome that provided easy selection of SNP tags for throughout genome coverage [43]. There is a correlation between DNA methylation and tissue-specific blocks of CpG sites, which can span up to 1000 bases [44]. Comprehensive understanding of these block structures in tissues and cell types will hopefully improve the accurate and detailed selection of CpG sites for EWAS, at least until the methylome map is completed. High-resolution methylome maps are currently available for fetal fibroblasts, human embryonic stem cells, and peripheral blood monocytes, which means that potentially satisfactory EWAS can be performed for early developmental disorders and blood-based diseases [45]. The International Human Epigenome Consortium aims to produce 1000 reference epigenomes and methylome maps, which will be specific for human tissues and cell types. These maps will be the epigenetic equivalent of the human haplotype map and will provide an outstanding resource for performing EWAS for most common human diseases.

Another challenge regarding EWAS will be the availability of relevant tissue samples. Establishment of a tissue bank center will allow researchers to search for a given sample and associated data that is relevant to their specific EWAS. Pertinent EWAS also require the development of technologies for the analysis and interpretation of data. Achieving this goal will require cooperation from biotechnologists, clinicians, sample providers, and funding resources.

18.5 EPIDEMIOLOGIC AND EPIGENETIC ANALYSIS

Genetic and epidemiologic studies have provided a better framework for understanding the combined effect of an individual's genotype and environmental factors on the risk of disease; however, adding epigenetic data will help to explain the functional basis underlying these combined effects by providing unbiased biologic measures that may change gene activity [46].

18.6 MODELING THE EPIGENETIC DATA

Epigenetics is considered as a relatively new field of study and seems to be a suitable approach to model the methylation information. Few studies have integrated data obtained about genetic, epigenetic, and environmental factors, and separate associations may exist between the genotype and phenotype, genotype and epigenotype, and epigenotype and phenotype. Various environmental exposures (i.e., smoking and alcohol) may affect the epigenotype and have independent effects on the phenotype. Therefore a suitable statistical model will be based upon the question of interest and knowledge about the relevant biologic pathways. Methods for epigenetic modeling, such as generalized estimating equations, mixed models, or the rudimentary approach of comparing means, have attempted to correlate outcomes with the epigenome. CpG sites with susceptibility to methylation often occur in clusters and may indicate coordinated methylation alterations, therefore analytic approaches may be required to summarize the correlated data [46,47]. Most epigenetic analysis approaches consider the methylation at individual CpG sites. The correlation between methylation and gene expression may be nonlinear and may differ from site to site; however, new statistical software, such as Stata's GLM command (Stata Corporation, College Station, TX), provide the modeling of proportional outcomes within the generalized linear framework. In the future, new modeling approaches need to be devised to improve the interpretation of epigenetic data.

18.7 FUTURE CHALLENGES IN COMPUTATIONAL EPIGENETICS

Technological advancements such as ChIP-on-chip and ChIP-seq have provided an opportunity to turn epigenetic studies into high-throughput outputs, hence the development of computational and bioinformatic tools is expected to make a dramatic difference to the field. In fact, pioneering research in computational epigenetics was only made possible by the rapid emergence of experimental data generation techniques in which sophisticated computational approaches were used as tools to determine the genomic distribution pattern of epigenetic marks. These subjects will remain as relevant areas of study, and the quantity of epigenetic data from epigenome projects will create a bioinformatic challenge, therefore these studies should be armed with computational studies as they are becoming significantly broader. Moreover, the analysis of epigenetic information will apply data from proteins that are involved in reading and writing the epigenetic data, as well as their regulatory networks and interaction partners. In fact, the reverse engineering of epigenetic regulatory patterns could result in quantitative models and sophisticated manipulation of the critical steps that determine cell fate [48]. Furthermore, epigenome mapping will become affordable, allowing quantitative analysis of epigenetic modifications in large-scale human populations. Twin studies have demonstrated that both environmental factors and genetic variations can affect epigenetic marks [47,49]. Therefore bioinformatic technology should be capable of assigning functional connections by combining epigenome information with genetic marks (i.e., gene expression data) and haplotype maps from large sample sizes and heterogeneous populations. Performing comprehensive epigenome mapping in multiple species will also contribute to evolutionary data. Early data demonstrated that orthologous regions harbor similar epigenetic marks in various mammals [50,51], as the DNA itself encodes its own epigenetic circumstance [52,53]. It should be noted that comparative epigenomics has the potential to significantly improve the ability to discover functionally important positions in the human genome. Alternately, developing theoretical modeling will help to deepen our mechanistic and quantitative knowledge of epigenetic regulations. Interactions between proteins that sign the epigenetic marks are necessary for maintaining the quality of an epigenetic change at the molecular level in a media with highly dynamic perturbations [54,55]. As a result, explanation of the circumstances during epigenetic regulation caused by the dynamic interactions of different epigenetic regulatory mechanisms is regarded as the benefit of computational modeling studies.

18.8 DEVELOPMENT OF EPIGENETIC BIOMARKERS

Epigenetic regulatory mechanisms control gene expression in a manner that is stably distributed over multiple cell divisions: however, it is also flexible enough to respond to environmental factors. This intermediate state of epigenetic change, between stability and flexibility, makes epigenetic data highly beneficial for cellular-state monitoring as it is rendered through personalized medicine. Aberrant epigenetic modifications are also known to cause common diseases such as autoimmune diseases. Epigenetic biomarker development aims to design experimental assays to produce relevant information for prognosis, diagnosis, and optimal therapy in routine clinical treatment as well as in drug discovery.

Several epigenetic regulatory mechanisms have been discovered to date, including DNA methylation, histone methylation, histone acetylation, and ncRNAs (i.e., miRNAs) [56]; however, most of the biomarker development studies have concentrated on the DNA methylation status. This is largely due to the stability of the methylation marks, the simplicity of measuring DNA methylation, and the well-known role of dysregulated DNA methylation in human diseases [57–59].

Epigenetic biomarker development on a genome-wide scale has been facilitated by progress in high-throughput sequencing approaches [11,60]. The expanded spectrum of investigation should help to identify novel regions with disease-specific epigenetic modifications outside the usual CpG islands and gene promoters [61]. Nonetheless, bioinformatic techniques should also be improved alongside the experimental discoveries by high-throughput sequencing data. Furthermore, challenges are faced when translating data of candidate biomarkers into diagnostic approaches in clinics. Therefore a systematic approach for the development of epigenetic biomarkers should have the potential to dominate the limitations and challenges. This approach should entail: (1) maximal genomic coverage through the initial evaluations; (2) the use of computational and bioinformatic techniques to identify the most suitable candidate biomarkers; and (3) the use of large-cohort evaluations to validate the performance of the biomarker. Due to the high cost of genome-wide methylation analysis, developing the epigenetic biomarkers occurs when individual experiments with different levels of genomic coverage provide sufficient data [15,22,62,63]. However, medium-scale validation experiments should be able to evaluate the status of DNA methylation in thousands of CpG sites in hundreds of cases, and techniques such as bisulfite sequencing and epigenotyping (Illumina Infinium HumanMethylation27) [64–66] afford these requirements.

After investigating the specificity and sensitivity of data obtained from the medium-scale validation step, highly favorably predictive regions are chosen as candidate biomarkers for further studies. An optimized test must then be developed to analyze the DNA methylation status at selected CpG sites. Evaluation of DNA methylation by several methods results in the measurement of a small number of CpG sites at low costs and provides an approach for biomarker confirmation that leads to well-suited clinical applications.

Bisulfite pyrosequencing [67], combined bisulfite restriction analysis [68], and methylation-sensitive single-nucleotide primer extension [69] confer quantitative DNA methylation information for each CpG site. Conversely, methylation-specific PCR [70] and MethyLight [71] techniques measure the methylation circumstance at many CpG sites at the same time and provide sensitive and specific detection of the methylation pattern. Ultimately, the clonal bisulfite sequencing approach, which is considered as the gold standard for evaluating DNA methylation [72], is beneficial for quality control. Despite being a toil for common use in clinics, this approach is beneficial for exploring well-suited CpG sites within a genomic region [73].

18.8.1 The Effects of Epigenetic Biomarkers in Treatment Approaches

The aim of clinical biomarkers is to arm physicians with beneficial data about the presence or absence of a specific disease, or to provide characteristics of the patient and disease that may affect the treatment strategy decision making. Epigenetic biomarkers are made up of two complementary sections: (1) an experimental or laboratory assay, which generates accurate information about epigenetic alterations in a patient at a single locus or multiple genomic regions; and (2) predicted disease criteria, which translates the experimental data into biomarker results.

Most of the mechanisms underlying epigenetic regulations have been recently identified and mainly include DNA methylation, histone methylation, histone acetylation, miRNAs, and other ncRNAs [74]. The development of epigenetic biomarkers has mainly concentrated on DNA methylation due to the simplicity of practical considerations [59,75] and the well-described role of DNA methylation in autoimmune diseases [56]. However, it is hoped that epigenetic biomarkers relating to histone modifications [76] and ncRNAs [77] may develop in the near future.

The development of epigenetic biomarkers for prognosis and optimal therapy meets two major elementary problems that are faced in the development of diagnostic biomarkers: (1) the biopsy of the disease tissue is commonly available after the diagnosis has been confirmed, whereas the diagnostic biomarkers should be present in more easily accessible tissues (i.e., blood), which may not provide sufficient information of the disease-involved pathways and mechanisms; and (2) the required specificity is lower for prognostic and therapy-optimization biomarkers than diagnostic biomarkers. This is because diagnostic biomarkers are designed for screening vast numbers of the healthy population, in which a relatively small false-positive rate can produce substantial numbers of unnecessary follow-up examinations. Considering this fact, it seems logical that most epigenetic biomarkers should focus on discovering disease subtypes alongside distinct clinical characteristics [78] instead of discriminating between disease-affected patients and healthy individuals.

18.8.2 Systems of Epigenetic Biomarker Development

Recent breakthroughs in high-throughput sequencing [11,60] have permitted the discovery of epigenetic biomarkers on a genome-wide scale. These cutting-edge techniques are likely to reveal many new genomic regions that harbor disease-specific epigenetic alterations, such as those found outside of well-known candidate regions (i.e., CpG islands and gene promoters) [61]. However, this fast-growing area demands significantly improved bioinformatic interpretation and the need to address the statistical aspect of a multiple-testing. More efforts are also needed to facilitate the rapid translation of biomarkers into diagnostic approaches that will provide physicians with sufficient information. A systematic approach to biomarker development has the potential to resolve most of these challenges. The major concepts of this approach are to maximize the genome coverage in the early steps of the examination, use computational methods for optimizing and discovering well-suited candidate biomarkers, and confirm the biomarker validation in large cohorts. This procedure can be used for any well-known disease in instances where a minimum of two high-quality case–control cohort studies are available. The first cohort study is used for biomarker development, and the second validation cohort study is used for the unbiased evaluation of biomarker function. Genome-scale analysis of DNA methylation is not yet economically feasible, therefore developing large-scale epigenetic biomarkers will only be possible when a number of experimental approaches with various spectrums of genome coverage are merged. The use of experimental methods with maximum genomic coverage is suggested in the initial step of the discovery phase [15,22,62,63], despite the number of samples that can be evaluated being limited by the high per-sample costs. The low validation rate resulting from the small sample size during the initial screening phase is compensated in the following steps by another medium-rate confirmation study. The experimental methods applied in this medium-scale confirmation study should be customizable and have the ability to analyze DNA methylation at thousands of CpG sites in a maximum of 100 samples. The estimated specificity and sensitivity outcomes from the medium-scale confirmatory step are used to define the highly predictive regions of the genome as candidate biomarkers. A well-optimized assay is designed for each of these epigenetically changed regions to analyze the DNA methylation status at a small number of representative CpG sites. This step is based on the assumption that the status of DNA methylation is correlated between adjacent CpG dinucleotides [44], hence evaluation of a number of carefully selected CpG sites provides accurate DNA methylation information of all the CpG sites in the CpG island or covered genome [79,80]. Some experimental techniques can accurately measure the DNA methylation status at a small number of CpG sites without high costs, suggesting that they may be beneficial tools for biomarker confirmation and subsequent clinical application.

18.8.3 Future Challenges of Developing Epigenetic Biomarkers

Most of the research performed on epigenetic biomarkers has focused on cancer; however, developing epigenetic biomarkers has also become attractive for researchers investigating common diseases (i.e., autoimmune disorders). The hypothesis is that epigenetic modifications may link environmental factors to disease susceptibility, hence producing a biochemical trace of the risk factor exposure record that cannot be interpreted based on genetic and environmental information alone. Advances in measurement techniques have made epigenome mapping easy and cost-effective, providing clinicians with the ability to monitor the epigenetic modifications in a given disease or patient. Research regarding epigenetic modifications and disease-specific biomarkers is likely to continue on a genome-wide scale, which is regarded as the pattern of GWAS and their related DNA sequence polymorphisms [81]. Some theoretical hypotheses have also suggested combining the results of epigenetic and genetic association studies on a united genome-wide scale [46,82,83]. Although the integrated results of the genetic and epigenetic investigations may improve the understanding of complex diseases, some setbacks still exist. Experience from the initial days of GWAS highlighted that the results from studies lacking sufficient reliability need to be replicated and that noted improvements in the sample size and statistical approaches should be performed before the GWAS could be assumed as the central approach in identifying disease-causing genes [84]. It should be noted that performing investigations on a small number of validated epigenetic biomarkers instead of studies on large amounts of differentially methylated regions in an unreplicated manner can identify more validated epigenetic biomarkers. Moreover, the validation of epigenetic biomarkers is a critically important issue that should be addressed in more than one cohort.

18.9 PROSPECTS FOR EPIGENETIC THERAPY IN AUTOIMMUNE DISEASES

Pharmacogenetics has almost failed to describe the differences in the individual response to medications; however, epigenomics appears to have the potential to explain the differences in the drug response both intra- and interpersonally [85]. Epigenetic-related factors, which affect the drug response quality, commonly impact on drug-metabolizing enzymes (i.e., cytochrome P450) [86], drug transporters (i.e., ATP-binding cassette [87] and soluble carrier transporter group) [88], and nuclear receptors (i.e., retinoic acid receptor family [89] and steroid hormone receptor estrogen receptor α [90]). One of the most awaited outcomes of the current Human Epigenome Project is the exhaustive study of pharmacogenomics. It is obvious that medications have failed to provide enough options to either prevent the initiation or ameliorate the perpetuation of autoimmune diseases. Nowadays, similar drugs are prescribed for all conditions that require an immunosuppressive treatment, regardless of the adverse effects.

Identifying the aberrant epigenetic changes underlying the pathogenesis of autoimmune diseases has improved vastly over recent years. This has attracted the attention of researchers wishing to find tissue- or cell-specific epigenetic clinical signatures for early diagnosis as well as those wishing to develop novel therapeutic approaches. "Epigenetic therapy" is characterized by drugs with the potential to alter methylation patterns on DNA molecules or change histones proteins. Several of the drugs approved for the treatment of cancer could also have beneficial effects for multiple autoimmune diseases. Among the studied epigenetic therapeutic agents for autoimmune diseases are drugs such as histone deacetylase (HDAC) inhibitors and DNA methyltransferase (DNMT) inhibitors. Studies of these drugs in several autoimmune diseases were initiated in animal models and in vitro trials and seem to be promising for future use in clinical trials [91]. The potency and safety of cutting-edge treatment procedures depends on validation in carefully designed control trials. The drugs currently endorsed for hematologic carcinomas, which are also applicable for clinical trials in solid malignancies such as breast cancers [92] and nonsmall cell lung cancers [93], comprise those that focus on epigenetic modifications (i.e., DNMT inhibitors and HDAC inhibitors) [94,95]. However, the ontogeny of many next-generation epigenetic medications is still in process [96]. It should be noted that studies have assessed the therapeutic potential of inhibitory agents by targeting the enzymes that play a role in the alteration of DNA methylation and acetylation and lysine residues in both histone and nonhistone proteins. However, it is mandatory to strengthen our knowledge to devise novel therapeutics in this field. Moreover, the availability of sophisticated bioinformatic tools will help to achieve the goal of personalized therapy in immune-mediated disorders [97,98], as has already been exploited in cancer [5].

18.10 CONCLUSION

The evidence relating epigenetic modifications to almost all autoimmune diseases has been achieved by a mounting number of investigations suggesting the potential application of epigenetic biomarkers in clinical practice. Epigenetic studies in autoimmune disorders have primarily concentrated on a particular subset of mechanisms. In most studies, DNA methylation and histone acetylation have been evaluated, as these mechanisms are regarded as well-characterized and measurable. These epigenetic mechanisms are also commonly studied regardless of the observation that histone modifications and DNA methylation do not act independently. Therefore it would be advantageous to simultaneously evaluate these different epigenetic changes to obtain correct insights into epigenetic interactions. We are at the initiation of a very promising journey toward the application of epigenetic data in new areas of therapies, such as personalized medicine. New developing techniques and data modeling and interpretation will hopefully contribute to autoimmune disease treatment in the future.

References

[1] Hamburg MA, Collins FS. The path to personalized medicine. N Engl J M 2010;363(4):301–4.

[2] Bruehl S, Apkarian AV, Ballantyne JC, Berger A, Borsook D, Chen WG, et al. Personalized medicine and opioid analgesic prescribing for chronic pain: opportunities and challenges. J Pain 2013;14(2):103–13.

[3] Hedrich CM. Epigenetics in autoimmunity. Encyclopedia of medical immunology. Springer; 2014. 399-408.

[4] Egger G, Liang G, Aparicio A, Jones PA. Epigenetics in human disease and prospects for epigenetic therapy. Nature 2004;429(6990):457–63.

[5] Esteller M. Epigenetics in cancer. N Engl J Med 2008;358(11):1148–59.

[6] Joss-Moore LA, Lane RH. Epigenetics and the developmental origins of disease: the key to unlocking the door of personalized medicine. Epigenomics 2012;4(5):471–3.

[7] Safar AM, Spencer H III, Su X, Cooney CA, Shwaiki A, Fan C-Y. Promoter hypermethylation for molecular nodal staging in non-small cell lung cancer. Arch Pathol Lab Med 2007;131(6):936–41.

[8] Wong NC, Morley R, Saffery R, Craig JM. Archived Guthrie blood spots as a novel source for quantitative DNA methylation analysis. Biotechniques 2008;45(4):423.

[9] Frommer M, McDonald LE, Millar DS, Collis CM, Watt F, Grigg GW, et al. A genomic sequencing protocol that yields a positive display of 5-methylcytosine residues in individual DNA strands. Proc Natl Acad Sci USA 1992;89(5): 1827–31.

[10] Schumacher A, Kapranov P, Kaminsky Z, Flanagan J, Assadzadeh A, Yau P, et al. Microarray-based DNA methylation profiling: technology and applications. Nucleic Acids Res 2006;34(2):528–42.
[11] Bernstein BE, Meissner A, Lander ES. The mammalian epigenome. Cell 2007;128(4):669–81.
[12] Epigenomics EV. Mapping the methylome. Cell Cycle 2006;5(2):155–8.
[13] Beck S, Rakyan VK. The methylome: approaches for global DNA methylation profiling. Trends Genet 2008;24(5):231–7.
[14] Rakyan VK, Down TA, Thorne NP, Flicek P, Kulesha E, Gräf S, et al. An integrated resource for genome-wide identification and analysis of human tissue-specific differentially methylated regions (tDMRs). Genome Res 2008;18(9):1518–29.
[15] Irizarry RA, Ladd-Acosta C, Carvalho B, Wu H, Brandenburg SA, Jeddeloh JA, et al. Comprehensive high-throughput arrays for relative methylation (CHARM). Genome Res 2008;18(5):780–90.
[16] Mill J, Tang T, Kaminsky Z, Khare T, Yazdanpanah S, Bouchard L, et al. Epigenomic profiling reveals DNA-methylation changes associated with major psychosis. Am J Hum Genet 2008;82(3):696–711.
[17] Ehrlich M. Cancer-linked DNA hypomethylation and its relationship to hypermethylation. DNA methylation: development genetic disease and cancer. Springer; 2006. p.251-74.
[18] Ehrlich M, Wang R. 5-Methylcytosine in eukaryotic DNA. Science 1981;212(4501):1350–7.
[19] Gama-Sosa MA, Slagel VA, Trewyn RW, Oxenhandler R, Kuo KC, Gehrke CW, et al. The 5-methylcytosine content of DNA from human tumors. Nucleic Acids Res 1983;11(19):6883–94.
[20] Gama-Sosa MA, Wang RY, Kuo KC, Gehrke CW, Ehrlich M. The 5-methylcytosine content of highly repeated sequences in human DNA. Nucleic Acids Res 1983;11(10):3087–95.
[21] Mohsen K, Johansson S, Ekström TJ. Using LUMA: a luminometric-based assay for global DNA-methylation. Epigenetics 2006;1(1):46–9.
[22] Meissner A, Mikkelsen TS, Gu H, Wernig M, Hanna J, Sivachenko A, et al. Genome-scale DNA methylation maps of pluripotent and differentiated cells. Nature 2008;454(7205):766.
[23] Taylor KH, Kramer RS, Davis JW, Guo J, Duff DJ, Xu D, et al. Ultradeep bisulfite sequencing analysis of DNA methylation patterns in multiple gene promoters by 454 sequencing. Cancer Res 2007;67(18):8511–8.
[24] Hindorff LA, Sethupathy P, Junkins HA, Ramos EM, Mehta JP, Collins FS, et al. Potential etiologic and functional implications of genome-wide association loci for human diseases and traits. Proc Natl Acad Sci USA 2009;106(23):9362–7.
[25] Feinberg AP, Irizarry RA. Stochastic epigenetic variation as a driving force of development, evolutionary adaptation, and disease. Proc Natl Acad Sci USA 2010;107(Suppl. 1):1757–64.
[26] Petronis A. Epigenetics as a unifying principle in the aetiology of complex traits and diseases. Nature 2010;465(7299):721–7.
[27] Das PM, Singal R. DNA methylation and cancer. J Clin Oncol 2004;22(22):4632–42.
[28] Ramsahoye BH, Biniszkiewicz D, Lyko F, Clark V, Bird AP, Jaenisch R. Non-CpG methylation is prevalent in embryonic stem cells and may be mediated by DNA methyltransferase 3a. Proc Natl Acad Sci USA 2000;97(10):5237–42.
[29] Lister R, Pelizzola M, Dowen RH, Hawkins RD, Hon G, Tonti-Filippini J, et al. Human DNA methylomes at base resolution show widespread epigenomic differences. Nature 2009;462(7271):315–22.
[30] Kriaucionis S, Heintz N. The nuclear DNA base 5-hydroxymethylcytosine is present in Purkinje neurons and the brain. Science 2009;324(5929):929–30.
[31] Tahiliani M, Koh KP, Shen Y, Pastor WA, Bandukwala H, Brudno Y, et al. Conversion of 5-methylcytosine to 5-hydroxymethylcytosine in mammalian DNA by MLL partner TET1. Science 2009;324(5929):930–5.
[32] Véron N, Peters AH. Epigenetics: Tet proteins in the limelight. Nature 2011;473(7347):293–4.
[33] Zaratiegui M, Irvine DV, Martienssen RA. Noncoding RNAs and gene silencing. Cell 2007;128(4):763–76.
[34] Rassoulzadegan M, Grandjean V, Gounon P, Cuzin F. Inheritance of an epigenetic change in the mouse: a new role for RNA. Portland Press Limited; 2007.
[35] Laird PW. Principles and challenges of genome-wide DNA methylation analysis. Nat Rev Genet 2010;11(3):191–203.
[36] Harris RA, Wang T, Coarfa C, Nagarajan RP, Hong C, Downey SL, et al. Comparison of sequencing-based methods to profile DNA methylation and identification of monoallelic epigenetic modifications. Nat Biotechnol 2010;28(10):1097–105.
[37] Bock C, Tomazou EM, Brinkman AB, Müller F, Simmer F, Gu H, et al. Quantitative comparison of genome-wide DNA methylation mapping technologies. Nat Biotechnol 2010;28(10):1106–14.
[38] Beck S. Taking the measure of the methylome: two comparative studies from the International Human Epigenome Project find high concordance between different methods for measuring genomic methylation. Nat Biotechnol 2010;28(10):1026–9.
[39] Pidsley R, Zotenko E, Peters TJ, Lawrence MG, Risbridger GP, Molloy P, et al. Critical evaluation of the Illumina MethylationEPIC BeadChip microarray for whole-genome DNA methylation profiling. Genome Biol 2016;17(1):208.

[39a] Zhang D, Cheng L, Badner JA, Chen C, Chen Q, Luo W, et al. Genetic control of individual differences in gene-specific methylation in human brain. Am J Hum Genet 2010;86(3):411–9.
[40] Bell JT, Pai AA, Pickrell JK, Gaffney DJ, Pique-Regi R, Degner JF, et al. DNA methylation patterns associate with genetic and gene expression variation in HapMap cell lines. Genome Biol 2011;12(1):R10.
[41] Bell CG, Finer S, Lindgren CM, Wilson GA, Rakyan VK, Teschendorff AE, et al. Integrated genetic and epigenetic analysis identifies haplotype-specific methylation in the FTO type 2 diabetes and obesity susceptibility locus. PLoS One 2010;5(11):e14040.
[42] Ernst J, Kellis M. Discovery and characterization of chromatin states for systematic annotation of the human genome. Nat Biotechnol 2010;28(8):817–25.
[43] International HapMap Consortium. A haplotype map of the human genome. Nature 2005;437(7063):1299–320.
[44] Eckhardt F, Lewin J, Cortese R, Rakyan VK, Attwood J, Burger M, et al. DNA methylation profiling of human chromosomes 6, 20 and 22. Nat Genet 2006;38(12):1378–85.
[45] Li Y, Zhu J, Tian G, Li N, Li Q, Ye M, et al. The DNA methylome of human peripheral blood mononuclear cells. PLoS Biol 2010;8(11):e1000533.
[46] Foley DL, Craig JM, Morley R, Olsson CJ, Dwyer T, Smith K, et al. Prospects for epigenetic epidemiology. Am J Epidemiol 2009;169(4):389–400.
[47] Heijmans BT, Kremer D, Tobi EW, Boomsma DI, Slagboom PE. Heritable rather than age-related environmental and stochastic factors dominate variation in DNA methylation of the human IGF2/H19 locus. Hum Mol Genet 2007;16(5):547–54.
[48] Boyer LA, Lee TI, Cole MF, Johnstone SE, Levine SS, Zucker JP, et al. Core transcriptional regulatory circuitry in human embryonic stem cells. Cell 2005;122(6):947–56.
[49] Fraga MF, Ballestar E, Paz MF, Ropero S, Setien F, Ballestar ML, et al. Epigenetic differences arise during the lifetime of monozygotic twins. Proc Natl Acad Sci USA 2005;102(30):10604–9.
[50] Bernstein BE, Kamal M, Lindblad-Toh K, Bekiranov S, Bailey DK, Huebert DJ, et al. Genomic maps and comparative analysis of histone modifications in human and mouse. Cell 2005;120(2):169–81.
[51] Enard W, Fassbender A, Model F, Adorján P, Pääbo S, Olek A. Differences in DNA methylation patterns between humans and chimpanzees. Curr Biol 2004;14(4):R148–9.
[52] Bock C, Walter J, Paulsen M, Lengauer T. CpG island mapping by epigenome prediction. PLoS Comput Biol 2007;3(6):e110.
[53] Segal E, Fondufe-Mittendorf Y, Chen L, Thåström A, Field Y, Moore IK, et al. A genomic code for nucleosome positioning. Nature 2006;442(7104):772–8.
[54] Dodd IB, Micheelsen MA, Sneppen K, Thon G. Theoretical analysis of epigenetic cell memory by nucleosome modification. Cell 2007;129(4):813–22.
[55] Sontag LB, Lorincz MC, Luebeck EG. Dynamics, stability and inheritance of somatic DNA methylation imprints. J Theor Biol 2006;242(4):890–9.
[56] Aslani S, Mahmoudi M, Karami J, Jamshidi AR, Malekshahi Z, Nicknam MH. Epigenetic alterations underlying autoimmune diseases. Autoimmunity 2016;49(2):69–83.
[57] Rodríguez-Paredes M, Esteller M. Cancer epigenetics reaches mainstream oncology. Nat Med 2011;330–9.
[58] Feinberg AP, Tycko B. The history of cancer epigenetics. Nat Rev Cancer 2004;4(2):143.
[59] Suzuki MM, Bird A. DNA methylation landscapes: provocative insights from epigenomics. Nat Rev Genet 2008;9(6):465–76.
[60] Jeddeloh JA, Greally JM, Rando OJ. Reduced-representation methylation mapping. Genome Biol 2008;9(8):231.
[61] Irizarry RA, Ladd-Acosta C, Wen B, Wu Z, Montano C, Onyango P, et al. The human colon cancer methylome shows similar hypo- and hypermethylation at conserved tissue-specific CpG island shores. Nat Genet 2009;41(2):178–86.
[62] Down TA, Rakyan VK, Turner DJ, Flicek P, Li H, Kulesha E, et al. A Bayesian deconvolution strategy for immunoprecipitation-based DNA methylome analysis. Nat Biotechnol 2008;26(7):779.
[63] Oda M, Glass JL, Thompson RF, Mo Y, Olivier EN, Figueroa ME, et al. High-resolution genome-wide cytosine methylation profiling with simultaneous copy number analysis and optimization for limited cell numbers. Nucleic Acids Res 2009;37(12):3829–39.
[64] Ball MP, Li JB, Gao Y, Lee J-H, LeProust EM, Park I-H, et al. Targeted and genome-scale strategies reveal gene-body methylation signatures in human cells. Nat Biotechnol 2009;27(4):361–8.
[65] Deng J, Shoemaker R, Xie B, Gore A, LeProust EM, Antosiewicz-Bourget J, et al. Targeted bisulfite sequencing reveals changes in DNA methylation associated with nuclear reprogramming. Nat Biotechnol 2009;27(4):353–60.

[66] Thirlwell C, Eymard M, Feber A, Teschendorff A, Pearce K, Lechner M, et al. Genome-wide DNA methylation analysis of archival formalin-fixed paraffin-embedded tissue using the Illumina Infinium HumanMethylation27 BeadChip. Methods 2010;52(3):248–54.

[67] Tost J, Gut IG. DNA methylation analysis by pyrosequencing. Nat Protoc 2007;2(9):2265.

[68] Brena RM, Auer H, Kornacker K, Plass C. Quantification of DNA methylation in electrofluidics chips (Bio-COBRA). Nat Protoc 2006;1(1):52.

[69] Gonzalgo ML, Liang G. Methylation-sensitive single-nucleotide primer extension (Ms-SNuPE) for quantitative measurement of DNA methylation. Nat Protoc 2007;2(8):1931.

[70] Herman JG, Graff JR, Myöhänen S, Nelkin BD, Baylin SB. Methylation-specific PCR: a novel PCR assay for methylation status of CpG islands. Proc Natl Acad Sci USA 1996;93(18):9821–6.

[71] Eads CA, Danenberg KD, Kawakami K, Saltz LB, Blake C, Shibata D, et al. MethyLight: a high-throughput assay to measure DNA methylation. Nucleic Acids Res 2000;28(8):e32.

[72] Zhang Y, Rohde C, Tierling S, Jurkowski TP, Bock C, Santacruz D, et al. DNA methylation analysis of chromosome 21 gene promoters at single base pair and single allele resolution. PLoS Genet 2009;5(3):e1000438.

[73] Schüffler P, Mikeska T, Waha A, Lengauer T, Bock C. MethMarker: user-friendly design and optimization of gene-specific DNA methylation assays. Genome Biol 2009;10(10):R105.

[74] Bird A. Perceptions of epigenetics. Nature 2007;447(7143):396–8.

[75] Esteller M. Cancer epigenomics: DNA methylomes and histone-modification maps. Nat Rev Genet 2007;8(4):286–98.

[76] Kondo Y, Shen L, Cheng AS, Ahmed S, Boumber Y, Charo C, et al. Gene silencing in cancer by histone H3 lysine 27 trimethylation independent of promoter DNA methylation. Nat Genet 2008;40(6):741–50.

[77] Thai T-H. MicroRNA in autoimmunity. Encyclopedia of medical immunology: autoimmune diseases. 2014;724-9.

[78] Feinberg AP. Phenotypic plasticity and the epigenetics of human disease. Nature 2007;447(7143):433–40.

[79] Mikeska T, Bock C, El-Maarri O, Hübner A, Ehrentraut D, Schramm J, et al. Optimization of quantitative MGMT promoter methylation analysis using pyrosequencing and combined bisulfite restriction analysis. J Mol Diagn 2007;9(3):368–81.

[80] Meng H, Murrelle EL, Li G. Identification of a small optimal subset of CpG sites as bio-markers from high-throughput DNA methylation profiles. BMC Bioinf 2008;9(1):457.

[81] Altshuler D, Daly MJ, Lander ES. Genetic mapping in human disease. Science 2008;322(5903):881–8.

[82] Johannes F, Colot V, Jansen RC. Epigenome dynamics: a quantitative genetics perspective. Nat Rev Genet 2008;9(11):883–90.

[83] Butcher LM, Beck S. Future impact of integrated high-throughput methylome analyses on human health and disease. J Genet Genomics 2008;35(7):391–401.

[84] McCarthy MI, Abecasis GR, Cardon LR, Goldstein DB, Little J, Ioannidis JP, et al. Genome-wide association studies for complex traits: consensus, uncertainty and challenges. Nat Rev Genet 2008;9(5):356–69.

[85] Baer-Dubowska W, Majchrzak-Celińska A, Cichocki M. Pharmocoepigenetics: a new approach to predicting individual drug responses and targeting new drugs. Pharmacol Rep 2011;63(2):293–304.

[86] Ingelman-Sundberg M, Sim SC, Gomez A, Rodriguez-Antona C. Influence of cytochrome P450 polymorphisms on drug therapies: pharmacogenetic, pharmacoepigenetic and clinical aspects. Pharmacol Ther 2007;116(3):496–526.

[87] Kusaba H, Nakayama M, Harada T, Nomoto M, Kohno K, Kuwano M, et al. Association of 5′ CpG demethylation and altered chromatin structure in the promoter region with transcriptional activation of the multidrug resistance 1 gene in human cancer cells. Eur J Biochem 1999;262(3):924–32.

[88] Park JY, Helm JF, Zheng W, Ly QP, Hodul PJ, Centeno BA, et al. Silencing of the candidate tumor suppressor gene solute carrier family 5 member 8 (SLC5A8) in human pancreatic cancer. Pancreas 2008;36(4):e32–9.

[89] Youssef EM, Lotan D, Issa J-P, Wakasa K, Fan Y-H, Mao L, et al. Hypermethylation of the retinoic acid receptor-β2 gene in head and neck carcinogenesis. Clin Cancer Res 2004;10(5):1733–42.

[90] Bovenzi V, Momparler RL. Antineoplastic action of 5-aza-2′-deoxycytidine and histone deacetylase inhibitor and their effect on the expression of retinoic acid receptor β and estrogen receptor α genes in breast carcinoma cells. Cancer Chemother Pharmacol 2001;48(1):71–6.

[91] Lu Q, Qiu X, Hu N, Wen H, Su Y, Richardson B. Epigenetics, disease, and therapeutic interventions. Ageing Res Rev 2006;5(4):449–67.

[92] Connolly R, Stearns V. Epigenetics as a therapeutic target in breast cancer. J Mammary Gland Biol Neoplasia 2012;17(3–4):191–204.

[93] Vendetti FP, Rudin CM. Epigenetic therapy in non-small-cell lung cancer: targeting DNA methyltransferases and histone deacetylases. Expert Opin Biol Ther 2013;13(9):1273–85.

[94] Dawson MA, Kouzarides T. Cancer epigenetics: from mechanism to therapy. Cell 2012;150(1):12–27.
[95] Højfeldt JW, Agger K, Helin K. Histone lysine demethylases as targets for anticancer therapy. Nat Rev Drug Discov 2013;12(12):917–30.
[96] Helin K, Dhanak D. Chromatin proteins and modifications as drug targets. Nature 2013;502(7472):480–8.
[97] Virgin HW, Todd JA. Metagenomics and personalized medicine. Cell 2011;147(1):44–56.
[98] Yan Q. Immunoinformatics and systems biology methods for personalized medicine. Systems biology in drug discovery and development: methods and protocols. 2010;203-20.

Index

C

D

E

F

G

FSC
www.fsc.org
MIX
Papier | Fördert
gute Waldnutzung
FSC® C083411

Kolibri 360 GmbH
Ferdinand-Jühlke-Straße 7
99095 Erfurt, Deutschland
produktsicherheit@kolibri360.de